THE BEHAVIOUR

of

DOMESTIC ANIMALS

THE BEHAVIOUR
of
DOMESTIC ANIMALS

Edited by

E. S. E. HAFEZ
Ph.D.(Cantab.)

Professor, Department of Animal Sciences
Washington State University, Pullman, Washington, U.S.A.

LONDON
BAILLIÈRE, TINDALL & CASSELL
7 AND 8 HENRIETTA STREET
1969

First published 1962

Second edition 1969

SBN 7020 0290 9

Published in the United States of America by
the Williams and Wilkins Company, Baltimore

MADE AND PRINTED IN GREAT BRITAIN

List of Contributors

BANKS, Edwin M., Departments of Zoology and Animal Science, University of Illinois, Urbana, Illinois 61801, U.S.A.

CAIRNS, Robert B., Department of Psychology, Indiana University, Bloomington, Indiana 47401, U.S.A.

DENENBERG, Victor H., Department of Psychology, Purdue University, Lafayette, Indiana 47907, U.S.A. (*also* Department of Biobehavioral Sciences, University of Connecticut Storrs, Connecticut, U.S.A.)

EWBANK, R., Sub-Department of Animal Husbandry, Faculty of Veterinary Science, University of Liverpool, England

FISCHER, Gloria J., Department of Psychology, Washington State University, Pullman, Washington 99163, U.S.A.

FOX, M. W., Department of Psychology, Washington University, St Louis, Missouri 63130, U.S.A.

FULLER, J. L., Roscoe B. Jackson Memorial Laboratory, Bar Harbor, Maine 04609, U.S.A.

GOY, Robert W., Department of Reproductive Physiology and Behavior, Oregon Regional Primate Research Center, Beaverton, Oregon 97005, U.S.A.

GUHL, A. M., Department of Zoology, Kansas State University, Manhattan, Kansas 66502, U.S.A.

HAFEZ, E. S. E., Reproduction Laboratory, Department of Animal Sciences, Washington State University, Pullman, Washington 99163, U.S.A.

HALE, E. B., Departments of Biology and Poultry Science, The Pennsylvania State University, University Park, Pennsylvania 16802, U.S.A.

HATTON, Glenn I., Department of Psychology, Michigan State University, East Lansing, Michigan 48823, U.S.A.

HULET, C. V., U.S. Sheep Experiment Station, Sheep and Fur Animal Research Branch, A.R.S., Dubois, Idaho 83423, U.S.A.

JOHNSON, Jr., John I., Departments of Biophysics, Psychology and Zoology, Michigan State University, East Lansing, Michigan 48823, U.S.A.

KLING, Arthur, Departments of Psychiatry and Physiology, University of Illinois College of Medicine, Chicago, Illinois 60680, U.S.A.

KOVACH, Joseph K., Menninger Foundation, Box 829, Topeka, Kansas 66601, U.S.A.

McKINNEY, Frank, Department of Ecology and Behavioral Biology, James Ford Bell Museum of Natural History, University of Minnesota, Minneapolis, Minnesota 55455, U.S.A.

ROSS, Sherman, Department of Psychology, Howard University, Washington, D.C. 20001, U.S.A.

SCHEIN, Martin W., Department of Biology, West Virginia University, Morgantown, West Virginia 26506, U.S.A.

SCHLEIDT, Wolfgang M., Department of Zoology, University of Maryland, College Park, Maryland 20742, U.S.A.

SCOTT, J. P., Center for Research on Social Behavior, Bowling Green State University, Bowling Green, Ohio 43402, U.S.A.

SIGNORET, J. P., Laboratoire de Physiologie de la Reproduction, Tours-L'Orfrasière, 37, Nouzilly, France

TUCKER, Thomas J., Neuropsychiatric Institute, University of Illinois College of Medicine, Chicago, Illinois 60680, U.S.A.

WIERZBOWSKI, S., Institute of Zootechnics, Department of Animal Reproduction and A.I., Balice/Krakow, Poland

WILLIAMS, Moyra, Department of Psychiatry, The United Cambridge Hospitals, Addenbrooke's Hospital, Cambridge, England

ZARROW, M. X., Department of Biological Sciences, Purdue University, Lafayette, Indiana 47907, U.S.A. (*also* Department of Biobehavioral Sciences, University of Connecticut Storrs, Connecticut, U.S.A.)

Preface

Extensive investigations have been carried out on the evolution, genetics, breeding, nutrition, endocrinology, physiology, and pharmacology of domestic animals, but further progress in many areas of research is hindered by insufficient knowledge of animal behaviour. Since the literature on animal behaviour is widely scattered in many different journals and known only to specialists, I thought that a useful purpose would be served if it could be summarized in a single volume readily available to students of behaviour.

During the revision of the previous edition it was apparent that more attention has been paid to feeding and sexual behaviour than to other behavioural patterns, and that some species have been studied more extensively than others. An attempt has therefore been made to establish a balance. Because our knowledge of the behaviour of domestic animals still contains many gaps, a number of the introductory chapters are based on studies of laboratory animals. This necessity is not altogether unfortunate, since most of these species are also domestic animals, but accentuates the need for further study of other domestic animals.

In addition to students of behaviour, this volume should also interest zoologists, animal breeders, physiologists, veterinarians, psychologists, philosophers, and sociologists. To the investigator who is experimenting with domestic animals, this book will provide the basic information on the habits and abilities of the species with which he must be familiar.

The first edition has been extensively revised to include recent information about the neural and endocrine mechanisms of behaviour with emphasis on the hypothalamus, sensory capacities, and early experience. Most chapters have been completely rewritten to include current concepts and recent references. The editor takes this opportunity to express his appreciation to all contributors for revising their chapters meticulously, and to Mr P. R. West and the staff of Baillière, Tindall and Cassell for their excellent co-operation.

It will be noticed that no attempt has been made to impose consistency in the use of Imperial and metric measurements. It is hoped that the conversion tables printed on p. 627 will be found useful.

E. S. E. HAFEZ

Pullman, Washington

Acknowledgements

Acknowledgements are due to the following who assisted the contributors in reviewing manuscripts, and in preparing or giving permission to use illustrations: D. K. Adams, J. O. Almquist, M. Altmann, R. P. Amann, James Anderson, G. A. Bartholomew, Margaret Bastock, F. R. Bell, V. R. Berliner, P. B. Bhattacharya, W. Bielanski, Helen H. Blauvelt, A. Brownlee, Jan Bruell, H. K. Buechner, J. F. Burger, S. J. Cowlishaw, T. J. Cunha, D. E. Davis, Almut Dettmers, Philip Dziuk, H. D. Fausch, M. Freer, J. C. Gill, J. G. Gordon, Sir John Hammond, James L. Hancock, K. W. Harker, H. Heitman, Jr., A. F. Hicks, H. A. Hochbaum, C. A. Hultnäs, R. F. Hunter, I. J. Inkster, W. T. James, H. Kalmus, George W. Kelley, Jr., J. J. Kiser, F. D. Klopfer, D. H. K. Lee, G. P. Lofgreen, R. E. McDowell, W. McGee, F. F. McKenzie, A. Manning, Jean Mayer, R. DuMesnil DuBuisson, U. Moore, Joan Munro, Y. Niwa, Y. Nishikawa, John M. Ogle, Lloyd Peterson, G. D. Phillips, W. M. Schleidt, R. Schloeth, H. L. Self, J. P. Signoret, A. W. Stokes, R. S. Summerhays, T. Y. Tanabe, W. R. Thompson, von E. Trumler, E. S. Valenstein, S. W. J. Van Rensberg, W. I. Welker, C. F. Winchester, D. G. M. Wood-Gush, M. X. Zarrow, H. P. Zeigler, and D. R. Zimmermann.

Permission for quotation or for the reproduction of plates and figures was kindly given by the following: *Acta Agriculturæ Scandinavica*, Sweden; Agricultural College in Krakow, Poland; American Psychological Association; *Animal Behaviour*, England; *Animal Breeding Abstract*, Scotland; Association for Research in Nervous & Mental Disease, U.S.A.; Baillière, Tindall & Cassell Ltd, England; *Behaviour*, Netherlands; British Society of Animal Production, England; Cambridge University Press, England; Charles C. Thomas, Springfield, Ill., U.S.A.; Comptes Rendus des Séances de L'Académie des Sciences, France; *Cornell Veterinarian*, U.S.A.; *Empire Journal of Experimental Agriculture*, England; Gaines Dog Research Center, New York, U.S.A.; Government Printer, Pretoria, South Africa; Grassland Research Institute, Hurley, Berkshire, England; Japan Racing Association, U.S.A.; *Journal of Comparative Psychology*, U.S.A.; *Journal of Agricultural Sciences*, England; *Journal of Animal Science*, U.S.A.; *Journal of Dairy Science*, U.S.A.; *Journal of Genetic Psychology*, U.S.A.; McGraw-Hill Book Company, U.S.A.; Macmillan Book Company, U.S.A.; Masson & Cie Éditeurs, France; Meldinger fra Norge Landbrukshogskola, Norway; Methuen & Co., England; Missouri Agric. Expt. Sta., Columbia, U.S.A.; National Academy of Sciences, Committee for Research in Problems of Sex, U.S.A.; National Institute of Agricultural Sciences, Chiba, Japan; *Nature*, England; Ohio Agric. Expt. Sta., Wooster, U.S.A.; *Onderstepoort Journal of Veterinary Research*, South Africa; Pennsylvania State Univ., University Park, U.S.A.; Roscoe B. Jackson Memorial Laboratory, Bar Harbor, Maine,

U.S.A.; *Poultry Science*, U.S.A.; Utah State Univ., Logan, U.S.A.; Washington Agric. Expt. Sta., Pullman, U.S.A.; *Zeitschrift für Tierpsychologie*, Germany; *Zeitschrift für vergleichende Physiologie*, Germany.

Contributors wish to make personal acknowledgements as follows:

Chapter 6. The preparation of the chapter was in part supported by the National Institute of Child Health and Human Development grant HD-02068, U.S. Public Health Service.

Chapter 7. The chapter was prepared while J. I. Johnson was recipient of a Career Development Award from the National Institute of Child Health and Human Development, U.S. Public Health Service. We wish also to acknowledge N.I.H. Grants to R. W. Goy (MH-08634) and to the Oregon Regional Primate Research Center (FR-00163).

Chapter 8. The preparation of this chapter was in part supported by a National Science Foundation grant GB-4996 for Edwin M. Banks.

Chapter 13. This chapter was supported in part by a research grant HD-02068 from the National Institute of Child Health and Human Development, U.S. Public Health Service.

Chapter 15. This chapter was written while the authors were recipients of U.S. Public Health Service grants 5 SO1 FR-05517-06 and HD-02277. This work was also supported in part by the Menninger Foundation.

Contents

Chapter

PART FOUR

BEHAVIOUR OF BIRDS

Part One

Behaviour and Domestication

Chapter 1

Introduction to Animal Behaviour

J. P. Scott

Why should prehistoric men have chosen certain animal species for domestication out of the hundreds and thousands which are available? Even today primitive peoples capture many young animals and keep them as pets for short times, but these never become truly domesticated. Again, why have not more animals been domesticated in historical times? The answer to these questions is an interesting scientific puzzle, and we can get some clues to its solution from studying the behaviour of our farm and household animals.

Considered as a group, our domestic mammals and birds are highly social, and their wild relatives form elaborately organized groups under natural conditions. The one possible exception is the domestic cat, and even in this animal strong social relationships develop between mother and off-spring, and neighbouring males show social dominance when they meet. Another general characteristic of domestic animals is their ability to adapt themselves to other conditions than those in which they normally live. It is possible that relatively few species have this capacity and thus have a native aptitude for domestication.

If the aptitude for domestication includes both a high degree of socia-bility and general adaptability, we should find general resemblances in the behaviour of all domestic species. To be understood, the behaviour of each must be systematically described, analysed, and compared with that of others. This chapter will provide a guide to the kinds and significance of information to be found in later chapters. These will reveal that much more is known about certain species than others, and that certain kinds of behaviour have been studied much more extensively than the rest. This introductory chapter will also provide a systematic guide to the kinds of information that will lead to a more complete understanding of our domestic animals.

I. PATTERNS OF BEHAVIOUR

Each animal species has characteristic ways of performing certain functions and rarely departs from them. For example, a female dog characteristic-ally cleans its puppies by licking them, incidentally stimulating reflexes of defaecation and urination in the young. One result of this behaviour is to keep the litter and its sleeping place clean, and it is correlated with the fact that wolves, the ancestors of dogs, lived in lairs. On the other hand, sheep and cattle, which live in the open and move from place to place, have no such special patterns of eliminative behaviour.

We can define a behaviour pattern as an organized segment of behaviour having a special function. Its nature is determined chiefly by heredity, but it can also be modified by training and learning. A dog, for example, normally travels on four legs, because the structure of its body makes this the simplest and most effective means of locomotion under most circumstances, but it can be taught to walk on its hind legs for a short time. Patterns of behaviour are related to the fundamental anatomy and life processes of animals and thus are extremely stable under conditions of domestication and even of intense selection. A dog, cow, or sheep must use its native behaviour patterns, no matter how unusual the environment into which it is placed.

Tinbergen (1951) has stressed the importance of making a complete inventory, or "ethogram", of the behaviour patterns for every species studied. This is the raw material of the science of animal behaviour and, as the following chapters will show, it has still not been completely collected for even our common domestic animals.

A. Causes of Behaviour

The primary function of behaviour is to enable an animal to adjust to some change in conditions, whether external or internal. Most animals have a variety of behaviour patterns which can be tried out in a given situation, and in this way they learn to apply one or the other according to that which produces the best adjustment. A cow placed in a milking stanchion can attempt to break loose or stand quietly until released. Since only the latter produces results, most animals choose it.

However, before an animal can learn the results of its behaviour there must first be something which calls forth a response. Each behaviour pattern has some sort of primary stimulus or releaser which elicits behaviour in the absence of any previous experience (Lorenz, 1935). For example, the "hiding" reaction of young turkeys is stimulated by the alarm call of the mother. This behaviour can also be elicited by any similar sound, such as tapping on a window with a metallic object. Alarm reactions in young puppies are produced by any loud noise. In general, birds react to primary elicitors which are much more specific than those which stimulate mammals, and they have less ability to modify their behaviour by learning. Understanding the nature of the primary stimuli peculiar to a species is a basic part of understanding an animal's behaviour and being able to control it in practical situations.

From these illustrations we can see that behaviour may have several general causes. One is the general hereditary organization of the species, which determines its behaviour patterns. Another is the presence or absence of the primary stimulation which produces the behaviour; there must be some sort of change in conditions either in the external environment or inside the body in order for behaviour to occur. Finally, animals organize their behaviour through the processes of learning and behave according to what they have learned by previous experience.

B. Daily and Seasonal Cycles of Behaviour

If we watch the behaviour of animals living under the fairly uniform conditions typical of domestication, we find that they often do the same thing each day at a regular time. Part of this is caused by habit formation, as when cows gather around the barn just before milking time. Part of it is also caused by regular changes in environmental conditions as day changes to night and back to day. Animals are likely to be most active at the time of greatest change, namely at dawn and dusk, and least active either in the middle of the day or middle of the night. Part is also caused by internal physiological rhythms that are partially independent of external events. These recur at approximately 24-hour intervals and hence are called "circadian rhythms" (Aschoff, 1965). Whatever factors are involved, most animals tend to live a highly regular existence from day to day.

Animals also change their behaviour from season to season. Part of this is a direct response to changes in weather conditions. Grazing animals are likely to be more active at night during hot weather and less so in cold weather. In addition, seasonal changes in behaviour may accompany seasonal breeding activity. Most wild animals have regular breeding seasons which, in turn, result in regular seasons for the birth and care of offspring. In many domestic animals the breeding season has been modified or extended by artificial selection in order to increase fertility, but most still show a regular seasonal round of behaviour.

C. The Physiological Basis of Behaviour

Since behaviour is activity, it necessarily involves physiological function. Any sort of behaviour involves the reception of stimuli through sense organs, the transformation of these stimuli into neural activity, the integrative action of the nervous system, and, finally, the activity of various motor organs, both internal and external. In addition to these general physiological mechanisms, each special pattern of behaviour may have its own special physiology activated by particular external stimuli which are transformed and transmitted over particular neural pathways so that integrated activity results.

D. Environmental Modification of Behaviour

There is a general trend in evolution towards a greater degree of behavioural adaptability. Indeed, such adaptability is highly characteristic of most of the species which have been successfully domesticated. It rests primarily on possessing a variety of alternate behaviour patterns and consequent lack of fixed responses to specific stimuli, together with the capacity to learn from experience, so that an animal may select the behaviour pattern which has proved successful and repeat it in a similar future situation. In vertebrate animals, the capacity for learning appears early in development and continues throughout life. There are two basic methods of studying learning. One is to consider behaviour as an effect of stimulation. In the experiments of Pavlov (1927), the technique is to associate the behavioural response to a primary stimulus with a previously neutral stimulus, so that the behaviour is now produced by a new cause. In the method of Skinner (1938), the behaviour is considered a cause which is modified according to the result it produces; i.e. behaviour is modified according to its contingencies. It seems likely that the same basic neural processes are associated with each, but the latter probably has wider application in the behaviour of animals acting under free situations. These two techniques are respectively called "classical conditioning" and "operant conditioning".

E. Developmental Changes in Behaviour

All vertebrate animals are born or hatched in an immature state and at first show patterns of behaviour which are suitable only for early life. Many domestic mammals and birds are quite precocious, but even animals like sheep, cattle, and chickens do not complete their development of behaviour until much later in life. Young animals come into the world with a few well-developed patterns of behaviour, such as pecking in chickens or nursing in mammals, but most other behaviour patterns develop under the influence of post-natal environmental stimulation and are greatly affected by later learning (Cruikshank, 1954). In some species the proper development of behaviour patterns depends upon the opportunity for play and social contact with others of their kind. Play behaviour usually consists of immature forms of adult patterns of behaviour. In such species as the monkey, playful behaviour is necessary in order to develop successful adult patterns of sexual behaviour (Mason, 1963). It also has the function of helping develop muscular strength through exercise. Closely confined animals never show the muscular development and motor skill developed by those which live under more free conditions.

It should be remembered that no behaviour is inherited as such. The only things which can be biologically inherited are the nuclei from the two parental germ cells plus the cytoplasm contained in the egg. Capacity for behaviour develops through the process of growth and is organized by hereditary factors and the process of learning. Thus, the old question of whether behaviour is acquired or inherited becomes meaningless; behaviour develops and is differentiated under the influence of genetic and environmental factors, neither one of which can act independently.

F. Genetic Differences in Behaviour

Since behaviour is often concerned with adjustment to changes in environmental conditions, a large part of behavioural variation is caused by differences in environmental stimulation. However, even when environmental stimulation is uniform, individuals still react differently from each other. Males act differently from females, and various breeds and strains of animals often differ from each other (Fuller & Thompson, 1960; Scott & Fuller, 1965). Game cocks are much more aggressive than the roosters of other breeds of chickens. Such effects of heredity on behaviour will be described in detail in Chapter 3.

Besides variation between individuals and breeds within a species, there are large differences in behaviour between species, and these are usually proportional to the relative taxonomic positions of the animals concerned. Birds are, of course, very different from mammals, and many of these differences are related to their primary methods of adaptation. With few exceptions, the structure of the whole class Aves is adapted for flight, and this is reflected in their sense organs, their body covering of feathers, and their small size and weight. Their behaviour patterns are likewise related to flight, as is the rapid development of such behaviour. Birds generally develop much more rapidly than mammals and consequently have less time to acquire learned behaviour.

Most mammals, on the other hand, are terrestrial animals, and many more of them are nocturnal in their habits than are birds. Their sense organs tend to emphasize odour as well as vision, and the body covering of hair permits much more behaviour involving direct body contact. While visual stimuli are highly important in birds, they are much less so in mammals.

The important domestic birds belong to two orders, the Anseriformes and the Galliformes. The former includes the family Anatidae, including the ducks and geese, belonging to related genera. Consequently, their behaviour is relatively similar. On the other hand, our other two important species of domestic fowl, the chicken and turkey, belong to two different families, and their behaviour is quite different.

The common domestic mammals belong to several different orders. Cats and dogs are both Carnivora, but belong to separate families. As is well known, their behaviour is quite dissimilar. Rabbits, which belong to the order Lagomorpha, are not closely related to other domestic animals. In the order Rodentia there are several domesticated forms. Rats and mice belong to the family Muridae, but guinea pigs come from a different family. The behaviour of the latter is distinctly different from that of the more ordinary rodents.

The domestic mammals which are currently most important commercially are all members of one order, Artiodactyla, the even-toed hoofed animals. Cattle, sheep, and goats all belong to one family, the Bovidae, and there are many similarities between their behaviour. Swine belong to a different family, the Suidae, and have distinctly different behaviour patterns. Horses are classified in a different order, the Perissodactyla, and have behaviour patterns strikingly unlike those of the other domestic ungulates. Such original differences in behaviour have continued to evolve under domestication, as will be described in Chapter 2.

II. SYSTEMS OF BEHAVIOUR

Each pattern of behaviour has a definite special adaptive function, which can generally be related to one of the nine general functions described below (Scott, 1958). A group of behaviour patterns with a common general function comprises a behavioural system. The organization of behavioural systems differs from species to species, being well or poorly developed, and similarities are closest between taxonomically closely related species. In most domestic animals only two systems of behaviour have been extensively studied, ingestive and sexual, since these are important commercially. Other systems are often less perfectly understood.

A. Ingestive Behaviour

Ingestive behaviour involves the consumption of food or nourishing substances, including both solids and liquids. Eating and drinking are ingestive behaviour, and each species has its own particular methods. Chickens and turkeys obtain their food by pecking, but with their broad, soft bills, ducks are more likely to nibble or scoop their food.

Cows, sheep, and goats have in common the behaviour pattern of rumination or chewing the cud. After eating, the animal usually lies down and repeatedly regurgitates mouthfuls of food, which it chews and re-swallows. The stomach is divided into sections which facilitate this behaviour by separating coarse and finely ground food.

The pattern of grazing in cattle and sheep is correlated with the lack of upper incisors. Cattle wrap their tongues around mouthfuls of grass, then jerk their heads forward so that the grass is cut by the lower teeth. The characteristic pattern of eating in pigs is rooting. A pig will thrust its nose into the ground and lift forward and upward, moving dirt out of the way and exposing earthworms, grubs, and roots. Horses eat in still another fashion. They graze, but bite their food with both upper and lower teeth, chew it more thoroughly, and do not ruminate.

All of the species mentioned so far are either herbivorous or omnivorous and spend many hours of the day eating. The same is true of domestic rabbits, which nibble their food with their sharp front incisors and then draw their heads back to chew it laterally many times before swallowing. The food habits of the domestic carnivores are quite different. Adult dogs and cats normally eat only one quick meal a day. Dogs are especially rapid eaters, gulping their food in large mouthfuls. Being meat eaters, their nourishment is highly concentrated and, since food is not constantly available, they are likely to eat large amounts of it at one time. Unlike the herbivorous animals, the teeth of carnivores do not permit lateral grinding movements. Soft food is cut into large chunks and swallowed, whereas bones are merely cracked and swallowed in pieces.

The patterns of ingestive behaviour are thus related to the anatomy and physiology of each species and the nature of its characteristic food. Because of its economic importance, this behaviour has been studied in detail in many species.

B. Eliminative Behaviour

This type of behaviour is closely related to the general ecology of wild ancestral species. It has little importance in the common birds and herbivorous mammals. In water birds, or those which live in trees, the disposal of faeces presents no problem, and the feeding habits of herbivorous animals force them to wander widely and thus to distribute their waste products.

Dogs and cats, whose ancestors lived in lairs, show elaborate patterns of elimination. Cats bury faeces and urine, whereas dogs have a tendency to deposit them at particular spots known as scent posts. Understanding these characteristic patterns of behaviour is the basis of housetraining.

C. Sexual Behaviour

Again, each species has special patterns of sexual behaviour. Among wild animals, such differences make mating between different species difficult. Among birds, the mating behaviour of turkeys, chickens, and ducks is distinctly different. Among mammals, members of the family Bovidae have similar patterns of sexual behaviour, and sheep and goats will readily mate with each other when penned together, although the matings are never fertile. Captivity has the effect of producing many such distortions of sexual behaviour. In zoos, some animals become hypersexual, and others are bred only with great difficulty. A species with a tendency toward hypersexuality would adjust well to domestication, and it is interesting that two of the examples given by Hediger (1950), the American bison and wild swine, are closely related to domesticated animals. Sexual behaviour is of great practical importance in animal breeding and will be described in detail in later chapters.

D. Care-giving (Epimeletic) Behaviour

This kind of behaviour is highly important to survival and is especially important to the young of social animals. In mammals, the care of the mother for her offspring is the most common type of epimeletic behaviour. In many birds, the male parent also assists in feeding the young, but male turkeys and chickens have little relationship to their offspring. Care-giving behaviour is therefore largely confined to females in our common domestic animals and is usually described as "maternal" (Rheingold, 1963).

All mammals allow their offspring to suckle. The minimum care of this type is provided by rabbits which visit their offspring only once per day and allow suckling for only a few minutes (Ross et al., 1963). In contrast, a mother dog or cat will spend almost 24 hours per day with her offspring during the first week or so of life. Besides nursing, the common patterns of care involve nest-building (rabbits), brooding (chickens and turkeys), and cleaning or grooming (dogs and cats).

As a part of animal husbandry, man himself exhibits a great many patterns of care-giving behaviour towards domestic animals, providing them with food, water, and shelter, cleaning and grooming them, and taking care of the sick. This has resulted in the relaxation of selection for good maternal behaviour, so that even more work is thrown on the human owner.

E. Care-soliciting (Et-epimeletic) Behaviour

Most young animals are unable to completely care for themselves. In situations which call for special patterns of behaviour in the adult, they substitute a general pattern of behaviour in the form of a call or signal for help.

1*

Young chickens have a loud, insistent chirp when distressed. Calves bawl, lambs bleat, and puppies whine or yelp. The type of distress call may be related to the situation producing it, although it is often difficult to distinguish a hungry animal from one which is lost or hurt. The loudness of the call is always a clue to the amount of distress involved.

Knowledge of such distress calls and their meaning is of considerable help in rearing young animals, particularly in cases of accident or attacks by predators. Such behaviour is not entirely confined to young animals. Adult cattle and sheep will call when hungry or separated from their kind. In this case, the distress call is often directed towards the human caretaker.

F. Agonistic Behaviour

This includes fighting, flight, and other related reactions associated with conflict. Fighting is an important practical problem in animal management, and it can be understood only in relation to its special function in a particular species. In the first place, fighting is more pronounced in the males of all domestic mammals and birds, being particularly associated with competition for mates. Farmers and ranchers have for centuries used castration to produce more docile males, particularly in horses, cattle, and swine, as these large animals can be extremely dangerous to human beings.

In the bovine mammals, fighting is a regular part of social behaviour, regulating the space between individuals and determining which males shall do the mating for the year. Each species has characteristic patterns. Bulls paw the ground and bellow, whereas sheep back off and charge headlong. Combats are usually individual, and combined attacks on a single animal are extremely rare.

The social fighting behaviour of cats and dogs differs from the normal patterns of attack on prey and is much less likely to lead to serious injury. Much of such behaviour is limited to ritualized expression of threats, and avoidance or submission. Cats keep the claws retracted in any but serious fights, and dogs use an inhibited bite in which the teeth are not firmly clamped together.

The control of undesirable fighting is necessary in domestication, both between the animals in a group and between them and human beings. This is normally done through the development of social relationships and social organization (Scott, 1958a).

G. Allelomimetic Behaviour

This involves two animals doing the same thing at the same time with some degree of mutual stimulation. Everyone who has seen sheep or chickens in flocks is familiar with this type of behaviour, but not all of us realize that it is important in all common domestic animals except rabbits and cats. Its most general function is to maintain a social group and to provide safety, for when one animal sees danger all become alerted by responding to his behaviour. In addition, this behaviour in dogs is important in hunting and group attacks on prey, since several animals working together are more likely to be successful than one alone.

Under most conditions, domestic animals are protected from predators and other dangerous situations, but allelomimetic behaviour still has important consequences. By stimulating each other, animals produce the phenomenon of "social facilitation", defined as any increment of activity resulting from the presence of another individual (Crawford, 1939; Simmel et al., 1968). When animals eat in groups they eat more than if they are fed separately. Furthermore, in groups they are likely to be less fearful and hence more contented, and so healthier and more productive. The common practice of milking cows in groups thus has a sound psychological basis.

Any animal which normally lives in a flock or herd will become either lonely and depressed or frightened and agitated if kept by itself. This forms a special problem with individual riding horses or milking cows and with such solitary pet animals as dogs which are left by themselves for any length of time. Human companionship will substitute in part, but in general two animals get along better than one.

Many commonly observed sorts of behaviour are understandable as patterns of allelomimetic behaviour. A chicken feeding by itself will run toward the flock when alarmed. Similarly, frightened sheep will at first bunch together and then move away in a direction initiated by the leader of the flock.

The allelomimetic behaviour of horses comes out strongly when they are used on a pack trip. A single horse will proceed very slowly and hesitantly into strange areas, but will follow briskly behind a group in order to avoid separation. The common practice is to ride in front with a confident and fast-stepping traveller. The rest will then keep up without urging.

H. Shelter-seeking Behaviour

All species of animals seek an environment in which they find themselves comfortable. In some cases, they obtain shelter from the bodies of other animals, and in others they find shelter within barns or near such natural objects as trees. Given the opportunity, most animals will select the most beneficial environment, but this is not always the case. The kind of shelter needed and used depends on the type of environment to which the original wild species was adapted. Turkeys are a particular problem, showing little tendency to seek shelter in rain and hail. It should be remembered that these animals originally lived in forests where shelter was readily available in any tree in which the birds could roost. Sheep may be a problem in large flocks because they sometimes mass together and smother in winter storms. This usually happens where there is no natural shelter.

I. Investigatory Behaviour

All animals have a tendency to explore their environment. Whenever an animal is introduced into a new place, its first reaction is to explore it, the type of behaviour depending on the sense organs. The large herd animals, whose original wild habitats were plains and deserts, do most of their investigation with their eyes, whereas dogs and rabbits use their noses a great deal. An important part of managing animals is to allow time for investigation, either of new quarters or of new individuals introduced into those quarters, before attempting work with them.

J. Behaviour Disorders

Maladaptive behaviour is rare in wild populations but occurs frequently in domestic animals. Beginning with Pavlov's (1927) and Liddell's (1944) studies of conditioned reflexes, there have been many attempts to produce experimental neuroses in domestic animals. When analysed, such experiments all have certain factors in common (Scott, 1958a). The animal must be highly motivated, either by training or by strong external or internal stimulation. It must be unable to escape from the source of stimulation and, finally, it must be unable to adapt to the situation. None of these three factors of high motivation, captivity, and inability to adapt will produce maladaptive behaviour by itself.

When all three factors are present, the animal will eventually show some sort of inappropriate behaviour whose nature depends on the kind of stimulation, the previous experience of the animal, and the objects and animals present. When examined, the behaviour usually turns out to be partially adaptive. In short, an animal placed in an impossible situation attempts to adapt as well as it can.

The conditions of close captivity under which many domestic animals are kept fulfil one of the prime requirements for maladaptive behaviour. Theoretically, any system of behaviour may be involved, but one of the commonest is sexual behaviour. Homosexual behaviour, masturbation, interspecific mating reactions, and other partially adaptive forms of sexual behaviour are commonplace in a barnyard. Isolation will produce many bizarre forms of behaviour in species which are strongly allelomimetic, such as the jungle fowl (ancestor of the domestic chicken, Kruijt, 1964) and dog (Fuller, 1967).

Maladaptive behaviour takes two general forms. In one case, the animal gives the appropriate response, but directs it toward another object or animal instead of reacting to the original stimulus. Thus, a cow threatened with a stick may hook her neighbour instead of the cowherd. This is called *displaced* behaviour, or, specifically, displaced aggression. In the second case, the animal responds inappropriately, as when a cock engaged in fighting will suddenly stop and begin pecking the ground as if feeding (Fraser, 1959). This is called *displacement* behaviour, or more properly, displacement *of* behaviour. Another example is the head shaking of closely caged chickens, as if tossing off invisible drops of water (Levy, 1954).

A common type of maladaptive behaviour in captive animals, whether wild or domestic, is stereotyped movement (Levy, *loc. cit.*). Horses weave in their stalls, and dogs spin or run in circles when excited. The analysis of any particular case of maladaptive behaviour often requires an inspired bit of clinical detection, although many cases are obvious. For example, a dog is reported to be chewing up rugs and furniture. Inquiry reveals the fact that the dog is left alone in an apartment for many hours each day. The animal is obviously lonesome, and as soon as the owners go out the dog attacks the door and later the furniture. As soon as they come back, their first thought is the dog, whose behaviour has the result of generating extra attention as well as working off tension. The situation can be helped either by giving the dog company, or by shutting him up where he can do no harm.

In addition to maladaptive behaviour induced by social experience and other environmental conditions, there are a large number of behavioural abnormalities caused by organic defects and disease. Lesions of the central nervous system, endocrine deficiencies, poisons, and many other conditions can produce aberrations in behaviour (Fox, 1965).

III. GROUP FORMATION

Each of the general systems of behaviour has at least some tendency to draw animals together, with the exception of agonistic behaviour, which has the effect of keeping animals at a distance or driving them apart. Shelter-seeking keeps animals together because they may find shelter in each other's bodies, or because they seek common shelter in a barn or under a tree. Ingestive and investigatory behaviour may draw animals apart if they are investigating the surrounding environment and searching for food. On the other hand, mutual investigation, a common source of food, or nursing may draw animals together in groups. In animals which form flocks and herds, allelomimetic behaviour is a strong and constant cohesive force. By following and mimicking each other they form tightly organized groups. Care-giving and care-soliciting behaviour are also powerful forces of attraction. Sexual behaviour causes strong attraction, particularly at certain times and seasons. Thus, most animal species will form social groups, even if not held together in pens.

Homogeneous Groups

Since the patterns of behaviour differ between the two sexes, and between young and adult animals, there is a natural tendency for animals to form groups of like individuals. For example, in wild mountain sheep and in many other ungulates, the males and females form separate groups except during the breeding season. Before the lambing season, females stay in a separate group, and after the lambs arrive there is a tendency for the young to form little groups of their own a short distance away from their mothers. The adult females spend most of their time grazing, but the young lambs receive much of their nourishment in a more concentrated form and hence have more spare time for play and other activities.

Groups of females are often formed artificially by the practices of animal husbandry, but they have a natural basis in many species. Groups of males are used more rarely, partly because of economy and partly because of the likelihood of severe fighting during the breeding season.

Heterogeneous Groups

Besides the three kinds of groups formed from like individuals, there are four possible combinations of unlike individuals. The female-young group is the most important and common one in domestic animals. In domestic species, the males pay little if any attention to the young, although males help care for their offspring in many species of wild birds and mammals. Even in the latter there is usually joint care by males and females, so that the male-young group is very rare in the higher animals, although it does occur in some fish, where the males guard nests and later guard the young fry. Male-female groups are naturally formed in many species during the breeding season, and similar groups are formed artificially in domestication.

The most elaborate heterogeneous group is that of males, females, and young. Among domestic mammals and birds, such groups are rarely permitted, and among their wild ancestors, wolves alone have this type of group. Male as well as female wolves help take care of the young, and attachments between the two sexes are long lasting.

In rearing domestic mammals and birds, we would expect that individuals would adjust best to each other in groups which are natural to the species concerned, and most poorly in groups not formed naturally.

IV. SOCIAL RELATIONSHIPS

Within social groups animals which are kept together quickly form habits of responding to each other. Their behaviour thus becomes regular and predictable, and such behaviour between two individuals is called a social relationship (Carpenter, 1934; Scott, 1958). Some of the more important social relationships are listed below.

A. Care-dependency Relationships

This is the usual relationship developed between mother and offspring. In sheep and goats, in which the young animal is allowed to stay with its mother for a long time, such a relationship becomes very strong and persists into adult life.

B. Dominance-subordination Relationships

When two strange adult animals meet for the first time, they are likely to respond by mild or severe fighting. As a result, one animal loses and the other wins. This behaviour is quickly reduced to habit, usually with the result that one animal, the dominant one, always attacks or threatens, while the subordinate animal submits or avoids contact. This relationship is a solution for the problem of conflict and competition, and usually results in relatively peaceful behaviour. However, some pairs simply form a habit of attacking each other whenever they meet. When young animals are reared together, there is a tendency to form habits of peaceful behaviour, and dominance may never appear. If it does appear between young animals, it is likely to develop without severe fighting. In the management of farm animals, placing like-sexed strange adults together often results in severe fighting, because dominance relationships have not yet been established.

In most groups a stable dominance organization gradually develops and reduces overt fighting. If strangers are introduced into such a group, social disorganization results in the outbreak of serious and injurious fighting. In a flock of hens, egg production and the rate of physical growth will decline (Guhl & Allee, 1944).

C. Sexual Relationships

These are formed between adult males and females and may be very weak and tenuous, as they are in the herd animals, or relatively strong, as they are in some carnivores. Fox breeders frequently have difficulty in getting males to mate with more than one individual. In female dogs, mating preferences often develop (Beach & LeBoeuf, 1967). No such difficulty is found in species where the males normally mate with many females, as in the large herd animals.

D. Leader-follower Relationships

These are important in sheep, goats, cattle, and horses. The young animals follow their mothers and later generalize to all older individuals. As a result, older animals tend to be the leaders. Leader-follower relationships also occur between young domestic birds and their mothers, and they are particularly prominent in geese and ducks.

The occurrence of social relationships depends upon the natural social behaviour patterns of the species concerned. Leader-follower relationships are very strong in sheep, where lambs follow their mothers from birth, but are quite weak in dogs, where the pups are not physically capable of following adults on extended trips until several months of age. Leader-follower relationships should be distinguished from dominance, in which one animal may drive or herd another rather than lead it.

E. Relationships between Different Species

Social relationships are normally formed between members of the same species but can, in special circumstances, be developed between two different species. In domestication this tendency has important consequences, partly because several species may be kept together in the same field, and partly because of the close association between man and domestic animals.

Such interspecific relationships can be produced artificially, as when a hen is given a setting of duck eggs. In addition, many bizarre relationships occur spontaneously in captive animals. In one case a cat with kittens was offered live domestic mice; she responded by taking them into her nest box and attempting to treat them like kittens until they escaped.

Social relationships can also be transferred to human beings (Hediger, 1950). An animal caretaker normally forms a care-dependency relationship with the animals under his charge. In order to do his job successfully, he must also form a dominance-subordination relationship in which he is the dominant individual. In a relationship with a dog, the master may form a leader-follower relationship, although this is sometimes difficult to do because of the native behaviour patterns of the species.

Many breeders of milk goats make a practice of removing a kid immediately after birth so that the first milking is done by the caretaker. The mother forms an affectionate relationship with the milker, making milking much easier, and in effect adopts the milker in place of her kid. In this case, the human takes the dependent role in a care-dependency relationship.

V. THE PROCESS OF SOCIALIZATION

Every species of social animal so far studied has a brief period, usually occurring early in life, when its first social relationships are formed. Normally these attachments are made to the parents and other members of its own species, but they may also be formed with human beings and members of other species as well. This process of forming a primary social relationship is very important in domestic animals, and indeed may be the way in which all mammals and birds were first domesticated.

A. Socialization in Birds: Imprinting

Lorenz (1935) was the first to call attention to the importance of this phenomenon, basing it on his experiences with wild geese and jackdaws, as well as those of the German ornithologist Oscar Heinroth, who had attempted to hand-raise many species of European wild birds. In general, hand-reared birds become attached to human beings and respond to them as they might normally do to their own species, even giving mating reactions. The hand-reared birds may still react to their own species, but prefer their human foster parents. The social bonds are formed so rapidly that Lorenz called the process "imprinting".

The same phenomenon occurs in domestic geese, ducks, chickens, and turkeys (Sluckin, 1965). If a newly hatched bird is exposed only to human beings during the first 2 or 3 days of life, it will become firmly attached to them in spite of subsequent opportunities to form attachments with its own species. It will also become attached to moving models, or even motionless objects to which it is exposed at this time (Gray, 1960).

B. Socialization in Domestic Mammals

Mammals show the same tendency early in life to form attachments to their caretakers and associates, whether these belong to the same or another species. In precocious animals like sheep, this takes place within the first 10 days of life. In dogs, which are born in an immature state, the formation of primary social relationships begins at a later age, and the critical period begins at approximately 3 weeks of age, reaches its peak at 5 to 7 weeks and declines slowly thereafter (Freedman et al., 1961). The capacity to form social relationships never entirely disappears, but is limited by the development of other behaviour patterns in both the young animal and its parents. Social mammals will also form attachments to animals other than man (Cairns, 1966). Young lambs, for example, become attached to dogs, and dogs to rabbits.

These findings are of great practical importance to persons who handle and breed domestic animals. Taking an animal away from its kind at an early age and hand-rearing it has the effect of transferring all its social relationships to human beings, with the result that it may become difficult to use as a breeder. Its care of young may also be disturbed, as sometimes happens in sheep raised on the bottle. On the other hand, contact with and handling of young animals at the sensitive age will make them much more amenable to later human handling. The young of most species of mammals have the capacity to form relationships simultaneously with their own species and with human handlers. If the handler postpones the attempt to form a social relationship to a later age, the usual result is a permanently timid animal. Horses, for example, which have been trained only as adults, are wilder and much less tractable than those with which human contacts have been established in early life.

Accompanying the process of socialization young animals also become attached to their physical surroundings, and may become greatly disturbed when, as older animals, they are taken away from a familiar environment. The tendency to form attachments, whether with other animals or with places, seems to be largely an internal process which takes place at the proper time in development, irrespective of the kind of external stimulation which the animal receives (Scott, 1967). The time at which socialization or imprinting normally takes place is thus a "critical period" in the life of an individual (Lorenz, 1935). Experience at that time determines the individuals which will be its close social relatives, and by extension determines its behaviour towards similar individuals in later life (Scott, 1962).

VI. LOCALITY AND BEHAVIOUR

As indicated above, almost all animals become attached to particular spots, usually those in which they are born or where they spend their youth. This is true even of animals which show little tendency to become attached to other members of the species. Such social animals as the dog become attached to a particular area during the critical period of primary socialization (Scott & Bronson, 1964).

A. Home Range

Having become attached to a particular locality, an animal continues to live nearby. The area in which it lives and wanders is called its "home range" (Burt, 1943). The range of domestic animals is usually limited artificially by fences, although not in the stock farms and ranges of the western United States and Australia. The Soay sheep, a feral domestic sheep which lives on islands off the Scottish coast, has a home range of many acres; this varies in size from season to season (Jewell, 1966). The principal practical importance of the phenomenon is that an animal forcibly removed from its home range becomes emotionally disturbed, a fact which is obvious in livestock sent to markets or transferred to a new farm.

B. Homing and Orientation

An animal removed from its home range makes every effort to return to it. Cats are particularly attached to their home localities and are capable of finding their way back over unfamiliar ground for many miles. The strong tendencies of horses to return to their home localities are well known to anyone who has ever ridden or driven a horse away from its home barn and back. The same tendency has a practical application in homing pigeons, although these birds are now primarily raised for sport rather than carrying messages. Homing races of pigeons are held over courses hundreds of miles in length.

Being able to return from long distances implies that an animal has some means of finding the proper direction for travel. How orientation is accomplished is still an interesting research problem, but most evidence shows that orientation is accomplished through the usual senses, and that there is no special "sense of direction". Animals may orient themselves by familiar landmarks, or even through the sun and stars in some species (Griffin, 1958). Mammals which live on the ground are much more limited than birds, and many of them seem to find their way simply by following their outward path, or wandering around until they reach familiar ground.

C. Migration

Many animals have two home ranges and regularly move from one to the other as the seasons pass. Several of our domestic species are closely related to migratory wild animals. Wild ducks and geese migrate hundreds of miles each year. Rocky Mountain sheep formerly migrated each year from mountain pastures to plains areas and back, before their range was restricted by fencing and hunting.

D. Territoriality

In addition to having a home range, some animals defend particular areas as *territories*. In the strict sense, a territory is an area with a fixed boundary, on one side of which the resident animal will attack a stranger, and on the other side will leave him alone. Some of the best examples are the breeding territories of song birds. Defending a territory successfully is a difficult task, and in most species the territory is much smaller than the home range. True territoriality has little importance for most domestic mammals, with the exception of dogs. Like their wolf relatives, dogs defend the small area immediately surrounding the den or home. This is usually a very small proportion of the area over which a dog will wander if permitted; and wolves sometimes have home ranges 20 miles or more across.

Territoriality has little importance in domestic birds, although a setting hen will defend the small area within reach of her beak. Cattle, sheep, and goats do not defend particular areas, although goats will attack other animals who get too close to them. In this case, the animal defends a "living space" wherever it goes, rather than any particular piece of ground. Horses are sometimes said to be territorial, but this conclusion has yet to be verified by studies on wild horses.

VII. PRACTICAL APPLICATION

The behaviour of animals can be systematically studied by describing the behaviour patterns involved in the nine principal types of adaptation or adjustment. Each species has its own peculiar basic patterns of behaviour which can be partially suppressed or exaggerated under conditions of domestication. However, selection and domestication have not produced new patterns, but have only modified the old.

Understanding the behaviour of a particular species involves not only knowledge of its basic behaviour patterns and their limitations, but also the primary stimuli or releasers which produce them. These primary reactions can be associated with other stimuli, with the result of modifying behaviour by the process of learning. Understanding the behaviour patterns peculiar to a species is an essential part of successful care and management. As indicated above, one basic principle is that the animals must behave according to their basic capacities and behavioural organization. This principle can be consciously used to increase productivity, as in using the phenomenon of social facilitation to increase feed consumption and therefore weight gain. A second basic principle is that the animals will adjust to a wide variety of new conditions if they are reared in similar conditions as young animals. The greatest changes can be produced if this is done beginning with the period of socialization.

The process of socialization (imprinting) is strongly developed in all domestic animals and probably represents the way in which primitive man originally tamed wild ancestors of our present household and farm animals. Young wild animals reared by hand become strongly attached to the persons who rear them and pay little attention to their native species. Hand-reared pets may have been the foundation stock for the first domestic flocks and herds.

In a similar way, most animals also become attached to particular areas, and most strongly to the places in which they grew up as young animals. When removed from familiar places and companions, animals therefore become emotionally disturbed. This becomes an important commercial factor in shipping or moving animals from place to place.

The behaviour of domestic animals has been altered by conscious and unconscious selection, as well as by improved feeding and changed environmental conditions. The most general effects are to produce early sexual maturity, hence an increased reproductive potential, and to decrease the amount of agonistic behaviour shown in either fighting or flight responses. Along with these desirable characteristics, selection may have produced some degenerative changes in behaviour, particularly among domestic animals used for pets. Human care has been substituted for the natural care of dog mothers for their offspring, so that the young of poor mothers can survive. In chickens there has been deliberate selection against maternal care or "broodiness", in order to increase egg production. The detailed history of domestication and its relationship to the general process of evolution will be discussed in the next chapter.

REFERENCES

ASCHOFF, J. (ed.) (1965). *Circadian Clocks: Proceedings of the Feldafing Summer School, 1964*. Amsterdam: North-Holland Publishing Co.

BEACH, F. A. & LEBOEUF, B. J. (1967). Coital behaviour in dogs. I. Preferential mating in the bitch. *Anim. Behav.*, **15**, 546–558.

BURT, W. H. (1943). Territoriality and home range concepts as applied to mammals. *J. Mammal.*, **24**, 346–352.

CAIRNS, R. B. (1966). Attachment behaviour of mammals. *Psychol. Rev.*, **73**, 409–426.

CARPENTER, C. R. (1934). A field study of the behavior and social relations of howling monkeys. *Comp. Psychol. Monogr.*, **10** (2), 1–168. Reprinted in: C. R. Carpenter (1964). *Naturalistic Behavior of Nonhuman Primates*. University Park, Penn.: Pennsylvania State University Press.

CRAWFORD, M. P. (1939). The social psychology of the vertebrates. *Psychol. Bull.*, **36**, 407–466.

CRUIKSHANK, R. M. (1954). Animal Infancy. Ch. 3 in: *Manual of Child Psychology*, 2nd edn. L. Carmichael (ed.). New York, N.Y.: Wiley.

FOX, M. W. (1965). *Canine Behavior*. Springfield, Illinois: Charles C. Thomes.

FRASER, A. F. (1959). Displacement activities in domestic animals. *Brit. vet. J.*, **115**, 1–16.

FREEDMAN, D., KING, J. A. & ELLIOT, O. (1961). Critical periods in the social development of dogs. *Science*, **133**, 1016–1017.

FULLER, J. L. (1967). Experiential deprivation and later behavior. *Science*, **158**, 1645–1652.

FULLER, J. L. & THOMPSON, W. R. (1960). *Behavior Genetics*. New York, N.Y.: Wiley.

GRAY, P. H. (1960). Evidence that retinal flicker is not a necessary condition of imprinting. *Science*, **132**, 1834.

GRIFFIN, D. R. (1958). *Listening in the Dark*. New Haven: Yale University Press.

GUHL, A. M. & ALLEE, W. C. (1944). Some measurable effects of social organization in flocks of hens. *Physiol. Zool.*, **17**, 320–347.

HEDIGER, H. (1950). *Wild Animals in Captivity*. London: Butterworths; New York, N.Y.: Academic Press.

JEWELL, P. A. (1966). The concept of home range in mammals. *Symp. zool. Soc. London*, No. **18**, 85–109.

KRUIJT, J. P. (1964). *Ontogeny of Social Behavior in Burmese Red Jungle Fowl (Gallus gallus spadiceus)*. Leiden: Brill.

LEVY, D. M. (1954). The relation of animal psychology to psychiatry. In: *Medicine and Science*, N.Y. Acad. Med. Lecture No. XVI, New York, N.Y.: International Universities Press.

LIDDELL, H. S. (1944). Conditioned reflex method and experimental neuroses. In: *Personality and Behavior Disorders*. J. McV. Hunt (ed.), New York, N.Y.: Ronald Press.

LORENZ, K. (1935). Der Kumpan in der Umwelt des Vogels. *Z. f. Ornithol.*, **83**, 137–213; 289–413. For English summary see: The companion in the bird's world. (1937). *Auk*, **54**, 245–273.

MASON, W. A. (1963). The effects of environmental restriction on the social development of rhesus monkeys. In: *Primate Social Behavior*, C. H. Southwick (ed.). New York, N.Y.: Van Nostrand.

PAVLOV, I. P. (1927). *Conditioned Reflexes*. London: Oxford University Press.

RHEINGOLD, H. L. (ed.) (1963). *Maternal Behavior in Mammals*. New York, N.Y.: Wiley.

ROSS, S., SAWIN, P. B., ZARROW, M. X. & DENENBERG, V. H. (1963). Maternal behavior in the rabbit. In: *Maternal Behavior in Mammals*. H. L. Rheingold (ed.). New York, N.Y.: Wiley.

SCOTT, J. P. (1958). *Animal Behavior*. Chicago: University of Chicago Press.
SCOTT, J. P. (1958a). *Aggression*. Chicago: University of Chicago Press.
SCOTT, J. P. (1962). Critical periods in behavioral development. *Science*, **138**, 949–958.
SCOTT, J. P. (1967). The development of social motivation. In: *Nebraska Symposium on Motivation*. D. Levine (ed.). Lincoln: University of Nebraska Press.
SCOTT, J. P. & BRONSON, F. H. (1964). Experimental exploration of the et-epimeletic or care-soliciting behavioral system. In: *Psychobiological Approaches to Social Behavior*. P. H. Leiderman & D. Shapiro (eds.). Stanford: Stanford University Press.
SCOTT, J. P. & FULLER, J. L. (1965). *Genetics and the Social Behavior of the Dog*. Chicago: University of Chicago Press.
SIMMEL, E. C., HOPPE, R. A. & MILTON, G. A. (eds.). (1968). *Social Facilitation and Imitative Behavior*. Boston: Allyn & Bacon.
SKINNER, B. F. (1938). *The Behavior of Organisms*. New York, N.Y.: Appleton-Century-Crofts.
SLUCKIN, W. (1965). *Imprinting and Early Learning*. Chicago: Aldine.
TINBERGEN, N. (1951). *The Study of Instinct*. Oxford: Oxford University Press.

Chapter 2

Domestication and the Evolution of Behaviour

E. B. Hale

Domestication of animals is a recent event in human history and is defined as that condition wherein the breeding, care, and feeding of animals are more or less controlled by man. Under modern husbandry and laboratory practices, complete control of breeding and maintenance is typical (Wood-Gush, 1961). Domestication involves some biological change (morphological, physiological, or behavioural) in the animal. Thus, domestication differs from *taming*, which is defined as the elimination of tendencies to flee from man (Hediger, 1950). *Wild* animals (species occupying natural habitats), like birds, lizards, and sea-lions of the Galapagos Islands (Eibl-Eibesfeldt, 1961) may be tame, whereas individuals of most domestic species require taming only if their association with man is limited during early life. Untamed domestic animals should be differentiated from *feral* animals, or domestic animals which have reverted to the wild state.

I. ORIGINS OF DOMESTICATION

The following discussion briefly outlines the general aspects of domestication, including the probable origins (Sauer, 1952; Zeuner, 1963; Braidwood, 1958; Reed, 1959) and generic distribution of domesticated forms. The more recent history of domestication is thoroughly documented elsewhere (Daubeny, 1857; Darwin, 1875; Lush, 1945; Wood-Gush, 1959).

We are dependent upon archaeology for an understanding of the early history of animal domestication. Archaeologists, however, have done little work in this area (Reed, *loc. cit.*). In a critical evaluation of the archaeological data, Reed points out that all too frequently the crucial bones were either discarded or identified carelessly. In addition, the presence of wild forms in the area may have suggested domestic animals when actually none were present. As Reed points out, one cannot think of domestication as happening independently of the geographical factors that determine animal distribution, or independently of the culture of the domesticators.

A. Process of Domestication

Many anthropologists believe that animals were not initially domesticated for economic reasons. Sauer (*loc. cit.*) suggests a ceremonial or religious basis for domestication and others believe that man may have only vaguely realized what was happening as a loose social bond developed between the animal and him. The economic potential was recognized later and planned domestications emerged with literate societies (Reed, 1959; Wood-Gush, 1959). Our knowledge of domestication is so inadequate that it permits little more than fanciful reconstructions. Lorenz (1955) has written one such imaginative account for the dog, although his suggestion that the dog was derived from the jackal is generally rejected.

1. Origin of Domestication

The propensity of women and children to keep pets has been suggested as the most probable spur to domestication (Sauer, 1952; Zeuner, 1963; Reed, loc. cit.). Wild forms are most easily caught and tamed as young animals, particularly if they are caught during critical periods of imprinting or socialization (Chapter 1). Sauer (loc. cit.) indicates that other cultures may have a more precise knowledge of such critical periods than our own. Darwin's observation that primitive peoples in all parts of the world easily tame and rear wild animals is indirect evidence of this. Providing milk for captive infant mammals might pose a problem. However, Sauer reports that there are still tribes in tropical America and southeast Asia whose women nurse pups, pigs, and kids. Domestication of one mammalian form would facilitate the domestication of additional ones, with lactating animals replacing the human nurse. Reed (loc. cit.) suggests that once sheep and goats became domesticated, milk was available for orphaned calves and colts, thus opening the way for domestication of these larger species.

It is also possible that a few adults from a wild species became attached to human settlements. Females loosely attached to man in this way might then provide milk for the young of other species. In any event, domestication of one mammal undoubtedly facilitated the domestication of others and an explosion of new domestic species occurred.

2. Unconscious and Directed Selection

Darwin (1875) proposed that domestic animals were modified through unconscious selection long before man selected for specific characteristics. Changes in the population occurred because man preserved the animals most useful or pleasing to him and destroyed or neglected the others without any conscious intention of altering the stock. Selection is given direction when man decides on a specific objective and directs the improvement of his animals along that line. The breeder of cocks for fighting, sheep for wool, cattle for milk, turkeys for meat, horses for speed, or dogs for herding gives direction to the subsequent evolution of the domestic type: only the most ideal individuals are selected for breeding.

Improvement of domestic types has followed an erratic history. During Roman times the husbandry and breeding of livestock and chickens had reached a high level of development (Daubeny, 1857; Wood-Gush, loc. cit.). Yet, with the decline of the Roman Empire the chicken reverted to the position of a scavenger in the barn yard and the flourishing livestock industry showed a parallel disintegration. Even today chickens may be kept as scavengers roaming about a settlement although they are more often maintained under conditions of extreme automation and intensive selective breeding. Domestic mammals, especially dogs and cats, may be similarly controlled by man, even in advanced societies. Although artificial selection has undoubtedly been practised for thousands of years following the realization that descendants resemble their ancestors, the advent of Mendelian genetics provided the first organic basis for hereditary transmission (Chapter 3).

B. Centres of Domestication

The Neolithic initiation of cultivation and domestication and the subsequent food-producing revolution was probably one of the three or four great cultural innovations (Childe, 1952). This revolution added new dimensions to man's biological and cultural existence and was a necessary prelude to urban civilization. The domestication of animals as a source of food apparently occurred somewhat later than plant cultivation. It was from this farming culture with livestock as an original element that later pastoral herdsmen were derived (Childe, 1951; Sauer, 1952). The dog is the only animal which was domesticated before the development of village-farming communities.

Domestication apparently began in three major centres (Fig. 1). The earliest centre arose in southeast Asia (Davenport, 1910; Sauer, *loc. cit.*), was associated with farming and involved household animals—chicken, duck, goose, dog, and pig. A new world centre included vegetative planters in South America and seed planters in Central America. The vegetative planters contributed llamas, alpacas, guinea pigs, and Muscovy ducks, while the seed planters contributed the turkey. The best documented and most extensive centre was the fertile crescent of southwest Asia, extending from Iran to Jordan, where such herd animals as goats, sheep, cattle, and swine were domesticated in association with seed farming. Zebu cattle and donkeys were also domesticated on the peripheries of this centre.

C. Generic Distribution of Domestic Animals

Perhaps the most remarkable aspect of early domestication is that it generally centred around a single order of mammals, the Artiodactyla (Table 1), and particularly around Bovids (sheep, goats, ordinary cattle, zebu, buffalo, yak, gayal, banteng). Indeed, there may be no bovid which cannot be domesticated (Reed, 1959). A similar but less extreme situation exists in birds with the order Galliformes contributing a large number of species including chickens, pheasants, peafowl, guinea fowl, and turkeys.

A listing of domestic animals, as in Table 1, poses several problems since the term domestic has been widely applied and frequently only means that, dead or alive, an animal is economically valuable as a source of raw materials or labour, and that its slaughter, castration, and if possible copulation, is controlled by man (Spurway, 1955). However, an attempt has been made to apply the definition adopted in the introduction. Recent domestications include the musk ox and such laboratory animals as the rat and drosophila.

Fig. 1. Proposed centres of origin of major domestic animals. (Adapted from Sauer, 1952; Reed, 1959.)

II. CHARACTERISTICS FAVOURING DOMESTICATION

The distribution of most domestic animals among relatively few orders suggests that such orders can adapt to a wide range of environmental conditions and have many potential uses. Darwin observed that "complete subjugation generally depends on an animal being social in its habits, and on receiving man as the chief of the herd or family. In order that an animal should be domesticated it must be fertile under changed conditions of life, and this is far from being always the case" (Darwin, 1875, chapter 28). An examination

Table 1. Generic Distribution of Major Domestic Species

Class Mammalia	Class Aves
Order Perissodactyla	**Order Anseriformes**
Family Equidae	Family Anatidae
Equus caballus—horse	*Anas platyrhynchos*—duck
Equus asinus—ass or donkey	*Cairina moschata*—Muscovy duck
Order Artiodactyla	*Anser anser*—goose
Family Suidae	*Branta canadensis*—Canada goose
Sus domesticus—swine	**Order Galliformes**
Family Camelidae	Family Phasianidae
Camelus bactrianus—Bactrian camel	*Gallus gallus*—chicken
Camelus dromedarius—Arabian camel	*Coturnix coturnix*—Japanese quail
Lama pacos—alpaca	*Phasianus colchicus*—ring-necked pheasant
Lama glama—llama	*Pavo cristatus*—peafowl
Family Cervidae	Family Numididae
Rangifer tarandus—reindeer	*Numida meleagris*—guinea fowl
Family Bovidae	Family Meleagrididae
Bos taurus—European cattle	*Meleagris gallopavo*—turkey
Bos indicus—zebu (humped) cattle	**Order Columbiformes**
Bos grunniens—yak	Family Columbidae
Bibos sondaicus—banteng	*Columba livia*—pigeon
Bibos frontalis—gayal	
Bos bubalus bubalis—Indian buffalo	
Ovibos moschatus—musk ox	
Ovis aries—sheep	
Capra hircus—goat	
Order Carnivora	
Family Canidae	
Canis familiaris—dog	
Family Felidae	
Felis catus—cat	
Order Rodentia	
Family Muridae	
Rattus norvegicus—rat	
Family Caviidae	
Cavia porcellus—guinea pig	
Order Lagomorpha	
Family Leporidae	
Oryctolagus cuniculus—rabbit	

of the characteristics which favour domestication, apart from usefulness to man, is essential for understanding the nature of behavioural changes or lack of change under domestication.

A. Behavioural Characteristics

Behavioural characteristics which conceivably facilitate domestication are summarized in Table 2. The progenitor of a domesticated species probably did not possess all the indicated characteristics, but it is unlikely that a species with all the unfavourable characteristics listed in the right-hand column of the table could be domesticated successfully. It is of special interest that many ungulates and gallinaceous birds have the characteristics listed as favouring domestication. As will be discussed subsequently,

Table 2. Behavioural Characteristics which favour Domestication and those which do not

Favourable Characteristics	Unfavourable Characteristics
1. *Group Structure:*	
a. Large social groups (flock, herd, pack), true leadership	a. Family groupings
b. Hierarchical group structure	b. Territorial structure
c. Males affiliated with female group	c. Males in separate groups
2. *Sexual Behaviour:*	
a. Promiscuous matings	a. Pair-bond matings
b. Males dominant over females	b. Male must establish dominance over or appease female
c. Sexual signals provided by movements or posture	c. Sexual signals provided by colour markings or morphological structures
3. *Parent-Young Interactions:*	
a. Critical period in development of species-bond (imprinting, etc.)	a. Species-bond established on basis of species characteristics
b. Female accepts other young soon after parturition or hatching	b. Young accepted on basis of species characteristics (e.g. colour patterns)
c. Precocial young	c. Altricial young
4. *Responses to Man:*	
a. Short flight distance to man	a. Extreme wariness and long flight distance
b. Little disturbed by man or sudden changes in environment	b. Easily disturbed by man or sudden changes in environment
5. *Other Behavioural Characteristics:*	
a. Omnivorous	a. Specialized dietary habits
b. Adapt to a wide range of environmental conditions	b. Require a specific habitat
c. Limited agility	c. Extreme agility

species which originally possesssed many unfavourable characteristics may tend to develop more favourable ones under the selection pressures of domestication.

1. Structure of Social Group

Species which live in large herds or flocks organized along hierarchical lines and including both sexes, at least during part of the year, are especially amenable to domestication. These forms can be readily kept in the numbers required to provide large quantities of produce (milk, eggs, meat, wool, hides, etc.). Such groups would be subjected to little social stress when confined because their social behaviour minimizes fighting (see Chapter 5). Since males of these species are typically more aggressive than females, castration of males would also reduce fighting behaviour.

Predominantly territorial forms would be more difficult to maintain in large numbers without undue fighting. It is of interest that territorial species have been domesticated primarily as family pets and man has taken advantage of territorial behaviour in developing such animals as the watch dog.

Social structure is of minor importance in species maintained for ornamental purposes since these forms are usually reared in small numbers and frequently maintained as individuals.

2. Sexual Behaviour

Promiscuous sexual behaviour in animals is an advantage in domesticating a species and also in carrying out a breeding programme based on the use of a few desirable sires. Because any female can be mated to any male, the chances of a suitable pairing are greatly increased over those possible when *pair-bonds* between male and female must first be established. In breeding foxes for fur, a male must remain with a female for no more than a few hours since a pair-bond would otherwise be established with that particular female and the male would not breed with other females thereafter (Enders, 1945).

Display functions in courtship to synchronize male and female sexual behaviour and to reduce aggression between male and female (Hinde & Tinbergen, 1958). Among ungulates and gallinaceous birds, males generally dominate females: this reduces the conflict between animals and facilitates mating.

Movements, colour patterns, vocalizations, and unique morphological structures serve as signals for social interactions like sexual and fighting behaviour. Colour markings and morphological structures commonly change under the conditions of domestication, and would result in failure to breed if specific patterns served as essential stimuli. Fortunately, movements or posture serve as major signals (Seitz, 1942; Manning, 1959; Tinbergen, 1959) and are more stable under domestication than are other characteristics. The importance of posture in eliciting sexual and fighting responses has been demonstrated in both domestic chickens and turkeys (Fisher & Hale, 1957; Hale, 1960). Domesticated mammals, and presumably their wild progenitors, generally lack colour markings or morphological characteristics similar to those serving as social signals in other species.

3. Parent-Young Interactions

Immediately after birth or hatching many *precocial* birds and mammals, able to move about almost immediately, react to a variety of objects as surrogate (substitute) mothers (Chapter 6). During this imprinting period the young normally establish a species-bond to their own species and as adults direct their sexual behaviour towards species members. Critical periods in the socialization of *altricial* mammals, with incomplete motor development at birth, may occur later, but have similar profound effects. If young birds and mammals are removed from their parents during these critical periods and cared for by humans, the bond is established to man. Thus an otherwise wild form comes readily under the control of man and acquires one of the characteristics essential for domestication.

Although imprinting brings wild forms under human control, complications may arise when man attempts to breed such individuals because imprinting modifies its conception of a sexual partner. Young pigeons of one wild species reared by adults of another species prefer to mate with birds like the foster parents (Craig, 1908). Among mammals, the sexual partner may be selected from the species which suckled the young (Spurway, 1955). Acceptance of man as the intraspecific sexual partner may disrupt courtship with true intraspecific partners. Male turkeys imprinted to man but subsequently reared with other turkeys prefer man as a sexual partner, but will respond to turkey hens if man is not present (Schein, 1960). Thus, an animal may respond sexually to a species to which it is not imprinted, if its experience is heterogeneous. Man can best take advantage of the imprintability of a species by rearing the young in groups away from their parents and under intensive human care. These animals imprint to their own species and as adults are tame, adapted to successful reproduction under confinement, and even though a few individuals may react sexually to man, all prefer their own species (Chapter 17).

If recognition of young is based on experience rather than species characteristics, young of a non-domesticated species may be reared by a domesticant to facilitate domestication of the new form. Several species of domestic birds accept the young of other birds, but later reject young differing from the brood even though they are of the same species as the mother (Ramsay, 1951). Recognition of young is established in sheep and goats soon after parturition and other young may be adopted at that time. However, continuous contact is important and females may reject their own lambs or kids if separated from them for more than a few hours immediately after parturition (Collias, 1956).

If recognition of young were based on colour or markings rather than experience, females might kill mutant young possessing altered down or coat patterns. Such behaviour has been observed among pheasants and ducks, although the selectivity could have been learned (Tinbergen, 1951).

4. Responses to Man

Wild animals do not always flee on seeing man. Only when a specific distance, which differs for each species, is overstepped does the flight response occur, at least until its specific escape distance exists once again (Hediger, 1950, 1955). By careful handling and approach, this flight distance may be reduced to zero during the process of taming. However, bonds established with man on the basis of taming are more labile than those formed by imprinting. Nevertheless, some species with otherwise few characteristics favouring domestication might be domesticated on this basis alone. The domestic cat may fall into this category.

5. Food and Habitat Characteristics

For complete domestication, a species must be able to live on the by-products of man's agriculture or on the food which he can provide. It is perhaps for this reason that most domestic animals are herbivores or scavengers with flexible dietary habits. Dogs, cats, and pigeons are typical scavengers and chickens and swine may also be kept as scavengers.

Since one of the inescapable effects of confinement is to prevent an animal from searching for its needs, those species without specialized habitat needs are most easily domesticated. Specialized habitat requirements may preclude domestication of large species with potential economic value. Man has routinely harvested seal populations, but would find it difficult to provide a controlled habitat. However, with small forms, such as aquarium fish, unique habitat requirements are not a problem.

Such unusually agile animals as antelopes might be difficult, if not impossible, to confine in the human environs in large numbers. If a species has morphological characters that can be altered so that it can be maintained in captivity, e.g. flight feathers which can be clipped, it may be possible to domesticate it.

B. Ability to Reproduce in Confinement

The failure of many captive wild animals to breed in confinement is well documented (Darwin, 1875; Hediger, 1950). Those species or individuals least disturbed under confinement are most likely to become parents of domestic forms. Strong flight and fright responses to man may be adaptive in nature, but in confinement these responses may be the source of agitated behaviour and possible injury. Apprehensive animals may not mate and disturbed mothers may desert or eat their young. Development of the reproductive organs may be inhibited. Richter (1954) has described differences in adrenal activity between domesticated and wild Norway rats which were related to differences in reproduction. Wild pintail ducks fail to breed in captivity because it inhibits normal gonadotrophin secretion in some manner unrelated to adrenal activity (Phillips & Tienhoven, 1960).

Courtship patterns may be disrupted under confinement if their performance requires a large amount of space. Dogs may adapt to a restricted space, but cats will not (Chapter 15). In general those species which can adapt their sexual patterns to a limited space with negligible endocrine dysfunction are most amenable to domestication.

III. DOMESTICATION AS AN EVOLUTIONARY PROCESS

Domestication is an evolutionary process in which emphasis has shifted from natural to artificial selection. Mathematical models which were originally developed to describe the evolution of natural populations have recently been successfully applied to populations of domestic animals (Lush, 1945; Lerner, 1950). Yet, there are important differences between evolution under natural and domestic conditions.

A. Process of Evolution

The principles of evolution, which were first discovered in terms of morphological changes, apply to behaviour as well. *Evolution* involves changes in the gene structure of a population and differential reproduction of genotypes.

1. EVOLUTIONARY SEQUENCES UNDER NATURAL AND DOMESTIC ENVIRONMENTS

Evolutionary progress is favoured by the availability of unoccupied ecological niches. The shift in human culture from food-collecting to plant cultivation created an unoccupied niche favouring rapid domestication of a variety of animal species. The events which followed appear in Table 3 where they are contrasted with evolutionary sequences under natural conditions.

The adaptive radiation of a species in areas useful to man includes several behavioural adaptations. Most conspicuous is the divergence of behavioural characteristics among dogs to include herding, bird hunting, trail hunting, guarding and draught purposes (Chapter 14). Horses have diverged into types selected for speed, and riding; chickens have been selected for fighting, as well as for meat and egg production. Breed formation based on non-functional characteristics has occurred at several stages in the sequence.

Examples of convergent as well as divergent evolution may be observed. A well-established breed may split by divergent selection as happened among Shorthorn cattle with one line selected for beef and the other for milk production (Hammond, 1941). As the two lines diverged from one another they converged with other breeds selected for the same functions. The beef Shorthorn now resembles the Angus breed more than it resembles the dairy Shorthorn, which now resembles the Friesian breed.

Behavioural display movements are another example of an evolutionary sequence. Display behaviour arose from three primary sources of movement (Hinde & Tinbergen, 1958; Tinbergen, 1959, 1960): (a) *intention movements,* the preparatory movements which often occur at the beginning of an activity (see Chapter 17); (b) *displacement activities,* which appear to be irrelevant when they occur, e.g. bill wiping and preening during courtship; and (c) *redirected activities,* behaviour redirected towards an object other than that which normally elicits it. These movements have been incorporated into complex behaviour patterns.

Table 3. Evolutionary Sequences in Natural and Domestic Environments

Under Natural Conditions	*Under Domestic Conditions*
1. Emergence of a new character or character complex making it possible to occupy previously unoccupied ecological niches.	1. A shift in human culture provides a new ecological niche for those species with favourably adapted characteristics.
2. Explosive adaptive radiation into all available ecological niches (carnivores, herbivores, etc.).	2. Explosive domestication of many forms to serve human culture (scavengers, milk producers, draught animals, etc.)
3. Selection between species.	3. Selection between species.
4. Adaptive radiation within species.	4. Divergence within domesticated species toward different specialized functions (e.g. among dogs, cattle, etc.).
5. Increasing specialization within the ecological niche.	5. Increasing improvement along a straight line for a specialized physiological or behavioural function (increased milk production, speed, etc.).
6. Speciation by minor or non-adaptive (neutral) specialization	6. Development of breeds or varieties by non-functional (morphological) diversification and imposed reproductive isolation.
7. Cessation of discernible evolutionary change—highly specialized in an unchanging environment (e.g. horseshoe crab); extinction if unable to adapt to the changing environment.	7. Some forms apparently do not respond to further selection: the condition is likely to be transient, or the stock replaced by a line which will continue to respond.

2. Factors producing Evolutionary Changes

Factors which alter the genetic makeup of populations include mutation, selection, migration, and isolation. Isolation disrupts the gene flow between populations of the same species and thereby increases the likelihood that the disconnected populations will diverge into separate species. Isolating mechanisms may be distinguished as geographical or reproductive (Dobzhansky, 1951).

Geographical isolation is provided by geographical barriers which physically prevent populations of the same species from interbreeding. Domestication is a form of geographical isolation imposed by man. Geographically separated (allopatric) populations may or may not interbreed if brought together artificially. If cross-fertilization is still possible, man can move animals across geographical barriers, circumvent the isolation occurring in nature, and add to the genetic variance of a domesticated form.

Reproductive isolation may involve a period of geographical separation during which populations diverge until they can no longer interbreed. Interbreeding is also prevented if parental forms living in the same area are prevented from meeting through *ecological isolation* (parents occupy different habitats in the same general region) or *seasonal isolation* (breeding periods fall at different times of the year). Man can overcome reproductive isolation based on these factors by providing a common habitat and by synchronizing breeding seasons, i.e. manipulating factors which regulate reproductive cycles (e.g. temperature, light).

Interbreeding may be excluded, even where parental forms meet, through *ethological isolation* (differences in behaviour patterns or releasing stimuli preclude synchronized courtship and mating) or *mechanical isolation* in which it is anatomically impossible for animals to breed. These types of reproductive isolation may be surmounted with artificial insemination. Behavioural isolation may sometimes be circumvented by providing stimulation from an intraspecific partner, but permitting physical contact with an interspecific partner only (Tinbergen, 1951). Effective use of imprinting to overcome behavioural isolation was used by Craig (1908) to cross wild species of pigeons. Young were reared with adults of another species and at maturity they preferred mates like their foster parents. Mule breeders similarly raise young jacks with mares to facilitate subsequent crossing of the two species (Harper, 1913).

Mutation and *selection* are major factors contributing to evolutionary progress. Mutation increases genetic variance and selection tends to reduce variance and give direction to evolutionary changes. Behavioural as well as morphological adaptations are shaped during evolution by a variety of environmental agents (selection). Natural selection is undirected in the sense that no goal is present, but a gradual increase occurs in the efficiency of structural, physiological, and behavioural adaptations. Natural selection continues to act under domestic conditions but the purposeful goals of artificial selection are added as additional agents shaping behaviour and phenotypes.

B. Effects induced by the Changed Environment and Man

Removing an animal from the ecological niche in which it evolved and placing it in a drastically different environment is likely to induce a new phenotypic expression of the same genotype (Chapter 3). Animals may undergo morphological changes in muscle and fat, skin, and bone proportions during growth (Spurway, 1955). Behavioural changes related to a more sedentary life, as well as nutritional changes, are contributing factors. The body conformation of modern pigs on a low plane of nutrition is similar to that of the wild boar (Hammond, 1941, 1958).

Sexual behaviour may develop earlier and continue longer under confinement since an abundance of high quality feed and protection from fluctuating climatic conditions promote more rapid growth, earlier sexual maturity, and a prolonged breeding season (Hammond, *loc. cit.*). Egg laying can be prolonged in many species of birds by simply removing eggs from the nest as soon as they are laid. Unselected jungle fowl hatched and reared under modern management conditions, lay an average of 62 eggs the first year. Although this is only a third of the number produced by Leghorns bred for increased production, it is considerably more than that expected of jungle fowl in a natural habitat (Hutt, 1949). In mammals, maternal care and lactation may be prolonged by providing new young at appropriate intervals or by more thorough milking, respectively (Leblond, 1940). In contrast, disturbances which activate the secretion of adrenaline interfere with milk let-down (Ely & Petersen, 1941).

IV. BEHAVIOURAL ADAPTATIONS UNDER DOMESTICATION

Different aspects of behaviour may be modified independently. Therefore, it is appropriate to discuss the consequences of domestication on specific aspects of behaviour. Three categories will be considered: (a) *patterns of behaviour*; (b) effective *stimuli* which elicit behaviour; and (c) the *level of response* as measured by the frequency at which appropriate stimuli elicit specific response patterns. The time when a behaviour appears during development may also shift. Changes in one type of behaviour may also cause changes to occur in other types of behaviour (Chapter 3). In passerine birds, for example, selection for territorial behaviour, distinctive song, sexual dimorphism in colour and behaviour, and suppression of male aggressiveness in courtship may all be linked so that a trend in the direction of one of them will influence all the others (Hinde & Tinbergen, 1958). Similarly under domestication, alterations of one aspect of social behaviour, for instance group structure, may alter the selective advantages of other aspects of social behaviour.

In general, behaviour patterns tend to be more *conservative* (less subject to change) than the stimuli which release them (Mayr, 1958). Least conservative and most readily modified are those aspects of behaviour related to level (frequency or intensity) of response. Genetic changes readily modify behaviour by changing thresholds (Manning, 1959; Fuller & Thompson, 1960) and numerous examples of quantitative changes in behaviour are noted in Chapters 3, 6 and 7.

A. Relative Stability of Behavioural Components

1. UNIQUE STABILITY OF PATTERNS OF BEHAVIOUR

Basic motor patterns of behaviour are clearly the most stable of the three major behavioural components. In some instances display patterns are sufficiently conservative to be used by taxonomists in systematic studies. Mayr (1958) cites a series of examples in which a species, a genus, or a group of genera was shifted from its original classification in a zoological system to a new one on the basis of specific behaviour patterns. The new arrangement was subsequently confirmed by new or re-evaluated morphological evidence.

An interesting example is that of the sand grouse (*Pterocletidae*), long considered to be a family of gallinaceous birds because the downy young greatly resembled young grouse. However, a few structural features suggested some relation with the pigeons. The limited anatomical evidence was strongly reinforced by the fact that only sand grouse and pigeons drink by sticking their bills into the water and pumping it up through the oesophagus.

Obviously, the anatomical characteristics of an animal influence its behaviour patterns. However, a variety of species-specific patterns occur within structurally related groups (examples cited by Mayr, *loc. cit.*). Sheep and goats are similar in gross morphology, but have markedly different patterns of fighting behaviour unrelated to structural differences (Scott, 1960). Turkeys and chickens exhibit marked differences in displays and vocalizations unrelated to differences in structure. Therefore we assume that, within the limits of the gross morphological structures, species-specific behaviour patterns reflect differential neuromotor organization.

Stability of behaviour patterns under domestication is not unexpected if we remember that the phenotypically uniform behaviour patterns of a wild species are relatively constant for the wide range of genotypes represented in the population (Spurway, 1955; Müntzing, 1959). Animals with divergent behaviour patterns (genotypes) would probably be rapidly removed from wild populations since aberrant displays would disrupt the synchronization of mating behaviour and reduce reproductive fitness. There is no evidence that the behaviour patterns of domestic ungulates, carnivores, and gallinaceous birds differ from those of their wild progenitors. With the exception of tail carriage, no behavioural traits have been observed in dogs that are not seen in their wild relatives (Scott, 1954). Careful observers now agree that wolves do bark, although less frequently than some breeds of dogs. Domestic dogs share distinctive patterns of mouse catching with the European fox and American coyote (Thorpe, 1956). Domestic rabbits placed in semi-natural conditions without prior digging experience, burrow and construct nests similar to wild forms (Deutsch, 1957). Banks (1956) found no basic difference in the social organization of red jungle fowl and domestic hens. Some differences in escape patterns of wild and domestic turkey poults in response to alarm calls have been observed (Leopold, 1944), but differences between wild races are even greater (Chapter 17).

The effects of domestication on patterns of behaviour cannot be thoroughly evaluated until more complete evidence on the variability of behaviour patterns in both wild and domestic forms is available.

2. STABILITY OF CERTAIN PERCEPTUAL CHARACTERISTICS

Perceptual patterns related to survival apparently remain highly stable during evolution. Tests of depth perception by means of a "visual cliff" with a glass-covered pit are most illuminating. Chicks, lambs, kids, piglets, kittens, and pups have been tested and in each case the reaction is clearly related to the role of the animal's vision in its survival (Gibson & Walk, 1960; Walk & Gibson, 1961). Survival requires that depth perception develop by the time the young begin independent locomotion, whether at one day (chick, goat) or three to four weeks (cat). Although provided with repeated experience of the tactual solidity of the glass, young goats never learned to remain on it without optical support as well (Walk & Gibson, *loc. cit.*). Results conform with the ecological niches in which the animals evolved and perceptual patterns have remained unchanged in the presumed absence of direct selection pressures under domestication.

Responses to alarm calls also appear to be stable, despite the protection provided under domestication. Collias (1960) found that the acoustical signals of birds and mammals are frequently characterized by certain common sound properties. In many species of birds continuous alarm notes signal the presence of an aerial predator and intermittent calls a ground predator. Domestic chickens, and wild and domestic turkeys respond similarly to various alarm calls (Chapter 17). Although presumably not exposed to predation for hundreds if not thousands of generations, captive finches from the Galapagos Islands likewise exhibited extreme escape responses when a variety of predatory birds came into view (Orr, 1945).

B. Behavioural Changes induced under Domestication

1. ADAPTATIONS BASED ON LEARNING OR PHYSIOLOGICAL ADJUSTMENTS

Although behaviour patterns terminating behavioural sequences (*consummatory acts* of eating, drinking, copulation, attack, etc.) tend to be fixed and little subject to modification, the more variable *appetitive behaviour* preceding the consummatory response can be modified through learning (Thorpe, 1956). Under confinement, animals rapidly develop new methods of locating feed, water, and potential mates.

Learned adjustments. New adjustments may simply involve learning to respond to new stimuli. Approach of the caretaker or slamming of an unseen door may signal feeding. Animals learn to anticipate feeding time by moving to the area where fed. The stimulus for milk let-down has shifted from the calf to man except for unruly milkers unaccustomed to human presence. Preparations for milking, including the rattling of milk pails, may elicit the release of milk prior to actual manipulation of the udder (Ely & Petersen, 1941).

Adjustments may also develop through *operant learning*, in which an animal must "operate" some aspect of the environment to obtain access to food or other animals. The *Skinner-box* in which an animal must press a bar or peck a key to obtain food is one such learning device used extensively in psychology laboratories. Most appetitive behaviour may be considered operant behaviour (Verplanck, 1957; most of the terms used in this section are defined in detail by Verplanck). Pigs lifting the cover on a self-feeder provide an excellent example of natural operant learning. The writer observed that a flock of chickens developed their own Skinner-box by pecking the siding of a building in which wheat was stored in order to jar grain loose. The flock used this home-made dispenser to such an extent that paint was removed and the wood worn away over a very extensive area. Thus, the environment may provide opportunity for adjustments through operant learning without the need for selective breeding.

Although operant learning may introduce additional movements into a behaviour pattern (Skinner, 1951), it more often involves a rearrangement of the same pattern. Dogs are commonly selected for their ability to learn specialized tasks (Scott, 1954). Again, this might be thought of as an elaboration of patterns, rather than the development of new ones. The morphology of some species under domestication is highly variable and this is probably one of the most important sources of variation in captivity (Spurway, 1955).

Physiological adjustments under domestication may also alter behaviour both in birds (Lehrman, 1959) and mammals (Beach, 1952). Changes in environmental conditions may, for example, modify physiological needs for energy and water and parallel changes in feeding behaviour follow. Increases in growth rate and egg production change the level of feed intake and the selection of specific nutrients. Both water and feed consumption increase during milk production.

Stimulus and threshold effects. Unique stimulus conditions may produce alterations under confinement. Suitable conditions for optimal nesting behaviour are typically absent in most human habitats and man may need, for example, to provide artificial nests for birds or nesting materials for pigs. Some birds find these nests inadequate and use them only if trained to do so by the caretaker (Chapter 17). Man's efforts to entice birds to use nests by placing eggs in them may attract hens to the nest, but little is gained since eggs also release incubation behaviour. Certain sounds in the domestic environment which resemble alarm calls and which Tinbergen (1951) calls *supernormal stimuli*, cause chickens and turkeys to pile in corners of pens and increase mortality.

Husbandry practices involving separation of the sexes may produce transient lowering of behaviour thresholds (Spurway, *loc. cit.*; Schein & Hale, 1959). Thus, a number of males with marginal levels of sexual response would superficially resemble those with higher sexual potential and make distinction of genotypes difficult unless increased demands were made on their sexual behaviour (Hale & Almquist, 1960). Males with exceptionally low thresholds for aggressive behaviour must be kept apart; castration of non-breeders reduces aggressiveness and makes maintenance in large groups possible.

2. CHANGES IN BEHAVIOUR PRODUCED BY SELECTIVE BREEDING

The most readily changed components of behaviour provide us with the best examples of directed evolution of behaviour under domestication. A few selected examples should serve to give a general picture of the role of selective breeding in shaping behaviour.

Patterns of behaviour and social structure. Modification of basic motor patterns under domestication is probably a rare event. Patterns of attack in fighting cocks may represent one of the few clear examples; Kentucky Dominique cocks fly high into the air before striking down towards the opponent, whereas Allen Roundhead cocks move forward on the ground and bring their legs and feet forward in a sweeping motion (Fennell, 1945). Hybrids with a combination of the two patterns may be prone to self-inflicted injuries. The presumed reduction of flight, swimming, and tree roosting in Muscovy ducks may also be the result of selection pressures under domestication.

Territorial patterns of behaviour tend to break down under confinement and selection pressures favour animals with more flexible territorial and social behaviour. By appropriately structuring the environment, green sunfish were shifted from hierarchical to territorial patterns (Greenberg, 1947). Fighting cocks in open areas show some territorial behaviour and each territory contains a single cock and a harem of females (Fennell, *loc. cit.*). The transitory nature of shifts in social structure under confinement is suggested by the rapid reversion of feral horses to groupings including a stallion and his harem. Wild mallards are monogamous in the natural habitat, but fail to develop pair-bonds with specific females in captivity and become polygamous or promiscuous.

Intraspecific social signals. The social signals of most successful domestic forms (particularly ungulates, carnivores, and gallinaceous birds) are not based on colour markings or unique morphological structures and such species were derived from animals without conspicuous external features. However, the distinctive plumage of wild mallard ducks is a specific social signal producing reproductive and flocking behaviour (Chapter 18). Lorenz (1940, 1943) studied some of the consequences of domestication among ducks. He found that in contrast to wild mallards, male domestic ducks forced copulation with females while the female was trying to escape.

Threshold changes. Changes in the frequency or intensity of behaviour are readily accomplished by selection which changes behavioural thresholds and represent the major changes obtained by selective breeding. Examples of successful selection for both lower thresholds (increased sexual and fighting behaviour) and higher threshold (decreased broody, sexual, fighting, and emotional behaviour) are presented in various chapters throughout the book and need not be considered in detail here.

Apparent elimination of a behaviour pattern by selective breeding may on analysis reveal instead an unusually high threshold to external stimuli. All females in a strain of turkeys established as "non-broody" and showing no broodiness under the management conditions in which the selective breeding was carried out, became broody and stopped laying if eggs were left on the nest continuously (Chapter 17). Although domestic rats spontaneously kill fewer mice than do wild Norway rats, certain brain lesions caused non-killers to start killing (Karli, 1957).

The sensitivity of neural structures to hormones may also change and modify levels of fighting and sexual behaviour (Hale, 1954; Riss & Young, 1954). A depression of thyroid activity may decrease neural functions and thereby produce decrements in various behaviours (Young, 1951).

Behavioural changes may occur indirectly when selective breeding shifts physiological thresholds. Selection for increased egg production in domestic birds has favoured activation of the anterior pituitary independent of day-length. Shifts of this type in the sensitivity of endocrine organs to extero-ceptive stimuli constitute a major factor in behavioural responses under domestication. Such changes depend in some instances, e.g. the annual breeding cycle of dogs, on only one or two gene pairs (Scott *et al.*, 1959).

3. Abnormal Behaviour related to Genetic Instabilities

Extreme selection pressure or hybridization may disrupt behavioural stability. Two effects typically occur: reduced precision of co-ordinated movements and the appearance of bizarre behaviour patterns.

Disrupted co-ordination of behaviour patterns. Extreme selection for economic characteristics may disrupt co-ordinated behaviour patterns unless accompanied by selection for the desired behaviour. In strains of turkeys selected for extremes in body conformation, some males typically tread the ground and exhibit copulatory movements, but rarely attempt to mount females. Natural selection tends to eliminate such aberrant patterns even under domestication, but artificial insemination combined with continued high selection pressure for extreme body type may produce a bird which cannot reproduce unaided.

Evaluation of disorganized behaviour patterns is complicated by our lack of information on the incidence of similar patterns in wild populations. That wild species do exhibit similar disorganized patterns is confirmed by observations of the Sage Grouse (*Centrocercus urophasianus*), in which strutting cocks frequently tread a pile of earth as if copulating (Simon, 1940). This behaviour is so common that prior to Simon's observation of actual matings, a prevalent myth held that the female was fertilized by eating the seminal fluid that males ejected on the ground.

Threshold changes with maladaptive consequences. Selection for a desired behavioural characteristic may inadvertently lead to undesirable changes in other behaviour. Behaviour of some commercial turkey strains indicates that attempts to select for more sexually active males have produced birds with exceptionally low imprinting thresholds to man. This situation results from selecting for breeding those males strutting most vigorously in the presence of the breeder (Chapter 18). Selection of bulls with refined characteristics has often nearly eliminated masculine secondary sexual characteristics. As a result, bulls have a feminine appearance, are very docile, and often show reduced sexual behaviour (Lagerlöf, 1957). Certain strains of turkeys, as well as individual females, may show such low thresholds to tactile stimulation that their sexual receptivity ends before the male is in position for copulation.

Behavioural phenodeviants. Certain types of behavioural abnormalities appear in inbred lines or in certain strain crosses of birds. Some turkey hybrids show such deviant behaviour that they practically denude many members of the flock, although neither parent line shows any inclination towards unusual feather picking (Chapter 17). A curious pattern of cannibalism in which female mice eat specific digits from their young while cleaning them suggests that parallel phenomena exist in mammals (Hauschka, 1952).

C. New Adaptive Peak achieved under Domestication

In nature, each animal species is more or less restricted to certain habitats within its geographical range. Such habitats represent a limited set of environmental conditions and adaptation of a species depends upon a multitude of unique biochemical, physiological, and behavioural adaptations shaped during evolution in that particular habitat.

Behavioural characteristics adapting a species to a specific habitat often include behavioural preferences leading to active selection of the appropriate habitat. Mice of the subspecies *Peromyscus maniculatus bairdii* are restricted to open-field prairie habitats, whereas *P. m. gracilis* commonly occurs in forested areas. Experiments on habitat selection have demonstrated positive preferences of *gracilis* for artificial habitats simulating forests and of *bairdii* for those simulating grasslands (Harris, 1952). That genetic characteristics play an important role is demonstrated by the fact that even laboratory-bred *Peromyscus* exhibit a preference for the habitat normally inhabited by the subspecies in nature.

As man transplants a species from its natural habitat to confinement, he is shifting a species from a natural adaptive peak to a new peak representing the domestic habitat. Man's success depends upon his ability to move the species across the intervening valley without loss of the reproductive fitness among the population. Species with many characteristics favouring domestication (Table 2) must cross a relatively shallow valley and are more readily domesticated than a species forced across a deep valley. Even species well adapted to domestication may show reduced reproductive fitness under selective breeding for specialized functions as man attempts to move the domesticated form to a somewhat different adaptive peak (e.g. specialization for increased milk, meat, or egg production). In order to maintain reproductive fitness it may be necessary to relax selection for the desired trait.

It has been shown in the present chapter that behavioural adaptations under domestication include non-genetic adjustments to the new environment and man based on learning, physiological adaptations, or stimulus changes which prolong some behaviours and terminate others. Species having many behavioural characteristics which favour domestication (e.g. many ungulates and gallinaceous birds) may exhibit few changed behaviour patterns under domestication. The intensity or frequency of most behaviours are modified by selective breeding. The totality of changes produced under domestication serve to shift a species to a new adaptive peak.

REFERENCES

BANKS, E. M. (1956) Social organization in Red Jungle Fowl Hens (*Gallus gallus* subsp.). *Ecology*, **37**, 239–248.

BEACH, F. A. (1952). "Psychosomatic" phenomena in animals. *Psychosom. Med.*, **14**, 261–276.

BRAIDWOOD, R. J. (1958). Near Eastern prehistory. *Science*, **127**, 1419–1430.

CHILDE, V. G. (1951). *Social Evolution*. New York, N.Y.: Schuman.

CHILDE, V. G. (1952). *Man Makes Himself*. 3rd edn. New York, N.Y.: Mentor.

COLLIAS, N. E. (1956). The analysis of socialization in sheep and goats. *Ecology*, **37**, 228–239.

COLLIAS, N. E. (1960). An ecological and functional classification of animal sounds. In: *Animal Sounds and Communication*. W. E. Lanyon & W. N. Tavolga (eds.). Washington, D.C.: Amer. Inst. Biol. Sci., pp. 368–391.

CRAIG, W. (1908). The voices of pigeons regarded as a means of social control. *Amer. J. Sociol.*, **14**, 86–100.

DARWIN, C. (1875). *The Variation of Animals and Plants under Domestication.* 2nd edn. 2 vols. London: John Murray.

DAUBENY, C. G. B. (1857). *Lectures on Roman Husbandry.* Oxford: Oxford Univ. Press.

DAVENPORT, E. (1910). *Domesticated Animals and Plants.* New York, N.Y.: Ginn.

DEUTSCH, J. A. (1957). Nest building behaviour of domestic rabbits under semi-natural conditions. *Brit. J. anim. Behav.*, 5, 53–54.

DOBZHANSKY, T. (1951). *Genetics and the Origin of Species*, 3rd edn. New York, N.Y.: Columbia University Press.

EIBL-EIBESFELDT, I. (1961). *Galapagos: The Noah's Ark of the Pacific.* (Trans. from the German by A. H. Brodrick). New York, N.Y.: Doubleday.

ELY, F. & PETERSEN, W. E. (1941). Factors involved in the ejection of milk. *J. Dairy Sci.*, 24, 211–223.

ENDERS, R. K. (1945). Induced changes in the breeding habits of foxes. *Sociometry*, 8, 53–55.

FENNELL, R. A. (1945). The relation between heredity, sexual activity and training to dominance-subordination in game cocks. *Amer. Nat.*, 79, 142–151.

FISHER, A. E. & HALE, E. B. (1957). Stimulus determinants of sexual and aggressive behavior in male domestic fowl. *Behaviour*, 10, 309–323.

FULLER, J. L. & THOMPSON, W. R. (1960). *Behavior Genetics.* New York, N.Y.: Wiley.

GIBSON, E. J. & WALK, R. D. (1960). The "visual cliff". *Sci. Amer.*, 202 (4), 64–71.

GREENBERG, B. (1947). Some relations between territory, social hierarchy, and leadership in the green sunfish (*Lepomis cyanellus*). *Physiol. Zool.*, 20, 267–299.

HALE, E. B. (1954). Androgen levels and breed differences in the fighting behavior of cocks. *Bull. Ecol. Soc. Amer.*, pp. 35, 71.

HALE, E. B. (1960). Role of head height in releasing sexual versus fighting behavior in turkeys. *Anat. Rec.*, 138, 354–355.

HALE, E. B. & ALMQUIST, J. O. (1960). Relation of sexual behavior to germ cell ouput in farm animals. 4th Biennial Sympos. Anim. Reprod., *J. Dairy Sci.*, 43 (Suppl.), 145–169.

HAMMOND, J. (1941). *Farm Animals.* London: Edward Arnold.

HAMMOND, J. (1958). Darwin and animal breeding. In: *A Century of Darwin.* S. A. Barnett (ed.). Cambridge: Harvard Univ. Press, pp. 85–101.

HARPER, M. W. (1913). *Management and Breeding of Horses.* London: Kegan Paul, Trench & Trubner.

HARRIS, V. T. (1952). An experimental study of habitat selection by prairie and forest races of deermouse, *Peromyscus maniculatus. Cont. Lab. Vert. Biol., Univ. Mich.*, 56, 1–53.

HAUSCHKA, T. S. (1952). Mutilation patterns and hereditary (?) cannibalism. *J. Hered.*, 43, 117–123.

HEDIGER, H. (1950). *Wild Animals in Captivity.* London: Butterworth.

HEDIGER, H. (1955). *Studies of the Psychology and Behaviour of Captive Animals in Zoos and Circuses.* London: Butterworth.

HINDE, R. A. & TINBERGEN, N. (1958). The comparative study of species-specific behavior. In: *Behavior and Evolution.* A. Roe & G. G. Simpson (eds.). New Haven: Yale Univ. Press, pp. 251–268.

HUTT, F. B. (1949). *Genetics of the Fowl.* New York, N.Y.: McGraw-Hill.

KARLI, P. (1957). The Norway rat's killing response to the white mouse: an experimental analysis. *Behaviour*, 10, 81–103.

LAGERLÖF, N. (1957). Biological aspects of infertility in male domestic animals. *Intern. J. Fertil.*, 2, 99–129.

LEBLOND, C. P. (1940). Nervous and hormonal factors in the maternal behavior of the mouse. *J. genet. Psychol.*, 57, 327–344.

LEHRMAN, D. S. (1959). Hormonal responses to external stimuli in birds. *Ibis*, 101, 478–496.

LEOPOLD, A. S. (1944). The nature of heritable wildness in turkeys. *Condor*, 46, 133–197.

LERNER, I. M. (1950). *Population Genetics and Animal Improvement.* Cambridge: Cambridge Univ. Press.

LORENZ, K. Z. (1940). Durch Domestikation verursachte Störungen arteigenen Verhaltens. *Z. angew. Psychol.*, 59, 2–82.

2*

LORENZ, K. Z. (1943). Die angeborenen Formen möglicher Erfahrung. Z. Tierpsychol.,
5, 235–409.

LORENZ, K. Z. (1955). Man Meets Dog. Boston: Houghton Mifflin.

LUSH, J. L. (1945). Animal Breeding Plans. 3rd edn. Ames: Iowa State Coll. Press.

MANNING, A. (1959). The sexual behaviour of two sibling Drosophila species. Behaviour,
15, 123–145.

MAYR, E. (1958). Behavior and systematics. In: Behavior and Evolution. A. Roe &
G. G. Simpson (eds.). New Haven: Yale Univ. Press, pp. 341–362.

MÜNTZING, A. (1959). Darwin's views on variation under domestication. Amer. Sci., 47,
314–325.

ORR, R. T. (1945). A study of captive Galapagos Finches of the genus Geospiza. Condor,
47, 177–201.

PHILLIPS, R. E. & TIENHOVEN, A. VAN (1960). Endocrine factors involved in the failure of
pintail ducks Anas acuta to reproduce in captivity. J. Endocr., 21, 253–261.

RAMSAY, A. O. (1951). Familial recognition in domestic birds. Auk, 68, 1–16.

REED, C. A. (1959). Animal domestication in the prehistoric Near East. Science, 130,
1629–1639.

RICHTER, C. P. (1954). The effects of domestication and selection on the behaviour of the
Norway rat. J. nat. Cancer Inst., 15, 727–738.

RISS, W. & YOUNG, W. C. (1954). The failure of large quantities of testosterone propionate
to activate low drive male guinea pigs. Endocrinology, 54, 232–235.

SAUER, C. O. (1952). Agricultural Origins and Dispersals. New York, N.Y.: Amer. Geo-
graph. Soc.

SCHEIN, M. W. (1960). Modification of sexual stimuli by imprinting in turkeys. Anat.
Rec., 137, 392.

SCHEIN, M. W. & HALE, E. B. (1959). The effect of early social experience on male sexual
behaviour of androgen injected turkeys. Anim. Behav., 7, 189–200.

SCOTT, J. P. (1954). The effects of selection and domestication upon the behavior of the
dog. J. nat. Cancer Inst., 15, 739–758.

SCOTT, J. P. (1960). Comparative social psychology. In: Principles of Comparative Psychology. R. H. Waters, D. A. Rethlingshafer & W. E. Caldwell (eds.). New York, N.Y.:
McGraw-Hill, pp. 250–288.

SCOTT, J. P., FULLER, J. L. & KING, J. A. (1959). The inheritance of annual breeding
cycles in hybrid basenji-cocker spaniel dogs. J. Hered., 50, 255–261.

SEITZ, A. (1942). Die Paarbildung bei einigen Cichliden. 2. Die Paarbildung bei Hemi-
chromis bimaculatus. Z. Tierpsychol., 5, 74–101.

SIMON, J. R. (1940). Mating performance of the Sage Grouse. Auk, 57, 467–471.

SKINNER, B. F. (1951). How to teach animals. Sci. Amer., 185 (6), 26–29.

SPURWAY, H. (1955). The causes of domestication: an attempt to integrate some ideas of
Konrad Lorenz with evolution theory. J. Genet., 53, 325–362.

THORPE, W. H. (1956). Learning and Instinct in Animals. Cambridge: Harvard Univ. Press.

TINBERGEN, N. (1951). The Study of Instinct. Oxford: Clarendon Press.

TINBERGEN, N. (1959). Comparative studies of the behaviour of gulls (Laridae): a progress
report. Behaviour, 15, 1–70.

TINBERGEN, N. (1960). The evolution of behavior in gulls. Sci. Amer., 203 (6), 118–130.

VERPLANCK, W. S. (1957). A glossary of some terms used in the objective science of behavior. Psychol. Rev., 64 (Suppl.), 1–42.

WALK, R. D. & GIBSON, E. J. (1961). A comparative and analytical study of visual depth
perception. Psychol. Monogr. No. 519, 44 pp.

WOOD-GUSH, D. G. M. (1959). A history of the domestic chicken from antiquity to the
19th century. Poult. Sci., 38, 321–326.

WOOD-GUSH, D. G. M. (1961). Domestication. In: The Dictionary of Birds. L. Thompson
(ed.). Brit. Ornith. Union.

YOUNG, W. C. (1951). Internal secretions and behavior. In: Comparative Psychology.
C. P. Stone (ed.). 3rd edn. New York, N.Y.: Prentice-Hall, pp. 110–136.

ZEUNER, F. E. (1954). Domestication of animals. In: A History of Technology. C. Singer,
E. J. Holmyard & A. R. Hall (eds.). Oxford: Oxford Univ. Press. vol. 1, pp. 327–352.

Part Two

Fundamentals of
Behaviour

Chapter 3

The Genetics of Behaviour

J. L. Fuller

Behavioural differences between animals depend basically upon different genetic endowments. Since, however, it is not generally possible to produce fertile interspecies crosses, genetics has been largely concerned with variation within rather than between species. New breeds have been developed for specialized purposes and old breeds have been steadily improved by means of artificial selection. Behavioural characters, like physical ones, are heritable and we shall show that behaviour responds to selection as readily as does structure. Actually, structure, physiological functions, and behaviour are not independent, but are different ways of looking at a living organism. Genetic changes affecting structure inevitably have physiological and behavioural consequences and the opposite relationships are also true. Some of these effects may be so small as to be unimportant practically, but they must always be considered as potentially significant. This chapter will deal with the nature of hereditary variation in behaviour, the transmission of behavioural traits, and problems encountered in selection based upon behaviour.

I. INDIVIDUAL AND STRAIN DIFFERENCES IN BEHAVIOUR

Different species of the same breed show differences in behaviour. One need only contrast the polo pony and the Percheron, the Airedale terrier and the Bulldog, to realize the truth of this statement. It is not the purpose of this section, however, to detail examples of strain differences among domestic animals. Some material of this sort will be included in the chapters dealing with the individual species.

A. Heritability of Behaviour

The term, genetics of behaviour, implies an extension of the meaning of the word phenotype to include transient response patterns in addition to its ordinary signification of form, colour, and chemical composition. If the view is accepted that a response is determined as much by the structure of the reacting organism as by the eliciting stimuli, the justification for the broadened meaning of phenotype is apparent. Some genes with major metabolic action have profound effects upon structure of the nervous system and thus upon behaviour. Such genes, in heterozygous or homozygous state, can be recognized by their anatomical correlates, usually regarded as abnormal variations, and the behavioural phenotype is clearly related to the observable structure. But much behavioural variation is not so obviously determined by heredity. Since behavioural characters are attributes of whole organisms, one would expect them to be inherited in a complex way. Although every genetic variation might, in certain circumstances, contribute to behavioural variation, it is likely that often the contribution of a single gene substitution to behavioural variation is so minute that it can be neglected for practical purposes. The cumulative effects of a number of such small substitutions can be, however, quantitatively important, and such effects are the basis of selection.

Briefly, heritability is defined as that proportion of total variance attributable to additive genetic factors. This may be expressed by the following formula:

$$h = \frac{\sigma_g^2}{\sigma_g^2 + \sigma_e^2}$$

where h = heritability, σ_g^2 = genetic variance, and σ_e^2 = environmental variance.

Total variance can be obtained from a set of observed measurements upon the population of interest. Members of kinships within the population share more genes in common than do members of the population at large. Computation of the variances, both within and between kinships, leads to estimates of σ_g^2 and σ_e^2, which can be used to determine heritability. The genetic variance can be computed by calculation based upon the degrees of relationship of the individuals in the population. For example, the coefficient of relationship between a parent and his offspring is 0·50, since the parent provides one-half the genes of his offspring. Similarly, the average genetic correlation between full siblings is also 0·50. This means that the average probability that any two sibs carry the same member of a gene pair, heterozygous in a parent, is one-half. If all variation in a trait were due to heredity, one could predict that the phenotypic correlation between siblings on the trait would also be 0·50. If the heritability were only one-half, the phenotypic correlation between siblings would be 0·25. Most of the techniques for studying inheritance of quantitative characters are based upon intraclass correlations within various categories of related individuals.

It should be emphasized that obtaining a correlation of 0·50 between siblings does not necessarily imply a heritability of 1·0. Individuals of a family share factors in addition to a common pool of genes; members of a mammalian litter have a common intra-uterine environment and share similar climatic, nutritional, and experiential factors as they develop. The combined action of such influences might well lead to a between-sibling correlation of close to 0·50 in a trait which had zero heritability. Thus the actual measurement of heritability of a behavioural character involves considerable experimental sophistication in genetics and in psychology.

A relatively simple experimental method of determining heritability is to compare the actual gain per generation of selection with the gain reached for. Suppose, for example, that the mean of a population on some behavioural measure is 50, with a range extending from 30 to 70. We select as breeders individuals whose mean score is 65; thus we are reaching for a gain of 15 points. The progeny, however, have a mean score of only 55 for an actual gain of 5 points. Heritability in this example would be five-fifteenths or 0·33. The basis of this method is set out in Fig. 2. This simple method of computing heritability is quite useful, as it tells the experimenter quickly whether his efforts have paid off. It must be recognized, however, that a single determination is subject to wide error. The gain of 5 points in our example might have been due to better conditions of rearing which had nothing to do with heredity, or it may have taken place in the face of some minor catastrophe

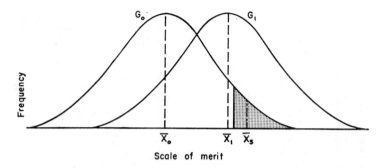

Scale of merit

Fig. 2. Curve G_0 is the distribution of a base population according to some scale of merit. The mean score of this population is $\bar{x}_0$, and from it a group of superior animals (shaded portion) with a mean score of $\bar{x}_s$ is selected for propagation. The first selected generation is represented in curve G_1 with a mean of $\bar{x}_1$. Heritability is estimated by the ratio $(\bar{x}_1 - \bar{x}_0)/(\bar{x}_s - \bar{x}_0)$, which is the proportion of superiority of the selected animals retained in their offspring.

which reduced the gain potentially derivable from selection. Thus, controls for environmental effects on non-selected populations are essential and heritability determinations should be carried out over several generations.

As an example from breeding practice, we shall describe the programme of selection for aggressiveness in White Leghorn chickens (Guhl *et al.*, 1960). The course of selection in males and females is shown in Fig. 3. The lines represent the mean percentage of paired encounters won in pairings between birds selected for high and low aggressiveness. It will be noted that the selected animals (shown by solid lines) are more extreme than the progeny (shown by dotted lines). Direct estimates of heritability were made by finding a ratio between the amount of effect produced by selection and that selected for.

B. Pathways of Gene Action

No reason exists to believe that special sets of genes are assigned to control behaviour. The present-day concept of the gene is that it is a complex nucleoprotein, which functions as a sort of template for the production of proteins of the cell. Genes replicate themselves in the process of mitosis and thus ensure the continuity of chemical organization. In very simple organisms most gene action can be described in terms of its effects upon specific chemical transformations or the synthesis of particular proteins. This is also true to some extent of higher organisms where genes have been found which produce abnormal types of haemoglobin, variations in plasma proteins, or other "metabolic errors". These aspects of gene action have been treated by Wagner & Mitchell (1964) and, in special relation to man, by Harris (1959). Some of these errors produce gross aberrations of behaviour, while others seem to be compatible with normal intellectual functions.

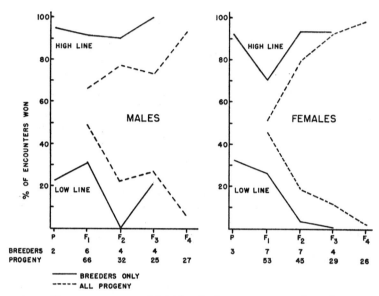

Fig. 3. Mean percentages of initial paired encounters won in each generation and by the individuals selected for breeding. (From Guhl *et al.*, 1960.)

Many effects of genes, however, have not yet been traced to unit chemical processes. Genes may interfere with histogenesis at a particular stage of development with consequent disruption of structural patterns. The deficiency in cartilage formation, achondroplasia, results in certain types of dwarf beef cattle, or in breeds such as the Dachshund dog and the Ancon sheep. Such variants are restricted in motor activity as a result of their heredity.

Genes are also known to affect hormone biosynthesis and tissue sensitivity to hormones. In most cases, such effects have not yet been related to a single chemical reaction, though we may hope that such discoveries will eventually come from the biochemical laboratories. Endocrine variations between strains which probably have effects upon behaviour have been found, for example, in the thyroid (Chai *et al.*, 1957), and adrenal gland (Wragg & Speirs, 1952) of the mouse, and in pituitary hormones of the fowl (Munro *et al.*, 1943).

There remains a great deal of behavioural variation which has been shown to be heritable, but which has not been correlated with organic characters. Possibly this is because psychologists and biologists have not looked in the right places. The fine structure of the nervous system and the distribution of enzyme activity between various neural centres has not been worked out thoroughly. We expect that more and more examples of inherited behaviour differences will be shown to depend upon chemical or physiological factors. Nevertheless, it is likely that for a long time behavioural tests will be the most efficient means of detecting certain kinds of genetic variation. In fact, the sensitivity of such tests for the detection of certain phenotypes will probably always exceed the capabilities of biochemical measurements. Genes could produce a major effect upon behaviour by acting upon a few cells whose volume in the brain is too small for accurate analysis and whose distribution is variable and indeterminate.

II. EVIDENCE FOR GENETIC VARIABILITY

Evidence for genetic variability in behaviour has, in general, been obtained from three sources. Perhaps the most extensive documentation comes from comparisons between strains or breeds within a species. Such differences, provided environmental causes can be ruled out, are presumptive evidence of heredity at work, but by themselves provide no insight either into modes of transmission or pathways between genes and behaviour. A second source of evidence is the success of selection procedures based upon behavioural criteria. Again such procedures afford presumptive evidence of hereditary factors and can lead to quantitative estimates of heritability, though few of the selection studies have utilized the presently available biometric techniques. Exceptions are studies of selection in the domestic fowl for sexual activity (Wood-Gush & Osborne, 1956) and for aggressiveness (Guhl et al., 1960). Finally, one may analyse mechanisms of genetic transmission by cross-breeding, followed by behavioural measurements of the offspring. Few studies of this nature have employed the larger domestic animals. Since large numbers of individuals are generally required, it is more convenient to utilize the small laboratory rodents, though one major study of the transmission of behavioural characters was conducted on the domestic dog (Scott & Fuller, 1951, 1965; Fuller & Scott, 1954).

In this section, we shall draw upon evidence from all three sources as applied to various types of behaviour of particular importance to animal husbandry. The overbalance of studies from the laboratory rodents simply means that, in general, these species are more economical to maintain than farm animals. Since the mechanisms of heredity and of behavioural development are very similar among mammals, one is justified in some degree of generalization to farm animals from experiments on rats and mice. Nevertheless, there are gaps in our knowledge which must be filled by studies of the larger domesticated species.

A. Variability in Sexual Behaviour

Different males of the same strain show different intensities of sexual behaviour. It might be hypothesized that such variation reflected differences in the supply of androgenic hormones, since it is well known that castration eliminates, or at least greatly decreases, sexual activity. Furthermore, when exogenous androgen is provided, as by implanting a hormone pellet subcutaneously, sexual activity is restored.

When high sex drive and low sex drive guinea pig males were castrated and supplied with adequate androgen, the restoration of function in each subject was only to the level found previous to castration (Grunt & Young, 1953). These results may be interpreted as indicating that an inherited capability to respond to androgen, rather than supply of androgen, was the limiting factor in determining sex drive.

In a similar way, significant strain differences which persist in spayed individuals given hormone replacement therapy have been found in the sexual behaviour of female guinea pigs (Goy & Young, 1957). Thus it appears that the sensitivity of the organism to hormones rather than its level of hormone production is the main channel through which heredity exerts its effects upon sexual behaviour.

In the domestic fowl also, sexual activity is a heritable trait (Wood-Gush & Osborne, 1956; Wood-Gush, 1958). In this species a negative correlation was found between success in mating and height of comb. Since comb height is a measure of androgen production, we may conclude that part of the hereditary control in this instance is exerted through rate of hormone production. It should be noted that measures of mating frequency were not good indices of fertility (Wood-Gush, 1960), and the so-called low sex drive cockerels actually had better semen production.

B. Variation in Parental Behaviour

Some domestic animals, for example Race X rabbits (Sawin & Curran, 1952), contain numerous females who fail to provide adequate maternal care to their offspring. Such deficiencies may in part be a result of the unnatural living conditions which do not provide the stimuli necessary for the elicitation of full-fledged parental behaviour. It is also possible that during the process of domestication, selection for characters of economic importance had a deleterious effect upon the biological requirements for parental behaviour.

In at least one important species, the domestic fowl, selection has been deliberately applied against broodiness, which is a form of parental or caretaking behaviour. Here the objective has been to keep birds actively laying eggs without taking time out for the care of a brood of chicks. Thus, much of our knowledge of the inheritance of parental behaviour comes from studies on broodiness in the domestic fowl. Results of a number of such investigations are summarized in Table 4 which compares the results of reciprocal crosses between broody and non-broody breeds of fowl. It will be noted that the proportion of broody offspring is, in general, greater when the sire rather than the dam is of the broody strain. This can be interpreted as evidence that the sex chromosome carries factors favourable for broodiness. It will be recalled that in fowl the males are homogametic ZZ and the females are heterogametic WZ. Thus, hens receive one Z chromosome from their sire and one W chromosome from their dam. If the Z chromosome carries important factors affecting broodiness, daughters will resemble the strain of their sire. Sex-linked transmission of broodiness in fowl has not been found in all investigations, and it is probable that the genetic mechanisms differ from breed to breed.

Table 4. Results of Reciprocal Crosses between Broody and Non-broody Breeds of Fowl. (From Fuller & Thompson, 1960.)

Dam	Sire	Per cent. Broody Offspring	Reference
Br. Leghorn	*Langshan*	29	}Punnett & Bailey, 1920
Langshan	Br. Leghorn	50	
Wh. Leghorn	*Cornish*	88	}Roberts & Card, 1934
Cornish	Wh. Leghorn	37	
R.I.R.	*Plymouth Rock*	40, 39[1]	}Knox & Olsen, 1938
Plymouth Rock	R.I.R.	12, 46[1]	
Wh. Leghorn	*Plymouth Rock*	42	}Knox & Olsen, 1938
Plymouth Rock	Wh. Leghorn	12	
Leghorn	*Greenleg*	78	}Kaufman, 1948
Greenleg	Leghorn	0	

Broody breeds identified by italics
[1] Results on replication of experiment.

C. Variation in Agonistic Behaviour

Breed differences in agonistic behaviour are fairly common. For example, bulls of the dairy breeds are considered less docile than those of the beef breeds; Wire-haired fox terriers score lower on submission than Cocker spaniels (Scott & Charles, 1954); Race *X* rabbit females defend their young more readily than Race *III* (Sawin & Curran, 1952; Sawin & Crary, 1953). In general, domestic breeds are more tractable than are related wild species and we may be sure that selection for docility has always been an objective of the practical animal breeder. It would be a mistake, however, to conclude that wild species actually fight more than domestic species. Naturalists (e.g. Murie, 1944) who have observed the wolf in its native habitat report little fighting between members of a pack. In contrast, certain dog breeds cannot be kept in groups of more than two or three, because of the danger of internal strife which may lead to death of the weaker individuals (Fuller, 1953).

Aggressiveness may be measured in terms of the incidence of fighting, as a proportion of encounters won, or indirectly by some measure of social dominance based upon priority of access to food or other rewards. These measures are not perfectly correlated and the genetics of agonistic behaviour will appear somewhat different depending upon the index chosen. In spite of these reservations, however, it is obvious that under specified conditions heredity plays an important part in determining the level of aggressive behaviour. We have already cited the successful selection of fowl for success in fighting (Guhl *et al.*, 1960). Rats selected for non-emotionality were found to initiate many more attacks than a strain selected for emotionality (Hall & Klein, 1942). Strain differences in aggression among mice have also been reported (Scott, 1942), and Lagerspetz (1964) successfully selected for fighting ability in the same species. Since aggressiveness is related to emotional responses in which endocrine glands participate, some attention has been given to the role hormones might play in the determination of strain differences in aggression. Richter (1954) has hypothesized that the process of domestication involves a reduction in the dependence of the animal upon adrenal cortical hormones and an increase in the behavioural effects of gonadal hormones. A considerable body of observation supports these views, which are based largely upon comparisons of domestic and wild rats. It would be desirable to obtain confirmation in other species.

In herd or pack animals unrestrained aggression would disrupt group organization. The phenomenon of social dominance which is widely found in the animal kingdom (Collias, 1944) serves to reduce overt fighting by setting up a system of priorities based upon social status. In general, priorities for food, mates, and the like follow successful battles with competitors. Studies of heterogeneous groups of dogs (James, 1951; Pawlowski & Scott, 1956) and of birds (Potter, 1949) have shown that regularly animals of certain breeds are socially dominant over other breeds. It is probable that dominance in genetically heterogeneous populations is in part determined by heredity.

D. Variation in Timidity

In one sense, aggressive behaviour and fearful behaviour stand at opposite poles. One involves attack and fighting, the other avoidance and fleeing. However, this is probably a false dichotomy, since a fearful animal may quickly turn from timidity to vicious attack. Either fearful behaviour or aggressive behaviour may at different times result from similar kinds of stimulation, such as sudden movements or forceful restraint. It follows that studies of the heritability of fearful and emotional behaviour must be interpreted with respect to the special circumstances of testing. Emotionality has frequently been measured in rats by the amount of defaecation when the animal is placed in an unfamiliar situation. In two separate studies it has been shown that selection for emotional defaecation is quite effective (Hall, 1938; Broadhurst, 1958). Both authors have shown that differences in emotionality are correlated with other behavioural traits such as fighting and learning, and larger adrenals and thyroids have been reported for the nonemotional rats (Yeakel & Rhoades, 1941). Urination and defaecation sometimes accompany emotional disturbances in other species, but, except in mice, have not been utilized for quantitative measurement of strain differences. Some doubt exists (Thompson & Bindra, 1952) as to the generality of the trait of emotional elimination, even in rats.

Among home-reared dogs, differences in fearful behaviour between breeds have been reported to override variations in environment and training (Mahut, 1958). In a large research colony particularly timid dogs were found to be descended from one unusually fearful female (Thorne, 1940). Crouching behaviour, which is the forerunner of fearful and subordination responses, varies characteristically between puppies of different breeds (Scott & Charles, 1954). Although evidence for genetic control of emotional responses is strong, the physiological channels through which this control is exerted have not been well defined. The endocrine system has been implicated, but we do not have experiments designed to show whether behaviour differences are primarily dependent upon the secretion rate of hormones or upon responses of target organs.

E. Variation in Learning Ability

A great deal of research on learning has been carried out on animals with the objective of finding general laws. Relatively few studies have dealt with individual differences in learning ability and their relation to genetic factors, although this question is of great importance for human beings. Enough has been done to prove that selection based upon good or poor performance in learning a maze is effective in producing "maze-bright" and "maze-dull" sublines (Tryon, 1940; Thompson, 1954). For the Tryon strains, at least, there is evidence that the maze-bright animals were not superior in learning in all situations, but had a combination of traits valuable in adapting to a complex automatic maze (Searle, 1949). Some of these traits were of an emotional nature; for example, a lesser tendency to be distracted by the noises of the automatic equipment. Other traits, more cognitive in nature, appeared to be related to space perception and the retention of goal orientation from different parts of the maze. For the analysis of the basic nature of hereditary differences in learning ability, complex tasks such as maze-running are unsuitable, since success depends upon so many factors. Perhaps we can fairly state that there is, as yet, no good evidence in animals for a general factor of intelligence which operates in all learning situations. In so far as some animals are superior to others in trainability on a number of tasks, an explanation can ordinarily be given in terms of adaptation to the testing situation and the animal handler. This statement should not be taken as implying that all members of a species are uniform in their ability to learn specific tasks. Great differences can be shown among dogs, for example, in the number of trials required for learning a simple visual discrimination, or in ability on the delayed response test (Scott & Fuller, 1965). Delayed response requires that the subject respond to a cue which is not present at the moment it must select its response. In a sense, the animal must remember its instructions and this is supposed to require some degree of concept formation.

III. GENETIC TRANSMISSION OF BEHAVIOURAL CHARACTERS

In the preceding section we have summarized the evidence for inheritance of differences in behaviour. It may be concluded that practically any quantitative behavioural character could be modified by changing the genotype and similarly that any large and some small (i.e. single locus) changes in the genotype probably will have some effect upon behaviour.

Geneticists, however, like to go beyond statements of strain differences to consider mechanisms of hereditary transmission. In some cases, waltzing in mice for example, a behavioural character can be shown to be dependent upon a single locus, and inheritance patterns are those of a simple recessive gene. The effects upon behaviour of substitutions at single loci range from near zero to major disturbances of sensory and motor functions. Much is still to be learned about pleiotropic behavioural effects of single genes affecting structural or biochemical characters, but most important variations between breeds of domestic animals are undoubtedly polygenic in origin.

To go beyond simple statements of breed differences requires breeding experiments. In laboratory experiments, summarized in Table 5, several in-

Table 5. Results of Crossing Strains or Breeds of Unlike Behaviour. (Adapted from Fuller, 1960.)

Behaviour Measure	Species	Cross	Results	Reference
Rate of running in activity wheel	Rat	Active × Inactive selected strains	F_1 active but less so than active parent. Genes for activity generally dominant in other crosses	Brody, 1942
Running speed	Mouse	Wild (fast) × Tame (slow)	F_1 slightly less fast than wild parent. F_2 and backcross data agrees with polygenic inheritance. Genes for speed generally dominant	Dawson, 1932
Numbers of errors in maze	Rat	Bright × Dull selected strains	F_1 intermediate to parents. F_2 similar to F_1 and about as variable	Tryon, 1940
Number of pellets hoarded	Rat	Brown-hooded (much hoarding) × Irish (little hoarding)	F_1 hoards actively, similar to B-h parent. Backcross to B-h is intermediate; possibly shows a bimodal distribution	Stamm, 1956
Errors in spatial orientation	Dog	Basenji (many errors) × Spaniel (few errors)	F_1, F_2, and backcrosses make equal or fewer errors than spaniels	
Docility rating	Dog	Basenji (low docility) × Spaniel (high docility)	Reciprocal F_1's differ. Young resemble mother. Other hybrids intermediate, but tend to be like mother	

vestigators have crossed divergent strains and studied the behaviour of the offspring. Some of the parental strains were obtained by selection in opposite directions (for example, high and low activity); others were inbred strains not intentionally selected for behaviour. In other instances, crosses were made between wild and domestic strains. It must be admitted that these experiments have yielded little precise information on genetic transmission mechanisms. Estimates of the number of loci producing the differences between parental strains are in the order of 3 to 6, but the formulae used for computation involve several unproved assumptions and no great significance can be attached to these values. In the author's view, measurements of heritability would be more useful, but few computations of this type have been made.

A few trends may be observed in the behavioural data. The F_1 hybrid of a cross between unlike breeds is usually intermediate to the parental stocks and often resembles the more active and aggressive parent. Backcrosses are generally intermediate between the F_1 and the respective parent, though sometimes the backcross to the more active parent is indistinguishable from the parental stock. As might be expected, F_2 hybrids are intermediate to the parents, and in some experiments the F_2 generation has shown an increase in variance over the F_1, predictable on the basis of segregation and independent assortment of Mendelian units. On the whole, however, behavioural variances do not correlate well with theoretical expectancies, even though averages fall in an orderly fashion. We should emphasize here that the variances (and to some extent, the means) can be modified by changing the scale of measurement. Frequently scale transformation is used to bring a set of data into conformity with a genetic hypothesis (Bruell, 1959). Such transformations are arbitrary, but they have proved useful in other areas of quantitative genetics.

It is too early to evaluate the contribution of quantitative genetic models to behaviour genetics. The difficulties which have been encountered in the application of classical biometrical genetics to behavioural data may be due to the long, complex pathways between genes and the phenomena measured. In classical systems of analysis it is ordinarily assumed that genes or groups of genes affect traits in the same direction in most environments and in association with most other genes. This is probably a fair approximation of the truth when one is dealing with traits close to primary gene action, and the assumption seems to give reasonably good results when applied to inheritance of body size or egg production.

It is quite conceivable, however, that behavioural traits are not related to their physiological substratum in a linear fashion. Learning, for example, may be most efficient in an animal of intermediate neural excitability. Genetic mechanisms operate on learning, let us say through biochemical reactions affecting excitability, while selection operates on problem solving ability. Let us imagine a group of genes, X_1, X_2, X_i, X_n, each of which increase neural excitability, with alleles x_1, x_2, x_i, and x_n, which decrease excitability. The effect on intelligence of X_i will depend upon the number of other X genes present. Where there are few, the possession of X_i is favourable to intelligent behaviour; where there are many, X_i is deleterious. In this model the scale of measurement of intelligence is not linear with respect to X genes, nor to their biochemical effects. Although no definite proof of this theory exists, it is based on reasonable assumptions and explains many experimental findings.

The points discussed above indicate a non-congruent relationship between genetics and behaviour (Fuller & Thompson, 1960). Non-congruence simply means that particular genes are not uniquely concerned with traits as defined by the animal behaviourist. The ramifications of a gene substitution may extend narrowly or widely over the behavioural repertoire of an animal depending upon the amount of modification of its physiological status. In this view, it is not really profitable to look for specific hereditary mechanisms for specific traits. Instead, we concern ourselves with heritability of characters in specified populations and environments. On the other hand gene substitutions may be used as treatments and their effects noted on a variety of behavioural characters. Merrell (1965) has argued that this method which starts with a known genotypic difference and moves outward to the behavioural phenotypes is superior to the more common method of starting with a phenotypic difference and seeking a genotypic explanation. Studies on the effect of the gene for albinism on the behaviour of mice illustrate this method. A discussion of the limitations of the approach and references to other studies is found in Fuller (1967).

The idea of non-congruence does not imply that there is no orderliness to the inheritance of behaviour. The fact that selection for behavioural criteria is so commonly successful is evidence that a certain number of genes do act in an additive fashion upon behaviour. The non-congruent view, however, does emphasize that a good part of individuality depends upon unique gene combinations which cannot readily be fixed by selection. Fixation of homozygous gene combinations is possible by inbreeding and combinations of selection and inbreeding could lead to the fixation of desired characters which are not inherited in an additive manner. The process will be long and expensive, however, since there is no way to control the particular genes which become homozygous.

IV. SELECTION FOR BEHAVIOURAL CHARACTERS

For centuries animal breeders have selected for desirable characteristics, but not until the twentieth century could the science of genetics provide scientific bases for guidance. From time immemorial behavioural characteristics must have been important criteria for selection. The greater docility of domestic species, their early maturity, and their adaptability to conditions of restraint can all be ascribed to selection in its pre-scientific days.

In individual selection the choice of sires and dams is based upon the characteristics of each prospective breeding animal. For example, in breeding Pointers, the animal with the highest record in field trial competitions would be chosen. Such a procedure implies that field trial success is a heritable character and that the selected individual's success is attributable to the particular group of genes which it possesses. Furthermore, the assumption is made that these genes act in a generally additive manner and that Pointers with a large proportion of genes from a champion will be superior animals. Family selection has a broader base than individual selection, for it takes into account the performance of close relatives. The breeder may have a choice between two prospective sires of good quality, one from a line which includes many other superior individuals, the other an exceptional individual from a mediocre line. Giving weight to the performance of relatives might actually result in selection of the individual with the lower score.

A. Examples of Selection for Behaviour

Selection for behaviour among laboratory animals has been described in considerable detail, but except for previously cited studies on the chicken, we have little information on the economically important domestic species. However, laboratory results provide good models to be followed in applied programmes. Selection for behaviour is reasonably rapid, and significant effects may be evident in one or two generations. Among the traits which have been used as bases for selection are activity in a rotating cage (Rundquist, 1933); maze learning (Tryon, 1940; Heron, 1941; Thompson, 1954); emotionality (Hall, 1938); and alcohol preference (Mardones *et al.*, 1953). A particularly interesting example is the selection of mice for sensitivity to audiogenic or sound-induced seizures (Frings & Frings, 1953). These investigators obtained a rapid response in only two generations of selection, but susceptibility, as measured by seizure latency, continued to increase for six or seven generations. Figure 4 illustrates the distribution of susceptibility for unselected animals and for those of the 2nd and 7th selected generations. An originally rare genotype became the standard for the strain within a relatively few generations.

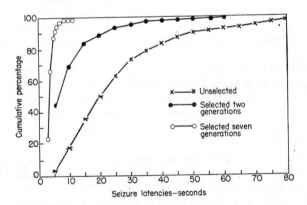

Fig. 4. Results of selection of mice for high susceptibility to audiogenic seizures. After two generations of selection the median latency is approximately equal to that of the most susceptible 15 per cent of the unselected population. After seven generations 90 per cent of the mice are as susceptible as the most susceptible 2 per cent of the original stock, and long latencies have completely disappeared. (From Frings & Frings, 1953.)

The cited studies demonstrate that a great variety of behavioural characters can be shifted by means of selection. However, none of the experimenters was interested in the process of selection itself, and they concentrated instead on the phenotypic variations obtained. Thus the published data do not permit a complete genetic analysis. Nevertheless, we conclude that a breeder of domestic animals who wishes to improve traits of behaviour by selection is likely to be successful if he invests the necessary care and effort. It is important to note that selection will not produce new traits, but it can greatly change the proportions of desired and non-desired phenotypes.

B. Selection for Multiple Characters

A problem encountered in most practical selection programmes is the need to consider several characters at once. The laboratory investigator who wishes to change the behaviour of his stock need not concern himself with the appearance or economic value of the animals he produces. For scientific purposes he may even deliberately select for characters such as extreme emotionality which are undesirable in ordinary laboratory practice. An animal breeder cannot afford to sacrifice economic qualities for the sake of a small improvement in behaviour, nor will he usually select on the basis of behaviour alone, as the experimentalist is free to do. This limitation is important, because selection for multiple uncorrelated characteristics is much less efficient than selection based upon a single criterion.

Various techniques have been devised for selecting for more than one character at a time (Hazel & Lush, 1942; Lerner, 1958). A general conclusion based upon both theory and practical experience is that best progress is made when animals are chosen for breeding by their standing on a selection index. Such an index gives points for each desirable economic character. If an attempt is made to improve behaviour this means that behavioural traits must be included along with physical characters. Point systems are familiar to animal breeders and it is not difficult to devise arbitrarily a rating system which includes characters of conformation, colour, growth, temperament, and trainability. The problem is in deciding what weight should be assigned to each character. In general, higher weights should be given to those characters of maximum economic importance, but with behavioural traits this may be a little difficult to judge. Weights should also be related to the heritability of the characters, for there is no sense to selecting on the basis of a trait whose variance is caused entirely by environmental factors. Since the heritability of behaviour patterns has not been quantitatively determined, it is impossible at this time to construct a precise selection index which includes behaviour. As guides, empirical studies are currently better than theory. However, selection experiments with a stronger theoretical orientation should yield even more useful results, and contribute both to basic behaviour genetics and practical breeding.

C. Correlated Variation

Selection for a single character is not only economically infeasible in most instances, but is an abstraction which does not correspond to biological reality. For example, let us consider a programme of selection for non-broodiness in fowl. Maternal nesting and brooding are unnecessary for survival of chicks in a modern hatchery, so it might seem that there are no barriers to total elimination of these patterns. From the viewpoint of an egg producer this would be highly desirable, since his hens would not be taking time off from their economic function of laying eggs. Suppose, however, that selection for non-broodiness modifies hormonal production in a direction which lowers egg production. Persistence in selection for non-broodiness could conceivably decrease fertility so much that the whole line might be extinguished. Long before this, however, the poultry raiser would probably have decided to tolerate a small residuum of broodiness rather than to reduce profits to the vanishing point.

Correlated changes such as those postulated between non-broodiness and egg laying are of general importance in selection for behaviour and other traits. The valuable economic attributes of our domesticated breeds, continuous egg production, high milk yields, and the like, have obviously been purchased at the cost of fitness, at least as measured under natural conditions. Selection modifies a total genotype and the total genotype controls constellations of characters, not single traits. Almost any change in a genotype which has been established through generations of natural selection must, in a sense, be a change for the worse from the animal's point of view. By providing shelter, feed, and care the animal husbandman counterbalances decreased fitness, though an alternative way of expressing the same facts might be to consider domestic animals as selected for fitness in the stable and the barnyard.

The relationship between behavioural and physiological phenotypic changes under selection works both ways. Selection for physical characters has been shown to affect behavioural traits (Castle, 1941; MacArthur, 1949; Lewis & Warwick, 1953), and behavioural consequences are certainly possible in any selection programme whatever may be its major objective. The occurrence of correlated changes in a selection programme can be explained by effects of loci closely linked with genes which are actively selected and by multiple phenotypic effects (pleiotropy) of selected genes. In the long run, the effects of pleiotropy are more important, since linkage groups gradually break up through crossing-over between homologous chromosomes.

Selection for extreme phenotypic characters ordinarily does result in some decrease of fitness. This is understandable from the fact that wild species and well-established domestic species are the outcome of many generations of selection. Their genotype has been balanced or co-adapted in the sense that diverse physiological and behavioural phenotypes operate at levels which maximize survival value. This balance is readily disturbed by too rapid and intense selection.

V. APPLICATIONS OF SELECTION

For the breeder of domestic animals the most important application of genetics is in the elimination of undesirable and increase of desirable traits.

Selection involves changing the average genotype of a population to produce more individuals with high quality phenotypes. The extension of the concept of phenotype to include behavioural characters does not introduce new genetic complications into selection, but with behavioural characters, the problem of quantitative measurement is more serious than with structural characters. In the following summary, we shall deal briefly with selection as a device for the elimination of undesirable traits, and the strengthening of desirable traits. In a sense, these are opposite sides of the same coin, but the economic significance of the two may be quite different. We shall assume that an undesirable trait is one so serious that it reduces profits to the vanishing point. Hence, its elimination may be considered critical to success of the breeding enterprise. Breeding for a desirable trait will be considered as an attempt to improve the economic value of a strain which can be managed profitably as it exists.

A. Eliminating Undesirable Traits

As a general principle, the possibilities of environmental control of undesirable traits should be exhaustively studied before any programme of genetic improvement is inaugurated. Early experience, mode of feeding and housing, and size of social group are powerful forces affecting the temperament and trainability of animals. Selection to eliminate deleterious behaviour is unlikely to be successful if the management system favours the development of these same traits.

If genetic means are employed to modify behaviour, the technique will, of course, depend upon the particular genetic mechanisms involved. A behavioural defect which is associated with a dominant gene with other visible effects can easily be eliminated by culling all "tagged" individuals. A trait associated in a similar manner with a recessive gene with visible effects would also be affected by selection against the structural phenotype. Since recessive genes in heterozygous condition are not exposed to selection, the process is less efficient than with dominants. Traits with visible correlates are apt to involve major neurological or endocrinological syndromes of which the behavioural traits form only a part.

Selection is more complex when multiple factors determine the trait of interest. Suppose that extreme timidity must be eliminated from a strain of horses which are otherwise desirable. Timidity should be given heavy weight in the culling of potential breeding stock. At the same time, care must be taken that other desirable characteristics are not lost. Selection on the basis of multiple criteria is slow. If n characters are considered, the rate of progress is $1/\sqrt{n}$ of the rate for a single trait (Hazel & Lush, 1942). New genes may be introduced by outcrossing to a "non-timid" strain and following this by repeated backcrossing of selected hybrids into the "timid" strain. By expanding this outcrossed stock, a new strain can be produced which has assimilated the "non-timidity" genes and retained other desirable characteristics. Such a programme is not to be undertaken lightly. It may be economically better to find another strain and make a new start.

B. Improving Desirable Traits

We shall now inquire into the prospects of improving the quality of animals which are currently paying their way. Is it possible by selective breeding to develop strains which are easier to manage because their behaviour is better adapted to modern systems of animal management? Wild species breed rather poorly in small cages and are much harder to handle even when reared in a laboratory from birth. In the laboratory, the rabbit, rat, and mouse strains used widely in research have been selected to be much more manageable than their wild ancestors. The development of non-broody fowl is another example with economic value. It is probable, however, that there are definite limits to our capacity for the synthesis of improved behaviour by selection. The domestic dog, for example, shows no behaviour patterns which are not found in its wild ancestor, the wolf (Scott, 1954). Selection has accentuated certain traits and reduced the intensity of others, but the patterns of feeding, mating, and group social organization are markedly alike in the Beagle dog and the Timber wolf.

Selection for extremes of one character will also modify other traits, perhaps in an undesirable direction. Hence, caution is advisable in trying to modify a trait which is already good enough to get by. These remarks are probably more applicable to selection for traits such as non-broodiness which run counter to requirements for survival under natural conditions than they are for traits such as intelligence, which is probably useful under both domestic or wild conditions. It seems unlikely that selecting a Pointer for high ability in field trials would impair other traits, since excellence in this particular area is dependent upon many different attributes, sensory acuity, physical vigour, and trainability. The whole subject requires more investigation in controlled laboratory experiments.

VI. PRODUCING EXCEPTIONAL ANIMALS

The production of exceptional animals will never be of much significance to the breeder of cattle, sheep, or poultry, where the essential economic worth of the enterprise depends on the average quality of his population. But the breeders of race horses or exhibition dogs gain more from a single outstanding individual than from a hundred near winners. Hence, breeding in these species can and should involve attention to the production of exceptional individuals, a field not widely studied by geneticists. It is probable that often superior performance can be obtained most readily in hybrids whose traits, derived from diverse parental stocks, "mesh" in a favourable fashion. The hybrids may be of various levels, between strains of a single breed, or between breeds. In a hybrid dominant genes from both parental strains are effective, and there is also the possibility of over-dominance, another name for superiority of the heterozygote. In general, dominant genes contribute more to fitness than do their recessive alleles, so that a hybrid benefits from either dominance or over-dominance.

As far as behaviour is concerned, the effects of heterozygosity and non-additive genetic interaction is an almost unexplored area of research (see

Mordkoff & Fuller, 1959). Laboratory studies are needed to serve as models for the breeding of exceptional animals and to increase our knowledge of behaviour genetics. (Cf. Fuller & Thompson, 1960; Hirsch, 1967; Parsons, 1967.)

REFERENCES

BROADHURST, P. L. (1958). Determinants of emotionality in the rat. III. Strain differences. *J. comp. physiol. Psychol.*, **51**, 55–59.

BRODY, E. G. (1942). Genetic basis of spontaneous activity in the albino rat. *Comp. Psychol. Monog.*, **17**, 1–24.

BRUELL, J. H. (1959). Dominance and segregation in the inheritance of quantitative behavior in mice. *Paper presented at Amer. Psychiatric Symposium, Roots of Behavior.* To be published in: *Roots of Behavior.* E. L. Bliss (ed.).

CASTLE, W. E. (1941). Influence of certain color mutations on body size in mice, rats and rabbits. *Genetics*, **26**, 177–191.

CHAI, C. K., AMIN, A. & REINEKE, E. P. (1957). Thyroidal iodine metabolism in inbred and F_1 hybrid mice. *Amer. J. Physiol.*, **188**, 499–502.

COLLIAS, N. E. (1944). Aggressive behavior among vertebrate animals. *Physiol. Zool.*, **27**, 83–123.

DAWSON, W. M. (1932). Inheritance of wildness and tameness in mice. *Genetics*, **17**, 296–326.

FRINGS, H. & FRINGS, M. (1953). The production of stocks of albino mice with predictable susceptibilities to audiogenic seizures. *Behaviour*, **5**, 305–319.

FULLER, J. L. (1953). Cross-sectional and longitudinal studies of adjustive behavior in dogs. *Ann. N.Y. Acad. Sci.*, **56**, 214–224.

FULLER, J. L. (1960). Genetics and Individual Differences. In: *Principles of Comparative Psychology.* Waters, Rethlingshafer & Caldwell (eds.). New York, N.Y.: McGraw-Hill, pp. 325–354.

FULLER, J. L. (1967). Effects of the albino gene upon behaviour of mice. *Anim. Behav.*, **15**, 467–470.

FULLER, J. L. & SCOTT, J. P. (1954). Heredity and learning ability in infrahuman mammals. *Eugen. Quart.*, **1**, 28–43.

FULLER, J. L. & THOMPSON, W. R. (1960). *Behavior Genetics.* New York, N.Y.: Wiley.

GOY, R. W. & YOUNG, W. C. (1957). Strain differences in the behavioural responses of female guinea pigs to alpha-estradiol benzoate and progesterone. *Behaviour*, **10**, 340–354.

GRUNT, J. A. & YOUNG, W. C. (1953). Consistency of sexual behavior patterns in individual male guinea pigs following castration and androgen therapy. *J. comp. physiol. Psychol.*, **46**, 138–144.

GUHL, A. M., CRAIG, J. Z. & MUELLER, C. D. (1960). Selective breeding for aggressiveness in chickens. *Poult. Sci.*, **39**, 970–980.

HALL, C. S. (1938). The inheritance of emotionality. *Sigma Xi Quart.*, **26**, 17–27.

HALL, C. S. & KLEIN, S. J. (1942). Individual differences in aggressiveness in rats. *J. comp. Psychol.*, **33**, 371–383.

HARRIS, H. (1959). *Human Biochemical Genetics.* Cambridge: Cambridge Univ. Press.

HAZEL, L. N. & LUSH, J. L. (1942). The efficiency of three methods of selection. *J. Hered.*, **33**, 393–399.

HERON, W. T. (1941). The inheritance of brightness and dullness in maze learning ability in the rat. *J. genet. Psychol.*, **59**, 41–49.

HIRSCH, J. (ed.) (1967). *Behavior-Genetic Analysis.* New York, N.Y.: McGraw-Hill.

JAMES, W. T. (1951). Social organization among dogs of different temperaments, terriers and beagles, raised together. *J. comp. physiol. Psychol.*, **44**, 71–77.

KAUFMAN, L. (1948). On the mode of inheritance of broodiness. *Proc. 8th World Poultry Congr.*, Copenhagen, 301–304.

KNOX, C. W. & OLSEN, M. W. (1938). A test of cross-bred chickens, Single Comb White Leghorns and Rhode Island Reds. *Poult. Sci.*, **17**, 193–199.

LAGERSPETZ, K. (1964). Studies on the aggressive behaviour of mice. *Ann. Acad. Sci. Fenn. B.*, **131**, 1–13.

LERNER, I. M. (1958). *The Genetic Basis of Selection.* New York, N.Y.: Wiley.

LEWIS, W. L. & WARWICK, E. J. (1953). Effectiveness of selection for body weight in mice. *J. Hered.*, **44**, 233–238.

MACARTHUR, J. W. (1949). Selection for small and large body size in the house mouse. *Genetics*, **34**, 194–209.

MAHUT, H. (1958). Breed differences in the dog's emotional behavior. *Canad. J. Psychol.*, **12**, 35–44.

MARDONES, R. J., SEGOVIA, N. M. & HEDERRA, A. D. (1953). Heredity of experimental alcohol preference in rats. II. Coefficient of heredity. *Quart. J. Stud. Alc.*, **14**, 1–2.

MERRELL, D. J. (1965). Methodology in behavior genetics. *J. Hered.*, **56**, 263–266.

MORDKOFF, A. M. & FULLER, J. L. (1959). Variability in activity within inbred and cross-bred mice: A study in behavior genetics. *J. Hered.*, **50**, 6–8.

MUNRO, S. S., KOSIN, I. L. & MACARTNEY, E. L. (1943). Quantitative genic hormone interactions in the fowl. I. Relative sensitivity of five breeds to an anterior pituitary extract possessing both thyrotropic and gonadotropic properties. *Amer. Naturalist*, **77**, 256–273.

MURIE, A. (1944). *The Wolves of Mt. McKinley.* U.S. Dept. Interior, Fauna Ser., No. 5, Washington, D.C. Government Printing Office.

PARSONS, P. A. (1967). *The Genetic Analysis of Behaviour.* London: Methuen & Co.; U.S. Distributor, Barnes & Noble.

PAWLOWSKI, A. A. & SCOTT, J. P. (1956). Hereditary differences in the development of dominance in litters of puppies. *J. comp. physiol. Psychol.*, **49**, 353–358.

POTTER, J. (1949). Dominance relations between different breeds of domestic hens. *Physiol. Zool.*, **22**, 261–280.

PUNNETT, R. C. & BAILEY, P. G. (1920). Genetic studies in poultry. II. Inheritance of egg colour and broodiness. *J. Genet.*, **10**, 277–292.

RICHTER, C. P. (1954). The effects of domestication and selection on the behavior of the Norway rat. *J. nat. Cancer Inst.*, **15**, 727–738.

ROBERTS, E. & CARD, L. E. (1934). Inheritance of broodiness in the domestic fowl. *Proc. 5th World Poult. Congr.*, **2**, 353–358.

RUNDQUIST, E. A. (1933). The inheritance of spontaneous activity in rats. *J. comp. Psychol.*, **16**, 415–438.

SAWIN, P. B. & CRARY, D. D. (1953). II. Some racial differences in the pattern of maternal behaviour. *Behaviour*, **8**, 43–53.

SAWIN, P. B. & CURRAN, R. H. (1952). Genetic and physiological background of repro-duction in the rabbit. I. The problem and its biological significance. *J. exp. Zool.*, **120**, 165–201.

SCOTT, J. P. (1942). Genetic differences in social behavior of inbred strains of mice. *J. Hered.*, **33**, 11–15.

SCOTT, J. P. (1954). The effects of selection and domestication upon the behavior of the dog. *J. nat. Cancer Inst.*, **15**, 739–758.

SCOTT, J. P. & CHARLES, M. S. (1954). Genetic differences in the behavior of dogs: a case of magnification by thresholds and by habit formation. *J. genet. Psychol.*, **84**, 175–188.

SCOTT, J. P. & FULLER, J. L. (1951). Research on genetics and social behavior. *J. Hered.*, **42**, 191–197.

SCOTT, J. P. & FULLER, J. L. (1965). *Genetics and the Social Behavior of the Dog.* Chicago, Ill.: University of Chicago Press.

SEARLE, L. V. (1949). The organization of hereditary maze-brightness and maze-dullness. *Genet. Psychol. Monogr.*, **39**, 279–325.

STAMM, J. S. (1956), Genetics of hoarding: II. Hoarding behavior of hybrid and back-crossed strains of rats. *J. comp. physiol. Psychol.*, **49**, 349–352.

THOMPSON, W. R. (1954). The inheritance and development of intelligence. *Proc. Assoc. Res. nerv. ment. Dis.*, **33**, 209–231.

THOMPSON, W. R. & BINDRA, D. (1952). Motivational and emotional characteristics of "bright" and "dull" rats. *Canad. J. Psychol.*, **6**, 116–122.

THORNE, F. C. (1940). Approach and withdrawal behavior in dogs. *J. genet. Psychol.*, **56**, 265–272.

TRYON, R. C. (1940). Genetic differences in maze-learning ability in rats. *39th Yearbook Nat. Soc. Stud. Educ. (Part 1)*, Bloomington, Ill.: Public School Pub. Co., pp. 111–119.

WAGNER, R. P. & MITCHELL, H. K. (1964). *Genetics and Metabolism*, 2nd edn. New York, N.Y.: Wiley.

WOOD-GUSH, D. G. M. (1958). Genetic and experimental factors affecting the libido of cockerels. *Proc. roy. Soc. Edinburgh*, **27**, 6–7.

WOOD-GUSH, D. G. M. (1960). A study of sex drive of two strains of cockerels through three generations. *Anim. Behav.*, **8**, 43–53.

WOOD-GUSH, D. G. M. & OSBORNE, R. (1956). A study of differences in the sex drive of cockerels. *Brit. J. anim. Behav.*, **4**, 102–110.

WRAGG, L. E. & SPIERS, R. S. (1952). Strain and sex differences in response of inbred mice to adrenal cortical hormones. *Proc. Soc. exp. Biol. Med. N.Y.*, **80**, 680–684.

YEAKEL, E. H. & RHOADES, R. P. (1941). A comparison of the body and endocrine gland (adrenal, thyroid, and pituitary) weights of emotional and non-emotional rats. *Endocrinology*, **28**, 337–340.

Chapter 4

The Physical Environment and Behaviour

M. W. Schein and E. S. E. Hafez

In its natural situation, each species occupies a reasonably well-defined ecological niche wherein individuals tolerate or readily adapt to most variations in the physical environment. The species can compensate for minor environmental fluctuations physiologically and/or behaviourally; major environmental changes demand major behavioural changes.

A warm-blooded animal, for example, responds physiologically to a hot summer day by reducing its metabolic heat production and using all available mechanisms for heat-loss. Behaviourally, the animal reflects the reduced metabolic rate by lethargy, reduced feed intake, and postures that maximize the opportunities for cooling. If the hot environmental conditions persist for some time (the usual seasonal shift in temperature), the behaviour of the animal gradually changes so as to minimize the physiological stress imposed on it by the environment: feeding and general activities are shifted to the cooler early morning, late evening, or night hours, whereas the hot midday is reserved for idling in the shade. If the environment changes markedly beyond the capacity for physiological adaptation, then major behavioural adjustments are required: the animal may be forced to migrate, hibernate, aestivate, become nocturnal, or take similar extreme steps.

Ecologists are well acquainted with the rules of Bergmann, Allen, and Gloger, each of which relates some aspect of animal morphology to the environment. According to Bergmann's law, the body mass of homoiotherms is inversely related to environmental temperature, i.e. birds and mammals from Arctic (Antarctic) regions are generally larger and therefore have less surface area per unit of body weight than their tropical relatives. Allen's rule states that the extremities (ears, legs, tail, etc.) tend to be shorter in colder climates, reducing the exposed surface area of the body across which heat may be lost (cf. Allee et al., 1949). Gloger's rule (cf. Hesse et al., 1937) relates coat colour to environmental temperature and humidity: species of birds and mammals living in cool, dry areas tend to have lighter coat colour than their relatives in warm, humid areas.

These rules were formulated in the nineteenth century and their status as laws, or even rules, is questionable at present. Both Bergmann's and Allen's rules, for example, ignore the fact that the body is differentially insulated and that all parts of the body's surface do not lose heat at the same rate. The racoon

3+

is only one example of a mammal which is smaller in the northern parts of its range than in the southern temperate parts. Furthermore, body mass is related to things other than temperature (e.g. general activity). There is also some question that the evolution of extremities depends primarily, if at all, on climate: it seems more reasonable that the evolution of appendages is related to their use (e.g. movement). Nor has it been demonstrated that heat radiation from animals with white coats is less than that from animals with darker coats. All Arctic animals are not white: the black raven, for example, occurs throughout the Arctic. Furthermore, a good case can be made that the relationship of coat colour to environment is one of cryptic coloration, especially since burrowing animals, which presumably would not be continually exposed to the surface environment, tend to be darker in warm, humid areas (where vegetation and soil is often dark) and lighter in cold, dry areas (where there is often less vegetation, exposed light-coloured soil, or snow). This is amply illustrated by subspecies of *Peromyscus* in Florida and Alabama (cf. Allee *et al.*, 1949, p. 627). *P. polionotus leucocephalus* lives on the white sand of the Florida coast and has a uniformly white coat; *P. p. polionotus* occurs on the dark soil of Alabama (59 miles away) and is uniformly dark in colour. Such differences in animals from the same general climatic area are probably better explained as adaptations to background colour than by Gloger's rule.

There are, however, examples, both from field and laboratory studies, in which the laws of Bergmann and Allen appear to hold. Mice reared at 31° to 33·5°C have longer tails and less stocky bodies than animals of the same strain reared at 15·5° to 20°C. The same appears to be true of domestic fowl: birds kept at 6°C when 3 to 4 months of age are shorter, have shorter tarsi and tails, and weigh more than flock mates retained at 21° to 24·5°C during the same period (cf. Allee *et al.*, 1949).

These ecological rules apply exceptionally, rather than generally (Gordon, 1968, p. 321; Irving, 1964, pp. 368–369). It is difficult, for example, to make a case for these laws when dealing with burrowing mammals and those which hibernate, and with migrating species of birds, since such animals physiologically and behaviourally are not directly exposed to the environment. What is perhaps more important is that such laws relegate behavioural and physiological adaptations to cold to a place of minor importance. Both types of adaptation are, of course, methods by which the animal avoids directly confronting the environment.

No formal rule correlating behaviour with climate has been postulated, except for the generally accepted inverse relationship between temperature and productive output (meat, milk, work, etc.): in tropical climates, lethargy and low work productivity are more the rule than in temperate or cooler climates. However, Lee (1948) points out that even this general rule may not be physiologically sound, since experience during World War II has shown that men from temperate zones can live healthily and work reasonably well in tropical climates. Even so, psychological fatigue and disturbances occur more easily in the tropics than in temperate climates; these conditions can be alleviated or minimized by providing additional mental stimulation (Lee, *loc. cit.*).

I. THERMOREGULATORY BEHAVIOUR

Efficient operation and continued activity of the homoiothermic animal depend upon its ability to maintain a reasonably constant body temperature. To do so, internal and external sensory receptors continually feed temperature information into the central nervous system, which in turn activates mechanisms that prevent significant changes in body temperature. The animal is essentially in thermal balance with its environment at *ca.* 70°F: the gradient from deep body (core) temperature to peripheral body (shell) temperature to environmental temperature permits a normal outward flow of continuously generated metabolic heat. Too steep a gradient (core temperature unchanged, environmental temperature below 70°F) increases metabolic heat loss, thereby endangering a stable core temperature. Efforts must be made to increase the internal production of heat, to decrease the outward flow of heat, or to reduce the difference between skin temperature and environmental temperature (i.e. moving to a warmer area). Each of these adaptations has behavioural connotations, and is used singularly or in combination to counter the stress of a low environmental temperature. Major behavioural adjustments must also be made when environmental temperature exceeds core temperature, in which case heat flows from the environment into the body.

A. Hot Environment

Short of physically moving to a cooler area, an animal confronted with high environmental temperatures must either reduce its internal heat production, enhance its heat dissipation mechanisms, or both. Heat production can be lowered by reducing motor activities and food intake, phenomena which involve energy transformation and therefore produce heat. Loss of appetite and increased degree of lethargy commonly accompany continuous exposure to high temperatures. However, this approach to countering high environmental temperatures has serious limitations: internal heat production cannot be reduced beyond basal levels, and even by aestivating, the animal cannot eliminate *all* muscular activity for extended periods of time. Furthermore, if environmental temperature exceeds body temperature, the animal will gain heat anyway, even if its own heat production could be reduced to zero. Thus, *dissipation* of heat by radiation and/or evaporation becomes a key factor in survival under hot conditions.

Heat loss by radiation is often limited, since it can occur only if part or all of the surrounding area is cooler than the animal itself. Thus, probably the most important physiological and behavioural heat-dissipating device available to the animal involves the evaporation of water from body surfaces: the animal's own body heat is used to evaporate water from the surface of the skin (sweat or insensible transpiration) or from the pulmonary and oral mucosae. However, the effectiveness of evaporative cooling depends on the moisture content of the surrounding air, which must be at less than saturation if the air is to take up any additional water from the animal. Behaviour is directed toward increasing possibilities for evaporation: body extensions maximize the surface area exposed to the air; increased respiratory rates bring more air into contact with the respiratory surfaces; activities, such as fanning, increase air movement which in effect removes the layer of moisture-saturated air from body surfaces. It is, of course, obvious that increased water loss necessitates increases in water intake; otherwise, the animal quickly becomes dehydrated.

A summary of the behavioural adjustments aimed at resisting increases in body temperature is presented in Table 6 and Fig. 5.

Table 6. Behavioural Adjustments of Birds and Mammals in Hot Environments

Body Temperature	Cooling Strategy	Behavioural Adjustment[1]
Normal	Reduced heat production	Decrease food consumption Decrease motor activities
	Thermal radiation	Avoid direct sunlight (solar heat load) Seek cooler surfaces Avoid warm surfaces (group dispersion)
	Evaporation	Increase water consumption Increase respiratory rate Wet body surfaces by licking, drooling, wallowing, etc. Increase body surface area by spreading, etc. Increase air movements over body surfaces by fanning, etc.
Moderate increase not critical	Reduced heat production	Anorexia Lethargy
	Thermal radiation	Same as above
	Evaporation	Respiration rate at maximum (panting and gasping) Other factors same as above
Critical increase	None	Anorexia, weakness and stupor, postural imbalance, "staggers", convulsions

[1] Assuming that escape to a cooler environment is not possible.

Mammals. The physiological responses of cattle to hot environments have been studied with the basic aim of improving milk and meat production in the tropics and subtropics. Zebu cattle (*Bos indicus*), which are tropically-evolved, can withstand hot climates better than European cattle (*Bos taurus*). The *comfort* or *thermoneutral zone* (environmental temperature range in which no apparent demands are made upon physiological thermoregulatory mechanisms) for European cattle is between 35° and 70°F, whereas for Zebu cattle it is between 50° and 80°F (summarized by Brody, 1956). Above 80°F, both breeds make physiological and behavioural adjustments, but European cattle must adjust more than Zebu cattle.

Bovine adjustments to severe short-term exposures (7 to 10 hours at 105° and 34 mm Hg vapour pressure) include an increased respiratory rate (panting), drooling, sweating and increased water consumption. However, the

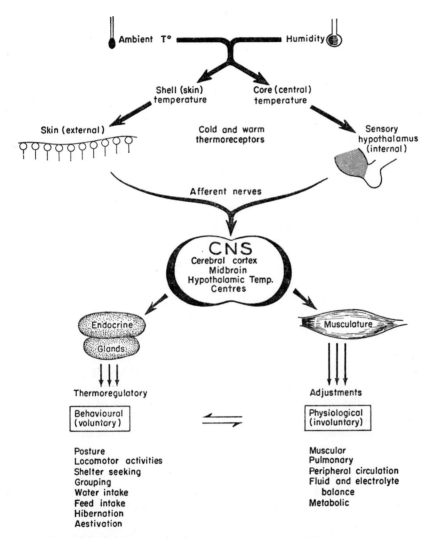

Fig. 5. Diagrammatic illustration of behavioural and physiological adjustments for thermoregulation. Arrows indicate the relationships between the environment and the body systems. (Adapted from Hafez, 1968.)

ventilation rate can increase only to a maximum associated with age and breed: about 130 respirations per minute for adult Jerseys and 110 respirations per minute for adult Zebu-Jersey crossbreeds (Schein *et al.*, 1957).

Drooling is also of little help in dissipating excess body heat. Consequently, the animal cannot compensate behaviourally for the increased heat load and body temperature rises.

Long-term exposures of several days or weeks to severe thermal stress in the field are quite unusual; temperatures in the hot-wet areas of the world rarely rise above 98° to 100°F for more than a few hours at a time, but commonly hover between 90° to 95°F for extended periods at vapour pressures of 25 to 30 mm Hg. Thus, since environmental temperatures are lower than body temperatures, heat flow is outward and behavioural adjustments are directed toward enhancing the core-to-environment gradient. In hot-wet areas, evaporative cooling is made difficult by the great amount of moisture in the surrounding air: water that does not evaporate, but rather drips off the body surface, does not help cool the animal. Hence, cooling by reduced heat production and increased radiation are more effective.

The hot-dry areas of the world have higher environmental temperatures than hot-wet areas, but are more compatible with homoiothermal life because the air contains little moisture. Temperatures above 110°F for prolonged periods of time are not uncommon in desert regions, and man and his domestic animals easily adapt to such conditions provided food and water are available. In hot-dry environments, an animal can successfully dissipate excess body heat by evaporative cooling; water consumption by both birds and mammals under these conditions rises markedly.

Radiation becomes a major factor in hot-dry environments, since outside temperatures exceed body temperatures. The heat input from surrounding bodies is not large, however, and is readily dissipated by evaporative cooling provided the animal is sufficiently hydrated. However, the heat load imposed by solar radiation is immense, and efforts are made to avoid direct exposure to sunlight. Sheep seek shade, especially for the head; if shade is unavailable, they face away from the sun. Camels do likewise, with the result that the smallest possible surface area is exposed to solar radiation. The surface area of sheep, for example, approaches 1 m²: half or less of this is exposed to the sun, depending on the time of day and the orientation of the animal (cf. Macfarlane, 1968).

Small desert mammals avoid severe heat stress by staying in cool burrows during the day. Some of them, the kangaroo rat (*Dipodomys*) in particular, do not need succulent foods or free drinking water. They utilize metabolic water with great physiological efficiency: body evaporation is reduced to minimum levels and urine is highly concentrated. Others, such as the pack rats (*Neotoma*), obtain water from succulent foods (summarized by Schmidt-Nielson & Schmidt-Nielson, 1952).

Large desert mammals, such as camels and gazelles, cannot easily hide from the sun and must therefore expend water for heat regulation. However, these animals have evolved highly efficient thermoregulatory mechanisms including insulating coats to protect them from solar radiation and the ability to withstand large fluctuations in body temperature and severe dehydration. They can consequently survive without free drinking water for extended periods of time (Schmidt-Nielson, 1959).

The heat which a large desert mammal absorbs or emits depends largely on the length, thickness, and colour of its coat (cf. Macfarlane, 1968). A white coat, for example, reflects about 40 per cent more light energy than a black one of the same length and texture. The importance of length and thickness is illustrated by three characteristic desert coats, each operating on a slightly different principle to protect the animal from overheating. The coat of the Merino sheep is a dense, fine fleece with about 70×10^6 fibres per sq. m of surface and a thickness approaching 5 cm. Its surface absorbs most solar radiation and radiates long-wave heat energy (as much as 150 to 300 kcal per hour in quiet air); its thick subsurface layer prevents heat from reaching the skin. In a wind (characteristic of deserts during the hot part of the day), convection can more than double heat loss from the surface of such a coat. The loose, long wool or hair-wool coats of Awassi sheep and desert goats are much thinner than those of Merino sheep, containing only 2 to 4×10^6 fibres per sq. m of surface. They permit convective ventilation and allow limited reflection and radiation. Hairier fleeces and the short, shiny hair coats of antelope, camels and cattle reflect solar radiation: the fibres of hair or fleece lie parallel to the skin, producing a smooth, reflecting surface. Cooling occurs principally by convection and short-wave reflection. This type of coat also readily permits sweat to evaporate from the surface of the skin.

In hot-wet areas, long-term exposure to *moderately* unfavourable climates is the rule. Under such conditions, the respiratory rates and body temperatures of cattle show a diurnal pattern, increasing during the hotter parts of the day and decreasing at night (Seath & Miller, 1946). The animals graze at times other than the hotter parts of the day (reviewed by Findlay & Beakley, 1954). During the summer in Louisiana (lat. 31°N), cattle graze less than 2 hours during the day and about $6\frac{1}{2}$ hours at night; however, on cool days daytime grazing averages $4\frac{1}{2}$ hours and night grazing decreases to about 5 hours. Total grazing time during the 24-hour period is about 1 hour longer on cooler days than on hotter days (Seath & Miller, *loc. cit.*). These findings have been confirmed by Payne et al. (1951), who report that cattle in Fiji (lat. 18°S) do 60 per cent of their grazing at night; in addition, total grazing time falls when mean day temperatures exceed 80°F. Seath & Miller (1947) note that on hot days Holstein cows frequent wet spots in the shade and often wallow in mud and water, whereas Jerseys show fewer such tendencies. This is not surprising since Jerseys can withstand hot-wet climates better than Holsteins (Seath & Miller, 1947; Worstell & Brody, 1953). Jerseys have more surface area per unit of mass than do Holsteins, and can therefore more readily dissipate excess body heat. Bare-skinned animals, e.g. swine and buffaloes, frequent shaded, wet spots and often wallow in mud, water, or areas wet with urine and faeces. If soil is accessible, pigs root up the ground and lie on the cooler subsoil. If a water pool is present, they root around it and wallow to cool themselves.

The behavioural adjustments of sheep to hot-wet environments resemble those of cattle and include panting and increased water intake (reviewed by Lee, 1950). In the subtropics, sheep also show a diurnal and a seasonal rhythm of rectal temperatures and respiration rates, with higher values occurring during hotter parts of the day and year (Hafez et al., 1956).

Similarly, the rectal temperatures and respiration rates of swine increase with increasing environmental temperatures, whereas food intake decreases (Heitman & Hughes, 1949). In addition, when air temperatures exceed 80°F, the animals become increasingly lethargic and lay stretched out on their sides (see Chapter 11). Areas moist with urine or faeces are used for wallowing, and hogs turn over from time to time, exposing their moist sides to the air.

Only limited information is available on the heat tolerance of other domestic animals (horses, rabbits, dogs, and cats), but there is little reason to expect large behavioural or physiological differences from those mentioned above (Table 6). Lee et al. (1941) investigated the rabbit's response to high environmental temperature and found essentially the same behavioural adjustments as those in cattle, except perhaps that rabbits depend more on respiratory evaporation for heat dissipation than do cattle. Dogs, of course, are exceptionally dependent on evaporation from the tongue and oral mucosa at high environmental temperatures (Robinson & Lee, 1941a); cats lick their body surfaces, thus facilitating body cooling by evaporation (Robinson & Lee, 1941b).

Birds. Birds respond to thermal stress as do mammals, with increased body temperatures and respiration rates, and decreased food consumption. Since birds normally have higher body temperatures than mammals, they are able to withstand correspondingly higher environmental temperatures. Evaporative cooling of the air sac system and oral mucosa (e.g. gular flutter) surfaces is a major means of dissipating heat, since evaporation from body surfaces is negligible because of the insulating character of feathers. However, the feather cover on the undersurface of the wing is generally sparser than that on the rest of the body, and on hot days birds often spread their wings as they crouch on the ground. This posture facilitates heat dissipation by radiation to cooler shaded areas and also favours some evaporative cooling.

Under moderate heat stress, chickens hold their wings away from the body so that air can circulate past the less insulated undersurface. They may also lie down with their head and neck stretched forward or sit on the soil. The thick insulation of the dorsal feathers and the still air above them impede the flow of heat into the bird, while the squashing of the breast feathers facilitates heat loss to the cool soil. Lethargy and open-mouth panting are also common. In hot weather, juvenile albatrosses often sit balanced on their heels with their feet off the soil (Howell & Bartholomew, 1961): this posture permits heat dissipation to the air from the vascularized foot webbing.

Lee et al. (1945) present a good description of the behavioural changes which occur during acute heat stress in chickens. With a 1° increase in rectal temperature (normally 107°F), respiration changes from closed-mouth breathing to open-mouth panting or tonguing; the hen becomes lethargic and stands with her feathers erect and wings partly outstretched. At a rectal temperature near 113°F, the hen is markedly distressed: her gait is unsteady, feathers are fluffed, and open-mouth panting gives way to slower deep sighs or gasps. The bird will collapse and die rapidly unless relief is given. Somewhat less severe heat exposure often causes the birds to enter a state resembling partial moult. Short-term exposures to moderately high temperatures are not noticeably detrimental to the sexual behaviour of birds:

Long & Godfrey (1952) found no evidence of reduced mating activity in New Hampshire cocks at environmental temperatures up to 88°F.

B. Cold Environments

Behavioural adjustments to cold environments tend to be the reverse of those made to hot environments. Efforts are made to increase internal heat production and decrease the steep thermal gradient between the body's core and the environment. Body surface is minimized to conserve heat; respiratory rates are reduced to decrease heat loss from the surface of the respiratory tract; motor activities are increased. However, if these efforts fail to maintain body temperature, they tend to cease. The consequence is death for all homoiotherms except those few which escape the rigours of cold environments by hibernating.

A summary of the behavioural adjustments aimed at resisting decreases in body temperature is presented in Table 7.

Table 7. Behavioural Adjustments of Birds and Mammals in Cold Environments

Body Temperature	Warming Strategy	Behavioural Adjustment[1]
Normal	Increased heat production	Increase food consumption Increase motor activities
	Thermal radiation	Seek warmer bodies (group huddling) Seek direct sunlight
	Evaporation	Decrease respiratory rate Decrease exposed body surface (flexure) Minimize air movement over body surfaces
Moderate decrease, not critical	Increased heat production	Hyperphagia Shivering Decrease general activity
	Thermal radiation	Same as above
	Evaporation	Increase body flexure to maximum Depress respiration
Critical decrease	None	Cessation of shivering, depressed metabolism, lethargy, reduced mental alertness

[1] Assuming that escape to a warmer environment is not possible.

Most domesticated animals are physiologically and behaviourally better equipped to deal with cold environments than with hot ones, since in general they have evolved from temperate-zone stock (see Chapter 2). The comfort or thermoneutral zone of European cattle extends to near freezing temperatures:

3*

under artificially controlled environmental conditions, temperatures near 0°F had no appreciable effect on the heat production of Holstein cows, but greatly increased the heat production and feed consumption of Zebu cattle. Smaller European cattle (Jerseys) than Holsteins exhibited moderate increases in heat production and feed consumption under similar conditions. Peterson & Young (1955) could not find a direct relationship between the level of metabolic activity and the intensity of the sex drive of male guinea pigs; nor was sex drive impaired by exposure to cold temperatures (ca. 32°F).

Behavioural responses to cold temperatures are well known, and are common to all our domestic animals: metabolic rate and therefore food consumption increases; muscle tonus is raised and shivering is apparent; animals seek protective shelter from wind and rain and huddle together, to conserve warmth (Brody, 1948). Seasonal trends in the quantity and quality of feed intake by turkeys have been demonstrated by Almquist (1953); feed intake declined approximately 30 per cent from January to July and rose sharply again as the temperature declined in autumn.

Fur-bearing mammals and birds are capable of mechanically altering the arrangement of their pelage to increase its insulative capacity and thus reduce heat loss. The piloerection of the fur in mammals and the ruffling of feathers in birds observed during cold stress traps a layer of still air close to the skin and within the pelage: this serves as additional insulation reducing heat loss.

In cold weather, chickens bunch up with their breast feathers fluffed out. If a bird becomes active its feathers lie flat again, but are fluffed out as soon as the bird stops. At night, when metabolism is at its lowest, the feathers are thoroughly fluffed and the bird sleeps much of the time with its head under its wing; this posture reduces heat loss considerably.

Weiss & Laties (1961) studied the interrelationships between operant conditioned behaviour and the subcutaneous body temperature of rats, using a Skinner-box: the animal could press a lever which activated an infra-red heat lamp for a short period of time. The device was placed in a cold environment (ca. 36°F) and the fur of the experimental animals was removed so that they could not possibly maintain normal body temperatures during the test without the external heat source (Fig. 6). The data revealed that initiation of a steady response rate, i.e. bar pressing for heat, depended on a relatively large decrease (about 14°F) in subcutaneous temperature, whereas the response rate, once initiated, was governed by much smaller fluctuations in skin temperature. When a rat is first placed in the cold, lever pressing apparently competes with other responses to the environment, such as shivering and huddling. When these mechanisms fail to maintain normal body temperatures, they are replaced by gross random motor activities, which by chance result in depression of the lever. The immediate reward of a small increment of heat from the lamp reinforces lever pressing and training is quickly accomplished. Once the rat has learned to manipulate the external source of heat, lever pressing replaces shivering and huddling as a response to peripheral cooling.

Pigs can also learn to press a panel-type switch with the snout in order to obtain bursts of radiant heat. The frequency of pressing increases as the ambient temperature falls. The animals also learn to press a switch which turns off a draught-creating fan. Baldwin & Ingram (1966) have shown that cooling the hypothalamus leads to increased pressing activity.

Severe hypothermia (deep body temperatures reduced to nearly 32°F) impairs the problem-solving abilities of rats, whereas moderate hypothermia

Fig. 6. The heat-reinforcement Skinner-box. Depression of the lever turns on an infra-red heat lamp. Rats adjust their rate of response to their needs. (After Weiss, 1957. *J comp. physiol. Psychol.* **50**, 481.)

(deep body temperature 55° to 65°F) has little effect (Russell, 1956). However, animals subjected to either severe or moderate hypothermia exhibit poorer retention abilities than non-exposed control animals.

Cold-acclimatized rats, trained in the cold, solve maze problems more rapidly and maintain higher rates of response than cold-acclimatized rats trained at optimal or higher temperatures (Hellmer, 1943; Moore, 1944). In non-acclimatized rats, the strength of an operant conditioned response learned in an optimum environment decreases when the animals are subjected to a cold environment. However, with continued exposure to the lower temperature (i.e. gradual acclimatization), the rate of response recovers to pre-exposure levels (Teichner & Kobrick, 1955). Teichner (1957, 1958) suggests that these effects depend on both the status of acclimatization and the degree of habitation (i.e., the amount of training and experience) of the individual.

C. Parental Behaviour and Thermoregulation

There are striking differences among various homoiotherms in the degree of physiological maturity at birth or hatching. The neonate of most species has poor thermal control; the time required to achieve true homoiothermy depends upon the species. Guinea pigs are relatively mature at birth and have more resistance to cold than the newborn cat or rat. In the human infant, homoiothermy becomes established at one to two years of age. Huddling for warmth is effective in polytocous species and in animals born in a nest. The inclination towards communal huddling persists to adulthood in pigs: pigs as heavy as 90 kg may lie on top of one another in sub-zero weather when shelter is limited. Monotocous species giving birth in the open suffer high neonatal mortality during cold seasons.

Thermoregulation by nest building in mammals and by incubation and broodiness in birds are striking examples of quantitative regulation of parental behaviour supplying heat until the young develop their own regulatory mechanisms. These behavioural patterns make survival possible in conditions in which young animals would otherwise die of exposure and exhaustion.

Nest building. The behavioural effects of cold environments and hypothermia on albino rats have been extensively investigated; these data should be generally applicable to larger domestic mammals and perhaps to domestic birds as well. Nest building by albino rats is thermoregulatory behaviour in that the activity is initiated by peripheral body cooling and inhibited by peripheral heating (Kinder, 1927). The behaviour occurs as early as the 20th day of age, appears in both sexes, and persists at a fairly constant level throughout life. An increase in relative humidity at low temperatures tends to increase nest building activity. At air temperatures above 80°F, nest building is inhibited in all rats except mothers suckling young. Adult animals continue to build nests until the ambient temperature drops to 40°F, but young animals cease to build nests at ambient temperatures below 56°F (presumably because of the increasing sluggishness caused by hypothermia). Nests constructed at low air temperatures (around 50°F) are compact and close-knit, whereas those built at high temperatures (around 80°F) are loose and scattered. The nests of wild animals serve as refuges from predators, living quarters during periods of diurnal inactivity, and nurseries for the young.

Incubation and broodiness. Careful regulation of the temperature and humidity of the immediate environment of avian eggs is necessary. The female usually selects a nest site which is shaded or in the soil. The Megapodes bury eggs in soil which is warmed by solar radiation, volcanic activity, or the decay of vegetative material. In the genus *Alectura*, the male digs holes in the mound, rams its head into the holes and then adds material (which raises the temperature) or removes material (which cools the mound) in order to maintain the desired temperature (cf. Lehrman, 1961). The time spent sitting on the eggs is greater at lower ambient temperatures than at higher ones. On very hot days, some birds cool the eggs by standing over them. The average mound and egg temperatures of various Arctic birds, measured in the nest, are about the same as those of species living in milder climates.

The transfer of heat from the bird to the eggs takes place across "brood patches", areas on the ventral side of the bird which become bare during the reproductive season and contain elaborate capillary circulatory systems (Fig. 7). Some birds possess one large brood patch, while others, such as the Herring

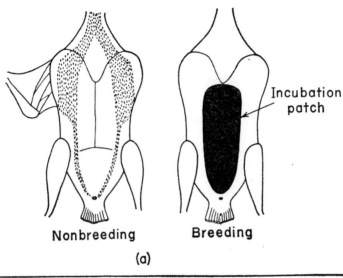

Nonbreeding   Breeding

(a)

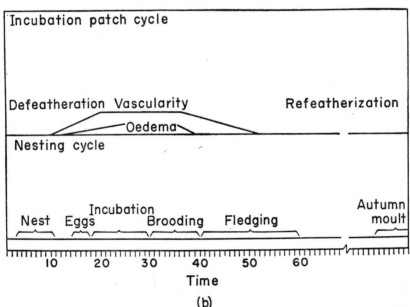

Incubation patch cycle

Defeatheration Vascularity   Refeatherization

Oedema

Nesting cycle

Incubation   Autumn
Nest   Eggs   Brooding   Fledging   moult

Time

(b)

Fig. 7. Incubation behaviour in the female white-crowned sparrow: (*a*) the location of the incubation brood patch; (*b*) stages in brood patch development. (From Marler & Hamilton, 1966. *Mechanisms of Animal Behavior*, New York: John Wiley; after Bailey 1952. *Condor,* **54,** 121.)

gull (*Larus argentatus*), have as many as three separate patches. When the female settles on her nest, all movements (ruffling the belly feathers, wagging, quivering, shifting the eggs) serve to maximize contact between the brood patch and the eggs, which are thereby insulated from the external environment.

At times, the danger to the eggs from overheating may be more important than the more common problem of keeping them warm. Desert birds, which place their nests on the ground, shade their eggs to avoid fatally overheating the embryo. Likewise, the nestlings must be kept cool and provided with water after hatching, which can be a major problem if food with a high water content is not readily available.

II. BIOLOGICAL RHYTHMS AND PHOTOPERIODIC RESPONSES

Solar radiation directly or indirectly exerts a profound effect on the behaviour of organisms. In the form of thermal radiation (short-wave or infrared), it provides heat which is sought in cold environments and avoided in hot environments. In the form of visible radiation (light), it provides the photoperiod which governs many, if not all, diurnal and seasonal activity patterns.

Many animals exhibit a precise rhythm of motor activity which appears to be controlled by an internal physiological clock. When a process repeats itself in more or less constant time intervals, the phenomenon is described as a "biological rhythm". Changes in biological rhythms may result from interactions with such external physical rhythms as day length, temperature, humidity, or barometric pressure (Pittendrigh, 1961; Aschoff, 1963, 1965; Bunning, 1964; Sollberger, 1965).

Many animals consistently exhibit specific behaviours each day at a regular time. Some of these are caused by habit, as when cows gather just before milking. Others are initiated by diurnal changes in environmental conditions. Most of the latter activities occur in cycles of about 24 hours and are called *circadian rhythms* from the Latin words *circa* and *dies*, meaning "about a day".

Circadian rhythms are defined as innate self-sustaining oscillations which can be entrained by certain environmental variables, particularly day length and temperature (Aschoff, *loc. cit.*; Hart, 1964). They involve neither learning nor imprinting. Animals of various species, when raised for several generations under constant conditions, still exhibit circadian rhythms. (This is, in fact, one definition of a circadian rhythm: a rhythm which continues under conditions of constant light and/or constant darkness.) Typical examples include the locomotor and food gathering activities of mammals, and the perching and egg-laying activities of birds. These measurable activities bear a fixed relationship to the day-night cycle. The precise temporal relationships between day-night cycles and the activity rhythms of domestic animals are unknown. There are, however, typical cases among invertebrates in which circadian rhythms are used as true "clocks". Honey bees visit particular flowers at certain times of the day coinciding with the period of nectar secretion. Such synchronization has adaptive significance for both species: bees gather maximum amounts of food with minimal effort, and plants are pollinated without the metabolic waste involved in continuous nectar secretion.

Photoperiodic responses. The synchronization of annual reproductive cycles with environmental factors which influence survival of the offspring involves, at least in large part, physiological responses to seasonally changing day length (photoperiod). The sexual activity of many mammals and birds is primarily controlled by the light-dark ratio. However, the nature and duration of the sexually active season of any species is also related to the extent of its domestication, its geographical origin, and its inherent reproductive biology. The seasonal incidence of oestrus in sheep is inversely related to day length and latitude (Fig. 6). At the equator, where the day length is constant, any seasonal sexual activity is probably conditioned by environmental temperature, rainfall, pasture condition and/or nutrition. The onset of oestrus in sheep can be experimentally modified by changing the photoperiod (Hafez, 1952). Thwaites (1965) found that Merino sheep lose their normal pattern of seasonal breeding after one year when exposed to artificial equatorial photoperiods: the occurrence of oestrus becomes sporadic and apparently unrelated to any definable external environmental factor.

An animal mediates some of its physiological responses to environmental stimuli by means of the hypothalamus and pituitary gland. Photoperiod, via the retina and optic nerves, stimulates the release of FSH and LH from the adenohypophysis. The pituitary's secretory cells are activated by a chemical substance (neurosecretory substance) discharged from certain hypothalamic neurons. Neurosecretory material passes down the axons and diffuses into the capillary loops of the hypophysical portal vessels, whence it passes into the sinusoids of the adenohypophysis. External stimuli may also influence hypothalamic-hypophysical function via numerous other exteroceptive pathways (Hafez, 1968).

The functional day (when light intensity is sufficient to initiate hypothalamico-hypophyseal secretions) begins 25 minutes before sunrise and ends 25 minutes after sunset (civil twilight), i.e. when the sun is about 6° below the horizon. Animals respond to light when there is sufficient light to see a large object clearly. Full moonlight also probably has some photoperiodic effect on sheep and cattle.

Future investigations, using controlled environmental chambers, are needed if we are to understand the relative importance of various environmental factors on the physiological responses of domestic species.

III. CHEMICAL AND PHYSICOCHEMICAL FACTORS

Chemical and physicochemical factors play relatively minor roles in the behaviour of domestic animals, since these factors are by and large adequately controlled by man or are not in critically short supply. Oxygen, nitrogen, and carbon dioxide are almost always present in adequate amounts, except perhaps at high altitudes. The decreased oxygen supply at high altitudes (10,000 to 12,000 feet) limits both the rate and the amount of motor activity which the non-acclimatized animal can perform. However, physiological adjustments are possible, enabling acclimatized animals to perform quite efficiently at high altitudes. Soil types generally affect behaviour only in so far as they do or do not support adequate vegetative growth: lush succulent pastures result in relatively short grazing times and distances, whereas sparse pastures increase the amount of work necessary to obtain an adequate daily ration (see Chapter 9).

Rainfall and air movements modify behaviour by imposing or reducing temperature stresses upon animals. Rainfall in a hot-dry environment and air movement in a hot-wet environment both help an animal dissipate excess body heat and are therefore not avoided. Rainfall in a hot-wet environment and air movement in a hot-dry environment are both undesirable, the first because it makes evaporative cooling more difficult, and the second because it prevents the efficient use of water for cooling purposes; these conditions, however, are not necessarily avoided by the animal unless environmental parameters are varied beyond the animal's range of thermal neutrality. Both rainfall and air movement are detrimental in cold environments, since they tend to work against the conservation of body heat; the animal therefore actively avoids rain and wind by shelter-seeking behaviour.

IV. AVAILABLE SPACE

The consideration of space as a factor in the physical environment is often neglected, despite its prominent role in animal and human societies. The concept of territories (see Chapters 1 and 5) is based on the relationship between social behaviour and space. The direct effects of limited space on the individual have been demonstrated in the following simple experiments: four sunfish (*Lepomis cyanellus*) grouped in an aquarium grew faster than did single individuals in one-fourth the total aquarium volume; also, single fish in a larger tank grew better than those in a smaller one (Allee *et al.*, 1948).

Levy (1944) examined the behavioural consequences of limited space and movement restrictions. He noted that chickens isolated in laying cages (about 1/2 sq. ft of floor space) displayed quick rotary spasms of the head (*head-shaking tic*) much more frequently than those roaming freely in pens. The *tic* was directly related to available space, since birds isolated in larger cages performed fewer head shakes per unit time than those in small laying cages. Large parrots (*Psittacus erithacus* and probably all *Amazona* species) exhibit a stereotyped pattern of headwiping when closely confined; this activity is much less frequently observed among birds in spacious quarters (Dilger, 1961).

In few places is the interaction between space and social organization more apparent than at the feed trough. If such space is in short supply, dominant animals do not hesitate to replace subordinate ones, which in turn are forced to chase away still more subordinate animals (Schein & Fohrman, 1955). The lowest ranking animals consequently get less to eat. The spatial distribution of food, as well as its total quantity, must also be considered when studying food as a factor in the ecology of a species (Orgain & Schein, 1953).

The adverse effects of high population densities, which result in decreased space per individual, as well as increased intensity of social interactions, have been amply demonstrated. Under high population pressures, the reproduction and survival rate of wild mammal populations decreases sharply, probably largely through endocrine and physiological disturbances of the individual (Christian, 1950, 1957; Christian & Davis, 1964). On the other hand, Bressler & Estep (1959) have been able to maintain domestic chickens at three times the "standard" population densities: 1 sq. ft of floor space per bird as opposed to 3 sq. ft per bird. They found no significant differences in egg production between crowded and uncrowded birds, if they supplied adequate amounts of feed, water, and nesting and perch space to each bird; however, there was a noticeably greater incidence of cannibalism among crowded birds, which again illustrates the inseparability of space from social interactions. Modern systems of animal husbandry, such as confinement rearing, aim at the most economical, rather than the maximum, utilization of available space.

V. PRACTICAL AND THEORETICAL CONSIDERATIONS

The husbandryman strives to maintain his animals in as favourable an environment as possible, since this course seems the most practical in terms of optimum production and reproduction. Deviations from optimum conditions result in physiological and behavioural adaptations which presumably are made at the expense of the animal's production output. Although man cannot control the overall weather pattern as yet, he can and does create small favourable climatic régimes for his animals. He does so by constructing barns, shelters, and shade pens, and installing fans, heaters, ventilators, lights, and/or sprinklers in them. Breeds, strains, and varieties of domestic animals have been developed for specific environments, such as Zebu cattle and lightweight breeds of fowl for tropical areas, as opposed to European cattle and heavy breeds of fowl for temperate regions.

Although favourable climatic regimes for domestic animals are desirable if only for humane (anthropomorphic) reasons, serious doubts have arisen concerning the direct effect of environmental stress on animal production (McDowell, 1959). Many early studies were so concerned with feed intake that the environmental conditions previously recommended as optimal may be quite unnecessary. For example, in hot-wet areas heat dissipation is aided by adequate shade, increased air movement, and adequate amounts of preferably cool water for drinking or wallowing. These conditions help to maintain normal body temperatures and therefore minimize or eliminate the behavioural consequences of hyperthermia. However, recent findings with dairy cattle and swine in Georgia, U.S.A. (lat. 32°N) make the value of shade questionable in hot humid environments: shading permits the animal to avoid making physiological adjustments to the environment so that when severe days occur, shaded animals are more markedly affected than those that have lived in the sun (McDowell, 1961). Unshaded cows take in slightly less feed, but produce as much milk as shaded animals, and are therefore more efficient to the husbandryman.

In the past, relationships between the physical environment and the behaviour of animals were often gleaned as by-products of physiologically oriented studies. With the emergence of animal behaviour as a respectable field of study, more and more direct investigations of the behavioural consequences of environmental fluctuations are being undertaken. These are made possible primarily by the development of adequate tools for measuring behaviour and by the establishment of a firm physiological framework with which to explain behavioural acts.

Environmental factors which are poorly understood or even as yet unrecognized may play a role in future behavioural investigations: ionization in the surrounding air, cosmic radiation, and coriolis forces, for example, may conceivably affect behaviour. As the effects of more environmental factors become apparent, electronic computers will undoubtedly be used as the most efficient means of accounting for the myriad of potentially important environmental variables. The result of such operations will help us to understand and therefore to predict more successfully the behaviour of an animal in complex dynamic circumstances.

REFERENCES

ALLEE, W. C., EMERSON, A. E., PARK, O., PARK, T. & SCHMIDT, K. P. (1949). *Principles of Animal Ecology*. Philadelphia: Saunders.

ALLEE, W. C., GREENBERG, B., ROSENTHAL, G. M. & FRANK, P. (1948). Some effects of social organization on growth in the green sunfish, *Lepomis cyanellus*. *J. exp. Zool.*, **108**, 1–20.

ALMQUIST, H. J. (1953). *Feed Requirements of Turkeys as Related to Time of Year*. The Grange Co., Modesto, Calif., 9 pp.

ASCHOFF, J. (1963). Comparative physiology: diurnal rhythms. *Ann. Rev. Physiol.*, **25**, 581–600.

ASCHOFF, J. (1965) (ed.). *Circadian Clocks*. Amsterdam: North-Holland Publ. Co.

BALDWIN, B. A. & INGRAM, D. L. (1966). Effects of cooling the hypothalamus in the pig. *J. Physiol.*, **186**, 72–73P.

BRESSLER, G. O. & ESTEP, R. D. (1959). Pullet performance in the solar house and a well insulated conventional house, 1956–57 and 1957–59. *Pa. Agr. Expt. Sta. Prog. Rept. No. 208*.

BRODY, S. (1948). Environmental physiology with special reference to domestic animals. I. Physiological backgrounds. *Mo. Agric. Expt. Stat. Res. Bull. No. 423*.

BRODY, S. (1956). Climatic physiology of cattle. *J. Dairy Sci.*, **39**, 715–725.

BUNNING, E. (1964). *The Physiological Clock*. New York: Academic Press Inc.

CHRISTIAN, J. J. (1950). The adreno-pituitary system and population cycles in mammals. *J. Mammal.*, **31**, 247–259.

CHRISTIAN, J. J. (1957). A review of the endocrine responses in rats and mice to increasing population size including delayed effects on off-spring. *Lect. & Rev. Ser. 57–2, Naval Med. Res. Inst.*, Bethesda, Md., pp. 445–462.

CHRISTIAN, J. J. & DAVIS, D. E. (1964). Endocrines, behavior and population. *Science*, **146**, 1550–1560.

DILGER, W. C. (1961). *Personal communication*. Laboratory of Ornithology, Cornell Univ., Ithaca, N.Y.

FINDLAY, J. D. & BEAKLEY, W. R. (1954). Environmental physiology of farm mammals. In: *Progress in the Physiology of Farm Animals*, J. Hammond (ed.), Vol. I, London: Butterworth.

GORDON, M. S. (1968). *Animal Function: Principles and Adaptations*. New York, N.Y.: The Macmillan Co.

HAFEZ, E. S. E. (1952). Studies on the breeding season and reproduction of the ewe. *J. agric. Sci.*, **42**, 232–233.

HAFEZ, E. S. E. (1968) (ed.). *Adaptation of Domestic Animals*. Philadelphia: Lea & Febiger.

HAFEZ, E. S. E., BADRELDIN, A. L. & SHARAFELDIN, M. A. (1956). Heat-tolerance studies of fat-tailed sheep in the subtropics. *J. agric. Sci.*, **47**, 280–286.

HART, J. S. (1964). Geography and season: mammals and birds. In: *Handbook of Physiology*, Sect. 4: *Adaptation to the Environment*. D. B. Dill, E. F. Adolph & C. G. Wilber (eds.). Washington, D.C.: American Physiological Society.

HEITMAN, H., JR. & HUGHES, E. H. (1949). The effects of air temperature and relative humidity on the physiological well being of swine. *J. anim. Sci.*, **8**, 171–181.

HELLMER, L. A. (1943). The effect of temperature on the behavior of the white rat. *Amer. J. Psychol.*, **56**, 408–421.

HESSE, R., ALLEE, W. C. & SCHMIDT, K. P. (1937). *Ecological Animal Geography*. New York, N.Y.: Wiley.

HOWELL, T. R. & BARTHOLOMEW, G. A. (1961). Temperature regulation in Lysan and black-footed Albatrosses. *Condor*, **63**, 185–197.

IRVING, L. (1964). Terrestrial animals in cold: birds and mammals, pp. 361–377. In: *Handbook of Physiology*, Sect. 4: *Adaptation to the Environment*. D. B. Dill, E. F. Adolph, and C. G. Wilber (eds.). Washington, D.C.: American Physiological Society.

KINDER, E. F. (1927). A study of the nest-building activity of the albino rat. *J. exp. Zool.*, **47**, 117–161.

LEE, D. H. K. (1948). Heat and cold. *Ann. rev. Physiol.*, **10**, 365–386.

LEE, D. H. K. (1950). Studies of heat regulation in the sheep with special reference to the Merino. *Austral. J. agric. Res.*, **1**, 200–216.

LEE, D. H. K., ROBINSON, K. & HINES, H. J. G. (1941). Reactions of the rabbit to hot atmospheres. *Proc. roy. Soc. Queensld.*, **53**, 129–144.

LEE, D. H. K., ROBINSON, K. W., YEATES, N. T. M. & SCOTT, M. I. R. (1945). Poultry husbandry in hot climates; Experimental enquiries. *Poult. Sci.*, **24**, 195–207.

LEHRMAN, D. S. (1961). Hormonal regulation of parental behaviour in birds and infrahuman mammals. In: *Sex and Internal Secretions*. W. C. Young (ed.). Ch. 21, vol. II. Baltimore: Williams & Wilkins Co.

LEVY, D. M. (1944). On the problem of movement restraint. *Amer. J. Orthopsychiatry*, **14**, 644–671.

LONG, E. & GODFREY, G. F. (1952). The effect of dubbing, environmental temperature, and social dominance on mating activity and fertility in the domestic fowl. *Poult. Sci.*, **31**, 665–673.

MACFARLANE, W. V. (1968). Adaptation of ruminants to tropics and deserts, pp. 164–182. In: *Adaptation of Domestic Animals*. E. S. E. Hafez (ed.). Philadelphia: Lea & Febiger.

McDOWELL, R. E. (1959). Adaptability and performance of cattle under the climatic conditions existing in the southern United States. *Proc. 15th Internat. Dairy Congr., London*, **1**, 333–340.

McDOWELL, R. E. (1961). *Personal communication*. Dairy Cattle Research Branch, U.S. Dept. Agr., Beltsville, Md.

MOORE, K. (1944). The effect of controlled temperature changes on the behavior of the white rat. *J. exp. Psychol.*, **34**, 70–79.

ORGAIN, H. & SCHEIN, M. W. (1953). A preliminary analysis of the physical environment of the Norway rat. *Ecology*, **34**, 467–473.

PAYNE, W. J. A., LAING, W. I. & RAIVOKA, E. N. (1951). Grazing behaviour of dairy cattle in the tropics. *Nature, Lond.*, **167**, 610–611.

PETERSON, R. R. & YOUNG, W. C. (1955). Prolonged cold, sex drive and metabolic responses in the male guinea pig. *Amer. J. Physiol.*, **180**, 535–538.

PITTENDRIGH, C. S. (1961). On temporal organization in living systems. In: *Harvey Lectures*, pp. 93–125. New York, N.Y.: Academic Press Inc.

ROBINSON, K. & LEE, D. H. K. (1941a). Reactions of the dog to hot atmospheres. *Proc. roy. Soc. Queensld.*, **53**, 171–188.

ROBINSON, K. & LEE, D. H. K. (1941b). Reactions of the cat to hot atmospheres. *Proc. roy. Soc. Queensld.*, **53**, 159–170.

RUSSELL, R. W. (1956). Some effects of severe hypothermia in behaviour. *Brit. J. anim. Behav.*, **4**, 75.

SCHEIN, M. W. & FOHRMAN, M. H. (1955). Social dominance relationships in a herd of dairy cattle. *Brit. J. anim. Behav.*, **3**, 45–55.

SCHEIN, M. W., McDOWELL, R. E., LEE, D. H. K. & HYDE, C. E. (1957). Heat tolerances of Jersey and Sindhi-Jersey crossbreds in Louisiana and Maryland. *J. Dairy Sci.*, **40**, 1405–1415.

SCHMIDT-NIELSON, K. (1959). The physiology of the camel. *Sci. Amer.*, **201**, 140–151.

SCHMIDT-NIELSON, K. & SCHMIDT-NIELSON, B. (1952). Water metabolism of desert mammals. *Physiol. Rev.*, **32**, 135–166.

SEATH, D. M. & MILLER, G. D. (1946). Effect of warm weather on grazing performance of milking cows. *J. Dairy Sci.*, **29**, 199–206.

SEATH, D. M. & MILLER, G. D. (1947). Heat tolerance comparisons between Jersey and Holstein cows. *J. anim. Sci.*, **6**, 24–34.

SOLLBERGER, A. (1965). *Biological Rhythm Research*. Amsterdam: Elsevier.

TEICHNER, W. H. (1957). Manual dexterity in the cold. *J. appl. Physiol.*, **11**, 333–338.

TEICHNER, W. H. (1958). Reaction time in the cold. *J. appl. Psychol.*, **42**, 54–59.

TEICHNER, W. H. & KOBRICK, J. L. (1955). Effects of prolonged exposure to low temperature on visual-motor performance. *J. exp. Psychol.*, **49**, 122–126.

THWAITES, C. J. (1965). Photoperiodic control of breeding activity in the Southdown ewe with particular reference to the effects of an equatorial light régime. *J. agric. Sci.*, **65**, 57–64.

WEISS, B. & LATIES, V. G. (1961). Behavioural thermoregulation. *Science*, **133**, 1338–1344.

WORSTELL, D. M. & BRODY, S. (1953). Environmental physiology and shelter engineering with special reference to domestic animals. XX. Comparative physiological reactions of European and Indian cattle to changing temperature. *Mo. Agric. Expt. Stat. Res. Bull. No. 515.*

Chapter 5

The Social Environment and Behaviour

A. M. Guhl

The term "social environment" refers to organizations of animals which occur in groups or colonies, in which one or both parents survive to co-operate with their young when they are mature, and in which a division of labour occurs (Michener, 1953). In this sense it applies only to man and certain social insects. Although such limitations can be justified, an evolutionary viewpoint (Allee, 1931, 1938, 1947) seeks simpler relationships on a subsocial, or even protosocial, level to develop a broader base for what is now known as sociobiology or psychobiology. According to Carpenter (1942) social behaviour refers to the reciprocal interactions of two or more animals and the resulting modifications of an individual's action. Usually the interactions considered are by individuals of the same species, although in some cases another species may be included (Scott, 1951).

In modern methods of animal husbandry man exerts considerable control over intraspecific relationships. The types of social behaviour within a species can be placed into broad categories such as social dominance, sexual behaviour, and parental behaviour. Each of these may include interactions between individuals of the same or opposite sex, between parent and young, young of different age classes, castrates and normals, and sometimes aberrant behaviour. The degree to which these social relationships are controlled varies greatly. For example, large breeding herds of sheep and beef cattle may include all possible relationships whereas "feeders" may approximate an age class of females and castrates. Matings are controlled in dairy herds and cows are moved from group to group according to state of lactation. Among domestic birds the turkeys and broilers are age classes regardless of sex, whereas layers among chickens are usually segregated according to age and sex. The incubator substitutes for parental behaviour among the fowls. Thus man simplifies some social relationships and complicates others as indicated in this and other chapters.

I. LIVING SPACE AND BEHAVIOUR

Man determines the space and available facilities for his domestic animals. The area may be restricted to a cage, to an area circumscribed by the radius of a chain, or to hundreds or thousands of acres for cattle.

The first concern of newcomers, whatever the species, is to investigate the strange area, much as residents do when any alteration or addition is made to their home area. Thus, if newcomers are placed with residents they are at some disadvantage, especially in competitive situations. Until an animal has familiarized itself with its surrounding sights, sounds, and odours, it shows some tenseness or alertness, ready to flee it knows not where. Since habits are formed during the first few days or weeks in a new situation, management practices should facilitate the newcomer's adjustments.

The facilities available to domestic animals need more attention in the light of behaviour. Management practices among some farm animals often have more consideration for the convenience (or cost per man per hour) of the caretaker than for the optimal productive efficiency of his charges. For example, the accessibility of feed may be more important to the animals than the amount of feed present. Thus a predetermined minimum amount of feeding space per individual, perhaps in several scattered feeding areas, may be most economical for optimum production although such management may be more demanding of time and equipment. This would be especially true of species or breeds which are highly competitive.

The accessibility of feed influences the degree of competition, the space available affects the extent of mobility during interactions and density as well as the population size determine how frequently any two individuals may meet. These factors determine to some extent the intensity of social interactions and the reinforcement of habitual social relationships.

Overcrowding may have deleterious effects which have been considered among causes of cyclic populations in some wild mammals. Christian (1950) developed a working hypothesis for the population crash in which he related the symptoms and conditions associated with "die-offs" to the *General Adaptive Syndrome* of Selye (1947). Subsequent experiments by Christian & Davis (1955), Christian (1956) and others noted that stresses associated with population densities resulted in enlargement of the adrenals. Somewhat similar reactions may occur during crowding of domestic species. Laying hens with a floor area of 1·33 sq. ft. per bird had significantly heavier adrenals than others penned at 4 sq. ft. per bird (Siegel, 1959a). Social competition among cocks might influence the weight and function of the adrenals (Siegel & Siegel, 1961). Males maintained in groups had heavier adrenals than those kept in individual cages, and these glands also increased in weight when several cocks were placed singly as strangers into a pen of acquainted males. The extent to which continued stressors from overcrowding or other factors may cause reactions beyond the stage of resistance is still to be explored. Crowded pullets reduced their clutch length and adrenals were hypertrophied but these results could not be correlated with egg production factors (Siegel, 1959b).

II. HETEROGENEITY OF THE GROUP

With the tendency towards specialization in agriculture and an attempt to increase efficiency, species are now segregated, although mixing of breeds among dairy or beef cattle may be common. Castrates are less aggressive than normal males and may be maintained with females, but the trend is towards unisexual grouping, except with horses where geldings may be run with mares. There are many factors, managerial and economic, that brought about these changes. An example of the effects of heterogeneity in groups may be given for chickens. When breeds and strains were reared, intermingled and compared with similar birds reared in pure strain groups, King & Bray (1959) found differences with respect to methods of rearing but considered them borderline in importance. However, in a very similar test, Tindell & Craig (1959) noted that interstrain competition seemed to result in higher performance levels for the more aggressive strains, whereas the less aggressive strains tended to mature later, feed less often, lay at a lower rate, and to have poorer livability. In flocks of less than 10 breed recognition and previous experience with an individual of a different breed results in a tendency for members of one breed to react to the other as an entity (Hale, 1957). In a herd of dairy cows certain differences between breeds may influence the ranks attained in the social order (Guhl & Atkeson, 1959).

III. HOMOGENEOUS GROUPS

Most domestic animals form some kind of social organization. If all the individuals in a group are reared together the social order forms during the period of growth and maturation (Guhl, 1958). If the individuals are assembled from different sources, i.e. are unacquainted, they establish social relations soon after they are brought together. There are other methods by which individuals enter a group, for example, in flocks of sheep (Scott, 1945) and breeding herds of beef cattle the young are born and reared within the group. In dairy herds some individuals may be removed temporarily (e.g. maternity cases) or permanently, and are replaced by younger animals (Schein & Fohrman, 1955; Guhl & Atkeson, loc. cit.). Whatever the manner of entry into the group, social relationships are formed immediately or shortly thereafter.

Social stress may develop when absentees are returned to the group or when small organized groups are combined. The length of the memory span of the species is a factor in the level of strife which may occur during the reassembly of former associates. In chickens an absence of 2 weeks may be sufficient to cause fighting when returned to the home flock (see Chapter 16). However, dairy cows re-enter the herd after calving with few complications.

Evidences of effects of seniority in a group were found in shifting flock memberships among chickens (Guhl & Allee, 1944). The longer the residence in the pen the higher the rank. When the complete assemblage of a large flock of pullets ranged over several weeks, it was found that those housed first ranked higher in the peck order than those added late (Guhl, 1953). In groups of dairy cows the newcomers are usually heifers which have calved recently, and here age is a factor associated with seniority. However, such factors as body weight, skill in encounters, and experience are closely related to age. Nevertheless, seniority was found to be an important factor in the attainment of high rank among cows (Schein & Fohrman, 1955; Guhl & Atkeson, loc. cit.). All of these results suggest that, if possible, the ideal situation would be a group in which the individuals were assembled at the same time, were uniform in age, of the same breed or strain and sex.

IV. SOCIAL ORGANIZATION OF THE GROUP

Levels of aggressiveness and types of agonistic behaviour vary among domestic species and even between breeds or strains within species. For most domestic animals the social order is based on some form of domination. Aggressive and submissive behaviour may be vocal signals, posture stances, gestures, in addition to manners of attacking an opponent. Schloeth (1956, 1958) describes the posture and signals for a number of bovine species. Dogs may growl, snap, or bite; cows bunt or butt; horses may kick, attempt to bite with bared teeth and ears laid back; chickens peck; and pigeons and turkeys slap an adversary with one wing as well as peck. Subordinate status is shown by some form of avoidance or flight, although other behaviour patterns such as crouching or even infantile behaviour may be evoked when under extreme harassment. These behaviour patterns usually occur between strangers when dominance relations are being established, and under competitive situations when they serve to reinforce the interindividual dominance relationships.

Table 8. Dominance Relationships in Small Flocks of Chickens

Each bird pecks those listed below it; deviations from a straight line peck order are indicated by arrows. The letters represent colours used for individual identification.

Number pecked	Peck order	Number pecked	Peck order	Number pecked	Peck order
9	G	10	BY	9	GV
8	Y	9	G	7	→G
7	RV	8	YG	6	→GB←
6	RG	6	→GV	6	VB←
5	RR	5	└GG←	5	→RR
4	V	4	YY┘	5	GY┘
3	B	3	RG	3	B
2	VV	3	YV	2	└VY←
1	RY	1	V	1	GR
0	R	0	B	1	BB┘

The types of social relations within a species determine the form of the social organization (Scott, 1956). Vertebrate species have societies with characteristic structures, but for convenience they may be classified into two main categories, dominance orders, or hierarchies, and territories. The hierarchies (Allee, 1952) may be based on unidirectional despotism, a "peck right" system, in which the individuals are ranked according to the number of individuals in the group that each may dominate without any retaliation. The order may be straightline or have a geometrical pattern composed of "pecking triangles" (see Table 8). Some species, e.g. pigeons and doves, show bidirectional dominance (formerly designated as "peck-dominance") in which individuals exchange "pecks" but one of each contact pair maintains an advantage by means of a higher frequency of attacks (Bennett, 1939). This social order is less stable than the "peck right" system.

The second major category, territory, has been defined in various ways (Hinde, 1956; Carpenter, 1958). Usually the members of a species are organized on a spatial pattern in which an individual (or mates) shows aggression towards any invaders of the same species. There are several kinds of territory and the functions ascribed to them are many. Territories may be relatively permanent or seasonal. Among domestic animals this type of organization is unusual although male cats show territorial behaviour (Green et al., 1957), and dogs may be intolerant of other dogs in or near their premises. Among domestic birds the ducks pair and establish territories, and group territorialism was found by McBride & Foenander (1962) in chickens. Some species maintain a territory within the home range. Burt (1940) defines the home range as the area which is traversed by an animal in its normal activities of searching for feed, mating, and caring for the young.

The evolutionary significance of aggressiveness still needs to be clarified. It is most common among vertebrates from fishes to man. Probably the tendency to be self-assertive contributed to reproductive and individual success in a highly competitive environment. But it also tends towards intraspecific strife and the dispersion of individuals as in territorialism. However, the ability to form a dominance order, made possible by the alternative submissive behaviour, permits the establishment of a within-group organization. The framework of the hierarchy channels the activities of the individuals and thereby reduces strife, promotes toleration by the formalization of agonistic behaviour, and conserves energy. Social life has a number of advantages to the individual and to the species (Allee, 1938, chapter 5; Tinbergen, 1953, chapter 3; Scott, 1958, chapter 11).

There is a trend towards maintaining some domestic animals in spatial isolation. The caged layer system is now rather common with chickens, and is being tried with other domestic animals. Reference is not to stalls or stanchions, but to an enclosure in which a single animal is confined. The type of isolation and the age (see Chapter 16) at which it is begun may affect an animal's behaviour. If the isolation of adults is essentially spatial, and it does not obscure others visually and audibly, the disadvantages may be minor. However, feed consumption may be reduced if there is any interference with social facilitation, and sexual behaviour may be modified when male and female are brought together. Among a number of managerial advantages, the one related to behaviour is the elimination of social stress for subordinate individuals. Problems related to social dominance are reintroduced in chickens with crowding two or more layers into small cages. Nevertheless it is claimed that multi-bird caging is more profitable despite a reduction in egg production and viability (Blount, 1965; Wilson et al., 1967).

V. INTEGRATION OF THE GROUP

Dominance orders establish sequences of precedence according to rank in certain competitive situations within the group. In themselves, they do not integrate the group *per se*, but by channelling interindividual behaviour the social orders do promote the formalization of agonistic behaviour and thereby reduce actual fighting. The rank order shows that members of the group not only recognize one another but also form appropriate habits of interaction with each member of the group. Although the frequency of encounters between dominant and subordinate members of a pair reinforce these habits, the submission or mild avoidance of the inferior is a weaker stimulus for aggression than is the normal posture. Therefore the intensity of aggression becomes reduced as does also that of avoidance, and the behaviour patterns shift to symbolic forms such as threats and other gestures. Thus toleration develops within the group, and social inertia promotes social stability. It is possible for the tolerance between some pairs of individuals to develop to the point of extinction of agonistic behaviour and the dominance-subordinate relationship. In such cases incidences may arise to renew the relationship or to cause a reversal of dominance (Guhl, 1968).

Groups of animals are integrated through various types of behavioural mechanisms. The stimuli may be visual, auditory, chemical, and tactile, and the relative importance of each is related to the sensory capacities of the species. The stimuli act as social signals which co-ordinate activities such as feeding, sexual behaviour, escape from enemies, resting, and roosting, to mention only a few. Visual and auditory signals often function at a distance in some domestic animals although odours dominate in perception in those such as dogs. Tactile stimuli function in close associations such as sexual contacts, parent-young relations, and mutual grooming. Learning is involved in these reciprocal stimulus-response relations which are reinforced by the interaction of drives. Thus it may be concluded that an organized group has an entity which is more than the summation of its parts.

Some of the benefits to the individuals, and indirectly to the husbandman, of the integrated flocks, herds, and packs of domestic animals are the result of social facilitation. Scott & Marston (1950) found that dogs running with unfamiliar animals are apt to run the course more slowly than when running alone. Strangeness enhances competition. Social facilitation is usually indicated by any increment of activity by an individual which results from the presence of another individual. The term needs to be distinguished from imitation and allelomimetic behaviour (Vogel *et al.*, 1950), although these two behaviours may lead to social facilitation. This concept is particularly applicable in the training and feeding of domestic animals (Bayer, 1929; Ross & Ross, 1949; Scott, 1945; Smith, 1957). Some such activities are readily observed, as for example, when one individual feeds others are stimulated to do likewise although they may have fed recently. Alarm and warning calls alert others which may join in the chorus. A strong escape reaction by one will cause others to follow, and sexual activities by one male may stimulate other males. Many of the parent-young behaviour patterns may be included in this category.

VI. SOCIAL ORGANIZATION AND SEXUAL BEHAVIOUR

Sexual behaviour may be considered as a category of social behaviour since two or more individuals are involved in more or less complex interactions. With a number of domestic species the breeding programme is controlled by separating the breeding pair from the group. In other species, as in poultry, sheep, and beef cattle, one or more males are permitted to run with the females of the breeding flock or herd. Since males are more aggressive than females and usually have a social order apart from that among the females (Schloeth, 1958; Guhl, 1953), the social ranks of the individuals influence the synchronization of the sexes and the success in mating (see Chapter 16). Briefly, among chickens (Guhl *et al.*, 1945) and turkeys high rank among males is associated with a high frequency of mating whereas among hens it interferes with sexual behaviour. The various behaviour patterns which act as signals appear to function best in an integrated group, one in which the individuals are conditioned to each other. Presumably animals do not recognize sex *per se*, and courting or display behaviour shifts the signal-response patterns from agonistic behaviour to the sexual sequence. As stated by Tinbergen (1953) mating behaviour involves the suppression of escape behaviour in the female and her reactions appease the male, i.e. divert his pugnacity. The type of social organization is only one of many factors that influence reproductive behaviour, including sexual and parental behaviour. The respective chapters for each domestic species present known effects of dominance relationships and territoriality to relative success in sexual behaviour. All activities of a social species interrelate to promote adaptation. All of the facilities and husbandry should conform to the species' potential adaptability.

VII. COMMUNICATION

Animal communication, or zoosemiotics, differs from human communication chiefly in the absence of abstract ideas. However, the sharing and/or conveying of information between members of a group is a common observation. Although vocalizations are the most apparent form of communication, postures and activities with or without accompanying sounds may alert others to the presence of food or enemies. Signal-response mechanisms may be the means of conveying specific information, as for examples, the alarm calls of fowls bring forth specific responses, and certain gaits among ungulates may be a signal which is followed by a stampede. Vocalizations have advantages over visual signals because reception from a distance is not limited to the forward direction, nor do vegetation, various objects, or land contour obscure the source of the information. A useful classification of animal language based on ecological background and functions is given by Collias (1960). Acoustic signals may function in (1) feeding, as in sounds of hunger by young, food finding and hunting cries; (2) warning and alarm calls which announce the approach or presence of an enemy and the all clear signal following the departure of a predator; (3) sexual behaviour and related fighting, such as territorial defence vocalizations and precoitional notes; (4) parent-young interrelations to establish contact and evoke care behaviour; and (5) maintaining the group in its movements and assembly.

Visual cues in communication include postural stances, certain movements associated with stances, and signals from various organs such as facial expressions, tail movements of dogs, raised hackles in chickens. Other sign stimuli may be chemical such as scent markings by male dogs, and identification of females in heat. Among domestic animals various social signals still need to be recognized and their functions determined.

VIII. MAN AND THE SOCIAL ENVIRONMENT

The relative importance of man in the social environment of domestic animals varies with the species and with the frequency and duration of the relationship. Household pets may be in continuous association and actually become a member of the family in many activities. Saddle and draught horses may be socially integrated with man to a lesser extent. Dairy cows have closer relations to man than beef cattle and swine because they are in daily contact with man during milking, feeding, and movements. Range cattle and sheep have the least frequent contact with man because they are born on the pasture and reared by their dams, whereas cats and dogs, as pets, are separated from their mothers post-weaning and become socialized during certain critical periods (see Chapter 6). Among poultry, the laying flocks of chickens and turkeys have the most human contact because eggs must be collected at regular intervals. The degree of association with man influences the level of tameness, as does also the manner in which one approaches and cares for the animals. The closer the relationship the greater the need to understand the behaviour of the species or breed.

All domestic animals make some adjustments to man, from the complex associations with dogs, cats, and horses known as "training", to the feeding and other care given to cattle, sheep, swine, and poultry. Social interaction with man varies from species to species, as for example, dogs and cats must be dominated in an appropriate fashion by their masters. Scott (1950) has stated that behaviour patterns in dogs and men are sufficiently similar so that the meaning of dog behaviour can be recognized by people. The social relationships between domestic species are based upon partial similarity in basic behavioural patterns. This is confirmed experimentally by Menzel & Menzel (1962) in obtaining interspecific social organizations. Although much may be learned from observing pets, scientific experiments in social behaviour with other domestic animals should avoid such intimate adjustments to man, because an animal sometimes considers man as one of its kind and treats him as a member of the same species (zoomorphism) and of its social organization (Hediger, 1950, 1965).

Man, all too often, treats his animals as property rather than as living, responsive organisms. He too frequently considers only the chores to be performed rather than the needs, drives, and responses of his animals. One may readily measure a man's understanding of animals by the manner in which he approaches them and moves among them. Some will startle them or put them to flight, whereas others evoke approach behaviour.

To be successful with domestic animals one should be in rapport with them. Observations have suggested that some persons have much difficulty in sensing the behaviour of their animals, whereas others soon have a complete understanding and can anticipate their reactions. Examples of the latter are found in the writings of Lorenz (1952) and Hediger (1955) with wild animals. For domestic animals the contributions in this book should be useful.

REFERENCES

ALLEE, W. C. (1931). *Animal Aggregations. A Study in General Sociology.* Chicago: University of Chicago Press.

ALLEE, W. C. (1938). *The Social Life of Animals.* New York, N.Y.: Norton.

ALLEE, W. C. (1947). *Animal Sociology.* Chicago: Encyclopedia Britannica.

ALLEE, W. C. (1952). Dominance and hierarchy in societies of Vertebrates. In: *Structure et Physiologie des Sociétés Animales.* Colloques Internationaux 34, P. P. Grasse (ed.), 157–181. Paris: Centre National de la Recherche Scientifique.

BAYER, E. (1929). Beiträge zur zwei Komponententheorie des Hungers. *Z. Psychol.,* **112,** 1–54.

BENNETT, M. (1939). The social hierarchy in ring doves. *Ecology,* **20,** 337–357.

BLOUNT, W. P. (1965). Egg production in multi-bird hen battery cages. *Vet. Rec.,* **77,** 825.

BURT, W. H. (1940). Territorial behavior and populations of some small mammals in southern Michigan. *Misc. Publ. Mus. Zool. Univ. Mich.,* **45,** 1–58.

CARPENTER, C. R. (1942). Characteristics of social behavior in non-human primates. *Trans. N.Y. Acad. Sci., Series II,* **4,** 248–258.

CARPENTER, C. R. (1958). Territoriality: A review of concepts and problems. In: *Behavior and Evolution,* A. Roe & G. G. Simpson (eds.), pp. 224–250. New Haven: Yale University Press.

CHRISTIAN, J. J. (1950). The adrenopituitary system and population cycles in mammals. *J. Mammal.,* **31,** 247–259.

CHRISTIAN, J. J. (1956). Adrenal and reproductive responses to population size in mice from freely growing populations. *Ecology,* **37,** 258–273.

CHRISTIAN, J. J. & DAVIS, D. E. (1955). Reduction of adrenal weight in rodents by reducing population size. In: *Trans. 20th N. Amer. Wildl. Confer.,* pp. 177–198.

COLLIAS, N. E. (1960). An ecological and functional classification of animal sounds. In: *Animal Sounds and Communication,* W. E. Lanyon & W. N. Tavolga (eds.), 368–391. Washington, D.C.: American Institute of Biological Sciences.

GREEN, J. D., CLEMENTE, C. D. & DEGROOT, J. (1957). Rhinencephalic lesions and behavior in cats. *J. comp. Neur.,* **108,** 505–536.

GUHL, A. M. (1953). Social Behaviour of the Domestic Fowl. *Kans. Agric. Expt. Stat. Tech. Bull. No. 73*

GUHL, A. M. (1958). The development of social organization in the domestic chick. *Anim. Behav.,* **6,** 92–111.

GUHL, A. M. (1968). Social inertia and social stability in chickens. *Anim. Behav.* **16,** 219–232.

GUHL, A. M. & ALLEE, W. C. (1944). Some measurable effects of social organization in flocks of hens. *Physiol. Zool.,* **17,** 320–347.

GUHL, A. M. & ATKESON, F. W. (1959). Social organization in a herd of dairy cattle. *Trans. Kans. Acad. Sci.,* **62,** 80–87.

GUHL, A. M., COLLIAS, N. E. & ALLEE, W. C. (1945). Mating behavior and the social hierarchy in small flocks of White Leghorns. *Physiol. Zool.,* **18,** 365–390.

HALE, E. B. (1957). Breed recognition in the social interactions of domestic fowl. *Behaviour,* **10,** 240–254.

HEDIGER, H. (1950). *Wild Animals in Captivity.* London: Butterworths.

HEDIGER, H. (1955). *Psychology and Behaviour of Captive Animals in Zoos and Circuses.* New York, N.Y.: Criterion Books.

HEDIGER, H. (1965). Man as a social partner of animals and vice versa. *Symp. zool. Soc. Lond.* **14,** 291–300.

HINDE, R. A. (1956). The biological significance of the territories of birds. *Ibis,* **98,** 340–369.

KING, S. C. & BRAY, D. F. (1959). Competition between strains of chickens in separate versus intermingled flocks. *Poult. Sci.,* **38,** 86–94.

LORENZ, K. Z. (1952). *King Solomon's Ring.* New York, N.Y.: Thomas Y. Crowell.

McBRIDE, G. & FOENANDER, F. (1962). Territorial behaviour in flocks of domestic fowls. *Nature, Lond.*, **194**, 102.

MENZEL, R. & MENZEL, R. (1962). Über Interferenzerscheinungen zwischen socialer und Rangordnung. *Z. Tierpsychol.*, **19**, 332–355.

MICHENER, C. D. (1953). Problems in the development of social behavior and communication among insects. *Trans. Kans. Acad. Sci.*, **56**, 1–15.

ROSS, S. & ROSS, J. G. (1949). Social facilitation of feeding of dogs. *J. genet. Psychol.*, **74**, 97–108.

SCHEIN, M. W. & FOHRMAN, M. H. (1955). Social dominance relationships in a herd of dairy cattle. *Brit. J. anim. Behav.*, **3**, 45–55.

SCHLOETH, R. (1956). Zur Psychologie der Begegnung zwischen Tieren. *Behaviour*, **10**, 1–80.

SCHLOETH, R. (1958). Cycle annuel et comportement social du Taureau de Carmargue. *Mammalia*, **22**, 121–139.

SCOTT, J. P. (1945). Social behavior, organization and leadership in a small flock of domestic sheep. *Comp. Psychol. Monogr.*, **18** (4), 1–29.

SCOTT, J. P. (1950). The social behavior of dogs and wolves: An illustration of sociobiological systematics. *Ann. N.Y. Acad. Sci.*, **51**, 1009–1021.

SCOTT, J. P. (1951). *Minutes of the Conference of the Effects of Early Experience on Mental Health.* R. B. Jackson Memorial Laboratory, Bar Harbor, Maine, U.S.A.

SCOTT, J. P. (1956). The analysis of social organization in animals. *Ecology*, **37**, 213–221.

SCOTT, J. P. (1958). *Animal Behavior.* Chicago: University of Chicago Press.

SCOTT, J. P. & MARSTON, M. (1950). Social facilitation and allelomimetic behavior in dogs. II. The effects of unfamiliarity. *Behaviour*, **2**, 135–143.

SELYE, H. (1947). The general-adaptive-syndrome and the diseases of adaptation. In: *Textbook of Endocrinology*, H. Selye (ed.), 837–866. Montreal: Acta Endocrinologica, Montreal University.

SIEGEL, H. S. (1959a). The relation between crowding and weight of adrenal glands in chickens. *Ecology*, **40**, 495–498.

SIEGEL, H. S. (1959b). Egg production characteristics and adrenal function in White Leghorns confined at different floor space levels. *Poult. Sci.*, **38**, 893–898.

SIEGEL, H. S. & SIEGEL, P. P. (1961). The relationship of social competition with endocrine weights and activity in male chickens. *Anim. Behav.*, **9**, 151-158.

SMITH, W. (1957). Social "learning" in domestic chicks. *Behaviour*, **11**, 40–55.

TINBERGEN, N. (1953). *Social Behaviour in Animals.* London: Methuen.

TINDELL, R. & CRAIG, J. V. (1959). Effects of social competition on laying house performance in the chicken. *Poult. Sci.*, **38**, 95–105.

VOGEL, H. H., JR., SCOTT, J. P. & MARSTON, M. (1950). Social facilitation and allelomimetic behavior in dogs. I. Social facilitation in a non-competitive situation. *Behaviour*, **2**, 121–134.

WILSON, H. R., JONES, J. E. & DORMINEY, R. W. (1967). Performance of layers under various cage régimes. *Poult. Sci.*, **46**, 422–425.

Chapter 6

The Effects of Early Experience

Victor H. Denenberg

Environmental forces, acting upon an immature organism, can drastically modify the future behavioural and physiological capabilities of that organism. An animal's ability to learn, to solve problems, to copulate, or to survive the effects of a severe stress are critically dependent upon the characteristics of the environment in which it has been reared during its very early life.

At one time it was thought that the major function of the environment was to "support" the developing organism by maintaining a *milieu* which would permit the natural unfolding of the organism's genetic potential. This extreme maturational viewpoint is no longer tenable. The environment may better be thought of as a dynamic force which actively interacts with the psychobiological processes of the living organism at each point in its developmental history. This interaction changes these processes. In turn, these changes affect the manner in which the environment interacts with the modified processes.

These environmental forces have broader and longer lasting effects when administered early in life than when administered to the adult. This is probably a function of the gross immaturity of the organism. There is much evidence from experimental embryology and experimental teratology (the study of monstrosities) to show that relatively small lesions can have widespread effects upon the developing foetus when the organism is still in an early stage of differentiation; the lesion can affect organs and structures which are still to be developed. An analogous relationship holds for behavioural development. Even though the animal, at birth, has developed to a point where it can maintain its physiological integrity in the new environment, it is still in the very early stages of differentiation with respect to most psychobiological processes. Thus, environmental stimulation at this time can have widespread effects. However, this stimulation, unlike the lesions in embryonic development, have generally been found to be beneficial to the future growth and development of the organism.

Very little systematic research on the effects of early experience upon subsequent behaviour was done before the 1940s. Beach & Jaynes (1954) have summarized the findings through the early 1950s, King (1958) has extended the summary somewhat in his discussion of relevant early experience parameters, and Denenberg (1962, 1967) and Levine (1962a, 1962b) have prepared more recent summaries of the literature (cf. Newton & Levine, 1968). Because of the relative newness of this field of investigation, very little work has been done using domestic animals. Thus, the literature summarized in this chapter will be based primarily upon laboratory animals.

The major objective of this chapter is to survey some of the relevant research in the field of early experience with the purpose of emphasizing the primary thesis of this chapter: *environmental factors have profound and drastic effects upon the immature organism, and one can literally change an animal's behavioural and physiological capabilities through the appropriate manipulation of environmental dynamics.*

The first topic to be discussed will be *critical periods in development* followed by a section on *imprinting*. These two topics will give the reader a broad survey of the field. The remainder of the chapter will be concerned with a more detailed analysis of research findings. There are enough consistencies among the procedures used by different researchers to order much of the work under three headings.

Time of stimulation. In general, researchers have administered stimulation before birth, after birth but before weaning, or after weaning. In some studies animals have been stimulated both before and after weaning.

Type of stimulation. In most instances animals have been stimulated in one of two manners. One method is to use *continuous* stimulation as the independent variable and the other method is to use *discontinuous* stimulation. Examples of the former are different environments (e.g. enriched, deprived, isolated) in which the animals live for some relatively long period of time. In the second procedure all animals live in the same environment (usually laboratory cages) but different groups are exposed daily to brief periods of stimulation of a more-or-less intense nature (e.g. handling, shock).

Adult behavioural measures. Included here are such psychological variables as learning, perception, and emotionality and such physiological variables as changes in body weight and degree of resistance to stress.

I. CRITICAL PERIODS

The age at which stimulation is administered in early life is a variable of major importance. This is part of the broader concept of critical periods in development which has been described and discussed in detail (Scott, 1958; Scott & Marston, 1950; Williams & Scott, 1953). The hypothesis states that stimulation administered during certain early periods of development will have a more marked and longer lasting effect than will the same stimulation administered either earlier or later in development. Observations of the development of the dog (Scott, 1958; Scott & Marston, 1950) and the mouse (Williams & Scott, 1953) have led to the abstractions of four major developmental periods: *neonatal*, characterized mainly by growth and development of behaviour patterns already present at birth; *transitional*, which is a period of rapid change and development of new behavioural patterns; *socialization*, which is marked by the occurrence of social interactions between mother and young, or young and young; and, *juvenile*, which begins at weaning and lasts until sexual maturity. From his theoretical analysis Scott concluded that the period of socialization was the most important in the development of behavioural patterns since the central nervous system should be mature enough for relatively permanent learning to occur, and the organism is learning to adjust to social living.

The procedure required to test the hypothesis of critical periods is to present to different groups of animals the same physical stimulation at different early ages and then measure their behaviour at some later date. If the age at which stimulation is administered is important, then the groups should differ in their behaviour. Research using this general methodology has clearly established that the age at which the animals are stimulated is critically important. Work with the *dog* has demonstrated differences in trainability as guide dogs for the blind (Pfaffenberger & Scott, 1959), ability to interact with a human handler (Freedman *et al.*, 1961), and the development of behavioural patterns (Scott & Marston, 1950; Scott *et al.*, 1951). For the *mouse* it has been found that stimulation at different ages will affect classical conditioning (Denenberg, 1958, 1960), avoidance learning (Denenberg & Bell, 1960), body weight and survival (Denenberg & Karas, 1959), aggressive and sexual behaviour (King, 1957), and the development of behavioural patterns (Williams & Scott, 1953). For the *rat* stimulation at different ages will differentially affect avoidance learning (Denenberg & Karas, 1960; Levine, 1956), body weight and survival (Denenberg & Karas, 1959, 1961; Levine & Otis, 1958), hoarding (Hunt, 1941; Seitz, 1954), competition for feed (Wolf, 1943), ascorbic acid depletion (Levine & Lewis, 1959), and blood sugar concentration (Bell *et al.*, 1961).

We may conclude that the age at which stimulation is administered in early life markedly affects later behaviour. Paradoxically, this evidence compels us to revise drastically the critical period hypothesis. First, it has been clearly established that stimulation prior to the period of socialization will significantly affect the adult behavioural patterns of several species. In addition, definite evidence of learning prior to the period of socialization has been found in the dog (Cornwell & Fuller, 1961). Denenberg & Karas (1960, p. 319) have concluded "that there are many continuous and rapid developmental changes occurring between birth and weaning and there may be as many 'critical periods' as there are quantitative and qualitative changes in developmental processes. Based upon the present state of knowledge the most reasonable hypothesis is that there are critical periods as defined by Scott in terms of *discrete* observable developmental changes such as the beginning of psychomotor co-ordination, opening of the eyes and ears, weaning, and sexual maturity; but, in addition, there are other less obvious *continuous* changes taking place which are also affected by external stimulation".

For a detailed discussion of the critical period concept and review of the literature with rodents see Denenberg (1968). Because of the extreme difficulty in demonstrating clear-cut critical period, many researchers, especially European ethologists, prefer to use the more descriptive term "sensitive period". In addition to being descriptive, this term also does not have the embryological connotations which the critical period concept has.

II. IMPRINTING

Directly related to the concept of sensitive or critical periods is the phenomenon of imprinting. Imprinting is the process whereby young birds develop social preferences, usually to their own species.

4+

Lorenz (1937), who introduced the term, called imprinting the ". . . process of acquiring the biologically 'right' object of social reactions by conditioning them, not to one individual fellow-member of the species, but to the species as such . . .". Imprinting has been primarily studied in birds although an analogous phenomenon has been described in other animals including mammals (Hess, 1959).

Imprinting has been called a special type of rapid learning which differs from associative learning in the following ways (Lorenz, 1937): (i) Imprinting occurs only during a very definite period early in the animal's life cycle, in many cases a period which is of short duration. (ii) Once imprinting has occurred, its effect is irreversible. (iii) Imprinting determines future adult behaviour; for example, turkeys, imprinted to humans, directed their courtship activity in adulthood towards humans rather than towards members of their own species. (iv) Imprinting can be transferred (generalized) from the specific stimulus object to other members of the class from which that stimulus object came. In other words, the imprinted animal will respond to the species (general) characteristics of the stimulus object rather than the individual (unique) characteristics of the object.

Imprinting is a "process" and, as such, cannot be measured directly. Behavioural referrents are needed which will permit inferences concerning the process of imprinting. The behaviours most commonly used to study imprinting are the "*following*" response and "*preference*" behaviour.

A. Following Behaviour

One apparatus used to measure following behaviour is shown in Fig. 8 (Hess, 1959). It consists primarily of a circular runway, 5 ft in diameter and 1 ft wide, around which a decoy duck, containing a loudspeaker, can be moved. The walls are made of clear Plexiglas. The speed of rotation and pattern of movement of the decoy is regulated from the control panel which is situated behind a one-way vision screen.

Hess (*loc. cit.*) has intensively studied the imprinting behaviour of the mallard duck and domestic chick in this apparatus. He has determined that the critical period for the imprinting phenomenon is within the first day of life. The largest imprinting score (85 per cent positive responses) was obtained with animals exposed to the moving decoy between the 13th and 16th hours of life; a score of 50 per cent occurred for animals exposed between the 1st and 4th hours of life, while no more than 20 per cent of positive responses were obtained with birds over 24 hours of age. Even though there are some negative findings (e.g. Hinde *et al.*, 1956), the presence of a critical or sensitive period early in life has been repeatedly confirmed and may be accepted as established.

However, the conclusion that imprinting is irreversible is highly questionable. Hinde *et al.* (1956) studied the following response of young coots (*Fulica atra*) and moorhens (*Gallinula chloropus*) and concluded that "the first two characteristics of the imprinting process given by Lorenz (the existence of a sensitive period and irreversibility . . .) are not valid for the following response of moorhens and coots". Moltz (1960), in his review of the experimental findings and theoretical significance of imprinting, has concluded that "the acquisition of the following response is neither rigidly restricted to a specific period in ontogeny nor irreversible once it is established".

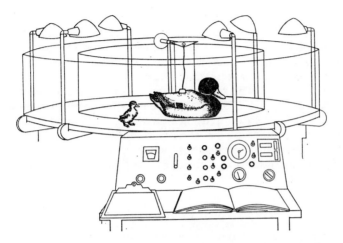

Fig. 8. Apparatus used to study following behaviour. The movement of the decoy duck is controlled by the panel in the foreground. A duckling is following the decoy in the drawing. (After Hess, 1959.)

B. Preference Behaviour

When one measures following behaviour, the emphasis is placed on the locomotor pattern of the bird. Bateson (1966) has suggested that too much attention has been paid to the motor response of the animal while too little attention has been paid to relevant sensory inputs, viz. visual and auditory contact with the imprinted object.

Bateson's own research has been concerned with the development of visual preferences (or visual discriminations) in chicks as a function of characteristics of the animal's rearing environment. For example, at hatching chicks were placed either into black and white pens or into yellow and red pens (Bateson, 1964c). At three days of age the animals were tested in an apparatus similar to the one used by Hess. Half of each group was tested with a moving black and white model while the other half was tested with a moving red and yellow model. The chicks tested on a model similar in colour to their home pen avoided the model for a shorter period of time and responded socially to it sooner than did chicks tested on a dissimilar model. Bateson concluded that the animals had learned the details of their isolation pens and were able to discriminate between them and dissimilar objects (see also Bateson, 1964a).

One of the stimulus parameters which aids in developing a visual preference appears to be the conspicuousness of the stimulus as defined by the human eye. Bateson has been able to demonstrate this finding in a series of studies. In the first one (Bateson, 1964b) chicks were reared in either black and white pens or grey pens. Both sets of pens reflected an equal amount of light. The activity of the animals in their pens was recorded for the first seven days of life. On each day those chicks housed in the grey pens were found to be more active than those housed in the black and white pens. Bateson suggested that this difference could have been brought about by the difference in complexity (or conspicuousness) of the two housing environments. This observation was followed up in a series of three experiments (Bateson, 1964d). In a choice situation a greater number of one-day-old chicks approached conspicuous static patterns in preference to inconspicuous ones. The subsequent two experiments established that the conspicuousness of a chick's rearing pen affected the extent to which the animal avoided a novel moving object: in both studies those animals reared with an inconspicuous pattern avoided the novel moving models for a shorter period than did those reared with conspicuous patterns. It was suggested that the effectiveness of a stimulus in imprinting experiments is related to its conspicuousness and that imprinting can be initiated by static objects.

C. Theories of Imprinting

Hess has suggested two developmental factors to account for the changes in imprinting effectiveness as a function of time. The onset of imprinting effectiveness is related to the development of locomotor ability; the lack of ability to move prevents the following response from occurring. The termination of the critical period is coincidental with the onset of fear; the occurrence of fear responses causes the bird to avoid rather than follow a moving object, thus effectively preventing the occurrence of imprinting. Hinde *et al.* (1956) have also emphasized the importance of the tendency to flee as a factor limiting imprinting effectiveness.

The observation that imprinting occurs most efficiently at a time when the bird does not exhibit any fear behaviour suggested to Moltz (1960) that the process underlying imprinting is the association of an attention-evoking object (the moving stimulus) and a state of low anxiety. This association is mediated by the principles of classical (Pavlovian) conditioning. When anxiety is subsequently aroused, any response on the part of the animal which will bring it near to or in contact with the stimulus object will be followed by anxiety reduction because the object had initially been associated with a low anxiety state. Thus, the following behaviour is instrumental in reducing anxiety; this behaviour is learned by the principles of instrumental (operant) learning. In two experiments Moltz & Rosenblum (1958) and Moltz *et al.* (1959) obtained data confirming the anxiety hypothesis; they found in the Peking duck that the strength of the following response was positively related to the degree of anxiety generated in the test situation (i.e. the greater the anxiety, the greater the following response).

A major difficulty in Moltz's position is that it does not account for the initial following of the bird; this first occurs at a time when anxiety is non-existent. To account for this Moltz modified his position in 1963. He pointed out that his precocial birds initially would only follow a retreating object but not an approaching one. He stated that there were classes of stimuli which would elicit approach responses (because they triggered off parasympathetic autonomic responses) or avoidance responses (brought about by sympathetic nervous system stimulation). Any stimulus causing a para-sympathetic response would increase the tendency of the young bird to approach that stimulus.

The most recent theoretical discussion of imprinting is that of Bateson. The reader is referred to his 1966 paper for a thoughtful and critical review of the imprinting literature as well as for a detailed explanation of his theoretical position. A shorter presentation of his theoretical position with some additional data will be found in Bateson (1969). In brief, his position may be summarized as follows.

The onset of imprinting behaviour is brought about by the rapid development of visual and auditory perceptions and locomotor ability in the precocious bird, one consequence of which is that the animals will interact with their environment as soon as they are capable of doing so (Bateson, 1966, 1969). Those stimuli which are conspicuous to the human eye are ones which will act as powerful reinforcers for chicks, and they will very quickly become familiar with such stimuli (and will follow them). The termination of the sensitive period (but not of imprinting) is brought about when the birds have become familiar with the characteristics of their environment and can discriminate between this environment and other forms of stimulation. Therefore, avoidance behaviour is not brought about by an endogenous development of fear (an assumption which cannot be tested), but, instead, may be regarded as a response to conspicuous but unfamiliar objects (cf. Hebb, 1946).

The advantage of both Moltz's and Bateson's positions concerning imprinting over those of Hess and Lorenz is that the former have used constructs and assumptions which are amenable to experimental tests (e.g. Moltz's notions of classical and instrumental conditioning and Bateson's concepts concerning familiarity) while Hess and Lorenz postulate endogenous processes which must be assumed to be genetically determined and which are not able to be experimentally evaluated. Furthermore, both Bateson and Moltz place the phenomenon of imprinting into the broader context of research on learning and novelty.

There has always been a mystique about imprinting when discussed by those steeped in the traditional concepts of ethology. Oddly enough, though such ethologists insisted that imprinting was a unique phenomenon, many people unjustifiably lifted the term from the context in which it was placed and rather glibly used it as an explanatory concept to explain the behaviour of non-avian species. This is a *non sequitur* for those who espouse the Lorenz or the Hess position. However, when viewing imprinting from the theoretical positions of either Moltz or Bateson it is more reasonable to attempt to generalize since neither attributes unique properties to imprinting but attempts to explain the phenomenon in broader psychological terms. However, one must still emphasize the need for caution when generalizing imprinting beyond avian behaviour.

D. Imprinting and Mammalian Behaviour

One of the major functions of the imprinting process appears to be that of establishing a social relationship between the young bird and others of its species. A similar phenomenon has also been observed with precocious mammals. Hess (1959) has reported the "following" response in sheep and guinea pigs. "Heeling" behaviour in the young moose-calf (*Alces alces shirasi*) has been described by Altman (1958); she states that it appears to be similar to imprinting though it lacks irreversibility and an exactly limited period of sensitivity.

In a natural setting imprinting is defined by the behaviour of the young towards the mother and other members of the species. There have been several reports in which the behaviour of the mother towards the young is the critical element. If a lamb or kid is removed from its mother at birth for a period of two to four and a half hours, the mother will probably reject the young upon its return (Collias, 1956). Bartholomew (1959) has reported that the pup of the Alaskan fur seal (*Callorhinus ursinus*) will attempt to nurse off any female but that the female will drive off all pups but her own. Scott (1945) has reported that a female lamb, which was bottle raised for the first three days of life, was driven off by ewes having lambs of their own when she was placed with the flock; in adulthood this female kept away from the flock until her first heat period and left it immediately afterwards. A male lamb, which had been separated on the 4th day of life for a period of 8 days, showed much more social behaviour in adulthood. Working with goats, Hersher et al. (1958) removed kids from their mothers 5 to 10 minutes after birth for periods of one-half to one hour. Mother-young interactions were studied at 2 and 3 months of age. Goats whose kids had been removed in infancy nursed all young indiscriminately while control mothers would only nurse their own young (see Chapter 10).

It is difficult to tell whether or not the above are examples of the mother being imprinted upon the young. Scott (1960) believes that the formation of mother-young relationships and imprinting are both examples of the process of the formation of primary social relationships. These primary relationships develop during the critical period of socialization. This attempt of Scott to integrate the phenomena of imprinting, critical periods, and socialization is an important theoretical contribution. Further research is required to determine the efficacy of this theory. The reader is referred to Chapter 1 for a further discussion of this matter.

Salzen (1966) has discussed imprinting in birds and primates and has shown that there are a number of common behaviours among these two major groups. However, it must be clearly emphasized that merely because different species of animals engage in similar behaviour (i.e. behaviour which seems to have the same biological functions), this in no way indicates that the mechanisms causing these behaviours are the same.

E. Effects of Imprinting upon Fear Behaviour and Vigilance

Tinbergen (1951) has reported that gallinaceous birds, ducks, and geese exhibited fear behaviour when a model of a hawk (characterized by its short neck) was "flown" overhead but these animals did not become upset when a long-necked model (goose) was "flown" overhead. Hirsch et al. (1955) attempted to study this phenomenon experimentally using the chicken; they failed to find any differential responses to models of a goose and hawk. Likewise, Rockett (1955) studied both chickens and ducks and found that short-neckedness was no more fear-producing than long-neckedness (see also Tinbergen's, 1957, comment on Hirsch et al., as well as Hirsch's, 1957, reply to Tinbergen).

Melzack et al. (1959) attempted to resolve these conflicting findings by systematically manipulating the past experience of mallard ducks. One group of ducklings was imprinted upon a hawk model within 20 hours after hatching; these animals were then given 120 presentations per day of the hawk model from the 4th to the 21st day of life. A second group received the identical experience using a goose model; controls were raised in isolation during this time. Tests for fear behaviour were made by alternately flying a model of a goose and a hawk overhead. Ducks having no prior experience with the models (controls) exhibited fear to both the hawk and goose model while birds with prior experience showed no fear to either the hawk or goose model. In one series of tests, controls exhibited more fear to the hawk than to the goose model but habituation to the models eliminated the fear responses after the third test day.

The discrepancies among these results appear to have been resolved by a further analysis by Melzack (1961) in which he examined the protocols gathered in his previous study (Melzack et al., loc. cit.) and scored the behaviour of the ducks in terms of a concept of "vigilance", defined as non-emotional orientating responses. Both experimental and control ducks had a high percentage of vigilance responses. Differential sensitivity to the hawk and goose figures was also found; a greater percentage of orientating responses was made to the hawk figure when moving directly overhead, but the ducks orientated more to the goose figure when it was moved at angles of 60° to 70° to their cage.

Thus, prior imprinting experience with either a hawk or a goose model served to bring about more adaptive behaviour. These animals, unlike the controls, did not exhibit emotional fear behaviour towards the flying models, but they did maintain a high degree of vigilance towards the moving objects. This maintenance of attention without expending energy in emotional behaviour would appear to be the most adaptive response the organism could make under the circumstances.

It is apparent that one could interpret the results described above within Bateson's framework of conspicuous and familiar stimuli. Such an interpretation would lead to an interesting set of experiments.

F. Conclusions

There is no dispute concerning the facts of imprinting. The existence of a sensitive period has been well established, the phenomenon of imprinting appears to be a stable one, and, at least in some species, there are long-term effects upon sexual behaviour. It is in the interpretation of these data where differences of opinion prevail. Imprinting appears to be a form of learning. It occurs very rapidly because of the precocial sensory and motor character-istics of birds (cf. one-trial learning), it generalizes to the class of stimuli of which the imprinted stimulus object is a member (based upon familiarity and stimulus generalization), and is reversible (extinction or unlearning). None of these events makes imprinting unique since there are many examples of one-trial learning in the psychological literature, and because the pheno-mena of stimulus generalization and extinction are also well documented (Kimble, 1961). Imprinting differs from our usual notions about learning in two ways. First, this phenomenon is most effective early in life; in this sense there is a sensitive period for imprinting. However, the age and range of imprinting varies as a function of species and environmental factors (see Bateson, 1966). Secondly, imprinting may effect the subsequent social (including sexual) behaviour of the animal. These two phenomena specify the unique features of imprinting behaviour and indicate that imprinting can be considered to be a form of early social learning.

III. PRENATAL EXPERIENCE

Thompson (1957) addressed himself to the question of whether "anxiety" on the part of the mother during pregnancy would affect the postnatal be-haviour of her offspring. Female rats were trained, prior to pregnancy, to associate the sound of a buzzer with strong electric shock; during pregnancy these rats were exposed to the buzzer without shock. Thompson assumed that the prior association of the buzzer with shock was so strong that the occur-rence of the buzzer alone would generate anxiety. The offspring were tested for emotionality and timidity between 30 and 40 days of age and then again at 130 to 140 days. At both ages it was found that the young from mothers which were made anxious during pregnancy were significantly more emotional and timid than young from control mothers.

A number of other investigators have replicated Thompson's general finding (Ader & Belfer, 1962; Doyle & Yule, 1959; Hockman, 1961; Morra, 1965).

A significant advance in the analysis of prenatal effects is seen in the study by Joffe (1965) who was able to separate out the effects of conditioning stress prior to mating from the stress induced during gestation. Joffe's findings indicate that it is the premating stress which is the cause of the change in the open field activity of the offspring. Joffe also found that the animal's geno-type interacted with the experimental treatments. Also, different patterns of results were obtained when different endpoints were evaluated. Since Joffe was not able to replicate Thompson's finding, it is not possible to assess these several experiments, though their implications are many. Interestingly, from clinical observations Sontag (1941) has suggested that the emotional life of the human during pregnancy may have effects upon the foetus.

Non-genetic maternal and grandmaternal effects. Joffe had found that stress imposed upon females immediately preceding pregnancy affected the subsequent behaviour of their offspring. Several other experiments have also shown that the experiences which females have prior to pregnancy can influence their offspring, and, indeed, their grandchildren as well. Denenberg & Whimbey (1963) evaluated the body weight and open field performance of rats whose mothers had either been handled or non-handled in their (i.e. the mothers') infancy, and found that the mothers' experiences significantly affected the behaviour and weight of their offspring. In a subsequent study Denenberg & Rosenberg (1967) were able to demonstrate that the grandpups of female rats which had received differential experiences in early life were significantly changed. Using a very different methodology Ginsburg & Hovda (1947) transplanted fertilized ova from one strain of mouse to another and studied the audiogenic seizures of the descendants. Though the percentage of animals seizing did not change, the incidence of deaths dropped over several generations.

Levine (1967) has also shown that the adrenocortical response of weanling rats is influenced by the handling experience which their mothers had when the mothers were infants.

IV. INFANTILE EXPERIENCE

We are concerned here with stimulation administered some time between birth and weaning. The terms "infancy" and "infantile experience" will be used only with reference to this time interval.

A. Continuous Stimulation

Seitz (1954) investigated the effects of litter sizes of 6 and 12 rat pups upon the behaviour of the mother as well as upon the subsequent behaviour of the pups themselves. He selected litter size as a variable because the rat has 12 teats, some of which are non-functional, thus providing a situation of feeding frustration for the pups from litters of 12. Mothers assigned litters of 6 were more maternal towards their young than were mothers with litters of 12. Rats from large litters hoarded more feed pellets and were more emotional as measured in an open field situation and a timidity test; males from large-sized litters weighed less and exhibited significantly more mating behaviour than those reared in small litters. In a second study Seitz (1958) used litters of 3, 6, 9, and 12 pups and confirmed his previous findings that the maternal behaviour of the female rat decreased as litter size increased.

The research of both Thompson and Seitz suggests that it may be possible to affect the behaviour of offspring through the experimental manipulation of the mother. Denenberg et al. (1962) investigated this hypothesis. Some rat mothers were shocked every day from the time their litter was born until weaning. Other females were paired on the basis of having litters born on the same day; these females were alternately rotated between their own litter and the other female's litter every 24 hours. The shock variable was used as a method of generating "anxiety" while the purpose of the rotation variable was to expose the young to a greater range of individual differences than they would get from one mother alone. Some litters were given one or the other of these experimental treatments, while other litters received both or neither (controls). Shocking the mother resulted in the young being

4*

emotional in adulthood as well as weighing less than rats reared by non-shocked mothers. The effects of rotation were to cause significant infantile mortality and reduced weaning weights of the surviving young as well as increased emotionality in adulthood. In an extension and partial replication of the paper by Denenberg et al., Ottinger et al. (1963) verified that rotation increased emotionality, and that the emotionality of the offspring was affected both by the prenatal (and presumably genetic) characteristic of the mother as well as by her postnatal behaviour toward the young.

The effect of enforced delayed weaning was studied by Levine (1958) who forced rat pups to nurse off the mother until 35 days of age by the procedure of having no feed in the nest cage. The mother was fed twice daily in a different cage. Since the late-weaned group had an inadequate diet, several control groups were used so that the effects of late weaning could be evaluated separately from the effects of inadequate diet. In adulthood the late-weaned animals required more trials to learn an avoidance response than did the normally weaned animals. The scores of the animals from the control groups were such that it could be concluded that major the treatment variable was the prolonged maternal contact rather than the inadequate diet.

Another approach to the study of social interactions in infancy is to foster the young of one species to a lactating mother from a different species. In this manner one is able to separate the genetic and prenatal effects upon behaviour from the postnatal influence. In a series of studies newborn mice have been fostered to lactating rat mothers (Denenberg et al., 1964, 1966; Hudgens et al., 1967, 1968). The characteristics of the peer group were also varied (no peers, mouse peers, rat peers). The mouse strain studied (C57BL/10J) is noted for its aggressive behaviour, and the most dramatic finding from these experiments is that the incidence of aggression can be reduced or eliminated if the mouse is reared by a rat mother. Independent of this finding, the absence of mouse peers during infancy results in an increase in aggressiveness.

Southwick (1968) has shown that aggressiveness is affected by intraspecies fostering as well. The aggressive behaviour of male mice from the passive A/J strain was increased by fostering at birth to mothers of an aggressive strain (CFW). However, aggression was not reduced when CFW pups were fostered to A/J mothers.

Social interactions between the cat and her kittens during infancy have been studied by Rosenblatt et al. (1961). They raised experimental animals in isolation in incubators which contained a feeder with a nipple attached. Different groups were reared in the incubators for different lengths of time and at different pre-weaning ages. All kittens were returned to their mothers and social-nursing behaviour was observed. Sucking after returning the kittens to the mother was reduced or eliminated, disturbance and withdrawal signs were noted when the young were returned to the mother, and orientation in the home cage was deficient. (See Chapter 14 for a more extensive discussion of this study.)

The effects of age of separation from the mother upon the adult behaviour of the cat have been described by Seitz (1959). One group of kittens was separated at 2 weeks of age, a second group was separated when they began

Table 9. Effects of Experimental Manipulations in Infancy upon Subsequent Behaviour.

Species	Experimental Variable	Main Findings	Author
Rat	Litters of 3, 6, 9, or 12 young	Maternal behaviour declined as litter size increased. Rats from litters of 12 hoarded more feed and were more emotional than rats from litters of 6	Seitz (1954, 1958)
Rat	Shocking mothers and/or rotating mothers between 2 litters	Shocking mothers led to young weighing less and being emotional in adulthood. Rotating mothers led to pre-weaning mortality of young, reduced weaning weights, and emotional behaviour in adulthood	Denenberg et al. (1962); Ottinger et al. (1963)
Rat	Delayed weaning	Late-weaned rats took longer to learn to make an avoidance response	Levine (1958)
Cat	Isolation rearing	Being reared in isolation interfered with sucking behaviour when returned to the mother and litter	Rosenblatt et al. (1961)
Cat	Age of separation from mother	Kittens weaned earlier than usual exhibited maladaptive behaviour in adulthood	Seitz (1959)
Rat	Visual or auditory deprivation	Temporary deprivation of a particular sense led to poorer performance in adulthood in a competitive situation where use of that sense was required	Wolf (1943); Gauron & Becker (1959)
Dog	Age of weaning	Puppies weaned before 19 days sucked more than controls. Puppies which gained weight at the slowest rate sucked the most	Scott et al. (1959)
Mouse	Rat mother; rat, mouse, or no peers	Rat mother causes reduction of aggressiveness. Absence of mouse peers increases aggressiveness	Denenberg et al. (1964, 1966) Hudgens et al. (1967, 1968)
Mouse	Different strain mouse mother	Aggression increased if reared by mother of an aggressive strain	Southwick (1968)

to lap milk spontaneously from a saucer (approximately 6 weeks of age), and a third group was forced to nurse from the the cat until 12 weeks of age by making no feed other than the mother's milk available to them. Behavioural tests after adulthood determined that the group weaned at 2 weeks of age had the most maladaptive behaviour.

Scott *et al.* (1959) weaned puppies at ages varying from 10 to 36 days and measured the daily amount of finger sucking. Puppies weaned before 10 days had higher sucking scores than controls; there were no differences between the groups at or after 19 days. For the animals weaned before 19 days a negative correlation between sucking scores and average weight gain after weaning indicated that the animals which gained at the slowest rate engaged in the most sucking behaviour. The authors concluded that sudden weaning produced an increase in non-nutritive sucking in puppies under 19 days and that this effect was inversely related to the previous state of nutrition.

The studies described above investigated the effects of social variables during infancy. A very different kind of study was carried out by Wolf (1943) who examined the role of sensory deprivation upon subsequent competitive behaviour. One group of rats was deprived of hearing from day 10 to day 25 (weaning); a second group was deprived of the use of vision from day 12 to day 25; and a control group was not deprived of any sense. In adulthood the animals were individually trained and then were placed in competition for food, in response to a visual signal and in response to an auditory signal. Auditorily deprived rats won more often under a visual signal; visually deprived rats won more often when an auditory signal was used. Thus, animals temporarily deprived of the use of a particular sense in infancy performed less well in adulthood in a competitive situation where use of that sense was required. Gauron & Becker (1959) were able to replicate the major findings in Wolf's experiment though they found it necessary to modify Wolf's technique rather extensively.

Table 9 summarizes the salient points of these studies. In nearly every case the experimental manipulations resulted in less adaptive behaviour on the part of the young. The studies on cross-fostering the mouse to the lactating rat mother are an exception to this statement, since aggressiveness could be either increased or decreased by the appropriate manipulation. Another exception is seen in Seitz's study of litter size; rats from small-sized litters consistently performed better than animals from large-sized litters.

B. Discontinuous Stimulation

The approach followed in most of the studies to be summarized in this section has been to expose the infant animal to some form of stimulation for a brief period of time on one or more days during infancy and then to test these animals as well as controls at some later date (usually in early adulthood).

Rats and mice have been most extensively studied. Several methods have been used to stimulate the young pups during infancy; a widely used method is the procedure of "handling" or "manipulation" (not to be confused with "gentling" as discussed later in the chapter). This consists of removing the young from the cage, and placing them individually into a container for a brief period of time (usually 3 minutes), after which they are returned to the nest cage. The mother is usually left in the nest cage during this period. This is done once a day for various days during infancy. The term "handling" or "handled" will be used to designate this general procedure in this chapter. Another common procedure has been to administer electric shock to the young.

The experimental operations of handling or shocking young laboratory rodents may be placed into the broader context of this chapter by considering these as methods of presenting differing amounts of extrinsic stimulation to the newborn animal. The research summarized below will show, for the rat and mouse, that the introduction of extrinsic stimulation during the period of infantile development will markedly change psychological and physiological proficiency in adulthood. The generality from the research is that there is an optimal range of stimulation in infancy which will lead to optimal behaviour in adulthood. Whether the amount of stimulation that the young of a particular species receives is at, below, or above this optimal is a matter for empirical determination.

For clarity the following discussion is classified under six categories: learning, emotionality, experimental neurosis, social behaviour, physiology, stress resistance.

1. LEARNING

The importance of handling as an experimental variable was first reported by Levine et al. (1956) who shocked one group of rats daily from birth until weaning (21 days), placed a second group on the unelectrified grid, and did not disturb a third group. The handled rats were found to be the best avoidance learners in adulthood while the controls had the poorest learning scores. In a second study Levine (1956) again found that animals handled for the first three weeks of life were superior to controls in learning while animals handled from 50 to 70 days of age were not significantly different from the controls.

Denenberg & Karas (1960) attempted to extend Levine's work by examining the relationship between the number of days of stimulation in infancy and avoidance learning in adulthood. Rat pups were handled during the first 10 days of life, the second 10 days, the first 20 days, or were not disturbed (controls). The groups handled for 10 days were better learners than the groups handled for 20 days or the controls. However, Denenberg & Karas (loc. cit.) were not able to repeat Levine's findings that the animals handled for 20 days will learn better than controls. Both sets of data are equally valid since Levine and Denenberg & Karas have been able to repeat their own findings. The hypothesis was put forth by Denenberg & Karas (loc. cit.) that the discrepancy between the two sets of data may be a function of strain differences. In essence, this hypothesis states that there is a major interaction between infantile stimulation and genetic complex. Such an interaction has been reported by King & Eleftheriou (1959) for the deermouse (Peromyscus maniculatus). Other examples of interaction between infantile stimulation and genotype may be found in Levine & Broadhurst (1963), Levine & Wetzel (1963), and Lindzey et al. (1963).

Working with the inbred house mouse Denenberg & Bell (1960) stimulated independent groups of animals at 2 to 3, 8 to 9, or 15 to 16 days of age with electric shock of 0·1, 0·3, or 0·5 ma. Other animals were placed on the grid but not shocked and still others were not disturbed at all until weaning. Those animals which had received 0·3 ma. of shock in infancy had the best learning scores in adulthood. In addition, the age at which the mice were stimulated in infancy was a significant determiner of adult learning. This work was an extension of previous work by Denenberg (1959) in which mice which had received an intermediate amount of electric shock at 25 days had the highest classical conditioning scores at 50 days.

The studies summarized above reveal that animals with an intermediate amount of stimulation in infancy (as determined either by intensity or number of days of stimulation) have had the best learning scores. However, this does not indicate that an inverted U function (i.e. a curvilinear function) will always describe the relationship between intensity of stimulation in infancy and adult learning. In this respect Karas & Denenberg (1961) were able to predict a linear relationship between infantile stimulation and learning for an "easy" discrimination learning problem (also, see Smith 1967 for similar results). The general conclusion from these studies is that a certain range of extrinsic stimulation in infancy is optimally beneficial for later learning.

2. EMOTIONALITY

Since the rat and mouse "freeze" and defaecate when exposed to different and/or noxious stimuli, these behaviours have been used to measure the construct of emotionality. One standardized method is to place the animal into a brightly lit open field which is an enclosure about 4 ft square marked off into smaller squares. The animal is observed for a constant period of time and records are kept of time required to move off the first square, total number of squares crossed, urination, and defaecation. The "non-emotional" animal is one that has a short latency to leave the first square, crosses a number of squares (i.e. is an active animal), and does not urinate or defaecate. For a fuller discussion of the concept of emotionality, methods for measuring the concept, and validity data see Denenberg (1967, 1969).

Rats handled for the first 20 days of life were found to be less emotional at weaning and in adulthood than undisturbed controls; female mice handled in infancy were less timid than non-handled controls, and handled male mice had shorter latencies to fight than did controls (Levine, 1956, 1959a, 1959b). The last effect may be interpreted as due to the lesser amount of emotionality on the part of the handled males.

The findings that a moderate amount of stimulation in infancy resulted in better adult learning suggested to Denenberg (1959) and Denenberg & Karas (1960) that emotionality decreased as the intensity of stimulation increased and that some intermediate amount of emotionality was optimal for learning. Denenberg & Karas (loc. cit.) found that rats handled for 20 days during infancy and non-handled controls both made poor scores on an avoidance learning task. But observation of the behaviour of these animals during the learning trials indicated that the two groups were performing poorly *for very different reasons*. The controls were quite agitated when the shock occurred and a number of them rushed about the apparatus in a wild manner. On the other hand the rats which had been handled for 20 days appeared to be more "casual" in their responses to the shock, and some of these animals were observed sitting on the electrified grid for a few seconds before moving at a slow run into the escape box. Thus, it appeared that one group was too upset by the shock stimulus to learn the task efficiently while the other group was not motivated enough by the shock to learn.

Several experiments have tested this hypothesis that emotionality increases as intensity of infantile stimulation increases. Denenberg *et al.* (1962) found that rats handled for 20 days in infancy defaecated the least in the open field test, followed by rats handled for 10 days in infancy while non-handled controls had the highest defaecation score. Denenberg & Smith (1963) reported that rats shocked in infancy had the smallest percentage of defaecation, followed by animals which had been handled in infancy, while the non-handled controls had the highest incidence of defaecation. Using the amount of adrenocorticosterone in the blood as a physiological indicator of degree of emotional upset, Denenberg & Haltmeyer (1967) showed a "dose-response" relationship between amount of handling and steroid level. They had groups which were not disturbed, or handled for 5, 10, or 20 days in infancy. A negative linear relationship was found with the non-handled group giving the greatest steroid response while the group handled for 20 days gave the least response. Denenberg (1964) has formalized and extended this hypothesis in a theoretical paper.

3. Exploratory Behaviour

Not only does handling affect learning and emotionality, but an organism's exploratory behaviour is also markedly modified. Rats handled in infancy spend more time exploring novel stimuli than do controls (Denenberg & Grota, 1964). Handled animals also increase their stimulus-seeking behaviour as the environment becomes more complex while non-handled controls decrease such behaviour as a function of increase in complexity of the environment (DeNelsky & Denenberg, 1967a, 1967b). (Other research has established that exploratory behaviour is independent of emotional reactivity; see Whimbey & Denenberg (1967) for more information.)

4. Experimental Neurosis

Research with sheep and goats has established that experimental neurosis can be precipitated in these animals by subjecting them to a rigid conditioning schedule of a 10-second conditioned stimulus (usually a light or a metronome) followed by electric shock to one leg (Liddell, 1954). The neurotic pattern includes violent struggles at the signal and the shock, jerky movements of the leg which received shock, urination, defaecation, vocalization, laboured breathing, and repeated movements of the head and ears. Three-week-old kids will exhibit this neurotic pattern if they are tested in isolation. However, if kids are tested in the presence of the mother, no neurotic pattern appears. Furthermore, this experience has long-term effects upon these animals. Eight goats which had undergone conditioning as infants were retested at two years of age. Four of these had been protected in infancy by the mother's presence while the other four had not. The four goats which had been initially conditioned in the presence of the mother gave no sign of experimental neurosis at 2 years of age while the other four goats exhibited the usual signs of neurosis (Liddell, *loc. cit.*).

5. Social Behaviour

The effects of handling upon aggressive behaviour in mice was studied by Levine (1959a). The handled mice started fighting more quickly than controls.

Hersher *et al.* (1958) separated 24 goat mothers from their kids 5 to 10 minutes after birth for periods from one-half to one hour; a control group followed the usual newborn caretaking pattern. Tests at 2 and 3 months of age found that control mothers would only nurse their own kids while experimental mothers nursed all kids indiscriminately; some of the control mothers rejected all kids, including their own. The authors concluded that the separation of some of the goats from their young had influenced the social structures of the herd as a whole. See Chapter 10 for a more extended discussion of this study, and see Moore (1968) for a broad review of the effects of modifying maternal care in the sheep and goat. Moore's general conclusion is that the experimental manipulations of (i) interrupting maternal care, (ii) having a foster mother adopt the young, and (iii) eliminating the mother completely by means of bottle-feeding all have the same general consequence of making the experimental animals appear to be chronologically younger. The responses given by the experimental animals during conditioning were very similar to those of normal animals who were conditioned immediately after parturition.

6. Physiology

Rats handled during infancy weighed more at weaning and at 70 days than controls; they also had smaller adrenals following a glucose injection and food and water deprivation (Levine, 1957). Infant rats were handled daily until sacrificed at 10, 12, or 14 days when the brains were analysed to determine cholesterol content (Levine & Alpert, 1959). The handled animals had larger amounts of cholesterol present than did the controls at 12 days of age. Rats handled until 12, 14, or 16 days of age, and then sacrificed, showed depletion of adrenal ascorbic acid while controls failed to show a depletion until 16 days of age (Levine *et al.*, 1958). In addition, handled rats evidenced a depletion approximating the level normally seen in adult animals.

A possible mechanism for the effects of infantile stimulation upon later behaviour and physiology has been suggested by the research on the effects of the sex hormone upon brain organization (Levine & Mullins, 1966; Young et al. 1964 for reviews of this literature). The general finding is that the injection of the male hormone, androgen, into a female organism at an appropriate time in early life will cause that female, when an adult, to have behavioural and physiological characteristics of a male rather than a female. There is a critical period for this phenomenon. In the rat the critical period is approximately the first 5 days after birth, while in the guinea pig and the monkey, both of which are precocial species, the critical period occurs prior to birth. The general conclusion is that the function of the sex steroids in early life is to organize the brain with respect to maleness or femaleness while, in adulthood, the steroids play an activating role.

Because of the findings that the sex steroids do have major effect upon brain organization during very early life, several groups of researchers began looking at other steroids which could have similar consequences for the effects of infantile stimulation. The hormone which has been investigated most intensively is corticosterone from the adrenal gland. In 1962 Levine reported that rats which were handled in infancy showed an increase in corticosterone 15 seconds after electric shock was applied, while non-handled controls did not show an elevation in steroid level until 5 minutes after shock. This finding was confirmed by Haltmeyer et al. (1967). However, when handled rats were exposed to a novel environment in adulthood (in contrast to the noxious effect of electric shock) their steroid response is markedly less than non-handled controls (Levine et al., 1967). Investigations then shifted to an analysis of corticosterone in the neonatal rat. Several studies have clearly shown that the newborn rat is capable of releasing corticosterone to stressors such as electric shock, heat, histamine, and ether (Haltmeyer et al., 1966; Levine, 1965; Levine et al., 1967; Levine & Mullins, 1966; Zarrow et al., 1966; Zarrow et al., 1967). Having established that intense stressors were capable of bringing about a release of corticosterone in the newborn rat, the next important question was whether the procedure of handling, which appears to be a much less stressful form of stimulation, is also capable of releasing a significant amount of corticosterone. It was established that the handling procedure itself is a sufficient condition to bring about a release of corticosterone in the 2-day old rat (Denenberg et al., 1967). Following the model from the sex steroid literature, once corticosterone is in the bloodstream the prediction was that some of it would localize in the brain, and in all likelihood the hypothalamus. Therefore, Zarrow et al. (1968) injected radioactive corticosterone into 2-day-old rats and examined the hypothalamus and the rest of the brain at varying intervals after the injection. A significant amount of corticosterone was taken up in the rat's brain and both at 10 and 30 minutes after injection there was a greater amount of corticosterone in the hypothalamus than in the rest of the brain. At the end of an hour most of the material had disappeared. Though these findings do not demonstrate that corticosterone is part of the mechanism mediating the effects of early stimulation, all the experiments to date have been consistent with such an hypothesis. Research over the next several years should establish whether or not this is a valid hypothesis.

7. Survival Ability

Rats handled daily from birth through weaning (21 days) had fewer mortalities following 120 hours of food and water deprivation than did undisturbed controls (Levine & Otis, 1958). Denenberg & Karas (1959) found that rats handled for the first 20 days of life lived longer on a total food and water deprivation than controls (thus confirming Levine & Otis's data), and they also found that 10 days of handling (either during the first 10 or second 10 days of life) resulted in even longer survival. Denenberg & Karas (*loc. cit.*) repeated the same experiment with mice and found that 20 days of handling led to more rapid death as compared to controls while 10 days of handling had very little effect. The mouse data were confirmed by Levine & Cohen (1959) who handled mice for the first 24 days of life and then injected leukaemia virus into experimentals and controls at about 50 days of age. Their handled mice died earlier than controls.

The stress of learning to avoid shock in adulthood will interact with infantile experience to complexly modify both body weight and survival time (Denenberg & Karas, 1961).

The same physical stimulation appears to have a more severe effect upon the mouse than the rat since mice handled for 20 or more days died sooner than controls under stress while the reverse was true for the rat. One of the main differences between these two species is that the mouse is a more rapidly developing organism; this suggests that the more rapid an organism's development, the greater the effect of infantile experience (Denenberg & Karas, 1959).

V. POST-WEANING EXPERIENCE

A. Continuous Stimulation

The question studied here has been whether the characteristics of the environment in which animals are reared after weaning affect their subsequent behaviour. Again, laboratory animals have been most thoroughly studied. We may distinguish three gross types of environments: restricted, normal, and enriched. The exact meanings of these terms vary from one experiment to another. In general, restricted environments are those which reduce sensory information and/or social contact. A normal environment is always with respect to the usual laboratory conditions; this varies considerably from one laboratory to another. Interestingly, there is often good agreement among researchers when it comes to setting up an enriched environment. Most researchers have followed Hebb's (1949) suggestions concerning the so-called free environment or some variation thereof (e.g. Hymovitch, 1952). The free environment is an enclosed box approximately 4 ft square in which rats are placed after weaning. The box contains tunnels, ramps, seesaws, and a number of small objects that can be manipulated by the animals. Feed and water are present in the apparatus.

1. Problem Solving and Learning

Problem-solving behaviour has usually been measured by the so-called Hebb-Williams maze (Hebb & Williams, 1946) as modified by Rabinovitch & Rosvold (1951). Although devised for the rat, the technique can be carried over to other species as Thompson & Heron (1954b) have shown with the dog. The rationale underlying the use of the maze is as follows. Hungry animals are first trained to go from a start box through an empty enclosed field to a goal box where they obtain food. After they have learned this, barriers are placed in the field between the start box and the goal box. Since the animals have already learned that food is present in the goal box, their task is to take the most direct path to the food. Whenever an animal takes a path which is not the most direct one to the food, it receives an error score. By varying the number and position of the barriers many different "problems" can be constructed. It is assumed that the animal with the least number of errors is the best problem solver and, hence, the most "intelligent". These assumptions have been questioned by Woods (1959) and Woods et al. (1960). The Hebb-Williams maze certainly needs more validation studies to establish exactly what is being measured, although the results to date indicate that some aspects of problem solving, among other things, are being measured.

The effects of an enriched environment upon problem-solving behaviour in the Hebb-Williams maze are unequivocal. Animals reared in the complex free environment are distinctly superior. This is true for the rat (Hebb, 1947; Forgays & Forgays, 1952; Hymovitch, 1952) and for the dog (Clarke et al., 1951; Thompson & Heron, 1954b). This generalization extends beyond the characteristics of the Hebb-Williams maze since similar findings have been obtained with other apparatus (Bingham & Griffiths, 1952; Luchins & Forgus, 1955).

A question of importance is: At what age in an animal's life should enriched experience be given to maximize problem-solving behaviour? Forgays & Read (1962) gave rats enrichment during weeks 1–3, 4–6, 7–9, 10–12, or 13–15 and tested the animals on the Hebb-Williams maze when they were about 18 weeks old. All groups except those receiving extra experience during the 13th to 15th week performed better than controls. The animals with the best scores were those which received enrichment for three weeks immediately after weaning. Denenberg et al. (1968) also found that enriched experience immediately after weaning had greater effects on problem solving than experience before weaning. The most important part of the study by Denenberg et al., however, was that they tested their animals 10 months after the termination of the experimental experience and still found differences, thus establishing the long-term effects of early enrichment.

A major interaction between genetic constitution and environmental complexity has been demonstrated by Cooper & Zubek (1958) in a study which has a number of important implications. Rats selectively bred for maze brightness (few errors) or maze dullness (many errors) were reared in enriched, normal, or restricted environments. Animals from the maze-dull group, when reared in the enriched environment, performed as well as maze-bright animals which were also reared in the enriched environment. Conversely, maze-bright rats when reared in a restricted environment made as many errors in the Hebb-Williams maze as did the maze-dull animals. Chapter 3 discusses the genetics of maze brightness and maze dullness.

2. PERCEPTION

To study the effects of early visual experience on later perception of form, Forgus (1956) placed cards containing the forms of a cross, a circle, a square, and a triangle into the living cages of rats from the 16th to the 41st day of life. No forms were placed into the cages of controls. The animals were then given a form discrimination task in which they had to learn to go to a triangle and to avoid a cross in order to obtain food. The experimental animals were superior to controls.

Meier & McGee (1959) have confirmed and extended Forgus's work by determining that tactual experience as well as visual experience contributes significantly to perceptual discriminations in adulthood. Riesen (1947, 1950) has found that chimpanzees reared in darkness are extremely slow to develop visual perceptions, and require extended periods of training to develop their perceptual capacities.

The sense of pain is of such fundamental importance for survival that it may intuitively be thought that this is one of nature's givens and that past experience is not of much importance. Melzack & Scott (1957) reared dogs in isolation from weaning to maturity in specially constructed cages which drastically reduced the sensory experiences of the animals. Control litter mates were reared normally as pets in homes or in the laboratory. Several tests were used to determine the perceptual capacity to pain. In one set of tests the dogs received electric shock which could be avoided by making the appropriate response. The dogs which had been restricted in early life required more time to learn to avoid the noxious stimuli than the controls. In another set of tests the dogs were subjected to nose-burning with matches and painful pin-pricks. The normally reared controls rapidly learned to avoid these noxious stimuli. The behaviour of the experimental dogs was markedly different; though their responses indicated that burning and pin-pricking were indeed painful to them, the experimental dogs did not appear to perceive that the flaming match or the pin were the causes of their pain. In fact, these animals spent more time in the vicinity of the experimenter after painful stimulation than before being stimulated.

That restriction of activity in early life can affect the perceptual processes of higher mammals has been shown by Nissen et al. (1951) who encased a chimpanzee's limbs in cardboard cylinders from the age of 4 weeks until 31 months. The animal was markedly deficient in tactual discrimination and in tactual motor tasks though he learned a visual discrimination problem as rapidly as controls.

The most extensive series of studies concerning the morphological, physiological, and behavioural effects of visual deprivation has been carried out (Hubel & Wiesel, 1963, 1965; Wiesel & Hubel, 1963a, 1963b, 1965a, 1965b) on the cat. Deprivation of visual experience for the first 3 months of life results in an animal which appears to be permanently impaired since, even with a recovery period of over a year, major defects were found at the behavioural, morphological, and physiological levels (Wiesel & Hubel, 1965b). Since 3 months of visual deprivation to an adult cat produced no detectable physiological abnormality (Wiesel & Hubel, 1963a, 1963b), it is apparent that the age at which deprivation occurs is critical. They conclude from their data that the proper connections between the eye and brain are present at birth and that visual deprivation brings about a disruption of these connections (Wiesel & Hubel, 1963b). Of equal importance is their finding that bilateral eye closure had much less disruptive effects than they had predicted from their monocular eye closure findings. They conclude that "It thus appears that at the cortical level the results of closing one eye depend upon whether the other eye is also closed. The damage produced by monocular closure may therefore not be caused simply by disuse, but may instead depend to a large extent on interaction of the two pathways" (Wiesel & Hubel, 1965a, p. 1040).

3. SOCIAL BEHAVIOUR

For clarity the following discussion is arranged according to species.

Chicken. Wood-Gush (1958) failed to find any clear-cut differences in the sexual behaviour of isolation-reared and normally reared cockerels. Similarly, Baron & Kish (1960) failed to obtain evidence of any long-term impairment of social behaviour in chickens reared in isolation, pairs, or in a flock. The isolation-reared chickens, at 4 weeks of age, spent less time near a stimulus animal than did the pair-reared and flock-reared animals; the latter two groups did not differ from each other. However, at 10 weeks of age no differences were found among the three groups.

Turkey. Rearing conditions have an effect on certain aspects of sexual behaviour in turkeys (Schein & Hale, 1959). Androgen-injected male and female turkeys were reared under group conditions or in visual isolation from other turkeys. A number of objects, including a poult head and a human hand, were used as sexual stimuli during testing at around 35 days of age. (The poult head had previously been found to be an adequate stimulus to elicit sexual behaviour in normally reared adult male turkeys.) Regardless of rearing conditions, the birds exhibited the normal pattern of sexual behaviour. However, the isolation-reared turkeys responded sexually to the human hand while group-reared animals favoured the poult head. Schein (1960) has reported that these differences continued into adulthood. Males were tested approximately 10 months later when they were sexually mature; those males which had been group-reared initially responded in the normal manner to sexually receptive females while the males initially reared in isolation ignored the females and displayed sexual behaviour towards humans. A year later the isolation-reared males mated with live female turkeys when humans were absent, but they ignored the females when an observer was present.

Rat. Beach (1942) studied the effects of social rearing upon sexual behaviour. After weaning at 21 days animals were raised in individual cages, in one large cage with females, or in one large cage with no females. The proportion of copulators in adulthood was highest among males reared in isolation, followed by the males raised with females, and lowest in males raised in segregation. In a later study Beach (1958) determined that rats isolated at 14 days of age were as efficient as group-reared animals in their sexual behaviour. Kagan & Beach (1953) reared male rats in individual cages and permitted them to play with another male or female once a week until testing. The rats which had remained continually in isolation had the highest frequency of ejaculation in adulthood while the group which had had early play experience with females had the lowest frequency. The authors attributed this to a conflict in behavioural tendencies. The males had formed habits of playful wrestling during their pre-pubertal experience with the females and this response tendency was strong enough to interfere with sexual behaviour.

Mouse. Mice were reared in isolation from weaning until testing at approximately 110 days (Group *I*); Group *M* animals were reared with fathers and brothers from weaning until 45 days of age, and then in isolation thereafter; and Group *F* mice were reared with mothers and sisters from weaning until 45 days and in isolation thereafter. Group *I* mice were less aggressive in adulthood than the *M* and *F* groups which did not differ from each other (King & Gurney, 1954). Hudgens *et al.* (1968) also found that mice raised in isolation after weaning were less aggressive than mice reared with other mice or with rats.

To determine whether social isolation also affected sexual behaviour King (1956) reared two groups of mice under the same conditions as the Group *I* and Group *M* animals had been raised in the previous study. No differences in sexual behaviour were found between the two groups except that mice raised with other mice showed twice as much sniffing of genitalia.

In further work King (1957) reported that a period of 10 days of social contact immediately after weaning was sufficient to produce the maximum amount of aggressiveness while 5 days was insufficient. He also found that social contact immediately after weaning was more critical than social contact at maturity. There was no relationship between sexual and aggressive behaviour suggesting that experience affecting adult aggression did not generalize to sexual behaviour.

Guinea Pig. Although isolation does not appear to affect sexual behaviour in the mouse or rat, it definitely does in the guinea pig. Valenstein & Young (1955) found that socially reared males engaged in more sexual behaviour in adulthood than isolation-reared animals. Following castration sex activity of both groups declined to the same baseline. When testosterone proprionate was injected, only those males which had previously shown organized patterns of sexual behaviour were able to copulate successfully after hormone treatment. Thus, the past history of the animal was critical in determining whether testosterone proprionate would be effective in eliciting sexual behaviour.

Genetic factors are also important. Valenstein & Young had used inbred Strain 2 guinea pigs. Valenstein *et al.* (1955) studied Strain 2, Strain 13, and genetically heterogeneous stock. Following weaning at 25 days socially reared animals from Strains 2 and 13 showed greater sexual behaviour than did the isolates. The same finding was obtained for the heterogeneous males weaned at 10 days. When the heterogeneous males were weaned at 25 days, no difference was found between socially reared and isolation-reared animals. This suggests that the critical period for isolation to affect social behaviour is earlier in the more rapidly growing heterogeneous animals. Castration at birth did not impair sexual behaviour for those guinea pigs receiving both testosterone and social experience, but the group which received testosterone without social experience exhibited reduced sexual behaviour; this confirms the work of Valenstein & Young (*loc. cit.*).

Dog. Clarke *et al.* (1951) reared three Scottish terriers as pets while three others from the same litter were reared together in a relatively small cage made of an opaque material. These dogs were never removed from the cage and their only contact with humans was during feeding. They were kept in this cage until $7\frac{1}{2}$ months of age and were then given the run of the laboratory. When tested the restricted dogs were inferior to those reared as pets in a competitive situation. The restricted animals tended to assume a subordinate role and were indifferent to the presence of new dogs. In a test of hunger frustration, the deprived dogs were much more passive and withdrawn.

In a follow-up experiment, Melzack & Thompson (1956) reared 21 Scottish terriers under varying kinds of restriction while 16 control dogs were reared as pets in homes or in the laboratory. In a feed competition test normal dogs had more wins than did the restricted dogs; restricted dogs shared their feed with each other. Also, older experimental dogs would have their feed taken away by younger normally reared dogs.

It is well known that strong social bonds can be established between dog and man. Melzack & Thompson (*loc. cit.*) used this information to devise a series of ingenious tests to measure the dog's reactions to human stimulus patterns. At different times an experimenter played the role of a *friendly* man, a *timid* man, and a *bold* man. Scoring was done for friendly behaviour, avoidance, diffuse emotional excitement, and aggressive stalking. Nearly all the control dogs showed friendly behaviour to the *friendly* man while none of the experimental did. The two groups responded in a similar fashion to the *timid* man. The restricted dogs were unable to avoid physical contact with the *bold* man as he strode towards them.

Monkey. Mason (1960) studied the effects of social restriction on the behaviour of rhesus monkeys. The restricted group consisted of 6 monkeys born in the laboratory and separated from their mothers before the end of the first month of life. They were housed in individual cages which prevented physical contact between monkeys but did allow them to see and hear other monkeys. The second group of animals consisted of 6 feral monkeys of comparable age captured in the field and housed in groups of two or more until separated one month prior to testing. Testing consisted of pairing each monkey with the other members of his own group and recording the social and sexual behaviour which ensued. The restricted monkeys showed more frequent and prolonged fighting and less prolonged grooming activities than feral pairs. The restricted monkeys were deficient in sexual behaviour, particularly the males. Because the females differed in their sexual experience, a subsequent study was run in which restricted and feral males were tested with the same socially experienced females. Gross differences in the organization of the male sexual pattern of the restricted monkeys were still present under these conditions.

A similar phenomenon of sexual incompetence has been reported by

Harlow (1960) for rhesus monkeys removed from their mothers at birth and reared by means of mother surrogates (Harlow, 1958). In addition, the maternal behaviour of those few females which became pregnant was markedly defective.

Species differences. Table 10 summarizes the main findings concerning the effects of rearing conditions upon subsequent social behaviour. Isolation or restriction after weaning has been found to have deleterious effects upon future social behaviour in 5 of the 7 species discussed. The data also suggest that restriction of social experiences has both a more profound effect and broader effects as one ascends the phylogenetic scale.

4. INVESTIGATORY BEHAVIOUR

Montgomery & Zimbardo (1957) were not able to find any difference in investigatory behaviour of rats which had been reared in normal laboratory cages, much smaller cages which restricted activity, or cages which restricted activity and sensory stimulation. However, when normal cage-reared rats were compared with rats reared in a Hebb free environment, it was found

Table 10. Effects of Rearing Conditions upon Subsequent Social Behaviour

Species	Rearing Conditions	Main Findings	Author
Chicken	Isolation; flock	No differences in sexual behaviour	Wood-Gush (1958)
Chicken	Isolation; pairs; flock	No long-term impairment of social behaviour	Baron & Kish (1960)
Turkey	Isolation; flock	Isolation-reared birds responded sexually to humans	Schein & Hale (1959); Schein (1960)
Rat	Isolation; with males; with females	Isolation rearing resulted in highest proportion of copulators. Lowest proportion found in male-reared group	Beach (1942, 1958)
Rat	Isolation; play experience with males; play experience with females	Isolation rearing resulted in highest frequency of ejaculation. Lowest frequency found in group that played with females	Kagan & Beach (1953)
Mouse	Isolation; with males; with females	Isolation-reared mice less aggressive than other 2 groups. No difference in sexual behaviour. Social contact immediately after weaning more important than social contact at maturity	King & Gurney (1954); King (1956, 1957); Hudgens et al. (1968)

Table 10. Continued

Species	Rearing Conditions	Main Findings	Author
Guinea Pig	Isolation; with females	Social rearing resulted in more sexual behaviour. Past history determined effectiveness of testosterone in eliciting sexual behaviour. Critical period for isolation to have effect much earlier with a heterogeneous strain than with 2 homogeneous strains	Valenstein & Young (1955); Valenstein *et al.* (1955)
Dog	Restriction; pets	Restricted dogs inferior in competition; more passive and withdrawn	Clarke *et al.* (1951)
Dog	Restriction; pets	Restricted dogs inferior in competition; shared their feed. Pets exhibited friendly behaviour towards *friendly* man; restricted dogs did not. Restricted dogs could not avoid contact with *bold* man	Melzack & Thompson (1956)
Monkey	Restriction; group	Restricted monkeys showed more fighting, less grooming, deficiencies in sexual behaviour	Mason (1960)
Monkey	Artificial mother	Inability to copulate or form social group. Defects in "personality". Inadequate maternal behaviour	Harlow (1960); Seay *et al.* (1964); Arling & Harlow (1967)

that the cage-reared group exhibited significantly more investigatory behaviour; this was attributed to the greater complexity or novelty of the test situation for the cage-reared animals (Zimbardo & Montgomery, 1957). Woods *et al.* (1960) obtained essentially the same findings as did Zimbardo & Montgomery.

Thompson & Heron (1954a) compared the investigatory activity of dogs raised as pets and those raised under varying degrees of restriction and found that the restricted animals investigated more.

B. Feed Deprivation

The group of experiments to be described next could have been discussed in the prior section since they involve a form of constant stimulation during the immediate post-weaning period. Since all of the studies have the common element of some form of feed deprivation, they are presented together in a separate section.

Hoarding. Hunt (1941) raised rats on a reduced feeding schedule for 15 days, starting either at 24 or 32 days of age. In adulthood the group first deprived at 24 days hoarded more feed pellets than their controls, but only when tested under conditions of feed deprivation. The same pattern of results was obtained by Hunt *et al.* (1947).

Learning. One explanation of the findings concerning infantile feeding deprivation and hoarding is that the early deprivation leads to greater motivation (a heightened drive level) in adulthood when the animal is again placed in a stressful situation. Denenberg & Naylor (1957) tested this general hypothesis by deducing that equal amounts of feed would not be equally rewarding to animals which have had differential food experiences in their early life. It was expected that feed would be more rewarding to animals which had been deprived in early life. This was tested by comparing the rate at which infantile deprived rats and controls learned a discrimination problem with feed as the reward. As predicted, the group which had experienced deprivation in infancy was superior to controls in learning. Mandler (1958) found that rats which had been on a feeding deprivation schedule in infancy acquired a bar-pressing response for feed reward faster than controls. Once the response had been acquired there was no difference in response rate until the groups were shifted to reinforcement ratios of 10:1 and 20:1 (i.e. 10 or 20 responses for one feed pellet). At these points the experimentals responded at a higher rate than did controls.

Social Behaviour. Shortly after weaning Frederickson (1951) had pairs of mice compete for one pellet of feed, deprived or not deprived; controls always had feed. In adulthood all animals were tested for fighting behaviour under non-deprived conditions. All the experimental animals exhibited fighting while none of the controls did.

Consummatory Behaviour. Elliott & King (1960) manipulated feed deprivation by maintaining seven puppies on a reduced diet from the 3rd to the 7th weeks while controls had feed *ad lib.* Tests at 42 and 57 days found that the experimental puppies ate faster than controls though they did not eat any more feed than controls. Mandler (*loc. cit.*) found that rats reared on a deprivation schedule after weaning ate sooner than controls and showed other feed-oriented behaviours though no differences were found between the groups on investigatory behaviour.

Thus, a reduced feeding schedule immediately or shortly after weaning will significantly affect adult hoarding behaviour, learning performance, fighting behaviour, and rate of eating. In view of the broad range of effects, it is surprising that more research on the effects of feeding schedules has not been carried out. This is an important research area and one in which laboratory and domestic animals can be used with equal efficiency.

C. Discontinuous Stimulation

The major method used to administer brief periods of stimulation to post-weaning rats is the procedure called "gentling" which was first reported by Bernstein (1952). The clearest description of the procedure is given by Weininger (1956): "Gentling consisted of holding the animal in [the experimenter's] left hand with the hand placed against [the experimenter's] chest. The back of the rat was stroked with the right thumb, from the head to the base of the tail, at approximately fifty strokes a minute." This procedure is usually administered once a day for a period of 10 minutes.

Gentling is similar to handling in that it is a method of introducing extrinsic stimulation to the animal. However, the gentling procedure is usually done with post-weaning animals which are at a more mature stage in development than animals which receive handling experience. Thus, the effects of gentling should be different from the effects of handling because of the differences in ontogeny. Gentling with rats may reflect, at least in part, the general phenomenon of taming animals.

Learning. Bernstein (1957) attempted to test the hypothesis that the formation of a "relationship" between the experimenter and the animal would facilitate learning. He found that rats gentled from weaning on learned a discrimination problem faster than either of two control groups. After learning the animals were tested for retention; rats which continued to be gentled during the retention trials made less errors than animals which were not gentled during this time. Unfortunately, the lack of appropriate control groups makes it impossible to accept or reject the hypothesis that the differences in learning and retention were brought about by the formation of a relationship between man and rat; there are several alternative interpretations which are more parsimonious. However, it may be concluded that the procedure of gentling did result in much better learning.

Classical buzzer-shock conditioning in early life will also affect later learning and retention (Denenberg, 1958). Different groups of mice received buzzer-shock conditioning between 16 and 40 days of age. Retention scores at 50 days (measured by an extinction criterion) found that retention was monotonically related to age of prior conditioning; animals conditioned at 16 days gave no indication of retention while those conditioned at 40 days had the highest retention scores. The animals were then reconditioned; mice initially conditioned between 20 and 40 days had essentially the same reconditioning score and were better than the group originally conditioned at 16 days and a control group. These data suggested that the shift from the critical period of socialization to the juvenile period occurred between the 16th and 20th day in the mouse.

Social Behaviour. Rosen (1958) gentled one group of rats from day 21 to day 41 while controls were not disturbed. The animals were then allowed to eat individually each day for 5 minutes from a metal cylinder which was only large enough to allow one animal at a time to eat. The animals received this experience in a special enclosure. After this training a hungry gentled and a hungry non-gentled rat were released simultaneously into the enclosure. They were allowed 5 minutes in the unit and the total time each animal spent eating from the cylinder was recorded. Rosen reports that there was vigorous fighting around the feeding cylinder and that the gentled animals spent more time at the cylinder than the controls. This finding was interpreted in terms of the gentled animals having greater resistance to the emotional disturbance aroused by the fighting.

Activity, Physiology, Body Weight. No consistent relationship has been found between gentling and measures of activity, physiology, and body weight (Denenberg, 1962).

The effects of gentling administered after weaning are not as consistent nor as broad as the effects of handling pre-weaning animals. In fact a comparison of handling and gentling both before and after weaning found that handling and gentling before weaning had the equivalent effect of increasing weight and survival rate while gentling after weaning was of no benefit, as compared to a control group (Levine & Otis, 1958).

In conclusion, we are just beginning to understand and appreciate the tremendous impact that environmental factors (some of them apparently very subtle) can have upon the developing organism. Much more work is needed to fill in major gaps in our knowledge. A greater number of species needs to be studied, more techniques must be tried out; parametric studies using existing techniques are needed to determine functional relationships.

REFERENCES

ADER, R. & BELFER, M. L. (1962). Prenatal maternal anxiety and offspring emotionality in the rat. *Psychol. Rep.*, **10**, 711–718.

ALTMAN, M. (1958). Social integration of the moose calf. *Anim. Behav.*, **6**, 155–159.

ARLING, G. L. & HARLOW, H. F. (1967). Effects of social deprivation on maternal behaviour of the rhesus monkey. *J. comp. physiol. Psychol.*, **64**, 371–377.

BARON, A. & KISH, G. B. (1960). Early social isolation as a determinant of aggregative behavior in the domestic chicken. *J. comp. physiol. Psychol.*, **53**, 459–463.

BARTHOLOMEW, G. A. (1959). Mother-young relations and the maturation of pup behaviour in the Alaskan fur seal. *Anim. Behav.*, **7**, 163–172.

BATESON, P. P. G. (1964a). Changes in chicks' responses to novel moving objects over the sensitive period for imprinting. *Anim. Behav.*, **12**, 479–489.

BATESON, P. P. G. (1964b). Changes in the activity of isolated chicks over the first week after hatching. *Anim. Behav.*, **12**, 490–492.

BATESON, P. P. G. (1964c). Effect of similarity between rearing and testing conditions on chicks' following an avoidance response. *J. comp. physiol. Psychol.*, **57**, 100–103.

BATESON, P. P. G. (1964d). Relation between conspicuousness of stimuli and their effectiveness in the imprinting situation. *J. comp. physiol. Psychol.*, **58**, 407–411.

BATESON, P. P. G. (1966). The characteristics and context of imprinting. *Biol. Rev.*, **41**, 177–220.

BATESON, P. P. G. (1969). Imprinting and the development of preferences. In: *The Functions of Stimulation in Early Postnatal Development*. J. A. Ambrose (ed.). New York, N.Y.,: Academic Press Inc. (in press).

BEACH, F. A. (1942). Comparison of copulatory behavior of male rats raised in isolation, cohabitation, and segregation. *J. genet. Psychol.*, **60**, 121–136.

BEACH, F. A. (1958). Normal sexual behavior in male rats isolated at fourteen days of age. *J. comp. physiol. Psychol.*, **51**, 37–38.

BEACH, F. A. & JAYNES, J. (1954). Effects of early experience upon the behavior of animals. *Psychol. Bull.*, **51**, 239–263.

BELL, R. W., REISNER, G. & LINN, T. (1961). Recovery from electroconvulsive shock as a function of infantile stimulation. *Science*, **133**, 1428.

BERNSTEIN, LEWIS. (1952). A note on Christie's "Experimental naivete and experiential naivete". *Psychol. Bull.*, **49**, 38–40.

BERNSTEIN, L. (1957). The effects of variations in handling upon learning and retention. *J. comp. physiol. Psychol.*, **50**, 162–167.

BINGHAM, W. E. & GRIFFITHS, W. J. JR. (1952). The effect of different environments during infancy on adult behavior in the rat. *J. comp. physiol. Psychol.*, **45**, 307–312.

CLARKE, R. S., HERON, W., FEATHERSTONHAUGH, M. L., FORGAYS, D. C. & HEBB, D. O. (1951). Individual differences in dogs: preliminary reports on the effects of early experience. *Canad. J. Psychol.*, **5**, 150–156.

COLLIAS, N. E. (1956). The analysis of socialization in sheep and goats. *Ecology*, **37**, 228–239.

COOPER, R. M. & ZUBEK, J. P. (1958). Effects of enriched and restricted early environments on the learning ability of bright and dull rats. *Canad. J. Psychol.*, **12**, 159–164.

CORNWELL, A. W. & FULLER, J. L. (1961). Conditioned responses in young puppies. *J. comp. physiol. Psychol.*, **54**, 13–15.

DeNELSKY, G. Y. & DENENBERG, V. H. (1967a). Infantile stimulation and adult exploratory behavior: Effects of handling upon tactual variation seeking. *J. comp. physiol. Psychol.*, **63**, 309–312.

DeNELSKY, G. Y. & DENENBERG, V. H. (1967b). Infantile stimulation and adult exploratory behaviour in the rat: Effects of handling upon visual variation-seeking. *Anim. Behav.*, **15**, 568–573.

DENENBERG, V. H. (1958). Effects of age and early experience upon conditioning in the C57BL/10 mouse. *J. Psychol.*, **46**, 211–226.

DENENBERG, V. H. (1959). The interactive effects of infantile and adult shock levels upon learning. *Psychol. Rep.*, **5**, 357–364.

DENENBERG, V. H. (1960). A test of the critical period hypothesis and a further study of the relationship between age and conditioning in the C57BL/10 mouse. *J. genet. Psychol.*, **97**, 379–384.

DENENBERG, V. H. (1962). The effects of early experience. In: *The Behaviour of Domestic Animals*. E. S. E. Hafez (ed.). London: Baillière, Tindall & Cox, pp. 109–138.

DENENBERG, V. H. (1964). Critical periods, stimulus input, and emotional reactivity: A theory of infantile stimulation. *Psychol. Rev.*, **71**, 335–351.

DENENBERG, V. H. (1967). Stimulation in infancy, emotional reactivity, and exploratory behavior. In: *Neurophysiology and Emotion*. D. C. Glass (ed.). New York: Rockefeller University Press and Russell Sage Foundation, pp. 161–190.

DENENBERG, V. H. (1968). A consideration of the usefulness of the critical period hypothesis as applied to the stimulation of rodents in infancy. In: *Early Experience and Behavior*. G. Newton and S. Levine (eds.). Springfield, Ill.: Thomas, pp. 142–167.

DENENBERG, V. H. (1969). Open-field behavior in the rat: What does it mean? *An. N.Y. Acad. Sci.* (in press).

DENENBERG, V. H. & BELL, R. W. (1960). Critical periods for the effects of infantile experience on adult learning. *Science*, **131**, 227–228.

DENENBERG, V. H. BRUMAGHIM, J. T., HALTMEYER, G. C. & ZARROW, M. X. (1967). Increased adrenocortical activity in the neonatal rat following handling. *Endocrinology*, **81**, 1047–1052.

DENENBERG, V. H. & GROTA, L. J. (1964). Social-seeking and novelty-seeking behavior as a function of differential rearing histories. *J. abn. soc. Psychol.*, **69**, 453–456.

DENENBERG, V. H. & HALTMEYER, G. C. (1967). Test of the monotonicity hypothesis concerning infantile stimulation and emotional reactivity. *J. comp. physiol. Psychol.*, **63**, 394–396.

DENENBERG, V. H., HUDGENS, G. A. & ZARROW, M. X. (1964). Mice reared with rats: Modification of behavior by early experience with another species. *Science*, **143**, 380–381.

DENENBERG, V. H., HUDGENS, G. A. & ZARROW, M. X. (1966). Mice reared with rats: Effects of mother on adult behaviour patterns. *Psychol. Rep.*, **18**, 451–456.

DENENBERG, V. H. & KARAS, G. G. (1959). Effects of differential infantile handling upon weight gain and mortality in the rat and mouse. *Science*, **130**, 629–630.

DENENBERG, V. H. & KARAS, G. G. (1960). Interactive effects of age and duration of infantile experience on adult learning. *Psychol. Rep.*, **7**, 313–322.

DENENBERG, V. H. & KARAS, G. G. (1961). Interactive effects of infantile and adult experience upon weight gain and mortality in the rat. *J. comp. physiol. Psychol.*, **54**, 170–174.

DENENBERG, V. H., MORTON, J. R. C., KLINE, N. J. & GROTA, L. J. (1962). Effects of duration of infantile stimulation upon emotionality. *Canad. J. Psychol.*, **16**, 72–76.

DENENBERG, V. H. & NAYLOR, J. C. (1957). The effects of early food deprivation upon adult learning. *Psychol. Rec.*, **7**, 75–77.

DENENBERG, V. H., OTTINGER, D. R. & STEPHENS, M. W. (1962). Effects of maternal factors upon growth and behavior of the rat. *Child Develop.*, **33**, 65–71.

DENENBERG, V. H. & ROSENBERG, K. M. (1967). Nongenetic transmission of information. *Nature, Lond.*, **216**, 549–550.

DENENBERG, V. H. & SMITH, S. A. (1963). Effects of infantile stimulation and age upon behavior. *J. comp. physiol. Psychol.*, **56**, 307–312.

DENENBERG, V. H. & WHIMBEY, A. E. (1963). Behavior of adult rats is modified by the experiences their mothers had as infants. *Science*, **142**, 1192–1193.

DENENBERG, V. H., WOODCOCK, J. M. & ROSENBERG, K. M. (1968). Long-term effects of preweaning and postweaning free-environment experience on the rat's problem solving behavior. *J. comp. physiol. Psychol.*, **66**, 533–535.

DOYLE, G. & YULE, E. P. (1959). Early experience and emotionality: I. The effects of pre-natal maternal anxiety on the emotionality of albino rats. *S. Afr. J. Soc. Res.*, **10**, 57–65.

ELLIOTT, O. & KING, J. A. (1960). Effect of early food deprivation upon later consumma-tory behavior in puppies. *Psychol. Rep.*, **6**, 391–400.

FORGAYS, D. G. & FORGAYS, J. W. (1952). The nature of the effect of free-environmental experience in the rat. *J. comp. physiol. Psychol.*, **45**, 322–328.

FORGAYS, D. G. & READ, J. M. (1962). Crucial periods for free-environment experience in the rat. *J. comp. physiol. Psychol.*, **55**, 816–818.

FORGUS, R. H. (1956). Advantage of early over late perceptual experience in improving form discrimination. *Canad. J. Psychol.*, **10**, 147–155.

FREDERICKSON, E. (1951). Competition: the effects of infantile experience upon adult behavior. *J. abn. soc. Psychol.*, **46**, 406–409.

FREEDMAN, D., KING, J. A. & ELLIOTT, O. (1961). Critical period in the social development of dogs. *Science*, **133**, 1016–1017.

GAURON, E. F. & BECKER, W. C. (1959). The effects of early sensory deprivation on adult rat behavior under competition stress: An attempt at replication of a study by Alex-ander Wolf. *J. comp. physiol. Psychol.*, **52**, 689–693.

GINSBURG, B. E. & HOVDA, R. B. (1947). On the physiology of gene controlled audiogenic seizures in mice. *Anat. Rec.*, **99**, 621–622.

HALTMEYER, G. C., DENENBERG, V. H., THATCHER, J. & ZARROW, M. X. (1966). Response of the adrenal cortex of the neonatal rat after subjection to stress. *Nature, Lond.*, **212**, 1371–1373.

HALTMEYER, G. C., DENENBERG, V. H. & ZARROW, M. X. (1967). Modification of the plasma corticosterone response as a function of infantile stimulation and electric shock parameters. *Physiol. Behav.*, **2**, 61–63.

HARLOW, H. F. (1958). The nature of love. *Amer. Psychol.*, **13**, 673–685.

HARLOW, H. F. (1960). The maternal and infantile affectional patterns. Salmon Lecture 2.

HEBB, D. O. (1946). On the nature of fear. *Psychol. Rev.*, **53**, 259–276.

HEBB, D. O. (1947). The effects of early experience on problem-solving at maturity. *Amer. Psychol.*, **2**, 306–307 (abst.).

HEBB, D. O. (1949). *The Organization of Behavior*. New York, N.Y.: Wiley.

HEBB, D. O. & WILLIAMS, K. (1946). A method of rating animal intelligence. *J. gen. Psychol.*, **34**, 59–65.

HERSHER, L., MOORE, A. U. & RICHMOND, J. B. (1958). Effect of post partum separation of mother and kid on maternal care in the domestic goat. *Science*, **128**, 1342–1343.

HESS, E. H. (1959). Imprinting. *Science*, **130**, 133–141.

HINDE, R. A., THORPE, W. H. & VINCE, M. A. (1956). The following response of young coots and moorhens. *Behaviour*, **9**, 214–242.

HIRSCH, J. (1957). Careful reporting and experimental analysis—a comment. *J. comp. physiol. Psychol.*, **50**, 415.

HIRSCH, J., LINDLEY, R. H. & TOLMAN, E. C. (1955). An experimental test of an alleged innate sign stimulus. *J. comp. physiol. Psychol.*, **48**, 278–280.

HOCKMAN, C. H. (1961). Prenatal maternal stress in the rat: Its effects on emotional reared with artificial squint. *J. Neurophysiol.*, **28**, 1041–1059.

HUBEL, D. H. & WIESEL, T. N. (1963). Receptive fields of cells in striate cortex of very behavior in the offspring. *J. comp. physiol. Psychol.*, **54**, 679–684.

HUBEL, D. H. & WIESEL, T. N. (1965). Binocular interaction in striate cortex of kittens young, visually inexperienced kittens. *J. Neurophysiol.*, **26**, 994–1002.

HUDGENS, G. A., DENENBERG, V. H. & ZARROW, M. X. (1967). Mice reared with rats: Relations between mothers' activity level and offspring's behavior. *J. comp. physiol. Psychol.*, **63**, 304–308.

HUDGENS, G. A., DENENBERG, V. H. & ZARROW, M. X. (1968). Mice reared with rats:

Effects of preweaning and postweaning social interactions upon adult behaviour. *Behaviour*, **30**, 259-274

HUNT, J. McV. (1941). The effects of infant feeding-frustration upon hoarding in the albino rat. *J. abnorm. soc. Psychol.*, **36**, 338–360.

HUNT, J. McV., SCHLOSBERG, H., SOLOMON, R. L. & STELLAR, E. (1947). Studies of the effects of infantile experience on adult behavior in rats. I. Effects of infantile feeding frustration on adult hoarding. *J. comp. physiol. Psychol.*, **40**, 291–304.

HYMOVITCH, B. (1952). The effects of experimental variations on problem solving in the rat. *J. comp. physiol. Psychol.*, **45**, 313–321.

JOFFE, J. M. (1965). Genotype and prenatal and premating stress interact to affect adult behavior in rats. *Science*, **150**, 1844–1845.

KAGAN, J. & BEACH, F. A. (1953). Effects of early experience on mating behavior in male rats. *J. comp. physiol. Psychol.*, **46**, 204–208.

KARAS, G. G. & DENENBERG, V. H. (1961). The effects of duration and distribution of infantile experience on adult learning. *J. comp. physiol. Psychol.*, **54**, 170–174.

KIMBLE, G. A. (1961). *Conditioning and Learning* (2nd ed.), New York, N.Y.: Appleton-Century.

KING, J. A. (1956). Sexual behavior of C57BL/10 mice and its relation to early social experience. *J. genet. Psychol.*, **88**, 223–229.

KING, J. A. (1957). Relationship between early social experience and adult aggressive behavior in inbred mice. *J. genet. Psychol.*, **90**, 151–166.

KING, J. A. (1958). Parameters relevant to determining the effects of early experience upon the adult behavior of animals. *Psychol. Bull.*, **55**, 46–58.

KING, J. A. & ELEFTHERIOU, B. E. (1959). Effects of early handling upon adult behavior in two subspecies of deer-mice, *Peromyscus maniculatus*. *J. comp. physiol. Psychol.*, **52**, 82–88.

KING, J. A. & GURNEY, N. L. (1954). Effects of early social experience on adult aggressive behavior in C57BL/10 mice. *J. comp. physiol. Psychol.*, **47**, 326–330.

LEVINE, S. (1956). A further study of infantile handling and adult avoidance learning. *J. Personality*, **25**, 70–80.

LEVINE, S. (1957). Infantile experience and resistance to physiological stress. *Science*, **126**, 405.

LEVINE, S. (1958). Effects of early deprivation and delayed weaning on avoidance learning in the albino rat. *Arch. Neurol. Psychiat.*, **79**, 211–213.

LEVINE, S. (1959a). Emotionality and aggressive behavior in the mouse as a function of infantile experience. *J. genet. Psychol.*, **94**, 77–83.

LEVINE, S. (1959b). The effects of differential infantile stimulation on emotionality at weaning. *Canad. J. Psychol.*, **13**, 243–247.

LEVINE, S. (1962). The psychophysiological effects of infantile stimulation. In: *Roots of Behavior*. E. L. Bliss (ed.). New York: Harper, pp. 246–253.

LEVINE, S. (1962a). Plasma-free corticosteroid response to electric shock in rats stimulated in infancy. *Science*, **135**, 795–796.

LEVINE, S. (1962b). The effects of infantile experience on adult behavior. In: *Experimental Foundations of Clinical Psychology*. A. J. Bachrach (ed.). New York: Basic Books, pp. 139–169.

LEVINE, S. (1965). Maturation of the neuroendocrine response to stress. VIth Pan American Congress of Endocrinology, *Excerpta Med. Internat. Congress Series* No. 99.

LEVINE, S. (1967). Maternal and environmental influences on the adrenocortical response to stress in weanling rats. *Science*, **156**, 258–260.

LEVINE, S. & ALPERT, M. (1959). Differential maturation of the central nervous system as a function of early experience. *A. M. A. Arch. gen. Psychiat.*, **1**, 403–405.

LEVINE, S., ALPERT, M. & LEWIS, G. W. (1958). Differential maturation of an adrenal response to cold stress in rats manipulated in infancy. *J. comp. physiol. Psychol.*, **51**, 774–777.

LEVINE, S. & BROADHURST, P. L. (1963). Genetic and ontogenetic determinants of adult behavior in the rat. *J. comp. physiol. Psychol.*, **56**, 2, 423–428.

LEVINE, S., CHEVALIER, J. A. & KORCHIN, S. J. (1956). The effects of early shock and handling on later avoidance learning. *J. Personality*, **24**, 475–493.

LEVINE, S. & COHEN, C. (1959). Differential survival to leukemia as a function of infantile stimulation in DBA/2 mice. *Proc. Soc. exp. Biol. Med., N.Y.*, **102**, 53–54.

LEVINE, S., GLICK, D. & NAKANE, P. K. (1967). Adrenal and plasma corticosterone and vitamin A in rat adrenal glands during postnatal development. *Endocrinology*, **80**, 910–914.

LEVINE, S., HALTMEYER, G. C., KARAS, G. G. & DENENBERG, V. H. (1967). Physiological and behavioral effects of infantile stimulation. *Physiol. Behav.*, **2**, 55–59.

LEVINE, S. & LEWIS, G. W. (1959). Critical period for the effects of infantile experience on the maturation of a stress response. *Science*, **129**, 42–43.

LEVINE, S. & MULLINS., R. F., Jr. (1966). Hormonal influence on brain organization in infant rats. *Science*, **152**, 1585–1592.

LEVINE, S. & OTIS, L. S. (1958). The effects of handling before and after weaning on the resistance of albino rats to later deprivation. *Canad. J. Psychol.*, **12**, 103–108.

LEVINE, S. & WETZEL, A. (1963). Infantile experiences, strain differences, and avoidance learning. *J. comp. physiol. Psychol.*, **56**, 879–881.

LIDDELL, H. (1954). Conditioning and emotions. *Scient. Amer.*, **190**, 48–57.

LINDZEY, G., WINSTON, H. D. & MANOSEVITZ, M. (1963). Early experience, genotype and temperament in *Mus Musculus*. *J. comp. physiol. Psychol.*, **56**, 622–629.

LORENZ, K. (1937). The companion in the bird's world. *Auk*, **54**, 245–273.

LUCHINS, A. S. & FORGUS, R. H. (1955). The effect of differential post-weaning environments on the rigidity of an animal's behaviour. *J. genet. Psychol.*, **86**, 51–58.

MANDLER, J. M. (1958). Effects of early food deprivation on adult behavior in the rat. *J. comp. physiol. Psychol.*, **51**, 513–517.

MASON, W. A. (1960). The effects of social restriction on the behavior of rhesus monkeys: I. Free social behavior. *J. comp. physiol. Psychol.*, **53**, 582–589.

MEIER, G. W. & McGEE, R. K. (1959). A re-evaluation of the effect of early perceptual experience on discrimination performance during adulthood. *J. comp. physiol. Psychol.*, **52**, 390–395.

MELZACK, R. (1961). On the survival of mallard ducks after "habituation" to the hawk-shaped figure. *Behaviour*, **17**, 9–16.

MELZACK, R., PENICK, E. & BECKETT, A. (1959). The problem of "innate fear" of the hawk form: An experimental study with mallard ducks. *J. comp. physiol. Psychol.*, **52**, 694–698.

MELZACK, R. A. & SCOTT, T. H. (1957). The effects of early experience on the response to pain. *J. comp. physiol. Psychol.*, **50**, 155–161.

MELZACK, R. & THOMPSON, W. R. (1956). Effects of early experience on social behavior. *Canad. J. Psychol.*, **10**, 82–90.

MOLTZ, H. (1960). Imprinting: Empirical basis and theoretical significance. *Psychol. Bull.*, **57**, 291–314.

MOLTZ, H. (1963). Imprinting: An epigenetic approach. *Psychol. Rev.*, **70**, 123–138.

MOLTZ, H. & ROSENBLUM, L. A. (1958). The relation between habituation and the stability of the following response. *J. comp. physiol. Psychol.*, **51**, 658–661.

MOLTZ, H., ROSENBLUM, L. A. & HALIKAS, N. (1959). Imprinting and level of anxiety. *J. comp. physiol. Psychol.*, **52**, 240–244.

MONTGOMERY, K. C. & ZIMBARDO, P. G. (1957). Effects of sensory and behavioral deprivation upon exploratory behavior in the rat. *Percept. Mot. Skills*, **7**, 223–229.

MOORE, A. U. (1968). Effects of modified maternal care in the sheep and goat. In: *Early Experience and Behavior*. G. Newton & S. Levine (eds.). Springfield, Ill.: Thomas, pp. 142–167.

MORRA, M. (1965). Level of maternal stress during two pregnancy periods on rat offspring behavior. *Psychonom. Sci.*, **3**, 7–8.

NEWTON., G. & LEVINE, S. (eds.) (1968). *Early Experience and Behavior*. Springfield, Ill.: Thomas.

NISSEN, H. W., CHOW, K. L. & SEMMES, J. (1951). Effects of restricted opportunity for tactual, kinesthetic, and manipulative experience on the behavior of a chimpanzee. *Amer. J. Psychol.*, **64**, 485–507.

OTTINGER, D. R., DENENBERG, V. H. & STEPHENS, M. W. (1963). Maternal emotionality, multiple mothering, and emotionality in maturity. *J. comp. physiol. Psychol.*, **56**, 313–317.

PFAFFENBERGER, C. J. & SCOTT, J. P. (1959). The relationship between delayed socialization and trainability in guide dogs. *J. genet. Psychol.*, **95**, 145–155.

RABINOVITCH, M. S. & ROSVOLD, H. E. (1951). A closed-field intelligence test for rats. *Canad. J. Psychol.*, **5**, 122–128.

RIESEN, A. H. (1947). Visual discriminations by chimpanzees after rearing in darkness. *Amer. Psychol.*, **2**, 307.

RIESEN, A. H. (1950, July). Arrested vision. *Scient. Amer.*, **183**, 16–19.

ROCKETT, F. C. (1955). A note on "An experimental test of an alleged innate sign stimulus" by Hirsch, Lindley, and Tolman. *Percept. Mot. Skills*, **5**, 155–156.

ROSEN, J. (1958). Dominance behaviour as a function of post-weaning gentling in the albino rat. *Canad. J. Psychol.*, **12**, 229–234.

ROSENBLATT, J. TURKEWITZ, G. & SCHNEIRLA, T. C. (1961). Early socialization in the domestic cat as based on feeding and other relationships between female and young. In: *Determinants of Infant Behaviour*. B. Foss (ed.)., London: Methuens.

SALZEN, E. A. (1966). Imprinting in birds and primates. *Behaviour*, **28**, 232–253.

SCHEIN, M. W. (1960). Modification of sexual stimuli by imprinting in turkeys. *Anat. Rec.*, **137**, 392 (abst.).

SCHEIN, M. W. & HALE, E. B. (1959). The effect of early social experience on male sexual behaviour of androgen injected turkeys. *Anim. Behav.*, **7**, 189–200.

SCOTT, J. P. (1945). Social behavior, organization, and leadership in a small flock of domestic sheep. *Comp. Psychol. Monogr.*, **18**, (4), 1–29.

SCOTT, J. P. (1958). Critical periods in the development of social behavior in puppies. *Psychosom. Med.*, **20**, 42–54.

SCOTT, J. P. (1960). Comparative social psychology. In: *Principles of Comparative Psychology*. R. H. Waters *et al.* (eds.), New York, N.Y.: McGraw-Hill.

SCOTT, J. P., FREDERICKSON, E. & FULLER, J. L. (1951). Experimental exploration of the critical period hypothesis. *Personality*, **1**, 162–183.

SCOTT, J. P. & MARSTON, M. V. (1950). Critical periods affecting the development of normal and maladjustive social behavior in puppies. *J. genet. Psychol.*, **77**, 25–60.

SCOTT, J. P., ROSS, S. & FISHER, A. E. (1959). The effects of early enforced weaning on sucking behavior of puppies. *J. genet. Psychol.*, **95**, 261–281.

SEAY, B., ALEXANDER, B. K. & HARLOW, H. F. (1964). Maternal behavior of socially deprived rhesus monkeys. *J. Abnor. soc. Psychol.*, **69**, 345-354.

SEITZ, P. F. D. (1954). The effects of infantile experiences upon adult behavior in animal subjects: I. Effects of litter size during infancy upon adult behavior in the rat. *Amer. J. Psychiat.*, **110**, 916–927.

SEITZ, P. F. D. (1958). The maternal instinct in animal subjects: I. *Psychosom. Med.*, **20**, 215–226.

SEITZ, P. F. D. (1959). Infantile experience and adult behavior in animal subjects: II. Age of separation from the mother and adult behavior in the cat. *Psychosom. Med.*, **21**, 353–378.

SMITH, A. M. (1967). Infantile stimulation and the Yerkes-Dodson law. *Canad. J. Psychol.*, **21**, 285–293.

SONTAG, L. W. (1941). The significance of fetal environmental differences. *Amer. J. Obstet. Gynec.*, **42**, 996–1003.

SOUTHWICK, C. H. (1968). Effect of maternal environment on aggressive behavior of inbred mice. *Communications in behavioral biology*, **1**, 129–132.

THOMPSON, W. R. (1957). Influence of prenatal maternal anxiety on emotionality in young rats. *Science*, **125**, 698–699.

THOMPSON, W. R. & HERON, W. (1954a). The effect of early restriction on activity in dogs. *J. comp. physiol. Psychol.*, **47**, 77–82.

THOMPSON, W. R. & HERON, W. (1954b). The effects of restricting early experience on the problem-solving capacity of dogs. *Canad. J. Psychol.*, **8**, 17–31.

TINBERGEN, N. (1951). *The Study of Instinct*. New York, N.Y.: Oxford University Press.

TINBERGEN, N. (1957). On anti-predator responses in certain birds—a reply. *J. comp. physiol. Psychol.*, **50**, 412–414.

VALENSTEIN, E. S., RISS, W. & YOUNG, W. C. (1955). Experiential and genetic factors in the organization of sexual behavior in male guinea pigs. *J. comp. physiol. Psychol.*, **48**, 397–403.

VALENSTEIN, E. S. & YOUNG, W. C. (1955). An experiential factor influencing the effectiveness of testosterone proprionate in eliciting sexual behavior in male guinea pigs. *Endocrinology*, **56**, 173–177.

WEININGER, O. (1956). The effects of early experience on behavior and growth characteristics. *J. comp. physiol. Psychol.*, **49**, 1–9.

WHIMBEY, A. E. & DENENBERG, V. H. (1967). Experimental programming of life histories: The factor structure underlying experimentally created individual differences. *Behaviour*, **29**, 296–314.

WIESEL, T. N. & HUBEL, D. H. (1963a). Effects of visual deprivation on morphology and physiology of cells in the cat's lateral geniculate body. *J. Neurophysiol.*, **26**, 978–993.

WIESEL, T. N. & HUBEL, D. H. (1963b). Single-cell responses in striate cortex of kittens deprived of vision in one eye. *J. Neurophysiol.*, **26**, 1003–1017.

WIESEL, T. N. & HUBEL, D. H. (1965a). Comparison of the effects of unilateral and bilateral eye closure on cortical unit responses in kittens. *J. Neurophysiol.*, **28**, 1029–1040.

WIESEL, T. N. & HUBEL, D. H. (1965b). Extent of recovery from the effects of visual deprivation in kittens. *J. Neurophysiol.*, **28**, 1060–1972.

WILLIAMS, E. & SCOTT, J. P. (1953). The development of social behavior patterns in the mouse, in relation to natural periods. *Behaviour*, **6**, 35–64.

WOLF, A. (1943). The dynamics of the selective inhibition of specific functions in neurosis: a preliminary report. *Psychosom. Med.*, **5**, 27–38.

WOOD-GUSH, D. G. M. (1958). The effect of experience on the mating behaviour of the domestic cock. *Anim. Behav.*, **6**, 68–71.

WOODS, P. J. (1959). The effects of free and restricted environmental experience on problem-solving behavior in the rat. *J. comp. physiol. Psychol.*, **52**, 399–402.

WOODS, P. J., RUCKELSHAUS, S. I. & BOWLING, D. M. (1960). Some effects of "free" and "restricted" environmental rearing conditions upon adult behavior in the rat. *Psychol. Rep.*, **6**, 191–200.

YOUNG, W. C., GOY, R. W. & PHOENIX, C. H. (1964). Hormones and sexual behavior. *Science*, **143**, 212–218.

ZARROW, M. X., DENENBERG, V. H., HALTMEYER, G. C. & BRUMAGHIM, J. T. (1967). Plasma and adrenal corticosterone levels following exposure of the two-day-old rat to various stressors. *Proc. Exp. Biol. Med.* **125**, 113–116.

ZARROW, M. X., HALTMAYER, G. C., DENENBERG, V. H. & THATCHER, J. (1966). Response of the infantile rat to stress. *Endocrinology*, **79**, 631–634.

ZARROW, M. X., PHILPOTT, J. E., DENENBERG, V. H. & O'CONNER, W. B. (1968). Localization of $^{14}C-4$–corticosterone in the 2-day old rat and a consideration of the mechanism involved in early handling. *Nature, Lond.*, **218**, 1264–1265.

ZIMBARDO, P. G. & MONTGOMERY, K. C. (1957). Effects of "free environment" rearing upon exploratory behavior. *Psychol. Rep.*, **3**, 589–594.

Chapter 7

The Physiological Analysis of Animal Behaviour

J. I. Johnson, G. I. Hatton and R. W. Goy

I. PHYSIOLOGICAL PROCESSES UNDERLYING BEHAVIOUR

The study of behaviour is different from other branches of natural science only in the level of organization of its subject matter. It differs from physiology, chemistry and physics in that its purview is the activity of the organism considered as a whole. But much of the understanding and explanation at each of these levels of scientific inquiry derives from study at the next more basic level. Thus, just as physics supplies explanations for chemical processes, and chemistry elucidates physiological activity, so physiology answers many questions posed by behavioural phenomena.

Books have been devoted to extensive treatment of the relationships between physiological processes and behaviour (e.g. Altman, 1966; Thompson, 1967). Here we shall be restricted to a sketch of this physiological analysis of behaviour, with a presentation of examples of such analyses that have elucidated some of the behaviours found among the domesticated animals.

We learn more about behaviour, the activity of whole animals, by considering individually the activities of parts of animals, the workings of organs and systems, especially as they relate to one another. Physiology is the study of the functions, processes, activities of these parts of animals. Behaviour can be viewed as the organismic integration of all the physiological processes taking place in an animal; these processes have been considered as the pieces, the parts, the fractions, the components, the foundations or the substrates of behaviour.

Behaviour can also be regarded as the complex of interactions of an organism with its environments; physiological analysis describes the role of individual organs and processes in those interactions. Foremost among these elements are the sense organs and processes relating the animal to environmental events, and the effectors by which the animal acts upon the environment. These constitute, along with neural and endocrine mechanisms for processing and programming both information and activity, the organic components basic to the behaviour of birds and mammals.

A. Sensory Processes

Detector organs gather information from the current environment of an animal. The phrase "internal environment" has been used to emphasize the fact that information is obtained from within an animal's body as well as from events and things outside of it. Information is transmitted from environment(s) to animals as a configuration of energy changes. We must emphasize that, in nearly all sensory and perceptual processes in the birds and mammals we are considering in this book, it is the configuration rather than the energy change itself that is important to behaviour. For examples, ". . . the higher animals and therefore their nerve cells want light mainly as a carrier of information about objects, directions, time, and so on" (Bullock, 1966, p. 8); and "The listener is unaware of the pressure changes acting upon his eardrum, and he does not perceive their transmission upon the sensory surface . . . yet he can understand the meaning of a spoken message" (Hess, 1967, pp. 1280–1281).

1. Energy Selectivity and Sensitivity

Animals are continuously subjected to energy fluctuations occurring in their environments. Some types of energy variations are obvious carriers of messages influencing behaviour (e.g. visible light changes, sound variations, etc.), while others are apparently either not at all related to behaviour or are less potent and dramatic in their influence (e.g. X-rays and extremely high and extremely low frequency pressure waves). Mechanisms have evolved which detect changes in the energy of the environment. These mechanisms are rather finely tuned, and the energy changes to which they respond are limited both quantitatively and qualitatively.

A first quantitative limitation is that detection capabilities are restricted to some fraction of the total possible extent to which the energy can fluctuate. Thus we humans cannot see ultraviolet "light" even though it is the same type of energy as visible light. Similarly, we cannot hear longitudinal pressure changes when they occur in the megacycle frequency range. However, it is not safe to assume because our receptor capacities are limited to some specific energy band, that other animals are identically limited. Other animals certainly differ from us as well as among themselves with respect to the ranges of variations to which they are selectively attuned. We all know that dogs hear frequencies that are above our audible range, and that bats use high frequency echolocation as a means of navigation. Nevertheless, large portions of the energy spectra go undetected by animals.

A second general quantitative limitation of animal mechanisms for detecting energy fluctuations is their degree of sensitivity. Even within the selected range of detectable energy levels, all variations in level are not detected. That is to say, these sensing systems have minimal-energy-change or threshold requirements, below which they are insensitive or hyposensitive. Sensory systems, then, have the property of being sensitive to energy fluctuations only if these are equal to or greater than some predetermined value (in part predetermined by the genetic and experiential history of the animal, and in part by the conditions existing immediately prior to the fluctuation).

In addition to these two quantitative properties, being tuned (1) to changes within a portion of the energy spectrum, and (2) to a predetermined level of intensities within that portion, individual sensing or detection systems are further characterized by the kind or quality of energy variations to which they are most sensitive. The structure of each sensing element is specialized for transduction of usually one or two types of environmental energy into coded neural information. The animal may, at any given time, be called upon to transduce and transmit information about changes in light, sound, odour, touch, pain, heat, cold, blood pressure, osmotic pressure, oxygen tension, etc. Animals have evolved sensing elements which constantly report on all of these important types of energy fluctuations.

These quantitative and qualitative characteristics are presumably related to the biological importance of the energies thereby selected, or more particularly, of the messages carried by means of the energies selected. It would be difficult to say, however, that selection is restricted to ranges, levels, and kinds that are important to the survival of individuals and species. This is a case of the general rule that if a mechanism is adaptive, it survives; but the fact of survival is not a demonstration of adaptive significance.

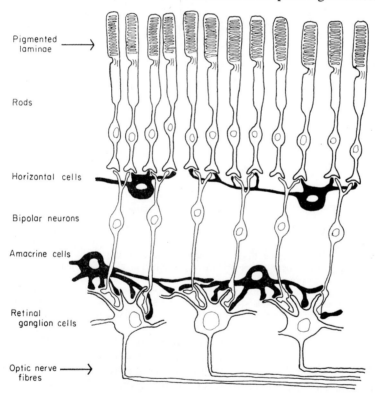

Pigmented laminae

Rods

Horizontal cells

Bipolar neurons

Amacrine cells

Retinal ganglion cells

Optic nerve fibres

Fig. 9. Schema of interrelationships of elements of a rod-dominated portion of mammalian retina. Light acting on pigmented laminae or rods presumably instigates neural activity in all elements. This activity constitutes processing whose end results are recorded as patterns of nerve impulses in the retinal ganglion cells, whose branches travel from the eye to the brain as fibres of the optic nerve. (Based on data from Sjöstrand, 1953a, b; Dowling & Boycott, 1966; Walls, 1942.)

2. RECEPTORS

The essential element of a detector in a sense organ is a cell, called a receptor cell, which has the capacity of responding to some change in the selected energy impinging upon it. For a first example, in the visual sense organs of vertebrates the receptor cells are photosensitive rods and cones in the retina. When a change within the selected range of the ambient photic energy impinges upon a rod, the rod reacts. The rod reaction appears to result in the generation of a nerve impulse in a neighbouring nerve cell (or neuron, see Fig. 9). This impulse in turn sets up (or prevents) impulses in other neighbouring neurons, and we have departed from the receptor component and entered the neural processing component of the behaving system.

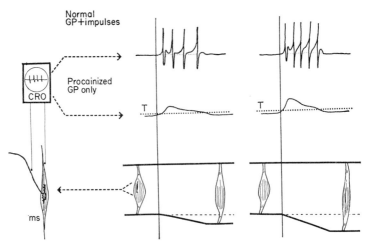

Fig. 10. A form of transduction and neural coding accomplished in a receptor ending of a nerve fibre. Stretching of muscle spindle *ms* produces impulses in the nerve fibre recorded on cathode ray oscilloscope *CRO*. Abolition of impulses by application of procaine makes it possible to visualize the generator potential *GP*. Impulses are generated when *GP* amplitude is above threshold *T*. Impulse frequency is a function of *GP* amplitude, which is proportional to muscle stretch. Relative amplitude scale for *GP* has been doubled to facilitate visualization. (Based on data from Katz, 1950).

For a second example, consider a receptor which is itself a neuron. Certain fibres in muscle tissue are arranged into a structure known as a muscle spindle. These spindles have a complex innervation; this includes one nerve ending which is specialized to react when the muscle fibres are stretched. Stretching of the muscle in some way stretches or bends the outer limiting membrane of the nerve end. The bent or stretched membrane apparently allows certain ions to pass through the membrane which otherwise could not. This ionic movement results in a change in the electrochemical balances on the two sides of the membrane; this change is known as a generator potential (see Fig. 10). It is proportional in amplitude to the degree of stretching of the membrane. If the amplitude of the change in potential is sufficiently large, a nerve impulse or a series of impulses proceeds from the receptor ending up the nerve fibre and on through all other branches of the nerve cell. These branches reach into the spinal cord; some even attain the base of the brain. Thus communication of the receptor with other parts of the body is accomplished, and neural processing is under way.

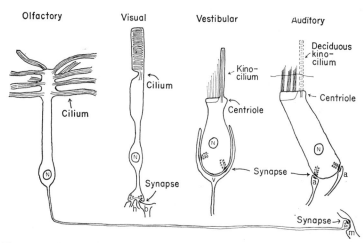

Fig. 11. Schematic illustration of vertebrate receptors which have a cilium and synapses at opposite ends of the cell. Olfactory receptors have many cilia, each too long to be shown here in entirety, and a very fine long axon which travels to the olfactory bulb where it synapses with a branch of a mitral cell *m* (based on data from De Lorenzo, 1960; Moulton & Beidler, 1967; Okano *et al.*, 1967). All that remains of the phylogenetically primordial cilium in the visual rod is the slender stalk connecting the inner and outer segments; the branch bearing synapses has the structure of an axon which establishes synaptic contact with horizontal cells *h* and bipolar cells *b* (based on data from Sjöstrand, 1953a, b; Dowling & Boycott, 1966; Eakin, 1966). Each vestibular receptor bears many sensory hairs, which seem responsible for activity resulting in impulses in vestibular nerve cell *v*, and one kinocilium whose function is speculative. During embryonic development, the auditory receptors along the basilar membrane also have sensory hairs and one kinocilium. The hairs are arranged in rows. The kinocilium disappears before the receptor becomes functional, leaving only its base as the centriole. Synapses of vestibular and auditory receptors with vestibular *v* and auditory *a* bipolar neurons have a characteristic synaptic bar which is surrounded by vesicles. (Based on data from Wersäll *et al.*, 1966.)

Other nerve endings serving as receptors are known to subserve the senses of touch, temperature, and pain. The transduction processes whereby physical energy is converted into neural activity are still quite mysterious. One interesting fact emerging from recent studies of the ultrastructure of receptors is the presence of cilia-like structures at the "receiving end", and synapse-like connections at the "sending end", of a wide variety of transducer cells, including those subserving vision, audition, olfaction, vestibular and labyrinthine sensation (Fig. 11).

3. CODING OF STIMULUS CHARACTERISTICS

Intensity coding. Consider again the activity of the muscle stretch receptor just described. The generator potential is directly proportional to the degree of stretching of the muscle. Furthermore, the number and frequency of the nerve impulses generated is proportional to the amplitude of the generator potential, and thus to the degree of stretch (Katz, 1950). In other words, stimulus intensity has been coded into frequency and duration of a train of nerve impulses (Fig. 10).

Intensity can also be coded simply by whether certain receptors are activated or not. If some of the receptors in a given region of the body are less sensitive than neighbouring receptors, then a stimulus which activates the less sensitive ("higher threshold") receptors will also activate the more sensitive ones. Coding is happening in two ways at the same time: the intense stimulus activates more receptors, and it activates some receptors which signal only intense stimuli. A second kind of recruiting also indicates intensity in those cases where more intense energies spread out in space. As the energy spreads it activates receptors over a wider area; once again a more intense stimulus activates a greater number of receptors.

Topographical coding. Besides these kinds of intensity-coding, another type of coding occurs in most, if not all, sensory systems. This is *topographical* coding: a receptor transmits information about the location of a stimulus by virtue of its own location in the body. Activity in a receptor in the foot signals that stimulation is impinging on the foot and not the head; a retinal cell in the lower part of the retina, in its every activation, indicates that photic events are taking place in the upper part of the visual field (which is "seen" with the lower retina). Auditory receptors not only signal location of sound by which of two ears is more strongly stimulated, they also may signal pitch by topographical coding. Going from the base to the apex of the spiral basilar membrane on which auditory receptors are located, cells in any region are activated most strongly by sounds of certain frequency. Cells near the base respond to high frequency waves, those near the apex respond to low frequency, those between respond to ordered intervening frequencies. In visual, auditory, and tactile systems this topographical specificity is maintained through several levels of neural processing in the brain.

B. Effector Processes

A simple mechanism occurs in the limbs of many vertebrates wherein activation of a receptor produces a movement in an animal. Activation of a mechanoreceptive distal ending of a nerve cell produces nerve impulses which travel centrally along the cell membrane to other specialized endings of branches of the cell. These endings contact other nerve cells, in specialized structures known as synapses, where the external limiting membranes of the two cells adhere to one another, with a space or cleft existing between the adhering membranes (Fig. 12). Arrival of the nerve impulse at a synapse brings about the release there of a chemical transmitter substance from the pre-synaptic membrane. This transmitter somehow changes the ionic permeability of the post-synaptic membrane, such as to change the electro-chemical balance across this membrane of the second cell. This change, very similar to the generator potential of the receptor, is called an excitatory post-synaptic potential (EPSP), since if it is of sufficiently large amplitude, it will excite or generate nerve impulses in the second, post-synaptic cell.

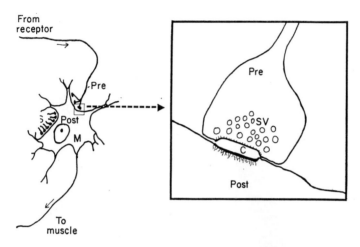

Fig. 12. Synapse as found in monosynaptic reflex connection on a motoneuron *M*. Relative density of synapses in moto-neuron surface is suggested by section near *s*. Enlarged view of synapse at right shows location of synaptic vesicles *sv* which are presumed to contain chemical transmitter substance. Upon arrival of nerve impulses, transmitter is released from pre-synaptic cell *PRE* into synaptic cleft *c*, where it acts on mem-brane of post-synaptic cell *POST*. Based on data from de Robertis (1964) and Eccles (1964).

In a *monosynaptic reflex* (Figs. 12 and 13), the second cell is a motoneuron, whose chief axon terminates on muscle fibres in special structures known as myoneural junctions or motor end-plates. As in the synapse, arrival of nerve impulses brings about the release of transmitter substance, which in this case is known to be acetylcholine, into the junction. The transmitter alters the electro-chemical balance of the next cell membrane, and impulses are set up that travel along the membranes of the muscle fibres. But the specialization of the muscle fibres is such that when the impulse passes along the surface membrane, the fibrils within the cells are structurally altered so that their components are drawn together and the muscle fibres shorten in length (Katz, 1966).

Virtually all observable behaviour of birds and mammals and many other animals consists in such contractions of muscle. The different behavioural entities are but different sets and organizations of a multitude of individual contracting muscle fibres, acting under the influence of their motoneurons. A monosynaptic reflex is one means whereby fibres contract in response to

5*

activity of a receptor. (In this reflex mechanism, many receptors activate many fibres, but only one synapse intervenes between any given receptor and the related motoneuron.) In context, this contraction forms a small part of the actions making up what we think of as a behavioural response. One actual monosynaptic reflex is involved in maintaining posture against the force of gravity in a jointed limb (Fig. 13).

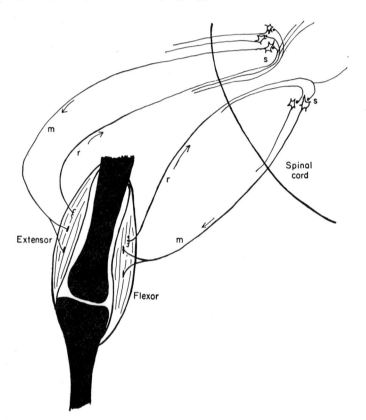

Fig. 13. The monosynaptic myotatic reflex. Contraction of the flexor muscle stretches the extensor. Fibres from stretch receptors *r* in the extensor through synapses *s*, excite the moto-neorons *m* serving the extensor, causing it to contract. This stretches the flexor, causing it to contract through a similar mechanism. Both muscles are thus maintained in a state of contraction, keeping the joint firmly but flexibly in posturally supportive position. The experimental reports demonstrating this reflex, by Sherrington, Lloyd and others are collected in Barnes & Kircher (1968).

C. Neural Processing and Programming

Behaviour is more than a set of responses to stimuli. For one thing responses are organized into relation with each other, and are influenced by each other, through neural interconnections. For example, walking movements of vertebrates proceed in a highly organized fashion with a minimum of sensory guidance and regulation. Proprioceptive impulses from a single limb were sufficient to maintain the orderly cycle of walking movements in all four limbs in toads (Gray, 1950). Artificial electrical stimulation at various locations in the nervous system has produced long sequences of finely co-ordinated movements (Hess, 1957). Responses are further influenced by the immediate and remote past experience of the animal. Also, varying levels of hormones in the circulatory system exert profound effects on the responding elements and their organization.

Even the responses to stimuli are rarely as simple and direct as in the reflexes; this is why reflexes are considered as special types of responses: they are uncomplicated and uniform stereotyped actions in response to a particular stimulation. Most responses, among the vertebrates, are not reflexes. They are occasioned by analysis and abstraction of sensory input which isolates significant features of stimulation received, relates these features to the phylogenetic experience of the species, the personal history of the animal, the current state of affairs within the animal, and only then is a programme of organized activity put into effect. The mechanisms, whereby this takes place, are largely unknown. Of those that are known, we present a few examples of the kinds of synthesizing processes that take place in the programming and production of behaviour.

1. Effector processing Mechanisms

Neural *facilitation* and *inhibition* have been identified as two basic processes, occurring at synapses throughout the nervous system, that are fundamental to synthesis and analysis of transmitted information. As a highly simplified case of facilitation, consider a motoneuron that is synaptically contacted by endings from two receptor cells. An impulse that arrives at either ending alone does not induce impulses in the motoneuron; a small EPSP is generated, but it does not reach the threshold amplitude necessary to fire an impulse. However, when impulses arrive in both endings at or near the same time, the EPSP of one *summates* with that of the other, and together their joint amplitude passes the impulse-generation threshold and the motoneuron fires. In this case the release of transmitter substance in the first ending is said to *facilitate* the transmitter effect from the other, and this facilitation is necessary for the activation of the motoneuron impulse. It is also said that the transmitter effects of the two endings *summate*; their influence on the motoneuron is in some ways additive. The mechanisms of *inhibition*, the prevention of neuronal impulse generation—and there are apparently several quite different mechanisms—are still being worked out (see Eccles, 1964 for details). One mechanism has been described in detail; this is the inhibitory post-synaptic potential or IPSP. This was first demonstrated in special cells of the spinal cord located near motoneurons and called Renshaw cells after the discoverer of their effects (Eccles, 1957). Renshaw cells synapse on motoneurons, and when a Renshaw cell is activated, a series of impulses arrives at the synapse, and as long as the impulse train persists, the motoneuron fires only with extreme levels of excitation (much summation of

EPSPs), if at all. The transmitter released by the Renshaw cell produces an IPSP which is opposite in polarity to EPSPs, and cancels the effect of the excitatory transmitter, thus preventing impulse generation. Another type of inhibitory ending is known in several locations in the nervous system; it acts on an excitatory ending directly and prevents the release of all or part of the excitatory transmitter.

Whether a motoneuron, or other neuron, fires or not is determined by the sum of all these synaptic effects operative at a given time upon the cell in question. When EPSPs are more numerous and powerful than IPSPs, the motoneuron fires and muscles contract, otherwise there is no activation transmitted.

A motoneuron is, of course, subject to influences other than reflex connections from receptors and inhibition by Renshaw cells. The surface of a cell body of a motoneuron, with its many short processes called dendrites, is contacted by as many as 10,000 synapses (Gelfan, 1964), both inhibitory and excitatory. These synapses receive impulses from many places in the spinal cord, and many are activated directly or indirectly from sources all over the brain. The motoneuron has thus become known as the "final common pathway" of neural influences affecting behavioural output; they are the ultimate controllers of activity, and all overt behavioural effects of sensory stimulation and experience must operate upon and through the action of motoneurons.

The massive and variegated input to motoneurons must be finely and discriminatively programmed to produce the co-ordinated behaviours that we observe. But beyond a few reflex mechanisms, our knowledge of the relevant neural programmes is quite sparse. Only recently has it been demonstrated that in cells of the cerebral cortex, whose branches reach to the region of the motoneurons, there is specific activity which always precedes certain voluntary muscular movements (Evarts, 1966). The functional role in the programming of muscular contractions, of many other regions of the brain known to effect motoneuron activity, remains to be determined.

Facilitation and inhibition are a means of organizing input to the nervous system, as well as its output in behaviour. The following, rather extended example of information processing in the mammalian visual system is the most advanced explanation yet derived for the handling of sensory data in the brain; and there is growing evidence that analogous processes take place in the other information-processing networks of the nervous system (Bishop, 1967, pp. 437, 444).

2. SENSORY PROCESSING MECHANISMS

Little is known about the generation of nerve impulses by the activity of rods and cones in the retina. These receptors are in contact with two types of neural cells, called *horizontal* and *bipolar* cells, respectively (Fig. 9). The bipolar cells in turn contact *amacrine* cells and *retinal ganglion* cells. The ganglion cells send processes into the brain, and these processes make up the optic nerve and serve as the "final common pathway" for all information coming to the rest of the nervous system from the eye. Of all these cell types, it has so far been possible to record the activity of single cells only from the retinal ganglion cells. And detailed study reveals that a great deal of processing of the information transduced by rods and cones has already taken place by the time the ganglion cells are activated.

Most vertebrate retinal ganglion cells so far studied, and virtually all in the cat where most have been studied, possess specially differentiated patterns of response to visual stimuli which depend upon the relative location of the stimulus in the "receptive field" of the ganglion cell (Kuffler, 1953; Rodieck

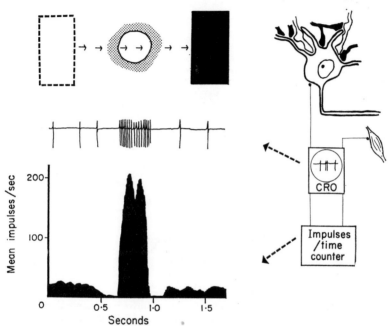

Fig. 14. Response of a retinal ganglion cell, with off-centre on-surround organization of receptive field, to movement of a black bar across the field (upper left). Impulses (left centre) generated in the ganglion cell (upper right) are recorded by a cathode-ray oscilloscope *CRO* and counted as a function of time intervals by counter (lower right) to produce a histogram of activation (lower left). Data in the histogram are from 30 passes of the stimulus. As the black bar passes (light off) into the surround, the cell is inhibited from firing; even the normal "non-stimulated" spikes cease. Then as the bar crosses the centre, a high-frequency burst of spikes is generated. As the bar passes through the surround on the other side, there is another "silent" period where the cell is inhibited from firing. (Data from Rodieck & Stone, 1965.)

& Stone, 1965). The receptive field for any neural unit is that region of the external world (in this case that small portion of the visual field) in which stimuli are effective in influencing the cell. To understand the response of the retinal ganglion cell, it is necessary to think of the receptive field as consisting of a roundish *centre*, with a peripheral wide border around it called the *surround* (Figs. 14 and 15). In one type of ganglion cell, onset of light, or the rapid succession of a dark object by a light one, when confined to the centre of the field, induces a train of impulses in the cell. The cessation of light, or succession of a light object by a dark one, in the centre of the field produces a suppression of all impulses, as if the cell were being actively inhibited—even the normal "spontaneous" firing of the cell ceases. With the same cell, onset of light in the peripheral *surround* suppresses impulses, while onset of dark causes a train of impulses, which is just the reverse of their effects in the *centre* of the field. This type of ganglion cell is known as an "on-centre, off-surround" cell, named for the types of stimulation which produce cell activity. About half the cells of the cat retinal

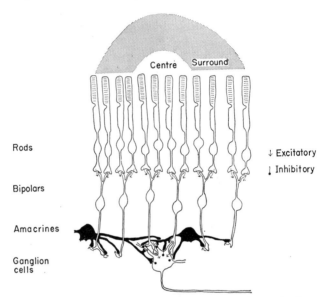

Fig. 15. Possible mechanism of centre-surround organization of the receptive field of a retinal ganglion cell, simplified from that suggested by Dowling & Boycott (1966). Rods activated by stimuli in the centre of the receptive field stimulate bipolar cells which excite impulses in the ganglion cell. Rods activated by stimuli in the surround stimulate bipolar cells, which excite amacrine cells, which inhibit the ganglion cell.

ganglion layer are of this type. The other half are nearly all of a complementary opposite "off-centre, on-surround" type, with onset of light in the centre inhibiting the cell, in the surround exciting it, etc. In both types of cell, a uniform field of brightness, or darkness, will induce equal excitatory and inhibitory effects and the cell will show no influence of stimulation. Strong contrasting edges in the field, particularly when they move across the surround and centre produce maximal activations and inhibitions, yielding sharply defined bursts of impulses in the cell set off by periods of "*absolute silence*" (Fig. 14). Thus these retinal ganglion cells are attuned to (or are programmed to) respond maximally (i.e. most definitively, with highest signal-to-noise ratio) to the movement of highly contrasting contours in the visual field. Apparently some combinations of activity in the rods and cones, and the horizontal, bipolar and amacrine cells exert the requisite combinations of summed excitatory and inhibitory effects on these retinal ganglion cells. Knowledge of their precise function awaits the development of adequate technology to record the activity of these cells. A hypothetical mechanism which accounts for the available data is presented in Fig. 15.

Still another type of retinal ganglion cell has been demonstrated in rabbits, pigeons, and some other animals. These cells respond only to contrast-contours moving in one certain specific direction (Fig. 16). Barlow and

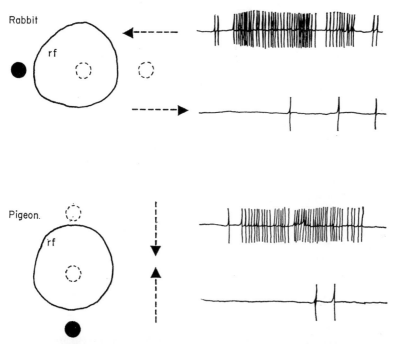

Fig. 16. Selectively directional responses of retinal ganglion cells in rabbit and pigeon retina. When a moving stimulus crosses the receptive field rf of the cell, impulses are generated in a continuous train as long as the stimulus is in the field and moves in the preferred direction. When the same stimulus is moved in the opposite direction, few if any impulses are generated. Different cells in a given retina respond to different preferred directions. Data from Maturana & Frenk (1963) and Barlow & Levick (1965). Such directional sensitivity occurs in a minority of cells of retinas thus far investigated.

Levick have proposed a mechanism whereby this highly selective response to an abstracted quality of the stimulus configuration is accomplished (Fig. 17). It is based on the known anatomical and physiological character-istics of the cells involved, and shows how, in theory, an assembly of nerve cells can synthesize a signal corresponding to an abstract or general class of external events out of the simple energy changes recorded by the receptors. The ecological significance of this class of signals for the rabbits, pigeons, etc., remains speculative, however.

Additional striking evidence of the capability of neuronal systems to abstract and identify pertinent aspects of environmental events is provided by studies of the activity of neural units in the cerebral cortex of cats (Hubel & Wiesel, 1962, 1965). Cells in the visual area of the cortex responded to photic stimulation in restricted parts of the visual field; thus they have receptive fields. Topographic coding is operative, even at this cortical level, such that cells lying posteriorly in visual cortex respond to events in the upper part

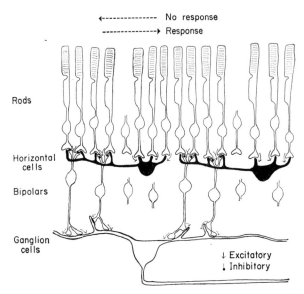

Fig. 17. A very simplified diagram of a hypothetical mechanism to account for the behaviour of directionally selective retinal ganglion cell, proposed by Barlow & Levick (1965). The ganglion cell is connected to relatively few bipolar cells, some at a distance from others. These bipolar cells are in contact with rods and inhibitory horizontal cells, such that a stimulus moving in the non-preferred direction (no response) activates rods which excite the horizontal cells which inhibit activity in the bipolar cells at the time when the rods connected to the bipolar cells are activated. A stimulus moving in the preferred direction (response) first activates rods which activate the bipolars and the ganglion cell, and *then* the rods which activate the horizontal cells; thus the bipolars are inhibited *after* the stimulus has passed and the ganglion cell has been activated.

of the visual field; cells lying anteriorly in cortex are activated from the lower part of the visual field. Cortical cells are generally more selective of stimuli than were retinal ganglion cells or the intervening cells of the lateral geniculate nucleus in the thalamus (the function of these thalamic cells is still obscure; their receptive field and response properties are in many ways similar to those of the retinal ganglion cells). Instead of the "centre-surround" type of functional organization of the receptive field seen in retinal and thalamic cells, many cortical neurons respond rather to the movement of straight linear contrast edges oriented at a particular specified angle across a specific locus in their receptive field. This activation can be explained by the hypothetical mechanisms proposed by Hubel & Wiesel (Fig. 18), wherein synaptic input from a number of neighbouring, similar retinal ganglion cells, after transmission through the thalamus, converges upon a single cortical cell. Summation of activity of this line of cells (oriented at the angle and locus

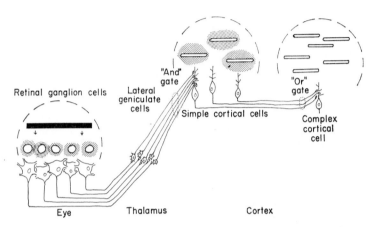

Fig. 18. Postulated connections of some visual cortical cells of cats (Hubel & Wiesel, 1962). A group of retinal ganglion cells, whose receptive fields are located in a linear array, are connected through lateral geniculate cells of the thalamus to a simple cell in the cerebral cortex. These connections function as an And-gate: all of the retinal and thalamic cells in this group must be simultaneously activated in order to excite the simple cortical cell. The simple cell thus responds only to linear stimuli which lie in the same location and angular orientation as the array of retinal receptive fields. Several of these line-detecting simple cells which detect lines of the same orientation but in different locations, are connected to a complex cortical cell; these connections function as an Or-gate: activity in any one or other of the simple cells will activate the complex cell. Thus any line of appropriate orientation in any of several locations in space will produce a response in the complex cell. The network may be said to abstract the properties of linearity and orientation from the stimulus manifold.

to which the cortical cell responds) as they are activated by a linear contrasting edge in the visual field, serves to excite the cortical cell. Summation of the activity of all the retinal and thalamic cells is necessary to produce the response in the cortical cell. (Electronic engineers will recognize this as an n-legged And-gate.)

A second type of cortical cell responds to linear edges, again oriented at a particular angle, but at any location in the receptive field rather than one specified exclusive locus. Hubel & Wiesel suggest that this type of cell is activated by excitatory input from several of the previous type cells having the same angle of effective linear stimulus but varying in its location in the field from one unit to another. Stimulation of any one of these contributing cells is sufficient to produce response in the cell in question; no summation is necessary. (This would function like an Or-gate.)

To describe a third type of cell briefly, we must oversimplify the evidence even more than we have been doing. These cells appear to receive input from two sorts of line-detecting cells, the lines of one set being oriented at right angles to those of the other. The combined input to this type of cell, requiring the summation of the inputs, enables the cell to detect and selectively respond to the presence of corners in its receptive field. Inhibition is called into play once again to explain why the cells respond to corners and not to crosses (the explanation can be found in Hubel & Wiesel—this example has already served its purpose).

The centre-surround pattern of excitation and inhibition has been reported for somesthetic (Mountcastle & Powell, 1959), auditory and taste (Békésy, 1967) receptors; in each case the mechanism results in an emphasis and heightening of the effect of contours between contrasting stimulus states. Further parallels to the state of organization in the visual system are found in studies of cortical auditory units, which selectively respond only to definite patterns of acoustic stimulation (Suga, 1965).

The organization of the visual cortex in cats is apparently determined by genetic rather than experiential factors; Hubel & Wiesel (1963) found units in the cortex of visually inexperienced infant kittens responding selectively to lines and corners just as in the adults. We can only speculate at this time whether similar mechanisms exist to account for apparently genetic selection of particular behaviour-governing stimuli. The apparent inborn knowledge of the patterns of stars in the night sky which guide migrating birds (Sauer, 1957; Emlen, 1967); the peculiar significance of certain shapes passing overhead for preyed-upon birds; and even the sensory qualities of preferred foods and mates may rely on genetically determined neural recognition mechanisms. Other mechanisms may be learned, although it may be that learning is a means of modifying existing genetic programmes rather than constructing entirely new ones.

II. BEHAVIOUR RELATING TO ENERGY REGULATION: MAINTENANCE OF THE INDIVIDUAL

Feeding, drinking and elimination are overt behaviours which are a part of the overall process of energy regulation. The physiological mechanisms which underlie these behaviours interact with one another and with other systems in an extremely complicated fashion.

These behaviours are steps in a dynamic process in time, and what happens at earlier stages modifies what occurs later in the process. Thus, it behooves us to employ an analysis-of-process approach, which recognizes these dependencies. One approach that has proved fruitful, and will be used here, is that suggested by O'Kelly (1963). He suggested that the sensory, effector and central neural processes involved in food and water regulation can be reclassified into the following analysable components:

(a) *System variables.* These correspond to the detectable energy fluctuations (in the internal and external environment) which are of critical importance to the regulatory process.

(b) *Detector components*. These are, for the most part, the internal sensing systems which monitor the changes in the system variables.

(c) *Orientation components*. Functionally, these enable the animal to orient to environmental factors so that ensuing action is appropriately directed.

(d) *Correctional components*. These are, in part, the overt motor acts of chewing, licking, sucking, swallowing, urination, defaecation; and, in part, internal (metabolic) adjustments which are an integral part of the process, but which have no overt behavioural manifestations.

(e) *Satiety detection components*. These are conceived of as similar to the detector components ((b) above) which initiate the regulatory process; but the satiety mechanisms serve to terminate correctional behaviour by signalling the return of the internal environment to some optimum state of balance.

A. Feeding Behaviour

System variables. There are two sets of conditions which seem at present the most likely candidates as causative factors in the onset of feeding behaviour. One of these is the arteriovenous (A-V) glucose difference (Mayer & Thomas, 1967). Even though the absolute glucose level may be high as in diabetic hyperglycaemia, this A-V glucose difference is small in states of hunger and is large after feeding, indicating glucose utilization by the cells.

The second set of conditions, this time hormonal, also seems to bear a consistent relationship with the onset of hunger and feeding (Kennedy, 1966). Plasma levels of growth hormone (which stimulates lipolysis when glucose levels are low) show a rapid rise which correlates with the time of onset of hunger; whereas this hormone is absent, or nearly so, after feeding. Present knowledge does not permit a decision in favour of either the A-V glucose or the growth hormone hypotheses, or for a combination of them. Nevertheless, the system variables seem to be blood borne factors which can affect some detector systems.

Detector components. It is quite likely that there are some peripheral mechanisms important in the detection of hunger states (among which stomach contractions should not be included; Davis *et al.*, 1959), but the bulk of evidence to date is on central neural detection. Neural elements in the ill-defined lateral hypothalamic area (Hla in Plate I) seem to play a major role in the instigation of feeding behaviour. It has been known for some time that artificial stimulation of this area causes eating, or at least typical feeding movements, in a variety of animals; in sheep and goats it has caused rumination; in cats it has produced voracious eating, and mouthing of objects (Akert, 1959). Bilateral lesions in Hla (see Fig. 83) have reliably led to aphagia or refusal to eat (Anand, 1961; Morgane, 1961); eating seems to resume with proper nursing (Teitelbaum and Epstein, 1962). Aphagia may result from interrupting the fibres from the *globus pallidus* (Plate II) which pass through Hla on their way to the midbrain and other places (Morgane, 1961; Gold, 1967).

In an elegant series of electrophysiological studies, Oomura *et al.* (1967) demonstrated in cats that activity in Hla was directly related to food-seeking behaviour.

Orientation components. All of the exteroceptive sense modalities, i.e. vision, olfaction, audition, etc., participate in orienting the animal to environmental factors which are appropriate to its need state. However, hunger turns out to be a complex need state which can be general, as in an animal which has simply depleted its energy stores; or it can be specific, as in an animal which has incurred a vitamin or mineral deficiency. A combination of these two states of imbalance is not rare. In the former case, any high caloric substance in adequate quantity will serve to correct the imbalance; in the latter case, a much narrower range of substances will suffice. Obviously, since what constitutes utilizable food for a carnivore will not meet the needs of herbivorous animals and vice versa, the ingested substance must be metabolizable by the animals' enzyme systems.

At this point in the process, taste, and probably olfaction, are the sensory modalities usually relied upon. There are individual as well as species differences in taste preference (Nachman, 1959) and, in evolution, tastes as inherited properties of species allow for selective survival of those individuals which perceive nutritious foods as preferable, whether their current need state is general or specific. The neural systems which subserve the orienting behaviours of hungry animals are most likely the same as those involved in attention and alertness to any other set of stimulus events. These are discussed in a later section of this chapter (pp. 175 ff.).

Correctional components. Once the animal has oriented to an appropriate food object, the actual feeding behaviour ensues. The net effect of this behaviour is, of course, to correct the original deficit or imbalance in the system variables which was sensed by the detector component. Eating is a relatively stereotyped behaviour, being highly similar for all members of a species. However, the patterns, and duration of eating can be modified by experience, current and past, as well as by the intensity of the physiological need.

A number of external discriminative stimuli have been shown to influence feeding. In chickens, the pecking response is a crucial component of the complex of behavioural acts associated with feeding. Although a wide variety of objects can elicit pecking, in newly hatched chicks, abundant evidence indicates that both colour and form act as discriminative cues for the manifestation of preferences (reviewed by Fantz, 1957). Feeding by adult hens is affected by a variety of external stimuli. More is eaten when a larger quantity of food is present; satiation appears to be determined not by the absolute amount consumed, but rather by the proportion of the available food that is eaten. The presence of other hens eating also stimulates increased food intake (summarized in Krech & Crutchfield, 1958, p. 295).

Albino rats ate according to a diurnal cycle, the time of day being more influential than food deprivation in instigating feeding (Bare, 1959). This cycle was evidently related to the light-dark cycle. Rats kept in a standard constant illumination ceased to follow the diurnal cycle. Thereupon food intake was directly proportional to the degree of deprivation. Two to four days after normal light and dark periods were reintroduced, the diurnal rhythm resumed its prepotency over deprivation (Siegel, 1961).

Infantile suckling may be an exceptional ingestion behaviour pattern, particularly responsive to external stimulus factors. In infant dogs, suckling was not initiated by hunger, nor inhibited by satiation, nor was it reinforced by obtaining milk. Rather, it appeared to be a reflex activated by external stimulation including contact and temperature components (James, 1957). Intensity and frequency of suckling movements are both affected, however, by hunger. Contact stimulation inside the mouth can be an effective instigator. Decerebrate animals who will not otherwise eat, will chew and swallow food placed in the mouth.

Some other factors known to affect the frequency of feeding behaviour, which in the long run alters the total intake of the animal, are: temperature, exercise, and caloric value of the food eaten. At high ambient temperatures food intake is drastically reduced, while low temperatures produce consistently elevated intakes. Excessively high ambient temperatures can result in the death of the animal due to self-imposed starvation, even though food is abundantly available. These facts demonstrate the interaction of food and temperature regulating systems in homeotherms (Brobeck, 1960). Direct heating or cooling of the preoptic area of the hypothalamus (Po in Plate I) similarly affects food intake in goats (Andersson et al., 1964). Elements in this area are known to be involved in the regulation of body temperature.

When the amount of exercise or activity engaged in by an animal increases, total food consumption increases, having the net effect of matching the energy output with equivalent input. The immediate effect of exercise, however, is to decrease or inhibit feeding. This seems to be due largely to inhibition of gut motility, increase in blood glucose due to glycolysis, and a rise in insulin titres, during exercise.

Although comparable data are not available for large animals, it has been demonstrated in young rats that the caloric value of the food is a strong determinant of amounts consumed (Adolph, 1947). When the diet was diluted with non-nutritive bulk, these animals ate more. In fact, they adjusted their intakes so that their caloric intake was the same as it had been on the normal diet. When the percentage of non-nutritive bulk in the diet became very large, however, the rats could not increase their intakes enough to compensate, and weight losses followed. These relationships are not quite so clear in older rats or animals with lesions in certain areas of the hypothalamus, mainly because palatability factors seem to gain some importance as determinants of intake, and diluents alter the palatability.

Satiety detection components. Feeding behaviour is usually terminated in advance of the animal's restoration to a state of balance. That is, in some way the amount of food necessary to make up an incurred deficit is anticipated, so that the animal avoids going from a state of deficiency to one of excess, either of these constituting imbalance. This part of the overall regulatory process is not quite so obvious in grazing animals, whose intake proceeds at a relatively slow rate, as it is in dogs and cats, where a meal is eaten in a short period of time. The anticipatory nature of the behaviour forces us to consider a mechanism for satiety which is different from merely turning off the detectors which were responsible for initiating the process which resulted in feeding.

The first candidate for a satiety signalling device is the stomach, or, more precisely, stomach filling or distension. Undoubtedly, the neural signals accompanying such distension travelling the vagus (Xth cranial) nerve to the brain are *part* of the satiety mechanisms of many animals. However, the gut stretch receptors cannot be the only satiety detectors because mere bulk in the gut has only a transient influence on food intake, whereas nutritive substances in the same amount have a lasting inhibitory effect on eating. Furthermore, vagotomy does not have an appreciable effect on feeding (reviewed by Mayer & Thomas, 1967). Kennedy (1966) has advanced the hypothesis that the relatively high levels of insulin associated with lipogenesis following eating are the stimuli to which a central satiety detector might respond. This is, of course, an alternative to Mayer's notion of arteriovenous glucose difference, outlined above under System Variables.

A great deal of research points to the ventromedial nucleus of the hypothalamus (Vem in Plate I) as the central neural detector of satiety. Electrolytic lesions in and around these nuclei lead to a syndrome which has been called hypothalamic hyperphagia (see Fig. 83). This syndrome includes a stage of immense appetite and rapid weight gain, followed by a stage of relatively normal food intake sufficient to maintain but not to increase the obesity produced in the earlier stage. That the Vem is involved in the satiety aspects of feeding is suggested by the observation that the lesioned animal eats larger amounts per meal, rather than eating more often (Teitelbaum & Campbell, 1958).

Hypothalamic hyperphagia and subsequent obesity have been demonstrated in mice, rats, cats, dogs, monkeys, and, recently, in goats (Baile & Mayer, 1966). In mice, hyperphagia and obesity can be produced by injections of gold thioglucose, a substance which can enter the brain cells and deposit the metal moiety, thereby destroying the affected cells. This substance rather selectively (although it is not completely selective) destroys the Vem of the mouse brain. Electrolytic lesions close to, but not actually damaging, the Vem in rats have resulted in the syndrome (Graff & Stellar, 1962). An overwhelming number of studies reporting successful lesions also report using female animals, a fact which may, in the last analysis, turn out to be important if not crucial. Matsuda (1966) found gonadal atrophy in some of the male hyperphagics, but did not find this in females. In both sexes he found some animals with adrenal hypertrophy. These findings suggest extensive and complex hormonal involvement in the syndrome.

One of the frequent, but not entirely consistent, symptoms of hypothalamic hyperphagia is "finickiness", or refusal to eat or drink adulterated foods in order to maintain energy balance (Graff & Stellar, *loc. cit.*; Corbit, 1965). When obesity was induced by an extended series of insulin injections prior to Vem destruction, the lesion caused little or no additional weight gain (Hoebel & Teitelbaum, 1966). This suggests some kind of regulation at the obese level in the same way that normal animals regulate at lean body weights.

The electrophysiological studies of Oomura *et al.* (1967) leave little doubt that the Vem and Hla are reciprocally related in their activity. Stimulation of Vem inhibited both feeding and activity in Hla; Hla stimulation produced feeding and a reduction in Vem activity. "Spontaneous unit discharges" were highest in one of these areas when they were lowest in the other.

All of these data suggest that feeding behaviour is controlled by elements of these two areas working antagonistically. A final word is in order, lest the impression be given that the animal is merely a mouth and a hypothalamus: the Vem and Hla receive and process a great deal of information from sources outside the hypothalamus, and it is this information which influences the moment-to-moment dominance in this continuous antagonism.

B. Drinking Behaviour

System variables. Animals living in an air environment are constantly losing water by a variety of routes. These losses constitute depletions of total body water and consequent increases in the concentration of the body fluids. Whatever the detection mechanisms that have evolved as instigators of drinking behaviour, they must be sensitive to: (a) decreases in body fluid volume, (b) increases in body fluid concentration, or (c) a combination of these two. It should be noted that the consequences of both of these events, separately or in concert, could be loss of intracellular water; this, in turn then, would be *the* thirst stimulus. The assumption here is that certain nerve cells, which shrink in volume due to water loss, respond to the resulting deformation of their membranes with a pattern of neural signals which starts the process that eventuates in drinking behaviour. It is equally logical, at the present state of our knowledge, to assume that these detector cells respond to the altered ionic concentrations which are a direct result of the water losses. These increases in concentration might then alter membrane excitability to the point where the response pattern mentioned above might occur.

In any case, these changes are induced by variations in the extracellular compartments, chiefly in the blood plasma. When blood volume depletion has been accomplished without increase in osmotic pressure, by haemorrhage, drinking has resulted. This has been observed in dogs and rats (reviewed by Greenleaf, 1966). However, the blood losses produced in these animals were very large and thus cannot be thought of as the stimuli which normally evoke drinking. When increases in osmotic pressure are produced locally in certain brain areas, without significant volume changes, drinking follows (Andersson, 1953). Here, again, the stimulus cannot be the normal one because it is outside the range of those which occur physiologically. Nonetheless, these techniques have been valuable in the dissection of this mechanism.

It seems, then, that volume alone, or osmotic pressure alone, or both together are adequate to trigger drinking responses. Normal drinking is probably a response to the simultaneous occurrence of these two changes whereby the detector systems are stimulated by cell water losses.

Detector components. While thirst has been associated with dry mouth in many animals, it seems, like the role of stomach contractions in feeding, to be of secondary importance in the initiation of water intake. Drinking behaviour persists when: (a) the oropharyngeal cavity is denervated, or (b) the oesophagus is fistulated so that the fluid ingested is not allowed to reach the stomach for subsequent absorption (reviewed by Fitzsimons, 1966). These two facts and the finding of Montemurro & Stevenson (1957), that lesions in the lateral hypothalamus can cause adipsia (refusal to drink) without aphagia, have directed the search for the detectors toward the central nervous system.

At this writing, there is no complete picture of a "drinking mechanism" and the neural systems which are apparently functional in such a mechanism are known only grossly. It is apparent from the work of Andersson and his associates (1955a & b, 1956, 1960) with dogs and goats, that the lateral hypothalamus contains many elements, the stimulation of which seems to cause drinking. Stimulation of the areas near the paraventricular nucleus of the hypothalamus (Pa in Plate II) evoked antidiuresis and milk ejection; slightly posterior to this, drinking and antidiuresis were often produced; and from the area between the fornix and the mammillothalamic tract, drinking with no antidiuresis was elicited. Stimulation of similar nature has also produced drinking responses in pigeons (Andersson et al., 1960). To date, however, no definable cell masses have been clearly associated with such evoked drinking responses.

There are several areas of the brain containing cells which respond to increases in plasma osmotic pressure, thus implicating themselves as possible components functioning in water balance. Changes in electrophysiological activity in the olfactory bulb of dogs (Sundsten & Sawyer, 1959) and rabbits (Sawyer & Gernandt, 1956) have been observed to result from systematically administered hypertonic fluids. Cross & Green (1959), working with rabbits, found osmotically sensitive cells in the medial and lateral preoptic hypothalamus, the supraoptic nucleus (Soa and Sot in Plate I) and the Pa (Plate II). In addition, a few similarly responding cells were found in the massa intermedia of the thalamus. The medulla oblongata of cats appears to contain some of these osmotically sensitive cells also (Clemente et al., 1957). While it is clear that osmosensitivity is a property of many brain areas, it is not at all clear what, if anything, these areas have to do with one another. Some of these (e.g. Soa and Sot) have demonstrable functions (to be described in the next section) in water regulation, while the others still challenge the imagination. It seems noteworthy that no electrophysiological studies of osmotic stimuli in the hypothalamus have found responses from the lateral areas, which, when ablated, cause adipsia. If these areas are indeed critical to the initiation of water intake, then they might also be expected to be sensitive to osmotic stimuli, directly or indirectly.

Orientation components. Once the animal has detected the state of imbalance occasioned by significant deviations in stimuli to which the detector components respond, appropriate action must be taken to assure that the animal obtains the corrective substance. It must orient to water or some source which will yield water in adequate quantity. Here, again, attentional and alerting mechanisms in the nervous system are called upon to sort out the relevant aspects of the environment. Sensing systems, which sample the external environment and discriminate on the basis of experience or genetic propensity, must accomplish this sorting process. Many free-living species of rodents show selective preference for foods with high water content, and in captivity will ignore (to the point of severe dehydration) foods with similar water stores if they are strange to the animal (reviewed by Chew, 1965).

Whether these preferences are genotypic or experiential remains to be determined.

In some of the cases of adipsia produced by Hla lesions, it appears as if this discrimination component, rather than the detectors, has been disrupted. The response patterns of these animals to water resembles that of normal animals when they are offered solutions which are extremely bitter. Thus Andersson's dogs, after sustaining such lesions, would avidly consume milk or broth while refusing to drink water (reviewed by O'Kelly, 1963).

Correctional components. As in feeding behaviour, the actual correctional mechanisms are those of ingestion, swallowing, etc., and are highly stereotyped within species. There are, however, large cross-species differences in the amount, rate and frequency of water intake (Chew, 1965). Amount is proportional to body size; rate of intake depends to some extent upon the water deficit at the time of drinking; and the frequency, as a cross-species comparison, is most directly related to the availability of water in the environment and the water content of the food. All of these have interesting facets, but it is not the province of this chapter to explore them.

Satiety detection components. Most mammals are able to approximate their degree of deficit in their drinking. Since this is done in advance of complete tissue rehydration, the mechanism of anticipation is of interest. The major factors involved in such a satiety mechanism are: (a) the head receptors, (b) distension of the gut, and (c) intracellular changes upon rehydration.

Drinking can be depressed by preloading the stomach with water or by inflating a balloon in the stomach. It is depressed even more if the same load is actually drunk. That is, some information about the water in the mouth and/or the act of ingestion is important in determining the final cutoff of drinking. There is some evidence to suggest that the brain receives small quantities of ingested substances before feedback from postingestion can occur (Maller *et al.*, 1967).

The stretch receptors in the gut have been shown to be important inhibitors of drinking behaviour in the dog and rabbit. The information from these receptors travels the vagus nerve to the brain stem (the effect of distension on drinking is eliminated by vagotomy). Exactly what neural systems, responsible for the cessation of correctional behaviour, receive and act on this information are not yet known. The inhibition of drinking by stomach distension is only temporary, however, and final satiation follows significant absorption by the intestine. It is conceivable that there exists a feedback system, whereby intestinal absorption rate is monitored by neural systems which relay information about impending satiety to the central nervous mechanisms that control drinking.

Lesions in the central nervous system have been shown to increase water consumption, but the data do not rule out the possibility that the primary effect is on elimination rather than on a "satiety mechanism". Harvey & Hunt (1965) demonstrated that lesions in the septal nuclei (Plates I and II) increase water intake. However, no measurement of water losses was made, leaving the question open as regards the primary nature of the effect. Fisher *et al.* (1938) demonstrated that water intake increased greatly with lesions in the supraoptic nuclei of cats, but this effect was due to excessive water losses via urination. The elevated intake was, thus, compensatory in nature. It is entirely possible that the septal lesions cause some increase in the rate of water loss, by urination, evaporative loss, etc., which is compensated for by increased water consumption.

C. Eliminative Behaviour

The elimination of fluid wastes by mammals provides instances of interrelated internal sensory, experiential, and humoral influences upon a behaviour pattern with a relatively fixed neural organization. Evidence indicates that the pre-behavioural physiology is governed by chemical receptors. The associated behaviour shows a typically standardized organization with dependence upon specific neural structures. The manifestation of this behaviour is occasioned and qualified by additional experiential, and in some cases hormonal, factors.

System variables and detectors. Osmotic pressure of the blood in the internal carotid artery is the chief governor of the rate of water excretion. In dogs it has been shown that osmoreceptors in the vascular bed of the artery are the mediators between the electrolytic tonicity of the blood and the secretion of an antidiuretic hormone by cells of the supraoptic nucleus of the hypothalamus (Verney, 1947). Increased osmotic pressure in the blood results in increased secretion of this hormone; decreased tonicity of the blood reduces the hormone flow. Hormone liberation has been secondarily stimulated by other factors which presumably act through the hypothalamus. These include emotional stress (Akert, 1959, p. 169), milking or vaginal stimulalation of oxytocin secretion in several species: rabbits (Cross, 1951), cows (Debackere & Peeters, 1960), humans (Harris & Pickles, 1953) and horses (see Chapter 12). Electrophysiological evidence indicates that the cells of the supraoptic nucleus respond to increases in the osmotic pressure of the plasma with changes in their firing rates as well as with antidiuretic hormone release. Recordings from single cells in this nucleus in rabbits and in cats have revealed a high degree of correspondence between increased rates of firing and antidiuretic response, as measured by reduced urine flow (reviewed by Cross & Silver, 1966).

Correctional components. The level of antidiuretic hormone, acting in the kidney, determines how much water is classified as waste to be released. Water-filled animals excrete large quantities of water; in dehydrated animals the water excretion is so scant and the hormone secretion so plentiful that the hormone itself is in excess and is excreted (Gilman & Goodman, 1937). As the water release collects in the urinary bladder, the increasing volume distends the bladder wall. This distension produces afferent impulses in fibres leading from the bladder, and the impulse frequency was found in cats to be patterned according to the relative increases in intrabladder pressure (Iggo, 1955). When the filling reaches a threshold, a complexity of reflex mechanisms is set in motion which results in the relaxation of the bladder sphincters, contraction of the bladder walls, and the expulsion of its contents. The characteristics of the micturition reflex system, and of its peripheral components in particular, have been studied in detail in cats

(Barrington, 1948). Whether all components of this system can be generalized to other animals, however, has been seriously questioned (Kuru, 1965). In any case, the mechanism of micturition in normal animals occurs in co-ordination with a complexity of other associated behaviours, and can be activated or delayed considerably by factors other than the need for elimination.

The neural organization of urinary behaviour was elucidated by the Hess method of brain stimulation. Micturition—often accompanied by defaecation—complete with preliminary exploration, site selection, co-ordinated posturing, and covering movements afterwards, was induced in cats by electrical stimulation of subcortical areas of the cerebrum. The primary effective areas were in the mid-basal forebrain (the perifornical septal nuclei (Plates I and II), and the bed-nucleus of the stria terminalis), and the response was also obtained from the lateral hypothalamus and from the grey matter near the rostral border of the midbrain (Akert, 1959). Similar results have been obtained with sheep (Andersson, 1951).

For the study of animal behaviour, a most significant characteristic of the micturition processes is their extreme susceptibility to what may be analogous to human conscious control. It has been suggested that the evolutionary significance of the urinary bladder consists in enabling its owner to choose the time and place of urinary deposit (Barrington, 1948). Thus an animal can keep the route of his travels obscured from the prying noses of other interested creatures, both predators and prey. He also possesses a means for olfactory signalling, marking, and stimulating (Martins & Valle, 1948; Ilse, 1955).

This voluntary control over urination is not evident in such incomplete specimens as infants and those with lesions in relevant central nervous structures. In these cases micturition appears to be governed by the state of the bladder. The cerebral cortex appears important in inhibition of reflex micturition. If there were anything in the bladder, cats urinated within a standard 5 to 10 minutes after decerebration (Barrington, 1921). The volumes urinated by decorticated cats were more standard, showing much less variability than did those of intact cats (Woolsey & Brooks, 1937). This may indicate increased responsiveness to a filling bladder threshold and less influence of more comprehensive factors of perception and experience (Root, 1956).

1. INFLUENCE OF SEX HORMONES

The behaviours associated with urination that are exhibited by adult male dogs—sniffing, choice of upright vertical targets, and leg lifting—have been shown to be a secondary sex character under hormonal influence (see Chapter 15). Absent in infants and females, these activities appear in males only after androgen has been supplied by pubertal secretion or artificial injection (Martins & Valle, 1948). It is not completely a hormonal effect, however. Females responded with the male pattern only if the hormone was introduced immediately after birth and continued until adulthood. The male pattern has been reported to depend upon a functioning olfactory epithelium (Freud & Uyldert, 1948). Evidence has been obtained from rats and humans that olfactory sensitivity to urinary steroids is altered by gonadal hormone levels (Sawyer, 1960). These aspects of behaviour could, it appears, be as well classified as sexual rather than as elimination behaviours.

III. REPRODUCTIVE AND OTHER SOCIAL BEHAVIOURS: MAINTENANCE OF THE SPECIES

To maintain existence beyond one generation, virtually all animals must interact with others of their own species. From a biological point of view, the goal of such interactions is the transfer of genetic materials to provide the gene-shuffling which allows selective survival and is necessary for evolution and adaptation to take place. Behaviours, as well as structures and physiological processes, have continually and progressively evolved to provide ever more efficient means of accomplishing this basic transfer. The behaviours most directly involved are sexual activities, but the general purpose of species perpetuation is also served by parental behaviours which provide care and protection to the offspring, and by other affiliative and agonistic behaviours which establish and regulate appropriately constituted groups of individuals within a species. (Agonistic actions can also serve for dealing with predators and thereby assist in survival.)

The physiological analysis of these interactional behaviours has proceeded farthest in the realm of sexual activities, where the behavioural influence of gonadal hormones has been shown to be remarkably pervasive.

A. Sexual Behaviour

The behavioural differences that distinguish males from females within a species are to a large degree determined through the activity of the gonadal hormones secreted by the testes or the ovaries. Young (1961) pointed out that the determining influence can take place in two different ways, and abundant evidence supports this dual-process view. Early in development, hormones can play a morphogenetic and organizational role in the establishment of the components (presumably neural in large part) of behaviour programmes. Secondly, they are involved in the activation of the programme at later ages.

In addition to delineating the roles of gonadal hormones, physiological analysis has striven to learn what factors govern the production of the hormones themselves, and (with somewhat less success) to elucidate the nature of the relevant neural programmes and the mode of hormone actions thereon.

1. Influence of Gonadal Hormones

An initial word of caution is in order: the character of sexual behaviour displayed can in no case be accounted for entirely by the type of hormone. The same hormone administered in exactly the same way contributes to performances which differ greatly between sexes, strains or even individuals. That is to say, hormones can be *necessary* for the appearance of certain behaviours, but in no case is a hormone *sufficient* explanation for an observed behaviour.

Role of early androgen in programme organization. From a chronological-ontogenetic point of view, a primary role of testicular hormones is their influence in the determination of the character of the behavioural substrate. Apparently these hormones directly establish a differential organization of the neural programmes underlying sexual and other behaviours typical of males (Phoenix *et al.*, 1967). Male patterns of behaviour, and of genital morphology, were produced in genetic female offspring of guinea pig and monkey mothers who were administered androgens during pregnancy. There was a general correlation between the degree of masculinization of genital morphology and of behaviour; but there were enough cases of non-correlation to indicate that the hormone was acting directly on some presumable neural substrate of behaviour without necessarily affecting genital tissue. Similar results were obtained administering androgen directly to newborn rats, who are in a stage of development corresponding to foetal stages in guinea pigs. When deprived of androgen during this developmental stage, by early castration or administration of chemical anti-androgens, the male offspring exhibited female behavioural characteristics in later life. The evidence indicates that normal masculinization of the behavioural programme is accomplished during this foetal or perinatal period of development by androgens normally secreted by the testes of male infants.

Thus early androgen appears to be a significant determining factor in the initial organization and development of sex-related behaviour programmes. If early androgen is present, the infant develops into a behavioural, and to some extent a morphological, male; if it is absent, the infant subsequently exhibits behavioural traits normally more characteristic of the genetic female. Oestrogen administration during this stage of development, in contrast, has consistently produced deleterious effects on the development of both male and female sexual behaviours in foetal or larval forms of either sex.

Activation of female programmes. In the absence of the ovary and its associated endocrine products, sexual behaviour is either not displayed at all or not displayed as regularly and frequently as by intact females. The extent to which display of sexual behaviours depends on the integrity of the gonad varies somewhat with the species. Among the higher primates, Young & Orbison (1944) reported that ovariectomized chimpanzees continued to display all of the elements of sexual behaviour except permitting copulation, although the frequency with which other elements were displayed declined greatly. For the female cat (Harris *et al.*, 1958) strict dependence of sexual behaviour on ovarian hormones has been noted. It may be generally concluded that for all vertebrate species, with the possible exception of the human female (Young, 1953), the characteristics of sexual behaviour typical of normal adults fails to appear in the absence of the ovarian hormones.

The sexual behaviour of most, if not all, female mammals is cyclical or episodic in character. Brief periods of sexual activity alternate with usually longer periods of either lowered sexual activity or total sexual inactivity. The interval of sexual inactivity cannot be accounted for by a corresponding absence of the ovarian hormones. Rather, as recent experiments suggest, the period is characterized by an active inhibition of sexual behaviour by progesterone (Goy, Phoenix & Young, 1966; Zucker, 1966). The sexually inhibiting actions of progesterone appear to apply to most mammalian species including those in which the same hormone acts synergistically with oestrogen to facilitate sexual behaviour.

Distinct species differences exist with respect to the hormonal factors essential to the restoration of normal sexual behaviour in ovariectomized females. Among rodents, normal receptivity does not occur unless a priming dose of oestrogen is followed by a small amount of progesterone. This hormonal requirement of sequential treatment with oestradiol and progesterone appears to duplicate the natural mechanism for spontaneous oestrus in rodents. Recent chemical studies of endogenous progesterone concentration have shown that an abrupt rise in the level of circulating gestagens occurs coincident with the appearance of the early oestrous reactions. In the intact female guinea pig (Feder, Resko & Goy, 1968a) the principal gestagen is progesterone. In the rat, both progesterone and 20-α-hydroxyprogesterone rise to very high concentrations at the onset of behavioural oestrus (Feder, Resko & Goy, 1968b). For the rabbit, a lagomorph, an oestrogen alone was considered sufficient for the induction of oestrus until the work of Sawyer and Everett (1959) demonstrated a facilitating effect of progesterone. More recently, the synthesis and release of 20-α-hydroxyprogesterone has been associated with the preovulatory oestrous condition in this species (Hilliard & Sawyer, 1964).

In other mammalian species the gestagens have not been implicated directly in the hormonal induction of oestrus. For the dog (Robson, 1938) and cat (Michael & Scott, 1957), oestrogen alone is sufficient for the induction of normal receptivity; and for the monkey, progesterone may be definitely inhibitory (Michael, Saayman & Zumpe, 1967). Whether or not the ewe proves to be representative of ungulates in general, an interesting relationship between oestrogen and progesterone in the production of sexual behaviour has been established (Robinson, 1955a, 1955b; Robinson et al., 1956). In this species the action of either endogenous or exogenous oestrogen is greatly facilitated by previous treatment with progesterone—a relationship exactly the reverse of that characteristic of rodents.

When amounts of hormone which are larger than necessary to induce behavioural changes are injected into ovariectomized females, augmentation of sexual behaviour does not necessarily result. Certain end-points, such as latency to heat and duration of heat, are unquestionably facilitated by suprathreshold doses (Peretz, 1968). Other end-points more representative of behavioural vigour are not influenced by the larger amounts of hormone. When spayed female guinea pigs were injected with 50, 100, 200, 400, and 800 I.U. of oestradiol benzoate followed by progesterone, the peak intensity of their oestrous reactions and the time of its appearance in the oestrous period were not facilitated (Goy & Young, 1957). The data suggest that the duration of oestrus is increased with the larger doses of oestradiol only by addition of the more feeble terminal responses.

The female cat (Michael & Scott, loc. cit.) and ewe (Moore & Robinson, 1957) are strikingly similar in the relationship which exists between exogenous hormones and certain sexual effects. In these species oestrous behaviour can be induced by the appropriate hormone treatment without complete cornification of the vaginal epithelium. Alterations in the treatment methods can be made so that oestrous behaviour and vaginal oestrus coincide. Presumably the physiological independence of these effects is characteristic of most or all female mammals, but other species have not been so extensively investigated.

The sensitivity to oestrogen characteristic of the tissues mediating sexual behaviour in the adult female mammal is attributed to some maturational effect. From 5 to 20 days of age, spayed female rats and guinea pigs either do not respond at all to injected oestrogen and progesterone, or require much larger amounts than the adult spayed female. By the 30th day of age, both rats and guinea pigs respond normally (Wilson & Young, 1941), but the presence of the ovary is not essential to the increase in sensitivity with age in either species.

Activation of male programmes. Among vertebrate males, dependence upon gonadal secretions for the manifestation of sexual behaviour appears to be as extensive as among females. However, relations between gonadal hormones and behaviour have not been carefully assessed for domestic animals and much controversy continues to exist. For carnivores such as the cat (Rosenblatt & Aronson, 1958a, b) and dog (Beach, 1960) there is little doubt that erection, intromission and even ejaculation may persist for a number of years after castration in mature and sexually experienced animals. The opinion is often expressed that the longer persistence of post-castration sexual behaviour in higher mammalian species is evidence for a lesser dependence on hormonal factors.

Among the many available synthetic androgens, testosterone (either in its free form or in one of its organic salts) is the most effective in restoring sexual behaviour in the castrated animal. Differences between species are not so marked among males as among females with respect to the nature of the hormonal requirements. In all species extensively studied, a single injection of testosterone has been ineffective in restoring the full pattern of sexual behaviour. Instead, repeated injections over a protracted period of time have been necessary, with components such as investigatory activities and mounting responding sooner and components such as intromission and ejaculation responding later to the treatments.

The amount of hormone injected is not closely related to the vigour of the behaviour displayed. Male rats may respond with increased vigour to higher dosages, but only up to some physiological maximum (Beach & Fowler, 1959). When restoration of the pattern of sexual behaviour characteristic of different groups of individuals is the end-point studied, increases in dosages beyond a threshold amount have not produced corresponding augmentations of the behaviour (Grunt & Young, 1952).

There is no substantial evidence that immature males require more hormone than mature ones to produce the same behavioural changes. Evidence does exist, however, for the role of maturation in determining the reactivity of the tissues mediating sexual behaviour. In two studies on the male guinea pig, large doses of testosterone propionate were injected daily into intact and castrate males from birth onwards (Riss *et al.*, 1955; Gerall, 1958). In both studies some acceleration of sexual behaviour was noted but the precocity was limited to the lower components.

Heterotypical sexual responses. While it is occasionally possible to induce, by means of hormonal alterations, behavioural manifestations appropriate to the opposite sex, it must be stated clearly that the injections of oestrogens into adult males and androgens into adult females have not been uniformly successful in producing complete heterotypical display. Young (1961) considered all 8 possible relationships between type of hormone, sex of recipient, and type of response. In general, it may be said that the quantity of hormone required to induce heterotypical display is large and the display less vigorous than that normally occurring in the proper sex. In addition, the responses of each sex to hormonal stimulation appear to be "predetermined", and there is a marked tendency for a particular pattern of behaviour to appear regardless of whether androgens or oestrogens are injected.

This predetermined character is probably ascribable to the early organization of behavioural programmes by foetal hormones described above, and it is one of the chief classes of evidence leading to the postulation of such programmes.

2. REGULATION OF GONADAL HORMONES

Neuro-hypophyseal direct control. Although the endocrine products of the gonad are directly instrumental in the production and maintenance of reproductive behaviours, the secretory activity of the gonads depends upon suitable chemical stimulation from the hypophysis (also known as the pituitary gland). Evidence now accumulating strongly suggests that both the rate of production and the rate of release of gonadotrophic hormones are controlled by nervous activities within the brain (Harris, 1960; Harris *et al.*, 1966).

A close anatomical relationship of the brain and pituitary gland is demonstrable throughout all vertebrate species. In all forms so far studied, the principal anatomical link between the brain and the anterior pituitary consists primarily of a vascular plexus of portal vessels which carry blood from ventral brain structures to the pituitary. The area believed to be the richest source of portal blood is the dense capillary network lying within the median eminence of the hypothalamus (Fig. 21). A large number of cells within the hypothalamus present the cytological appearance of both neural and secretory cells, and the axonal processes of many of these terminate on the capillaries within the median eminence. Nervous activity within the hypothalamus and other parts of the central nervous system is believed to cause the discharge of secretory products from the neurosecretory cells into the capillaries and portal vasculature from which the products reach the sinusoids of the anterior lobe proper. A variety of experiments illustrate that the integrity of the vascular link between the hypothalamus and the pituitary is essential to the maintenance of gonadal function. Other experiments demonstrate a gonadotrophic response to electrical stimulation of specific hypothalamic loci and a loss of gonadotrophic function after destruction of various hypothalamic areas (Harris, 1955, 1960; Sawyer, 1959). In addition, hypothalamic extracts have proved effective in activating the gonad-stimulating properties of the isolated pituitary (Harris *et al.*, 1966).

Photomicrographs of parasagittal sections through corresponding region of hypothalamus of rat (*lower left*), cat (*lower right*) and sheep (*upper right*). Line drawings indicate location of section in the respective brains; the brains are pictured as if divided in the parasagittal plane, the farther portion displaced downward to reveal the photographed area enclosed in a rectangle. Nomenclature after Bleier (1961).

bv	blood vessel
Db	nucleus of the diagonal band of Broca
fd	descending column of the fornix
hi	habenulo-interpenduncular tract
Hla	lateral hypothalamic area
Mnc	mammillary nuclear complex
mt	mammillo-thalamic tract
oc	optic chiasm
Po	preoptic area
Soa	supraoptic nucleus, anterior component
Sot	supraoptic nucleus, tuberal component
Sp	septal nuclear complex
St	bed-nucleus of the stria terminalis
tt	taenia thalami
Vem	ventromedial nucleus

* This section is through the edge of this structure and does not represent its relative size in this animal.

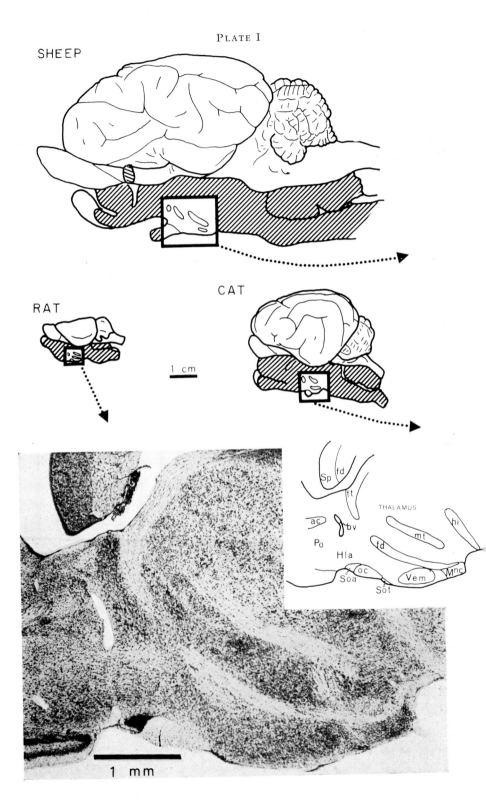

PLATE I

SHEEP

RAT

CAT

1 cm

1 mm

Sp fd

tt

THALAMUS

ac

bv

hi

Po

fd

mt

Hla

oc

Soa

Vem

Mnc

Sot

PLATE I

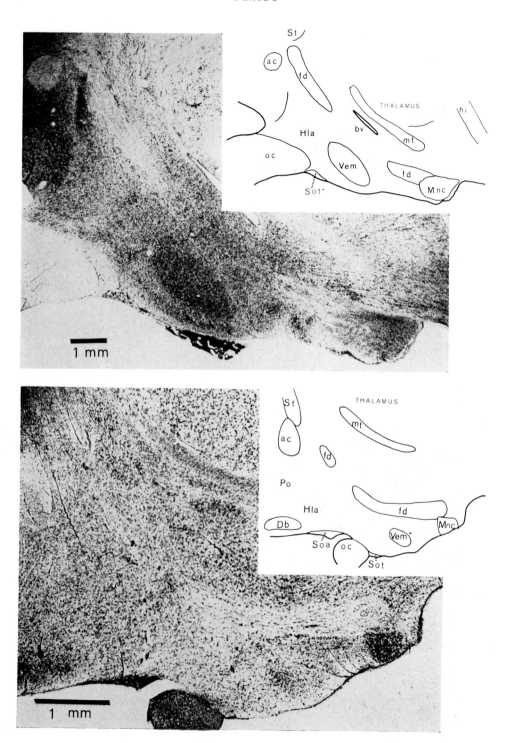

PLATE II

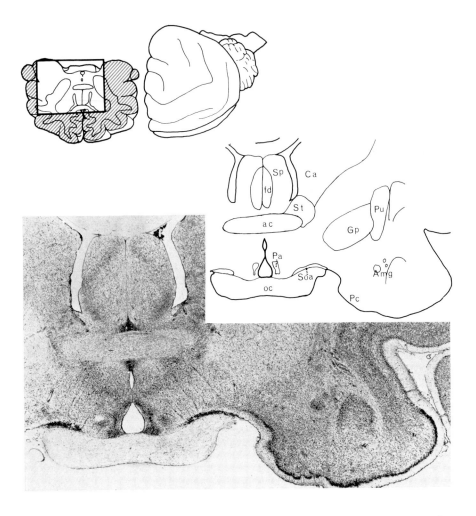

Coronal section through the anterior hypothalamic region of cat. Nomenclature after Bleier (1961).

Amg	amygdaloid complex
Ca	caudate nucleus
fd	descending column of the fornix
Gp	globus pallidus
cc	optic chiasm
Pa	paraventricular nucleus
Pc	pyriform cortex or periamygdaloid cortex
Pu	putamen
Soa	supraoptic nucleus, anterior component
Sp	septal nuclear complex
St	bed-nucleus of the stria terminalis

It is clear that the hypothalamic-pituitary-gonadal circuit functions as a feed-back system (Fig. 19). Gonadal hormones suppress the pituitary-stimulating properties of cells in the median eminence of the hypothalamus in both sexes. The evidence for this relationship is reviewed by Davidson & Smith (1967), who further present a case for considering the essential change

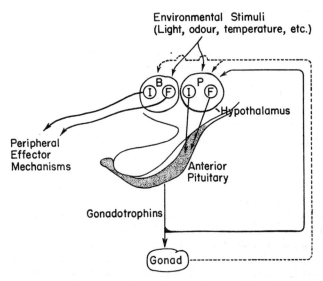

Fig. 19. A diagram to depict both facilitative (F) and inhibitory (I) hypothalamic mechanisms, influencing sexual behaviour (B), as well as pituitary function (P). These mechanisms serve independently as target areas for the gonadal, and possibly for the gonadotrophic, hormones in the production of sexual behaviour and in the feed-back system regulating the anterior pituitary. Independent neural loci are believed to exist for the behavioural and pituitary mechanisms. The particular regions involved vary greatly among species.

at puberty to be a reduction in hypothalamic sensitivity to this inhibition by circulating gonadal hormones.

In addition to this "long-loop" feedback from gonadal hormones, there exist "short-loop" feedback effects wherein pituitary hormones inhibit the production of their hypothalamic releasing factors (Fig. 19; Brown-Grant, 1966).

A differential organization of male and female hypothalamic mechanisms is suggested by the experiments of Barraclough & Gorski (1961). They found that the blockade of ovulation occurring in female rats subjected to early androgen treatment was attributable to an alteration of hypothalamic mechanisms controlling release of ovulating hormone. That the alteration may have been a change in neural organization in the masculine direction seems likely. This is a further indication that distinctive sexual characteristics are based upon programmes determined by early presence or absence of androgen.

Higher-level controls: regulation of neuro-hypophyseal activity. In addition to the feedback control exerted by gonadal hormones, there is increasing evidence that environmental events, acting through sensory exteroceptors, affect the activity of the hypothalamic-pituitary system. One frequent effect of these influences is to co-ordinate reproductive cycles with environmental and social conditions so as to ensure the production of offspring when these conditions are optimal.

In mice and guinea pigs, daily alternation of light and dark is a stimulus to pituitary secretion of ovulation-inducing hormones, such that ovulation usually occurs in all females in a group near the same time of night (Lehrman, 1964). The reproductive activity of several mammals is correlated with seasonal changes in the relative length of days and nights. Alteration of day length has correspondingly altered reproductive cycles in voles, mice, rabbits, ferrets, raccoons and sheep; closely related animals, however, were un- affected by similar manipulations of photoperiods (Thibault *et al.*, 1966; Wurtman, 1967). In some of these cases, changing the photoperiod was shown to affect the levels of circulating pituitary hormones, but in general the action of photic effects through the hypothalamus remains presumptive.

Ovulation in rabbits, cats, and ferrets is induced by reflex stimulation of pituitary release of ovulation-inducing hormones. The effective stimulus is the mechanical stimulation of the vagina in real or simulated copulation.

The assumption that olfactory stimuli affect hypothalamic-pituitary mechanisms makes it easier to explain some phenomena of oestrous cycling and gestation in mice. The onset of regular oestrous cycling depends upon the odours of males; and odour of strange males terminates pregnancy. Reported effects of olfactory stimuli from males on oestrous cycles in sheep, pigs, and goats may prove to be additional cases of exteroceptive regulation of hormonal cycles (Bruce, 1966).

Table 11. Relationships and Interdependence among Factors Influencing Repro- ductive Behaviour of Ring Dove Females

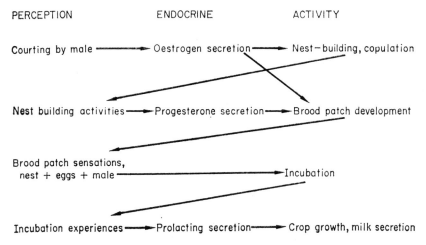

PERCEPTION   ENDOCRINE   ACTIVITY

Courting by male ⟶ Oestrogen secretion ⟶ Nest-building, copulation

Nest building activities ⟶ Progesterone secretion ⟶ Brood patch development

Brood patch sensations, nest + eggs + male ⟶ Incubation

Incubation experiences ⟶ Prolactin secretion ⟶ Crop growth, milk secretion

Sensory influences upon hypothalamic and hypophyseal activity are strongly exhibited in birds. A variety of photoperiodic effects on gonadal cycling have been demonstrated in a multitude of temperate-zone species where day-length is a function of the season of the year (Farner & Follett, 1966). The most precise correlations among complex exteroceptive stimuli, hormonal regulation, and reproductive behaviours have been demonstrated in the ring dove by Lehrman and his associates. An extensive series of experiments indicates that specific stimuli, largely visual, from perception of the activity of the mate, nesting materials, eggs, etc., activate successive pituitary and gonadal secretions, and elicit appropriate reproductive behaviours. The sequence of events for the females is presented in Table 11. Each perception, hormone, and activity in sequence is affected by its predecessors, and the three classes of phenomena are interlaced into an interdependent functional system (Lehrman, 1961, 1964).

3. IN SEARCH OF NEURAL PROGRAMMES

Peripheral processes. As to where and how the gonadal hormones exert their influence in activating behaviours, there are a number of candidates for neural target structures. We could begin with peripheral sensory systems: gonadal hormones have demonstrably altered olfactory thresholds in humans to several odorous substances, including musklike and urinoid odours (Moulton & Beidler, 1967). The site and mode of hormone action have not been determined.

Spinal and brain stem processes. Moving centrally in the nervous system, there are a number of spinal and brain stem reflexive mechanisms that act directly on the motoneurons responsible for the actions constituting sexual behaviours. Excepting the recent results of Hart (1967b), evidence so far indicates that these elements are not directly influenced by hormones (Beach, 1967).

Although final effector systems exist within the spinal cord for both male and female patterns of sexual behaviours, the response elicitable from spinal or decerebrate animals bears little resemblance to the typical pattern of behaviour. Elicitation of sexual activities in animals with spinal transections results in manifestations of individual components from the behavioural repertoire, and there can be little doubt that the highly organized picture of sexual behaviour characteristic of intact mature animals depends extensively on the participation of higher neural centres. Nevertheless, a surprising degree of temporal organization of the sexual response pattern appears to be mediated by purely spinal mechanisms in the male dog (Hart, 1967a) and rat (Hart, 1967b).

Forebrain processes. Contributions to the regulation of sexual behaviour from forebrain structures have proven to be both facilitating and inhibitory in nature. In both rodents and carnivores the removal of large amounts of neocortex from the dorsal brain surface results in a decrease in the frequency and vigour with which the masculine elements of sexual behaviour are displayed (Beach, 1940; Beach *et al.*, 1956). It is possible that the extent to which neocortical structures participate in mating behaviour reflects the complexity and need for plasticity imposed by the conditions under which mating is normally displayed. In support of such a possibility are the observations that neocortical structures are not as important to rodents as to carnivores and are generally more important to the male than to the female pattern of behaviour (Beach, 1942).

Recent work with rabbits has indicated that the action of progesterone (or oestrogen priming followed by progesterone) may be quite widespread in terms of altering the thresholds for neuronal activity at many different brain loci (Kawakami & Sawyer, 1959a). The threshold for the response of cortical arousal to stimulation of the reticular formation may decrease more than 50 per cent after suitable treatment with ovarian hormones. The action of injected progesterone is biphasic, causing first a decrease in threshold to electrical activation followed by a marked increase which can be maintained by continued injections of progesterone.

The same investigators have also identified a behavioural after-reaction to coitus or vaginal stimulation in the rabbit which is correlated with marked alterations in the neural activity in hippocampal, amygdaloid, septal, and ventrohypothalamic areas (Kawakami & Sawyer, 1959b; Sawyer & Kawakami, 1959). Electroencephalographic (EEG) characteristics of the after-reaction appear more reliable than the behavioural signs and consist nearly always of an early phase of "sleep spindles" followed by a later appearance of "hippocampal hyperactivity". These two phases constitute the EEG after-reaction and it can be induced by electrical stimulation of various regions (particularly the ventromedial hypothalamus) and by administration of pituitary or placental hormones. Vaginal stimulation or coitus produces the EEG after-reaction in oestrous females given ovarian hormones, but not in untreated ovariectomized females.

The thresholds for cortical arousal on the one hand and for the EEG after-reaction on the other hand are mutually antagonistic, since the former is related to behavioural arousal whereas the latter is related to a sleep-like depression. Nevertheless, changes in the two thresholds parallel one another after suitablehor monal treatment, and Sawyer & Kawakami suggest that the arousal threshold is related to oestrous behaviour and the after-reaction threshold to activation of the pituitary with consequent release of ovulatory gonadotrophin. It is important to distinguish between changes in the threshold for the EEG after-reaction and the reaction per se—particularly the phase of "hippocampal hyperactivity". This part of the EEG after-reaction is believed to be induced by the release of ovulatory hormone because of (a) the time at which it normally occurs following coitus, and (b) its induction by administration of gonadotrophins and oxytocin.

In contrast to the facilitating role of neocortex, the phylogenetically older pyriform cortex (Pc in Plate II and Fig. 21) of the ventral brain surface exerts strongly inhibitory control over the manifestation of sexual behaviour (Green et al., 1957). Removal or destruction of pyriform cortex and perhaps some of the associated amygdaloid complex (Amg in Plate II and Fig. 21) results in hypersexuality in carnivores and primates (Klüver & Bucy, 1939; Schreiner & Kling, 1956, see Chapter 15). The nature of the hypersexuality induced by such lesions has been most carefully assessed by Green et al.

(1957) and according to those investigators involves two fundamental post-operative changes in the character of the sexual behaviour displayed by male cats: (a) increased frequency of mating outside the established territory, and (b) increased frequency of mating attempts towards unsuitable or unadaptive sexual objects (e.g. inanimate objects or animals from other species).

Three things about the nature of the behavioural change induced by destruction of the piriform-amygdaloid complex are important to emphasize. First, no increase in the vigour of sexual behaviour has been documented. The effects of these lesions on such end-points as frequency of ejaculation, the post-ejaculatory refractory period, and number of ejaculations to sexual exhaustion have not yet been investigated. Second, the inhibitory control exercised by the piriform-amygdaloid complex appears to be specific to male mating behaviour and no convincing demonstration of changes in female receptivity has been published. Third, the hypersexual effect, although persistent, does not survive castration and therefore the piriform cortex must exert its inhibitory influence over lower centres which remain normally responsive to the presence or absence of testicular hormones. That the lower centres involved are contained within the hypothalamus was suggested by Schreiner & Kling (1953). A bilateral lesion involving the ventromedial nuclei of the hypothalamus was produced in one of their cats previously made hypersexual. Immediately after the operation, all sexual behaviour disappeared.

Hypothalamic processes. A great deal of evidence points to hypothalamic structures and processes as necessary components of organized sexual activity, and as special target tissues mediating hormonal influence on behaviour independently of the pituitary control mechanisms (Fig. 19). Hypothalamic lesions can eliminate all sexual behaviour despite adequate amounts of gonadal hormone supplied either by a normally functioning gonad or by injection. Studies demonstrating effective hypothalamic sites for lesions eliminating sexual responsiveness have been carried out on the female cat and rabbit (Sawyer & Robinson, 1956), the male cat (Schreiner & Kling, 1954), the ewe (Clegg *et al.*, 1958) and rodents (Phoenix, 1961). These lesions apparently remove some facilitating process necessary for sexual activity.

Supporting evidence comes from studies of oestrogen implantation and oestrogen uptake (Sawyer, 1967). In the anterior lateral hypothalamic region and the medial preoptic regions of cats, where lesions are effective in elimi-nating sexual behaviours in female cats, implantation of oestradiol produce behavioural oestrus with no stimulating effects on accessory genital tissues. Furthermore, tracing of radioactive oestrogens showed these anterior hypothalamic regions to be favoured sites for hormone uptake. These findings of Michael in cats (Sawyer, *loc. cit.*) have been confirmed in rats by Kato & Villee (1967) who also found a secondary target for oestrogen uptake in the median eminence where gonadal hormones exert feedback inhibition of pituitary stimulation. Thus they provide further evidence of two separate hypothalamic locations—the anterior and that in the median eminence—which are involved in the two respective gonadal functions—behavioural and endocrine—in cats and rats (see also Fig. 85, p. 497).

Also among males, implants of testosterone in the medial preoptic region in castrated male rats were found to induce restoration of male sexual behaviour without restoration of accessory genital tissues (Davidson & Smith, 1967). Thus in both sexes there appears to be a process whereby gonadal hormones act on or through elements in the preoptic or anterior hypothalamic region to facilitate sexual behaviour, and these processes are independent of other general hormonal effects on body tissues.

Studies of rabbits have also found regions wherein lesions eliminate, and hormone implants facilitate, sexual behaviour. However, in rabbits these regions are not in the anterior hypothalamic regions where similar effects were localized in cats and rats; they are rather in the neighbourhood of the mammillary nuclear complex (Mnc in Plate I) in the posterior hypothalamus (Sawyer, 1967). In the face of this contrast, it is manifestly unwise to generalize across species concerning structure and function of these hypothalamic processes.

In addition to facilitating mechanisms, there is also some evidence for hypothalamic mechanisms which inhibit sexual behaviours.

The work of Law & Meagher (1958) describes an effect of hypothalamic lesions which was until very recently unique. After certain lesions in the midventral hypothalamus, female rats with intact ovaries displayed receptive behaviour in the absence of vaginal oestrus. Moreover, spayed female rats displayed receptive behaviour even though they were untreated with exogenous hormones. Goy & Phoenix (1963) have confirmed this type of effect using spayed female guinea pigs. Midventral lesions in this species permit the "continuous" display of receptive behaviour in the complete absence of exogenous hormone even though the vaginal membrane remains continuously closed. The display of the behaviour is not transitory, but persists indefinitely and appears within 24 hours of operation. The existence of both facilitative and inhibitory mechanisms for female sexual behaviour within the hypothalamus bears a striking resemblance to the hypothalamic mechanisms regulating the release of FSH from the pituitary. Harris (1959) has emphasized that mechanisms exist for both stimulation and inhibition of the release of that hormone.

Still to be explained are the mechanisms whereby these small, nondescript hypothalamic regions play their role in the mediation of a great number and variety of activities which are organized into consistent and distinctive sequences of sexual behaviour. The perplexity is compounded when we consider the many other kinds of behaviours that can be affected by alterations introduced into some of these same regions of the hypothalamus, such as the preoptic and ventromedial (see preceding and following sections on Feeding, Agonistic Behaviour and Sleep), and when we consider the possibility that the same locus may be involved in the mediation of both male and female patterns of sexual behaviour.

B. Parental Behaviours

The mode of production of young differs greatly in birds and mammals. In birds, production and incubation of eggs follows closely upon the sexual activities culminating in copulation and fertilization. In mammals, the long period of uterine gestation intervenes between sexual and parental activities.

The impending production of mammalian young is evidenced behaviourally by gradual changes in general activity level and social responsiveness of expectant mothers. Their sexual cycles cease, they often become less active as weight increases, and they often display increased isolation and hostility—aggression or defensive behaviour (see later chapters). Many of these changes may be related to the drastic alteration in the endocrine system of the mother, due to the secretions of the corpora lutea in the ovary and the growth of the placenta and its hormonal secretions.

1. Nest Building

In both birds and mammals the first overt act directly relating to production and nurture of young is usually the construction of some sort of nest in which to shelter the eggs or young. In an exhaustive review of the influences of hormones on parental behaviour, Lehrman (1961) summarizes considerable evidence that nest-building by female birds coincides with the receptive phase of the oestrous cycle during the period of maximal follicle growth and oestrogen secretion. Thus the nest is built within a few days prior to egg-laying, and there is a strong suggestion that oestrogens induce nest-building. Nest-building by some male birds correspondingly occurs at a peak of testicular hormone activity. (As usual in avian hormone systems there are numerous specific exceptions: some females build nests when stimulated by androgens; some males do so when stimulated by oestrogens; many birds do not respond to hormones at all in this respect.)

Nest-building in mammals may also be related to hormonal secretion. It is associated with pre-parturitional hair loosening, which hair is utilized in nest construction in rabbits (see Chapter 14), lions and squirrels (Lehrman, 1961). Administration of oestradiol, progesterone and prolactin to spayed rabbits for 8 weeks induced both nest-building and hair-loosening. Shorter-term hormone administration resulted in nest-building but no hair-loosening (Farooq et al., 1963). Thus both nest-building and hair-loosening appear to be affected by endocrine factors, although through different mechanisms.

Hormonal secretion itself can be markedly affected by nest-building, or by the perception of nest-building. This is a further example of perceptual and behavioural determinations of endocrine activity which have been demonstrated by Lehrman and his associates (Table 11).

2. Parturition and/or Incubation

The mammalian birth process, in which the foetus is expelled as the neonatal young, is instigated by hormonal developments (Zarrow, 1961). It is accompanied, in most species, by maternal actions assisting the expulsion (see later chapters). Many mothers consume the ejected placenta, membranes, and fluids, thus cleaning the young (and this includes normally herbivorous types, as related in later chapters). Experiments with cats, monkeys, and rats indicate that this appetite is limited to a very brief time after parturition and is closely related to the temporary hormonal state of the mother (Lehrman, *loc. cit.*). Schneirla, Rosenblatt, Tobach (1963) present data which indicate that the parturitional licking of cats is directed to the attractiveness of the fluids rather than of the young. This transient appetite may be an example of the appetitive influence of a metabolic need or deficiency related to the sudden loss of placental secretions in the circulatory system of the mother. A specific appetite for metabolites may play a role in bringing about the cleaning of young by the mother. It has been shown, for example, that iodine, provided to the young in the mother's milk, is returned to the mother when she ingests the urine of the young (Samel & Caputa, 1965: Beltz & Reineke, 1968).

In birds, the egg is formed and delivered, then embryonic and foetal growth takes place. The occurrence and spacing of egg delivery is under hormonal control, but the hormonal production itself is considerably influenced by the perceptual experience of the bird with such stimuli as courting, nest-building and the presence of eggs already laid (Lehrman, 1961). The subsequent development of the young depends upon incubation, a behavioural process, on the part of the parent. To accomplish this, the eggs are contacted by a featherless, specially vascularized area of the skin of the parent, known as the incubation patch or brood patch. Ducks and pelicans are notable in that they do not form such an incubation patch. A combination of oestrogen and prolactin secretion or injection was necessary for the full development of the patch of sparrows and finches, both males and females (Lehrman, 1961). Lehrman hypothesizes that irritating stimuli inherent in the patch are quieted by contact with the eggs, leading the parent to sit on the eggs. Some similar cooling effect, of the eggs absorbing heat from the parent, may underlie the necessity for duck eggs to be wet (see Chapter 18) to ensure proper incubation by the patchless parent. Independently of patch formation, some birds can be induced to sit on eggs by administration of the pituitary hormone, prolactin, which also induces crop-milk production in pigeons and lactation in mammals. Furthermore, in many birds the secretion of prolactin is stimulated and maintained by the process of sitting on eggs, thus securing the continuation of incubation until the eggs hatch (Lehrman, 1961).

3. LACTATION AND NURSING

Reduction of mammary tension has been suggested as the basic maternal goal in nursing (e.g. Rosenblatt & Lehrman, 1963). But Moltz, Geller, & Levin (1967) found that rats persisted in attempting to nurse their young following and despite complete mammectomy. Once the young do begin suckling, the stimulus this provides to the mother's nipples not only maintains continued lactation by inducing the long-term secretion of prolactin and other pituitary hormones, it also causes expulsion of the milk on a short-term basis by means of the well-documented milk-ejection reflex (Cowie & Folley, 1961). In this neuro-hormonal reflex, sucking stimulates receptors in the nipple, which provide a neural message to the hypothalamus, which results in the neurosecretion there of the oxytocic hormone. The oxytocin travels in the circulatory system to the mammary glands, where it acts on the alveolar myoepithelium and causes the expulsion of milk.

4. INDICATIONS OF A NEURAL PROGRAMME

There is evidence of special interest concerning the role of elements of the cerebral cortex in programmes of parental behaviour in rats. Complete destruction of the neocortex was found to eliminate nursing (but not lactation), cleaning, and retrieving of the newborn rats by their mothers (Beach, 1937; Stone, 1938; Davis, 1939). Following restricted bilateral

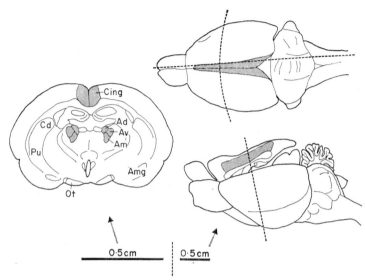

Fig. 20. Location of the lesions in cingulate cortex and of correlated degeneration in anterior thalamic nuclei associated with disorders in retrieving of young and other sequential motor patterns in rats. (Data from Slotnick, 1967.)

Ad	anterodorsal nucleus of the thalamus	Av	anteroventral nucleus of the thalamus
Am	anteromedial nucleus of the thalamus	Cd	caudate nucleus
		Cing	cingulate cortex
Amg	amygdaloid complex	Ot	optic tract
		Pu	putamen

lesions in the cingulate cortex on the medial surface (see Fig. 20) of the cerebral hemispheres (Stamm, 1955), nest construction was absent or inferior and while pups were carried, they were dropped about randomly rather than gathered. Cleaning of the newborn, however, was performed normally. Slotnick (1967) confirmed these results, and further found that the degree of behavioural disorder was correlated with the degree of retrograde degeneration produced in the cells of the anterior thalamic nuclei (Fig. 20) which send their fibres to the cingulate cortex. He further determined that the disorder lasted only 3 to 4 days post-partum, but could be reinstated at later times by placing the mother in a strange or novel environment. It has been suggested that the nature of the cortical deficit consists in an inability to integrate smaller behavioural units, e.g. carrying, into larger patterns such as collecting infants and nest-building (Beach, 1938; Pribram, 1960). In this regard it is of interest that the cingulate-lesions also interfered with food-

6*

gathering and hoarding (Stamm, 1954), and with the temporal integration of learned sequences of behaviour (Barker & Thomas, 1965). It should be emphasized, however, that the neural involvement in maternal activity is not restricted to cingulate cortex. In fact, cingulate cortex appears somewhat less extensively involved than dorsal hippocampus. (Hc in Fig. 21; lies between Cing and Ad in Fig. 20.) Kimble, Rogers & Hendrickson (1967) recently reported that lesions of dorsal hippocampus produced severe disruption of maternal activities without measurably impairing sexual behaviour. Moreover, the disruption was more severe than in the case of cingulate lesions (Stamm, 1955; Wilsoncroft, 1963; Slotnick, *op. cit.*).

C. Other Forms of Social Behaviour

In addition to sexual activities and caring for infants, there are many other sequences of social responses that subserve individual and species preservation effectively, if not quite so obviously and directly. Responses to and interactions with other animals are here regarded as social behaviours. These can be sorted, for survey purposes, into two types of activity. The first encompasses actions leading to aggregation of animals into groups, and such processes we will call affiliative. The other class consists in those activities which drive animals apart, and includes defence and aggression, flight and retreat. Dominance relationships among animals are a stabilized state resulting from such activities (Chapter 1). Scott's term *agonistic* will be used to designate the common, separative, characteristic of these behaviours.

1. AFFILIATION

Selection of companions and keeping in contact with them involves appropriate responding to the stimulus characteristics of other animals. There is some evidence that these responses become associated with stimuli provided by the companions of an animal's early experience. Thus, except after unusual conditions of rearing, most animals come to associate with those perceived as similar to parents and siblings, forming groups of the same species (Scott, 1960).

Physiological analysis of infant affiliation can be undertaken by a sufficiently detailed study of behaviour sequences and their effective stimuli, as in the following case (Welker, 1959a). When newborn puppies were cold or hot, they exhibited restless behaviour, with cries at each respiration, and right-left turning movements of the head, but little locomotion. When they were optimally warm, they remained quiet in sleeping posture, even when food-deprived. Deviations from optimal temperature conditions evidently lowered the threshold for activation of specific act sequences by stimulation of somatic receptors in the snout. The specific response pattern activated was determined by the direction of the deviation. In cold animals, bilateral warm contact with the snout, exclusively, resulted in forward progression by a consistent sequence of leg movements. Cold snout contact produced withdrawal of the head and extension of the forelegs. In hot animals, on the other hand, both warm and cold contact induced the withdrawal movements. Food deprivation also brought about forward progress and sucking movements in response to warm snout contact, regardless of the animal's temperature.

These mechanisms can obviously serve to bring puppies together and to the mother when they are cold, to the food supply when they are hungry, and can disperse them when they are warm. The contribution to puppy survival is evident. The heat-conservation value of social behaviour has also been demonstrated in mice (King & Connon, 1955).

Components of programmes serving group maintenance include the calls and postures used by many species to effect auditory and visual contact (e.g. see Chapters 14, 16, 17). By means of various glandular secretions, group and territory identification is accomplished using olfactory cues by primates (Hill, 1956; Evans & Goy, 1968), marsupials (Schultze-Westrum, 1965), and dogs (Chapter 14).

2. AGONISTIC BEHAVIOUR

The hierarchical neural organization of the various component activities into a pattern of behaviour subserving defensive or aggressive purposes has

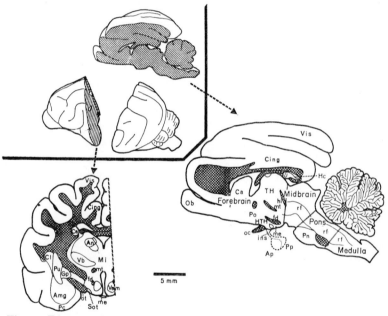

Fig. 21. Diagram of cat brain, with parasagittal and coronal sections showing locations of major divisions of the brain and certain subdivisions.

Amg	amygdaloid complex	Ob	olfactory bulb
An	anterior thalamic nuclear group	oc	optic chiasm
Ap	anterior pituitary or adenohypo-physis	ot	optic tract
		Pc	pyriform or periamygdaloid cortex
Ca	caudate nucleus	Pn	pontine nuclei
Cing	cingulate cortex	Po	preoptic area
Cl	claustrum	Pp	posterior pituitary or neurophypo-physis
Fd	descending column of the fornix		
Gp	globus pallidus	Pu	putamen
Hc	hippocampus, retrosplenial portion	rf	reticular formations
hi	habenulo-interpeduncular tract	TH	thalamus
HTH	hypothalamus	Vb	ventrobasal nuclear complex of the thalamus
ins	infundibular stalk		
me	median eminence of the hypothalamus	Vem	ventromedial nucleus of the hypo-thalamus
Mi	massa intermedia of the thalamus		
mt	mammillo-thalamic tract	Vis	visual cortex

been extensively studied in cats. A detailed description of findings is presented in Chapter 15. Plates I and II and Fig. 21 depict the locations of relevant neural regions in the cat brain. Animals with the entire midbrain and forebrain removed reacted to noxious stimulation only when it was quite severe. Growling, protrusion of claws, running movements by the legs (these animals were unable to stand or walk), pilo-erection, increased respiratory and cardiac rates, and righting of the head and shoulders constituted the response. Presence of the intact midbrain augmented the response to include vocalization, pupil dilation, hissing, tail-lashing, and running away. When midbrain and hypothalamus were left intact, head lowering, arching of the back, retraction of the ears, striking with the paws, and biting were integrated into the pattern (Bard & Macht, 1958). The hypothalamus appears to be the centre necessary for the normal defensive pattern in response to painful or threatening stimulation (Brady, 1960).

In intact cats, electrical stimulation of the dorsal hypothalamus around the *descending column of the fornix* resulted in well-directed attacks at the slightest external provocation (Akert, 1959). Included in the reaction were arousal, angry vocalization, unsheathing the claws, lashing the tail, crouching, arching the back, spitting, hissing, retraction of ears—the whole pattern of feline defence or aggression.

Similar results were obtained following lesions of the *ventromedial* nucleus of the hypothalamus (Wheatley, 1944), suggesting that this nucleus exerts some inhibiting influence over the activities induced in the dorsally neighbouring region. Attack or flight reactions were also reported to follow lesions, in the septal area in rats, cats, and dogs (King & Meyer, 1958; Spiegel *et al.*, 1940).

The forebrain region most implicated in aggressive-defensive activities includes the *amygdalar complex* and the neighbouring *pyriform cortex*. Removal of this region resulted in general absence of reaction to aversive stimulation, docility, submissiveness, and other symptoms (see sections on activity and sexual behaviour). Similar effects were observed in monkeys (Klüver & Bucy, 1939), cats (Schreiner & Kling, 1953) and dogs (Fuller *et al.*, 1957). Lesions of the lateral and basal portions of the amygdalar complex also eliminated the exaggerated defensive responsiveness of rats, which followed septal lesions (King & Meyer, *loc. cit.*). In monkeys with the amygdalar complex preserved intact, however, the lack of aggression and responsiveness was reported to follow lesions confined to the periamygdalar cerebral cortex (Pinto Hamuy *et al.*, 1957).

Electrical stimulation of central and medial amygdalar regions, or the dorsal hypothalamus, or the reticular formation in the midbrain, or the fibre tracts connecting these regions, produced growling, hissing, and attack or flight in cats (Fernandes de Molina & Hunsperger, 1959). Stimulation of two neighbouring amygdalar areas produced response patterns somewhat different from each other (Ursin & Kaada, 1960). The more rostral and medial area gave glancing movements of the head, wide dilation of pupils, cringing, followed by running away and hiding after choosing a suitable place. The area more posterior and lateral gave growling, retraction of ears, pawing with claws extended, but no further attack.

The mechanisms by which the amygdalar-cortical processes participate in defensive-aggressive activities is far from clear. Their influence may operate through the ventromedial nucleus of the hypothalamus, since destruction of this body eliminates amygdalar-cortical effects and leaves the typical ventromedial lesion syndrome of hyperaggressive behaviour (Gloor, 1960). Electrical stimulation in lateral and basal parts of the amygdala of cats suppressed attacks elicited by hypothalamic stimulation, while stimulation in a more posterior amygdalar region facilitated these attacks (Egger & Flynn, 1967).

Influence of gonadal hormones. Many birds and mammals show increased aggression and defensive reactions while caring for their young (Beach, 1948, p. 78). Winter flocks of birds disintegrate with the coming of spring, as aggressive displays drive individuals apart (Emlen, 1952). Males of several species of birds and mammals commonly engage in fighting during the breeding season. Such temporal correlation with hormonal cycles led to investigations of hormonal influences on aggressive activity.

Androgen administration has resulted in increased aggression in such diverse species as canaries (Shoemaker, 1939) and rats (Beach, 1948, p. 101). Orchidectomy, correspondingly, reduced defensive and aggressive behaviour in a wide range of vertebrate species (see Beach, *op. cit.* p. 99). In contrast, female hormones have suppressed aggressive tendencies. The female short-tailed shrew (*Blarina brevicauda*) attacks males regularly except during oestrus. Female hamsters have similar habits. Castrated female hamsters showed increased aggression when treated with oestrogen, but when progesterone treatment followed, oestrus was induced and the aggressive responses ceased (Kislak & Beach, 1955).

However, a simple rule that androgens increase and female hormones decrease aggressiveness cannot be generalized to all species. The female marten (*Martes americana*) is reported to fight with males only during oestrus (Beach, 1948, p. 98) and the aggressive behaviour of other species does not appear markedly affected in any direction by hormonal alterations (Mirsky, 1955).

IV. ACTIVITY

One basic pattern occurs, in normal situations, in almost any activity of intact animals. It can be called an arousal sequence, wherein a placid animal responds to stimulation by alerting, orientating, observing, then perhaps exploring or fleeing or resuming his placid state. These activities may form part of a larger pattern, leading into any of the behavioural sequences described in the previous sections. Or the cycle may occur without further complication as a simple series of events in the ordinary, but necessary, moment-to-moment negotiation with the environment. Arousal is most marked when it supplants sleep; and the neurophysiological correlates of sleep and arousal, and their influence on other activities, have attracted widespread interest in recent years.

A. Sleep

Most, if not all, birds and mammals exhibit periodic decreases in, or cessation of, activity and interaction with the environment. We have come to view these cyclic retreats as analogous to our own regular altered states of consciousness which we call sleep. On closer examination, details of these periods of rest often differ greatly from animal to animal (Kleitman, 1967)—indeed, a case can be made, for example, that cattle do not sleep at all in every sense that the term usually connotes (see Chapter 9; also Koella, 1967). Various cycles of resting have evolved in different creatures, and these are most frequently attuned to cycles of nature (Murray, 1965). This correspondence can be established by habit and repetition or by genetic selection (Kleitman, 1967), which can result in some sort of internal physiological clock. Richter (1967) presents evidence that such a clock, in the hypothalamus of rats, governs not only the cycles of sleep and waking, but the rhythms of other activities as well. These factors result in animals that are diurnal and sleep at night, nocturnal and sleep in the day, summer kinds that sleep in winter, etc., according to the adaptations required for each species.

A few of the physiological factors producing sleep are coming to light, but the remarkable truth is that neither the causes nor the functions of sleep have as yet been discovered. But deprivation of sleep, or severe or prolonged interruption of the established cycle of rest periods, produces metabolic disorders and energy deficits which result in serious organismic dysfunction (evidence reviewed in Murray, 1965).

A number of physiological indicators bear some degree of correlation with states of sleep. Many, but not all, postural and locomotory muscles are relaxed, and responsiveness to stimuli, metabolic activity, blood pressure, heart rate, lung ventilation, and body and brain temperature are all reduced during sleep (Altman, 1966, p. 423). A most useful sign, closely correlated with sleep, is the electrical activity recorded from the surface of the brain in the electroencephalogram (EEG). The EEG has proved a reliable indicator even of different stages of sleep; as many as 5 different stages, ranging from sensory drift or "floating", through light to deep sleep, have been discriminated by varying EEG patterns (Koella, 1967). Two major categories of sleeping states have been established on the basis of EEG recordings. In the first category are the progressive stages just mentioned, wherein EEG patterns are correlated with behavioural signs. The second category consists of cases where the EEG pattern is not distinguishable from the waking pattern, yet by many other indicators the animal is asleep. This state has been called "paradoxical sleep", and is characterized by tension in neck and other muscles, conjugated rapid eye movements, twitchings of vibrissae and legs, irregular and rapid respiration, and by dreaming according to reports by human subjects. Around 30–40 per cent of behavioural sleep is of this paradoxical type, and the percentage is higher in young animals (Loizzo & Longo, 1968).

Selective lesions in various portions of the nervous system have produced a pathological sleeping state, presumably by damaging processes necessary to maintaining waking, alert and active states (Koella, 1967; Sprague, 1967). These regions crucial for "sleep-prevention" include lateral hypothalamic areas, where lesions in cats, monkeys and rats have produced a somnolence lasting several days, and the reticular formation of the midbrain, where sufficiently large lesions result in somnolence lasting weeks or months or permanently. As in all lesion studies, rather than assuming that all the behavioural programmes or controls are localized in the lesioned region, it must be remembered that the functions affected may well be those of regions lying above or below the damaged area, particularly of regions which communicate through the altered region.

Koella (1967) has assembled impressive evidence that sleep is an actively induced state, not a simple absence of activity or a lapse in wakefulness. Reduced activity is a result of active synaptic inhibition of reflexes and neural action mechanisms. (Sleep also involves some cases of *deinhibition*, where inhibitory cells are prevented from firing, allowing activity during sleep which is normally suppressed in the waking state). Most important, sleep can actually be produced by electrical and chemical stimulation of neural structures.

In freely moving cats, dogs and rabbits, slow rhythmic electrical stimulation in the medial thalamus (between the mammillothalamic and habenulo-interpeduncular tracts; Plate I) induced behavioural signs of impending sleep, which were followed by actual sleeping without further stimulation. The sleep produced has all the behavioural and EEG signs of normal sleep. Stimulation of the preoptic area in the anterior hypothalamus (Plate I) and various locations among the reticular formations have also been reported to produce sleeping states. Stimulation in or near the pontine nuclei (Fig. 21) and the neighbouring pontine reticular formation has produced—and lesions in this area have precluded—paradoxical sleep in animals already sleeping. In waking animals, heightened arousal is produced by such stimulation.

Of special interest are recent experiments (Monnier & Hösli, 1965) wherein electrical stimulation of the medial thalamus was used to induce sleep in rabbits. Then, either blood or blood-borne substances from these stimulated-sleeping rabbits was introduced into the circulatory system of waking rabbits, who thereupon reliably fell asleep. Similar transfusion from aroused rabbits woke sleeping rabbits. Thus a promising case is laid for the existence of some neuro-humoural circulatory factor in the governance of sleep cycles.

B. Arousal, Attention and Orientation

A considerable body of evidence argues for the existence of a neural programme of general arousal. Electrical stimulation of the central core of the brain-stem, in the area designated for convenience as the *reticular formation* of the midbrain and lower brain-stem, which extends into the posterior hypothalamus, was found to produce the same pattern of activity in the cerebral cortex as was elicited by strong sensory stimulation. Furthermore, with either type of stimulation, sensory or electrical, the cortical activity was accompanied by behavioural arousal, and by awakening if the animal was sleeping (Magoun, 1958). Cutting the rostral connections of the reticular formation abolished wakefulness, leaving the animal permanently somnolent. In a variety of animals, sectioning the reticular connections with the cortex, leaving sensory connections intact, eliminated wakefulness, and sensory stimulation would no longer arouse the subjects. Sectioning the external sensory pathways to the cortex, on the other hand, leaving reticulocortical connections intact, did not interfere with the production of arousal by stimulating the reticular formation (Lindsley, 1960). It should be cautioned that *reticular formation* is a label of convenience rather than an anatomic or physiological entity. The region so designated consists of large numbers of small diversified nuclear groups greater in variety of cell types contained than is any other portion of the central nervous system (Olszewski, 1954).

All sensory stimulation is not equally effective in producing arousal. Somatic afferents from the face are more potent than other sensory activators (French, 1960). The reticular region receives connections *from* the cerebral cortex, and many other central nervous structures, as well as from sensory pathways. Thus reticular arousal may result from, or be influenced by, events in the brain as well as by external stimulation.

Reticular activation has produced effects in sensory systems which resemble those accompanying changes in attention. For example, improved resolution of detail in the visual system resulted from concurrent reticular stimulation (Lindsley, 1960). Reticular activation has also been observed to inhibit responses evoked by visual and auditory stimulation, in the same manner as does distraction by extraneous sensory stimulation (Livingston, 1960).

The behavioural aspect of attention most readily observed is orientation: moving the receptors into a position to better perceive a stimulus. As most receptors are gathered there, this usually consists in movement of the head. Pavlov identified this head movement as part of a larger, organized sequence of physiological and behavioural events termed the orienting reflex (Sokolov, 1963). These include an increase in cerebral blood flow, increase in sensitivity of many receptors, EEG of lower amplitude and higher frequency, and respiration of greater amplitude and decreased frequency (Lynn, 1966). The reaction is elicited by stimuli possessing unusual novelty, intensity, or a special meaningfulness to the subject that has been established by previous learning and association.

Sokolov has proposed the following mechanism for orientation (Lynn, 1966). Sensory stimulation activates the reticular arousal programme, and also the analysing programme in the cerebral cortex (e.g. Fig. 18). These cortical analysers in some way compare the incoming patterns of information with pre-formed "expectations" derived from past experience, individual or genetic. When the current effective stimuli do not match the expectations, cortical influences to the reticular arousal elements summate with the sensory activity already functioning there, and the programme of orienting activities is set in motion. If cortical "matching" occurs, arousal is not activated, and may even be inhibited. Some experimental support is claimed for this highly speculative scheme. Reticular activation by cortical influence has been shown experimentally (Lynn, 1966). Cortical units have been found which respond to stimuli only when the animal is orienting (Hubel *et al.*, 1959), and others which respond only when the stimulus is novel (Bettinger *et al.*, 1967). Lesions in cerebral cortex deleteriously affected orienting movements to auditory stimuli (Thompson & Welker, 1963)—but, contrary to the theory, some orienting still took place in the absence of cortex.

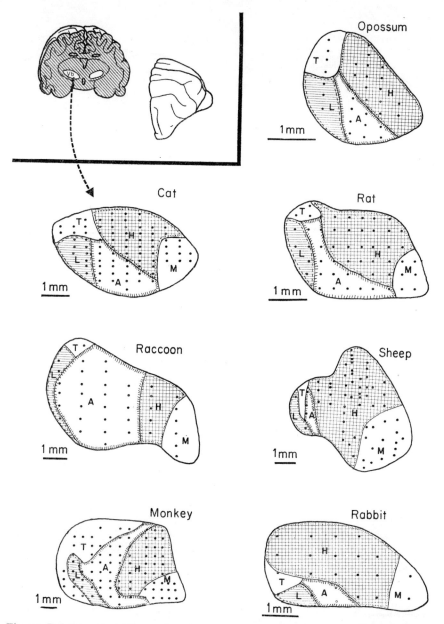

Fig. 22. Relative extent of somatic sensory projections from different body parts to the ventrobasal nuclear complex of the thalamus. Inset at upper left shows location of this region in cat brain. Dots represent locations at which electrophysiological responses were obtained in response to mechanical stimulation of the leg or foot (L); arm, hand, or fore-paw (A); trunk or tail (T); head (H); or tissues inside the mouth—gums, teeth, tongue or palate (M). Representative coronal planes, or combinations of planes, have been selected from studies of opossum (Pubols & Pubols, 1966, fig. 6); cat (Mountcastle & Henneman, 1949, figs. 3 and 4); rat (Emmers, 1965, figs. 2 and 3); raccoon (Welker & Johnson, 1965, fig. 2 plane 9 A); sheep (Cabral & Johnson, 1967); monkey (Mountcastle & Henneman, 1952, as cited in Rose & Mountcastle, 1959, fig. 3); and rabbit (Rose & Mountcastle, 1952, figs. 6 and 10). Relative hyperdevelopment of thalamic projection regions, related to behavioural specialization in tactile investigation, is notable in forelimb (A) projections in raccoons, in external head (H) projections in sheep and rabbit, and in intrabuccal (M) projections in sheep.

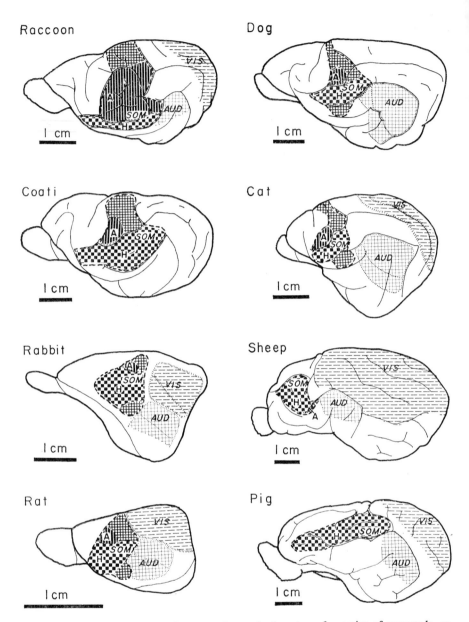

Fig. 23. Main sensory projection areas in cerebral cortex of a series of mammals, as determined by electrophysiological mapping. Electrical activity, in direct response to receptor activation, is evoked in areas AUD by auditory, SOM by somatic sensory, and VIS by visual stimulation. Boundaries of these projection fields have been established only in those cases represented with a boundary line around the shaded region. Shaded areas without boundaries represent a minimal portion of the projection region. Projections from the head (H) and arm and hand or forepaw (A) are differentially indicated in somatic sensory projection regions. Differential distribution of projections in areal extent can be seen, and is usually related to preferential use of sensory modalities in investigative activities. Pig and coatimundi engage in tactile exploration with the nose while raccoons use their hands. Cats and dogs can be expected to emphasize auditory information; sheep

C. Investigative Behaviour

Extension and elaboration of orienting responses grade into exploratory or investigative behaviour sequences. Attention, orientation, and investigative activities are similarly responsive to certain determinants, chief among which are novelty and complexity of the stimulation (Dember & Earl, 1957; Glanzer, 1958; Welker, 1961). Such activities occurred and developed in young animals independently of states of physiological need, although in later life exploration is often need-related. The sudden appearance of distinctive, attentive responses and exploratory activities, at a definite point in the animals' development, suggests that they are dependent upon the maturation of a specific set of neural connections (Welker, 1959b).

Following initial attentiveness, if animals further react to novel stimulation, they do so in one of two mutually exclusive types of behaviour (Welker, 1959c). Either fear, sometimes resulting in flight, ensues, or approach and investigation proceed. In this regard the effects of ablating the amygdalar-pyriform region of the forebrain are again of interest. Removal of these areas in cats (Schreiner & Kling, 1953) and in monkeys (Klüver, 1951) resulted in persistent and exaggerated investigative activity. The animals would approach and closely investigate any and every object encountered. Oral examination, including biting, chewing, smelling, and touching with the lips, was prominent. They paid excessive attention and reacted to all visual stimulation. Another prominent symptom following these lesions, it will be recalled, was the loss of responsiveness to aversive stimulation, with great decrease in aggressiveness and defensive reactions. It may thus be suggested that pyriform-amygdalar processes play some role in determining whether exploration of withdrawal follows orientation to novel stimulation.

Ablation of the cortex of the frontal cerebral poles has also produced unusual behaviour suggesting defects in attention and related behaviour. Frontal ablations in cats led to prolonged orienting and following to visual, auditory, and tactile stimulation (Hageman et al., 1959). Primates with frontal pole lesions have commonly exhibited almost constant locomotion in hyperreaction to external stimuli (French, 1959).

The particular direction taken by an animal's exploratory activity is apt to reflect his relative neural endowment in sensory processing systems. Wide variations in extent of brain regions receiving sensory information have been demonstrated in different mammals (Figs. 22 and 23). Variations exist within sense modalities, as in raccoon and coatimundi (Welker & Campos, 1963; Fig. 23), as well as between sense modalities, as in raccoon and sheep (Fig. 23). These figures provide considerable explanation as to why sheep and

---

have a highly developed visual afferent system; and raccoons and coatimundis have well developed somatic sensory processes. Data for raccoons are from Welker & Seidenstein (1959), visual and auditory determinations are incomplete; for dog from Tunturi (1944, 1950) and Pinto Hamuy et al. (1956), visual projections were not investigated; for coatimundi from Welker & Campos (1963), visual and auditory projections were not studied; for cat and rabbit from Woolsey (1958); for sheep from Adrian (1943), Woolsey & Fairman (1946), Hatton & Rubel (1967) and unpublished observations of Johnson, Hatton & Rubel; for rat from Woolsey (1952, 1958); for pig from Adrian (1943) and Woolsey & Fairman (loc. cit.), determinations of all sensory projections were incomplete.

coatimundi explore tactually with the snout, while raccoons use their hands. Sheep can also be expected to do more looking than feeling in most of life's investigations, at least those not immediately concerned with the ingestion of food at hand (or, rather, mouth).

D. Habituation

In general, attention, orientation, fear responses, and exploration are decreased by repetition of a particular stimulus pattern, which thus increases familiarity through experience (Glanzer, 1958; Welker, 1961). Physiological factors in habituation are suggested, though not demonstrated, in several studies of recording of central neural activity.

One example is the investigation of the electrical activity patterns in the *hippocampus* which are simultaneous with and related to cortical arousal (Green & Arduini, 1954). The patterns occur after stimulation of the mid-brain reticular formation, and are normally produced by external sensory stimulation. The patterns adapted (gradually disappeared) with repetition of a particular external sensory stimulus, but were reactivated by new or strange stimulation. For such activation, olfactory and tactile stimulation were most effective in rabbits, while auditory stimuli were powerful in cats. As another example, habituation was observed in correlated cortical arousal, reticular activation, and behavioural awakening, produced by auditory stimulation of sleeping cats (Sharpless & Jasper, 1956). Complete arousal, cortical, reticular, and behavioural, was observed with the first presentation of a tone. Succeeding repetitions of the same tone were progressively less effective in activating the animal, and finally sleep was undisturbed. Introduction of another tone of different pitch, however, provoked full-fledged arousal, and then repetition of this stimulus produced habituation.

Sokolov's scheme for the neural mechanism of the orientation reaction, described above, also serves as a model neuronal basis for the habituation of orientation: when the incoming stimulus configuration matches the "expectation" in the cortical analyser, cortical inhibition or lack of facilitation acting on reticular formation arousal results in habituation—a decrease in arousal. On the grounds that the time intervals of synaptic events are far too brief for synaptic inhibition and facilitation to serve as adequate mechanisms for habituation, Thompson & Spencer (1966) have proposed a somewhat simpler alternative scheme. In their view, habituation results from continued or repeated activity in a specific neural network. Only when a set of neurons not previously active is called into play, by some novelty introduced into the stimulus complex, are arousal and orientation set in motion.

This mechanism of neuronal desensitization is much like that advanced long ago by Pavlov, who proposed that repeated stimulation of a cortical analyser causes its neurons to become exhausted and prevents them from activating the orientation mechanisms. This theory had been rejected on the basis of evidence that, after habituation to strong or lengthy stimulation, animals would immediately orient to a weaker or shorter presentation of the otherwise same stimulation; hence the neurons were in no way fatigued or less sensitive (Lynn, 1966). However, Thompson & Spencer cite studies showing that weaker or shorter stimuli actually activate different neurons than do stronger or longer ones. This can happen in a neural network similar to that described in the retina in Fig. 14: if the receptive field organization of the cell were such that the surround is more influential than the centre, then a weak stimulus, activating only the centre, would stimulate the cell; a strong stimulus activating both centre and surround would inhibit the cell. The cell would thus respond selectively to weak stimuli only.

This revival of the Pavlovian hypothesis is more intriguing when we consider Pavlov's extension of the basic premise: in the process of exhaustion, the cells give off a substance which spreads and inhibits activity in other regions, thus explaining how constant or repetitive stimulation eventually puts animals to sleep. This reminds us of the findings of Monnier & Hösli where repetitive electrical stimulation of the medial thalamus does indeed appear to cause the production of some sleep-inducing substance into the circulation.

V. INTEGRATION OF BEHAVIOURAL PATTERNS

We hope to have illustrated, to some degree, the role of physiological components in behaviour patterns. Perhaps the most evident principle to be derived is that the chains of physiological events that make up a behavioural sequence are neither simple nor very well known. Through processes still for the most part undetermined, activations produced by particular stimuli are collated and organized, modulated by subtle effects from the past, and influenced by chemical states; individual muscular movements are arranged into consistent and effective series of actions. This integration is accomplished in the nervous system, the details of whose operations must be elucidated in order to fully understand behaviour.

For this to come about it will be necessary to abandon the all-too-human habit of considering as a unitary behavioural entity what is actually a diversity of response sequences, each with its own delicate and complex organization. Thus many quite different patterns of movement are lumped together, because of some common characteristic apparent to a hominid abstracter who sees, perhaps, purpose in events rather than seeing events. The activities are then classified according to their purpose, and all the various chains of activities become the X response, controlled by the Y centre, important in Z behaviour. Then, investigators are all too easily trapped into analysing labels rather than realities.

Once relatively simple (yet already quite complex) patterns have been identified and analysed, their integration into more comprehensive behavioural patterns must be analysed. Then the influence of sensory, hormonal and experiential factors can be assessed. Only then shall we approach a physiology of behaviour.

Detailed and persistent research is beginning to reveal something of the delicately ordered, fine interlacings of stimulus factors, physiological reactions, experiences, and response patterns that have developed in the evolution of viable species. We may hope that continuation and propagation of similar programmes will reveal equally detailed, complicated, and smooth-working organizations in a great variety of animal activities.

Simple stimulus and response explanations, and the parcelling of behaviours and their components and causes into hasty categories are at best crude beginnings. The anatomy of behaviour can be expected to surpass that of bodily structure in richness of detail and intricacy of design.

REFERENCES

ADOLPH, E. F. (1947). Urges to eat and drink in rats. *Am. J. Physiol.*, **151**, 110–125.

ADRIAN, E. D. (1943). Afferent areas in the brain of ungulates. *Brain*, **66**, 89–103.

AKERT, K. (1959). Physiology and pathophysiology of the hypothalamus. In: *Introduction to Stereotaxis with an Atlas of the Human Brain*. G. Schaltenbrand and P. Bailey (eds.). Stuttgart: Thieme, pp. 152–230.

ALTMAN, J. (1966). *Organic Foundations of Animal Behavior*. New York, N.Y.: Holt, Rinehart & Winston.

ANAND, B. K. (1961). Nervous regulation of food intake. *Physiol. Rev.*, **41**, 677–703.

ANDERSSON, B. (1951). The effect and localisation of electrical stimulation of certain parts of the brain stem in sheep and goats. *Acta physiol. scand.*, **23**, 8–23.

ANDERSSON, B. (1953). The effect of injections of hypertonic NaCl solutions into different parts of the hypothalamus of goats. *Acta physiol. scand.*, **28**, 188–211.

ANDERSSON, B., GALE, C. C. & SUNDSTEN, J. W. (1964). Preoptic influences on water intake. In: *Thirst*. M. J. Wayner (ed.). Oxford: Pergamon Press, pp. 361–379

ANDERSSON, B., LARSSON, S. & PERSSON, N. (1960). Some characteristics of the hypothalamic "drinking centre" in the goat as shown by the use of permanent electrodes. *Acta physiol. scand.*, **50**, 140–152.

ANDERSSON, B. & McCANN, S. M. (1955a). A further study of polydipsia evoked by hypothalamic stimulation in the goat. *Acta physiol. scand.*, **33**, 333–346.

ANDERSSON, B. & McCANN, S. M. (1955b). Drinking, antidiuresis and milk ejection from electrical stimulation within the hypothalamus of the goat. *Acta physiol. scand.*, **35**, 191–201.

ANDERSSON, B. & McCANN, S. M. (1956). The effect of hypothalamic lesions on the water intake of the dog. *Acta physiol. scand.*, **35**, 312–320.

BAILE, C. A. & MAYER, J. (1966). Hyperphagia in ruminants induced by a depressant. *Science*, **151**, 458–459.

BARD, P. & MACHT, M. B. (1958). In: *CIBA Foundation Symposium on the Neurological Basis of Behaviour*. Boston: Little, Brown, p. 55.

BARE, J. K. (1959). Hunger, deprivation and the day-night cycle. *J. comp. physiol. Psychol.*, **52**, 129–131.

BARKER, D. J. & THOMAS, G. J. (1965). Ablation of cingulate cortex in rats impairs alternation learning and retention. *J. comp. physiol. Psychol.* **60**, 353–359.

BARLOW, H. B. & LEVICK, W. R. (1965). The mechanism of directionally selective units in rabbit's retina. *J. Physiol., Lond.*, **178**, 477–504.

BARNES, C. D. & KIRCHER, C. (1968). *Readings in Neurophysiology*. New York, N.Y.: Wiley.

BARRACLOUGH, C. A. & GORSKI, R. A. (1961). Evidence that the hypothalamus is responsible for androgen-induced sterility in the female rat. *Endocrinology*, **68**, 68–79.

BARRINGTON, F. J. F. (1921). The relation of the hind brain to micturition. *Brain*, **44**, 23–42.

BARRINGTON, F. J. F. (1948). The surgical anatomy of the bladder and the physiology of micturition. In: *Textbook of Genito-Urinary Surgery*. H. P. Winsbury-White (ed.). Edinburgh: Livingstone, Ch. 18.

BEACH, F. A. (1937). The neural basis of innate behavior. I. Effects of cortical lesions upon the maternal behavior pattern in the rat. *J. comp. Psychol.*, **24**, 393–440.

BEACH, F. A. (1938). The neural basis of innate behavior. II. Relative effects of partial decortication in adulthood and infancy upon the maternal behavior of the primiparous rat. *J. genet. Psychol.*, **53**, 109–148.

BEACH, F. A. (1940). Effects of cortical lesions upon the copulatory behaviour of male rats. *J. comp. Psychol.*, **29**, 193–245.

BEACH, F. A. (1942). Central nervous mechanisms involved in the reproductive behavior of vertebrates. *Psychol. Bull.*, **39**, 200–226.

BEACH, F. A. (1948). *Hormones and Behavior.* New York, N.Y.: Paul Hoeber.

BEACH, F. A. (1960). Experimental investigations of species-specific behavior. *Am. Psychol.*, **15**, 1–18.

BEACH, F. A. (1967). Cerebral and hormonal control of reflexive mechanisms involved in copulatory behavior. *Physiol. Rev.*, **47**, 289–316.

BEACH, F. A. & FOWLER, H. (1959). Individual differences in the response of male rats to androgen. *J. comp. physiol. Psychol.*, **52**, 50–52.

BEACH, F. A., ZITRIN, A. & JAYNES, J. (1956). Neural mediation of mating in male cats. I. Effects of unilateral and bilateral removal of the neocortex. *J. comp. physiol. Psychol.*, **49**, 321–327.

BÉKÉSY, G. V. (1967). Mach band type lateral inhibition in different sense organs. *J. gen. Physiol.*, **50**, 519–532.

BELTZ, A. D. and REINEKE, E. P. (1968). Thyroid secretion rate in the neonatal rat. *Gen. comp. Endocrinol.*, **10**, 103–108.

BETTINGER, L. A., DAVIS, J. L., MEIKLE, M. B., BIRCH, H., KOPP, R., SMITH, H. E. & THOMPSON, R. F. (1967). "Novelty" cells in association cortex of cat. *Psychonom. Sci.*, **9**, 421–422.

BISHOP, P. O. (1967). Central nervous system: afferent mechanisms and perception. *A. Rev. Physiol.*, **29**, 427–484.

BLEIER, R. (1961). *The hypothalamus of the cat. A cytoarchitectonic atlas in the Horsley-Clarke co-ordinate system.* Baltimore, Md.: The Johns Hopkins Press.

BRADY, J. V. (1960). Emotional behavior. In: *Handbook of Physiology.* J. Field (ed.). Washington, D.C.: Amer. Physiol. Soc. Ch. 63. Section I. Neurophysiology, Vol. III.

BROBECK, J. R. (1960). Regulation of feeding and drinking. In: *Handbook of Physiology.* J. Field (ed.). Washington, D.C.; American Physiol. Soc. Ch. 47. Section I. Neurophysiology, Vol. II.

BROWN-GRANT, K. (1966). The action of hormones on the hypothalamus. *Br. med. Bull.*, **22**, 273–277.

BRUCE, H. M. (1966). Smell as an exteroceptive factor. In: *Environmental Influences on Reproductive Processes.* The Seventh Bienn. Symp. Anim. Reproduction, pp. 83–87. *J. Anim. Sci. Suppl.*, Vol. **25**.

BULLOCK, T. H. (1966). Strategies for blind physiologists with elephantine problems. In: *Nervous and hormonal mechanisms of integration.* Symposium Soc. exp. Biol. No. XX. G. M. Hughes (ed.). New York, N.Y.: Academic Press, pp. 1–10.

CABRAL, R. J. & JOHNSON, J. I. (1967). Somatic sensory projections in thalamus of sheep. *Am. Zoologist*, **7**, 792 (Abstract).

CHEW, R. M. (1965). Water metabolism of mammals. In: *Physiological Mammalogy,* Vol. 2: *Mammalian reactions to stressful environments.* W. V. Mayer & R. G. Van Gelder (eds.). New York: Academic Press, pp. 43–178.

CLEGG, M. T., SANTOLUCITO, J. A., SMITH, J. D. & GANONG, W. F. (1958). The effect of hypothalamic lesions on sexual behavior and estrous cycles in the ewe. *Endocrinology*, **62**, 790–797.

CLEMENTE, C. D., SUTIN, J. & SILVERSTONE, J. T. (1957). Changes in electrical activity of the medulla on the intravenous injection of hypertonic solutions. *Am. J. Physiol.*, **188**, 193–198.

CORBIT, J. D. (1965). Hyperphagic reactivity to adulteration of drinking water with quinine HCl. *J. comp. physiol. Psychol.*, **60**, 123–124.

COWIE, A. T. & FOLLEY, S. J. (1961). The mammary gland and lactation. In: *Sex and Internal Secretions.* W. C. Young (ed.). Baltimore, Md.: Williams & Wilkins, pp. 590–642.

CROSS, B. A. (1951). Suckling antidiuresis in rabbits. *J. Physiol., Lond.* **114**, 447 (Abstract).

CROSS, B. A. & GREEN, J. D. (1959). Activity of single neurons in the hypothalamus: effect of osmotic and other stimuli. *J. Physiol., Lond.*, **148**, 554–569.

CROSS, B. A. & SILVER, I. A. (1966). Electrophysiological studies on hypothalamus. *Br. med. Bull.*, **22**, 254–260.

DAVIDSON, J. M. & SMITH, E. R. (1967). Testosterone feedback in the control of somatic and behavioral aspects of male reproduction. In: *Proc. Second Internat. Congr. Hormonal Steroids*. Internat. Congr. Series No. 132. L. Martini, F. Fraschini & M. Motta (eds.). Amsterdam: Excerpta Medica Foundation, pp. 805–813.

DAVIS, C. D. (1939). Effect of ablations in neocortex on mating, maternal behavior, and the production of pseudopregnancy in the female rat and on copulatory activity in the male. *Am. J. Physiol.*, **127**, 374–380.

DAVIS, R. C., GARAFOLO, L. & KVEIM, K. (1959). Conditions associated with gastro-intestinal activity. *J. comp. physiol. Psychol.*, **52**, 466–475.

DEBACKERE, M. & PEETERS, G. (1960). The influence of vaginal distension on milk ejection and diuresis in the lactating cow. *Archs int. Pharmacodyn. Thér.*, **123**, 462–471.

DE LORENZO, A. J. (1960). Electron microscopy of the olfactory and gustatory pathways. *Ann. Otol-Rhinol. Lar.* **69**, 410–420.

DEMBER, W. N. & EARL, R. W. (1957). Analysis of exploratory, manipulatory, and curiosity behaviors. *Psychol. Rev.*, **64**, 91–96.

DE ROBERTIS, E. (1964). *Histophysiology of Synapses and Neurosecretion*. New York, N.Y.: Macmillan Co.

DOWLING, J. E. & BOYCOTT, B. K. (1966). Organization of the primate retina; electron microscopy. *Proc. R. Soc. Ser. B.*, **166**, 80–111.

EAKIN, R. M. (1966). Evolution of photoreceptors. In: *Cold Spring Harbor Symposia on Quantitative Biology*. L. Frisch (ed.). Cold Spring Harbor, L.I., N.Y, pp. 363–370, Vol. XXX. *Sensory Receptors*.

ECCLES, J. C. (1957). *The Physiology of Nerve Cells*. London: Oxford University Press.

ECCLES, J. C. (1964). *The Physiology of Synapses*. Berlin: Springer-Verlag.

EGGER, M. D. & FLYNN, J. P. (1967). Further studies on the effects of amygdaloid stimulation and ablation on hypothalamically elicited attack behavior in rats. In: *Progress in Brain Research*. Vol. 27. *Structure and Function of the Limbic System*. W. R. Adey & T. Tokizane (eds.). Amsterdam: Elsevier, pp. 165–182.

EMLEN, J. T. (1952). Flocking behavior in birds. *Auk*, **69**, 160–170.

EMLEN, S. T. (1967). Migratory orientation in the indigo bunting, *Passerina cyanea*. Part 1: Evidence for use of celestial cues. *Auk*, **84**, 309–342.

EMMERS, R. (1965). Organization of the first and second somesthetic regions (SI and SII) in the rat thalamus. *J. comp. Neurol.*, **124**, 215–229.

EVANS, C. S. & GOY, R. W. (1968). Social behaviour and reproductive cycles in captive Ring-tailed lemurs (*Lemur catta*), *J. Zool., Lond.*, **156**, 181–197.

EVARTS, E. V. (1966). Pyramidal tract activity associated with a conditioned hand movement in the monkey. *J. Neurophysiol.*, **29**, 1011–1027.

FANTZ, R. L. (1957). Form preferences in newly hatched chicks. *J. comp. physiol. Psychol.*, **40**, 422–430.

FARNER, D. S. & FOLLET, B. K. (1966). Light and other factors affecting avian reproduction. In: *Environmental Influences on Reproductive Processes*. Seventh Bienn. Symp. Anim. Reproduction; pp. 90–115. *J. Anim. Sci. Suppl.*, Vol. **25**.

FAROOQ, A., DENENBERG, V. H., SAWIN, P. B. & ZARROW, M. X. (1963). Maternal behavior in the rabbit: endocrine factors involved in hair-loosening. *Am. J. Physiol.*, **204**, 271–274.

FEDER, H. H., RESKO, J. A. & GOY, R. W. (1968a). Progesterone concentrations in the arterial plasma of guinea-pigs during the oestrous cycle. *J. Endocr.*, **40**, 505–513.

FEDER, H. H., RESKO, J. A. & GOY, R. W. (1968b). Progesterone: levels in the arterial plasma of pre-ovulatory and ovariectomized rats. *J. Endocr.*, **41**, 563–569.

FERNANDES DE MOLINA, A. & HUNSPERGER, R. W. (1959). Central representation of affective reactions in forebrain and brain stem: electrical stimulation of amygdala, stria terminalis and adjacent structures. *J. Physiol., Lond.*, **145**, 251–265.

FISHER, C., INGRAM, W. B. & RANSON, S. W. (1938). *Diabetes insipidus and the neurohormonal control of water balance*. Ann. Arbor, Mich.: Edward Bros.

FITZSIMONS, J. T. (1966). The hypothalamus and drinking. *Br. med. Bull.*, **22**, 232–237.

FRENCH, G. M. (1959). Locomotor effect of regional ablations of frontal cortex in rhesus monkeys. *J. comp. physiol. Psychol.*, **52**, 18–24.

FRENCH, J. D. (1960). The reticular formation. In: *Handbook of Physiology*. J. Field (ed.). Washington, D.C.: Amer. Physiol. Soc. Ch. 52. Section I. Neurophysiology, Vol. II.

FREUD, J. & UYLDERT, I. E. (1948). Micturition and copulation behaviour patterns in dogs. *Acta. brev. neerl. Physiol.*, **16**, 49–53.

FULLER, J. L., ROSVOLD, H. E. & PRIBRAM, K. H. (1957). The effect on affective and cognitive behavior in the dog of lesions of the pyriform-amygdala-hippocampal complex. *J. comp. physiol. Psychol.*, **50**, 89–96.

GELFAN, S. (1964). Neuronal interdependence. In: *Progress in Brain Research*. Vol. XI. *Organization of the Spinal Cord*. Amsterdam: Elsevier, pp. 238–260.

GERALL, A. A. (1958). An attempt to induce precocious sexual behavior in male guinea pigs by injections of testosterone propionate. *Endocrinology*, **63**, 280–284.

GILMAN, A. & GOODMAN, L. (1937). The secretory response of the posterior pituitary to the need for water conservation. *J. Physiol.*, **90**, 113–124.

GLANZER, M. (1958). Curiosity, exploratory drive, and stimulus satiation. *Psychol. Bull.*, **55**, 302–315.

GLOOR, P. (1960). Amygdala. In: *Handbook of Physiology*. J. Field (ed.). Washington, D.C.: Amer. Physiol. Soc. Ch. 58. Section I. Neurophysiology, Vol. II.

GOLD, R. M. (1967). Aphagia and adipsia following unilateral and bilaterally asymmetrical lesions in rats. *Physiol. Behav.*, **2**, 211–220.

GOY, R. W. & PHOENIX, C. H. (1963). Hypothalamic regulation of female sexual behaviour; establishment of behavioural oestrus in spayed guinea-pigs following hypothalamic lesions. *J. Reprod. Fert.*, **5**, 23–40.

GOY, R. W., PHOENIX, C. H. & YOUNG, W. C. (1966). Inhibitory actions of the corpus luteum on the hormonal induction of estrous behaviour in the guinea pig. *Gen. Comp. Endocr.*, **6**, 267–275.

GOY, R. W. & YOUNG, W. C. (1957). Strain differences in the responses of female guinea pigs to alpha-estradiol benzoate and progesterone. *Behavior*, **10**, 340–354.

GRAFF, H. & STELLAR, E. (1962). Hyperphagia, obesity and finickiness. *J. comp. physiol. Psychol.*, **55**, 418–424.

GRAY, J. (1950). The role of peripheral sense organs during locomotion in the vertebrates. In: *Symposia of the Society for Experimental Biology*. No. IV. *Physiological Mechanisms in Animal Behaviour*. Cambridge: Cambridge University Press, pp. 112–126.

GREEN, J. D. & ARDUINI, A. A. (1954). Hippocampal electrical activity in arousal. *J. Neurophysiol.*, **17**, 533–537.

GREEN, J. D., CLEMENTE, C. D. & DE GROOT, J. (1957). Rhinencephalic lesions and behavior in cats. *J. comp. Neurol.*, **108**, 505–536.

GREENLEAF, J. E. (1966). *Involuntary hypohydration in man and animals: a review*. Washington, D.C.: Nat. Aeronaut. Space Admin.

GRUNT, J. & YOUNG, W. C. (1952). Differential reactivity of individuals and the response of the male guinea pig to testosterone propionate. *Endocrinology*, **51**, 237–248.

HAGEMAN, W. D., LANCE, E. M. & UNGEWITTER, L. H. (1959). Increased responsiveness to stimuli following lesions of the forebrain. *Anat. Rec.*, **133**, 387–388.

HARRIS, G. W. (1955). *Neural Control of the Pituitary Gland*. London: Arnold.

HARRIS, G. W. (1959). The nervous system-follicular ripening, ovulation and estrous behavior. In: *Recent Progress in the Endocrinology of Reproduction*. C. W. Lloyd (ed.). New York, N.Y.: Academic Press, pp. 21–44.

HARRIS, G. W. (1960). Central control of pituitary secretion. In: *Handbook of Physiology*. J. Field (ed.). Washington, D.C.: Amer. Physiol. Soc. Section I. Neurophysiology, Vol. II.

HARRIS, G. W., MICHAEL, R. P. & SCOTT, P. P. (1958). Neurological site of action of stilbestrol in eliciting sexual behavior. In: *Ciba Foundation Symposium on the Neurological Basis of Behavior*. Boston: Little, Brown, pp. 236–251.

HARRIS, G. W. & PICKLES, V. R. (1953). Reflex stimulation of the neurohypophysis (posterior pituitary gland) and the nature of the posterior pituitary hormone(s). *Nature, Lond.*, **172**, 1049.

HARRIS, G. W., REED, M. & FAWCETT, C. P. (1966). Hypothalamic releasing factors. *Br. med. Bull.*, **22**, 266–272.

HART, B. L. (1967a). Sexual reflexes and mating behavior in the male dog. *J. comp. physiol. Psychol.*, **64**, 388–399.

HART, B. L. (1967b). Testosterone regulation of sexual reflexes in spinal male rats. *Science*, **155**, 1283–1284.

HARVEY, J. A. & HUNT, H. F. (1965). Effect of septal lesions on thirst in the rat as indicated by water consumption and operant responding for water reward. *J. comp. physiol. Psychol.*, **59**, 49–56.

HATTON, G. I. & RUBEL, E. W. (1967). Somatic sensory projections in cerebral cortex of sheep. *Anat. Rec.*, **157**, 256 (Abstract).

HESS, W. R. (1957). *The Functional Organization of the Diencephalon.* New York, N.Y.: Grune & Stratton.

HESS, W. R. (1967). Causality, consciousness, and cerebral organization. *Science*, **158**, 1279–1283.

HILL, W. C. O. (1956). Behaviour and adaptations of the primates. *Proc. R. Soc. Edinb.*, **66B**, 94–110.

HILLIARD, J. & SAWYER, C. H. (1964). Synthesis and release of progestin by rabbit ovary *in vivo*. In: *Proc. First Intl. Congr. Horm. Ster.*, Vol. 1. New York: Academic Press, pp. 263–272.

HOEBEL, B. G. & TIETELBAUM, P. (1966). Weight regulation in normal and hypothalamic hyperphagic rats. *J. comp. physiol. Psychol.*, **61**, 189–193.

HUBEL, D. H., HENSON, C. O., RUPERT, A. & GALAMBOS, R. (1959). "Attention" units in the auditory cortex. *Science*, **129**, 1279–1280.

HUBEL, D. H. & WIESEL, T. N. (1962). Receptive fields, binocular interaction, and functional architecture in the cat's visual cortex. *J. Physiol., Lond.*, **160**, 106–154.

HUBEL, D. H. & WIESEL, T. N. (1963). Receptive fields of cells in striate cortex of very young, visually inexperienced kittens. *J. Neurophysiol.*, **26**, 994–1002.

HUBEL, D. H. & WIESEL, T. N. (1965). Receptive fields and functional architecture in two nonstriate visual areas (18 and 19) of the cat. *J. Neurophysiol.*, **28**, 229–289.

IGGO, A. (1955). Tension receptors in the stomach and urinary bladder. *J. Physiol., Lond.*, **128**, 593–607.

ILSE, D. R. (1955). Olfactory marking of territory in two young male loris, *Loris tardigradus lydekkerianus*, kept in captivity in Poona. *Br. J. Anim. Behav.*, **3**, 118–120.

JAMES, W. T. (1957). The effect of satiation of the suckling response in puppies. *J. comp. physiol. Psychol.*, **51**, 375–378.

KATO, J. & VILLEE, C. A. (1967). Preferential uptake of estradiol by the anterior hypothalamus of the rat. *Endocrinology*, **81**, 567–575.

KATZ, B. (1950). Depolarization of sensory terminals and the initiation of impulses in the muscle spindle. *J. Physiol., London.*, **111**, 261–262.

KATZ, B. (1966). *Nerve, Muscle, and Synapse.* New York, N.Y.: McGraw-Hill.

KAWAKAMI, M. & SAWYER, C. H. (1959a). Neuroendocrine correlates of changes in brain activity thresholds by sex steroids and pituitary hormones. *Endocrinology*, **65**, 652–668.

KAWAKAMI, M. & SAWYER, C. H. (1959b). Induction of behavioral and electroencephalographic changes in the rabbit by hormone administration or brain stimulation. *Endocrinology*, **65**, 631–643.

KENNEDY, G. C. (1966). Food intake, energy balance and growth. *Br. med. Bull.*, **22**, 216–220.

KIMBLE, D. P., ROGERS, L. & HENDRICKSON, C. W. (1967). Hippocampal lesions disrupt maternal, not sexual, behavior in the albino rat. *J. comp. physiol. Psychol.*, **63**, 401–407.

KING, J. A. & CONNON, H. (1955). Effects of social relationships upon mortality in C57BL/10 mice. *Physiol. Zool.* **28**, 233–239.

KING, F. A. & MEYER, P. M. (1958). Effects of amygdaloid lesions upon septal hyperemotionality in the rat. *Science*, **128**, 655–656.

KISLAK, J. W. & BEACH, F. A. (1955). Inhibition of aggressiveness by ovarian hormones. *Endocrinology*, **56**, 684–692.

KLEITMAN, N. (1967). Phylogenetic, ontogenetic and environmental determinants in the evolution of sleep-wakefulness cycles. In: *Sleep and Altered States of Consciousness.* Proc. Assn. Res. Nerv. Ment. Dis. Vol. 45. S. S. Kety, E. V. Evarts, & H. L. Williams (eds.). Baltimore, Md.: Williams & Wilkins, pp. 30–38.

KLÜVER, H. (1951). Functional differences between the occipital and temporal lobes with special reference to the interrelations of behavior and extracerebral mechanisms. In: *Cerebral Mechanisms in Behavior.* The Hixon Symposium. L. A. Jeffress (ed.). New York, N.Y.: Wiley, pp. 147-199.

KLÜVER, H. & BUCY, P. C. (1939). Preliminary analysis of functions of the temporal lobes in monkeys. *Archs Neurol. Psychiat., Chicago,* **42**, 979-1000.

KOELLA, W. P. (1967). *Sleep, its Nature and Physiological Organization.* Springfield, Ill.: Charles C. Thomas.

KRECH, D. & CRUTCHFIELD, R. S. (1958.) *Elements of Psychology.* New York, N.Y.: Alfred A. Knopf.

KUFFLER, S. W. (1953). Discharge patterns and functional organisation of mammalian retina. *J. Neurophysiol.,* **16**, 37-68.

KURU, M. (1965). Nervous control of micturition. *Physiol. Rev.,* **45**, 425-494.

LAW, T. & MEAGHER, W. (1958). Hypothalamic lesions and sexual behavior in the female rat. *Science,* **128**, 1626-1627.

LEHRMAN, D. S. (1961). Gonadal hormones and parental behavior in birds and infrahuman mammals. In: *Sex and Internal Secretions.* W. C. Young (ed.). Baltimore, Md.: Williams & Wilkins, pp. 1268-1382.

LEHRMAN, D. S. (1964). Control of behavior cycles in reproduction. In: *Social Behavior and Organization among Vertebrates.* W. Etkin (ed.). Chicago, Ill.: University of Chicago Press, pp. 143-166.

LINDSLEY, D. B. (1960). Attention, consciousness, sleep and wakefulness. In: *Handbook of Physiology.* J. Field (ed.). Washington, D.C.: American Physiol. Soc. Ch. 64. Section I. Neurophysiology, Vol. III.

LIVINGSTON, R. B. (1960). Central control of receptors and sensory transmission systems. In: *Handbook of Physiology.* J. Field (ed.). Washington, D.C.: Amer. Physiol. Soc. Ch. 31. Section I. Neurophysiology, Vol. I.

LOIZZO, A. & LONGO, V. G. (1968). A pharmacological approach to paradoxical sleep. *Physiol. Behav.,* **3**, 91-97.

LYNN, R. (1966). *Attention, arousal, and the orientation reaction.* Oxford: Pergamon Press.

MAGOUN, H. W. (1958). *The Waking Brain.* Springfield, Ill.: Thomas.

MALLER, O., KARE, M. R., WELT, M. & BEHRMAN, H. (1967). Movement of glucose and sodium chloride from oropharyngeal cavity to the brain. *Nature, Lond.,* **213**, 713-714.

MARTINS, T. & VALLE, J. R. (1948). Hormonal regulation of the micturition behavior of the dog. *J. comp. physiol. Psychol.,* **41**, 301-311.

MATSUDA, Y. (1966). Effects of some centrally acting drugs on food intake of normal and hypothalamus-lesioned rats. *Jap. J. Pharmac.,* **16**, 276-286.

MATURANA, H. R. & FRENK, S. (1963). Directional movement and horizontal edge detectors in pigeon retina. *Science,* **142**, 977-979.

MAYER, J. & THOMAS, E. W. (1967). Regulation of food intake and obesity. *Science,* **156**, 328-337.

MICHAEL, R. P., SAAYMAN, G. S. & ZUMPE, D. (1967). Inhibition of sexual receptivity by progesterone in rhesus monkeys. *J. Endocr.,* **39**, 309-310.

MICHAEL, R. P. & SCOTT, P. P. (1957). Quantitative studies on mating behaviour of spayed female cats stimulated by treatment with oestrogens. *J. Physiol.,* **138**, 46-47 P.

MIRSKY, A. (1955). The influence of sex hormones on social behavior in monkeys. *J. comp. physiol. Psychol.,* **48**, 327-335.

MOLTZ, H., GELLER, D. & LEVIN, R. (1967). Maternal behavior in the totally mammectomized rat. *J. comp. physiol. Psychol.,* **64**, 225-229.

MONNIER, M. & HÖSLI, L. (1965). Humoral regulation of sleep and wakefulness by hypnogenic and activating dialysable factors. In: *Progress in Brain Research.* Vol. 18. *Sleep Mechanisms.* K. Akert, C. Bally & J. P. Schadé (eds.). Amsterdam: Elsevier, pp. 118-123.

MONTEMURRO, D. G. & STEVENSON, J. A. F. (1957). Adipsia produced by hypothalamic lesions in the rat. *Can. J. Biochem. Physiol.,* **35**, 31-37.

MOORE, N. W. & ROBINSON, T. J. (1957). The behavioural and vaginal response of the spayed ewe to oestrogen injected at various times relative to the injection of progesterone. *J. Endocr.,* **15**, 360-365.

MORGANE, P. J. (1961). Alterations in feeding and drinking behavior of rats with lesions in globi pallidi. *Am. J. Physiol.*, **201**, 420–428.

MOULTON, D. G. & BEIDLER, L. M. (1967). Structure and function in the peripheral olfactory system. *Physiol. Rev.*, **47**, 1–52.

MOUNTCASTLE, V. B. & HENNEMAN, E. (1949). Pattern of tactile representation in the thalamus of cat. *J. Neurophysiol.*, **12**, 85–100.

MOUNTCASTLE, V. B. & HENNEMAN, E. (1952). The representation of tactile sensibility in the thalamus of the monkey. *J. comp. Neurol.*, **97**, 409–440.

MOUNTCASTLE, V. B. & POWELL, T. P. S. (1959). Neural mechanisms subserving cutaneous sensibility, with special reference to the role of afferent inhibition in sensory perception and discrimination. *Bull. Johns Hopkins Hosp.*, **105**, 201–232.

MURRAY, E. J. (1965). *Sleep, Dreams and Arousal*. New York, N.Y.: Appleton-Century-Crofts.

NACHMAN, M. (1959). The inheritance of saccharine preference. *J. comp. physiol. Psychol.*, **52**, 451–457.

OKANO, M., WEBER, A. F. & FROMMES, S. P. (1967). Electron microscopic studies of the distal border of the canine olfactory epithelium. *J. Ultrastruct. Res.*, **17**, 487–502.

O'KELLY, L. I. (1963). The psychophysiology of motivation. *A. Rev. Psychol.*, **14**, 57–92.

OLSZEWSKI, J. (1954). The cytoarchitecture of the human reticular formation. In: *Brain Mechanisms and Consciousness*. J. F. Delafresnaye (ed.). Oxford: Blackwell Scientific, pp. 54–80.

OOMURA, Y., OOYAMA, H., YAMAMOTO, T. & NAKA, F. (1967). Reciprocal relationship of the lateral and ventromedial hypothalamus in the regulation of food intake. *Physiol. Behav.*, **2**, 97–116.

PERETZ, E. (1968). Estrogen dose and the duration of the mating period in cats. *Physiol. Behav.*, **3**, 41–43.

PHOENIX, C. H. (1961). Hypothalamus regulation of sexual behavior in male guinea pigs. *J. comp. physiol. Psychol.*, **54**, 72–77.

PHOENIX, C. H., GOY, R. W. & YOUNG, W. C. (1967). Sexual behavior: general aspects. In: *Neuroendocrinology*, Vol. 2. L. Martini & W. F. Ganong (eds.). New York, N.Y.: Academic Press, Ch. 21.

PINTO HAMUY, T., BROMILEY, R. B. & WOOLSEY, C. N. (1956). Somatic afferent areas I and II of dog's cerebral cortex. *J. Neurophysiol.*, **19**, 485–499.

PINTO HAMUY, T., SANTIBANEZ, G., GONZALES, C. & VICENCIO, E. (1957). Changes in behavior and visual discrimination performance after selective ablations of the temporal cortex. *J. comp. physiol. Psychol.*, **50**, 379–385.

PRIBRAM, K. H. (1960). A review of theory in physiological psychology. *A. Rev. Psychol.*, **11**, 1–40.

PUBOLS, B. H. & PUBOLS, L. M. (1966). Somatic sensory representation in the thalamic ventrobasal complex of the Virginia opossum. *J. comp. Neurol.*, **127**, 19–34.

RICHTER, C. P. (1967). Sleep and activity: their relation to the 24-hour clock. In: *Sleep and Altered States of Consciousness*. Proc. Assn. Res. Nerv. Ment. Dis., Vol. 45. S. S. Kety, E. V. Evarts & H. L. Williams (eds.). Baltimore, Md.: Williams & Wilkins, pp. 8–29.

RISS, W., VALENSTEIN, E. S., SINKS, J. & YOUNG, W. C. (1955). Development of sexual behavior in male guinea pigs from genetically different stocks under controlled conditions of androgen treatment and caging. *Endocrinology*, **57**, 139–146.

ROBINSON, T. J. (1955a). Quantitative studies on the hormonal induction of oestrus in spayed ewes. *J. Endocr.*, **12**, 163–173.

ROBINSON, T. J. (1955b). Endocrine relationships in the induction of oestrus and ovulation in the anoestrous ewe. *J. agric. Sci., Camb.*, **46**, 37–43.

ROBINSON, T. J., MOORE, N. W. & BINET, F. E. (1956). The effect of the duration of progesterone pretreatment on the response of the spayed ewe to oestrogen. *J. Endocr.*, **14**, 1–7.

ROBSON, J. M. (1938). Induction of oestrous changes in the monkey and bitch by triphenyl ethylene. *Proc. Soc. exp. Biol. Med.*, **38**, 153–157.

RODIECK, R. W. & STONE, J. (1965). Response of cat retinal ganglion cells to moving visual patterns. *J. Neurophysiol.*, **28**, 819–832.

ROOT, W. S. (1956). The physiology of micturition: a specific autonomic function. In: *Medical Physiology*. P. Bard (ed.). St. Louis: Mosby, pp. 1121–1127.

ROSE, J. E. & MOUNTCASTLE, V. B. (1952). The thalamic tactile region in rabbit and cat. *J. comp. Neurol.*, **97**, 441–490.

ROSE, J. E. & MOUNTCASTLE, V. B. (1959). Touch and kinesthesis. In: *Handbook of Physiology*. J. Field (ed.). Washington, D.C.: Amer. Physiol. Soc. Ch. 17. Section I. Neurophysiology, Vol. I.

ROSENBLATT, J. S. & ARONSON, L. R. (1958a). The decline of sexual behaviour in male cats after castration with special reference to the role of prior sexual experience. *Behaviour*, **12**, 285–338.

ROSENBLATT, J. S. & ARONSON, L. R. (1958b). The influence of experience on the behavioural effects of androgen in prepuberally castrated male cats. *Anim. Behav.*, **6**, 171–182.

ROSENBLATT, J. S. & LEHRMAN, D. S. (1963). Maternal behavior of the laboratory rat. In: *Maternal Behavior in Mammals*. H. L. Rheingold (ed.). New York, N.Y.: Wiley, pp. 8–57.

SAMEL, M. & CAPUTA, A. (1965). The role of the mother in I^{131} metabolism of suckling and weanling rats. *Can. J. Physiol. Pharmacol.*, **43**, 431–436.

SAUER, F. (1957). Die Sternorienterung nächtlich Ziehender Grasmücken. (*Sylvia atricapilla, borin und curruca* L.). *Z. Tierpsychol.*, **14**, 29–70.

SAWYER, C. H. (1959). Nervous control of ovulation. In: *Recent Progress in the Endocrinology of Reproduction*. C. W. Lloyd (ed.). New York: N.Y.: Academic Press, p. 1.

SAWYER, C. H. (1960). Reproductive Behavior. In: *Handbook of Physiology*. J. Field (ed.). Washington, D.C.: Amer. Physiol. Soc. Ch. 49. Section I. Neurophysiology.

SAWYER, C. H. (1967). Effects of hormonal steroids on certain mechanisms in the adult brain. In: *Proc. Second Internat. Congr. Hormonal Steroids*. Internat. Congr. Ser. No. 132. L. Martini, F. Fraschini & M. Motta (eds.). Amsterdam: Excerpta Medica Foundation, pp. 123–135.

SAWYER, C. H. & EVERETT, J. W. (1959). Stimulatory and inhibitory effects of progesterone on the release of pituitary ovulating hormone in the rabbit. *Endocrinology*, **65**, 644–651.

SAWYER, C. H. & GERNANDT, B. E. (1956). Effects of intracarotid and intraventricular injections of hypertonic solutions on electrical activity of the rabbit brain. *Am. J. Physiol.*, **185**, 209–216.

SAWYER, C. H. & KAWAKAMI, M. (1959). Characteristics of behavioral and electroencephalographic after-reactions to copulation and vaginal stimulation in the female rabbit. *Endocrinology*, **65**, 622–630.

SAWYER, C. H. & ROBINSON, B. L. (1956). Separate hypothalamic areas controlling pituitary gonadotropic functions and mating behavior in female cats and rabbits. *J. clin. Endocr. Metab.*, **16**, 914–915.

SCHNEIRLA, T. C., ROSENBLATT, J. S. & TOBACH, E. (1963). Maternal behavior in the cat. In: *Maternal Behavior in Mammals*. H. L. Rheingold (ed.). New York, N.Y.: Wiley, pp. 122–168.

SCHREINER, L. & KLING, A. (1953). Behavioral changes following rhinencephalic injury in cat. *J. Neurophysiol.*, **16**, 643–659.

SCHREINER, L. & KLING, A. (1954). Effects of castration on hypersexual behavior induced by rhinencephalic injury in cat. *Arch. Neurol. Psychiat.*, **72**, 180–186.

SCHREINER, L. & KLING, A. (1956). Rhinencephalon and behaviour. *Am. J. Physiol.*, **184**, 486–490.

SCHULTZE-WESTRUM, T. (1965). Innerartliche Verständigung durch düfte beim Gleitbeutler *Petaurus breviceps papuanus* Thomas (Marsupialia, Phalangeridae). *Z. vergl. Physiol.*, **50**, 151–220.

SCOTT, J. P. (1960). Comparative social psychology. In: *Principles of Comparative Psychology*. R. H. Waters, D. A. Rethlingshafer & W. E. Caldwell (eds.). New York, N.Y.: McGraw-Hill, Ch. 9.

SHARPLESS, S. & JASPER, H. (1956). Habituation of the arousal reaction. *Brain*, **79**, 655–680.

SHOEMAKER, H. H. (1939). Effect of testosterone propionate on behaviour of the female canary. *Proc. Soc. exp. Biol. Med.*, **41**, 299–302.

SIEGEL, P. S. (1961). Food intake in the rat in relation to the light-dark cycle. *J. comp. physiol. Psychol.*, **54**, 294–301.

SJÖSTRAND, F. S. (1953a). The ultrastructure of the outer segments of rods and cone of the eye as revealed by electron microscopy. *J. cell. comp. Physiol.*, **42**, 15–44.

SJÖSTRAND, F. S. (1953b). The ultrastructure of the inner segments of the retinal rods of the guinea pig eye as revealed by electron microscopy. *J. cell. comp. Physiol.*, **42**, 45–70.

SLOTNICK, B. M. (1967). Disturbances of maternal behavior in rats following lesions of the cingulate cortex. *Behaviour*, **29**, 204–236.

SOKOLOV, E. N. (1963). Higher nervous functions: the orienting reflex. *A. Rev. Physiol.*, **25**, 545–580.

SPIEGEL, E. A., MILLER, H. R. & OPPENHEIMER, M. J. (1940). Forebrain and rage reactions. *J. Neurophysiol.*, **3**, 538–548.

SPRAGUE, J. M. (1967). The effects of chronic brainstem lesions on wakefulness, sleep and behavior. In: *Sleep and Altered States of Consciousness*. Proc. Assn. Res. Nerv. Ment. Dis., Vol. 45. S. S. Kety, E. V. Evarts & H. L. Williams (eds.). Baltimore, Md.: Williams & Wilkins, pp. 148–194.

STAMM, J. S. (1954). Control of hoarding activity in rats by median cerebral cortex. *J. comp. physiol. Psychol.*, **47**, 21–27.

STAMM, J. S. (1955). Function of median cerebral cortex in maternal behavior in rats. *J. comp. physiol. Psychol.*, **48**, 347–356.

STONE, C. P. (1938). Effects of cortical destruction on reproductive behavior and maze learning in albino rats. *J. comp. Psychol.*, **26**, 217–236.

SUGA, N. (1965). Analysis of frequency-modulated sounds by auditory neurons of echo-locating bats. *J. Physiol., London*, **179**, 26–53.

SUNDSTEN, J. W. & SAWYER, C. H. (1959). Electroencephalographic evidence of osmo-receptive elements in olfactory bulb of dog brain. *Proc. Soc. Exp. Biol. Med.*, **101**, 524–527.

TEITELBAUM, P. & CAMPBELL, B. A. (1958). Ingestion patterns in normal and hyperphagic rats. *J. comp. physiol. Psychol.*, **51**, 135–141.

TEITELBAUM, P. & EPSTEIN, A. N. (1962). The lateral hypothalamic syndrome: recovery of feeding and drinking after lateral hypothalamic lesions. *Psychol. Rev.*, **69**, 74–90.

THIBAULT, C., COUROT, M., MARTINET, L., MAULEON, L., DU MESNIL DU BUISSON, F., ORTAVANT, R., PELLETIER, J. & SIGNORET, J. P. (1966). Regulation of breeding season and estrous cycles by light and external stimuli in some mammals. In: *Environmental Influences on Reproductive Processes*. Seventh Bienn. Symp. Anim. Reproduction, pp. 119–139. *J. Anim. Sci. Suppl.*, Vol. **25**.

THOMPSON, R. F. (1967). *Foundations of Physiological Psychology*. New York, N.Y.: Harper & Row.

THOMPSON, R. F. & SPENCER, W. A. (1966). Habituation: a model phenomenon for the study of neuronal substrates of behavior. *Psychol. Rev.*, **173**, 16–43.

THOMPSON, R. F. & WELKER, W. I. (1963). Role of auditory cortex in reflex head orientation by cats to auditory stimuli. *J. comp. physiol. Psychol.*, **56**, 996–1002.

TUNTURI, A. R. (1944). Audio frequency localization in the acoustic cortex of the dog. *Am. J. Physiol.*, **141**, 397–403.

TUNTURI, A. R. (1950). Physiological determination of the boundary of the acoustic area in the cerebral cortex of the dog. *Am. J. Physiol.*, **160**, 395–401.

URSIN, H. & KAADA, B. R. (1960). Functional localization within the amygdaloid complex in the cat. *Electroenceph. clin. Neurophysiol.*, **12**, 1–120.

VERNEY, E. B. (1947). Antidiuretic hormone and the factors which determine its release. *Proc. R. Soc. Ser. B.*, **135**, 25–106.

WALLS, G. L. (1942). *The Vertebrate Eye and Its Adaptive Radiation*. Bloomfield Hills, Mich.: Cranbook Inst. Sci. Reprinted 1963, New York, N.Y.: Hafner.

WELKER, W. I. (1959a). Factors influencing aggregation of neonatal puppies. *J. comp. physiol. Psychol.*, **52**, 376–380.

WELKER, W. I. (1959b). Genesis of exploratory and play behavior in infant raccoons. *Psychol. Rep.*, **5**, 764.

WELKER, W. I. (1959c). Escape, exploratory, and food-seeking responses of rats in a novel situation. *J. comp. physiol. Psychol.*, **52**, 106–111.

WELKER, W. I. (1961). An analysis of play and exploratory behavior in animals. In: *Functions of Varied Experience*. D. W. Fiske & S. R. Maddi (eds.). Homewood, Ill.: Dorsey, Ch. 7.

WELKER, W. I. & CAMPOS, G. B. (1963). Physiological significance of sulci in somatic sensory cerebral cortex in mammals of the family Procyonidae. *J. comp. Neurol.*, **120**, 19–36.

WELKER, W. I. & JOHNSON, J. I. (1965). Correlation between nuclear morphology and somatotopic organization in ventrobasal complex of the raccoon's thalamus. *J. Anat.*, **99**, 761–790.

WELKER, W. I. & SEIDENSTEIN, S. (1959). Somatic sensory representation in the cerebral cortex of the raccoon (*Procyon lotor*). *J. comp. Neurol.*, **111**, 469–502.

WERSÄLL, J., FLOCK, Å. & LUNDQUIST, P. G. (1966). Structural basis for directional sensitivity in cochlear and vestibular sensory receptors. In: *Cold Spring Harbor Symposia on Quantitative Biology*. L. Frisch (ed.). Cold Spring Harbor, L.I., N.Y.: Cold Spring Harbor Lab. Quant. Biol. Vol. xxx. *Sensory Receptors*, pp. 115–132.

WHEATLEY, M. D. (1944). The hypothalamus and affective behavior in cats. A study of the effects of experimental lesions with anatomic correlations. *Archs Neurol. Psychiat.*, *Chicago*, **52**, 298–316.

WILSON, J. G. & YOUNG, W. C. (1941). Sensitivity to estrogen studied by means of experimentally induced mating responses in the female guinea pig and rat. *Endocrinology*, **29**, 779–783.

WILSONCROFT, W. E. (1963). Effects of median cortex lesions on the maternal behavior of the rat. *Psychol. Rep.*, **13**, 835–838.

WOOLSEY, C. N. (1952). Patterns of localization in sensory and motor areas of the cerebral cortex. In *The Biology of Mental Health and Disease*. New York, N.Y.: Paul B. Hoeber, Ch. 14.

WOOLSEY, C. N. (1958). Organization of somatic sensory and motor areas of the cerebral cortex. In: *Biological and Biochemical Bases of Behavior*. H. F. Harlow & C. N. Woolsey (eds.). Madison, Wis.: Univ. Wis. Press, pp. 63–81.

WOOLSEY, C. N. & BROOKS, C. McC. (1937). Factors influencing micturition volume in the unanesthetized cat. *Am. J. Physiol.*, **119**, 423 (Abstract).

WOOLSEY, C. N. & FAIRMAN, D. (1946). Contralateral, ipsilateral and bilateral representation of cutaneous receptors in somatic areas I and II of the cerebral cortex of pig, sheep, and other mammals. *Surgery*, **19**, 684–702.

WURTMAN, R. J. (1967). Effects of light and visual stimuli on endocrine function. In: *Neuroendocrinology*. Vol. 2. L. Martini & W. F. Ganong (eds.). New York, N.Y.: Academic Press, Ch. 18.

YOUNG, W. C. (1953). Gamete-age at the time of fertilization and the course of gestation in mammals. In: *Pregnancy and Wastage*. E. T. Engle (ed.). Springfield, Ill: C. C. Thomas, pp. 38–50.

YOUNG, W. C. (1961). The hormones and mating behavior. In: *Sex and Internal Secretions*. W. C. Young (ed.). Baltimore, Md.: Williams & Wilkins, pp. 1173–1239.

YOUNG, W. C. & ORBISON, W. D. (1944). Changes in selected features of behavior in pairs of oppositely sexed chimpanzees during the sexual cycle and after ovariectomy. *J. comp. Psychol.*, **37**, 107–143.

ZARROW, M. X. (1961). Gestation. In: *Sex and Internal Secretions*. W. C. Young (ed.). Baltimore, Md.: Williams & Wilkins, pp. 958–1034.

ZUCKER, I. (1966). Facilitatory and inhibitory effects of progesterone on sexual responses of spayed guinea pigs. *J. comp. physiol. Psychol.*, **62**, 376–381.

Chapter 8

Techniques of Measurement and Evaluation

Victor H. Denenberg and Edwin M. Banks

The two fundamental objectives of any science are to describe and predict the subject matter under investigation. The subject matter of behaviour may be broadly defined as the *activity of an intact organism as it interacts with its environment*. The *description* of behaviour is very broad, ranging from rather primitive verbal statements to more sophisticated quantitative statements. Likewise, *predictions* of behaviour range from complete inability to make any estimates of future performance, through simple verbal statements, to more exact quantitative statements. Adequate description and prediction can be achieved only when techniques have been developed for *measuring* behaviour; the absence of appropriate measuring techniques limits or restricts scientific progress. Measurement means assigning numbers to objects or events according to certain rules. The purpose of this chapter is to present the major principles and procedures used in the measurement and analysis of behaviour and to describe specific methods used to measure behaviour in domestic animals.

I. THE MEASUREMENT OF BEHAVIOUR

Why measure behaviour? There are many simpler and more convenient characteristics of animals which can be investigated such as body weight, hormonal activity, or number of young born. Many considerations underlie the selection of an end point or dependent variable. A behavioural end point is the logical choice if one is interested in studying the integrated activity of a functioning organism. Non-behavioural variables (e.g. body weight) measure only certain fractionated aspects of the organism, though these may be extremely important. In the final analysis those who work in the life sciences are ultimately concerned with the intact organism though they may be working at a much more molecular level.

There are instances when behavioural measures are not the best, most efficient, or most appropriate ones to take. For example, in studying genetic mechanism, it is necessary to have a distinct phenotype. If the objective of the research is to study gene action rather than the genetics of behaviour *per se*, it is often much more convenient to select a morphological phenotype (e.g. coat colour) which can be classified visually, or is readily measured with little error, than it is to use a behavioural measure.

The behaviour which is finally selected for study and analysis may be only a very small aspect of the total activity in which the animal engages, or it may cover a relatively broad band on the activity spectrum. For example, in studying learning in the laboratory, the researcher may only observe and measure one specific response, such as leg flexion. On the other hand, the study of maternal behaviour spans the age range from birth through weaning and requires measuring a much wider range of activities.

A. Conceptual Meaning of Behavioural Measures

It is emphasized that there *must be some meaningful interpretation of the behaviour which is being measured.* The interpretation may be based upon a theoretical rationale or upon prior empirical findings. It is not enough to use a convenient semantic label as the basis for developing a conceptual system.

Not all behaviour which can be measured is necessarily important. Unless we can attach meaning to the behaviour, there is very little, if any, scientific merit in measuring it. It can be argued that "general activity" is being studied and innumerable references can be cited to show that activity has been of major interest to psychologists, ethologists, and other behavioural scientists. However, it is known that different measures of activity are often not significantly interrelated (Eayrs, 1954). This is obvious when one considers the various factors which can bring about changes in activity. Some of these factors are: temperature variation, hunger, thirst, sexual desire, introduction of a stranger or a novel stimulus into the group, changes in sensory stimulation in the immediate environment, and placing the animal into a strange environment. In other words, "general activity" does not refer to a basic unitary dimension or conceptual system.

Perhaps this point can be made clearer by contrasting the disciplines of physiology and behaviour. The meaning of a particular set of measurements is usually quite apparent in physiology, but the techniques required to obtain these measurements may be exceedingly difficult to learn and may require a high level of technical skill. In contrast, behavioural measurements are usually less difficult to obtain than physiological measurements. Anyone can be a "quantitative behavioural scientist" merely by using a microswitch, counter, or stopwatch. Since it is easy to obtain behavioural data many people have never stopped to ask just *what* it is that they are measuring. That is the critical question and unless some rational answer can be given to that question, the data are useless.

A behavioural classification scheme would be helpful. Unfortunately behaviourists are not in agreement as to a taxonomy of behaviour (discussed by King & Nichols, 1960).

B. Scales of Measurement

After deciding upon the behaviour to measure, the next problem is the degree of refinement of the measurement. This is the problem of the *scale of measurement* which will be used. In general, there are four scales of measurement: nominal, ordinal, interval, and ratio (Stevens, 1951). These scales differ in the assumptions underlying their use, the amount of information yielded, and the uses to which the numerical data can be put.

1. NOMINAL SCALE

This is the simplest type of scale; it is concerned only with the naming or classification of behaviour. Dichotomous data are probably the most common

7+

type but classification into three or more groups is also seen. Examples include aggressive versus non-aggressive, emotional versus non-emotional, and phylogenetic classification. The only requirement for the use of the nominal scale is that the behaviour be such that it can be placed into one of a number of categories. The categories should be mutually exclusive and totally inclusive. That is, a particular behaviour can only be assigned to one category, and all behaviours must be assigned to some category. There is no implication with the nominal scale that one category is "better" than another. These are merely qualitatively different behaviours. Since observations are merely classified into different groups, the usual numerical analyses are based upon percentages or frequencies within a group. Thus, relatively little information is obtained with the nominal scale. However, this scale is extensively used as, for example, in studying behaviour genetics.

2. ORDINAL SCALE

This scale assumes that the behaviour which is being measured can be ordered on a meaningful dimension in terms of such statements as "more of", "poorer than", "better than". A well-known ordinal scale in behavioural research is the "pecking order" of animals in their social interactions. Other examples include the classification of maternal behaviour on a scale such as excellent-good-fair-poor, and the ordering of aggressiveness by the "round robin" technique in which each animal is compared with every other animal. The numerical characteristics of this scale are such as to permit the use of ordinal (or rank) comparisons of the data.

No assumption is made concerning the *distance* between scale categories. For example, if three fowl are ranked A, B, and C in pecking order, all we can say is that A is more dominant than B and that B is more dominant than C. We *cannot* say that the distance between A and B is the same as the distance between B and C or that the distance between A and B is twice as great as the distance between B and C. To make such statements the interval scale must be used.

3. INTERVAL SCALE

With the ordinal scale we assign the ordinal numbers first, second, third, etc., to our observed behaviour. Thus, we are able to place the observed behaviours in their proper positions relative to each other on a meaningful dimension but we cannot assign scale values to the observed behaviours. If specific numerical scale values can be assigned to the observations so that the *distance* between two scale values is meaningful, this results in an interval scale. Formally, this is called the assumption of additivity. This means that the distances between observed behaviours can be added together in the same manner that we can add together the distance from one post to a second post, and from the second to a third and obtain the total distance from the first to the third (assuming that the three posts are in a straight line). Unfortunately, one cannot make such a simple check on the assumption of additivity with behavioural data as one can with the physical distance example where direct measurement is possible. Even in the physical world the assumption of additivity must be taken on faith more often than not. There are methods whereby one can determine whether the additivity assumption holds for behavioural data (Stevens, 1951; Guilford, 1954).

Several of the scales which measure intelligence are presumed to possess the requisite properties as are many psychophysical scales developed to measure sensory capacities (Stevens, *loc. cit.*). Usually an investigator more or less assumes that the procedure whereby he assigns numerical scores to the observed behaviour is compatible with the mathematical assumptions underlying the interval scale.

The numerical manipulations of addition, subtraction, multiplication, and division are permissible for purposes of obtaining measurement data with the interval scale. This order of scale will usually suffice for most research objectives. However, it is not possible to make statements of the following nature: A is four times as good as B on this task; C took twice as long to learn the problems as D. Such statements are *ratio* statements and require that the absolute zero of the scale be known (in the same manner, it is incorrect to state that 30°C is twice as hot as 15°C because the zero on the centigrade scale is an arbitrary zero point, not an absolute zero point).

4. RATIO SCALE

If we do have an absolute zero, then ratio statements are permissible; by absolute zero is meant the complete absence of the phenomenon we are measuring. For example, the absolute zero for temperature is the complete absence of molecular motion. This is 273° below the arbitrary zero of the centigrade scale and it can be seen that 30°C is actually 1·0521 times as hot as 15°C [that is, $(273 + 30)/(273 + 15)$]. The ratio scale is not often found in behavioural research. For example, the failure of an animal to solve even the simplest problem presented by a researcher does not mean that the animal is completely lacking in ability to learn (an absolute zero) since a problem could be constructed which the animal should be able to learn. When studying sensory phenomena, the absolute threshold (which is that

Table 12. The Four Scales used to Measure Behaviour Patterns

Scale	Measurement Requirements	Behaviour Patterns
Nominal	Mutually exclusive and totally inclusive categories.	Presence or absence of aggressiveness, emotionality, etc.
Ordinal	A unidimensional scale.	Peck order. Maternal behaviour.
Interval	A unidimensional scale possessing additivity.	Intelligence. Learning.
Ratio	A unidimensional scale possessing additivity and absolute zero.	Psychophysical scales of loudness, brightness, pitch.

amount of physical energy that can be perceived correctly 50 per cent of the time) may be taken to be an absolute zero. The measurement of sensory events is part of the field of psychophysics, where the majority of behavioural ratio scales are to be found (Stevens, 1951; Guilford, 1954). The numerical properties of the ratio scale are such that we are able to divide one number into another and come up with accurate and meaningful statements concerning ratios. Table 12 summarizes the measurement requirements of the four scales and gives examples.

C. Selection of Appropriate Scale

The behavioural researcher must first decide what type of measurement data he wishes to obtain in his experiment (nominal, ordinal, etc.) and then must set up the appropriate experimental procedures to attempt to obtain this type of information. When the experimental operations are perfectly compatible (isomorphic) with the mathematical assumptions for a particular scale of measurement, then the numbers assigned to the observations may be manipulated and interpreted meaningfully within the context of the scale of measurement. Whenever the experimental operations are not completely compatible with the mathematical assumptions, there is loss of information, and interpretations will be in error to some degree. As previously indicated, it is difficult to determine whether the experimental operations are completely compatible with the assumptions underlying the various scales of measurement. In most cases it is safe to say that our operations only approximate the assumptions.

D. Reliability of Measurement

We must now concern ourselves with ascertaining the *reliability* of our scores; by reliability is meant reproducibility, or consistency, of the observation under some other set of circumstances. These circumstances vary from one research problem to another and may be concerned with consistency of the behaviour at different times, at different places, or with respect to the judgments made by different individuals.

The importance of reliable measurements cannot be overemphasized. What we are asking, in essence, is whether we are measuring a behavioural phenomenon which is stable enough so that we have at least a fair degree of confidence that we will be able to reproduce this event. If we cannot do this, then we are measuring nothing more than "random" events which happened to occur at the particular moment that our observations were obtained.

The problem of reliability is probably more acute with the behavioural sciences than with the physical sciences. Behaviour is a dynamic process which is continually undergoing change as a function of internal and external events. In many instances we have to be certain that the characteristic we are measuring is relatively stable within this dynamic flux. Even if the behaviour is stable, our measurements may still be unreliable because of the instrument we use to obtain our measurements.

The general procedure used in ascertaining the degree of reliability of either the measuring instrument or the behaviour under investigation is to obtain at least two independent sets of measurements of the same phenomenon upon a random sample of animals. Thus, for each animal there will be at least two measures of its behaviour obtained independently. A *Pearson product-moment correlation coefficient* is then computed, and this statistic indicates the degree of stability of the characteristic being measured. The computational procedures for obtaining a correlation coefficient are discussed by Snedecor (1956) and Edwards (1954). The correlation coefficient will indicate the degree of agreement in the rank order of the animals for the two independent measurements. If the rank order is the same, then the correlation coefficient will be numerically high. If the rank order for one set of measurements has no relationship to the rank order for the second set of measurements then the correlation coefficient will be near zero.

1. Determining Reliability of the Measuring Instrument

To determine the stability or reliability of the measuring instrument it is necessary that the behaviour does not change during the time of measurement. If this happens, the failure to obtain satisfactory reliability may be due to the changes in the behaviour rather than the ineffectiveness of the instrument. Thus, the measurements are usually taken at very short time intervals, from instantaneous to about 24 to 48 hours.

One method, which is quite common when using check lists or rating scales, is to have two or more different experimenters make independent ratings of the behaviour to be measured. For example, it is often convenient to evaluate the nest built by a female for her young by rating the quality of the nest. A behavioural scale is constructed of the following general form: no nest, poor nest, fair nest, good nest, and excellent nest. Numerical values are assigned to these five scale positions. A description of what is an excellent nest, good nest, etc. is written and often sketches are drawn to illustrate the characteristics of the different types of nest (Sawin & Crary, 1953). Two experimenters would examine a number of nests and rate each one independently. The two sets of ratings would be compared via the correlation coefficient. If there is substantial agreement (a high coefficient), then it may be concluded that the behavioural rating scale is satisfactory as a measuring instrument. If there is poor agreement, then something is wrong with the measuring scale.

Often we measure behaviour by means of instruments rather than human observers. At times it is worth while to determine the reliability of the instruments we are using. One procedure is for the experimenter to observe the animal, make his own recordings, and see how well these correspond to the data obtained from the instruments. Failure to obtain close correspondence may indicate open circuits in the wiring, non-functional equipment, or lack of sensitivity of the instrument.

If the behaviour does not change within a relatively restricted time interval, then another common method of determining the reliability of the measuring instrument is to give the animal repeated testings on a short-term basis such as hour-to-hour or day-to-day. For example, the amount of feed ingested in successive 24-hour periods could be obtained for a random sample of animals. A high correlation coefficient would mean that the technique for recording feed intake was stable and discriminated among animals. A failure to obtain a high correlation would suggest that our measurement technique was not satisfactory.

2. DETERMINING RELIABILITY OF THE BEHAVIOUR

We have to distinguish between short-term behaviour changes and long-term changes. In general, short-term changes are those taking place within a few minutes to a few weeks (for mammals) while long-term changes extend beyond this time. Usually, behaviour is expected to be stable on a short-term basis. If no correspondence was found between two sets of scores obtained a few days apart, then no meaning could be attached to the scores. However, we must be aware of one important point: namely, the operations used in obtaining the measurements may have changed the behaviour so that the animals were dynamically different the second time they were tested. If so, then it is not too meaningful to correlate the first and second set of scores since, in a sense, different behaviours are being measured.

With respect to long-term changes, the behaviour may be stable in some instances but not in others. Feeding behaviour is usually constant but may change in old age or when the female gets pregnant. Learning, by definition, is a change in performance as a function of practice or prior experience. The correlation between performance during early learning trials and during later learning trials is often quite low. This does *not* mean that the behaviour is unstable; it means that the behaviour has changed between early and late learning and that the training has markedly changed the rank order of the animals between these two time intervals.

E. Validity of Measurement

A particular test (set of measurements) is said to be valid if the test actually measures what it is supposed to measure. Thus, validity is directly related to the point made previously that there must be some meaningful interpretation of the behaviour being measured. In some instances there is no problem of validity; in other cases it is a major problem.

There is no problem of validity when the concern is with studying specified behaviours such as grazing, maternal, sexual, and fighting. In studying maternal behaviour in the rabbit tests have to be developed to measure nest building, cannibalism, and scattering of the young, nursing of the young, protection of the young, etc. (Sawin & Crary, 1953). These are behaviours which occur when the doe has a litter and, thus, are valid measures of the complex phenomenon, maternal behaviour.

The problem of validity is pertinent when studying behavioural patterns which cannot be directly observed such as learning, emotionality, motivation, and perception. These terms denote "constructs", and, as such, can be measured only by indirection. The problem of validity is whether our indirect measurements of these constructs are valid measures of the construct itself. Usually, criteria are developed which determine whether a set of experimental operations are valid measures of a construct. For example, psychologists have developed special methods and criteria to study learning.

A related construct, "intelligence", requires developing a set of experimental operations which will yield numerical scores that are valid measures of that construct. Merely because a researcher states that his procedure measures "intelligence" does not make the test valid; the assignment of a semantic label is a far cry from the objective of constructing a test which is conceptually and empirically valid. A start towards the development of measures of "intelligence" in animals has been made by Hebb & Williams (1946) and Rabinovitch & Rosvold (1951). "Instinctive behaviour" is another example. To date no satisfactory set of experimental operations has been specified for measuring this construct. Indeed, it is doubtful if this is a very meaningful construct (Ross & Denenberg, 1960).

A non-behavioural example may help clarify this discussion. In genetics research the *gene* is not directly observed or measured but the *gene action* can be obtained by the operations of studying phenotypes and performing certain types of breeding experiments, the outcome of which may permit deductions concerning the presence or absence of genes.

II. GENERAL METHODS OF BEHAVIOURAL ANALYSIS

The development of a measuring instrument permits the *description* of certain aspects of behaviour with a known degree of accuracy. Our second objective is to *predict* the behaviour we have measured. This requires knowledge of antecedent events which are related, directly or indirectly, to the measured behaviour. The analysis of behaviour is concerned with the problem of prediction in the sense that the objective is to isolate the antecedent conditions which affect, modify, or change the behaviour. When successful in determining these antecedent conditions, laws of behaviour and mechanisms underlying behavioural patterns can be specified. Three of the major methods used in the analysis of behaviour will be described in this section. These are the naturalistic, experimental, and statistical methods.

A. Naturalistic Observations

Naturalistic or field observation, as the name implies, is the systematic observation of animal behaviour in a natural or semi-natural environment. Observations may be in terms of notes, check lists of behavioural patterns, motion pictures, or some combination of these. The researcher may or may not "interfere" with the behaviour he is observing (e.g. introducing a strange animal into a well-organized social group). The researcher is able to *describe* the behaviour of the animals through his systematic observations. If he is able to relate certain antecedent conditions with certain consequences (e.g. the introduction of a stranger leads to aggressive behaviour), he is then able to *predict* certain behavioural events.

Ethologists are probably the major group of researchers who have combined naturalistic observation and experimental study (reviewed by Thorpe, 1956). The observations of Lorenz & Tinbergen (Tinbergen, 1951) suggested that certain stimuli (releasers) caused certain behavioural patterns to be initiated; these hypotheses were later tested experimentally in the field and laboratory. A different example using the method of naturalistic observation is Scott's (1960) work with goats and sheep. Field observations lead to knowledge of the behavioural repertoire of the animal. In turn, this often leads to the isolation of variables which can be *experimentally* studied in the field or in the laboratory, or both.

B. Experimental Method

The experimental method can be briefly characterized as the systematic manipulation of one or more *independent variables* to study their effects upon some pertinent behaviour, or *dependent variable*, while *controlling*, by holding constant or randomizing, the effects of other variables which are likely to affect the behaviour being measured.

1. INDEPENDENT VARIABLES

These are variables which the researcher believes will affect the behaviour under study. They are "independent" in the sense that the particular variables and the particular values of these variables which are chosen for study are at the discretion of the experimenter; they are independent of any prior experimental condition. Any independent variable must have at least two values (levels) but may have more. Commonly, presence and absence of the variable are the two levels used.

2. FACTORIAL EXPERIMENTS

Two or more variables may be studied simultaneously in the same experiment. We could determine whether animals ate more when in isolation, or in groups of four or eight, and, at the same time, also determine whether the social facilitation phenomenon was the same when the animals were given a highly preferred or a less preferred feed. The size of the social grouping is one independent variable and the quality of feed is a second independent variable. The first variable is at three levels and the feed quality variable is at two levels, requiring $3 \times 2 = 6$ experimental groups to obtain all possible combinations of these two variables (Table 13). Some animals which ate alone ("groups" of one) would be given the highly preferred feed while others would be given the less preferred feed; these same two feed quality conditions would be repeated for groups of four and eight. The measurement of feed consumption would determine (a) the

Table 13. Experimental Layout showing how Two Independent Variables can each be Separately Evaluated in One Experiment and also showing the Interactive Effects of the Two Variables. Numbers are Hypothetical Means of Number of Feed Units Consumed.

Feed Preference	Size of Social Group			Mean
	I	4	8	
Highly preferred	13[1]	15[1]	15[1]	14·33[3]
Less preferred	3[1]	12[1]	15[1]	10·0 [3]
Mean	8·0[2]	13·5[2]	15·0[2]	

[1] These scores assess the joint effects of feed quality and social groupings.
[2] These means assess the effects of social grouping independent of feed quality.
[3] These means assess the effects of feed quality independent of social grouping.

effects of social grouping upon amount consumed, (b) the effects of feed quality upon amount consumed, and (c) the joint effects of social grouping and feed quality upon amount consumed. In Table 13 each set of marginal totals is independent of the effects of the variable which is perpendicular to it since the effect of that variable is balanced out. However, the scores within the table are not balanced out and they represent the joint effects of both variables acting in unison.

An experiment in which all combinations of two or more independent variables are studied at the same time is called a *factorial experiment*, and the joint effect of the two variables is called an *interaction*. Interactions can only be evaluated in a factorial design and are often of critical importance in the analysis and interpretation of experimental results. Suppose, in this example, that the researcher found that the animals ate the same amount regardless of social grouping when the feed was highly preferred, but, with the less preferred feed, they ate more as the social group increased in size. This is an example of an interaction and the researcher now has much more information for his experimental investment than if he had studied the effects of social grouping using only one type of feed.

3. DEPENDENT VARIABLE

This has been discussed in detail already. The variable is called "dependent" because it is assumed to be affected by (dependent upon) the particular levels of the independent variables. The dependent variable is the one measured in the experiment and it is the one analysed in the sense of determining the antecedent conditions (independent variables) which control its manifestation.

4. CONTROLS

In designing and conducting an experiment there are a number of variables which are not of interest but which may affect the behaviour being measured. The possible effects of these *extraneous variables* must be controlled; otherwise, the interpretation of the data will be ambiguous. If the caloric value of the highly preferred and less preferred feeds differed, then the amount consumed might be a function of this rather than the feed quality. Similarly, the results would be meaningless if the feeder space was inadequate in the group situation.

A pilot or preliminary study will often help isolate biasing variables. Those variables which are known to be potential sources of bias are held constant at some predetermined level.

5. RANDOMIZATION

Randomization is an integral part of an experiment. It is a necessary technique for ensuring unbiased results; randomization will not control bias by holding the effects of a variable constant but will do so by distributing its effects randomly among all groups. In general it may be stated that whenever a condition cannot be held constant its effects should be randomized. Factors such as assignment of animals to experimental groups, the sequence in which the animals are tested, the time of day of testing, and other potential sources of bias should always be randomized (Fisher, 1945).

7*

C. Statistical Analyses

On many occasions we are not able to conduct experimental research nor carry out systematic naturalistic observations but we are able to obtain data on one or more behavioural events. If these data can be quantified into one of the four scales of measurement, the method of statistical analysis can be used.

Statistical analysis will permit us to *describe* certain behavioural events in terms of averages (e.g. means, percentages) and variability (usually the standard deviation). Examples of this include statistical descriptions of migratory behaviour and seasonal hibernation. The descriptive data may also permit us to make *predictions*. For example, it is known that the density of animals may be related to geographical conditions (latitude, longitude, altitude), climate, and other ecological aspects. Suppose we had data showing relative population densities of a species for a period of 10 years and a systematic shift out of one region into another was observed. We might be able to relate this shift to certain ecological factors and thus be able to predict that when certain aspects of the ecology changed there would be a movement out of a particular region. In other words, descriptive statistical data are used to help draw inferences concerning antecedent conditions which brought about the change in the statistical values. This is obviously a circular process of reasoning but one which can be validated by finding another situation in which the hypothesized antecedent conditions are operating and determining whether the predicted behavioural changes do indeed occur.

1. Correlational Procedures

The correlational method is extremely useful for purposes of prediction. If two or more variables are significantly correlated, then knowing the scores on one variable we are able to predict, at least to some extent, the scores on the other variable. For example, Denenberg *et al.* (1958) found significant positive correlations between the quality of maternal nest built by rabbits and the per cent of young nursed by the doe. Knowing either the nest quality score or the per cent of young nursed, it is possible to predict the behaviour on the other variable. The danger inherent in using correlational techniques comes in interpreting the data. A significant correlation does not imply causation, *per se*. Quite often both measures are obtained at the same time or in close temporal sequence. Even when a considerable amount of time elapses between the measurement of the two variables this cannot be used alone as a logical argument to support the concept of causality. Quite often a significant correlation is brought about by a third variable which affects both of the variables which have been measured. Denenberg *et al.* (*loc. cit.*) suggested that the relationship between better nest quality and greater per cent of young nursed may have been brought about by the amount of prolactin secreted during pregnancy.

Even though correlation only indicates that two variables are related, it may be possible to make inferences concerning antecedent events and consequences. An example of this type of inference (and the arguments which ensue) is seen in the interpretations of the finding that a significant relationship has been obtained between smoking and lung cancer.

2. Two Uses of Statistics

The procedure of statistical analysis has been discussed as one of the major methods used in the analysis of behavioural phenomena together with the experimental method and naturalistic observations. This may be confusing to the reader since statistical procedures are quite often used to analyse experimental and observational data. There is a distinction between these two uses of statistics. When used in conjunction with experimental or observational research, statistics has the role of a powerful tool which helps in focusing the attention of the researcher upon the relevant bits of information among his mass of data. In this sense statistics is like a microscope or any other instrument used by a researcher to aid him in his research endeavours. It should be remembered, however, that excellent experimental and observational research were done before the advent of statistical theory (and there are still many researchers who use little or no statistics in their work), and excellent experimental and observational research were done before the microscope was invented. This concept of statistics is different from the use of statistical procedures as a major analytical method in its own right. This is of relatively recent origin and it is in this sense that we have listed statistical analyses as a separate method.

Table 13a summarizes the advantages and disadvantages of the three methods.

Table 13a. Advantages and Disadvantages of the Major Methods used in Behavioural Analysis

Method	Advantages	Disadvantages
Naturalistic	Best way to gain knowledge concerning natural behaviour patterns, both individual and social. Variables for future experimental analysis may be isolated.	Lack of controls may lead to ambiguity in interpretation.
Experimental	Rigid control over environmental events. Ability to manipulate variables in systematic manner. Excellent for isolation of relevant antecedent conditions ("causes").	Artificiality of laboratory conditions may affect generality of findings.
Statistical analyses	Excellent for population descriptions. Relationships among many measures can be determined by correlational techniques.	Lack of controls leads to difficulty in interpretation (correlation does not imply causation).

III. SPECIAL METHODS OF BEHAVIOURAL ANALYSIS

Researchers concerned with the analysis of behaviour are often interested in comparative behaviour, developmental functions relating age and behaviour, and techniques for reducing variability. Comparative analyses and age-behaviour functions can be investigated using any of the three major methods described above; techniques for reducing variability have generally been restricted to the experimental method.

A. Comparative Method

The objective of a comparative analysis of behaviour is to determine similarities and differences in behaviour patterns among different species and to relate these patterns to changes in the phylogenetic scale. Even though a great many species have been studied, very little work of a truly comparative nature has been carried out. The great majority of work has been concentrated upon a small number of animals (primarily rats and primates). Even when other animals have been used, little effort has been made to obtain comparable conditions of testing and measurement. The status of comparative psychology has been reviewed by Beach (1950).

One of the reasons for the lack of adequate comparative data can be seen if we refer back to our distinction between the measurement of observable specified behaviours and the measurement of behavioural processes which cannot be directly observed. Most animal research, at least by psychologists, has been concerned with behavioural processes like learning, drive, perception, etc. These processes are inferred from the behaviour of the animal under some set of artificial conditions (e.g. mazes, discrimination boxes). If we try to carry out comparative studies on these behavioural processes, a number of difficulties immediately arise. First, it is unlikely that the same apparatus and stimulus conditions have the same "meaning" to different species because of differences in sensory, motor, and neurological mechanisms. Secondly, there is no guarantee that the same laws underlie similar behavioural processes in different species (is learning in the earthworm controlled by the same behavioural and/or neurological principles as learning in the monkey?).

A truly comparative science of behaviour is more likely to develop through the systematic study of observable specified behaviours. These are behavioural patterns which are within the natural repertoire of the species. Comparative investigations of similarities and differences among these types of behaviour appear to be a very fruitful approach. Beach (1960) has made a similar point in his argument that behavioural researchers should study species-specific behaviour.

B. Age-Behaviour Relationships

The study of the development of behavioural phenomena can be an invaluable aid in the analysis of behavioural patterns. The investigation of development requires measuring animals at different ages. This can be done by studying the same animals at different times, the *longitudinal* approach. Another method is to study independent groups of animals which differ in age, the *cross-sectional* method. Each of these methods has both advantages and disadvantages.

I. LONGITUDINAL METHOD

The longitudinal method will result in less variability in the data because the same group of animals is measured repeatedly. This procedure of repeated measurements is of considerable importance because it is often possible to relate certain unique behavioural patterns at one age to known events which occurred earlier in time. This approach can be used in field studies and statistical analyses if the researcher can find some method of tagging or marking the animals so that they can be individually identified.

There are several disadvantages to the longitudinal method. First, a considerable time investment is required if the animals have a slow rate of development. Secondly, if there are major changes in the environment (e.g. seasonal), these changes are confounded with age changes, and it is not possible to determine whether a particular behaviour was brought about by changes in maturation or changes in the environment. Thirdly, the process of measuring the animal may change it so much that it is a different organism thereafter (e.g. once learning has taken place the animal is uniquely different from what it was before the learning occurred).

2. Cross-sectional Method

The cross-sectional approach avoids these difficulties by using animals which differ markedly in their age, measuring them only once, and measuring all the animals within a relatively narrow time span. The disadvantages of this method are the advantages of the longitudinal method. There is greater sampling error in using independent groups and the past history of the animals is less well known.

These two methods can be combined into one experimental design by studying one group of animals repeatedly while other groups of different ages are measured only once. However, this requires more time and effort.

C. Techniques for Reducing Variability

A basic principle of experimental design is to minimize the variability among the animals within the different treatment groups (Fisher, 1945). Variability among animals treated alike is used as the "error" term, or reference point, by which the experimental treatments are evaluated. If the variability among the different experimental treatments is not significantly greater than the variability within the treatment groups (error), we conclude that there is no evidence that the experimental treatments were effective. Therefore, it is apparent that by reducing variability within groups the chances of finding significant treatment effects are enhanced.

1. Each Animal as Its Own Control

Using each animal as its own control is one method of reducing variability; that is, each animal is tested under all treatment conditions. Intra-individual variability is usually less than inter-individual variability; thus, when each animal is used as its own control, the error term will generally be numerically less than the error term based upon independent groups. This procedure requires that the sequence in which the animals receive the treatments be randomized unless the researcher is specifically interested in studying sequence effects (as in learning). The major disadvantage of this procedure is that the experience of undergoing one of the treatments may change the animal so much that it cannot be tested again under another treatment. If, for example, we wish to determine the effects of group size upon the aggressive behaviour of animals reared in isolation, it would be best to test each animal only once.

2. Matching

The procedure of matching avoids the difficulty of not being able to use an animal for more than one treatment condition and retains many of the advantages of using each animal as its own control. This procedure requires that the experimenter have information about his animals on a variable which is significantly correlated with the dependent variable. Animals which have the same or similar scores on this variable are then paired (if there are two treatment conditions) and are randomly assigned to the two treatments. This is repeated for each of the pairs. It should be emphasized that the animals must be matched on an individual basis, not a group basis. The latter procedure actually results in a decreased likelihood of finding significant differences if the information about individual animals is not utilized.

An example of this procedure is a study by Pawlowski *et al.* (1961) investigating the effects of alcohol upon the retention of a learned response. Rats were trained to make an escape response to electric shock until their speed of running levelled off. Animals were then paired on their speed scores and one animal in each pair was randomly assigned to the experimental treatment of drinking tap water which contained 10 per cent alcohol. Controls drank tap water alone. A test of retention 100 days later found that the control group had retained the habit better than the alcohol group. This difference could not have been due to differences in initial learning since the groups had been matched on that basis.

This discussion is intended merely to introduce the reader to certain experimental procedures for reducing variability. Special statistical analyses are required for data obtained by the procedures described above (Fisher, 1945; Edwards, 1950; Snedecor, 1956). Parenthetically, the method of *covariance* is another method for reducing variability.

3. SPLIT-LITTER TECHNIQUE

A procedure which is often used with litters of animals and which resembles the matching procedure to some extent is the split-litter technique. Usually a number of litters are used, and animals from the same litter are randomly assigned to the different treatment conditions. Since animals from the same litter generally share the same maternal and physical environment, they should be more alike than animals selected strictly at random. The technique is quite useful if the experimenter suspects or knows that maternal factors or other uncontrolled elements in the environment in which the animals have been reared are significantly related to the dependent variable. It has also been assumed that the split-litter technique will equate the different treatment groups with respect to genetic factors since animals from the same parents are present in each of the treatment conditions. Ross *et al.* (1957) have discussed the split-litter technique in detail and have shown that the procedure will not eliminate genetic bias when using partially inbred lines of animals. In fact, the procedure is liable to lead to biased estimates and loss of precision. The authors do conclude that the method is useful as a control over certain environmental events.

IV. MEASUREMENT OF BEHAVIOUR IN DOMESTIC ANIMALS

Having presented the reader with a theoretical framework regarding the problem of measurement and evaluation of behaviour, we turn now to a consideration of some of the most widely used techniques. Our aim will be to delineate the objectives of a variety of techniques and to indicate the appropriateness of a given analytical method for obtaining meaningful behavioural information.

A. Description of Behavioural Patterns

It is virtually impossible to analyse the parameters of any kind of behavioural phenomenon without first having in hand a thorough description of the behaviour pattern of which it is a part. Moreover, it is equally important that the investigator have some insight into the entire behaviour repertoire of the species with which he is working. All too frequently untenable assumptions made in the design of an experiment lead to results which are ambiguous and difficult to interpret. The first task facing the behavioural scientist is that of preparing a detailed description of the behaviour under study.

1. Written Descriptions

A hand-written description of a behaviour pattern, made while observing animals, is the oldest and simplest method. After a series of preliminary observations, during which the major components of the behaviour are defined, a code or check-list can be devised. A properly constructed check-list can be used to record the frequency of occurrence of the various components of a behaviour pattern. One disadvantage of a hand-written record is the possibility of overlooking segments of a sequence when attention is diverted to write the data. Having two observers recording the activity simultaneously provides a margin of safety with regard to accurate identification of individuals and the notation of such parameters of the behaviour as frequency, etc. Statistical tests can then be employed to assess observer variability (Hultnäs, 1959).

2. Tape Recorder Descriptions

The availability of inexpensive, battery-operated tape recorders has simplified the task of recording rapidly occurring events. Such an instrument, when constructed of light-weight materials, can be usefully employed in studies in which the observation of free-ranging animals is required. When used in a laboratory situation care must be exercised to insure that the background noise produced by most tape recorders and the sound of the observer's voice does not distract the animals under study. One disadvantage associated with the use of tape-recorded data is the necessity of reducing the voice record to one which can be analysed. The frequency with which a given behaviour is displayed in a timed observation session can be readily obtained with this recording device. On the other hand, measurements of the duration of a given bout of behaviour are more difficult to record in this manner.

3. Recording Polygraph

Once the components of a behaviour pattern have been defined, it is possible, and usually desirable, to record the frequency and duration of interactions between two or more subjects. One device which serves this purpose admirably is the recording polygraph, an instrument consisting basically of a moving paper tape on which up to 20 recording pens may be fitted. Deflection from the baseline by the pens is activated by pressure-sensitive keys fitted into a keyboard. The keys may be coded according to the kinds of behaviour being studied; recorders of this type are usually provided with a system to regulate the speed with which the paper tape is moved. Hence, highly accurate measurements of frequency and duration can be made. One precaution is worth noting with regard to the use of the recording polygraph. It is possible for a trained observer to collect voluminous quantities of data by this means. Serious thought should be given to recording only those data which appear to be most relevant to the problem. Otherwise, the procedure of reducing the recorded information to manageable proportions for purposes of statistical evaluation becomes a major obstacle. This points up once more the importance to the investigator of having a thorough knowledge of the behaviour he is attempting to analyse.

In the event that the particular problem being attacked requires that records be obtained of every motor component in a complex interaction pattern, the investigator is well advised to turn to a technique which records the data in a form that can be directly processed by a programmed electronic computer. There are a number of electronic firms specializing in the design and manufacture of data acquisition systems. A quite sophisticated system that starts with an observer-activated keyboard can accumulate time and frequency scores and store these on magnetic tape for later analysis by computer. These observer-to-computer systems, although relatively costly, provide rapid acquisition and analysis of complex behavioural information. They can also reduce the drudgery of data reduction and analysis where behavioural measures recorded automatically rather than by observer are studied.

4. CINEMATOGRAPHY AND VIDEOTAPE RECORDING

There are two invaluable recording techniques which fix behaviour in a unique manner, motion picture film and videotape. The use of cinematography in behavioural studies is recommended for several reasons. Such records are permanent, may be re-examined many times, and when studied with a time and motion projector provide a means for detecting subtle postural adjustments that might otherwise pass unnoticed. When long-term observational surveillance is desired, it is possible with relatively inexpensive equipment to modify a movie camera to take single frame exposures at predetermined intervals. For example, in a pilot study of nursing/suckling behaviour of swine, a camera was set to expose a single frame of 16 mm film every 10 seconds during the first week after birth of the litter to obtain time scores of suckling, preference and competition for teat location, feeding behaviour of the sow, etc. (Banks et al., 1961). There are, in addition, a number of high-speed 16-mm cameras available with exposure speeds ranging from 100 to 2,000 frames per second. Such cameras, when provided with appropriate timing lights, provide the ultimate in slow-motion cinematography. Films so produced, when analysed with sophisticated time and motion equipment, permit extremely fine grain analysis of motor patterns. There are obvious limitations to the use of cinematography however, not the least of which is the relatively high cost of recording data in this way.

Recent advances in the development of vidicon cameras and videotape recorders and the appearance on the market of relatively inexpensive equipment provide another means of recording behavioural information. These systems have several distinct advantages over the movie camera, e.g. in contrast to movie film, videotape can be re-used many times; even the least costly system comes provided with audio pickup so that sound can be recorded simultaneously with action; vidicon tubes especially sensitive to the red end of the colour spectrum can be obtained so that animals on normal light-dark régimes can be monitored via closed-circuit television without modifying their photoperiod. Videotaping systems which provide slow-motion and stop-action replay as well as colour will no doubt become an economic reality to investigators of animal behaviour in due course. For analytical purposes, the problems of data reduction are much the same with the aforementioned techniques as was noted for the polygraph.

5. RADIOTRACKING AND TELEMETERING

The advent of miniature electronic components has opened up a new approach to the study of animal movements and the monitoring of physiological events coincident with ongoing behaviour of unrestrained animals. Radiotracking employs miniature radio transmitters attached to free-moving animals which emit signals of specified frequency. These signals provide input to appropriate receiver instruments and animal movements can be mapped with considerable accuracy (Sanderson, 1966). This technique has obvious advantages in studies of home range and territory of wild animals. It may also prove a more accurate means of measuring grazing behaviour in domestic mammals.

Small, implantable radio transmitters have also been designed to telemeter a variety of physiological measures such as body temperature and a heart rate in animals whose ongoing behaviour is under surveillance (Harris & Siegel, 1967). Techniques permitting remote stimulation of specified brain areas in free-moving animals have been demonstrated in behavioural studies of primates (Delgado, 1964).

B. Evaluation of Stimulus Characteristics

An adequate description of a behaviour pattern requires not only an inventory of the motor components involved, but also the isolation of the stimulus configuration which evokes the behaviour. Techniques used to uncover the stimuli which precipitate complex, species-specific behaviour in feral animals have been developed by Lorenz, Tinbergen, Thorpe, Baerends and others (cf. Tinbergen, 1951).

1. THE USE OF MODELS OR DUMMIES

One very useful method for studying the stimulus characteristics of a particular behaviour pattern is by the presentation of models or dummies. Taxidermic dummies of cocks and hens mounted in a variety of postures were used to delimit the contributions of posture and sex dimorphism in the evocation of various male sexual responses (Fisher & Hale, 1957). The stimuli eliciting sexual responses in male turkeys have been similarly investigated (Schein & Hale, 1957). In a further study of this phenomenon (Schein &

Hale, 1959), it was found that androgen-injected poults would display sexual responses to the disembodied head and neck of females of the same age.

One of the important attributes of the model technique is that its use may enable the investigator to study systematically the stimulus valence or value of various body regions, colour patterns, etc. The artificial stimulus need not be a taxidermic model, or portion thereof of the species being studied. In many studies on feral animals, two-dimensional models constructed of cardboard or other materials have provided important insights into stimulus characteristics such as body contours, size, and colour patterns.

2. MODIFICATIONS OF STIMULUS ANIMALS

Another means of evaluating stimuli which evoke specific responses is that of altering or modifying the appearance of stimulus animals and noting response aberrations in an experimental subject. An example of this approach was a study that investigated which morphological features were responsible for individual recognition in the domestic fowl (Guhl & Ortman, 1953). After having established the social organization in flocks of hens, the appearance of selected birds was altered by adding feathers or denuding certain body areas, and by changing the shape and colour of the comb. It was found that alterations involving the head and neck regions were particularly effective in bringing about loss of recognition of the modified bird by its flock-mates.

Cues other than those visually received may also be altered in the evaluation of stimulus characteristics. Among domestic animals the mammals are most likely to reveal species-specific responses to olfactory stimuli, particularly in reference to sexual behaviour. This is an area which has received too little attention, owing perhaps to an overly uncritical attitude. A male which appears to be guided to a receptive female primarily through olfactory cues may, in fact, be attending to information of quite a different character. Experiments designed to test the influence of odorants on complex behaviour patterns of domestic animals are needed, e.g. the role of olfactory cues in the establishment of mother-neonate responses in goats (Klopfer & Gamble, 1966), and the influence of olfactory cues on mating behaviour in sheep (Banks *et al.*, 1963).

That auditory stimuli play an important role in mother-neonate relationships of sheep and goats (Blauvelt, 1955; Collias, 1956) and in the parental behaviour displayed by hens (Collias & Joos, 1953) has been amply documented. The technique of recording vocalizations of animals under specified conditions and playing such recordings back in a controlled setting could be managed in such a way as to constitute an alteration experiment similar to that involving morphological modifications (see application in a study of responses to vocalization in crows by Frings *et al.*, 1958).

3. ELIMINATION OF SENSORY MODALITIES

The obverse approach to the alteration of the stimulus animal is that of creating changes in the sensory capacities of test animals. Lacking sensory input of a specified type, can the animal respond adaptively to a normal stimulus pattern? In a study of the neural mediation of sexual behaviour in the cat (Zitrin *et al.*, 1956) it was shown that ablations of the neocortex which eliminated vision interfered drastically with certain parameters of male behaviour. However, visually deprived male cats were able to perform copulation if placed in physical contact with an oestrous female. Visual stimuli, therefore, played an important role in the appetitive components of courtship, but were not essential for the consummatory responses. Another interpretation could be placed upon these findings, namely that olfactory cues presumably emanating from the oestrous female were not sufficient to guide the blinded males. The surgical elimination of the olfactory bulbs of the male rabbit did not result in aberrant sexual behaviour (Stone, 1925). Behavioural effects of the elimination of sensory modalities must be viewed with considerable care. Although a given type of sensory input may be unnecessary to the evocation of some behaviour pattern as indicated by deprivation experiments, it cannot be assumed that the eliminated modality plays no role at all in the intact animal.

4. STIMULUS SPECIFICITY AND SATIATION

Central to a study of the stimulus characteristics of a behaviour pattern is the demonstration of stimulus specificity. Involved in this concept is the idea that interacting systems of complex behaviour follow a non-random course. Our ability to predict the appearance of a given segment of behaviour in an interaction between two or more animals implies a degree of fixity in the display pattern. Whether such fixed patterns are exclusively a function of the heredity of the organism, depend upon experiential factors, or represent an interaction between these two causal bases is unimportant for the present discussion. What is important is that there does appear to be a considerable degree of stimulus specificity in the manifestation of complex behaviour. The behavioural scientist should be able to select those features which appear to exert greatest guidance over the behaviour. Many techniques used to study stimulus specificity have been treated above.

Further insight into other aspects of stimulus specificity and satiation may be gained by considering the sexual behaviour of bulls (Hale & Almquist, 1960). Various parameters (latency of ejaculation, number of ejaculations in an exhaustion test, etc.) were measured during the collection of semen from donor bulls. Either a cow or another bull provided ample stimulation during the early phases of collection from each donor bull, but as time passed, the response of the donors waned. To check for the possibility that the decline in sexual performance was the result of physiological fatigue, a change in the stimulus configuration was effected. The stimulus animal was replaced or the donor bull was moved back and forth in the collection frame. These modifications led to a resumption of high-level performance. The important consideration for the present discussion is the demonstration of stimulus satiation and the consequent reduction in responsiveness. The simple expedient of changing the stimulus animal was sufficient to effect a renewal of the behaviour being measured. Similar results have been obtained with respect to gobbling in turkeys (Schleidt, 1955). It would be worth while to bear in mind that cases of apparent physiological fatigue bringing about a reduction in behaviour may be due to stimulus satiation.

C. Quantitative Descriptions of Behaviour

Many of the items discussed previously have been concerned to some extent with methods that yield quantitative information. In this section techniques used to obtain data of this nature will be discussed in more detail.

1. FREQUENCY

A statistic of great usefulness in the analysis of behaviour is that of frequency, i.e. the number of times a given display is exhibited during a period of specified duration. The most expeditious means of recording frequency will depend to a large extent upon the kind of behaviour being studied. Thus, in a study of suckling behaviour of young lambs (Munro, 1956), the frequency of suckling was recorded by direct observation during the hours of daylight. A procedure of this type requiring close observation for extended periods is apt to be quite fatiguing for a single observer. It is sometimes possible, as it later proved to be in the above-cited example, to record short time samples over an extended number of days to arrive at suitably accurate measurements of frequency. When the behaviour is such that it can be recorded automatically, highly accurate measurements can be obtained. In experiments designed to test the rat's rate of drinking as a function of water deprivation, fluid intake was measured electronically (Stellar & Hill, 1952). Each time the subject licked the tube leading from the supply bottle, an electric circuit was completed that activated a counting device. It is suggested that this technique, suitably modified for larger domestic animals, would provide more quantitative data on fluid and solid food intake than is currently available.

As noted previously, measurement of frequencies is readily obtained from filmed data. If the animals are small and the behaviour of such nature that subjects will perform in a relatively restricted area such as a glass-walled container, a stopwatch may be suspended in the field of view for an accurate time base. This method was used in a study of pre-fighting behaviour of mice and allowed very accurate counts to be made of the behaviour components being studied. Films of behaviour displayed by large, free-ranging animals are also amenable to frequency determinations by means of a time and motion projector equipped with a frame counter. The number of frames during which the behaviour occurs can thus be easily counted, and knowing the speed at which the film was exposed, frequencies can be calculated. One precaution to be noted for this method is to be sure that the time exposure setting of the camera, usually given in terms of frames per second, is accurate.

2. LATENCY

Response latency measurements represent another valuable statistic for the analysis of behaviour. Typically, this parameter measures the time interval elapsing between presentation of a stimulus or stimuli and the elicitation of the response(s). One of the most universally measured aspects of sexual behaviour is that of response latencies of various types (see, for example, Rosenblatt & Aronson, 1958, on the cat and Hale & Almquist,

1960, on cattle). This parameter of behaviour may reveal important information regarding the motivational state of the animal, the adequacy of the stimulus configuration, and the influence of experimental manipulations of both or either of the foregoing. Latency measurements are most readily obtained in experimental situations, but may also be recorded when the study is one of a purely descriptive nature with free-ranging animals. The instrument of choice is the stopwatch or a 0·01 second timer without which quantitative studies of animal behaviour would be seriously impaired.

3. INTENSITY

Another aspect of behaviour subject to quantification is that of response intensity, also known as the dimension of magnitude or amplitude. By these terms we mean the apparent level and form of the performance. There are many objective measures of intensity, depending upon the particular behaviour being studied. Pavlov recorded the number of drops of saliva secreted by his dogs as a measure of the magnitude of conditioning. The amount of time an animal spends in performing an act may be considered another aspect of response intensity. In studies of maze learning, the speed of the response, obtained by the ratio of distance travelled to time, is a measure of intensity. There are other somewhat less objective indices of response intensity. For example, the terms "high-drive" and "low-drive" have frequently been used to classify animals with regard to the display of a given type of behaviour. In this context, the term "drive" is usually understood in a purely operational sense, with no necessary assumptions regarding motivation being implied. In a study of the genetic basis of sexual behaviour in cockerels, Wood-Gush (1960) selected breeding males on the basis of tests that separated the original population into two groups. The intensity of sexual behaviour exhibited by the two lines was measured in terms of frequency of mating activity during a series of test periods. Variations in the intensity of vocalization patterns have been recorded and studied in a large variety of animals. (For a review of this rapidly expanding field see Lanyon & Tavolga, 1960; Busnel, 1963.) The investigator should be careful to distinguish clearly between the intensity of the overt behaviour and inferences regarding its internal concomitants.

4. RESPONSE DECREMENT

The waning of a response in the face of a previously sufficient stimulus pattern may occur as the result of a number of factors such as general physiological fatigue, sensory adaptation, stimulus satiation, habituation, loss of supporting hormones, etc. The explanation for the decrement will obviously differ in different situations and further analysis is usually required before a causal relationship can be established. However, our concern here is not with causality, but rather with methods which have been used to obtain measures of this attribute of behaviour. Measurements of response decrement are obtained principally in the form of frequency or latency scores, and in a quantitative study of a behaviour pattern, constitute part of the essential data to be collected. There are numbers of apparently spontaneous response decrement phenomena in domestic animals that appear to be pathological in nature, e.g. balking in horses, the "downer" cow, etc.

(Fraser, 1960). In studies of domestic animals response decrements of the complex patterns involved in inter-individual encounters have been obtained, e.g. in sexual, agonistic, and parental behaviour. By way of example, the oestrous behaviour displayed by ewes undergoes a distinct decrement that can be correlated with changing levels of gonadal hormones (McKenzie & Phillips, 1930; Inkster, 1957). The causal relationships between the behaviour and hormones is firmly established in this instance and the decrement though gradual, proceeds to a zero level of response. In agonistic interactions among hens in a newly organized flock, the frequencies of pecks and threats reveal a response decrement that can be correlated with the duration of time in which the flock remains intact (see Chapter 16). The reason for this reduction in the display of agonistic behaviour is not nearly so clear as is that of the previous example.

Decrements in response to physical factors normally present in the environment such as photoperiod, feed, water, or experimentally imposed stimuli such as electric shock and sound have also been measured. The technique usually encountered entails frequency determinations of the response over stated periods of time. For a general discussion of changes in responsiveness to a constant stimulus, the reader is referred to Hinde (1954).

5. Response Recovery

Associated closely with the waning of a response is the phenomenon of response recovery. Once response decrements have set in, under what conditions will a return of the response take place? Here again, one must first establish that a recovery will occur and then proceed to delimit the causal factors for the recovery. The measure for recovery may be simply a reappearance of the behaviour after an intervening period during which no response is elicited, or the recovery may be measured in terms of an increase in frequency or intensity of response following a period of low-level activity.

6. Indices of Behaviour

When it is desirable to compare the performance of a sample of animals all of which have been exposed to the same stimulus situation, it is frequently useful to devise a weighted scoring system. An index number, computed from the scores of each animal can be applied to characterize its behaviour. Such indices have been used extensively in studies of sexual behaviour. After systematic observation, the male guinea pig's sexual pattern was fractionated into several components: nuzzling and sniffing at anogenital region, mounting, pelvic thrusts, circling, pursuit, etc. (Young & Grunt, 1951). Each component was assigned a numerical value corresponding to its position in the sequence from the onset of male-female interactions to post-copulatory activities and an index of sexual behaviour for each male was computed. Similarly conceived indices have been used in the study of sexual behaviour in cats (Rosenblatt & Aronson, 1958) and cattle (Hultnäs, 1959). Other types of behaviour are equally amenable to this scoring technique, e.g. the evaluation of temperament in different breeds of cattle (Tulloh, 1961). The index applied to each animal in this manner is usually an arbitrarily defined score and therefore should be interpreted with great caution. A criticism of this technique is the possibility that standard scores are not always used to obtain specific weightings. A more valid procedure would be to compute intercorrelations among the various scores to determine how they interact. If the data warrant it, a factor analysis can be carried out to determine the minimal number of factors to account for the observed scores. These factor scores can then be used in further analyses.

7. OTHER QUANTITATIVE MEASURES

Domestic animals have been used extensively in experimental studies of learning and conditioning. Intelligence in sheep was measured by the number of trials required by the subject to find its way through a maze without a mistake three times in succession (Liddell, 1954). Number of errors is a commonly used measure in maze-learning experiments. In instrumental conditioning studies, the number or per cent of successful responses is frequently used as a measure of learning, e.g. learning is measured in terms of successful avoidance responses in studies of instrumental avoidance conditioning. The reader is referred to a competent psychology text for further examples of quantitative measures used to study this kind of behaviour.

D. Role of some Major Factors in the Evocation of Behaviour

A causal analysis of a behaviour pattern which has been fully described in terms of major motor components and quantified stimulus-response parameters must proceed on the basis of testable hypotheses. To describe some of the experimental approaches which have been used to test hypotheses concerning causation of the behaviour of domestic animals, three major factors will be briefly examined, namely the role of the nervous system, of hormones, and of experiential factors.

1. NERVOUS SYSTEM INVOLVEMENT

For an extensive review of the literature regarding this subject the reader is referred to Field (1960). However, mention will be made here of the usual approaches used in studies of central nervous system-overt behaviour interrelations (see also Chapter 7 and Figs. 9 and 10).

Ablation experiments. The approach of removing large portions of the brain and noting change or the absence of change in the display of a given behaviour pattern has been followed. Thus Beach *et al.* (1956) studied the effects of unilateral and bilateral removal of the neocortex on sexual behaviour of male cats. The results of such massive insults to the organism are frequently difficult to interpret, particularly when the behaviour pattern being observed has a heavy motor component. The animal may not be able to perform the behaviour simply because it now lacks the motor co-ordination for proper orientation movements.

Placement of small lesions. Once specific areas of the brain have been implicated in the display of a given type of behaviour by means of gross ablations, a more refined approach can be used to identify precisely localized regions (nuclei, fibre tracts, areas). In studies of sexual behaviour in sheep, small electrolytic lesions in the anterior hypothalamus have been found to eliminate completely the manifestations of oestrus (Clegg *et al.*, 1958). The role of the hypothalamus in many other kinds of behaviour (feeding, drinking) has been studied in this manner.

Electrical recording. Still another approach, which complements that of the removal of brain tissue, is to study the electrical activity of various portions of the intact brain. The rationale here is an attempt to correlate patterns of electrical activity and the display of specified behaviour patterns. Bell (1960) recorded cortical electroencephalogram patterns (EEG) during somnolence and rumination in unrestrained goats. EEG recordings from the hypothalamus have implicated this structure in mating behaviour of cats (Porter *et al.*, 1957).

Electrical stimulation. Chronic implantation of stimulating electrodes can provide further information regarding the localization of neural areas that mediate overt behaviour. In studies on the chicken, von Holst (1960) has succeeded in evoking the display of complex behaviour by stimulating various areas of the brain.

Chemical stimulation. Another avenue used particularly in efforts to study the interrelationships between behaviour, hormones, and the nervous system, is the technique of chemical stimulation of selected sites. Fisher (1958) introduced minute quantities of sex hormone salts and other agents into localized brain regions to study their effects on sexual behaviour. A complementary approach is the study of the effects on overt behaviour of the introduction of minute quantities of enzyme inhibitors (Ginsburg, 1961).

2. The Role of Hormones in Behaviour

Although we still lack an understanding of the mode of action of hormones on the molecular level, many studies have been made relating hormones and behaviour. As in the preceding discussion, we will stress the various methods used to investigate hormone-behaviour interrelations.

Removal and replacement therapy. Following the classical approach of descriptive endocrinology, it is possible to study hormone effects by surgically removing the source of the hormone and observing the change or lack of change in the behaviour display. After a stated period, exogenous hormone is supplied to the animal and the effect is noted in the behaviour. The necessity for obtaining good pre-operative measurements of the behaviour cannot be over-emphasized. This procedure has been used extensively in studies of sexual and agonistic behaviour (Stone, 1939; Beach & Holz, 1946; Beeman, 1947; Rosenblatt & Aronson, 1958).

Treatment of immature animals with exogenous hormones. Once it has been established that a given behaviour pattern is hormone supported by using the previously described technique, it may be useful to investigate the influence of the hormone on pre-adult animals. In the case of sexual behaviour, on which most of this kind of experimentation has been done, intact pre-puberal animals are given exogenous hormone and effects are noted on the behaviour. Conclusions regarding features such as maturation of neuromuscular coordination, responsiveness to adequate stimulation, and levels of response in immature animals can often be reached by means of this technique. One example in which this approach was combined with several others was a study of the effect of early experience on sexual behaviour in turkeys (Schein & Hale, 1959; see also Chapter 17).

Precautions to be observed in the use of hormones. Perhaps the most important general observation to be made regarding hormone-behaviour studies is to guard against casualness in regard to the hormone dosage being used. Although techniques for evaluating circulating levels in blood of many hormones are just being perfected, a competent source should be consulted for information on what constitutes a physiological dose of a given hormone for a particular species. Failure to obtain this datum may lead to results which are quite ambiguous. Other aspects of hormone treatments which must be carefully considered are (i) the use of impure hormone preparations, (ii) the injection of hormones prepared from one species into animals of a different species, (iii) the mode of administration, e.g. injection, implants, and (iv) frequency of administration.

3. The Role of Experiential Factors in Behaviour

Intimately woven into the fabric of species-specific behaviour patterns are the day-to-day experiences of the animal. Although the role played by experience in the development and execution of a given behaviour may be difficult to measure, an exhaustive analysis of causality will suffer if this contribution is ignored. That early experiences profoundly influence many kinds of adult behaviours has been amply demonstrated (see Chapter 6).

Deprivation techniques. A simple approach to this problem is to interfere with the normal social environment of the developing animal. Raising animals from birth, hatching, or weaning age in complete sensory isolation from conspecifics is a generally used technique. At a stipulated age, the animal is confronted with the stimulus situation under study and its responses are noted. Sensory isolation from conspecifics in birds is not too difficult to achieve. In the production of specific-pathogen-free pigs, litters are delivered by hysterectomy and hand-reared in complete isolation. If accompanied by auditory isolation, such a procedure provides complete sensory isolation from conspecifics. However, the criticism is often made of such experiments that the subject can never be isolated from itself and therefore complete isolation is impossible. Another method is to modulate the degree of isolation by raising together unisexual size-controlled groups of litter-mates or birds hatched at the same time and hence control environmental impoverishment. Measurements of the adult behaviour may then be relatable to the number and sex of the animals encountered during early development.

Other methods. Another way of assessing the influence of early experience is to impose a treatment of some kind on young animals at one age or at a sequence of ages. The effect of the treatment on some measurable aspect of adult behaviour can then be evaluated (see Chapter 6). Training procedures for the performance of certain tasks are frequently imposed on adult animals. Methods are usually quite specific for the given task and species and the reader is referred to the species chapters for further information.

E. Specific Types of Behaviour

The remainder of this chapter is devoted to a summary in tabular form of a few examples of methods used to study behaviour of domestic animals.

Although not comprehensive in treatment, it should serve as a point of departure for readers interested in learning more about specific techniques. In each category of behaviour, the number of aspects which can be studied and the number of measurements which can be made are almost infinite. The selection by the investigator of the most appropriate kind of stimulus situations to obtain the most meaningful measurements is required for a satisfactory analysis. The measurement must obviously bear some realistic relation to the normal expression of the behaviour, i.e. whereas recording the number of intromissions yields useful information in the study of sexual behaviour in a rat, such a parameter has no meaning in bovine or avian sexual behaviour. When appropriate the samples will present the objectives of the study, the test situation favouring elicitation of the behaviour, the stimulus situation provided, the method of obtaining the measurement, and special advantages and limitations of the test.

1. INGESTIVE BEHAVIOUR

Table 14 summarizes a number of studies on ingestive behaviour. These investigations were designed primarily to describe the various components of grazing, ranging, and rumination in animals on pasture, water intake measurements, feed preference and taste discrimination tests.

2. SEXUAL BEHAVIOUR, MALE

Investigations of sexual behaviour in male domestic animals have been many and varied (Table 15). In addition to purely theoretical considerations, many studies have been made in response to questions raised by the practice of artificial insemination in the larger domestic animals. Interestingly enough, some of the techniques developed to select males for use as semen donors have yielded basic information regarding sex drive in the male.

3. SEXUAL BEHAVIOUR, FEMALE

One feature which appears as a common characteristic of the animals which have acceded to domestication is that of vigorous reproductive activity. It is, therefore, not surprising that the study of sexual behaviour in domestic animals has been intense.

The classical approach to the study of oestrous behaviour in the mammal was that of the determination of oestrus in the guinea pig (Young et al.). By systematically observing female guinea pigs throughout the course of their oestrous cycles, a method was arrived at by means of which the phase or state of oestrus could be ascertained using behavioural criteria. The method involved the palpation of the females and noting the responses to such stimulation, and of observing inter-female behaviour for further clues to oestrous condition.

In many domestic mammals it has been possible to determine the oestrous condition of females by noting such behaviour as postural adjustments made in response to stimulation by another female, a male, or by some experimental manipulation such as the vaginal or cervical smear technique (Table 16). Other criteria include gross motor activity, vocalizations, and emission of odoriferous substances (Parkes, 1960).

4. Parental Behaviour

Both mammals and birds display a considerable amount of parental behaviour (Table 17). In domestic mammals attention has focused on neonate nursing reactions, responses of mother, formation of social relationships between mother and offspring. In domestic poultry, the phenomenon of broody behaviour has been studied, but there is an urgent need for more analytical studies.

5. Social Behaviour

Social behaviours, other than sexual and parental, form a significant portion of the behaviour repertoire of all domestic animals (see Chapters 5 and 16). One of the most readily observed expressions of social interactions among individuals in a flock or herd is that of agonistic behaviour, including both aggressive and submissive activities. Studies of dominance orders based upon agonistic interactions have been conducted on most domestic animals. Techniques used to measure aggressiveness include the direct observation of dominance-subordinance encounters leading to group organization. Relative aggressiveness is sometimes studied by staging a contest between two animals in an area unfamiliar to either. In addition to recording the winner, other parameters have been used, e.g. accumulated attacking time, number of attacks, fighting response latency (Catlett, 1961). Under flock or herd conditions it is sometimes necessary to supply a conflict situation in order to enhance close contact and thus increase agonistic behaviour. A simple expedient is to provide feed in a restricted area with insufficient space to accommodate all individuals. This serves a twofold purpose of drawing the subjects into one area and providing a reward, so to speak, for which the animals may compete. Table 18 summarizes a few specific examples of agonistic behaviour studies and concludes with several concerned with various other aspects of social behaviour.

Table 14. Studies of Ingestive Behaviour

Behaviour	Animal	Aspect Measured	Technique	Comments	Source
	Cattle and sheep	Preference for individual plants or plant groups; rate of feed ingestion.	Test conditions: response to test feed presentation and continuous weighing of feed. Field condition: eye estimates of amounts and kinds of forage plants eaten from plots at daily or twice daily intervals.	Study represents a combination of field observation and experimental procedures. Conditions of field study prevent weighing of feed consumed.	Cowlishaw & Alder (1960)
Feeding	Sheep	Selective grazing.	Chemical composition of forage obtained by hand-clipping; quantitative evidence of selection by means of oesophageal-fistulated animals.	Hand-clipped forage analyses provided incomplete data for estimates of selective grazing.	Weir & Torell (1959)
	Cattle	Taste discrimination.	Qualitative measurements of threshold of discrimination using pairs of monozygotic twins.	Differing thresholds of taste discrimination for quinine found to have genetic basis.	Bell & Williams (1959)
	Cattle	Water intake rate.	Calibrated 12-gal. buckets examined at hourly intervals.	Crude method; requires frequent visitation to test area.	Rollinson et al. (1955)
Drinking	Cattle	Water intake; to determine whether ambient temperature-dry matter consumption-water intake relationships can be used to estimate water intake.	Statistical evaluation of data from many sources.	Careful consideration of variables determining water-intake rates indirectly.	Winchester & Morris (1956)

	Animal	Measurement	Method	Remarks	Reference
	Cattle	Frequency and duration of standing, lying down, grazing, chewing, ruminating. Total distances travelled.	Direct observation for 24-hr periods. Stopwatch readings for frequency and duration. Distance travelled measured by placing numbered stakes in field and recording positions of animals and paths travelled during observation period.	Possible fatigue of observer; distance measurement somewhat subjective.	Johnstone-Wallace & Kennedy (1944) (see Chapter 10)
Grazing	Sheep	Distances travelled	Rangemeter readings.	Provides objective distance measurement for free-ranging animals.	Cresswell & Harris (1959)
	Cattle	Comparison of continuous vs. intermittent observation of components of grazing behaviour.	One observer/animal. One lot studied continuously for 24 hr; another lot observed at 15, 30, and 60 min. intervals/24 hr	Comparison revealed 30-min. intervals adequate for measurement of major components of grazing. Method reduced fatigue of observer.	Hull et al. (1960)
	Cattle	Frequency of walking, lying down, jaw and head movements in unrestrained animals.	Recording device.	Automatic recording of basic components of grazing. Requires period of adaptation to harness.	Canaway et al. (1955)
Rumination	Goats	Rumination time/24 hr; variation in cycle length.	Recording device to measure jaw movements by air manometer system.	Animals partially restrained in indoor pen.	Bell & Lawn (1957) (see Chapter 11)
	Sheep	Number of chews/bolus; time in sec/bolus; rate of chewing, etc.	Recording of jaw movements for 7 days using 4 different diets; ink-writing continuous recorder.	684 hr data with diets ranging from 100% hay to 100% concentrate meal.	Gordon (1958)

Table 15. Studies of Male Sexual Behaviour

Animal	Aspect Measured	Technique	Comments	Source
Guinea pig	Description of motor components.	Direct observation of behaviour using spayed females brought into oestrus with hormones.	Stimulus constancy provided by hormone-treated females. Grading system established to provide an index value characterizing each male.	Young & Grunt (1951)
Sheep	Latency of successive ejaculates; total number of ejaculates as a measure of sex drive.	Libido measured by time required to produce successive ejaculates and by number of ejaculates/30-min. period; rams run with ewes in panel pens or small pastures.	Stress on non-behavioural aspects, e.g. semen analyses, lambing records; no component study of ram behaviour.	Wiggins et al. (1953)
Cattle	Latency to ejaculation; frequency; stimulus satiation, recovery period; reaction to new stimulus situation.	Standardized test periods; artificial vagina; constant stimulus animals. Responses measured in time and frequency scores; new stimulus animals.	Animals must be trained to collection room procedures. Concept of stimulus pressure is presented.	Hale & Almquist (1960)
Cattle and sheep	Ejaculatory response.	Artificial ejaculation using transparent artificial vagina; electroejaculators.	For study of physiology of ejaculation; measurement of thrust and penis lengthening; drug effects on response, etc.	Marden (1954); Rowson & Murdoch (1954); Hafez (1960)
Cattle	Variability in sexual performance; age effects;	Sex drive index devised on basis of performance of components of	Inbreeding coefficient determinations for 3 or 4 generations to	Hultnäs (1959)

222

Species		Method	Findings	Reference
	heritability; effect of inbreeding.	sexual pattern. Bulls exposed and measured 3 times with artificial vagina, 3 times with natural mating.	measure influence of inbreeding and heritability of performance. Selection of measurable features of behaviour to study cause of variation.	
Horse	Erection reflex time; effect of elimination of vision and suppression of olfaction on erection and leap reflexes; reactions to dummy.	Comparison of young and experienced "old" stallions in tests with blindfold; blindfold and nose mask; blindfold, nose mask, and odoriferous substances.	Young stallions apathetic to dummy presentations, whereas older stallions responded with normal sexual behaviour.	Wierzbowski (1959) (see Chapter 12)
	Stimulus determinants.	Taxidermic models of cocks and hens mounted in variety of postures.	Male display analysed in terms of stimulus properties of models and live hens.	Fisher & Hale (1957)
	Relation between aggressiveness and sexual drive in cockerels.	5-wk old cockerels caged singly; injected daily with androgen for 7 days and tested for aggressive drive; on 8th day killed female chick in crouching posture used to measure sex drive.	Male sexual responses graded in terms of form and frequency of sexual behaviour displayed.	Wood-Gush (1957)
Chicken	Male dominance over female and mating success.	Gonadectomized males treated with oestrogen and placed in flocks of hens.	Oestrogens increased male sexual responses without increasing aggressive behaviour. Common male-female dominance order.	Guhl (1949)
	Stimulus determinants: response levels measured by assigning weighted scores to displayed components of male pattern.	Control stimulus object: fully crouched model; experimental: headless, tailless body; bodyless head with tail; headless, bodyless tail; bodyless, tailless head.	Head found to provide orientation cues but little arousal value; body with high arousal value but no orientation cue.	Carbaugh & Schein (1961)

Table 16. Studies of Female Sexual Behaviour

Animal	Aspect Measured	Technique	Comments	Source
Sheep	Determination of oestrus; incidence.	Examination of vulva; exposure to sterile ram whose brisket has been ochred with oily paint; colour changed every 12 days.	Behavioural response of acceptance indicated by colour markings on rump of ewe. Under field conditions can be used to establish individual cycle patterns of large flocks. Components of female behaviour not indicated.	McKenzie & Phillips (1930); Hafez (1952)
	Characteristics of oestrous behaviour.	Direct observation of ewes presented with painted sterile ram in small pens.	Established a grading system for sexual drive based on differential response of ewes to ram. Responses described in qualitative terms; no time-scores obtained; procedure adequate for preliminary study.	Inkster (1957)
	Influence of ram on incidence of oestrus in relation to onset of breeding season.	Comparison of 2 flocks, one run with ram continuously, other for 1 month only. Oestrus detected with painted ram.	Variable of presence of ram well controlled.	Riches & Watson (1954).
	Gross activity during oestrus.	Pedometer attached to leg recorded number of steps taken before, during, and after oestrus.	Objective method of measuring gross activity as defined by number of steps in unrestrained animals.	Farris (1954)

Animal	Problem	Method	Comments	Reference
Cattle	Repeatability and heritability of degree of expression of oestrus, and relationship of latter to conception.	Degree of expression of oestrus evaluated on subjective scale ranging from 1–4 (vague to supranormal "heat" symptoms). Conception scored on "yes" or "no" basis.	Careful use of statistical techniques indicated subjective ratings for degree of expression of "heat" were of little or no value in selecting for increased conception rate.	Rottensten & Touchberry (1957)
Swine	Description of motor components; measurement of gross activity; several physiological parameters.	Direct observation; pedometer readings, cervical and nasal smears; bio-electric body potentials.	Integrated study of behavioural and physiological aspects.	Altmann (1941)
	Evaluation of auditory and visual stimuli which elicit mating stance.	Presentation of male odour and vocalization under several conditions. Response criterion was assumption of mating stance.	Rhythm of male vocalization pattern changed artificially; chemical nature of olfactory stimulus not identified.	Signoret & Du Mesnil du Buisson (1961); (see Chapter 11)
Chicken	Influence of social rank on sexual receptivity.	Introduction of cocks singly and in rotating order for timed test period. Frequency scores by direct observation of courting, avoidance, and crouching.	Study of interaction of relative aggressive levels measured by rank in social order and sexual receptivity.	Guhl (1950) (see Chapter 16)
	Role of gonadal hormones.	Injection of androgens and oestrogens into ovariectomized poulards. Stimulus situations included normal cock, normal hen, taxidermic hen dummy mounted in receptive posture.	Responses measured in reference to hormone treatments. Control of stimulus pattern added data on specificity, etc.	Davis & Domm (1943)

Table 17. Studies of Parental Behaviour

Animal	Aspect Measured	Technique	Comments	Source
Sheep	Suckling.	Time scores of suckling; number of sucks from right or left teat; attitude and willingness of ewe.	A descriptive study of major parameters of nursing-suckling interactions.	Munro (1955)
	Effect of suckling frequency on quality of milk.	Measurement of lamb weights before and after suckling bouts; comparison of amount suckled at 1 and 4 hr intervals.	Milk quality inferred from weight gains.	Munro & Inkson (1957)
Sheep and goat	Comparative study of main components of mother-neonate reactions.	Direct observation of initial responses at birth; eating and licking of birth membranes; vocalization patterns; maternal defence and nursing.	Such comparative studies are helpful in laying foundation for analytical work. Whether the taxonomic relationship of the two species is close enough to provide information for an evaluation of the evolution of parental behaviour is questionable.	Collias (1956)
	Mother-neonate relationships.	Data filmed and analysed with time and motion projector. Particular attention to development of social bonds.	Technique well suited to study visually received cues important in parental behaviour.	Blauvelt (1955)
	Mother-neonate relationships	Experiment with goats to determine contribution of odorants in establishing goat-kid bond.	Use of temporarily anosmic goats clarifies phenomenon of "olfactory imprinting".	Klopfer & Gamble (1966)

Swine			
Suckling behaviour.	Time/frequency scores obtained by direct observation of position of sow during nursing; teat selection and preferences; time spent in massage of udder, duration of milk flow, milk consumption/piglet/suckling period measured by weight changes.	Establishment of quantitative data on major parameters of nursing behaviour.	Gill & Thomson (1956)
Rabbit			
Effect of repeated litter production on quality of nest.	Quality rated on scale based on degree of hollowing out, arrangement, and covering.	Does provided with nesting box 2 days prior to parturition; found significant improvement in quality over first 4 litters.	Ross *et al.* (1956)
Nest building under semi-natural conditions.	Direct observation of entire nest building procedure.	Under semi-natural conditions, no improvement in nest construction was noted.	Deutsch (1957)
Chicken			
Broody behaviour.	Direct observation of responses of hens to chicks up to 4 wks of age.	A descriptive study providing data for fractionating broodiness into 4 stages.	Ramsay (1953)
Hormonal induction of broodiness in cockerels.	Injections of prolactin into cockerels and direct observation of responses to chicks and eggs.	Stimulus characteristics of chicks inducing display of broodiness in injected cockerels were not analysed.	Nalbandov *et al.* (1945)

Table 18. Studies of Social Behaviour

Animal	Aspect Measured	Technique	Comments	Source
Cattle	Dominance order.	Direct observation of different groups of the herd 1 hr/wk, and recording of agonistic interactions.	2 observers used to reduce errors in identification and judgment of the outcomes of interactions.	Schein & Fohrman (1955)
	Dominance order modification.	Dehorning.	Induced changes in rank order of dehorned cows.	Woodbury (1941)
	Play activities.	Direct observation: introduction of various manipulanda such as bales of straw, wooden trestle; rubbing neck or rump of animal.	11 play patterns observed; discussion of play releasers and proposal of a play instinct.	Brownlee (1954)
Horse	Dominance order of 3 females and 7 castrate male grade horses.	Direct observation; criteria of agonistic interactions: kick, bite, head bump, passive reaction.	Linear dominance order with 1 triangular relationship. Body weight appeared an important factor influencing rank.	Montgomery (1957)

	Level of aggressiveness.	Initial paired encounters of hens in an empty pen, unfamiliar to both; round robin sequence.	Comparison of rank in home flock organization and in paired encounter situation.	Allee et al. (1939) (see Chapter 16)
	Effect of debeaking hens on aggressive manifestations.	Comparison of flock formation by normal and debeaked hens; debeaking of alpha hens of organized flocks; paired encounters between normal and debeaked hens.	Higher pecking frequencies found initially in debeaked flocks than in normal; subordinate hens tended to ignore pecks from debeaked birds.	Hale (1948)
Chicken	Effect of hormones on level of aggressiveness.	Injections of testosterone propionate into low-ranking hens.	Upward social mobility demonstrated as a function of male hormone treatment.	Allee et al. (1939)
	Individual recognition.	Modification of head furnishings and feathers to study cues used in individual recognition.	Features of head and neck implicated as major stimuli for individual recognition.	Guhl & Ortman (1953)
	Stimuli evoking aggressive behaviour in cocks.	Taxidermic models of both sexes mounted in sexual crouch, aggressive posture, and a female in normal standing posture.	Designed to define the stimulus characteristics for the release of fighting.	Fisher & Hale (1957)
	Social communication.	Sound signals produced by hen and chicks recorded and analysed on acoustic spectrograph.	Identification of the characteristics of sound stimuli attractive or aversive to chicks.	Collias & Joos (1953)

REFERENCES

ALLEE, W. C., COLLIAS, N. E. & LUTHERMAN, C. Z. (1939). Modifications of the social order in flocks of hens by the injection of testosterone propionate. *Physiol. Zool.*, **12**, 412–440.

ALTMANN, M. (1941). Interrelations of the sex cycle and the behavior of the sow. *J. comp. Psychol.*, **31**, 481–489.

BANKS, E. M., BECKER, H. E. & RASMUSSEN, O. (1961). Unpublished data. University of Illinois: Urbana, Ill.

BANKS, E. M., BISHOP, R. & NORTON, H. W. (1963). The effect of temporary anosmia on courtship in the ram (*Ovis aries*). *Proc. XVI International Zoological Congress*, **2**, 25.

BEACH, F. A. (1950) The snark was a boojum. *Am. Psychol.*, **5**, 115–124.

BEACH, F. A. (1960). Experimental investigations of species-specific behavior. *Am. Psychol.*, **15**, 1–18.

BEACH, F. A. & HOLZ, A. M. (1946). Mating behavior in male rats castrated at various ages and injected with androgen. *J. exp. Zool.*, **101**, 91–142.

BEACH, F. A., ZITRIN, A. & JAYNES, J. (1956). Neural mediation of mating in male cats: I. Effects of unilateral and bilateral removal of the neocortex. *J. comp. physiol. Psychol.*, **49**, 321–327.

BEEMAN, E. A. (1947). Aggressive behavior of normal, castrate, and androgen-treated castrate C57 black and Bagg albino mice. *Physiol. Zool.*, **20**, 373–405.

BELL, F. R. (1960). The electroencephalogram of goats during somnolence and rumination. *Anim. Behav.*, **8**, 39–42.

BELL, F. R. & LAWN, A. M. (1957). The pattern of rumination behaviour in housed goats. *Brit. J. Anim. Behav.*, **5**, 85–89.

BELL, F. R. & WILLIAMS, H. L. (1959). Threshold values for taste discrimination in monozygotic twin calves. *Nature, Lond.*, **183**, 345–346.

BLAUVELT, H. (1955). Neonate-mother relationships in goat and man. In: "*Group Processes*," Trans. Second Conf., pp. 94–140. Josiah Macy, Jr. Foundation, New York, N.Y.

BROWNLEE, A. (1954). Play in domestic cattle in Britain: an analysis of its nature. *Brit. vet. J.*, **110**, 48–68.

BUSNEL, R. G. (1963). *Acoustic Behaviour of Animals*. Amsterdam: Elsevier Publishing Co..

CANAWAY, R. J., RAYMOND, W. F. & TAYLOR, J. C. (1955). The automatic recording of animal behaviour in the field. *Electron. Engng.*, **27**, 102.

CARBAUGH, B. T. & SCHEIN, M. W. (1961). Sexual response of roosters to full and partial models. *Amer. Zool.*, **1**, in press (Abstract).

CATLETT, R. H. (1961). An evaluation of methods for measuring fighting behaviour with special reference to *Mus musculus*. *Anim. Behav.*, **9**, 8–10.

CLEGG, W. T., SANTOLUCITO, J. A., SMITH, J. D. & GANONG, W. F. (1958). The effect of hypothalamic lesions on sexual behaviour and estrous cycles in the ewe. *Endocrinology*, **62**, 790–797.

COLLIAS, N. E. (1956). The analysis of socialization in sheep and goats. *Ecology*, **37**, 228–239.

COLLIAS, N. E. & JOOS, M. (1953). The spectographic analysis of sound signals of the domestic fowl. *Behaviour*, **5**, 175–188.

COWLISHAW, S. J. & ALDER, F. E. (1960). The grazing preferences of cattle and sheep. *J. agric. Sci.*, **54**, 257–265.

CRESSWELL, E. & HARRIS, L. E. (1959). An improved rangemeter for sheep. *J. Anim. Sci.*, **18**, 1447–1451.

DAVIS, D. E. & DOMM, L. V. (1943). The influence of hormones on the sexual behavior of domestic fowl. In: "*Essays in Biology*", Univ. of Calif. Press, Berkeley, pp. 171–181

DELGADO, J. M. R. (1964). Free behavior and brain stimulation. In: *Int. Rev. Neurobiol.*, **6**, 349–449.

DENENBERG, V. H., SAWIN, P. B., FROMMER, G. P. & ROSS, S. (1958). Genetic, physiological, and behavioural background of reproduction in the rabbbit: IV. An analysis of maternal behaviour at successive parturitions. *Behaviour*, **13**, 131–142.

DEUTSCH, J. A. (1957). Nest building behaviour of domestic rabbits under semi-natural conditions. *Brit. J. Anim. Behav.*, **5**, 53–54.

EAYRS, J. T. (1954). Spontaneous activity in the rat. *Brit. J. Anim. Behav.*, **2**, 25–30.

EDWARDS, A. L. (1950). *Experimental Design in Psychological Research.* New York, N.Y.: Rinehart.

EDWARDS, A. L. (1954). *Statistical Methods for the Behavioral Sciences.* New York, N.Y.: Rinehart.

FARRIS, E. J. (1954). Activity of dairy cows during estrus. *J. Am. vet. med. Ass.*, **125**, 117.

FIELD, J. (ed.) (1960). *Handbook of Physiology*, Section 1. Neurophysiology. Washington, D.C.: American Physiological Society.

FISHER, A. (1958). Induced sexual behaviour by intracranial chemical stimulation. *Anim. Behav.*, **6**, 122.

FISHER, A. E. & HALE, E. B. (1957). Stimulus determinants of sexual and aggressive behaviour in male domestic fowl. *Behaviour*, **10**, 309–323.

FISHER, R. A. (1945). *The Design of Experiments.* (4th ed.). Edinburgh: Oliver & Boyd.

FRASER, A. F. (1960). Spontaneously occurring forms of "tonic immobility" in farm animals. *Canad. J. comp. Med.*, **24**, 330–333.

FRINGS, H. M., JUMBER, J., BUSNEL, R., GEBAN, J. & GRAMET, P. (1958). Reactions of American and French species of *Corvus* and *Larus* to recorded communication signals tested reciprocally. *Ecology*, **39**, 126–131.

GILL, J. C. & THOMSON, W. (1956). Observations on the behaviour of suckling pigs. *Brit. J. Anim. Behav.*, **4**, 46–51.

GINSBURG, B. E. (1961). Personal communication. Univ. of Chicago.

GORDON, J. G. (1958). The act of rumination. *J. agric. Sci.*, **50**, 34–42.

GUHL, A. M. (1949). Heterosexual dominance and mating behaviour in chickens. *Behaviour*, **2**, 106–120.

GUHL, A. M. (1950). Social dominance and receptivity in the domestic fowl. *Physiol. Zool.*, **23**, 361–366.

GUHL, A. M. & ORTMAN, L. L. (1953). Visual patterns in the recognition of individuals among chickens. *Condor*, **55**, 287–298.

GUILFORD, J. P. (1954). *Psychometric Methods.* (2nd ed.) New York, N.Y.: McGraw-Hill.

HAFEZ, E. S. E. (1952). Studies on the breeding season and reproduction of the ewe. *J. agric. Sci.*, **42**, 189–265.

HAFEZ, E. S. E. (1960). Analysis of ejaculatory reflexes and sex drive in the bull. *Cornell Vet.*, **50**, 384–411.

HALE, E. B. (1948). Observations on the social behavior of hens following debeaking. *Poult. Sci.*, **27**, 591–592.

HALE, E. B. & ALMQUIST, J. O. (1960). Relation of sexual behavior to germ cell output in farm animals. *J. Dairy Sci.*, **43**, Suppl., 145–169.

HARRIS, C. L. & SIEGEL, P. B. (1967). An implantable telemeter for determining body temperature and heart rate. *J. appl. Physiol.*, **22**, 846–849.

HEBB, D. O. & WILLIAMS, K. (1946). A method of rating animal intelligence. *J. gen. Psychol.*, **34**, 59–65.

HINDE, R. A. (1954). Changes in responsiveness to a constant stimulus. *Brit. J. Anim. Behav.*, **2**, 41–55.

HOLST, E. VON (1960). Von Wirkungsgefüge der Triebe. *Naturwissenschaften*, **47**, 409–422.

HULL, J. L., LOFGREEN, G. P. & MEYER, J. H. (1960). Continuous versus intermittent observations in behavior studies with grazing cattle. *J. Anim. Sci.*, **19**, 1204–1207.

HULTNÄS, C. A. (1959). Studies on variation in mating behaviour and semen picture in young bulls of the Swedish Red-and-White breed and on the causes of this variation. *Acta agric. scand.*, **6**, Suppl., 1–82.

INKSTER, I. J. (1957). The breeding behaviour of two-tooth ewes. *Proc. N.Z. Soc. Anim. Prod.*, **17**, 72–76.

JOHNSTONE-WALLACE, D. B. & KENNEDY, K. (1944). Grazing management practices and their relationship to the behaviour and grazing habits of cattle. *J. agric. Sci., Camb.*, **34**, 190.

KING, J. A. & NICHOLS, J. W. (1960). Problems of classification. In: *Principles of Comparative Psychology.* R. H. Waters, D. A. Rethlingshafer & W. E. Caldwell (eds.). New York, N.Y.: McGraw-Hill.

KLOPFER, P. H. & GAMBLE, J. (1966). Maternal "imprinting" in goats: the role of chemical senses. *Z. Tierpsychol.*, **23**, 588–592.

LANYON, W. E. & TAVOLGA, W. N. (eds.) (1960). *Animal sounds and communication.* Washington, D.C.: American Institute of Biological Sciences.

LIDDELL, H. S. (1954). Conditioning and emotions. *Scient. Am.,* **190,** 48–57.

MARDEN, W. G. R. (1954). New advances in the electroejaculation of the bull. *J. Dairy Sci.,* **37,** 556–561.

MCKENZIE, F. F. & PHILLIPS, R. W. (1930). Some observations on the estrual cycle in sheep. *Rec. Proc. Am. Soc. Anim. Prod.,* **23,** 138–141.

MONTGOMERY, G. G. (1957). Some aspects of the sociality of the domestic horse. *Trans. Kans. Acad. Sci.,* **60,** 419–424.

MUNRO, J. (1955). Observations on the suckling and grazing behaviour of lambs. *Agric. Progr.,* **30,** 129–131.

MUNRO, J. (1956). Observations on the suckling behaviour of young lambs. *Brit. J. Anim. Behav.,* **4,** 34–36.

MUNRO, J. & INKSON, R. H. E. (1957). The effects of different suckling frequencies on the quality of milk consumed by young lambs. *J. agric. Sci.,* **49,** 169–173.

NALBANDOV, A. V., HOCHHOUSER, M. & DUGAS, M. (1945). A study of the effect of pro-lactin on broodiness and on cock testes. *Endocrinology,* **36,** 251–258.

PARKES, A. S. (1960). The role of odorous substances in mammalian reproduction. *J. Reprod. Fert.,* **1,** 312–314.

PAWLOWSKI, A. A., DENENBERG, V. H. & ZARROW, M. X. (1961). Prolonged alcohol con-sumption in the rat: 2. Acquisition and extinction of an escape response. *Quart. Jl. Stud. Alcohol,* **22,** 232–240.

PORTER, R. W., CAVANAUGH, E. B., CRITCHLOW, B. V. & SAWYER, C. H. (1957). Localized changes in electrical activity of the hypothalamus in estrous cats following vaginal stimulation. *Am. J. Physiol.,* **189,** 145–151.

RABINOVITCH, M. S. & ROSVOLD, H. E. (1951). A closed-field intelligence test for rats. *Canad. J. Psychol.,* **5,** 122–128.

RAMSAY, A. O. (1953). Variations in the development of broodiness in fowl. *Behaviour,* **5,** 51–57.

RICHES, J. H. & WATSON, R. H. (1954). The influence of the introduction of rams on the incidence of oestrus in Merino ewes. *Aust. J. agric. Res.,* **5,** 141–147.

ROLLINSON, D. H. L., HARKER, K. W. & TAYLOR, J. I. (1955). Studies on the habits of Zebu cattle. III. Water consumption of Zebu cattle. *J. agric. Sci.,* **46,** 123–129.

ROSENBLATT, J. S. & ARONSON, L. R. (1958). The decline of sexual behaviour in male cats after castration with special reference to the role of prior sexual experience. *Behaviour,* **12,** 285–338.

ROSS, S. & DENENBERG, V. H. (1960). Innate behaviour: the organism in its environment. In: *Principles of Comparative Psychology.* R. H. Waters, D. A. Rethlingshafer & W. E. Caldwell (eds.). New York, N.Y.: McGraw-Hill.

ROSS, S., DENENBERG, V. H., SAWIN, P. B. & MEYERS, P. (1956). Changes in nest building behaviour in multiparous rabbits. *Brit. J. anim. Behav.,* **4,** 69–74.

ROSS, S., GINSBURG, B. E. & DENENBERG, V. H. (1957). The use of the split-litter tech-nique in psychological research. *Psychol. Bull.,* **54,** 145–151.

ROTTENSTEN, K. & TOUCHBERRY, R. W. (1957). Observations on the degree of expression of estrus in cattle. *J. Dairy Sci.,* **40,** 1457–1465.

ROWSON, L. E. A. & MURDOCH, M. I. (1954). Electrical ejaculation in the bull. *Vet. Res.,* **66,** 326–327.

SANDERSON, G. C. (1966). The study of mammal movements—a review. *J. Wildl. Mgmt,* **30,** 215–235.

SAWIN, P. B. & CRARY, D. D. (1953). Genetic and physiological background of reproduc-tion in the rabbit: II. Some racial differences in the pattern of maternal behaviour. *Behaviour,* **8,** 128–146.

SCHEIN, M. W. & FOHRMAN, M. H. (1955). Social dominance relationships in a herd of dairy cattle. *Brit. J. Anim. Behav.,* **3,** 45–55.

SCHEIN, M. W. & HALE, E. B. (1957). The head as a stimulus for orientation and arousal of sexual behavior of male turkeys. *Anat. Rec.,* **128,** 617 (Abstract).

SCHEIN, M. W. & HALE, E. B. (1959). The effect of early social experience on male sexual behaviour of androgen-injected turkeys. *Anim. Behav.,* **7,** 189–200.

SCHLEIDT, M. (1955). Untersuchungen über die Auslösung des Kollerns beim Truthahn (*Meleagris gallopavo*). *Z. Tierpsychol.*, **11**, 417–435.

SCOTT, J. P. (1960). Comparative social psychology. In: *Principles of Comparative Psychology*. R. H. Waters, D. A. Rethlingshafer & W. E. Caldwell (eds.). New York, N.Y.: McGraw-Hill.

SIGNORET, J. P. & DU MESNIL DU BUISSON, F. (1961). Etude du comportement de la truie en oestrus. *Proc. 4th Int. Congr. Anim. Reprod.* (The Hague).

SNEDECOR, G. W. (1956). *Statistical Methods.* (5th ed.) Ames, Iowa: State College Press.

STELLAR, E. & HILL, J. H. (1952). The rat's rate of drinking as a function of water deprivation. *J. comp. physiol. Psychol.*, **45**, 92–102.

STEVENS, S. S. (1951). Mathematics, measurement, and psycho-physics. In: *Handbook of Experimental Psychology.* S. S. Stevens (ed.). New York, N.Y.: Wiley.

STONE, C. P. (1925). The effects of cerebral destruction on the sexual behavior of male rabbits. I. The olfactory bulbs. *Am. J. Physiol.*, **71**, 430–435.

STONE, C. P. (1939). Copulatory activity in adult male rats following castration and injections of testosterone propionate. *Endocrinology*, **24**, 165–174.

THORPE, W. H. (1956). *Learning and Instinct in Animals.* London: Methuen.

TINBERGEN, N. (1951). *The Study of Instinct.* Oxford: Clarendon Press.

TULLOH, N. M. (1961). Behaviour of cattle in yards. II. A study of temperament. *Anim. Behav.*, **9**, 25–30.

WEIR, W. C. & TORELL, D. T. (1959). Selective grazing by sheep as shown by a comparison of the chemical composition of range and pasture forage obtained by hand clipping and that collected by oesophageal-fistulated sheep. *J. Anim. Sci.*, **18**, 641–649.

WIERZBOWSKI, S. (1959). Odruchy płciowe ogierów. *Rozcniki nauk rolniczych*, **73**, 753–788. (English summary.)

WIGGINS, E. L., TERRILL, C. E. & EMIK, L. O. (1953). Relationships between libido, semen characteristics and fertility in range rams. *J. Anim. Sci.*, **12**, 684–693.

WINCHESTER, C. F. & MORRIS, M. J. (1956). Water intake rates of cattle. *J. Anim. Sci.*, **15**, 722–740.

WOODBURY, A. M. (1941). Changing the "hook-order" in cows. *Ecology*, **22**, 410–411.

WOOD-GUSH, D. G. M. (1957). Aggression and sexual activity in the Brown Leghorn cock. *Brit. J. Anim. Behav.*, **5**, 1–5.

WOOD-GUSH, D. G. M. (1960). A study of sex drive of two strains of cockerels through three generations. *Anim. Behav.*, **8**, 43–53.

YOUNG, W. C., DEMPSEY, E. W., HAGQUIST, C. W. & BOLING, J. L. (1937). The determination of heat in the guinea pig. *J. Lab. clin. Med.*, **23**, 300–303.

YOUNG, W. C. & GRUNT, J. A. (1951). The pattern and measurement of sexual behavior in the male guinea pig. *J. comp. physiol. Psychol.*, **44**, 492–500.

ZITRIN, A., JAYNES, J. & BEACH, F. A. (1956). Neural mediation of mating in male cats. III. Contributions of occipital, parietal, and temporal cortex. *J. comp. Neurol.*, **105**, 111–125.

Part Three

Behaviour of Mammals

Chapter 9

The Behaviour of Cattle

E. S. E. Hafez, M. W. Schein and R. Ewbank

The genus *Bos*, which includes all types of cattle, is commonly separated into four major categories: the taurine group (domestic cattle), the bibovine group (the gaur, gayal, and banteng), the bisontine group (the yak and the bison), and the bubaline group (the Eurasian buffaloes) (Rice *et al.*, 1957). The taurine group includes the two domesticated species with which we are chiefly concerned: *B. taurus*, cattle of the temperate and warm temperate zones, commonly called "European" cattle; and *B. indicus*, the Zebu or humped cattle of India. The European cattle, *B. taurus*, are probably descended from the wild Aurochs (*B. primigenius*) and Celtic Shorthorns (*B. longifrons*), while the Zebu cattle, *B. indicus*, are thought, by some, to descend from the Malayan banteng (see discussion in Zeuner, 1963). Despite their different origins, the behaviour patterns of European and Zebu cattle are quite similar, and the two groups will be discussed together unless the data clearly indicate species differences.

I. INGESTIVE BEHAVIOUR

Ingestive behaviour refers to the sequence of events leading to and culminating in the intake of feed or water (Table 19). The descriptive terms, *grazing, browsing, feeding, suckling*, and *drinking* refer to eating fresh herbage (usually grass) in the field, eating parts of shrubs and trees, eating prepared foods (concentrates or roughages), taking milk directly from the mother or from an artificial teat device, and taking in water, in that order. Each activity will be discussed in terms of the *patterns* of behaviour involved, the *intensity* or level of behaviour, and, insofar as is known, the *stimuli* which elicit the behavioural components which make up each complete pattern or sequence.

A. Grazing Behaviour

Modern studies of the grazing behaviour of cattle began with Cory in the 1920s followed during the 1940s by Johnstone-Wallace, and during the 1950s and early 1960s by the extensive studies of the New Zealand and British investigators. These investigations were based on the behaviour of individual animals, as well as of herds (reviewed by Hancock, 1953; Tribe, 1955; Waite, 1963).

1. PATTERNS OF GRAZING

The pattern of grazing behaviour of each herd member is relatively stereotyped. Slowly moving across the pasture, each animal keeps its muzzle close to the ground as it bites and tears off mouthfuls of grass which it swallows without much chewing. Cattle generally stand to graze, although very young calves will sometimes graze from the lying position (Roy *et al.*, 1955). Occasionally, an adult animal will kneel on its forelegs to reach some desirable herbage under a fence or barrier. As the animal consumes the preferred herbage immediately under its muzzle it shifts slightly to the right, left, or forwards to select the next bite. Thus, the head describes an arc of about 60 to 90 degrees, parallel to the ground, and on a uniform sward the animal mows a path perhaps twice its body width as it progresses slowly through the field. On mixed, and especially on rough pastures, the animals move about much more and no clear path is created.

Cattle differ from horses in the way they take in grass. Horses have both upper and lower incisors, and grasp the grass between their teeth to tear off each mouthful. Cattle, on the other hand, have no upper incisors and therefore use their highly mobile tongues as prehensile organs. The tongue emerges from the mouth, encircles a small stand of grass, and draws it into the mouth. Pinching action between the tongue and lower teeth then binds the grass so that it can be torn off. The structure of the lower jaw makes it impossible for cattle to graze closer than $\frac{1}{2}$ inch from the soil (Voisin, 1959), while sheep can graze virtually at soil level.

Table 19. A Summary of Ingestive Behaviour in Cattle. (Averages of data from Johnstone-Wallace & Kennedy, 1944; Hancock, 1950a, b; Tribe, 1950, 1955; Schmidt *et al.*, 1951; Castle & Halley, 1953; Corbett, 1953; Wardrop, 1953; Harker & Rollinson, 1961.)

Behaviour		*Values*[1] (over 24-hr period)
Grazing	Grazing time (hours)	4–9[2]
	Total number of bites	24,000
	Rate of grazing (bites/minute)	50–80
	Amount grazed (fresh herbage)	10% of animal weight
	Amount grazed (dry-matter, lb)	13–27
	Grazing distance (miles)	2–3
Rumination	Rumination time (hours)	4–9
	No. of rumination periods	15–20
	No. of boluses	360
	No. of bites/bolus	48
Drinking	No. of drinks (per day)	1–4
Activity	Time spent lying down (hours)	9–12
	Time spent loafing[3] (hours)	8–9

[1] These values and other values presented in this chapter represent average figures reported in the literature. Such values are subject to variations according to type, breed and age of animal, climate, condition of pasture, and managerial factors.

[2] Wagnon (1963) reports that beef cows under range conditions graze up to 14 hours a day and loaf for as little as 2 hours.

[3] Standing, but not grazing (can be ruminating).

The distance travelled during grazing (Table 19) averages about 2½ miles per day (reviewed by Hancock, 1953) and for the most part is covered in daylight. The grazing distance increases in warm, wet, or windy weather and where flies and external parasites prevail. Cattle grazing on old pastures travel twice as far as those on new pastures.

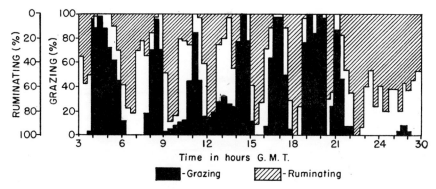

Fig. 24. Periodicity of grazing habits of 8 steers over a period of 24 hours; blank areas represent idling times.[1] (Adapted from Hughes & Reid, 1951.)

Individuals within a group of grazing cattle are usually orientated in the same direction, whereas in other situations, e.g. "idling[1]", body axes are randomly orientated. A suggested explanation for this phenomenon is presented below in the section on Leadership-Followership Patterns and Movement-Orders.

Grazing cycles. Cattle graze during 4 or 5 periods each 24 hours (Wardrop, 1953). The major periods of grazing during the day occur just before dawn, during mid-morning, in the early afternoon, and near sundown. The most continuous grazing periods occur in the early morning and near sundown (Fig. 24). Thus in a given locality, the commencement of early morning grazing is correlated with the season of the year (Fig. 25). Hughes & Reid (1951) reported that no definite pattern of grazing could be recognized in the daytime between the two primary periods. In beef cattle, intense grazing activity occurs for several hours from just before daybreak onwards, and again from late afternoon until darkness (Sheppard *et al.*, 1957; Wagnon, 1963). Grazing "peaks" during the middle of the day are variable in number and extent from day to day (Tayler, 1953).

Under some conditions, 1 or 2 grazing periods occur during the night, but they are less sharply defined than the day periods. Night grazing may be more frequent during the summer and under tropical conditions (European cattle), since in hot weather the animals prefer to graze during the cooler mornings and evenings; in the hot midday, they idle, rest, or ruminate. Smith (1959) reports, however, that high day temperatures do not cause Zebu cattle to increase the proportion of night grazing. On the other hand, European cattle under tropical conditions cut down the hours spent grazing

[1] Standing or lying and neither eating nor ruminating.

in a day during the hot months of the year (Payne *et al.*, 1951). European cattle in Louisiana grazed 1·9 hours in the day and 6·5 hours at night when the day temperature was 85°F, and 4·5 hours in the day and 4·7 hours at night when the day temperature was 72°F (Seath & Miller, 1946).

Subtle changes in patterns are evident during the course of each grazing period. Initially, grazing is intermittent and selectivity is low: grass stems

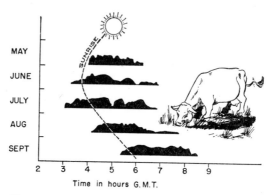

Fig. 25. Seasonal variation in the time of first daylight grazing. Note the close relationship with sunrise. The height of the black curves represents the intensity of herd grazing. (Adapted from Hughes & Reid, 1951.)

are ingested along with the leafy pasture. As the period progresses, grazing becomes steady and selectivity gradually increases. Towards the end, grazing again becomes intermittent and selectivity is marked with grass stems being completely avoided.

Periods of grazing alternate with periods of exploring, idling, and ruminating. In general, the herd functions as a unit: all members engage in the same activity at the same time, whether it be grazing, idling, or ruminating (Hancock, 1950a; Taylor *et al.*, 1955). Yates & Larkin (1965) suggest that, for least error, grazing animals should be weighed at the time of minimum fill and that this should be determined by studying their patterns of feeding activity.

Factors affecting grazing patterns. Breed differences in grazing patterns are related to the climatic adaptability of the animals (see Chapter 4) and to the capacity of the digestive tract. The bison is adapted to severe cold climates, European breeds to temperate climates, and Zebu breeds to the tropics and subtropics. During the winter in Canada, cattalo (hybrid between domestic cattle and bison) graze more frequently than do European breeds (Smoliak & Peters, 1955). It is generally held that in the tropics and subtropics, the Afrikander and Zebu breeds graze longer hours and travel greater distances than the European breeds (Rhoad, 1938; Miller *et al.*, 1951; Bonsma & Roux, 1953). This may be related to the fact that the capacity of the Zebu digestive system is considerably smaller than that of the European

(Swett *et al.*, 1961). Hence, because of limited capacity, the Zebu is forced to graze more often and to take less per grazing period than European cattle. There is one report, however (Lampkin & Quarterman, 1962), which indicates that in a mixed herd run in a hot humid climate, Zebu and European type cattle grazed for equal lengths of time each day and that the Zebu ruminated for a shorter period than the European cattle. This latter point may be partly explained by Phillips's suggestion (1961) that Zebus digest organic matter more efficiently than European cattle.

Differences in grazing patterns exist even among the European breeds. Holstein and Jersey cattle graze at the same rate at comfortable temperatures,

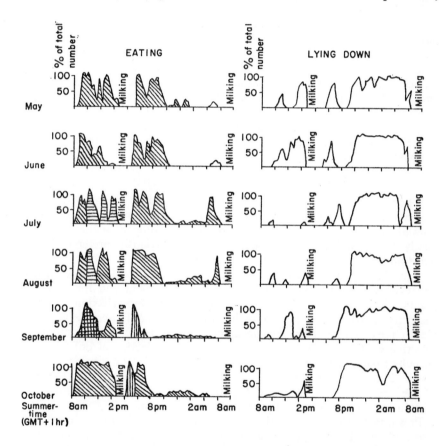

Fig. 26. Feeding behaviour in dairy cattle. The number of cows is expressed as a percentage of the total at each particular observation. There is a grazing or eating peak immediately after the cattle enter the field, followed by a gradual decline in the numbers grazing until the afternoon milking. Similar double or single grazing peaks occur after the afternoon milking but steadily lessen until 10 p.m. Lying-down time occurs mostly between the two milkings during the day. There is no apparent monthly variation in these patterns. (Diagonal hatching: grass; horizontal hatching: alfalfa lucerne; vertical hatching: hay; square hatching: silage.) (From Castle & Halley, 1953.)

but when the environmental temperature rises above the comfort zone Jerseys graze more than Holsteins; at extremely high temperatures there is again little difference (Miller et al., 1951).

Newborn calves do not graze. However, after a few days of age they take bites off grass, swallow after imperfect mastication, and rechew thoroughly during rumination (Brownlee, 1950). Periodic suckling is followed by intensive grazing periods. Calves are highly selective, "muzzling" herbage at length before eating it.

In dairy herds, there is a grazing or eating peak immediately after morning milking, followed by a gradual decline in the numbers grazing until the afternoon milking (Fig. 26). Harker et al. (1956) report that Zebu cattle in the tropics prefer to graze during daylight, but will graze at night on good pasture. In New Zealand, set-stocked calves graze 2 hours longer than those on rotational grazing (McArthur, 1952). The difference in grazing time may reflect more careful selection of suitable herbage. Very inclement weather reduces grazing time: in strong winds and driving rains, cattle cease grazing and stand motionless, or seek available shelter. Intensive grazing occurs between storms or showers, regardless of the time of day or night (Hancock, 1953). Ambient temperature has a strong influence on group structure: at high temperatures, distances between individuals are greater than at low temperatures (Voisin, 1959).

2. STIMULI ELICITING GRAZING BEHAVIOUR

When a field is surveyed after it has been grazed by a herd of cattle, the viewer is impressed by the fact that the field does not have a smooth evenly-mowed appearance. Some patches are closely cropped, while others are barely touched, indicating that the animals obviously preferred certain items and definitely rejected others. Cattle exhibit preferences not only for certain plant species, but for the same species at different stages of growth, and even for the various parts of an individual plant and for individual plants within a species. The phenomenon of specific selection is termed "selective grazing" (see general review by Fontenot & Blaser, 1965). As an aid to range and pasture management, Voisin (1959) further subdivides selective grazing into *progressive defoliation*, the selection of only the most succulent parts of the part, and *creaming*, the selection of only certain plant species in the pasture. The degree of selectivity is inversely related to the age of the animal: young calves may strip single leaves from stalky plants, whereas adults are more likely to take the entire plant.

Presumably through experience, cows learn visually to distinguish between plants. They tend to select leafy herbage of about 5 inches or less in height, which may, in part, reflect the lesser physical effort required to gather young grass with a low fibre content than is required with highly fibrous material. Young animals probably learn that certain plant materials are easily bitten off and require little chewing before swallowing, and it is only coincidence that such material is at the same time easily digestible.

It seems almost superfluous to point out that the hungrier the animal, the less fastidious it is about what it will eat. Somewhat less obvious and still debatable is whether or not the animal actively selects and prefers those items which it specifically needs ("nutritional wisdom"). For example, phosphorus-deficient cattle show the characteristic behaviour of osteophagia and/or allotriophagia: bones, cinders, sacks, and tins are readily eaten by these animals, but ignored by normal ones (Theiler *et al.*, 1924; Green, 1925). However, when cattle on phosphorus-deficient pastures were offered a choice between a calcium supplement and a calcium-phosphorus supplement, they failed to show preference for the latter (Gordon *et al.*, 1954). Voisin (1954) suggests that preferences for certain weed species may be associated with their high antibiotic content. Lofgreen *et al.* (1956) indicate that the net energy of total digestible nutrients (TDN) in forage selected by grazing steers may be greater than that in harvested fresh forage or hay made from the same crop. Although these observations are interesting, the authors feel that the definitive experiment on "nutritional wisdom" remains to be done.

With respect to the myriad of stimuli impinging on an animal from its environment, it must be able to ignore the vast majority which are of no immediate concern and then selectively distinguish between the few stimuli which are important at the moment. The ability to distinguish between two stimuli, of course, is basic to the exhibition of a preference. Since information from the environment reaches the animal by way of its sensory receptors, it is desirable to determine how these might be involved in eliciting grazing behaviour.

Gustatory stimuli. The taste of ingested material probably provides the ultimate basis upon which the animal forms a decision as to whether the material is highly preferred, merely tolerated, or completely rejected. Other information regarding the material, either olfactory or visual, may then be learned as a secondary reinforcer, but the primary test of a material is its taste.

Taste has been studied by electrically stimulating taste receptors (Liljestrand & Zotterman, 1954) and by measuring thresholds of discrimination. Bell & Williams (1959, 1960) measured the threshold for acceptance and rejection of a variety of sapid substances. Cattle react more negatively to bitter tastes than to salt or sour solutions, and accept fairly high concentrations of glucose (Table 20). In several hundred tests with 8- to 16-week old calves, Kare (1961a, b) found only a few that did not respond positively to 1 per cent sucrose; glucose and fructose were also preferred by the calves, but higher concentrations were necessary. Xylose, which is extremely offensive to chickens, was significantly preferred in concentrations of 1 per cent and higher by calves. Maltose at 1 per cent was not appealing to the calf, nor in general was saccharin. Calves have an unusual tolerance for acids: HCl and H_2SO_4, at pH3, were preferred over distilled water during 6-day trials (Kare, 1961a).

Bell & Williams (1960) induced sodium deficiency in one member of a pair of monozygotic twin calves by exteriorizing a parotid duct: the continuous loss of saliva caused sodium depletion. The calf accepted a higher concentration of sodium chloride (up to 5 per cent) and sodium bicarbonate (up to 8 per cent) than the normal calf (0·03 per cent, 0·25 per cent, respectively). When allowed continuous access to either 1 per cent NaCl or 2 per cent $NaHCO_3$, the calf took enough to maintain normal sodium

Table 20. Gustatory Thresholds of Calves to the Four Taste Modalities as Measured[1] by Taste Rejection and Acceptance. (Adapted from Bell & Williams, 1959.)

Taste Modality	Test Substance	Threshold Amount (moles)	
		Rejection[2]	Acceptance[3]
Bitter	Quinine dihydrochloride	0·000097	0·000024
Sour	Acetic acid	0·026	0·0083
Salt	Sodium chloride	0·42	0·105
Sweet	Glucose	No apparent rejection	1·11

[1] For each taste modality, the calf was allowed to choose between 2 buckets, one filled with tap water (control) and one filled with varying concentrations of the test substance.
[2] Consumed less than 20% of solution.
[3] Consumed 50% of test solution.

metabolism as indicated by a normal salivary Na/K ratio. Some evidence suggests that the threshold of taste discrimination is genetically controlled. Cattle and goats are markedly different in the threshold for bitter taste (see Chapter 12). Using four pairs of monozygotic twin calves, Bell & Williams (1959) found that the threshold of taste discrimination for quinine was similar within each pair and different among pairs.

Preference lists for different types of herbage reported by Ivins (1952), Voisin (1959), Cowlishaw & Alder (1960), and others are conflicting. It seems difficult to establish cattle preferences for herbage within the range of digestible species without taking into account the effects of early experience, local environmental variation, and individual variation. Kare (1961b) reported that for any one food, variability may be extreme: 1 per cent fructose was accepted with indifference by some calves and highly preferred by others. On the other hand, most calves show a uniformly high preference (over water) for a 1 per cent solution of sucrose.

Some plants develop chemical materials with a strong taste at late stages of growth and are then rejected by cattle. A high coumarin content is usually unacceptable, at least until animals have become accustomed to it. While the acceptability of herbage is not usually associated with the herbage's mineral content or nutritive value, it is sometimes affected by factors like soil, fertilizers, season, and stage of growth (reviewed by Ivins, 1955). Wavy hair grass (*Aira coepitosa*) is eaten in quantity by cattle in mountain regions, but is not touched at all in the plains, because at lower altitudes it is harder and tougher (Mott, cited by Voisin, 1959). Acceptability of a pasture may be increased by phosphate or potash fertilizers (Klapp, cited by Voisin, 1959); this may be caused by a modification of the flora or by a particular taste conferred by the fertilizer on the plant. The cutting and wilting of herbage usually changes its palatability: shrubs or tufted herbage which have been rejected are often eaten after they have been cut. Acceptability of herbage may also be increased by the use of common salt, urea, and molasses (Harker & Rollinson, 1961).

Olfactory and tactile stimuli. During grazing, cattle constantly "sniff" the herbage. Whether certain grass species are preferred or rejected on the basis of odour alone is a moot question. The odour of an exotic grass species or of an area contaminated with faeces may influence grazing selectivity: herbage contaminated with excreta is usually rejected, but if the pasture is widely contaminated then the herbage is eaten (Tribe, 1955).

The sense of touch plays a major role in determining which items are rejected and which preferred. For example, harshness and hairiness tend to render a species less acceptable: thistles and their relatives are generally avoided except in cases of extreme deprivation. On a very bare pasture, Brownlee (1951) watched hungry cattle eat stinging nettles; while doing so they licked their muzzles or rubbed them with their hind feet, presumably because of the smarting from the stings.

3. Grazing Intake

Time spent grazing. Cattle spend roughly 4 to 9 hours (in exceptional cases up to 15 hours—see Wagnon, 1963) a day grazing (Table 19), with an additional 2 hours per day going to or searching for suitable grazing sites. The grazing time differs significantly from day to day, with less variation between consecutive days than among separated days (Sheppard et al., 1957). In beef cattle, the grazing time ranged from 6·7 hours on orchard grass-redtop Korean lespedeza to 7·9 hours on orchard grass-ladino clover. Fescue is less palatable than orchard grass, whether grown alone or with clover. However, even with these palatability differences, cattle graze about the same length of time on each grass alone or in a mixture. Due to the open nature of swards, grazing time is one hour longer on cocksfoot (*Dactylis glomerata*) pasture than on a dense permanent pasture (Waite et al., 1951). In addition, when dry matter of herbage per acre falls from 1·5 to 0·3 ton, the grazing time increases from 5¾ to 7½ hours (Halley, 1953). If the cattle are given feed supplements, grazing time can be reduced by 7 to 10 per cent (Hardison et al., 1956; Holder, 1960). In the subtropics, Holstein cattle graze an average of 1·5 hours in the sunshine and 0·4 hours in the shade during the day, and 4·7 hours at night (Seath & Miller, 1946). In such areas, the use of sprinklers on cattle may cause an increase in grazing time (Frye et al., 1950).

Rate of grazing. The number of bites per minute and the size of each bite determine the feed intake. The rate of normal grazing in cattle ranges from 50 to 70 bites per minute (Hancock, 1952) but under very favourable grazing conditions increases to as many as 80 bites per minute (Table 19). Higher eating rates usually occur when herbage of low fibre and low dry matter content is being consumed (Duckworth & Shirlaw, 1958a). The head is raised between series of bites and ½ minute is taken to complete prehension and swallowing. When cattle are fed on clippings taken from a 4-inch pasture sward, eating may continue for as long as one hour without any head-raising pauses. The grazing rate starts at a maximum and gradually declines toward the end of the grazing cycle. Hancock (1953) observed the grazing rate of 10 pairs of identical twins. The fastest and slowest pair of twins showed grazing rates of 48½ and 37 bites per minute, respectively, while the average difference within twins was 2 bites per minute, thus demonstrating the effect of genetic factors on rate of grazing. There are no indications of diurnal variations in the rate of grazing.

Amount grazed. A fundamental factor determining the amount grazed by a cow is the height of the herbage (under conditions of equal density). In view of the structure of the cow's jaw, the average height of grass allowing the maximum harvest is about 6 inches.

Considerable information has been published about factors which influence the voluntary intake of grass by ruminants (see reviews by Balch & Campling, 1962; Campling, 1964; Fontenot & Blaser, 1965). Regulation of roughage intake is thought to be controlled mainly by the amount of material in the reticulum and rumen. The higher the digestibility of a roughage, the quicker it will move out of the reticulum and rumen, and thus the greater the intake possible per unit time.

Schmidt *et al.* (1951) found that in three dairy breeds (Fleckvieh, Braunvieh, and Hinterwalder, weighing 12·4, 11·4, and 8·5 cwt respectively), the daily quantity of fresh grass consumed was 10 per cent of the live weight. Estimates of the daily dry-matter intake of herbage are 13 lb for beef cattle (Brannon *et al.*, 1954) and 18 to 27 lb for dairy cattle (Waite *et al.*, 1950; Kane *et al.*, 1953) (Table 19). The average daily dry-matter intake for every 100 lb of live weight is 2·7 lb for young animals and 1·7 lb for older ones. Ittner *et al.* (1954) found that yearling Hereford steers ate 60 lb of green silage or 35 lb of wilted silage daily; the daily dry-matter intake was 15 lb for both types of silage. Lactating Guernsey cows ate 25 lb of dry matter daily on permanent fescue-rye grass and crimson clover (McCullough, 1953).

The often stated opinion that bloat occurs because cows eat greedily on "bloat-inducing" pastures has not been substantiated. Several "bloat-preventive" grazing methods have been attempted by Hancock (1955): "break" grazing involved restriction of the area available to cows and "breaks" varied from 7 to 11 per day. Regardless of the number of breaks, grazing with or without restriction gave no effective control of bloat. Cows on potentially dangerous pasture exhibit shorter grazing times with fewer periods of intense grazing.

Factors affecting grazing intake. Inherent individuality is an important factor in grazing. Individual differences may be related to such anatomical characteristics as length or width of the lower jaw, which may result in differences in bites per minute, intake per bite, or efficiency of selective grazing. The differences in grazing times between identical twins were very small, averaging only 7½ minutes, while the range among sets of twins was as great as 2¼ hours (Hancock, 1950b). In general, identical twins tend to graze together and behave similarly while at pasture (Joubert, 1952; Petersen, 1957), although it has also been shown (Ewbank, 1967) that single born calves, which have been reared together in pairs, are usually found together when they are turned out with a grazing herd.

As mentioned above, newborn calves do not graze. None the less, some grazing activity becomes evident within a few days post-partum although the actual intake of grass is still low. Bucket-fed calves learn to graze rapidly, whereas single-suckled animals running with their dams make no serious effort to graze until they are about 6 weeks old (Chambers, 1959). By 8 weeks of age the grazing rate is 15 bites per minute and by 5 months of age the rate reaches the adult level of 50 bites per minute (Table 19).

In the tropics, cattle reduce their heat production by voluntary anorexia (Chapter 4). This reduction in feed consumption as a means of reducing the heat load is reflected in the grazing behaviour. Feed intake and the muscular activity involved in foraging are both reduced by grazing less (Payne et al., 1951; Findlay & Beakley, 1954; Larkin, 1954).

B. Browsing Behaviour

Cattle will sometimes eat parts of shrubs and trees. Under carefully controlled grazing systems, as practised in the dairy areas of the world, browsing does not contribute significantly to the animals' food requirements. However, it becomes important when damage is done to hedges or poisonous plants are eaten. Under range conditions, browsing may contribute a large part of the animals' daily food needs. Innes & Mabey (1964), for example, report that cattle in African plain country may spend 40 per cent of their feeding time browsing, and it has been shown by Payne & MacFarlane (1963) that cattle do not browse indiscriminately, but select mainly browse plants high in nutritive value.

C. Feeding Behaviour

Cattle, especially dairy cows, often obtain an important part of their total feed intake in the barn in the form of hay, silage, prepared grain concentrates, or green chopped forage. In this section we shall be concerned with such extra-grazing intakes, all of which are dependent upon domestication.

1. Patterns of Feeding

The patterns of feeding vary somewhat, according to the physical consistency of the ration. Grains and ground mixtures are usually offered to cows or calves while they are in a stanchion or stall. Feed is poured into a trough, from which the animal eats by gathering up scoopfuls with its tongue and sucking it into the mouth. Since the feed is in the form of moderately small particles, no biting action is necessary, although chewing movements are evident. Protein cakes and pellets are also scooped into the mouth by tongue and lip movements and then vigorously chewed.

The prehension of hay and straw involves the use of the tongue, lips, teeth, head, and neck. A bite of hay is shredded away from the massed entanglement by pulling motions of the head and neck; after separation, the portions of stems still outside the mouth are drawn in by the tongue. The prehension of whole hay is much slower and the chewing more vigorous than with processed hay. Cattle quickly learn how to select the leaves from hay fed in a digestibility trial: they hold the leaves in their mouth and shake the stalks away (Bredon, cited by Harker & Rollinson, 1961). Silage and chopped forages are pulled into the mouth by tongue and lip movements, much like those used in eating grain; mastication follows. Since such feeds are normally chopped into short lengths, there is no entangled mass and pulling and shredding movements of the head are unnecessary. Small roots are eaten in much the same fashion as chopped forage, i.e. drawn into the mouth several at a time by lip and tongue movements. On the other hand, cattle bite and eat large tubers such as sugar beets much as we would eat an apple. Roots and tubers are chewed vigorously. Salt blocks are common on farms and feed-lots; these are licked by cattle and the salt-laden saliva is swallowed. Earth eating (*geophagia*) is widespread in certain parts of Africa. It has been assumed that cattle go to "earth licks" for salt, because the sodium content of herbage is often low; this hypothesis is not supported by chemical analysis (Harker & Rollinson, 1961).

2. SPEED AND DURATION OF FEEDING

There seems to be significant variation between cows in their basic rates of eating. Large cows tend to eat faster than small ones (Burt, 1957). Jersey cattle appear to eat concentrates more slowly than either Guernseys or Friesians, and cows of all breeds take concentrate foods faster in milking parlours than in tie-up cowsheds (Stallcup et al., 1959). Jones et al. (1966) report that (a) within the range of 2–8 lb, the larger the ration, the quicker the cow consumes it, and (b) cubed foods (1·18 min/lb) are eaten more swiftly than meals (1·75 min/lb).

The time spent in the mastication of different rations varies directly with the total number of chews made. Whole alfalfa requires four times as many chews as whole corn; the number of chews required for ground corn is 25 per cent greater than for shelled corn. Duckworth & Shirlaw (1958b) report that wetter diets are bulky and require more bites per pound of dry matter consumed than an equivalent ration in drier form. This supports the suggestion that eating speed depends not only on weight, but also on volume of feed (Weidlich & Wulff, 1961). The percentage of the daily time spent in mastication varies from 2 to 11 per cent, depending on the fineness of the ration (Kick et al., 1937). Younger animals spend more time masticating their feed than do older animals. The average number of jaw movements per minute varies from 60 to 90 according to type of feed. It is estimated that cattle make an average of 42,000 jaw movements daily during mastication and rumination. Typical eating speeds are 3·45 min/lb for silage (Stallcup et al., 1959); 5·0 to 7·1 min/lb for hay (Burt, 1957); and 1·5 to 2·5 min/lb for kale (Burt, 1957).

When dairy cattle are offered hay and silage *ad libitum*, they spend a longer time eating silage than hay: 4·0 hours at the silage face compared with 2·2 hours at the hay (Webb *et al.*, 1963). Lewis & Johnson (1954) found that when cattle fed on hay alone (*ad lib.*), they ate for 5·2 hours daily; later they also fed for 5·2 hours per day on a silage ration (again, *ad lib.*). Many loose housed cattle are kept on self-feed silage and rationed hay. Typical figures quoted for time spent feeding under these conditions are 3·5 to 4·6 hours per day for silage, and 1·4 to 2·25 hours per day for hay (Walker-Love & Laird, 1965; Small, 1966). Cows will not all be present at the silage face at any one time, possibly due to its limited width (Pinfold, 1965). Silage consumption under *ad libitum* conditions seems to be highest just after milkings; a fair amount of eating is also done at night (Small, 1966).

3. Stimuli Releasing Feeding

Since much of the discussion of stimuli releasing grazing behaviour also applies here, we need concern ourselves only with a few words on preferences and acceptability. Nevens (1927) found that, in free-choice feeding, not all cows showed preferences for the same feed and that some feeds were consumed almost exclusively; the preferences of several cows changed suddenly. In a study of feed preferences among dairy cattle, Blaxter (1944) delineated the following generalized order of preference: young herbage, excellent quality hay, certain protein cakes and cattle pellets, green fodders and roots, certain cereal meals and protein meals, average quality hay, and lastly, cereal chaff and straw. Calves offered various grains (corn, milo, and oats) of different degrees of fineness exhibited no preference between grains, but consumed smaller quantities of finely ground preparations (Ray & Drake, 1959). Brown *et al.* (1952) report that ground pellets are less acceptable than whole pellets; that pelleted milo is less acceptable than medium rolled milo (although the grain intake differed only slightly); that ¼-inch pellets were preferred to ½-inch pellets and hard pellets to soft ones. Putnam *et al.* (1967) showed that steers seem to prefer a coarsely ground 25 per cent hay ration to either the same material in a pellet form or to a coarsely ground or pelleted 89 per cent hay food.

4. Feed Intake (Non-grazing)

The amount of grains and ground mixtures consumed is largely determined by the quantity offered. However, cattle on silage eat larger quantities of less nutritious feed than when on pasture, presumably because of the preclusion of selective choosing (Meyer *et al.*, 1956). Dry-matter consumption is higher when animals are fed hay than when they are fed silage or pasture. There is some evidence that cattle eat less hay when fed individually than when group fed (Kidwell *et al.*, 1954), possibly because of the increased anxiety exhibited by temporarily isolated animals.

Feeding behaviour is affected by environmental temperature, condition of teeth, age of cattle, and nature and kind of feed. In general, feed consumption (in a controlled-climate laboratory) is depressed by increasing environmental temperature (Ragsdale *et al.*, 1948). Flies, mosquitoes, dogs, interference by herd-mates, or any condition which may frighten or disturb the animal may lead to a cessation of feeding.

D. Rumination

Rumination is the act of regurgitating, remasticating, and reswallowing previously ingested feed. Large quantities of slightly chewed herbage are consumed and stored in the rumen during grazing. More than simple storage is involved, since the herbage undergoes fermentation in the rumen. Boluses of herbage are later regurgitated from the rumen, thoroughly chewed, and reswallowed. The adaptive significance of this mode of ingestion has often been pointed out: grazing permits the animal to gather a maximum amount of feed with a minimum amount of exposure in an open field situation; rumination permits the animal to finish eating the gathered feed at its leisure, perhaps in a less vulnerable environment.

1. PATTERNS OF RUMINATING BEHAVIOUR

The basic patterns of rumination are quite simple: a bolus of feed is regurgitated and rechewed with lateral grinding jaw movements. Body positions during rumination are many, but typically the animal lies with its forelegs bent under its chest and its hind limbs brought forward so as to lie partly under the body. Cattle show a marked preference for rumination while lying down (65 to 80 per cent of the total rumination time), although the process can and does occur while the animal is either standing quietly or walking slowly. In wet periods, such as after heavy rainstorms, animals may spend much more time standing; thus, there are more "standing" rumination periods. The transition from grazing to ruminating is slow and deliberate as the animals become quiet and "settle down". On the other hand, the transition from ruminating to grazing is abrupt, and the latter behaviour often begins before the last bolus has been reswallowed.

2. LEVELS OF RUMINATION

Rumination time. Rumination time includes the time spent in regurgitation, mastication, swallowing of ruminal ingesta, and the short time intervals between boluses. Although ruminating behaviour is evidenced before the calf reaches 3 weeks of age, rumination time does not reach adult levels until the age of 6 to 8 months (Swanson & Harris, 1958). The average daily rumination time ranges from 4 to 9 hours (Table 19), with large daily variations both within and among animals, possibly related to both quantity and quality of herbage consumed. Rumination is divided into 15 to 20 periods scattered throughout the 24-hour day. The length of each rumination period varies from 2 minutes to 1 hour or more. The total number of boluses regurgitated parallels the total number of minutes of rumination. Some 50 to 60 seconds are devoted to the remastication and reinsalivation of each bolus; 4 to 5 seconds are spent in swallowing and regurgitation (Fuller, 1928; Dukes, 1955). The pause at the end of a cycle is 3 to 4 seconds.

The height of ruminating activity is generally reached soon after nightfall and is followed by a gradual continuous decline associated with shortened periods of rumination. Kick *et al.* (1937) report that cutting the herbage into ¼-inch pieces has no effect on rumination time, the number of rumination periods, the number of regurgitated boluses, the number of chews made, or the time spent on each bolus. However, rumination time is reduced when cattle are fed ground hay or concentrates instead of long hay (Balch, 1952).

Rumination : grazing ratio. The time devoted each day to ruminating is approximately three-quarters of that spent in grazing. The relationship of rumination time to grazing time is often expressed as the R:G ratio; it is useful only if grazing is not restricted by management, and is influenced by the abundance of pasture and climate. High quality pasture, abundant per cow as well as per acre, results in a comparatively long grazing time coupled with a short ruminating time, and therefore a low R:G value. Long grazing times occur on sparse pastures. If the herbage is succulent, ruminating time is short and the R:G value is low; if the herbage is poor and fibrous, ruminating time is longer and the R:G value is high (Tribe, 1955). Grazing time is only slightly in excess of ruminating time in mid-summer, but exceeds ruminating time by about 100 per cent in the spring and autumn. Thus, the R:G value varies seasonally from nearly 1·0:1 to approximately 0·5:1 under normal conditions. Such differences reflect seasonal changes in quantity and quality of herbage.

Harker *et al.* (1954) maintain that the R:G value is lower for Zebu than European cattle, i.e. that Zebu cattle must graze longer than European cattle for the same period of cud-chewing. Wilson (1961) disputes this, but the work of Lampkin & Quarterman (1962) (see section on Factors Affecting Grazing Patterns, p. 239) appears, at least in part, to support Harker's view.

Factors affecting rumination time. Hancock (1953) measured the relationship between rumination time and dry-matter content of herbage by offering cows equal quantities of machine-dried grass of different qualities. Using two grasses of different crude fibre content (29 per cent and 18 per cent), he found a direct relationship between dry-matter intake and average rumination time (Fig. 27). In addition, the grass of high fibre content required much

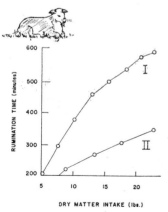

Fig. 27. The relationship between dry-matter intake and rumination times using two grasses of varying crude fibre content. The crude fibre content of grass I was 29 per cent, and of grass II, 17·5 per cent. (Data adapted from Hancock, 1953.)

longer rumination time per pound of either dry matter or crude fibre than the grass of low fibre content. Thus rumination time depends on both the quantity and quality of herbage.

When the ratio of ruminating time to eating time is plotted against total digestible nutrients (TDN), a difference is noted between grass species (Fig. 28). Lofgreen *et al.* (1957) showed that with alfalfa there is a marked change in the ratio of ruminating to eating when the TDN content of the forage changes. This agrees with Wardrop's report (1953) that rumination time is positively correlated with a low moisture and a high fibre content in the consumed herbage.

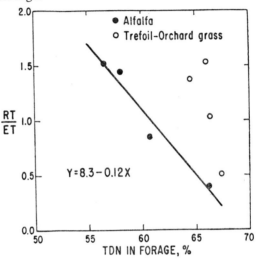

Fig. 28. The relationship between the rumination time (RT) to eating time (ET) ratio and the total digestible nutrients (TDN) in the diet selected by cattle on alfalfa pasture. (From Lofgreen *et al.*, 1957.)

3. STIMULI AFFECTING RUMINATION BEHAVIOUR

Since the nature of the interoceptive stimuli which induce rumination behaviour is unclear, it is considerably easier to discuss the stimuli which stop or interrupt rumination behaviour than to discuss those which initiate it.

Young cattle are usually very timid: any slight disturbance may cause them to stop ruminating. Rumination can also be terminated by a variety of circumstances including hunger, fear, pain, curiosity, or maternal anxiety. A cow deprived of its calf may interrupt rumination to bellow from time to time. Cattle stop ruminating momentarily while listening to an unusual sound and start again if nothing important arises (Brownlee, 1950). Rumination decreases, but does not cease, during oestrus. Cows do not ruminate during certain illnesses, during the final stages of parturition, and in the ensuing period spent licking the calf. (However, some bulls may continue to ruminate during electro-ejaculation.) It is believed that the longer rumination is delayed, the more difficult it is to resume, since grass may become dry and closely packed in the rumen and cause local irritation.

E. Drinking Behaviour

Drinking behaviour refers primarily to the ingestion of water. This behaviour is controlled by interoceptive receptors (thirst) and exteroceptive stimuli (the sight of water). Masai cattle in Africa are reputed to smell water over a hill (Harker & Rollinson, 1961). Social facilitation may occasionally induce some non-thirsty animals to drink.

Cattle drink by dipping their muzzles into the water and sucking the fluid into the mouth. The tongue plays only a passive role in drinking. Only the mouth is submerged while drinking: the nostrils are almost never under water. Water taken into the mouth is swallowed and passes into the rumen. The head need not be raised, for the liquid is actively sucked and flow does not depend on gravity. Cattle pastured on grass drink from 1 to 4 times daily (Table 19); more often in hot seasons or when grazing old pasture (Hancock, 1953). The frequency tends to increase with supplementary feeding, and the highest rates are shown by housed cattle (MacLusky, 1959). Grazing cattle usually drink in the forenoon, late afternoon, and evening; they seldom drink at night or in the early morning. Dairy cattle tend to drink after they have been milked (Small, 1966). MacLusky (1959) reports that the drinking activity of a grazing herd was highest after the evening milking and before sunset. In the tropics, Zebu cattle drink very little at night, but on most days some water is consumed during each hour throughout the day (Rollinson *et al.*, 1955). Dairy cattle sometimes drink between the afternoon milking and dusk, whereas those on pasture drink between grazing cycles.

Water intake. Water intake is influenced by many factors, including breed, age, dry-matter intake, ambient temperature, protein and salt content of the ration, pregnancy, and lactation. Water intake increases during late pregnancy and lactation.

Water intake has been measured in the tropics (Rollinson *et al.*, 1955; Lampkin *et al.*, 1958), subtropics (Ittner *et al.*, 1951, 1954; Kelly *et al.*, 1955), and in psychometric rooms (Ragsdale *et al.*, 1951, 1953; Casady *et al.*, 1956). It has also been measured in relation to changes in environmental temperature and live weight (Bailey & Broster, 1958), and type of production (Winchester & Morris, 1956). In general, hay-fed steers consume 1·5 U.S. gallons[1] per 100 lb of body weight in the subtropics (Ittner *et al.*, 1954); water consumption increases by 0·05 gallons per 1 °F rise in ambient temperature, and by 0·008 gallons per pound increase in live weight. In milk-fed calves, the average daily water consumption, including the water in feed, is 10 lb for every 100 lb of live weight; this does not include the water contained in the milk.

The amount of water consumed by cattle of various ages and physiological states in temperate climates is summarized in Table 21.

Factors affecting water intake. Winchester & Morris (1956) estimated the total water intake of cattle of various classes and sizes. While their estimates, expressed as "total water intake", include the water contained in the feed, for practical purposes they also represent the water that cattle drink when

---

[1] 1 U.S. gallon ≃ 0·8 Imperial gallon.

fed such dry rations as hay and grain (Fig. 29). At any given environmental temperature, European breeds and grade cattle consume more water than Zebu breeds (Lampkin *et al.*, 1958). In the Imperial Valley of California Hereford cattle drank 16 gallons of water daily, whereas Zebu drank only

Table 21. The Water Consumption of Cattle in Temperate Climates. (Adapted from Sykes, 1955.)

Class of Cattle	Conditions	Water Consumption (Pounds per day)
Holstein calves (liquid milk or dried milk and water supplied)	4 weeks of age	10–12
	8 weeks of age	13
	12 weeks of age	18–20
	16 weeks of age	25–28
	20 weeks of age	32–36
	26 weeks of age	33–48
Dairy heifers	Pregnant	60–70
Steers	Maintenance ration	35
	Fattening ration	70
Range cattle	—	35–70
Jersey cows	Milk production 5–30 lb/day	60–102
Holstein cows	Milk production 20–50 lb/day	65–182
	Milk production 80 lb/day	190
	Dry	90

10 gallons (Ittner *et al.*, 1951). The large difference in water consumption, which is even more marked at high ambient temperatures, is related to differences in physiological adaptability, as well as body size.

The relationship between water and feed intakes is of great practical importance in arid pastoral areas of the tropics and subtropics, where the availability of drinking water is the limiting factor for productivity in cattle. Water intake is a function of dry-matter ingestion so that a constant ratio of water intake to dry-matter intake is maintained, all other factors being constant. The restriction of water causes a reduction in dry-matter intake and a reduction in the ratio of water intake to dry-matter intake (Balch *et al.*, 1953; French, 1956). Restriction of water causes a larger decrease in the feed intake of Zebu breeds than in that of European breeds (Phillips, 1960). Water deprivation also causes a steady increase in the inorganic phosphorus level of the blood; when water is again made available this level falls rapidly to a subnormal value, then to normal (Rollinson & Bredon, 1960).

Cattle on high protein rations drink 26 per cent more water than animals on low protein (Ritzman & Benedict, 1924). Salt added to the ration causes a marked increase in water consumption. As might be expected, milk production directly reflects the level of water intake, since milk is approximately 88 per cent water.

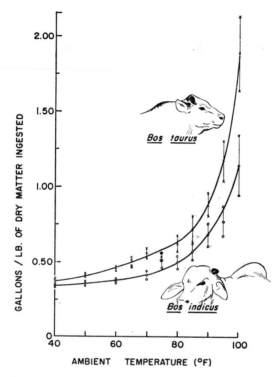

Fig. 29. Water intake expressed as a function of dry-matter consumption and ambient temperature; vertical lines represent standard deviations. European breeds at similar ambient temperatures and consuming feed with the same dry-matter content drink more than the Zebu breeds. (Adapted from Winchester & Morris, 1956.)

F. Suckling Behaviour

Newborn calves begin to suckle within 2 to 5 hours after birth. The neonate's first movements, after standing, are generally directed toward the mother and it finds a teat with seemingly random searching and nuzzling movements.

Until a teat is located, the calf readily mouths and sucks any protuberance on the mother's body. The mother seems to help the calf find a teat by positioning her body appropriately and by licking, nuzzling, and nudging the calf. Typically, the calf stands alongside the mother facing caudally (Plate III), but occasionally a calf's body will be at right angles to that of the mother. The mother must be standing for the calf to suckle: attempts to suckle from a reclining mother are generally unsuccessful. The calf suckles from right or left sides and occasionally from the rear of the cow, with no apparent preference for front or hind teats. When the cow's udder is small (as in heifers), the calf may be able to reach all four teats from one side.

1. PATTERNS OF SUCKLING BEHAVIOUR

The calf grasps the teat with its mouth and sucks vigorously without biting. It develops a negative pressure as great as, and more often in excess of, that required in a milking machine to milk the most hard-milking cow. By wrapping its tongue around the teat, it forms an air-tight compartment in the oral cavity and in this way develops the negative pressures necessary for milk flow. If a tube is inserted into the calf's mouth to prevent the development of a negative pressure, no milk can be obtained (Smith & Petersen, 1945). Negative pressure is produced by enlarging the oral cavity or the laryngeal cavity. Negative pressures were first measured by Krzwanek & Bruggemann (1930) with specially constructed nipples. The maximum negative pressure varies from 10 to 16 inches of Hg (Smith & Petersen, 1945). During suckling, the negative pressure is maximal when direct pressure on the teat is minimal; the reverse is true during swallowing. Individual calves range from those with high direct and low negative air pressure to those with the reverse combination (Martjugin, 1944).

Contrary to earlier opinions, mechanical movements and the position of the calf's head and neck are not significant in determining the course of milk flow through the stomach. Milk is prevented from entering the two fore compartments of the stomach by the closed oesophageal groove. From anatomical evidence, Wise et al. (1942) found that physiological factors regulate the opening and closing of the groove.

Since a calf is usually taller than the cow's udder, it must bend its neck downward and its head upward in order to suckle. By lowering the shoulders, the calf also has better access to the teats on the far side of the udder. This is probably one reason why older and stronger calves suckle far teats more often than weaker and younger animals do.

While suckling, the calf vigorously butts the mother's udder with its head. Presumably, these movements are elicited by a reduced milk flow and the subsequent jarring of the udder may help increase the flow. Peaceful suckling is also frequently accompanied by "wagging" movements of the tail. No obvious explanation is available for this phenomenon, which seems to be characteristic of several other species. However, Selman et al. (1967) suggest that vigorous tail-wagging, together with butting, are due to frustration and only occur when calves cannot find a teat or when no milk is available. It appears that a calf stays on each teat until it is dry, then moves to another.

Nursing involves both mother and calf. The cow initially helps the calf locate her teats. While the calf suckles, she licks its perineal and preputial areas stimulating it to urinate and/or defaecate. Interestingly, cows without calves or those which have recently calved, but no longer possess young, often show little interest in calves which may be present. They may, in fact, be hostile toward them. Nonetheless, Hafez & Lineweaver (1968), for example, report that a single calf suckled its dam and the mother of a pair of twins, whereas the twins never suckled the single calf's dam, even when she encouraged them to nurse. Subtle taste differences in the milks of the two cows could explain this.

In several countries, suckling calves are milk-fed by nipple-feeders or buckets. Calves normally refuse to drink very cold or very hot milk (above body temperature). Conditioning the calf for bucket feed requires physical effort to force the head down into the milk, even when the calf has already tasted the milk in the bucket. Reluctance to drink from the bucket may be related to the posture of the head during normal suckling from the mother. Once they are bucket-trained, certain calves thrust their noses to the bottom of the bucket, whereas others drink slowly from the surface (Walker, 1950).

Non-nutritional sucking behaviour occurs when a calf licks himself, another calf, or a nearby object. Selected licking areas include the penis sheath, scrotum, udder, and ears. No breed differences appear to exist in such activity (Hafez & Lineweaver, 1968). It is most likely to occur in herds where calves are bucket-fed and are grouped together shortly after birth (Wood *et al.*, 1967). Donald & Anderson (1967) have shown that twin calves tend to suck each other more frequently than single calves penned together. They suggest that the predisposition for sucking is either hereditary, comes from early pre- or post-natal environmental influences, or is due to a combination of these factors.

Non-nutritional sucking increases markedly in calves fed low-energy, low-protein diets (Wiltbank *et al.*, 1965), suggesting that it is related to the diet. Swanson (1956) estimates that calves suckling their dams drink more than bucket-fed animals, in which the incidence of non-nutritional sucking is high. Sucking also promotes salivation and the mixing of saliva and milk (Wise *et al.*, 1940). The agitation of milk during suckling also probably increases lipolysis.

However, non-nutritional sucking may occur because a teat is not at hand, since calves apparently spend some time suckling their dams after the udder is empty. Calves on Nurs-ettes suck for 16 to 27 minutes per 24 hours, whereas calves on their dams suckle 37 to 57 minutes per 24 hours (Hafez & Lineweaver, 1968). The fact that the calf changes teats more rapidly at the end of a natural suckling period may also indicate that the mother is dry. Such activity probably not only satisfies a need for sucking, but promotes the digestion of milk, as mentioned above.

From a herdsman's standpoint, non-nutritional sucking has important consequences, particularly if it continues into adult life (Harker & Rollinson, 1961). Such activity markedly decreases the dry matter consumption of calves and retards growth (Hafez & Lineweaver, 1968). Hair balls commonly occur in the rumina of calves which exhibit non-nutritional licking behaviour. These may attain a size of 3,788 g (Wiltbank *et al.*, 1965), and may be fatal if they block the entrance to the rumen and prevent eructation.

2. RATE OF SUCKLING

The rate of suckling and the amount of milk consumed by the young are related to age and size of the calf, breed (beef or dairy), method of suckling (natural, nipple-feeder, or bucket), easiness of milk let-down by the mother, and persistence of the calf during suckling.

After the first feeding, by which time the calf is familiar with the location

9+

of the teats, the calf satisfies its appetite in 10 to 15 minutes. The number of pulsations during suckling by Hereford calves ranges from 57 to 102 with an average of 74 per minute; the pulsation rate decreases near the end of the suckling period and as the calf grows older (Hafez, 1961).

Except for studies in which calves were weighed before and after suckling few accurate estimates of milk consumption exist for calves. The figures given by Walker (1950) represent the amount of milk required by the calf to meet its nutritional needs, rather than the amount actually consumed. Swanson (1956), using identical twin pairs, estimated that suckling calves drink more milk than bucket-fed animals. Calves seem to prefer nipple-feeding in the presence of an attendant to drinking milk from an "iron udder", presumably because of the attention and mothering given by the attendant (Brownlee, 1950). Sham-fed dairy calves consume milk 4 to 6 times more rapidly from a bucket than from a nipple-feeder (Wise et al., 1947). Hafez & Lineweaver (1968), however, have compared the suckling behaviour of Hereford and Holstein calves on dams and automatic milk dispensers (Nurs-ettes). Marked differences were found in the milk consumption not only between the two breeds, but among individual calves of each breed. Calves on their dams suckled an average of 37 to 57 minutes per 24 hours, whereas calves on Nurs-ettes averaged 16 to 42 minutes per 24 hours. These differences were apparently due to the composition of the milk, particularly the percentage of total solids. Calves on an artificial ration containing 19·5 per cent solids, for example, always suckled less than calves on a ration containing only 6·5 per cent solids. The average number of suckling cycles (700 ml milk per cycle) also decreased as the percentage of solids in the milk increased: Hereford and Holstein calves on 6·5 per cent solids averaged 26·9 and 28·7 suckling cycles per 24 hours, respectively; on a ration containing 19·5 per cent solids, they averaged 17·3 and 23·4 suckling cycles per 24 hours. Holstein calves on Nurs-ettes consumed 700 ml of milk in 0·88 to 2·19 minutes, and suckled for 22 to 44 minutes per day; Herefords required 1·19 to 1·49 minutes and suckled for 16 to 17 minutes a day. Calves with their dams suckled 8 to 10 minutes per feeding, and 37–57 minutes per day.

3. FREQUENCY OF SUCKLING

Newborn calves suckle their dams 5 to 8 times in each 24 hours (Walker, 1950), whereas calves on Nurs-ettes suckle as frequently as 55 times a day (Fig. 30), perhaps because the milk in Nurs-ettes is more dilute than normal cow's milk (Hafez & Lineweaver, 1968). Walker (1962) reports that older animals suckle less frequently: beef heifers nurse their 2- to 24-week-old calves only 3 to 5 times each day. Zebu calves suckle an average of 9·5 times per 24 hours at 1 month of age and 5·6 times at 6 months of age (Hutchison et al., 1962). However, Hafez & Lineweaver (1968) report that the suckling frequency of Hereford calves increases with age, probably because of the low milk production of beef cattle. Twin Hereford calves, for example, suckle 11 times a day, whereas single calves suckle only 4 to 6 times each day. If calves on natural photoperiods are placed in constant light, the day-night suckling pattern (Fig. 30) disappears, suggesting that no innate rhythm of suckling activity exists.

PLATE III

a. Dog-sitting posture: Hereford and Aberdeen Angus bulls sometimes rest in a dog-sitting posture with their forequarters supported by their extended forelimbs and their hindquarters resting on the ground. This appears to be a normal posture for bulls of some heavy beef breeds.

b. Nursing-suckling interactions: the cow orientates its body to facilitate nursing and in turn milk-ejection is maintained by the tactile stimulus or suckling.

c. (*left*) Pawing by Camargue bulls: this action sometimes precedes an attack, but is also frequently seen during the play activities of young individuals. (*Right*) Horning by a Camargue bull: a horn is used to dig into and churn the earth. (Photographs by R. Schloeth.)

PLATE IV

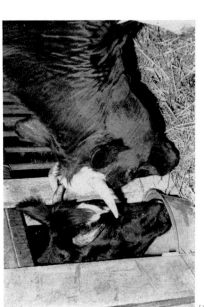

Social behaviour in dairy cattle based on competition for food. *a.* In a control test, the dominant cow (right) feeds and the subordinate cow (left) looks on timidly. *b.* A competitive situation between two blindfolded cows. Note the reaction of the dominant cow (right) when the subordinate one tries to feed. *c.* When the cows are prevented from interacting physically, both dominant and subordinate animals feed from the bucket. (Photographs by M.-F. Boussiou and J. P. Signoret.)

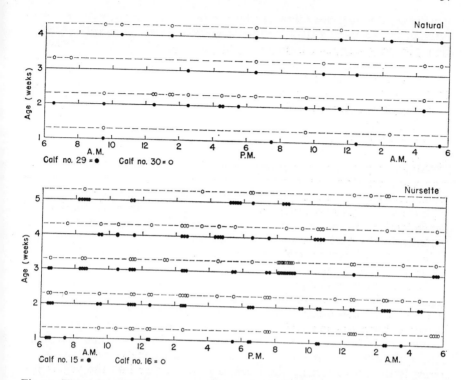

Fig. 30. The daily suckling patterns of Hereford calves reared in pairs, on their dam (*top*) and Nurs-ettes (*bottom*). Circles represent the number of suckling cycles per hour. (From Hafez & Lineweaver, 1968.)

II. SEXUAL BEHAVIOUR

Qualitative differences between the two sexes in sexual motor patterns and in stimuli releasing these patterns are common, indeed typical, in sexually dimorphic species. These differences preclude a comparison of the levels (intensities) of sexual behaviour between sexes. Although it is convenient to discuss the sexual behaviour of the two sexes separately, one must remember that they normally act in concert: the different patterns and stimuli of one sex are highly co-ordinated with those of the other, leading to successful insemination, reproduction, and thus survival of the species.

A recent general account of reproductive behaviour in cattle is to be found in Fraser (1968).

A. Sexual Behaviour of Males

The bull's penis is of the fibro-elastic type, relatively small in diameter, and rigid when non-erect. The amount of contractile tissue is limited, and although the penis becomes more rigid upon erection, it undergoes little enlargement. Protrusion is effected mainly by a straightening of the S-shaped sigmoid flexure and by relaxation of the retractor penis muscles. Copulatory patterns of the male are controlled by the autonomic nervous system. Both erection and ejaculation involve muscular contractions and cortical co-ordination primarily triggered through sacral parasympathetic nerves (Walton, 1955, 1960).

Pre-pubertally castrated bulls show some components of sexual behaviour (Folman & Volcani, 1966) and post-pubertally castrated animals may retain a high degree of sexual activity (Sokolowskyj, 1952).

1. Patterns of Male Sexual Behaviour

Normal copulation encompasses a sequence of behavioural elements: courtship, erection and protrusion, mounting, intromission, ejaculatory thrust and ejaculation, and dismounting. These elements are regarded as typical male behaviour, but for the most part are neither sex specific nor dependent on the possession of male organs. The manifestation of these patterns in the bull varies from full to no expression, typically by the following degrees: the normal complete chain of events including ejaculation; the complete chain of events without ejaculation; courtship and mounting without erection and/or protrusion; failure to mount, although the vulvar region and hindquarters of the oestrous cow are licked; no apparent interest in the oestrous cow other than standing near her or following her; failure to detect an oestrous cow; avoidance of or disinterest in an oestrous cow.

Intromission is the act of inserting the penis into the vagina. *Copulation* (*coitus*) is intromission culminating in ejaculation of semen. *Ejaculatory thrust* is the act of forcing the inserted penis further into the vagina just before ejaculation.

Courtship is more in evidence on open range than under the restricted conditions on small farms. The bull detects the pro-oestrous cow two days before oestrus and remains in her general vicinity. Schloeth (1961) terms this activity "guarding" (*hüten*) of the female by the male. In guarding, the body axes of the pair are parallel and body orientations are head to tail. Guarding encompasses more time than any other form of sexual behaviour in cattle on free range, and occurs both in and out of the breeding season (Schloeth, *loc. cit.*). A condensed (in time) version of this behaviour is shown by confined cattle. During guarding, the bull may attempt several mounts, showing partial erection and protrusion and perhaps the dribbling of accessory fluid. These mounts are usually unsuccessful because the female does not "stand".

As the cow reaches oestrus proper, the bull becomes intensely excited and follows her closely; he frequently mounts her or licks and smells around her external genitalia. Licking and smelling are often followed by a curling of the upper lip so as to expose the upper gum region while the head is held horizontal with the neck extended and nares distended; this posture has been termed *flehmen* (Schneider, 1931) or *la moue* (Schloeth, 1958a). The adaptive significance of *flehmen* is unknown, although it commonly appears in the pre-copulatory patterns of many other ungulates. The bull frequently displays what is termed masculinity by pawing and horning the ground (Plate III), throwing dirt over his back and withers, and rubbing his head and neck on the ground (described by Schloeth, 1958a, b). He may also snort with head lowered and nostrils distended, thereby threatening and chasing away young bulls or cows not in oestrus (Kerruish, 1955).

Chin-resting often appears before mounting: the bull orientates himself behind the cow and raises his head so that the chin and throat are in contact with the cow's rump; indeed, a mild forward pressure is exerted by the chin, throat, and chest against the female. Such behaviour superficially resembles and may be evolutionarily related to the *neck-thrusting* movements of other

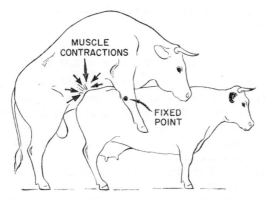

Fig. 31. Contraction of abdomen muscles in the bull during copulation. (Adapted from Hultnäs, 1961.)

ungulates, such as the Uganda kob (*Adenota kob*) (Buechner, 1961) and the Indian blackbuck (*Antelope cervicapra*) (Walther, 1958; Etkin & Schein, 1960): in these species such movements serve to drive oestrous females to specific mating territories or arenas. Non-receptive animals, male or female, respond to chin-resting by escape and avoidance. Oestrous cows respond to chin-resting pressure by "standing" or even exerting some back pressure; the bull then mounts.

Copulation in cattle is brief when compared to that in horses and swine. In mounting, the bull quickly shifts his weight to the hind legs, lifts his shoulders and forelegs off the ground, and moves forward to straddle the cow near the middle of her back. The penis is near the vaginal orifice; its free end sways to and fro until the *glans* reaches and penetrates the vagina. After further intromission, the cow's vulvar sphincter contracts around the penis. As a result of a sudden contraction of the bull's abdominal muscles (particularly the *rectus abdominis* muscle), while the forelegs are "fixed" on the cow's pelvis, the bull's pelvic region is quickly brought into direct apposition to the cow's genitalia and maximum intromission leading to ejaculation is achieved (Fig. 31). The force of muscular contraction is so strong that the bull's hind legs are often lifted completely off the ground, giving the appearance of an active leap. That no active leap occurs can be demonstrated by not permitting the bull a fixed point for his forelegs: in this case, the abdominal contraction draws the forelegs backwards while the hindlegs remain in place (Hultnäs, 1961). Thus, a cow with a narrow pelvis may be a poor choice for use as a teaser in an artificial insemination centre, other conditions being equal, since the "fix point" for the bull may be inadequate (Hulnäs, *loc. cit.*). Intromission is performed quickly, the ejaculatory thrust is given with maximum vigour and semen is ejaculated as a single violent gush near the *os cervix*. The abdominal muscles then relax and the bull dismounts slowly. In the process, he usually draws the lower surface of his jaw across the cow's back. This action has been used as a basis for marking cattle which have been mated. A crayon block is attached to the lower part of a strong leather headcollar. As the bull dismounts he marks the back of the cow (Yeates, 1965).

2. STIMULI ELICITING MALE SEXUAL BEHAVIOUR

In order to assess and interpret the stimuli which release any behaviour, it is important to first recognize which sensory modalities are involved. Certainly, the recognition of and response to a particular stimulus implies that the stimulus is perceived as a quality or entity distinguishable from background. We need to be aware of the sensory modalities involved, if only to maintain more efficiently the stimulus pressure so important to the adequate expression of the behaviour. We must also recognize that each sensory modality is capable of conveying information which might temporarily repress sexual excitation, either through the mechanism of negative conditioning or by inducing fear.

Sensory modalities. Of the major modalities, there is little to suggest that auditory or gustatory receptors convey sexually exciting information to the male, except in so far as secondary conditioning may be concerned. It would be relatively easy for a bull in an artificial insemination centre to learn to associate specific sounds in the collection room with the imminence of sexual stimulation; similarly, a free-ranging bull might quickly associate the smell or taste of urine from an oestrous cow with impending sexual activity. De Vuyst *et al.* (1964) have demonstrated that bulls can be sexually stimulated by playing recordings of certain noises made by cows. It is not certain whether these calls are specifically sexual in nature or just act as a general situation stimulant.

It has been often assumed that the sense of smell plays an important, if not primary, role in sexual stimulation of a bull. The assumption stems perhaps from generalization of the situation in dogs and perhaps from numerous casual observations of the bull sniffing the vulvar region of oestrous cows. Indeed, the studies of Hart *et al.* (1946) indicate that olfactory cues are all-important in the arousal of sexual behaviour in cattle. However, examination of these data reveals that the results obtained may be as easily attributed to visual cues as to olfactory ones. Attempts to elicit male sexual behaviour from bulls with olfactory stimuli while holding visual stimuli constant have proven unsuccessful (Hale, 1966).

The association of smell with sexual excitation is at best secondary conditioning: males provide the same degree of sexual stimulation as females (Almquist & Hale, 1956), and restrained anoestrous females are as stimulating as oestrous animals (Prabhu, 1956). It is common practice to use intact or castrated males and anoestrous females as teaser animals for semen collection.

Bonadonna (1956) reports that tactile stimulation of the penis is not the primary stimulus for erection. He induced insensitivity to needle pricks by the application of 15 per cent Novocain to the penis surface and found no detectable change in erection or sexual interest of the bull. None the less, the ejaculatory thrust normally occurs when the erect penis contacts the moist warm surface of the vagina and it is possible that association learning again is involved with respect to tactile stimulation: the bull associates stimulation of the penis with the subsequent "reward" of ejaculation.

Studies with blindfolded and permanently blinded bulls (Hale, 1966) indicate that sight is not essential for the development of the bull's reproductive function. Temporary or permanent absence of vision, however, does greatly reduce the probability that a bull will identify a sexual situation and initiate a sexual response. This is especially true when the animal is tested in unusual surroundings or under novel stimulation conditions.

Visual information received by a bull provides it with a large share of the stimulation needed to release sexual patterns. A visual stimulus may consist of some discrete part or configuration presented by the stimulus animal, or of the entire stimulus animal plus a particular physical setting. Present evidence points to an inverted U shape as the visual configuration most probably involved in releasing mounting behaviour. This particular shape is presented by any standing herd member, by bent-over people, and also by many inanimate objects, for example, which may act as sexual stimuli so long as they do not induce fear. Brownlee (1954) describes the mounting of a padded sawhorse by heifers. Wierzbowski & Janasz (1965) have shown that any three-dimensional object resembling a cow's trunk may act as a sexual stimulus and that movement of the object increases the stimulating effect. No doubt size is important, since a very small or very large object would probably not elicit mounting. The entire physical setting also plays an important role in visually stimulating the bull. Almquist & Hale (1956) report that a bull which has become satiated to a stimulus animal in a particular setting will often respond as if to an entirely new stimulus situation if the stimulus animal is moved but a few feet, or if some minor physical change (to us, at least) is made in the setting.

Masturbation is commonly practised by bulls. Pelvic movements while the back is arched serve to pass the penis in and out of the preputial orifice, which in turn provides the tactile stimulation which eventually causes ejaculation (Hultnäs, 1961). In some cases, contraction of the abdominal muscles is so forceful that the front and hind legs cross each other and the bull loses his balance. Masturbation is even more commonly observed among bulls on high protein diets, such as are used when preparing bulls for sale or exhibition. Hultnäs (*loc. cit.*) suggests the mucosa of the penis becomes more sensitive to tactile stimulation as a result of these diets.

Stimulus pressure. Hale & Almquist (1960) use the concept of *stimulus pressure* to explain the sum total sexual stimulation impinging on an animal over an extended period of time. Individual stimuli are presumably additive in nature: the stimulation provided by the sight of a sexual object may be enhanced by the additional association-learned stimulation of oestrous urine. Thus, high stimulus pressures ensure high levels of behavioural response, at least to some maximum potential. Conversely, low stimulus pressures result in low levels of behavioural response to the extent that inadequate stimulation results in no expression of behaviour.

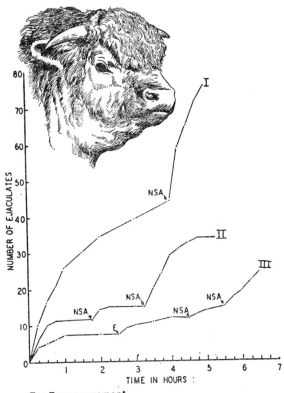

E = Encouragement
NSA = New Stimulus Animal

Fig. 32. Typical curves illustrating the effect of stimulation upon the sexual behaviour of bulls. Note that Bull III is markedly inferior to Bull I, in spite of the high stimulus pressure. (Adapted from Almquist & Hale, 1956.)

High stimulus pressures are maintained either by placing a new animal in an old setting or by placing the old stimulus in a new setting (Fig. 32). This technique results in the expression of high levels of sexual behaviour as measured by numbers of ejaculations per unit time (Milovanov & Smirnov-Ugrjumov, 1940; Smirnov-Ugrjumov, 1945; Almquist & Hale, 1956).

Factors contributing to high stimulus pressure. The nature of the phenomenon of recovery from sexual satiation is poorly understood. The amount of time necessary to recover from satiation to a particular stimulus animal or setting varies greatly among individual bulls. Thus, a bull which recovers completely in a week will show a constant reaction time for ejaculation during successive weekly trials in the same stimulus situation. If the recovery from satiation is slow, the bull shows steadily increasing reaction times and may eventually fail to react at all for a prolonged time (Fig. 33). There is no correlation between the initial sexual response to a stimulus situation and the rate of recovery after stimulus satiation (Almquist & Hale, 1956).

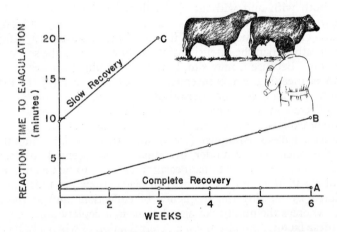

Fig. 33. Individual differences in the recovery of sexual responses following satiation in dairy bulls. The lines represent the reaction time to ejaculation when semen was collected once a week for 6 successive weeks, using an artificial vagina; the same female was used as a stimulus animal. The circles represent smoothed values at weekly intervals. A represents a bull with complete recovery between ejaculates, B one with incomplete recovery, and C one with poor recovery. (Adapted from Hale & Almquist, 1960.)

It seems reasonable to assume that the bull exhibits preferences for certain stimulus situations over others. Preferences may easily be established by the device of conditioning or association learning. Trautwein *et al.* (1958) report that bulls of Spotted breed tend to reject small cows of the Black-and-White or Red-and-White breeds, and Harker & Rollinson (1961) claim to have difficulties getting Boran bulls to mount Ankole cows. These phenomena may reflect previous social experience, for there is little reason to suppose that sexual preferences are innately determined. Conditional stimuli associated with sexual responses of animals are easily and often willingly established. The sound of a gate opening, the odour of a particular barn attendant, scratching on the neck, or the presence of a coloured light often contribute secondarily to the enhancement of stimulus pressure.

9*

3. Sex Drive (Libido)

Measurement. Indices designed to gauge the intensity of a bull's sexual behaviour (under farm or range conditions) vary from dependable values to almost irrelevant numbers. Under range conditions, the number of cows mated per unit time depends upon a host of interrelated factors: the oestrous cycles of the cows, the number of cows per bull, and aggressive interactions between bulls. Thus, the observer would merely measure the *expression* rather than the *potential intensity* of the behaviour.

In a controlled setting, such as when a bull is placed in an enclosure with an oestrous cow, sexual intensity may be measured as the total number of mounts, both with and without ejaculation, or the latency of ejaculation (*reaction time*). Each of these indices requires cautious interpretation, the first (total number of mounts) because it more likely measures a deficiency in expression rather than the intensity of the behaviour, and the second (latency of ejaculation) because it may simply reflect stimulus pressure.

The most meaningful measurements of sex drive are (a) the number of ejaculations during a constant period of time or until no more ejaculations can be collected in a reasonable period of time with all available stimuli (depletion or exhaustion test), and (b) the latency of ejaculation. In discussing these indices, Hale & Almquist (1960) state that "insight into the workings of the phenomenon under investigation is essential in determining when and under what conditions measurements are to be taken". Thus, the latency of ejaculation and the number of ejaculations per unit time may be manipulated by altering the frequency of collections and/or the stimulus pressure, whereas the number of ejaculations in a depletion test depends not only on these factors, but also upon a determination of when a bull is sexually satiated rather than physically depleted.

Levels of sexual behaviour can be gauged by measuring reaction times and the number of ejaculations per unit time only if stimulus pressures are maintained at an adequate constant level. At the same time, measurement of recovery time after stimulus satiation provides critically important information relating to the level of behaviour.

Number of ejaculations. Although the common practice in artificial insemination centres is to limit each bull to only one or two collections per week, Hale & Almquist (1960) report that they have collected six or more ejaculates per week from 29 bulls for 22 or more weeks with no deleterious effects. In another series of tests, they collected an average of 70 ejaculates per week for 6 weeks from a mature bull (Almquist *et al.*, 1958); they have recorded as many as 77 ejaculations from one animal in a 6-hour period (Almquist & Hale, 1956).

Factors affecting sex drive. Each bull has a characteristic, probably genetically controlled, level of sexual behaviour, as measured by the latency of ejaculation or the number of ejaculations in a given unit of time when subjected to a constant stimulus pressure (Fig. 32). Levels are fairly constant for each bull: latency of ejaculation and number of ejaculations are highly repeatable, provided adequate time is allowed for recovery between test periods. Evidence for the genetic control of levels of sexual behaviour stems from work with monozygotic twins and triplets (Olson & Petersen, 1951; Bane, 1954) and from comparisons between sires and sons (Hultnäs, 1959): great similarities in levels of expression between identical twin brothers, as well as large differences among twin pairs, are commonly reported (Figure 34).

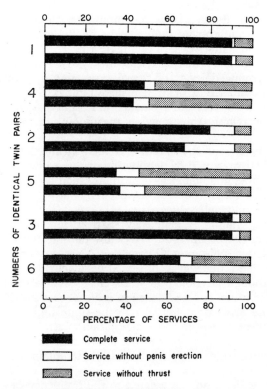

Fig. 34. Ejaculatory behaviour in identical twin dairy bulls in Sweden, including the percentage distribution of complete services, mounts without penis erection, and mounts without thrust. Note the similarities in the ejaculatory pattern of the twin brothers and the variability between the twin pairs. (Adapted from Bane, 1954.)

Bulls of European breeds readily mount anoestrous cows or other bulls, whereas those of the Zebu breed exhibit a well marked sexual sluggishness and mount only cows in full oestrus (Anderson, 1948, 1949). Ittner *et al.* (1954b) report that more Hereford calves than Brahman-Hereford or Short-ham-Hereford calves were born when equal numbers of equal-aged Brahman, Hereford, and Shortham bulls were run with a herd of Hereford range cows: however, Donaldson (1962) found that most calves came from the European-type bull, when equal numbers of Hereford and Brahman cross bulls were turned out with a Hereford herd. Smith (1951) suggests that differences occur in the sex drive of European breeds at semen collection: Shorthorn and Guernsey bulls react more slowly than Friesians. Strain differences have also been observed in the sex drive of Holstein bulls imported into Italy from the United States, Holland, and Germany (Bonadonna, 1953).

Young, inexperienced bulls used for semen collection for the first time are usually awkward to handle. They approach the cow hesitantly, spend a long time exploring the genitalia, and mount repeatedly without erection. The hesitancy of young novice bulls may be related to the fact that on open range, the bull must first socially dominate the females before being able to successfully complete a mating (Schloeth, 1961). It is also possible that observations of reduced levels of sexual expression in young bulls reflect timidity and unfamiliarity with strange surroundings. Schein & Hale (1959) have demonstrated that fear and avoidance, at least in turkeys, can completely mask sexual behaviour, thus making meaningful measures of intensity impossible.

Male sexual behaviour is markedly reduced in intensity during periods of physiological stress caused by debilitating diseases, low plane of nutrition, or climatic extremes. Pathological conditions, such as inflammation of the hooves or joints, eczema, tuberculosis, trichomoniasis, and the like, as well as direct physical injury, all reduce sexual expression (Bane, 1954; and others). Similarly, severe vitamin A deficiency (Erb et al., 1947; Csukas, 1949), protein deficiency (Meacham et al., 1963), low phosphorus diets (Sandstedt, 1949), molybdenum poisoning (Thomas & Moss, 1951), and social rank (Schloeth, 1961) all reduce levels of sexual behaviour.

Note that although physiological stress clearly reduces the expression of sexual behaviour, there is no evidence to indicate that sex drive *per se* is at all affected. The drive or potential may be quite high, despite its being masked by the general physiological adaptation to stress. Potential levels of sexual behaviour, i.e. sex drive, can be meaningfully measured only under optimal conditions, which are quite impossible to obtain if the individual is under stress.

Sex drive *per se* is in no way related to the frequency of copulation. By maintaining adequate stimulus pressure, Almquist & Hale (1956) have been able to collect as many as 77 ejaculates in a 6-hour period. Were it possible to collect from a bull until the animal was physically exhausted, the exhaustion itself would be a physiological stress inhibiting the further expression of behaviour; sex drive would be unaffected. More commonly, stimulus satiation occurs long before any degree of physical exhaustion is reached, and consequently the number of ejaculations per unit time usually declines as the number of copulations increases (Baker et al., 1955; Hafez & Darwish, 1956).

B. Sexual Behaviour of Females

As in most mammals, sexual behaviour of adult female cattle is overtly expressed only during oestrus or heat periods. Thus, a discussion of female sexual behaviour is concerned mainly with the brief oestrous period which occurs every few weeks.

1. Patterns of Female Sexual Behaviour

Normal patterns of oestrous behaviour. There are few, if any, observable differences between the sexual patterns of maiden heifers and adult cows during oestrus. From the onset of oestrus, the animal becomes hyper-reactive: it responds to environmental stimuli which ordinarily would be ignored. Relative social hierarchy positions are temporarily ignored as the animal indiscriminately approaches both dominant and subordinant herd-mates. Thus, agonistic interactions involving oestrous cows are relatively more common than those involving anoestrous cattle (Schein & Fohrman, 1955).

Grazing and feeding behaviours are frequently interrupted, rumination time is reduced, and milk yield declines in oestrous cows. Instead of grazing, the oestrous cow roams extensively (Farris, 1954) and attempts to mount or solicits mounts from other cows regardless of social rank. Tail switching and tail raising are common, and the frequency of urination is increased, although the frequency of defaecation is not.

Although non-oestrous animals generally refuse mounting attempts by oestrous cows, they themselves often mount the latter. In such situations, the oestrous cow makes no effort to escape; instead, it either stands quietly or leans caudally into the mount. Mylrea & Beilharz (1964) have shown that some, but not all, the animals in a group of heifers mount oestrous females. Mounting by female cattle, both in and out of heat, resembles the typical mounting behaviour of males even with respect to pelvic oscillations. The vulva of a receptive cow is often sniffed by the other cows in the herd. It seems curious that during a 13-month study of a heterosexual herd of free-ranging Camargue cattle, Schloeth (1961) never once observed mounting by adult females, except when the animals were being driven from one locality to another by humans; on the other hand males frequently mounted females.

Heinemann (1958) reported that some cows exhibit an "orgasm-like" reaction after copulation. This observation may stem from the fact that some females maintain a "urination posture" for several seconds after the ejaculatory thrust of the bull. Cordts (1953) measured the *psycho-galvanic reflex*, more commonly termed the *galvanic skin response* (GSR), of cows during copulation and found that the electrical resistance of the skin dropped suddenly at the time of ejaculation. The GSR is a sensitive indication of human emotional and autonomic responses in many situations (Morgan & Stellar, 1950), and it is not surprising to find it in cows as well. The presence of a GSR in cows during copulation is interesting, but does not conclusively prove that an orgasm has occurred.

Deviations from normal oestrous behaviour. Normal oestrous behaviour varies in frequency of occurrence and intensity of expression. In some cases, *silent heat* (ovulation without oestrus) cannot be distinguished on the basis of behaviour from *anoestrus* (absence of ovulation and oestrus). The intensity of oestrus, which does not easily lend itself to objective measurement, is considered "normal", if at least some oestrous behaviour is evident. How-ever, irregularities in the frequency of oestrus are usually considered abnormal, and are often related to pathological conditions of the ovaries (reviewed by Hafez, 1956). Cows which show intense oestrus-like behaviour persistently or at frequent irregular intervals are termed *nymphomaniacs*. Nymphomaniacal cows frequently are as sexually aggressive as bulls in seeking out and attempting to mount oestrous cows, but rarely do they themselves stand for mounting by other cows (Roberts, 1956); in this respect, their behaviour differs from cows in true oestrus. The nymphomaniac paws, bellows, and otherwise exhibits male-like behaviour; if the condition is not corrected, in time the voice and general body formation also tend to become male-like. The predisposition for nymphomania may be inherited (Garm, 1949), and the condition is more prevalent in dairy cattle than in beef cattle. Although nymphomania affects cows of all ages, it most commonly occurs at 4 to 6 years of age, after the cow has calved 2 or 3 times. Nymphomania is closely associated with ovarian cysts: 75 per cent of cows with cystic ovaries exhibit nymphomaniacal behaviour, whereas the other 25 per cent exhibit persistent anoestrus (Roberts, *loc. cit.*).

Onset, cessation, and duration of oestrus. It has been suggested that oestrus begins more often during the night and early morning than during the day. However, when reports from the literature (reviewed by Asdell, 1946, and others) are assessed, no definite trend is evident for the time of onset of oestrus. None the less, oestrus does end more often in the afternoon and early evening than during the night. The interpretation of these data is complicated by the effect of season on the duration of oestrus, the gradual cessation of oestrus, the frequency with which oestrus was observed, and the method used to differentiate between pro-oestrus, oestrus, and met-oestrus.

The duration of normal oestrus ranges from 9 to 28 hours and depends upon geographical locality, breed, and age of cattle. In general, oestrus is shorter in the tropics and subtropics than in the temperate zone. The duration of oestrus in Zebu cattle and their crosses is shorter than that of European breeds. In both types of cattle and in all locations, the oestrus of heifers is shorter than that of cows.

2. INTENSITY OF OESTRUS

The intensity of oestrus is usually subjectively measured in terms of how "excitable" the cow seems to be, or how much mounting and mount-soliciting she does. Since clear objective measurements are lacking, it is possible only to designate intensities as *strong*, *medium*, or *weak*, depending upon the degree of restlessness of the cow, congestion and turgidity of the vulva, appearance (palm leaf reaction upon drying of mucus) and volume of mucus, and conspicuous mounting behaviour on other cows. Such criteria provide a reasonably accurate measure of relative differences between breeds, ages, and individuals.

Simmental cows in Switzerland, Telemark cows in Norway, and the Swedish-Highland breed show strong oestrous behaviour, whereas the Swedish-Red breed exhibits weak oestrous behaviour (Lagerlöf, 1951). It seems that intensity of oestrus is less marked in the Brown Swiss than in other breeds (Brakel *et al.*, 1952) and is weaker in red, roan, and white cattle than in black cattle (Donald & Anderson, 1953). The relative importance of coat colour *per se* and genetic factors is not known. There are large differences in the intensity of oestrus between daughters of the same bull, and between progeny groups of the same herd in the same year and in different years. Estimates of the heritability (21 per cent) and repeatability (26 to 29 per cent) of the intensity of oestrus are very low (Rottensten & Touchberry, 1957).

C. Sexual Behaviour and Fertility

Sex drive is not generally related to fertility, i.e. to the quality and/or quantity of sperm per ejaculation. Of course, an animal with low sex drive will produce fewer ejaculations than one exhibiting more sexual activity. Sperm output in an ejaculation is correlated with the amount of sexual preparation. Increasing the amount of pre-coital stimulation has resulted in increased semen volume (Almquist et al., 1958), sperm and fructose concentration (Collins et al., 1951; Branton et al., 1952), sperm motility, percentage of live sperm (Crombach et al., 1956), and conception rate (M.M.B., 1954). These data are adequately summarized by Hale & Almquist (1960), who agree with Prabhu (1956) that bulls with the lowest sperm output before preparation usually show the greatest improvement with preparation.

Crombach (1961) and Signoret (1962) report that extra stimulation, such as a period of restraint and allowing false mounts before the collection of semen, increases sexual excitement and improves the quality of the ejaculate; further stimulation (presence of another bull, changing the teaser animal) continues to increase sexual excitement but does not improve semen characteristics.

Some bulls become apprehensive about sudden changes in the environment, such as changing the farm, the barn, the herdsman, or the locality of collection. Since fear and apprehension are inimical to sexual expression, the intensity of sexual behaviour will decline until the bull becomes accustomed to the new situation. The period required for adaptation varies with breed and age, and is longer in old animals than in young ones. Inhibition may develop as a result of repeated frustration, faulty management, wrong techniques during semen collection, distraction during coitus, and too rapid withdrawal of the teaser animal after copulation. Inhibition is characterized by refusal to copulate, incomplete erection, or incomplete ejaculation. The relationship between the sexual behaviour of bulls and management practices in artificial insemination centres is discussed by Trautwein et al. (1958), Hafez (1960), and Crombach (1961).

The determination of ovulation time, as gauged by oestrous behaviour, is complicated by the occurrence of "silent" and false heats. In a study of 500 oestrous periods in dairy cows, 19 per cent were silent heats and 15 per cent false oestrus (oestrous behaviour without ovulation or luteinization) (Trimberger, 1956). Consequently, ovulation may easily pass unnoticed or animals may be artificially inseminated at the wrong phase of oestrus. The careful studies by Rottensten & Touchberry (1957) indicate that the intensity of oestrous behaviour is not directly related to conception, despite some scattered reports to the contrary. These latter reports, from artificial insemination records, are often heavily biased by the herdsman's judgment of oestrus intensity.

The vagina of the cow contracts when she is approached by bulls or other oestrous cows, and when she is placed in a stand preparatory to artificial insemination (Parshutin, 1956). Such conditioned reflexes are developed by association learning and may have an effect on sperm passage to the site of fertilization and consequently on conception rate. It may be possible to improve the conception rate by preparing the cow for insemination in a manner similar to the preparation of bulls. Kirillov (1943), for example, reported an increase in breeding efficiency by pairing cows with vasecto-mized bulls. On the other hand, increased restlessness during oestrus may provide an unfavourable endocrine environment for the travel of sperm in the female tract. Heinemann (1958) reported that such restlessness can be curtailed by frequent mating at short intervals. Nervous cows conceive less frequently than relatively quiet ones when artificially inseminated (Pounden & Firebaugh, 1956). This, again, may be due to an unfavourable endocrine environment in the female tract or to mechanical difficulties in the insemina-tion of struggling cows.

III. SOCIAL BEHAVIOUR

An organism does not live by itself in a vacuum. This fact is especially pertinent to those gregarious animals which live in herds, flocks, and packs: the behaviour of each individual is strongly modified by the presence of others in the group (see Chapter 5). Cattle normally roam in groups varying in size from but a few individuals to tremendous herds. However, this section concerns primarily the normal farm operations, involving only a few to several hundred cattle.

A. Pair Interactions

1. MOTHER-YOUNG INTERACTIONS

A cow temporarily ceases to function as a member of the herd shortly before and immediately after parturition (Schein & Fohrman, 1955). She may wander off alone or stay behind as the herd moves away.

Few actual descriptions of parturient behaviour exist for cattle. Hence, the following account is based mainly on the work of Arthur (1961, 1964). As calving time approaches, the cow becomes restless and may show sudden pauses during eating or ruminating. During the first stage of labour, she will often get up and lie down repeatedly and may show some slight straining. Extensive bouts of straining occur during the second stage of labour and the cow will either adopt the normal resting position or lie on her side. Her upper legs may actually swing clear of the ground if she strains while lying on her side. Once the calf is born, the cow rests for a variable length of time and then gets up and licks the foetal membranes and fluids from the calf. She usually eats the placenta, and sometimes the bedding contaminated by foetal and placental fluids as well. The normally herbivorous animal suddenly and briefly exhibits a carnivorous appetite.

The neonate calf attempts to stand shortly post-partum, and although unsteady at first, quickly gains control over its legs, and is ambulatory within a few hours. During this time, the mother and calf stay close together, and the calf feeds for the first time (see Suckling Behaviour). Tactile stimulation of the udder during suckling plays a role in milk let-down. Visual and auditory cues provided by the suckling calf may quickly become associated with the tactile stimulation and thereafter serve as conditional stimuli for milk ejection. The pathways by which these stimuli reach the hypothalamus are not clearly understood; afferent stimuli from the udder may reach the hypothalamus via the medial lemniscus (reviewed by Cowie, 1957; Cross, 1961).

The neonate calf spends a considerable amount of time sleeping, while the mother grazes nearby. A day or two post-partum, the mother wanders more extensively with the calf at her side, and soon rejoins the herd. Thereafter, the calf gradually integrates with the herd and its relationship with the mother weakens. After weaning, the calf no longer depends upon the mother for defence: it assumes its own place in the herd dominance order and its ties with the mother are irrevocably broken. In a given herd, the calves tend to form "calf-groups" and "cow-with-calf" subgroups with their mothers. If the calf is separated from the herd, it may exhibit thigmotropic behaviour toward the attendant.

Recognition between the calf and its mother is based on olfactory, visual, and auditory cues (Schloeth, 1958c). Cows usually sniff their calves when they return from prolonged periods of grazing, and the calf recognizes the sounds characteristic of its mother. In the American buffalo (B. bison), cohesion between the calf and its mother was effected by the initiation or intensification of grunting between the two in the following circumstances: (a) by a mother when its calf strayed away; (b) by a mother during moments of danger; (c) by a calf when its mother strayed away; (d) by either calf or mother when one was separated artificially from the other; (e) by calves or mothers during herd movement (a distinct increase in grunting accompanied movement); (f) by either in answer to the other; or (g) by either prior to nursing (McHugh, 1958). Berloiz (1933) reported that buffalo (B. bison) cows bred to domestic (B. taurus) bulls often fail to suckle their hybrid calves: neglect of the calf may reflect impaired recognition of the hybrid.

The attachment of the mother to the neonate calf is very strong. Separation of the two results in bellowing and lowing by the mother for several days; she does not function as a herd member during this time. After a few days, the agitation passes and she rejoins the herd. The neonate calf, on the other hand, accepts the separation with less evident distress: it readily adapts and often attaches itself to other animate objects, such as neighbouring calves or human handlers. Calves which have been removed from their mothers just after birth and kept in isolated stalls rarely utter a sound. However, if separation from the mother occurs after strong pair-bond relationships are established (several days post-partum), then the separation may have severe effects on both the calf and the cow.

2. PLAY BEHAVIOUR

The term "play behaviour" has been used in ethological literature (Carpenter, 1934; Scott, 1945; Vogel *et al.*, 1950) to describe simulated adult behaviour which serves the adaptive purpose of developing and sharpening those motor patterns which may be of critical importance later in life. Brownlee (1954) proposes that "play" in cattle is an innate entity in itself, with its own drive, releasers, emotion, consummatory phase, and goal. A slightly different concept has been proposed by Schein (1954), who defines play as "an activity engaged in solely for the sake of the activity itself, and not for the normal end result of the activity. Thus, play fighting would be engaged in solely for the fight itself and not for the purpose of defeating the other animal."

Although play behaviour is not limited to the young, it is exhibited far more frequently by young than adult animals. A play-call, a play-specific tail position of calves, and specific play areas of cattle on free range are described by Schloeth (1961). Cattle play by gambolling (Scott, 1945), bucking, kicking, prancing, butting, vocalizing, head-shaking, snorting, goring, pawing (see Plate III), and mounting (Bownlee, *loc. cit.*). *Play-* or *mock-fighting* commonly occurs principally among calves, but not infrequently among adults as well. The gentle butting, pushing and horning of the two "combatants" does not lead to a decisive outcome, and is easily terminated by any mild distraction. In this manner, mock-fighting differs markedly from true agonistic interactions.

Well-fed healthy animals play more often than sick and poorly fed ones, and more often during good weather than in cold wet weather (Brownlee, *loc. cit.*). Adults often play upon release after a long period of confinement. Calves do so shortly after feeding, in new (but not strange) terrain, and if they have access to movable familiar objects (swinging doors, wheel barrows, etc.). Common to all these situations is the idea of a *new, but not unfamiliar*, object or field.

3. SEXUAL PAIRING

We have already described in detail both male and female sexual behaviour. We need only add a few words here concerning sexual pairing under free-roaming conditions. Cattle are promiscuous breeders: pair-bonds are very loosely established during oestrous periods and depend solely upon the ability of the dominant bull to keep other bulls away from the female. The bull mates freely with any female, and the oestrous female in turn accepts any dominant male. By $2\frac{1}{2}$ years of age, young bulls dominate all the cows in a herd and are therefore acceptable as sex partners. The partner, either a bull or a nymphomaniacal cow, tends to keep others away with varying degrees of success (see section on Courtship in Patterns of Male Sexual Behaviour). It is not known how often an oestrous cow is mated under range conditions; however, the 5 copulations observed by Schloeth (1961) during 13 months of observation were all performed by the two highest ranking males. The loose pair-bond is dissolved at the termination of oestrus and the animals drift apart.

4. AGONISTIC BEHAVIOUR

Aggressive behaviour is defined by King & Gurney (1954) as the initiation of a fight, whereas agonistic behaviour encompasses all behaviour associated with conflict, including escape and submissive behaviour (Scott, 1956).

Patterns of agonistic behaviour. Schein & Fohrman (1955) categorize the sequential patterns of agonistic behaviour in cattle as *approach, threat*, and *physical contact*. The approach may be passive or active: a passive approach should perhaps more properly be termed a "chance meeting", and is followed by a mild threat by one participant and immediate submission by the other; the action occurs so rapidly and so subtly that it often passes unnoticed even by a trained observer. If the threatened animal is slow to submit or fails to notice the threat (as sometimes happens when it is approached from the rear), the dominant animal *butts*. A butt (also called a *bunt*) is a blow with the forehead and is usually directed at the opponent's side or rump (Fig. 35). By virtue of the upswinging motion of the head during butting, the victim can be seriously injured if the attacker has horns. A subordinate animal is quick to retreat when butted or if a butt is imminent.

The active approach is purposeful, in that the movement of one individual is clearly directed toward another. Threats are usually exhibited by one or both participants when they are about 5 feet from each other. The threat posture of females (Schein & Fohrman, *loc. cit.*) resembles the "fight or flight" posture of males (Fraser, 1957a): head lowered, eyes directed at the opponent, hind legs drawn forward, forehead perpendicular to the ground; the head and horns, if present, are directed toward the opponent (Fig. 35). In other situations, an animal threatens by pawing, rubbing its head and neck on the ground, and digging its horns into the earth (Plate III), while still some distance from the other animal. Thereafter, the antagonists slowly approach each other and stand several feet apart in lateral threat positions, head to tail, with body axes parallel (Schoeth, 1958a, 1961). In the appeasement posture, the head is also lowered, but the forehead is nearly parallel to the ground and the neck is extended (Fraser, 1957b).

The threat may be followed by submission and avoidance behaviour on the part of the recipient animal, or by a threat in return. In the latter case, fighting occurs. Fighting among cattle, in fact among most or all ruminants, is "head-to-head": the animals push against each other with foreheads and horns (Fig. 35). There is much jockeying for position as each combatant strives for a flank rather than a frontal attack. A combatant which is flanked makes every effort to force the aggressor to a frontal attack: it even retreats and back-tracks if necessary to escape from a flank attack. If such attempts are unsuccessful, the animal submits by fleeing and avoiding the victor, who in turn may pursue the loser for a short distance. If neither combatant succeeds in flanking the other, they push against each other head-to-head until one submits. In the event that the two are evenly matched, the fight may be prolonged to the point of physical exhaustion of both participants. A fight may consist of one or more encounters; the interval between encounters varies from a few second to 10 minutes, during which time the antagonists remain either in the threat position or pursue threatening activities, pawing and horning the ground.

An interesting pattern of behaviour, the *clinch*, occurs in prolonged fights between females: one participant allows the opponent to gain a flank advantage, but at the same time it slips alongside the opponent and pushes its muzzle between the opponent's hind leg and udder (Schein & Fohrman, 1955). Neither cow can attack from such a position (Fig. 35): should one

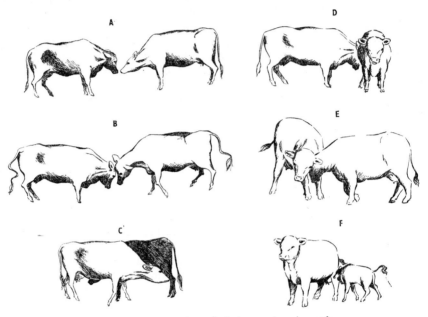

Fig. 35. Patterns of agonistic interactions in cattle.

A. Cows meeting after an *active approach*. The one on the left is *threatening*, whereas the one on the right has assumed a *submissive* posture.

B. Physical combat: a fight. The cows push against each other head to head, each striving for a flank position.

C. The *clinch*. One contestant of an evenly matched pair slips alongside the other; the head of the former is pushed between the hind leg and the udder of the latter. In unusually prolonged contests, the combatants rest briefly in the clinch between bouts of head-sparring.

D. Flank attack. The animal that gains a flank position is at a decided advantage over the other. The flanked animal either submits and flees, or strives to regain the head-to-head position.

E. The *butt*. A dominant animal directs an attack against the neck, shoulders, flank, or rump of a subordinate, who in turn submits and avoids the aggressor.

F. Play fighting. Butting of the mother by the calf.

(Drawn from photographs by Schein (A, B), D. E. Davis (C), and Hafez (D, E, F).)

attempt to turn for an attack, the other merely rides along by maintaining body contact. Thus, the fight cannot be resumed until both participants are ready. The clinch seems to be an ideal way for combatants to rest safely during prolonged bouts and is not seen often, since the majority of fights are decided quickly.

It should perhaps be noted that the *alpha*, or most dominant, animal in a herd is not necessarily the most aggressive one, but rather the one which bunts every other animal and is bunted by none in return (Boussiou, 1965).

Agonistic behaviour and management. Aside from injuries, agonistic behaviour can have detrimental effects when subordinate individuals are excluded from limited feeding areas (Appleman & Ronning, 1955; Wagnon, 1965). However, Lees (1962) reports that there is less bullying during feeding when cattle are housed on slatted floors, perhaps because all cows have much less self-confidence on slats than on other types of floor surfaces. Schein *et al.* (1955) have shown that the psychological disturbances engendered by poor management routines may result in as much as a 5 per cent average decline in milk yield. In this case, poor management routines consisted of increasing agonistic interactions by introducing unfamiliar animals to the milking herd.

The introduction of new animals into a small herd causes considerable excitement and occasional fighting as the newcomers establish their places in the over-all dominance rank. Agonistic behaviour is less marked in young calves than in older animals, and efforts should be made to introduce new herd members when they are young. Older animals should be introduced to the herd gradually, and kept in an adjoining field until they become familiar to the herd.

B. Group Interactions

1. Dominance Order

The outcome of an agonistic interaction determines the dominance-subordination relationship of two animals usually for as long as they remain in the same herd. Once the relationship is firmly established, it is recognized by each and further combat is superfluous: the subordinate animal retreats from the dominant at the slightest threat. The dominance order[1] of a herd may only be constructed if the relationship between every pair of animals is known (see Chapter 5). The dominance order has great survival value for the species: once established, it minimizes severe agonistic interactions between herd members.

When cattle are penned or pastured together, they immediately establish a hierarchy which remains stable for long periods (Boussiou, 1965). True fights rarely occur during this process. Simple gestures (e.g. movements of the head, or the "threat" posture described by Schein & Fohrman, 1955) apparently suffice to establish the social ladder of the group. The first encounters between the cattle in the group determine their long-term social status. The changes, which sometimes occur within the social structure of the group, appear suddenly, as suddenly as the initial relationships were set up.

When a strange animal is introduced into an established herd, violent fights are infrequent: the animal places himself at some appropriate level in the hierarchy, which is not related to his rank in previous groups. It appears then that the social hierarchy of cattle is determined mainly by visual means

---

[1] Also called hook order, social rank, social hierarchy, or rank order.

and that cattle are able to directly perceive whether the social rank of their neighbours is superior or inferior to their own.

Dominance orders among cattle have been described in the literature (Woodbury, 1941; Gyr, 1946) and studied in detail (Schein & Fohrman, 1955; Schloeth, 1958a; Guhl & Atkeson, 1959; Boussiou, 1965). Cattle establish social hierarchies which are *linear*, *linear-tending*, or *complex* (Fig. 36) (Boussiou, *loc. cit.*). In a linear hierarchy, the alpha animal domin-

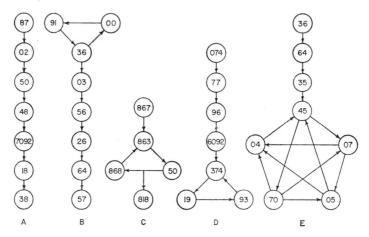

Fig. 36. The social hierarchies of domestic cattle. (*a*) Linear hierarchy; (*b*)–(*d*) linear-tending hierarchies with alpha (*b*), intermediate (*c*), and omega (*d*) triangles; (*e*) complex hierarchy. The alpha, or most dominant, animal is topmost in each diagram; the omega, or most subordinate, animal(s) is lowest. (Adapted from Boussiou, 1965.)

ates all others in the group, the beta animal dominates all but the alpha animal, and so on to the omega animal which is dominated by all others. A linear-tending hierarchy contains a "triangular relationship". For example, the alpha animal shown in Fig. 36*a* is sometimes replaced by the "alpha triangle" in Fig. 36*b*, in which animal 00 dominates 91, which dominates 36 which in turn dominates 00. The triangle may occur at the opposite end of the hierarchy (the "omega triangle" in Fig. 36*d*), or at any intermediate position (Fig. 36*c*). In complex hierarchies, the relationships between the most dominant animals may be linear, but those between less dominant ones are not clearcut (Fig. 36*e*).

Boussiou (*loc. cit.*) found that simple hierarchies occur most frequently in small groups of cattle: eleven of 15 groups of 4 to 8 cattle each set up linear or linear-tending hierarchies. Complex hierarchies, on the other hand, are particularly prevalent in large herds of cattle. One must be cautious in extrapolating these findings to herds of several thousand or more. The dominance order type of organization may tend to break down in large groups, or may exist as a series of dominance orders: an individual may simultaneously be a member of two or three different orders involving different animals. It is essential to consider the relationship of each animal to every other animal in the group in order to adequately assess the social hierarchy which they have set up among themselves. In complex hierarchies, it is perhaps easiest to classify individuals on the basis of the number of other animals which they dominate.

Several rank orders are evident in heterosexual herds of cattle on free range: one among the adult males, another among the adult females, and still a third among the juveniles. All adult males dominate all adult females, who in turn dominate all juveniles. However, at about $1\frac{1}{2}$ years of age, young males begin to fight with adult females and by $2\frac{1}{2}$ years of age they dominate all the females, and therefore join the adult male rank order (Schloeth, 1961). In general, agonistic interactions are much more common between neighbours in social rank (up to 3 rank positions apart) than between animals widely separated on the social scale (Schein & Fohrman, 1954; Schloeth, 1961).

Factors contributing to dominance order. Schein & Fohrman (1955) found significant correlations between dominance order and both age and weight in a closed dairy herd. They interpreted these correlations as coincidental, rather than causal, and suggest that *seniority* (which is also correlated with both age and weight) is the crucial factor in closed herds (those in which new members are added almost exclusively by being born into the herd). The data of Guhl & Atkeson (1959) also show a high correlation between social rank and seniority, while those of Dickson et al. (1967) show a significant correlation between dominance order and weight. Chest girth may influence rank order (Beilharz & Mylrea, 1963a), although height at the withers may be more important than either weight or chest girth (McPhee et al., 1964). In an open herd, where new individuals meet for the first time in unfamiliar surroundings, the rank order is based on weight, agility, and strength.

Schein & Fohrman (1955) could not find a relationship between breed (Jerseys compared with Zebu–Jersey crosses) and social rank. On the other hand, Guhl & Atkeson (1959) report a tendency for Holsteins and Ayrshires to dominate Jerseys (no trend was found between Holsteins and Ayrshires themselves), and Wagnon et al. (1966) indicate that Angus tend to dominate Shorthorns, which in their turn tend to dominate Herefords. These differences in breeds can be partly explained by differences in body size, except in the case of Wagnon et al. (*loc. cit.*) where dominance of Angus over Shorthorns, and Shorthorns over Herefords, was not due to size alone but also appeared to result from breed behavioural characteristics.

No relationship between rank order and milk production was found in the herds studied by Schein & Fohrman (*loc. cit*), Beilharz et al. (1965), and Dickson et al. (1967). According to Boussiou (1965), the type of hierarchy which is set up does not necessarily depend on the length of time the group has existed, or on variations in race, body weight, and age within it.

In studies with monozygotic twins, Hancock (1950b) demonstrated that the dominance order of 10 pairs of twins had only 10 positions: each twin shared a single position in the rank order with its sibling. He suggests therefore that the hereditary disposition of an animal may be a factor in determining rank position. However, it must also be recognized that a rank order based on age, weight, seniority, or past experience would inevitably link each animal with its twin.

2. LEADERSHIP-FOLLOWERSHIP PATTERNS AND MOVEMENT ORDERS

Leadership is a phenomenon quite distinct from dominance-subordination (Gyr, 1946). Although it is often assumed that the dominant animal in the herd is also the leader, this is often quite contrary to fact. It is perhaps important here to distinguish between the lead animal in a forced movement (e.g. cattle being driven into a weight crate or along a road) and the lead animal in a voluntary movement (e.g. grazing, or among dairy cattle entering a milking parlour).

Tulloh (1961a) found that the order in which young Hereford cattle entered a weight crate from a yard (forced movement) was significantly constant and unrelated to sex, weight, or the presence of horns. McPhee *et al.* (1964) similarly state that entry order is not related to the social rank of the animal. Such apparent non-random orders of entry may be a form of territorialism[1] in cattle yards so that the same animals will usually be found in the same part of the yard (Beilharz & Mylrea, 1963a). Donaldson & James (1963) have noted that pregnant cows are usually near the rear when animals enter a crush (squeeze). In a comparison of forced (cattle being driven along a fence line) and voluntary cattle grazing movement, Beilharz & Mylrea (1963b) found that the leaders of forced movements were low in the dominance order, whereas the leaders of voluntary movements tended to have high social rank. Cows in the middle of the dominance order tend to be in advance in herd voluntary movements, whereas the top dominance cows are usually in the middle of the moving group and low dominance animals at the rear (Kilgour & Scott, 1959). However, the entrance order of cows into a milking parlour (voluntary movement) is consistent, but apparently unrelated to body weight or to age, i.e. probably not related to dominance order (Dietrich *et al.*, 1965). Dickson *et al.* (1967) suggest that at least three social structures exist within an established herd of dairy cows: a milking order, a dominance hierarchy, and a leadership-followership pattern.

A leader is an individual which frequently is at the head of a moving column and often seems to initiate a new activity. However, the emphasis in this situation should be on *followership*, rather than *leadership*. Indeed, *active followership*, equated to *passive leadership*, explains why cattle often tend to have their body axes oriented in the same direction while grazing.

Two opposing "forces" are constantly brought to bear, with perhaps slightly different intensities, on the individual herd members. The first is a cohesive tendency, which draws the animals together and prevents scattering and dispersal; each animal tends to stay near another (Hancock, 1950b; Schein & Fohrman, 1955). The opposing force drives the animals apart,

---

1 "Home range" might be a better term, since "territoriality" implies defended area. The only reference to true territorialism in domesticated cattle seems to be that of Hunter & Edwards (1964), who reported a form of defended group territory in bulls kept at an A.I. bull holding farm. Dalton *et al.* (1967), however, did not observe this in their study of bull groups.

and prevents them from being too close to each other. A balance between these two opposing tendencies results in the relatively even spacing between individuals of the group.

The intensity of cohesion varies with a number of environmental and physiological, possibly also genetic, factors. A cow which has been raised among familiar herd-mates tends to be highly cohesive, whereas one that was introduced to the herd later in life is less so; the older the cow at introduction, the less it integrates into the group. Variations in cohesiveness exist within the herd (Schein & Fohrman, *loc. cit.*).

As the grazing group progresses slowly through a field, individuals with lower cohesive tendencies gradually move away from their neighbours. The neighbours, on the other hand, having high cohesive tendencies, will alter their orientations so as to follow those that are drifting apart. Should two less cohesive cows begin to drift in very different directions, they tend to adjust to each other (again by virtue of their own cohesive tendencies, even if somewhat reduced) so that the entire herd slowly progresses in an orderly manner in one direction. Should the two "leaders" fail to adjust to each other, they will pull the herd apart: some individuals will follow each of them. If all individuals follow only one of the two drifters, then the other will eventually be forced to alter its course because of the high positive correlation between the intensity of cohesiveness and the distance between animals. The result is that the herd moves as a unit, and since body axes are orientated in the direction of movement, they will all be parallel.

IV. OTHER BEHAVIOURS

In this section we will discuss eliminative behaviour, grooming and wallowing behaviour, resting and sleeping behaviour, temperament, investigatory behaviour, complex learning situations, and behaviour under restraint. Ranging behaviour, i.e. walking and loitering, has not been as adequately studied in cattle as in sheep (see Chapter 10).

A. Eliminative Behaviour

Defaecation

Cattle deposit their excreta haphazardly and make little or no effort to avoid walking through or lying in soiled areas, except in so far as a freshly wet area may be cold, and therefore uncomfortable. Cattle tend to bunch together at night; consequently, the faeces tend to be concentrated in a few areas. Cattle exposed to bad weather will stay in sheltered areas for long periods and will consequently defaecate much there. Powell (1967), in fact, suggests that dung-pat counts can be used as an indirect method of estimating shelter use. The counting of faecal pellet groups has also been put forward as a method for determining the graze use of range country (Julander, 1955). Cattle cannot apparently voluntarily control the passage of waste material, although Parsons (1966) suggests that an allelomimetic component sometimes causes a large proportion of milking herd to defaecate if one cow does. On the other hand, physical barriers which prevent an animal from assuming a proper stance tend to inhibit the relaxation of anal sphincter muscles, at least temporarily. Thus, an electrically-charged wire which prevents tail raising may, in turn, inhibit the passage of faeces.

Although defaecation may occur while the animal is walking or lying down, it more commonly happens while the animal is standing. Typically, a special stance is assumed: the base of the tail is raised and arched away from the body, the hind legs are placed slightly forward and apart, and the back is arched. The degree of arching varies among individuals. A special defaecation posture is assumed by both males and females during pawing and ground-rubbing behaviour; this stance is part of the general threat pattern. However, cattle attach no social significance to excreta (Schloeth, 1958b).

The defaecation stance assumed by the animal tends to minimize contamination of the body by faeces. An animal under duress, such as during illness, excitement, or exposure to high temperatures, frequently defaecates without arching its back and therefore soils itself. Healthy calves avoid skin contamination more carefully than do adults, possibly because of the irritating nature of the faeces resulting from a milk diet (Brownlee, 1950). It is interesting to note that during suckling, the cow licks (and cleans) the calf's perianal region.

Urination

Females do not urinate while walking and only rarely while lying down. The animal typically stands in a position much like that of defaecation, with perhaps greater arching of the back. The urine in expressed with some force so that contamination of the skin is even less likely than during de-defaecation. Brownlee (1950) noted that the complete micturition posture was assumed during the first urination (3 to 4 hours post-partum).

Males urinate from the normal standing position, with little change in posture except perhaps for a slight spreading of the hind legs. Unlike females, they can and often do urinate while walking, the urine dribbling from the sheath. Males pass urine very slowly and under not nearly as much pressure as do females.

Frequency of defaecation and urination

The frequency of elimination is affected by the quantity and quality of feed, ambient temperature, relative humidity, milk yield, and the individuality of the animal. Normal healthy cattle urinate an average of 9 times and defaecate 12 to 18 times in a 24-hour period. The daily number of urinations and defaecations tends to be higher in milking cows than in dry ones (Fuller, 1928). Although the output volume of urine and faeces per cow varies with the breed, the number of urinations and defaecations is almost the same (Table 22).

McDowell (1961) found a direct relationship between relative humidity and both the volume and frequency of urination: in hot dry environments (20 per cent R.H.), Holstein and Jersey cows averaged 3·2 urinations per 24-hour period; in hot wet environments (80 per cent R.H.), the same cows averaged 12·4 urinations per 24-hour period. Mullick *et al.* (1952) found that the daily urine volume of Kumaune Hill steers in India measured about 5 lb regardless of the season, although water intake varied seasonally from 21 to 34 lb per day. Harker (1960) reported that the average daily rate of faeces production in Zebu cattle was a constant 19 to 28 lb per animal, regardless of food intake.

Table 22. Daily Eliminative Patterns of Two Breeds of Dairy Cattle in New Hampshire, U.S.A. The figures are averages for 4 cows during a 3-day period. (Adapted from Fuller, 1928.)

	Breed	
	Holstein	*Jersey*
Water consumption (lb)	163	84
Milk yield (lb)	44	13
Live weight (lb)	1,400	1,000
No. of urinations	8	9
No. of defaecations	17	18
Urine output (lb)	32	27
Faeces output (lb)	88	61

B. Investigatory Behaviour

Investigatory behaviour involves all the perceptual facilities of the animal (see Chapters 1 and 7). Primary information regarding the presence of an object of curiosity is conveyed to the animal through visual or auditory pathways. If fear is not induced, the animal approaches the strange object cautiously with forehead parallel to the ground, ears up and aimed forward, and eyes fixed on the object. The investigatory posture resembles the submissive posture (Fig. 35, A), except that when an animal is investigating it sniffs and its nostrils quiver. When the object is reached, sniffing is replaced by licking. If the object is small and/or pliable, the animal may take it up and chew it, perhaps even swallowing it. Thus, olfactory, gustatory, and perhaps tactile perception play prominent roles in investigatory behaviour (Schein, 1954). Curiosity-inducing stimuli have two major characteristics: they do *not* induce fear and they are unfamiliar (either new objects in an old setting or old objects in a new setting). The object itself may be large or small, animate or inanimate, moving or stationary; it matters little, so long as fear is not induced.

The expression of investigatory behaviour is inversely related to age. The inverse relationship may be based on several factors: (a) older animals are familiar with most of the objects in their environment; therefore, the stimulus pressure releasing investigatory behaviour is low; (b) the potential level of investigatory behaviour is indeed lower in older animals; (c) investigatory behaviour patterns become more subtle through maturation and learning and are often not recognized as such. At the present time, we have no way of knowing which of these alternatives is correct.

C. Resting Behaviour

Cattle typically lie on the sternum and one side, with the fore limbs flexed under the body; one hind limb is flexed forward, under the body, while the other is stretched out (Brownlee, 1950). On sloping ground, the long axis of the cow's body is either along the slope or running slightly uphill with the forequarters placed a little higher than the hindquarters. The legs are downslope to the body in both positions. The head is usually held erect because other head positions tend to interfere with the eructation of gases from the rumen and with swallowing. Shifting of position from one side to the other is frequently observed. None the less, Nottbohm (1928) has suggested that cows lie on their left sides, and a study of a herd of 61 dairy cows (Ewbank, 1966) showed that 22 of the animals appeared to lie mainly (more than 75 per cent of their total lying time) on one particular side: 13 of these cows favoured the right side and nine the left. In contrast to horses, healthy adult cattle rarely lie fully outstretched on their sides and when they do it is only for short periods.

Hughes & Reid (1951) estimate that Hereford bullocks spend roughly 50 per cent of the daylight hours lying down, during most of which time the animals are ruminating. Zebu cattle in the tropics lie down mostly at night (Harker et al., 1961). In warm climates, European cattle lie down more if shade is available than if left exposed to the direct sunlight (McDaniel & Roark, 1956). In general, cattle spend 9 to 12 hours per day lying down (Table 19); these figures are subject to seasonal, breed, and individual variation. Furthermore, Turner (1961) suggests that old cows lie down for longer periods than young ones.

Hereford and Aberdeen Angus bulls are sometimes seen resting in a dog-sitting posture (Ewbank, 1964). Such animals spend lengthy periods of time with their forequarters raised up on their extended forelimbs and their hindquarters resting on the ground (Plate III). This appears to be a normal posture for bulls of some heavy beef breeds.

D. Sleeping Behaviour

Brownlee (1950) and Balch (1952) do not believe that adult cattle sleep. This may be so, but it has recently been demonstrated electrocardiographically (Klemm, 1966) that goats exhibit both orthodoxical (true) and paradoxical sleep, while lying in sternal recumbancy with their heads sometimes resting on their flanks and with their eyes closed. This position is certainly seen in cattle, although it only lasts for a short time on each occasion (Brownlee, 1950). If adult cattle do sleep, then it is possibly by many of these short periods. Calves commonly spend up to $\frac{1}{2}$ hour in the head-turned-back-to-upper-flank position and have been recorded as "sleeping" for as much as 3·2 hours a day (Roy *et al.*, 1955).

E. Temperament

The loosely defined concept of temperament[1] has been dealt with more commonly in popular writings than in scientific literature, primarily because it eludes firm definition and objective measurement. Moreover, the evaluation of temperament is complicated by several factors, such as age × genetic interaction and animal × man interaction. It is well recognized that temperament varies among breeds and individuals.

Attempts to evaluate temperament by using the latency of ejaculation or the time required to fasten an animal to a post (*restraint time*) are of dubious value: latency of ejaculation is more dependent upon factors unrelated to temperament (see Sexual Behaviour of Males), and restraint time may reflect learning and habituation more than temperament, to say nothing of the dexterity and experience of the human handler.

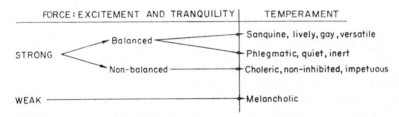

Fig. 37. The Pavlovian concept of temperament in terms of the force of nervous processes. (Adapted from Schein, 1961; Barȳshnikov & Kokorina, 1964.)

In an attempt to define temperament, Pavlov and his disciples tried to separate out and study its component parts. According to the Pavlovian concept (Fig. 37 and Table 23) (as described by Gantt, 1958; Schein, 1961; Barȳshnikov & Kokorina, 1964), animals can be divided according to the strength of the two main nervous processes, *excitement* and *tranquillity*. When these processes exist in great force, the animal is of the *strong* type; if they do not exist in great force, the animal is of the *weak* type. If both excitement and tranquillity forces are nearly equal, the animal is said to be *balanced*; if they are very unequal, the animal is *non-balanced*. Strong, balanced animals are either *sanguine* (lively) or *phlegmatic* (quiet), depending upon which primary force has the edge. They are well-adjusted animals, capable of readily adapting to new situations with the least amount of psychological trauma and are not susceptible to nervous breakdowns. Strong, non-balanced animals are *choleric* (non-inhibited) and are subject to fits of rage and expressions of extreme emotion. Weak animals are *melancholic* and, like choleric animals, are susceptible to nervous breakdowns. Melancholic animals are also the poor, unadaptable misfits in animal societies.

[1] Often called nervousness or skittishness.

Table 23. Relationships between Pavlovian Concepts of Temperament and Economic Attributes of Dairy Cows. (Adapted from Kudryavtzev, 1962.)

Temperament	Reaction to changes in Environment	Body Build	Milking Potential	Frequency of occurrence in Herds
Sanguine, lively, gay	Good adaptation	Strongly built	Highest	High in improved herds
Phlegmatic, quiet	Slow adaptation	Strongly built	Good	High in improved herds
Choleric, non-inhibited	Drastic drops in milk yields	Strongly built	Seldom high	Not high
Melancholic	Adapts with great difficulty	Poorly built	Low	Low

An extensive review of this essentially Russian work is given by Barȳshnikov & Kokorina (1964); in addition Kudryavtzev (1962) has indicated the manner in which it relates to the husbandry of dairy cattle.

Parez (1953) and Fraser (1957c) have attempted to classify the temperament of bulls by means of a modified Pavlovian/sexual activity approach and Tulloh (1961b) has studied the temperament of (yard) cattle by recording their behaviour when being handled for weighing.

F. Grooming and Wallowing Behaviour

Cattle spend a considerable amount of time licking themselves and each other. An individual licks any part of its body it can reach, including the hind limbs and tail (Brownlee, 1950), and often uses low branches and fence staves as scratching posts for the inaccessible parts of the body. Calves spend an average of 52 minutes a day licking and scratching themselves (Roy *et al.*, 1955); adult beef cattle average 152 licking periods and 28 scratching periods each day (Schake & Riggs, 1966). Individuals commonly submit to licking by herd-mates, typically standing with the head lowered and eyes half closed (Brownlee, *loc. cit.*). Tame cattle often submit to, and even solicit, scratching by humans. In a heterosexual herd of Camargue cattle on free range, licking interactions between males were more frequently observed than between females or juveniles. In 95 per cent of the cases, only the head, neck, and shoulder of the herd-mate were licked. Licking interactions in both sexes involved animals less than 3 rank positions apart on the social scale; most commonly, a subordinate animal licked its immediate rank superior (Schloeth, 1961).

Tail-switching is part of grooming behaviour in that it is commonly used to brush flies and other irritants off the skin. It may also be used to express intense emotion. Restrained individuals in a fearful situation tail-switch much more than usual as they struggle against the restraint; similarly, as already mentioned, young calves characteristically switch their tails while nursing.

The adaptive value of grooming is apparent, since it results in the removal of noxious contaminants such as faeces, urine, mud, and possibly some external parasites. What induces social grooming is less apparent, but we are tempted to speculate that salt or some other desirable chemical compound is being licked from the neighbouring animal.

Ferguson (1966) reports that European type dairy cows will, under heat stress, sometimes wallow in soft mud. This appears to be a method of thermoregulation.

G. Complex Learning Situations

Although the simple reflex type of learning has been extensively studied by the Russians (see Barȳshnikov & Kokorina, 1964), little seems to have been done on more complex types of learning. Brownlee (1957) has described some examples of learning and memory in cattle, and Albright et al. (1966) have shown that cows can, to a limited extent, be trained to a noise stimulus to enter a milking parlour in a specific order or to come to a particular place in a field. It has also been reported by Broadbent (1967) that cattle, wearing individually tuned coils around their necks, soon learn to unlock the electronically controlled doors of their individual feeding boxes. Young cattle appear easier to train to this device than older animals.

H. Behaviour under Restraint

Domesticated cattle are, from time to time, submitted to various handling procedures which restrict or confine them in some way. Stockmen over the ages have, of necessity, acquired an extensive knowledge of the behaviour of cattle under restraint. Unfortunately, much of this information is handed down by word of mouth and thus remains unrecorded in any formal sense. Exceptions to this are the reports of Bisschop (1961) and Sutton et al. (1967), who have recorded the behaviour of cattle being transported by rail; Tulloh (1961a, b), who has studied the behaviour of animals in collecting yards and weighing crates; and Ewbank (1961), who has described the various behaviour patterns seen when cattle are handled by means of a crush (squeeze).

A general account of the behaviour of animals (including cattle) under restraint can be found in Ewbank (1968).

REFERENCES

ALBRIGHT, J. L., GORDON, W. P., BLACK, W. C., DIETRICH, J. P., SNYDER, W. W. & MEADOWS, C. E. (1966). Behavioral responses of cows to auditory training. *J. Dairy Sci.*, **49**, 104–106.

ALMQUIST, J. O. & HALE, E. B. (1956). An approach to the measurement of sexual behaviour and semen production of dairy bulls. *Proc. 3rd Int. Congr. anim. Reprod.* (Cambridge), Plenary Papers, pp. 50–59.

ALMQUIST, J. O., HALE, E. B. & AMANN, R. P. (1958). Sperm production and fertility of dairy bulls at higher collection frequencies with varying degrees of sexual preparation. *J. Dairy Sci.*, **41**, 733 (Abstract).

ANDERSON, J. (1948). Improvement of cattle in East Africa by artificial insemination. *Proc. 1st Int. Congr. Physiol. Path. anim. Reprod. Artif. Insem.* (Milan), pp. 14–15.

ANDERSON, J. (1949). Artificial insemination in cattle breeding in Kenya. *E. Afr. agric. J.*, **14**, 148–150.

APPLEMAN, R. D. & RONNING, M. (1955). Cow feeding behavior in a self-feeding trench silo. *Proc. Ass. Southern agric. Wkrs. 52nd convention* (Louisville, Ky.), pp. 67–68 (Abstract).

ARTHUR, G. H. (1961). Some observations on the behaviour of parturient farm animals with particular reference to cattle. *Vet. Rev. May & Baker*, **12**, 75–84.

ARTHUR, G. H. (1964). *Wright's Veterinary Obstetrics*. London: Baillière, Tindall & Cassell, p. 131.

ASDELL, S. A. (1946). *Patterns of Mammalian Reproduction*. Ithaca, N.Y.: Comstock, p. 347.

BAILEY, G. L. & BROSTER, W. H. (1958). Experiments on the nutrition of the growing dairy heifer. III. Effects of air temperature and live weight on water consumption. *J. agric. Sci.*, **50**, 8–11.

BAKER, F. N., VANDEMARK, N. L. & SALISBURY, G. W. (1955). The effect of frequency of ejaculation on the semen production, seminal characteristics, and libido of bulls during the first post-puberal year. *J. Dairy Sci.*, **38**, 1000–1005.

BALCH, C. C. (1952). Factors affecting the utilization of food by dairy cows. 6. The rate of contractions of the reticulum. *Brit. J. Nutr.*, **6**, 366–375. (Also references by Balch *et al.*, published in the same journal.)

BALCH, C. C., BALCH, D. A., JOHNSON, V. W. & TURNER, J. (1953). Factors affecting the utilization of food by dairy cows. 7. The effect of limited water intake on the digestibility and rate of passage of hay. *Brit. J. Nutr.*, **7**, 212–224.

BALCH, C. C. & CAMPLING, R. C. (1962). Regulation of voluntary food intake in ruminants. *Nutr. Abstr. Rev.*, **32**, 669–686.

BANE, A. (1954). Studies on monozygous cattle twins. XV. Sexual functions of bulls in relation to heredity, rearing intensity and somatic conditions. *Acta agric. scand.*, **4**, 95–208.

BARYSHNIKOV, I. A. & KOKORINA, E. P. (1964). Higher nervous activity of cattle. *Dairy Sci. Abstr.*, **26**, 97–115.

BEILHARZ, R. G. & MYLREA, P. J. (1963a). Social position and behaviour of dairy heifers in yards. *Anim. Behav.*, **11**, 522–528.

BEILHARZ, R. G. & MYLREA, P. J. (1963b). Social position and movement orders of dairy heifers, *Anim. Behav.*, **11**, 529–533.

BEILHARZ, R. G., BUTCHER, D. F. & FREEMAN, A. E. (1965). Social dominance and milk production in Holsteins. *J. Dairy Sci.*, **48**, 795 (Abstract).

BELL, F. R. & WILLIAMS, H. H. (1959). Threshold values for taste in monozygotic twin calves. *Nature, Lond.*, **183**, 345–346.

BELL, F. R. & WILLIAMS, H. H. (1960). The effect of sodium depletion on the taste threshold of calves. *J. Physiol.*, **151**, 42–43.

BERLOIZ, J. (1933). L'élevage et l'hybridation du Bison au Canada. *Bull. Soc. Acclim. Fr.*, **80**, 47–53.

BISSCHOP, J. H. R. (1961). Transportation of animals by rail. I. The behaviour of cattle during transportation by rail. *Jl S. Afr. vet. med. Ass.*, **32**, 235–268.

BLAXTER, K. L. (1944). Food preferences and habits in dairy cows. *Proc. Br. Soc. Anim. Prod.*, 85–94.

BONADONNA, T. (1953). Problemi biologici e technologici della fecondazione artificiale. *Coll. tec.-sci. 'L. Spallanzani'*, No. 7, 507 pp. (*Anim. Breed. Abstr.* **21**, 406, 1953.)

BONADONNA, T. (1956). On some biological and non-biological factors that may affect the collection and quality of the semen. *Proc. 3rd Int. Congr. anim. Reprod.* (Cambridge), Plenary Papers, pp. 105–112.

BONSMA, J. C. & ROUX, J. D. (1953). Influence of environment on the grazing habits of cattle. *Fmg S. Afr.*, **28**, 43–47.

BOUSSIOU, M-F. (1965). Observations sur la hiérarchie sociale chez les bovins domestiques. *Annls. Biol. anim. Biochim. Biophys.*, **5**, 327–339.

BRAKEL, W. J., RIFE, D. C. & SALISBURY, S. M. (1952). Factors associated with the duration of gestation in dairy cattle. *J. Dairy Sci.*, **35**, 179–194.

BRANNON, W. F., REID, J. T. & MILLER, J. I. (1954). The influence of certain factors upon the digestibility and intake of pasture herbage by beef steers. *J. Anim. Sci.*, **13**, 535–542.

BRANTON, C., D'ARENSBOURG, G. & JOHNSTON, J. E. (1952). Semen production, fructose content of semen and fertility of dairy bulls as related to sexual excitement. *J. Dairy Sci.*, **35**, 801–807.

BROADBENT, P. J. (1967). Individual feeding device for livestock. *Fm Bldg Prog.*, No. 12, 7–8.

BROWN, D. C., WILLARD, H. S., HAMILTON, J. W. & READ, J. (1952). Alfalfa pellets-roughages or concentrates. *Wyo. Agric. exp. Sta. Bull.*, No. 321.

BROWNLEE, A. (1950). Studies on the behaviour of domestic cattle in Britain. *Bull. Anim. Behav.*, **1** (8), 11–20.

BROWNLEE, A. (1951). Studies in the behaviour of domestic cattle. (1) skin hygiene, (2) curiosity, (3) play. *Vet. Rec.*, **63**, 443 (Abstract).

BROWNLEE, A. (1954). Play in domestic cattle in Britain: an analysis of its nature. *Brit. vet. J.*, **110**, 48–68.

BROWNLEE, A. (1957). Higher nervous activity in domestic cattle. *Br. vet. J.*, **113**, 407–416.

BUECHNER, H. K. (1961). Territorial behavior in Uganda Kob. 16 mm. film, *A-V Aids Library*, Washington State Univ., Pullman, U.S.A.

BURT, A. W. A. (1957). The effect of variations in nutrient intake upon the yield and composition of milk. II. Factors affecting rate of eating roughage and responses to an increase in the amount of concentrates fed. *J. Dairy Res.*, **24**, 296–315.

CAMPLING, R. C. (1964). Factors affecting the voluntary intake of grass. *Proc. Nutr. Soc.*, **23**, 80–88.

CARPENTER, C. R. (1934). A field study of the behavior and social relations of howling monkeys. *Comp. Psychol. Monogr.*, **10**, 168 pp.

CASADY, R. B., LEGATES, J. E. & MYERS, R. M. (1956). Correlation between ambient temperatures varying from 60°–95°F. and certain physiological responses in young dairy bulls. *J. Anim. Sci.*, **15**, 141–152.

CASTLE, M. E. & HALLEY, R. J. (1953). The grazing behaviour of dairy cattle at the National Institute for Research in Dairying. *Brit. J. Anim. Behav.*, **1**, 139–143.

CHAMBERS, D. T. (1959). Grazing behaviour of calves reared at pasture. *J. agric. Sci.*, **53**, 417–424.

COLLINS, W. J., BRATTON, R. W. & HENDERSON, C. R. (1951). The relationship of semen production to sexual excitement of dairy bulls. *J. Dairy Sci.*, **34**, 224–227.

CORBETT, J. L. (1953). Grazing behaviour in New Zealand. *Brit. J. Anim. Behav.*, **1**, 67–71.

CORDTS, H. (1953). Zur Kenntnis der sexuellen Erregung bei Haustieren. *Z. Tierzücht ZüchtBiol.* **61**, 305–356.

COWIE, A. T. (1957). Mammary development and lactation. In: *Progress in the Physiology of Farm Animals*, J. Hammond (ed.), **3**, 907, London: Butterworth & Co.

COWLISHAW, S. J. & ALDER, F. E. (1960). The grazing preferences of cattle and sheep. *J. agric. Sci.*, **54**, 257–265.

CROMBACH, J. J. M. L. (1961). Some aspects of the behaviour of dairy bulls. The effect of stimulation on the ejaculate. *Z. Tierzücht ZüchtBiol.*, **75**, 331–391.

CROMBACH, J. J. M. L., DeROVER, W. & DE GROOT, B. (1956). The influence of preparation of dairy bulls on sperm production and fertility. *Proc. 3rd Int. Congr. anim. Reprod.* (Cambridge), Sec. 3, pp. 80–82.

CROSS, B. A. (1961). Neural control of lactation. In: *Milk: The Mammary Gland and its Secretion.* S. K. Kon & A. T. Cowie (eds.), **1**, 229. New York and London: Academic Press.

CSUKAS, Z. (1949). A karotin szerepe az emlösök szaporodásának szakaszosságában. (The role of carotene in the periodicity of reproduction of mammals.) *Magy. allatorv. Lap.* (14), 15 pp. (*Anim. Breed. Abstr.*, **18**, 165, 1950.)

DALTON, D. C., PEARSON, M. E. & SHEARD, M. (1967). The behaviour of dairy bulls kept in groups. *Anim. Prod.*, **9**, 1–5.

10+

DE VUYST, A., THINÈS, G., HENRIET, L. & SOFFIÉ, M. (1964). Influence des stimulations auditives sur le comportement sexual du taureau. *Experientia*, **20**, 648–553.

DICKSON, D. P., BARR, G. R. & WIECKERT, D. A. (1967). Social relationship of dairy cows in a feed lot. *Behaviour*, **29**, 195–203.

DIETRICH, J. P., SNYDER, W. W., MEADOWS, C. E. & ALBRIGHT, J. L. (1965). Rank order in dairy cows. *Am. Zool.*, **5**, 713 (Abstract).

DONALD, H. P. & ANDERSON, D. (1953). A study of variation in twin cattle. II. Fertility. *J. Dairy Res.*, **20**, 361–369.

DONALD, H. P. & ANDERSON, D. W. (1967). A study of behaviour in twin cattle. *Report of the Agricultural Research Council's Animal Breeding Research Organization.* Edinburgh, Jan. 1967, 18–25.

DONALDSON, L. E. (1962). Competition between Brahman cross bulls and Hereford bulls in mating. *Aust. vet. J.*, **38**, 520.

DONALDSON, L. E. & JAMES, J. W. (1963). A connection between pregnancy and crush order in cows. *Anim. Behav.*, **11**, 286.

DUCKWORTH, J. E. & SHIRLAW, D. W. (1958a). The value of animal behaviour records in pasture evaluation studies. *Anim. Behav.*, **6**, 139–146.

DUCKWORTH, J. E. & SHIRLAW, D. W. (1958b). A study of factors affecting feed intake and the eating behaviour of cattle. *Anim. Behav.*, **6**, 147–154.

DUKES, H. H. (1955). *The Physiology of Domestic Animals.* 7th ed. Ithaca, N.Y.: Comstock.

ERB, R. E., ANDREWS, F. N., HAUGE, S. M. & KING, W. A. (1947). Observations on vitamin A deficiency in young dairy bulls. *J. Dairy Sci.*, **30**, 687–702.

ETKIN, W. & SCHEIN, M. W. (1960). Social dominance in the male Black Buck. 16 mm. film, 9 min., PCR-117, *Psychol. Cinema Register*, Penn State Univ., University Park, U.S.A.

EWBANK, R. (1961). The behaviour of cattle in crushes. *Vet. Rec.*, **73**, 853–856.

EWBANK, R. (1964). The "dog-sitting" posture in cattle and sheep. *Vet. Rec.*, **76**, 388–393.

EWBANK, R. (1966). A possible correlation, in one herd, between certain aspects of the lying behaviour of tied-up dairy cows and the distribution of subclinical mastitis among the quarters of their udders. *Vet. Rec.*, **78**, 299–303.

EWBANK, R. (1967). Behavior of twin cattle. *J. Dairy Sci.*, **50**, 1510–1512.

EWBANK, R. (1968). The behaviour of animals in restraint. In: *Abnormal Behaviour in Animals*, M. W. Fox (ed.), Chapter 10. Philadelphia: Saunders.

FARRIS, E. J. (1954). Activity of dairy cows during estrus. *J. Am. vet. med. Ass.*, **125**, 117–120.

FERGUSON, W. (1966). Wallowing by dairy cattle. *Proc. Soc. vet. Ethology*, **1** (2), 6–7.

FINDLAY, J. D. & BEAKLEY, W. R. (1954). Environmental physiology of farm mammals. In: *Progress in the Physiology of Farm Animals.* J. Hammond (ed.) **1**, 252–298. London: Butterworth.

FOLMAN, Y. & VOLCANI, R. (1966). Copulatory behaviour of the prepubertally castrated bull. *Anim. Behav.*, **14**, 572–573.

FONTENOT, J. P. & BLASER, R. E. (1965). Symposium on factors influencing the voluntary intake of herbage by ruminants: selection and intake by grazing animals. *J. Anim. Sci.*, **24**, 1202–1208.

FRASER, A. F. (1957a). The state of fight or flight in the bull. *Brit. J. Anim. Behav.*, **5**, 48–49.

FRASER, A. F. (1957b). The state of submission in cattle. *Br. vet. J.*, **113**, 167–168.

FRASER, A. F. (1957c). The disposition of the bull. *Brit. J. Anim. Behav.*, **5**, 110–115.

FRASER, A. F. (1968). *Reproductive Behaviour in Ungulates.* London: Academic Press.

FRENCH, M. H. (1956). The effect of infrequent water intake on the consumption and digestibility of hay by Zebu cattle. *Emp. J. exp. Agric.*, **24**, 128–136.

FRYE, J. B., JR., MILLER, G. D. & BURCH, B. J. (1950). The effect of sprinkling dairy cows on midsummer milk production. *Proc. Assoc. Southern agric. Wkrs. 47th Ann. Conv.*, pp. 81–82.

FULLER, J. M. (1928). Some physical and physiological activities of dairy cows under conditions of modern herd management. *New Hamps. Agric. exp. Sta. Tech. Bull.*, No. 35.

GANTT, W. H. (1958). Pavlov. In: *The Central Nervous System and Behavior*, M. A. B. Brazier (ed.), New York, N.Y.: Josiah Macy, Jr. Foundation, pp. 163–186.

GARM, O. (1949). A study of bovine nymphomania with special reference to etiology and pathogenesis. *Acta endocr., Kopenh.*, **2** (Suppl. 3), 144 pp. (B).

GORDON, J. G., TRIBE, D. E. & GRAHAM, T. C. (1954). The feeding behaviour of phosphorus-deficient cattle and sheep. *Brit. J. Anim. Behav.*, **2**, 72–74.

GREEN, H. H. (1925). Perverted appetites. *Physiol. Rev.*, **5**, 336–348.

GUHL, A. M. & ATKESON, F. W. (1959). Social organization in a herd of dairy cows. *Trans. Kans. Acad. Sci.*, **62**, 80–87.

GYR, W. (1946). Die Kuhkämpfe im Val d'Anniviers. *Schweiz. Arch. Volkskunde*, **43**, 176–209.

HAFEZ, E. S. E. (1956). Nymphomania in cattle. *Vlaams diergeneesk. Tijdschr.*, **25**, 267–282.

HAFEZ, E. S. E. (1960). Analysis of ejaculatory reflexes and sex drive in the bull. *Cornell Vet.*, **50**, 384–411.

HAFEZ, E. S. E. (1961). Unpublished data. Observations on: Behaviour of Hereford calves during suckling; Comparative behaviour of Hereford bulls during semen collection by the electro-ejaculator or the artificial vagina; Manifestation of œstrus and facial expression of Hereford heifers before and after injection of gonadotropic and other hormones. Dept. Anim. Sci., Wash. State Univ., Pullman, U.S.A.

HAFEZ, E. S. E. & DARWISH, Y. H. (1956). Effect of successive ejaculation on semen characteristics in the buffalo. *J. agric. Sci.*, **47**, 191–195.

HAFEZ, E. S. E. & LINEWEAVER, J. A. (1968). Suckling behaviour in natural and artificially fed neonate calves. *Z. Tierpsychol.*, **25**, 187–198.

HALE, E. B. (1966). Visual stimuli and reproductive behavior in bulls. *J. Anim. Sci.*, **25** (Suppl.), 36–44.

HALE, E. B. & ALMQUIST, J. O. (1960). Relation of sexual behavior to germ cell output in farm animals. *J. Dairy Sci.*, **43** (Suppl.), 145–169.

HALLEY, R. J. (1953). The grazing behaviour of South Devon cattle under experimental conditions. *Br. J. Anim. Behav.*, **1**, 156–157.

HANCOCK, J. (1950a). Grazing habits of dairy cows in New Zealand. *Emp. J. exp. Agric.*, **18**, 249–263.

HANCOCK, J. (1950b). Studies in monozygotic twins. IV. Uniformity trials: Grazing behaviour. *N.Z. J. Sci. Tech.*, **32A**, 22–59.

HANCOCK, J. (1952). Grazing behaviour of identical twins in relation to pasture type, intake and production of dairy cattle. *Proc. 6th Int. Grassland Congr.* (Pa., U.S.A.), 1399–1407.

HANCOCK, J. (1953). Grazing behaviour of cattle. *Anim. Breed. Abstr.*, **21**, 1–13.

HANCOCK, J. (1955). Studies in grazing behaviour of dairy cattle. II. Bloat in relation to grazing behaviour. *J. agric. Sci.*, **45**, 80–95.

HARDISON, W. A., FISHER, H. L., GRAF, G. C. & THOMPSON, N. R. (1956). Some observations on the behavior of grazing lactating cows. *J. Dairy Sci.*, **39**, 1735–1741.

HARKER, K. W. (1960). Defaecating habits of a herd of Zebu cattle. *Tropical Agric.*, **37**, 193–200.

HARKER, K. W. & ROLLINSON, D. H. L. (1961). Personal communication. Animal Health Res. Centre, Entebbe, Uganda.

HARKER, K. W., TAYLOR, J. I. & ROLLINSON, D. H. L. (1954). Studies on the habits of Zebu cattle. I. Preliminary observations on grazing habits. *J. agric. Sci.*, **44**, 193–198.

HARKER, K. W., TAYLOR, J. I. & ROLLINSON, D. H. L. (1956). Studies on the habits of Zebu cattle. V. Night paddocking and its effect on the animal. *J. agric. Sci.*, **47**, 44–49.

HARKER, K. W., ROLLINSON, D. H. L., TAYLOR, J. I., GOURLAY, R. N. & NUNN, W. R. (1961). Studies on the habits of Zebu cattle. VI. The results on different pastures. *J. agric. Sci.*, **56**, 137–141.

HART, G. H., MEAD, S. W. & REGAN, W. M. (1946). Stimulating the sex drive of bovine males in artificial insemination. *Endocrinology*, **39**, 221–228.

HEINEMANN, F. (1958). Stille Brunst und Orgasmus weiblicher Rinder als verhaltens-biologische und züchterische Probleme. *Zuchthyg. FortpflStör. Besam. Haustiere.*, **2**, 235–265.

HOLDER, J. M. (1960). Observations on the grazing behaviour of lactating dairy cows in a sub-tropical environment. *J. agric. Sci.*, **55**, 261–267.

HUGHES, G. P. & REID, D. (1951). Studies on the behaviour of cattle and sheep in relation to the utilization of grass. *J. agric. Sci.*, **41**, 350–366.

HULTNÄS, C. A. (1959). Studies on variation in mating behaviour and semen picture in young bulls of the Swedish Red-and-White breed and on causes of this variation. *Acta agric. scand.* (Suppl.) 6, 82 pp. (B).

HULTNÄS, C. A. (1961). Personal communication. National Ass. for A. I., Hallsta, Sweden.

HUNTER, W. K. & EDWARDS, J. (1964). The maintenance of an A.I. stud in an inactive state. *Proc. 5th Int. Congr. anim. Reprod. Artif. Insem. (Trento)*, **4**, 341–347.

HUTCHISON, H. G., WOOF, R., MABON, R. M., SALEHE, I. & ROBB, J. M. (1962). A study of the habits of Zebu cattle in Tanganyika. *J. agric. Sci.*, **59**, 301–317.

INNES, R. R. & MABEY, G. L. (1964). Studies on browse plants in Ghana. III. Brouse/grass ingestion ratios (a) Determination of the free-choice Griffonia/Grass ingestion ratios for West African Shorthorn Cattle on the Accra plains using the "simulated shrub" technique. *Emp. J. exp. Agric.*, **32**, 180–190.

ITTNER, N. R., KELLY, C. F. & GUILBERT, H. R. (1951). Water consumption of Hereford and Brahman cattle and the effect of cooled drinking water in a hot climate. *J. Anim. Sci.*, **10**, 742–751.

ITTNER, N. R., BOND, T. E. & KELLY, C. F. (1954a). Increasing summer gains of livestock with cool water, concentrate roughage, wire corrals and adequate shades. *J. Anim. Sci.*, **13**, 867–877.

ITTNER, N. R., GUILBERT, H. R. & CARROLL, F. D. (1954b). Adaption of beef and dairy cattle to the irrigated desert. *Calif. agric. exp. Sta. Bull.*, No. 745.

IVINS, J. D. (1952). The relative palatability of herbage plants. *J. Br. Grassld Soc.*, **7**, 43–54.

IVINS, J. D. (1955). The palatability of herbage. *Herb. Abstr.*, **25**, 75–79.

JOHNSTONE-WALLACE, D. B. & KENNEDY, K. (1944). Grazing management practices and their relationship to the behaviour and grazing habits of cattle. *J. agric. Sci.*, **34**, 190–197.

JONES, C. G., MADDEVER, K. D., COURT, D. L. & PHILLIPS, M. (1966). The time taken by cows to eat concentrates. *Anim. Prod.*, **8**, 489–497.

JOUBERT, D. M. (1952). Twin births among cattle in South Africa. *Fmg S. Afr.*, **27**, 9–14.

JULANDER, O. (1955). Determining grazing use by cow-chip counts. *J. Range Mgmt*, **8**, 182.

KANE, E. A., JACOBSON, W. D., ELY, R. E. & MOORE, L. A. (1953). The estimation of dry-matter consumption of grazing animals by ratio techniques. *J. Dairy Sci.*, **36**, 637–644.

KARE, M. R. (1961a). Personal communication. Dept. Zoology, North Carolina State College, Raleigh, N.C., U.S.A.

KARE, M. R. (1961b). Comparative aspects of the sense of taste. In: *The Physiological and Behavioral Aspects of Taste*, M. R. Kare & B. P. Halpern (eds.), Chicago, Ill.: Univ. of Chicago Press.

KELLY, C. F., BOND, T. E. & ITTNER, N. R. (1955). Water cooling for livestock in hot climates. *Agric. Engng.*, **36**, 173–180.

KERRUISH, B. M. (1955). The effect of sexual stimulation prior to service on the behaviour and conception rate of bulls. *Brit. J. Anim. Behav.*, **3**, 125–130.

KICK, C. H., GERLAUGH, P., SCHALK, A. F. & SILVER, E. A. (1937). The effect of mechanical processing of feeds on the mastication and rumination of steers. *J. agric. Res.*, **55**, 587–597.

KIDWELL, J. F., BOHMAN, V. R. & HUNTER, J. E. (1954). Individual and group feeding of experimental beef cattle as influenced by hay maturity. *J. Anim. Sci.*, **13**, 543–547.

KILGOUR, R. & SCOTT, T. H. (1959). Leadership in a herd of dairy cows. *Proc. N.Z. Soc. Anim. Prod.*, **19**, 36–43.

KING, J. A. & GURNEY, N. L. (1954). Effect of early social experience on adult aggressive behavior in C57BL/10 mice. *J. comp. physiol. Psychol.*, **47**, 326–330.

KIRILLOV, V. (1943). Značenie steriljnyh samcov pri osemenii životnyh. (The importance of sterilized males in the insemination of animals.) *Sovhoz. Proizvod.*, No. 9, 39–40. (*Anim. Breed. Abstr.*, **13**, 20, 1945.)

KLEMM, W. R. (1966). Sleep and paradoxical sleep in ruminants. *Proc. Soc. exp. Biol. Med.*, **121**, 635–638.

KRZWANEK, W. & BRUGGEMANN, H. (1930). Modellversuche zur Physiologie des Saugaktes. *Tierärztl. Wschr.*, **20**, 710–713.

KUDRYAVTZEV, A. A. (1962). Higher nervous activity and physiology of the senses in lactating cows. *Proc. 16th Int. Dairy Congr. (Copenhagen)*, Volume D, 565–572.

LAGERLÖF, N. (1951). Hereditary forms of sterility in Swedish cattle breeds. *Fertil. Steril.*, **2**, 230–239.

LAMPKIN, G. H., QUARTERMAN, J. & KIDNER, M. (1958). Observations on the grazing habits of grade and Zebu steers in a high altitude temperate climate. *J. agric. Sci.*, **50**, 211–218.

LAMPKIN, G. H. & QUARTERMAN, J. (1962). Observations on the grazing habits of Grade and Zebu cattle. II. Their behaviour under favourable conditions in the tropics. *J. agric. Sci.*, **59**, 119–123.

LARKIN, R. M. (1954). Observations on the grazing behaviour of beef cattle in tropical Queensland. *Qd J. agric. Sci.*, **11**, 115–141.

LEES, J. L. (1962). Dairy cows on slats. *Agriculture, Lond.*, **69**, 226–229.

LEWIS, R. C. & JOHNSON, J. D. (1954). Observations of dairy cow activities in loose-housing. *J. dairy Sci.*, **37**, 269–275.

LILJESTRAND, G. & ZOTTERMAN, Y. (1954). The water taste of mammals. *Acta physiol. scand.*, **32**, 291–303.

LOFGREEN, G. P., MEYER, J. H. & PETERSON, M. L. (1956). Nutrient consumption and utilization from alfalfa pasture, soilage and hay. *J. Anim. Sci.*, **15**, 1158–1165.

LOFGREEN, G. P., MEYER, J. H. & HULL, J. L. (1957). Behavior patterns of sheep and cattle being fed pasture or soilage. *J. Anim. Sci.*, **16**, 773–780.

MARTJUGIN, D. D. (1944). Izučenie akta sosanija u teljat. Soobščenie 1-e. Soobščenie 2-e. (A study of the act of suckling in calves. Parts I & II.) *Trud. Timirjazev. seljskohoz Akad.*, **31**, 149–172, 173–184. (*Anim. Breed. Abstr.*, **14**, 18, 1946.)

MCARTHUR, A. T. G. (1952). The effect of management on the grazing behaviour of calves. *Proc. N.Z. Soc. anim. Prod.*, *11th Ann. Conf.*, p. 87.

MCCULLOUGH, M. E. (1953). The use of indication methods in measuring the contribution of two forages to the total ration of dairy cows. *J. Dairy Sci.*, **36**, 445–449.

MCDANIEL, A. H. & ROARK, C. B. (1956). Performance and grazing habits of Hereford and Aberdeen-Angus cows and calves on improved pasture as related to type of shade. *J. Anim. Sci.*, **15**, 59–63.

MCDOWELL, R. E. (1961). Personal communication. Dairy Husbandry Research Branch, U.S.D.A., Beltsville, Md., U.S.A.

MCHUGH, T. (1958). Social behavior of the American buffalo (*Bison bison bison*). *Zoologica*, N.Y. Zool. Soc., **43**, 1–40.

MACLUSKY, D. S. (1959). Drinking habits of grazing cows. *Agriculture, Lond.*, **66**, 383–386.

MCPHEE, C. P., MCBRIDE, G. & JAMES, J. W. (1964). Social behaviour of domestic animals. III. Steers in small yards. *Anim. Prod.*, **6**, 9–15.

MEACHAM, T. N., CUNHA, T. J., WARNICK, A. C., HENTGES, J. F. & HARGROVE, D. D. (1963). Influence of low protein rations on growth and semen characteristics of young beef bulls. *J. Anim. Sci.*, **22**, 115–120.

MEYER, J. H., LOFGREEN, G. P. & ITTNER, N. R. (1956). Further studies on the utilization of alfalfa by beef steers. *J. Anim. Sci.*, **15**, 64–75.

MILLER, G. D., FRYE, J. B., JR., BURCH, B. J., JR., HENDERSON, P. J. & RUSOFF, L. L. (1951). The effect of sprinkling on the respiration rate, body temperature, grazing performance and milk production of dairy cattle. *J. Anim. Sci.*, **10**, 961–968.

MILOVANOV, V. K. & SMIRNOV-UGRJUMOV, D. V. (1940). Problema racionaljnogo ispoljzovanija plemennyh proizvoditelei v svete učenija akad I. P. Pavlova. (Problems of the rational use of stud animals in the light of Pavlov's teachings.) *Vestn. seljskohoz. Nauki Zivotn*; No. 5:138. (*Anim. Breed. Abstr.*, **11**, 96, 1943.)

M.M.B. (1954). The importance of sex-reflex stimulation prior to service. *Milk Marketing Board Rep. Prod. Div.*, No. 5, pp. 84–87, Thames Ditton, Surrey, England.

MORGAN, C. T. & STELLAR, E. (1950). *Physiological Psychology*. 2nd ed., New York, N.Y.: McGraw-Hill.

MULLICK, D. N., MURTY, V. N. & KEHAR, N. D. (1952). Seasonal variation in the feed and water intake of cattle. *J. Anim. Sci.*, **11**, 42–49.

MYLREA, P. J. & BEILHARZ, R. G. (1964). The manifestation and detection of oestrus in heifers. *Anim. Behav.*, **12**, 25–30.

NEVENS, W. B. (1927). Experiments in the self-feeding of dairy cattle. *Illinois Agric. exp. Sta. Bull.*, No. 289.

NOTTBOHM, F. E. (1928). Cited by Munch-Petersen, E. (1938). *Bovine Mastitis.* Imperial Bureau of Animal Health, Weybridge, p. 10.

OLSON, H. H. & PETERSEN, W. E. (1951). Uniformity of semen production and behavior in monozygous triplet bulls. *J. Dairy Sci.*, **34**, 489–490.

PAREZ, M. (1953). *Contribution à l'étude de l'insémination artificielle. Psychologie du taurea. Conservation de la semence à trés basse température.* Theses, Paris (Alfort).

PARSHUTIN, G. V. (1956). The role of the nervous system in the reproduction of farm animals. *Proc. 3rd Int. Congr. Anim. Reprod.* (Cambridge), Plenary Papers, pp. 45–50.

PARSONS, L. M. (1966). *A Commentary and Summary of available Published Information on Cow Behaviour.* Agricultural Land Service, Min. of Agric., London (Mimeograph).

PAYNE, W. J. A., LAING, W. I. & RAIVOKA, E. N. (1951). Grazing behaviour of dairy cattle in the tropics. *Nature, Lond.*, **167**, 610.

PAYNE, W. J. A. & MACFARLANE, J. S. (1963). A brief study of cattle browsing behaviour in a semi-arid area of Tanganyika. *E. Afr. agric. For. J.*, **29**, 131–133.

PETERSEN, B. (1957). Uber das Verhalten eineüger Rinderzwillinge. *Zool. Anz.*, **20** (Suppl.), 91–95.

PHILLIPS, G. D. (1960). The relationship between water and food intake of European and Zebu types of cattle. *J. agric. Sci.*, **54**, 231–234.

PHILLIPS, G. D. (1961). Physiological comparisons of European and Zebu steers. I. Digestibility and retention times of food and rate of fermentation of rumen contents. *Res. vet. Sci.*, **2**, 202–208.

PINFOLD, W. J. (1965). Why did the cow cross the yard? *Fmr Stk Breed.*, **79** (3920), 83–84.

POUNDEN, W. D. & FIREBAUGH, J. G. (1956). Effects of nervousness on conception during artificial insemination. *Vet. Med.*, **51**, 469–470.

POWELL, T. L. (1967). A note on relationships between components of weather and the location of defaecation in out-wintered cattle. *Anim. Prod.*, **9**, 413–416.

PRABHU, S. S. (1956). Influence of factors affecting sex-drive on semen production of buffaloes. II. *Indian J. vet. Sci.*, **26**, 21–33.

PUTNAM, P. A., LEHMANN, R. & DAVIS, R. E. (1967). Ration selection and feeding patterns of steers fed in drylot. *J. Anim. Sci.*, **26**, 647–650.

RAGSDALE, A. C., BRODY, S., THOMPSON, H. J. & WORSTELL, D. M. (1948). Environmental physiology with special reference to domestic animals. II. Influence of temperature 50° to 105°F. on milk production and feed consumption in dairy cattle. *Missouri Agric. exp. Sta. Bull.*, No. 425.

RAGSDALE, A. C., THOMPSON, H. J., WORSTELL, D. M. & BRODY, S. (1951). Influence of temperature 40° to 105° on milk production in Brown Swiss cows, and on feed and water consumption and body weight in Brown Swiss and Brahman cows and heifers. *Missouri Agric. exp. Sta. Res. Bull.*, No. 471.

RAGSDALE, A. C., THOMPSON, H. J., WORSTELL, D. M. & BRODY, S. (1953). The effect of humidity on milk production and composition, feed and water consumption, and body weight in cattle. *Missouri Agric. exp. Sta. Res. Bull.*, No. 521.

RAY, M. L. & DRAKE, C. L. (1959). Effects of grain preparation on preferences shown by beef cattle. *J. anim. Sci.*, **18**, 1333–1338.

RHOAD, A. O. (1938). Some observations on the response of purebred *Bos taurus* and *Bos indicus* cattle and their crossbred types to certain conditions of the environment. *Proc. Amer. Soc. anim. Prod.*, *31st Ann. Meet.*, pp. 284–295.

RICE, V. A., ANDREWS, F. N., WARWICK, E. J. & LEGATES, J. E. (1957). *Breeding and Improvement of Farm Animals.* 5th ed., New York, N.Y.: McGraw-Hill.

RITZMAN, E. G. & BENEDICT, F. G. (1924). The effect of varying feed levels on the physiological economy of steers. *New Hamp. agric. Sta. Tech. Bull.*, No. 26.

ROBERTS, S. J. (1956). *Veterinary Obstetrics and Genital Diseases.* Ithaca, N.Y.: S. J. Roberts; Ann Arbor, Michigan: Edwards Bros. Inc.

ROLLINSON, D. H. L., HARKER, K. W. & TAYLOR, J. I. (1955). Studies on the habits of Zebu cattle. III. Water consumption of Zebu cattle. *J. agric. Sci.*, **46**, 123–129.

ROLLINSON, D. H. L. & BREDON, R. M. (1960). Factors causing alterations of the level of inorganic phosphorus in the blood of Zebu cattle. *J. agric. Sci.*, **54**, 235–242.

ROTTENSTEN, K. & TOUCHBERRY, R. W. (1957). Observations on the degree of expression of estrus in cattle. *J. Dairy Sci.*, **40**, 1457–1465.

ROY, J. H. B., SHILLAM, K. W. G. & PALMER, J. (1955). The outdoor rearing of calves on grass with special reference to growth rate and grazing behaviour. *J. Dairy Res.*, **22**, 252–269.

SANDSTEDT, H. (1949). Impotenssymtom hos tjur vid fosfatfattig utfodring. (Impotence in bulls on a low phosphate diet.) *Nord. Vet.-Med.*, **1**, 455–464. (*Anim. Breed. Abstr.*, **18**, 168, 1950.)

SCHAKE, L. M. & RIGGS, J. K. (1966). Diurnal and nocturnal activities of lactating beef cows in confinement. *J. Anim. Sci.*, **25**, 254 (Abstract).

SCHEIN, M. W. (1954). *Group Behavior Patterns in Dairy Cattle and Their Effect on Production.* Sc.D. Dissertation, Johns Hopkins University, Baltimore, Md., U.S.A.

SCHEIN, M. W. (1961). Report on a visit to the Soviet Union, with special reference to information on agricultural, biological and behavioral teaching and research. *Mimeographed ms., unpublished.*

SCHEIN, M. W. & FOHRMAN, M. H. (1954). The effect of rank separation on the intensity of contests in dairy cows. *Bull. ecol. Soc. Am.*, **35**, 74 (Abstract).

SCHEIN, M. W. & FOHRMAN, M. H. (1955). Social dominance relationships in a herd of dairy cattle. *Brit. J. Anim. Behav.*, **3**, 45–55.

SCHEIN, M. W. & HALE, E. B. (1959). The effect of early social experience on male sexual behaviour of androgen injected turkeys. *Anim. Behav.*, **7**, 189–200.

SCHEIN, M. W., HYDE, C. E. & FOHRMAN, M. H. (1955). The effect of psychological disturbances on milk production of dairy cattle. *Proc. Ass. Southern agric. Wkrs. 52nd Convention* (Louisville, Ky.), pp. 79–80 (Abstract).

SCHLOETH, R. (1958a). Cycle annuel et comportement social du taureau de Camargue. *Mammalia* (Paris), **22**, 121–139.

SCHLOETH, R. (1958b). Das Scharren bei rind und pferd. *Z. Saugetierk.*, **23**, 139–148.

SCHLOETH, R. (1958c). Über die Mutter-Kind-Beziehungen beim halbwilder Camargue-rind. *Saugetierkundliche Mitteilungen*, **6**, 145–150.

SCHLOETH, R. (1961). Das Sozialleben des Camargue-rindes. Qualitative und quantitative Untersuchungen über die sozialen Beziehungen—insbesondere die soziale Rangordnung—des halbwilden französischen Kampfrindes. *Z. Tierpsychol.*, **18**, 574–627.

SCHMIDT, J., NEHNER, A. & PIEL, H. (1951). Untersuchungen über Weideerträge zweier Betriebe, sowie über Verzehr und Leistung auf der Weide bei Milchkühen dreier Rassen. *Zuchtungskunde*, **23**, 110–122.

SCHNEIDER, K. M. (1931). Das Flehmen. *Zool. Gart*, **4**, 349–362.

SCOTT, J. P. (1945). Social behavior, organization and leadership in a small flock of domestic sheep. *Comp. Psychol. Monogr.*, Ser. 96, 22 pp.

SCOTT, J. P. (1956). The analysis of social organization in animals. *Ecology*, **37**, 213–221.

SEATH, D. M. & MILLER, G. D. (1946). Effect of warm weather on grazing performance of milking cows. *J. Dairy Sci.*, **29**, 199–206.

SELMAN, I. E., FISHER, E. W. & McEWAN, A. D. (1967). Observations of the early postpartum behaviour of cattle. *Proc. Soc. vet. Ethology*, **1** (3), 7–8.

SHEPPARD, A. J., BLASER, R. E. & KINCAID, C. M. (1957). The grazing habits of beef cattle on pasture. *J. Anim. Sci.*, **16**, 681–687.

SIGNORET, J. P. (1962). Étude de l'influence de divers éléments du compartement sexuel du taureau sur les caractéristiques du sperme. *Ann. Zootech.*, **11**, 93–101.

SMALL, C. A. (1966). Behaviour of a herd of cows with 24 hour access to self feed silage. *N.A.A.S. q. Rev.*, No. 72, 178–182.

SMIRNOV-UGRJUMOV, D. V. (1945). Racionaljnoe ispoljzovanie bykov—proizvoditelei učenie akademika I. P. Pavlova. (Itogi četyrëhletnih issledovanii.) (Rational exploitation of bulls and the doctrines of Pavlov (Summary of Four Years' Research).) *Trud. Lab. iskusst. Osemen. Zivotn.* (Moscow), **2**, 55–79. (*Anim. Breed. Abstr.*, **13**, 202, 1945.)

SMITH, C. A. (1959). Studies on the Northern Rhodesia Hyparrhenia veld. Part I. The grazing behaviour of indigenous cattle grazed at light and heavy stocking rates. *J. agric. Sci.*, **52**, 369–375.

SMITH, G. F. (1951). The behaviour of bulls at artificial insemination centres. *Proc. Brit. Soc. anim. Prod.*, *13th & 14th Meet.*, pp. 25–35.

SMITH, V. R. & PETERSEN, W. E. (1945). Negative pressure and nursing by calves. *J. Dairy Sci.*, **28**, 431–433.

SMOLIAK, S. & PETERS, H. F. (1955). Climatic effects of foraging performance of beef cows on winter range. *Canad. J. agric. Sci.*, **35**, 213–216.

SOKOLOWSKYJ, W. (1952). Experimental contribution to the sex-physiology of cattle. *Rep. 2nd Int. Congr. Physiol. Path. anim. Reprod. Artif. Insem. (Copenhagen)*, **1**, 50–52.

STALLCUP, O. T., BOSTAIN, D. V. & BIERWORTH, K. P. (1959). Time required for dairy cows to eat concentrates and silage. *Arkansas Agric. Expt. Sta. Bulletin*, No. 609.

SUTTON, G. D., FOURIE, P. D. & RETIEF, J. S. (1967). The behaviour of cattle in transit by rail. *Jl S. Afr. vet. med. Ass.*, **38**, 153–156.

SWANSON, E. W. (1956). The effect of nursing calves on milk production of identical twin heifers. *J. Dairy Sci.*, **39**, 73–80.

SWANSON, E. W. & HARRIS, J. D., JR. (1958). Development of rumination in the young calf. *J. Dairy Sci.*, **41**, 1768–1776.

SWETT, W. W., MATTHEWS, C. A. & McDOWELL, R. E. (1961). Sindhi-Jersey and Sindhi-Holstein crosses. *Tech. Bull. 1236*, U.S. Dept. Agr., 26 pp.

SYKES, J. F. (1955). Animals and Fowl and Water. In: *Water, The Yearbook of Agriculture*, U.S.D.A., pp. 14–18.

TAYLER, J. C. (1953). The grazing behaviour of bullocks under two methods of management. *Brit. J. Anim. Behav.*, **1**, 72–77.

TAYLOR, J. I., ROLLINSON, D. H. L. & HARKER, K. W. (1955). Studies on the habits of Zebu cattle. II. Individual and group variation within a herd. *J. agric. Sci.*, **45**, 257–263.

THEILER, A., GREEN, H. H. & DuTOIT, P. J. (1924). Phosphorus in the livestock industry *J. Dept. Agric. S. Afr.*, 8, 450–504.

THOMAS, J. W. & MOSS, S. (1951). The effect of orally administered molybdenum on growth, spermatogenesis and testes histology of young dairy bulls. *J. Dairy Sci.*, 34, 929–934.

TRAUTWEIN, V. K., BAUER, H. & FLUHR, F. (1958). Beobachtungen zur Psychologie der Bullen, speziell zum Deckverhalten. *Zuchthyg. FortpflStör. Besam. Haustiere*, **2**, 217–234.

TRIBE, D. E. (1950). The behaviour of the grazing animal. A critical review of present knowledge. *J. Br. Grassld Soc.*, **5**, 209–224.

TRIBE, D. E. (1955). The behaviour of grazing animals. In: *Progress in the Physiology of Farm Animals*. J. Hammond (ed.), **2**, 585. London: Butterworths.

TRIMBERGER, G. W. (1956). Ovarian functions, intervals between estrus and conception rates in dairy cattle. *J. Dairy Sci.*, **39**, 448–455.

TULLOH, N. M. (1961a). Behaviour of cattle in yards. I. Weighing order and behaviour before entering scales. *Anim. Behav.*, **9**, 20–24.

TULLOH, N. M. (1961b). Behaviour of cattle in yards. II. A study of temperament. *Anim. Behav.*, **9**, 25–30.

TURNER, R. R. (1961). Silage self feeding. *Vet. Rec.*, **73**, 1432–1436.

VOISIN, A. (1954). Biochemical aspects of ensilage. In: *La Conservation des Fourrages*, Paris.

VOISIN, A. (1959). *Grass Productivity*. English translation by C. T. M. Herriot. New York, N.Y.: Philosophical Library.

VOGEL, H. H., SCOTT, J. P. & MARSTON. M. (1950). Social facilitation and allelomimetic behaviour in dogs; I. Social facilitation in a non-competitive situation. *Behaviour*, **2**, 121–134.

WAGNON, K. A. (1963). Behavior of beef cows on a California range. *Calif. agric. Expt. Sta. Bull.*, No. 799.

WAGNON, K. A. (1965). Social dominance in range cows and its effect on supplemental feeding. *Calif. agric. Exp. Stn Bull.*, No. 819.

WAGNON, K. A., LOY, R. G., ROLLINS, W. C. & CARROLL, F. D. (1966). Social dominance in a herd of Angus, Hereford, and Shorthorn cows. *Anim. Behav.*, **14**, 474–479.

WAITE, R. (1963). Grazing behaviour. In: *Animal Health, Production and Pasture*. A. N. Worden, K. C. Sellers, D. E. Tribe (eds.), Ch. 9. London: Longmans.

WAITE, R., HOLMES, W., CAMPBELL, J. I. & FERGUSSON, D. L. (1950). Studies in grazing management. II. The amount and chemical composition of herbage eaten by cattle under close-folding and rotational methods of grazing. *J. agric. Sci.*, 40, 392–402.

WAITE, R., MACDONALD, W. B. & HOLMES, W. (1951). Studies in grazing management. III. The behaviour of dairy cows grazed under the close-folding and rotational system of management. *J. agric. Sci.*, 41, 163–173.

WALKER, D. E. (K) (1962). Suckling and grazing behaviour of beef heifers and calves. *N.Z. Jl. agric. Res.*, 5, 331–338.

WALKER, D. M. (1950). Observations on behaviour in young calves. *Bull. Anim. Behav.*, 1 (8), 5–10.

WALKER-LOVE, J. & LAIRD, R. (1965). Behaviour of dairy cows in loose housing with cubicles. *Fm Bldgs*, No. 7, 85 (Abstract).

WALTHER, VON F. (1958). Zum Kampf-und Paarungsverhalten einiger Antilopen. *Z. Tierpsychol.*, 15, 340–380.

WALTON, A. (1955). Sexual Behaviour. In: *Progress in the Physiology of Farm Animals*. J. Hammond (ed.), 2, 603–616. London: Butterworth & Co.

WALTON, A. (1960). Copulation and natural insemination. In: *Marshall's Physiology of Reproduction*, A. S. Parkes (ed.), Ch. 8, Vol. 1, Part 2. London: Longmans.

WARDROP, J. C. (1953). Studies in the behaviour of dairy cows at pasture. *Brit. J. Anim. Behav.*, 1, 23–31.

WEBB, F. M., COLENBRANDER, V. F., BLOSSER, T. H. & WALDERN, D. E. (1963). Eating habits of dairy cows under drylot conditions. *J. Dairy Sci.*, 46, 1433–1435.

WEIDLICH, C. & WULFF, C. (1961). Beobachtungen über Fresszeiten bei Kühen. (Time spent by cows on eating.) *Dt. Landwirt.*, 12, 40–42 (*Nut. Abstr. Rev.*, 31, 1030, 1961).

WIERZBOWSKI, S. & JANASZ, M. (1965). The effect of visual stimuli on sexual excitement in bulls. *Roczn. Nauk roln.* Ser. B, 86, 53–61. (*Anim. Breed. Abstr.*, 34, 509.)

WILSON, P. N. (1961). The grazing behaviour and free-water intake of East African short-horned Zebu heifers at Serene, Uganda. *J. agric. Sci.*, 56, 351–364.

WILTBANK, J. N., BOND, J. & WARWICK, E. J. (1965). Influence of total feed and protein intake on reproductive performance of the beef female through second calving. U.S.D.A. Tech. Bull., No. 1314, Washington, D.C.

WINCHESTER, C. F. & MORRIS, M. J. (1956). Water intake rates of cattle. *J. Anim. Sci.*, 15, 722–740.

WISE, G. H., ANDERSON, G. W. & MILLER, P. G. (1942). Factors affecting the passage of liquids into the rumen of the dairy calf. II. Elevation of the head as milk is consumed. *J. Dairy Sci.*, 25, 529–536.

WISE, G. H., MILLER, P. G. & ANDERSON, G. W. (1947). Changes in milk products "sham fed" to calves. I. Effects of volume of milk fed. *J. Dairy Sci.*, 30, 499–506.

WOOD, P. D. P., SMITH, G. F. & LISLE, M. F. (1967). A survey of intersucking in dairy herds in England and Wales. *Vet. Rec.*, 81, 396–398.

WOODBURY, A. M. (1941). Changing the "hook order" in cows. *Ecology*, 22, 410–411.

YATES, R. J. & LARKIN, P. J. (1965). Minimal error weight measurements in cattle experiments. *E. Afr. agric. For. J.*, 30, 263–264.

YEATES, N. T. M. (1965). *Modern Aspects of Animal Production*. London: Butterworths, p. 34.

ZEUNER, F. E. (1963). *A History of Domesticated Animals*. London: Hutchinson.

Chapter 10

The Behaviour of Sheep and Goats

E. S. E. Hafez, R. B. Cairns, C. V. Hulet and J. P. Scott

Sheep and goats belong to the family Bovidae, which includes cattle, buffaloes, antelopes, and other hollow-horned, ruminating animals. Sheep belong to the genus *Ovis* and the domestic variety is known as *Ovis aries*, while goats belong to the genus *Capra*. Wild sheep and goats are typically inhabitants of high mountain or plateau regions and have a somewhat similar distribution throughout the world, except that all species of wild goats live only in Europe and Asia. There are two North American species of wild sheep, *O. canadensis*, the bighorn, and *O. dalli*, the white sheep of Alaska and British Columbia, in addition to *O. ammon*, the wild sheep of the Old World. The Rocky Mountain goat belongs to a different genus, *Oreamnos*, and is a mountain antelope rather than a true goat.

The domestication of sheep and goats probably took place somewhere in the fertile crescent of land lying at the foot of the mountains surrounding Mesopotamia in the Middle East. According to Reed (1959) the bones of domestic goats have been definitely identified at Jericho and dated 6500 B.C. There is only one wild goat in the region, *C. hircus aegagrus*, and it is presumed to be the wild ancestor. Sheep, on the other hand, were not commonly found in the Middle East until about 4000 B.C., although they may have been domesticated as early as goats. There is more controversy about the ancestry of sheep, since there are several possible candidates in the Near Eastern and Mediterranean regions. The three subspecies of *O. ammon* found respectively in Sardinia, Cyprus, and the Near East are all like most breeds of domestic sheep in that the females are hornless. The adult females of the wild Rocky Mountain sheep, in contrast, have short, thin horns. Certain British domestic breeds also have horned females.

Domestic goats have changed relatively little from the wild species except in such superficial characteristics as the hair. The only way that an unknown goat skeleton can be definitely identified as a domestic or wild type is to examine the shape of the horn core. On the other hand, most domestic sheep are different from all wild varieties in having a long tail.[1] This argues for a single[2] origin of domestic sheep, with an early occurrence of a long-tailed

---

[1] In some areas, tails of sheep are docked early in life. Some domestic breeds of sheep have fat tails or fat-rumps.

[2] Contrary evidence is provided by the North Ronaldshay (Tribe & Tribe, 1949) and Shetland breeds, which are short-tailed domestic breeds of the old Celtic type (Gordon, 1961a) and may be explained either by a reverse mutation to the wild type, or by separate origin.

mutant which was used to distinguish domestic sheep from wild ones, just as the curly tail probably distinguished the first dogs from wolves. Domestic sheep have also completely lost the long guard hairs in the coats of their wild ancestors, retaining only the woolly undercoat.

I. GENERAL DESCRIPTION OF BEHAVIOUR

Commercial sheep flocks are maintained in a variety of conditions, depending on the nature of the available grassland. They vary in size from a few sheep kept on a small pasture to the large flocks of several thousand individuals kept on the open ranges in Australia, New Zealand, South America, and the western part of the United States. Studies of behaviour under these conditions have most frequently dealt with ingestive and sexual behaviour as these are directly related to the commercial success of the sheep industry.

Working on the hypothesis that the complete behavioural capacities of a species are likely to appear only under natural conditions, Scott (1945, 1960) made an extensive study of the behaviour of a small flock of sheep formed under semi-natural conditions. The flock was started with a pair of pregnant ewes and allowed to increase naturally, the numbers being restricted only when the flock began to exceed the capacity of the pasture. The Soay sheep (a feral domestic breed) of St Kilda have been observed extensively under even less restricted conditions (Boyd et al., 1964). These animals have been maintained in closed-island communities, and have not been managed by man in any way in recent years. Their social cohesion, range behaviour, habitat preferences, and population distribution have been studied in a series of recent investigations. Studies of the general behaviour of sheep maintained under normal flock conditions are also available (Hunter & Milner, 1963).

A. The Daily Cycle of Behaviour

During the autumn and spring months in a temperate climate, flocks of sheep will leave the barn or bedding ground in which they have spent the night and go out to graze. The flock moves along as a unit, spreading out as it goes, but following a regular route around the pasture or range. After two hours or so, the sheep stop, lie down, and start chewing the cud. Three or four such alternations of grazing and rumination take place, depending on the weather and season of the year.

The distance travelled while grazing is affected by both genetic and environmental factors. The Cheviot breed travels much farther than the Romney Marsh breed when both are kept in hill country, but only a little farther when both are allowed to graze in flat lands (Cresswell, 1960). The distance travelled under these conditions is constant from week to week and only changes prior to the breeding season, when the flocks travel farther. Thus the breed differences in behaviour are related to the habitat in which the breeds were developed.

The distance travelled by grazing sheep ranges from 5 to 10 miles per day (Cresswell, loc. cit.). Tribe (1949b) and England (1954) reported a shorter distance of 2 to 3 miles for sheep. Cory (1927) observed that range sheep travelled 3·8 miles per day and goats 6·0 miles. The increase in the distance travelled is associated with an increase in the total grazing time. In temperate climates sheep walk at night more in the summer than in the winter (Tribe, 1949b). Rambouillet ewes travel about 25 per cent farther than Hampshires; and Columbia ewes travel slightly farther than Hampshires in the same size pasture (Cresswell & Harris, 1959). When the field size is increased, there is a temporary increase in the daily activity.

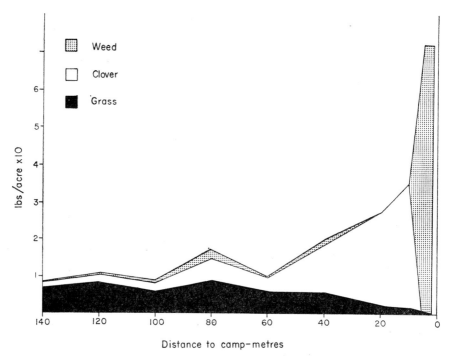

Fig. 38. Effect of "camping behaviour" in sheep on chemical composition of pasture. Note the abundance of weeds near the camp and the abundance of grass away from the camp. (Data by E. J. Hilder.)

Even when permitted unrestricted movement, sheep tend to remain in a limited area ("home range") while grazing. Among Soay ewes, the home range normally extends from $\frac{1}{3}$ to $\frac{2}{3}$ of a mile in length (Grubb & Jewell, 1966); in hill sheep, home ranges are usually not more than 120 acres in extent (Hunter & Davies, 1963). The adherence by sheep to such territories seems to be remarkably stable: ewe groups tend to remain in a given area throughout the lifetime of the constituent members. Adjacent home ranges do not, however, have sharply defined boundaries. Neighbouring groups of sheep frequently share contiguous grazing areas. Although rams also tend to remain in a given locale, it is not unusual for ram groups to alter their ranges, especially at the beginning of the breeding season. In general, ram groups have less clearly defined range areas and less stable membership than do ewe groups.

Camping behaviour is common in sheep. Certain breeds of sheep develop definite "bedding habits" known as camping behaviour. Sheep camp on high ground during cold weather and near water or under shade during hot weather. Camping sites vary with the topography of the pasture. High ground is preferred in hilly areas, whereas trees are used on the plains. Habit can have a dominating influence on camping site. With small numbers of sheep on adjacent small pastures of less than 10 acres, camping is usually along a common boundary or in contiguous corners. Camping behaviour apparently varies with breed; for example, Merinos have more definite camping areas than Dorset Horns.

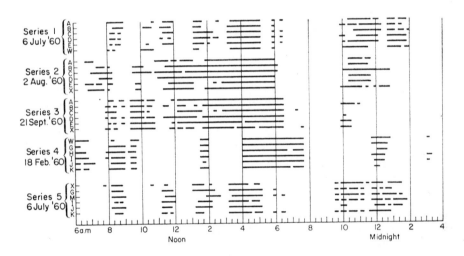

Fig. 39. Patterns of grazing for 2 groups of 6 Border Leicester × Merino sheep (three 24-hour observations on one group, and two 24-hour observations on the other group). Note that the beginning and cessation of major grazing periods varies little between sheep at any one observation. (From Arnold, 1962.)

As a result of camping behaviour, a large proportion of the faeces is dropped on the camping sites and there is a considerable transfer of plant nutrients into them from the main grazing area of the pasture. The productivity of the pasture is affected by this process and thus influences fertilizer requirements. In addition, this behaviour influences the distribution of plants and internal parasite eggs within the pasture. Weeds grow predominantly in areas closest to the camp, then clover grows in areas adjacent to the camp, whereas grasses grow at some distance from the camp (Fig. 38).

Another salient feature of sheep behaviour is that almost all activity is normally carried on within a flock and is highly co-ordinated. Sheep thus show considerable allelomimetic behaviour and social facilitation. This point is illustrated in Arnold's (1962) observations of grazing behaviour in two groups of sheep over a total of five 24-hour periods (Fig. 39). The beginning and the end of the major grazing periods were nearly simultaneous. Individual differences among sheep in grazing time were brought about by different lengths of non-grazing intervals during major grazing periods and by short-term subperiods of grazing. There was no evidence of particular sheep initiating either the onset or cessation of grazing. That is, it did not appear that any one of the sheep consistently determined the behaviour of the group.

Unlike cattle, sheep do not groom or lick each other. While resting, sheep sometimes lick the wool just above the hoof and then rub their faces with the licked wool; this pattern of self grooming commonly occurs after the animal has been lying down for some time and just before it gets up again (Keck, 1961).

B. The Seasonal Cycle of Behaviour

The above behaviour may be found at any season of the year, except that during the extremely hot weather sheep may graze more frequently at night than in the daytime. Certain other aspects of behaviour are definitely seasonal in their occurrence: during the autumn months there is an increase in sexual behaviour, as well as fighting (agonistic behaviour) between males.

Fig. 40. A. Et-epimeletic (care-soliciting) behaviour. Lambs "baa" when hungry or separated from the mother.

B. Allelomimetic behaviour in Bighorn rams (similar to domestic sheep). Animals do what other animals in flock do. One result of allelomimetic behaviour is the formation of footpaths as the sheep follow one another.

C. Agonistic behaviour, showing type of butting between rams during breeding season. (Redrawn from motion picture frames, Scott, 1945.)

Table 24. Behaviour Patterns of Sheep and Goats

Behaviours	Characteristic Pattern
Ingestive	Grazing, browsing, ruminating, licking salt, drinking.
	Suckling—nudging udder with nose, sucking, wiggling tail.
Shelter-seeking	Moving under trees, into barns.
	Huddling together to keep off flies.
	Crowding together in extreme cold weather.
	Pawing ground and lying down.
Investigatory	Raising head, directing eyes, ears, and nose towards disturbance.
	Nosing an object or another sheep.
Allelomimetic	Walking, running, grazing, and bedding down together.
	Following one another.
	Bouncing stiff-legged past an obstacle together (sheep).
Agonistic	Pawing; butting; shoving with shoulders.
	Running together and butting (rearing and butting, in goats).
	Bunching and running.
	Lying immobile.
	"Freezing" (in young kids only).
	Snort and stamp foot ("sneeze", in goats).
Eliminative	Urination posture:
	squatting of females;
	arching back, bending legs (male lambs).
	Defaecation:
	wiggling tails.
Care-giving	Licking and nibbling placental membranes and young.
(Epimeletic)	Arching back to permit nursing, nosing lamb at base of tail.
(females only)	Circling newborn lamb.
	Baaing when separated from lamb.
Care-soliciting	Baaing (distress vocalization by lamb when separated, hungry,
(Et-epimeletic)	hurt, or caught).
	Baaing of adults when separated from flock.
Sexual (Male)	Courtship: following female; pawing female; hoarse baa or grumble;
	nosing genital region of female;
	sniffing female urine, extending neck with upcurled lip;
	running tongue in and out;
	rubbing against side of female; biting wool of female;
	herding or pushing female away from other males.
	Copulation: wiggling tail (rare); mounting, thrusting movements of hind quarters.
Sexual (Female)	Courtship: rubbing neck and body against male;
	mounting male (rare).
	Copulation—standing still to receive male.
Play	Sexual: mounting (by either sex).
	Agonistic: playful butting.
	Allelomimetic: running together;
	"gambolling" (bouncing stiff-legged and turning in air) together.
	"Game playing"—jumping off and on rock or log together.

C. Basic Patterns of Behaviour

The behaviour of sheep and goats thus consists of a large number of special behaviour patterns, the most common ones being listed in Table 24. These behaviour patterns are related to nine basic functions and each group may be thought of as a behavioural system (see Chapter 1), the various patterns being organized and co-ordinated both by heredity and by the process of learning. As indicated above, the two important basic systems of behaviour in sheep and goats are ingestive behaviour and allelomimetic behaviour (Fig. 40).

II. INGESTIVE BEHAVIOUR

A. Grazing

The common general features of behaviour patterns, level of responses, and types of stimuli of grazing in ruminants have already been discussed under the behaviour of cattle (Chapter 9). However, certain special characteristic patterns of grazing specific to sheep and goats will be mentioned here.

Patterns of grazing. Sheep have a cleft upper lip which, though not prehensile, permits very close grazing. The lips, the lower incisor teeth, and the dental pad are the principal prehensile structures; the tongue does not protrude during grazing as in cattle. Since there are no upper incisors, leaves and stems have to be severed by the lower incisors against the dental pad as the animal jerks its head slightly forwards and upwards. In sheep, the jaws work very close to the ground and provide an opening of some 3 cm in diameter. For mechanical reasons, easily torn grasses are selected more often. The time spent grazing and the amount eaten are summarized in Table 25.

When lambs are two days of age, they begin picking at herbage, suck a blade of grass, and then let it go again. The blades of grass which have been sucked in this way are very slightly bruised but remain unbroken (Munro, 1955). At the age of 2 weeks the lamb begins to eat the grass blades.

Large flocks are not likely to graze together but split into subgroups occupying separate areas; it is not known whether the subgroups are based on families in the flock. Different breeds vary in their tendency to move and flock together: some breeds tend to stay within a part of the available grazing areas; others split into small groups occupying the area around a patch of good grazing. Gregarious breeds are not adapted to pastures in which patches of herbage are widely scattered; such breeds may be well suited to uniform and abundant pastures (Hunter, 1960).

Time of grazing. Sheep never graze continuously but rather in cycles interrupted by periods of rumination, rest, and idling. In general, intensive grazing begins around sunrise and stops about sundown; the longest periods of grazing occur in the early morning and between late afternoon and dusk

Table 25. Daily Ingestive and Eliminative Behaviour in Sheep. (Adapted from Tribe, 1949a; Schneider *et al.*, 1953; England, 1954; Sykes, 1955; Gordon, 1958b.)

	Pattern		Average Values Per Day
Grazing and feeding	Number of grazing periods		4–7
	Total grazing time (hr)		9–11
	Consumption of fresh herbage on permanent pasture (g)	Lambs	1,700–1,900
		Adults	1,300–5,000
	Dry matter consumption on permanent pasture (g)	Lambs	480–830
		Adults	520–1,300
Rumination	No. rumination periods		15
	Total rumination time (hr)		8–10
	No. of chews/rumination		39,000
	Rate of chews/min		91
	Duration of a rumination period (min)		1–120
	No. of boli regurgitated		500
	No. of chews/bolus		78
Drinking water (lb)	On range or dry pasture		5–13
	On hay and concentrates		0·3–6·0
Ranging	Distance travelled (miles)		1–8
Elimination	No. of urinations		9–13
	No. of defaecations		6–8

(Fig. 41). The incidence of night grazing depends on the temperature and the prevalence of flies. Observations of Cheviot sheep in Scotland indicate that grazing time in the night hours between 7 p.m. and 7 a.m. is considerably longer in summer than in winter (Tribe, 1949b). Contrasting findings with respect to night grazing have been reported by Arnold (1962). His observations of Border Leicester × Merino sheep in Canberra, Australia, indicate that the proportion of grazing time between 6 p.m. and 6 a.m. was greater in autumn and winter than in spring and summer. The much wider seasonal differences in daylight in Scotland might possibly account for the discrepancy.

The total grazing time of sheep ranges from 9 to 11 hours; the time an individual sheep grazes can differ markedly from the flock's average. Sheep graze more hours per day than do cattle.

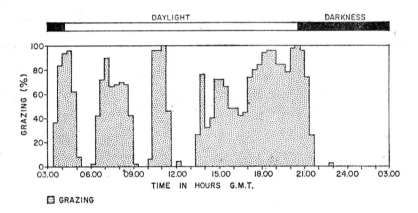

Fig. 41. Grazing periodicity over a 24-hour period. Ten sheep (Oxford Down half-breds and Clun Forest), England, July. Note two primary grazing periods, during the early morning and late afternoon to dusk. (After Hughes & Reid, 1951.)

Grazing intake. Grazing intake is partly a function of metabolic rate and body size. The daily intake of herbage (usually measured as dry-matter consumption of herbage) varies with the breed and age of sheep and the type and stage of growth of herbage. The daily average dry-matter consumption of orchard grass (*Dactylis glomerata*) is 0·8, 1·7, and 2·2 lb per 100 lb body weight for adult rams, yearlings, and lambs respectively (Schneider *et al.*, 1953). Rams of 250 lb weight will readily eat 3½ to 4 lb of dry matter per day (Cowlishaw, 1961).

Factors affecting grazing. In the pasture as a whole, the location of grazing may vary according to available grazing, climate, nutritional requirement, and the presence or absence of lambs. The effects of exteroceptive stimuli and of sensory capacities on grazing have been discussed previously (Chapter 9). Tribe (1950) observed two groups of grazing sheep; one group was supplemented with concentrates. When the two groups grazed separately, the supplemented group spent less time in grazing than the unsupplemented one. When the two groups grazed together, the grazing and resting times for both groups were similar, thus demonstrating the effect of allelomimetic behaviour on grazing behaviour.

Selective grazing. Sheep definitely prefer certain species of herbage, stages of growth in a given species, and particular parts of individual plants. Several psychological, physiological, and mechanical factors are potentially involved in selective grazing. Selective grazing was studied by comparing the chemical composition of forage samples obtained by harvesting and that by oesophageal-fistulated sheep on a wide variety of range (Weir &

PLATE V

a. A ewe licking the foetal membranes around the lamb a few seconds after birth.

b. A lamb reaching for the mother's body, 1 hour after lambing; after a few unsuccessful attempts to suckle between the front legs the lamb reaches to the udder.

c. One-day-old lamb suckling from a standing position. Note position of lamb in relation to that of ewe. Note tail wiggling.

Torrell, 1959). These sheep had their oesophageal fistulae opening to the outside, so that the food they ate could be collected in a container for identification. The sheep consistently selected forage higher in protein and lower in crude fibre than that obtained by harvesting. In general, selectivity is directly proportional to the amount of herbage available; thus the shorter the supply of feed, the less discriminating the animal becomes.

Sheep ordinarily reject plants contaminated with the odour of sheep urine and faeces. However, if sheep are in a pasture which is widely contaminated with excreta, they adapt to the odour of faeces and may eat the contaminated herbage (Tribe, 1949a, 1955); this behaviour contributes to over-grazing. Sheep and goats can distinguish between different strains of the same plant species so similar that a botanist might find the same task difficult. In Western Australia, Nichols (1944) found that sheep were able to distinguish between different varieties of mulga plants (*Acacia aneura*).

Cook et al. (1948) analysed the vegetation before and after being grazed by sheep. The range consisted primarily of summer-range browse, serviceberry (*Amelanchier* sp.), sweet anise (*Osmorphizia occidentalis*), wild rye (*Elymus* sp.), summer-range semi-bunch grass, mountain brome (*Bromus carinatus*), aspen peavine (*Lathyrus leucanthus*), and western chokecherry (*Prunus melanocarpa*). The sheep's grazed diet contained more mountain brome grass and aspen peavine than sweet anise and western chokecherry (Table 26). Except for one species not eaten (wild rye) there was a high correlation between the amount of food plant present in the pasture and the amount eaten.

Table 26. Selective Grazing in Sheep under Range Conditions. (Adapted from Cook et al., 1948.)

Name of Plant	Wt. of Plants per 100 sq. ft (a)	Wt. of Plant Consumed (b)	Wt. of Plant Consumed as Total of Diet	Wt. of Plants Consumed to wt. of plant available (c)
	(g)	(g)	%	%
Serviceberry	96	42	11	44
Sweet anise	48	18	5	28
Wild rye	1085	Nil	Nil	Nil
Mountain brome	631	134	34	21
Meadow rye	27	15	4	56
Aspen peavine	332	155	40	47
Western chokecherry	55	30	8	55
			102	

Cowlishaw & Alder (1960) observed the grazing preference of sheep and cattle on pastures made of strips of perennial rye-grass, cocksfoot, meadow fescue, tall fescue, red fescue, timothy, and reed canary grass. The authors concluded that sheep are relatively more fond of cocksfoot (*Dactylis glomerata*) and less fond of meadow fescue (*Festuca pratensis*) than are cattle.

Some species of herbage are rejected because of their gross morphological characteristics: sheep may reject woolly or hairy plants such as the mullein (*Verbascum* sp.) and species with greasy texture such as alpine butterwort (*Pinguicula alpina*) (Tribe, 1955). Other species may be rejected because of chemical composition. The same species of herbage grown in different environments may differ in taste and palatability; this may be attributed to climatic factors, altitude, soil properties, level of the water table, and the application of fertilizers.

Grazing behaviour and management. The type of forage also affects the transmission of parasites. Because of its particular conformation the trifoliate leaf of clover is likely to carry many strongylid larvae, while the upper leaves of grasses are relatively free. By preferentially selecting the clover leaves, sheep may collect most of the nematode larvae (Taylor, 1954) more than goats and other animals. Immunological studies, however, have shown that the lambs, during their first attempts at grazing, normally develop resistance by taking in small number of larvae of common parasites such as the whipworm (*Trichuris ovis*), brown hair worm (*Ostertagia* spp.), barber's pole worm (*Haemonchus contortus*), etc. (Brown, 1959). Further studies on the hygienic adaptive mechanisms of grazing are required.

Left to themselves on a sufficient area, sheep graze selectively. When pasturage is restricted, they crop grass very closely and will eat leaves of trees. The latter tendency can be controlled through training procedures (Goot, 1962). Merino ewes, fitted with a harness which restricted vertical movement of the head to about 20 inches above the ground, were permitted to graze on the sward growing between trees in orange groves. After 4 weeks of such experience, the harnesses were removed from some of the sheep for a 5-week period. All of the ewes and their lambs continued to confine their grazing to the sward. The extent to which these effects gradually extinguished was not determined. The early experience of the animals, age of training, availability of palatable grass, and behaviour of other members of the grazing group may influence the durability of the effect.

B. Feeding

Both feeding and rumination stimulate the secretion of saliva; and some investigators regard the active secretion of the saliva as an indicator of appetite.

Quantitative regulation of feed intake. The animal adapts its feed intake to environmental temperature, as if appetite is a thermoregulatory mechanism (see Chapter 4). In goats, the eating time and the mastication rate are increased when the ambient temperature is decreased, but below 10°C eating activity slows down (Fig. 42). The animals eat more in cool weather, but are inhibited by extreme cold. On the other hand, the voluntary intake of chopped hay by housed sheep is greater in the winter than in the summer when a constant level of concentrates is given (Gordon, 1961a). This may be explicable in terms of coat length in that the animals are shorn in the summer and that the heavy winter coat may render the animals independent of environmental temperature during cold weather.

Extensive studies were carried out to place on a quantitative basis the generalization that voluntary feed intake of ruminants increases with the quality of feed they are given. Blaxter *et al.* (1961) reported that voluntary intake of hay was related to the digestibility of the hay. The giving of concentrated feed resulted in a drop in the voluntary intake of hay: with high

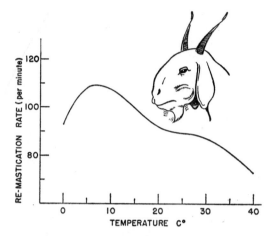

Fig. 42. Effect of ambient temperature on the re-mastication rate of Nubian goats. The curve represents smoothed values from data on 4 goats. At 20°C the average mastication rate was 90 per minute. At 5 to 10°C, the rate was 110 per minute. A further decrease in temperature to 0°C depressed remastication rate to one comparable with that at 20°C. (Data from Appleman & Delouche, 1958.)

quality hay 100 g concentrates replaced 100 g hay; with poor quality hay, 100 g concentrates replaced 47 g of hay.

Qualitative regulation of food intake. The status of the hypothesis of "nutritional wisdom"—that given a free choice of feed, deficient animals can select and balance their diet and thus correct the deficiency—is still controversial. On the one hand, Gordon & Tribe (1951) and Gordon *et al.* (1954) reported that sheep on a self-selection system did not eat to meet their nutritional requirements when the animals were bearing twins or suffering from oestophagia and allotriophagia (low blood phosphorus). Sheep will persist in diets that are ultimately injurious to their health or well-being. On the other hand, there have been some unambiguous instances reported where sheep have rectified quite precisely a biochemical deficiency by the intake of the appropriate materials. The case has been most clearly made for the intake of Na^+ solutions by Na^+ deficient animals. Not only do sheep show a very fine discrimination of variations in Na$^+$, but they adjust to changing concentrations of sodium in a solution by consuming more or less of it (Denton & Sabine, 1961). Arnold (1964) reports a similar effect for Na$^+$ preference, and Bell (1961) has shown that sodium-depleted cattle can select sodium bicarbonate out of four other choices (see Chapter 9).

Possibly some of the difficulty that has been encountered in the demonstration of nutritional wisdom over a long-term period is due to the fact that sheep normally shift in their preference for a substance if they have been maintained on that substance over an extended period. Peirce (1959) reports, for example, that although sheep are responsive to abrupt changes in salt concentration, animals which are gradually shifted to higher salt concentrations fail to show a decrease in water consumption. Further, if sheep are forced to graze exclusively on a generally disliked grass species for at least a month, that species becomes more acceptable (Arnold, 1964).

Other factors such as palatability and digestibility may operate against the selection of an unknown or an upsetting material. Brown & Caveness (1959) allowed sheep a free choice of different grains (corn, oats, milo, and wheat) of different preparations (whole, finely ground, crimped, and pellets). Preparational differences but not grain differences were statistically significant. It was also possible to alter feed preference through blindfolding or through inhibiting the sense of smell. The sheep were obviously selecting their feed, at least in part, by appearance and odour rather than content.

The information obtained from exteroceptive sensory stimuli does not appear to be essential, however, for the long-term maintenance of an adequate diet. Impairment of three primary senses (smell, taste, touch), singly or in combination, does not significantly alter the productivity of sheep over a period of one year (Arnold, 1966b). Similarly, preventing sheep from seeing has remarkably little effect on their total feed intake and body weight (Arnold, 1966a).

Taste discrimination. Electrophysiological studies have demonstrated the presence of specific taste receptors in the buccal cavity; the topographical distribution of the receptors on the tongue varies with the species (Pfaffman, 1941). Taste discrimination has also been evaluated by means of preference tests in some ruminants. Bell (1959) has shown that goats can distinguish various solutions: quinine dihydrochloride, sodium chloride, glucose, and acetic acid, representative of the four normal taste modalities of bitter, salt, sweet, and sour respectively. The animals have a definite threshold for each substance.

Goats have a high threshold for bitter taste as compared with cattle (but probably a lower threshold than camels). Bell (*loc. cit.*) suggests that the difference between the thresholds of rejections between goats and cattle may be explained by the contrast in grazing behaviour, natural habitat, and feed of these two species. Goats, which are browsing animals, feed on shoots of shrubs which are normally bitter; thus they have developed bitter taste receptors which may not be readily stimulated to a threshold of rejection. In goats, the destruction of the posteromedial ventral thalamic nuclei appears to abolish the sense of taste (Andersson, 1951; Andersson & Jewell, 1957).

C. Rumination

Sheep ruminate some 8 times at irregular intervals throughout the day and night (Fig. 43), the total rumination time being approximately 8 hours. (Table 25).

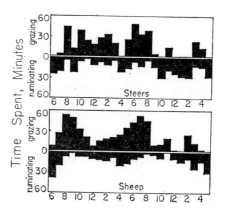

Fig. 43. Comparative study of the time spent eating and ruminating during the 24 hours by sheep and steers while on pasture. Sheep and cattle graze and ruminate throughout the day and night. (Lofgreen et al., 1957.)

Patterns. Chewing the bolus during rumination is characterized by regular intervals of short pauses for swallowing and regurgitation; chewing during eating, however, is *irregular* (Fig. 44). When fed on ground roughage, chewing is markedly reduced.

Gordon (1958a) studied the rumination behaviour of one caged sheep given four diet combinations ranging from 100 per cent hay to 100 per cent concentrates. He showed that the number of chews and not the chewing time is the most accurate way of measuring rumination quantitatively. "The

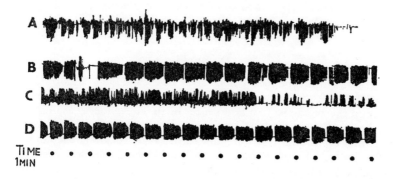

Fig. 44. Record of jaw movements during feeding (one feed of concentrates per day) and rumination. A rubber air cushion was held beneath lower jaw (English castrated male goats, housed). A and C show the irregular pattern of jaw movement associated with feeding. B and D show the regular period of chewing characteristic of rumination. (From Bell & Lawn, 1957.)

duration of the intervals between boli was constant at 15 per cent of total rumination time, but the rate of chewing varied from 83 to 99 chews per minute, being slower during midnight to noon, when most rumination occurred. Rumination periods last from under 1 minute up to about 2 hours. Tiring during a rumination period was reflected by an increase in the intervals between boli, but the number of chews per bolus remained unchanged" (Gordon, 1958b).

Stimuli. The stimuli which induce the onset and cessation of rumination have not been ascertained; physical factors in the rumen and other neural stimuli seem to be involved. Chopped hay evokes more rumination than the long hay, although the ground material produces much less rumination than either (Gordon, 1958a). Gordon (1961b) attempted to relate rumination to the effect of giving the same total amount of food in small frequent feeds as opposed to giving it in a single feed. He found that the rate of regurgitation of boli was more rapid when sheep were frequently fed, and he postulated that this may reflect a general increase of motility of the foregut, of saliva secretion, and of digestion.

Bell (1960) recorded electroencephalographs during rumination of unrestrained goats at various degrees of alertness. When the animal was in a profound somnolent state, the rumination periods were regular. In animals alerted by auditory stimuli, the rumination cycles were irregular and the periods between swallowing and regurgitation were prolonged. When the goats became quite alert, rumination ceased.

Andersson *et al.* (1959) electrically stimulated different sites in the caudal part of the medulla oblongata of conscious goats. The stimulation of certain centres caused rumination to start. It is postulated that there is a region in the brain, called *reticuloruminal motor centre*, caudal to the intercollicular plane, which can maintain co-ordinated activity of the reticulum and the rumen (Iggo, 1951). The importance of the vagal nerves and medullary reflex centres in the neural control of the reticulum, rumen, and oesophagus has often been demonstrated (Duncan, 1953; Bell & Lawn, 1955). Activity in the rumen and reticulum can be classically conditioned (Dedashev, 1959a).

D. Suckling

Patterns. Within the first $\frac{1}{2}$ hour after birth, the lamb is normally able to stand (Plate V). Suckling commences early, and the lamb begins to gain weight on the average 2 to 3 hours after birth (Alexander & Williams, 1964). Between sucks, the lamb pushes suddenly with its head, almost like butting, except that the mouth is directly forward. While suckling, the lamb rapidly wiggles its tail. Usually single lambs suckle both nipples, taking 2 to 3 sucking periods of 20 to 50 seconds before changing from one to the other. Among twins, about half of the pairs have a definite preference by the fifth week of life as to which teat each lamb will suckle (Ewbank, 1964, 1967). The suckling position of the lamb may be standing or kneeling. Suckling is disrupted by movement of the mother and the lamb then grazes, or returns to whatever its activity was before attempting to suckle. Little research has been done on the suckling behaviour of goats, and the following discussion is restricted to sheep.

Onset of suckling. Suckling behaviour is a prepotent activity in neonatal lambs. The lamb's initial attempts to suckle are inept; it tends to put its head under any large object or projecting surface (Smith, 1965). If the newborn lamb is prevented from gaining access to the ewe, it will mouth and suck other objects that are available (Smith *et al.*, 1966). If such non-nutritive sucking persists, it is likely to interfere with the performance of appropriate suckling behaviours. Blauvelt (1955) and others have suggested that maternal activity in the immediate *post-partum* period is instrumental in eliciting and maintaining suckling in the neonate. Experimental support for this proposition has been obtained in the work of Alexander & Williams (1964). Maternal ewes were restricted so that they were unable to either groom or orient their offspring. During the critical first 12 hours following birth, the lambs of restricted ewes gained less weight and showed less teat-seeking activity than did the lambs of ewes maintained under normal conditions. While maternal activity apparently facilitated appropriate suckling behaviour, it was not essential. Most lambs eventually began to suckle in the absence of maternal direction and stimulation.

Despite the strong tendency to suckle, newborn lambs frequently starve in natural settings. The problem of why starvation occurs has been the focus of further investigations by Alexander, Williams and their collaborators (e.g. 1966a, b). Early learning experiences play an important role in the development of suckling behaviour. For instance, newborn animals which are denied successful suckling experiences during the first hours after birth show a progressive decline in teat-seeking activity (Alexander & Williams, 1966a). Additional experiments demonstrated that the decline in teat-seeking activity cannot be attributed simply to fatigue or to the depletion of energy reserves (Alexander & Williams, 1966b). Though the suckling tendency is strong, apparently it can be extinguished. As Alexander & Williams (1966a) suggest, there are various circumstances which can postpone successful suckling for an extended period thus diminishing the chances that the lamb will ever demonstrate appropriate suckling patterns. Such natural conditions include factors which inhibit teat-seeking activity (e.g. activity levels are lowered when the lamb is cold and wet) and factors which make the nipples less accessible (e.g. primiparous ewes frequently avoid the approach of their offspring in the first *post-partum* hours).

Frequency and duration of suckling. After parturition, the mother usually allows suckling at any time of the day or night and for any length of time. During the first week, the interval between sucklings may be one hour or even less. Twin lambs were observed to suckle 33 times during the night and 45 times during the day, making a total of 78 times in 24 hours (Sojetado, 1952). Somewhat lower frequencies of suckling in twin lambs have been reported by Ewbank (1967) and Munro (1955). The times at which the lamb suckles alternate with periods of rest. After the first week, the mother does not go to the lamb for nursing but the lamb comes to the ewe who, in turn, limits the time spent in suckling by walking away. Lambs 4 to 6 weeks of age suckle only 6 times daily (Barnicoat *et al.*, 1949; Brown, 1959).

The duration of suckling from each of the two nipples of the ewe is illustrated in relation to age of a lamb (Fig. 45). In the first 2 weeks after lambing, each suckle may last 10 minutes; the lamb suckles some 5 times from each nipple (Munro, 1956). The duration of each suckling is shorter for older lambs than younger ones and shorter at night than during the day.

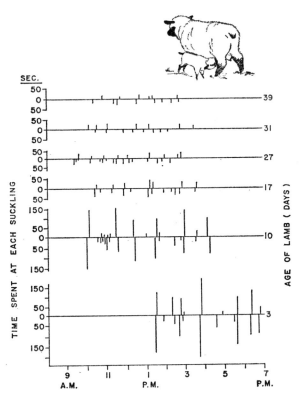

Fig. 45. Suckling behaviour of a single Blackface lamb from each of the two nipples of the ewe throughout 10 hours at different ages. The values above the lines represent one teat and values below represent the other teat. Three-day-old lambs were allowed to suckle, by the mother, at any time and for any length of time: the observation on the 1st day of experiment started in the p.m. With advancing age, the lambs spent less time in suckling. (After Munro, 1956.)

Suckling behaviour and management. The estimation of milk yield in the ewe has been based on lamb weight increases when suckling occurred at 3- and 4-hour intervals under experimental conditions (Wallace, 1948; Barnicoat *et al.*, 1949). Munro & Inkson (1957) showed that the milk consumption of lambs suckling at 4-hour intervals was similar to that of lambs suckling at 1-hour intervals.

When there is no "creep" feeding (supplemental feeding of lambs), the pre-weaning growth rate of the lambs is related to the amount of milk drunk during suckling; when the creep feeding is present, the relationship declines. This relationship is of great importance in twin-producing breeds and in areas of limited pasture, since milk yield is reduced as a result of a low level of nutrition during late pregnancy. Ewes suckling twins yield considerably more milk that those with single lambs, especially when the plane of nutrition is high. The external stimulus of suckling influences milk yield rather than any possible prenatal influences, since twin-bearing ewes rearing single lambs yielded no more milk than single-bearing ewes (Barnicoat et al., loc. cit.). When one lamb of twins or triplets is weaned several weeks after birth milk production drops. Even so, the presence of two or three lambs for the first few weeks seems to stimulate milking capacity for the whole lactation (Cowlishaw, 1961). Twin lambs which have been shown to have a fixed sucking preference will use both sides of the ewe's udder if one member of the twin pair is removed (Ewbank & Mason, 1967).

E. Drinking

Sheep make paths to drinking places, salt licks, or to shade, even though such places are nearby and quite visible across the field; the tracks are usually narrow, often only a foot across, and have sharply defined margins. Sheep tracks made by Shetland sheep are 7 to 8 inches wide (Gordon, 1961a). Under free range in Australia, sheep preferred to drink at an accustomed well and it was difficult to change them to new water, even in severe drought (Nichols, 1944). The amount of water drunk varies with the breed, climate, conditions of pasture, and reproductive phase (Table 25).

Four-month-old lambs which have been deprived of water show systematic increases in vigour and rate of sucking for water as a direct function of deprivation level (Cairns, unpublished data). Ten Hampshire lambs were fed ad lib. but were not permitted access to water over a 72-hour period. At various intervals the rate of consumption of 200 ml of water was determined. After the eighth hour of deprivation, rate of water intake increased systematically with length of deprivation. General activity level and rate of vocalization also increased with thirst (Fig. 46). Gordon (1955) has reported that a sheep deprived of water for 5 days refused to drink and had to be given over 3 litres per fistula before it drank voluntarily the next morning.

III. SEXUAL BEHAVIOUR

Unlike domestic cattle and swine which breed throughout the year, domestic sheep and goats breed during a restricted season, mainly during late summer, autumn and early winter in the temperate zone. The breeding season of sheep has been studied extensively under natural conditions at different latitudes and under artificial light treatments (reviewed by Hafez, 1952; Yeates, 1954). The effect of environmental temperature on the initiation of the breeding season has been demonstrated by Dutt & Bush (1955).

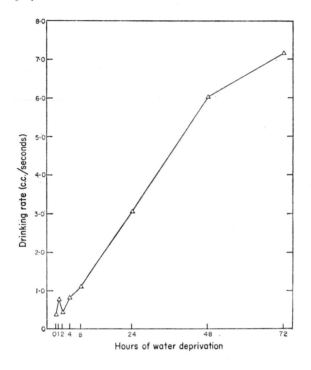

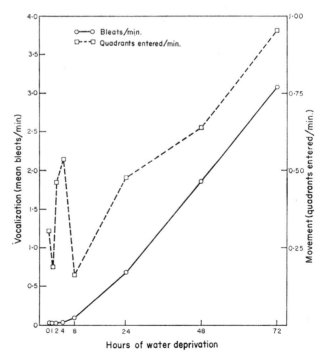

Fig. 46. The drinking rate (a), activity level, and vocalization rate (b) as a function of hours of water deprivation in 10 Hampshire Down lambs (4-5 months of age). (Unpublished data and Cairns, 1966b.)

A. Sexual Behaviour of Males

1. BREEDING SEASON

Unlike the female of the species, the male generally has no marked non-breeding season. This was not established until recent years when breeding has been initiated in the ewe during the normal anoestrous season by the use of progestagens and gonadotrophic hormones. Though there appears to be a moderate decline in semen quality in some rams during the spring and summer months (Nelson, 1958; Simpson *et al.*, 1959; Symington, 1961) and some decline in libido (Pepelko & Clegg, 1965), yet a large majority of rams have satisfactory libido and are fertile in the temperate zone at this time of year when the ewe does not normally show oestrus. Given an equal opportunity to copulate each month of the year rams produced more ejaculations during the autumn months. The greatest number of mounts per ejaculate occur during the winter period and the least in summer. In India the reaction time (time required to accomplish ejaculation) did not show any definite seasonal trend (Shukla & Bhattacharya, 1953). Certain of the mutton breeds may show a greater decline in sexual activity during the summer months than do the wool breeds. This needs to be demonstrated in a carefully conducted study.

2. COURTSHIP AND COPULATION

In sheep the display of sexual behaviour of the male is more elaborate than that of the female. The male typically responds to the urination of a female in oestrus by sniffing the urine, then extending his neck, with lips up-curled. The male moves his tongue in and out of the mouth as he follows the female, nosing her external genitalia and rubbing along her side biting her wool. A characteristic part of the sexual display or teasing is raising and lowering of one front leg in a stiff-legged striking motion (Fig. 47).

If the female is receptive, the male mounts making thrusting movements with the hind quarters. When the sensitive tip of the glans penis touches the warm mucous membrane of the vulva a reflex reaction occurs in which the penis is thrust deeply into the vagina with ejaculation occurring simultaneously. The ejaculatory thrust is easily recognized when observed. Immediately following successful copulation the ram usually appears depressed with head lowered. Temporarily he has no further sexual interest. The stimulating influence of the female has an important influence on the length of this sexually passive period. This will be discussed in greater detail in the section under sexual responses and sex drive.

If the ewe is not completely receptive, i.e. she will not stand when the ram attempts to mount, the ram continues to follow her, teasing and making repeated attempts to copulate.

In the breeding season a ram is more active than usual but apparently the energy requirements for copulation are not excessive; the amount of energy expended for one mating is estimated by Tomme & Odinec (1940) to be small compared to that required for maintenance. If this is true then the rapid loss in physical condition of breeding rams on the range might be due to lack of feed consumption due to the strong drive to copulate. However, the amount of time spent eating in small breeding pens did not have a significant negative correlation with the number of ewes in oestrus nor number of ewes mated (Hulet *et al.*, 1962a). The time spent lying down was negatively correlated with the number of ewes in oestrus.

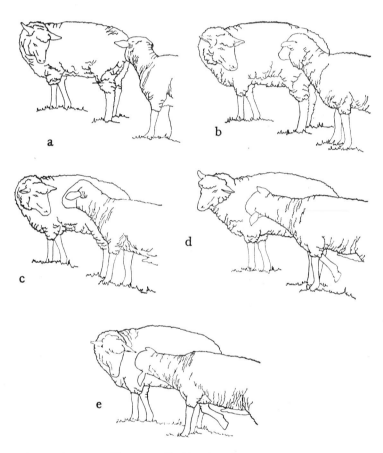

Fig. 47. Sequence illustrates display of nudging. Note partial erection which sometimes occurs with nudging. (Banks, 1965.)

The penis of sheep and goats is fibro-elastic in structure and does not enlarge during erection, as does the vascular penis of the stallion. The amount of precoital courtship is relatively shorter and the frequency of copulation is higher in rams and buck-goats than in boars and in stallions. The shape of the penis is peculiar, being slightly flexed in an S-shape in the sagittal plane. Its lower curvature is maintained by the retractor muscles, and the urethral process extends beyond the glans as a twisted filament structure. Using a short artificial vagina, Rodin (1940) has shown that the tip of the glans is the sensitive zone of the ram's penis. Stimulation of this portion alone results in ejaculation.

The number of ejaculations per day as a result of copulation ranged from about 12 to 48 over a 3-day period (Hulet, 1966), but it is greatly affected by individual, climate and the time at which the rams are introduced into breeding and the number of ewes in oestrus. Certain breeds tend to copulate more frequently than other breeds. Lambourne (1956) states that Southdown rams can copulate more often than Romney Marsh rams of the same age. No significant effect of breed was observed by Hulet et al. (1962a), although there was a tendency for Rambouillet rams to be more active breeders than Targhee and Columbia rams.

3. SEXUAL RESPONSES AND SEX DRIVE

The latency of ejaculation (reaction time for ejaculation) has been used to measure sex drive in rams (Wiggins et al., 1953). Another measure of sex drive closely related to reaction time is the number of ejaculates per hour (Hulet et al., 1962a; Pepelko & Clegg, 1964). Table 27 shows average latency periods, numbers of ejaculates per hour and factors affecting these measures of sex drive.

The expression of male behaviour is modified by exteroceptive factors, mainly the excitatory value of the sexual object, the female. Such factors may be determined by visual and olfactory cues such as size, appearance of wool, odour and colour of the fleece and especially the odour emanating from the vulva. Moreover, these factors seem to vary in their relative importance in modifying the expression of sexual behaviour of the male (Hafez, 1951). The two most important factors are how recently a ewe has exhibited the first manifestations of oestrus and the number of ewes in oestrus. Hulet et al.

Table 27. Sexual Behaviour of the Ram. (Data from Wiggins et al., 1953; Lambourne, 1956.)

Pattern		Value		Breed Observed
		Range	Mean	
Latency of	1st	1–242	7	Columbia
ejaculation	2nd	2–312	17	Corriedale
(min)[1]	3rd	4–367	28	Rambouillet
No. ejaculates/30 min		1–6	3	Targhee
No. ejaculates/day		12–15	14	Romney Marsh

[1] Reaction time to ejaculation.

(1962a) observed that ewes were mated 3·9 times during the first half of oestrus as compared to 2·4 times during the second half of oestrus. Pelpelko & Clegg (1964) measured sex drive in each of 10 rams. This was defined as the total number of ejaculations during a period of exposure to an oestrous female until sexual exhaustion occurred (a period of 20 minutes with no mounts). When, after exhaustion, a new recently unmated oestrous female was exchanged there was a 95 per cent recovery in sexual drive. If a new but recently mated oestrous female was exchanged, recovery was 39 per cent; whereas the recovery following the removal and reintroduction of the initial test female was only 18 per cent. By varying the length of time before reintroduction of the initial test female the following degrees of recovery were found: 30 minutes, 31 per cent; 2 hours, 57 per cent; 6 hours, 63 per cent; 24 hours, 79 per cent; 2 weeks, 73 per cent. These findings indicate that there is a reduction of sexual drive associated with habituation to the same partner and the normal stimulus may not be restored completely after 2 weeks. Also recent prior copulation with the female tended to reduce the normal male response.

Sexual experience may also modify the response of the male to the female. Banks (1965) studied the ontogenesis of male courtship behaviour in eight rams, four of which were entire, two castrated, and two deprived of sexual contact with females from weaning until well past the age of sexual maturity. None of the entire rams displayed a complete copulatory response during the first few exposures to anoestrous ewes, and all were well past physiological sexual maturity by the time of their first intromission and ejaculation. Of the two castrated rams, one exhibited the first complete copulatory response at 396 days of age, on the twelfth exposure to a stimulus female; the other failed to copulate. Of the two rams deprived of contact with females, the Rambouillet had not shown complete copulatory behaviour by 676 days of age, while the Hampshire displayed a complete sexual response upon its first exposure to a stimulus female at 419 days of age. Hulet et al. (1962a) observed no difference in initial mating response between 16 rams introduced to ewes for the first time since they were weaned as lambs and 16 rams used in the previous breeding season.

Rams at the U.S. Sheep Experiment Station are kept together in large monosexual groups under semi-wild or range conditions during the major part of their lives. This management practice promotes homosexuality in some rams and timidity in other rams because of older dominating rams in the same group. This environmental setting appears to condition heterosexual inhibition in some rams. Lohle (1954) reported that non-oestrous females may be used to stimulate buck-goats to deposit semen in an artificial vagina as long as the males have never encountered oestrous females; otherwise the males show no sexual interest unless the female is oestrous. This observation has not been verified by other investigators and is not true for the ram.

It is possible that olfactory stimuli as related to conditioning and sexual experience are more important in some species than in others and that the conspicuous odour of male goats is of some specific significance (see Chapters 11 and 15). Lindsay (1965) studied the contributions of olfactory stimuli in the ram to sexual behaviour. Two rams, subjected to olfactory ablation were compared with two normal rams in their ability to detect oestrus in two groups of equal numbers of oestrous and non-oestrous ewes. Identical tests were performed in separate years using different animals. Many typical signs of precopulatory behaviour were absent in the treated males. They approached females at random compared with normal males, but detected females in oestrus by nudging and attempting to mount. Foreplay was modified, but actual copulation appeared to be successful.

Soviet scientists studied sexual behaviour in relation to conditioned responses. In one study, Sokolova (1940) introduced the ewe to the ram without allowing intromission (by using a lattice screen on the vulva or tying the ewe's tail down). The ram was then permitted to mount a dummy. Within 10 days the ram became indifferent to ewes and ceased to react to the dummy. A shortlived revival of the sexual response was restored for 2 days when another sexual object, a castrated ram with different coat colour, was used behind the lattice screen. These observations show that visual cues and sexual experience are involved in conditioning and de-conditioning sexual responses.

In another experiment, a sexual object was introduced to the ram when a high-pitched bell was rung; a low-pitched bell was rung with no introduction of sexual object (Sokolova, *loc. cit.*). The rams could discriminate between the two electric bells after 10 days; they displayed sexual reactions when a high-pitched bell was rung and failed to respond to a low-pitched bell. On the 54th day, in the absence of the customary sexual object (a castrate ram), erection of the ram's penis was observed when the high-pitched bell was rung. Semen volume was considerably less when the high-pitched bell was not rung during copulation than when it was rung.

B. Sexual Behaviour of Females

1. Breeding Season

The wild sheep and the Grecian wild goat have a breeding season of a short duration (reviewed by Hafez, 1952). The improved breeds of domestic sheep are polyoestrus and, in the tropics and subtropics, sexual activity continues throughout the year (Hafez, 1953). Oestrus is associated with the ovarian cycle and limited by the onset and cessation of the breeding season which, in turn, are influenced by daylength and environmental temperature. Radford (1961) subjected Merino ewes to continuous light for three years. Despite an initial suppression of oestrus, most ewes under continuous light exhibited sexual activity which was seasonal in nature and little different from that exhibited by the control ewes. These results cannot be explained by any of the existing hypotheses of photoperiodic control of sexual activity in Merino ewes.

The Dorst Horn breed has a longer breeding season than the Border Leicester, Scottish Blackface, and Welsh Mountain breeds (Hafez, 1952). The length of the breeding season is related to the geographical origin:

11+

breeds originated at high latitudes or high altitudes have a shorter breeding season than others.

The she-goat is a seasonally polyoestrous animal which reaches her peak of oestrus in the autumn. The characteristic number of oestrous cycles during the year exhibited by the various types and breeds appear to be related, as in the ewe, to the severity of the environment: rhythmic breeding activity is under photoperiodic control (reviewed by Robinson, 1959). In general goats may be less restricted to seasonal breeding than sheep.

2. PATTERNS OF OESTRUS

The patterns of sheep female behaviour during oestrus are simple and consist of rubbing the neck and body against that of the male or putting her nose under his flank, fanning her tail vigorously and standing still to receive him. There is a strong tendency for ewes to stay with the ram and follow him about while they are in oestrus. When two or more ewes are in oestrus simultaneously, certain ewes frequently butt or crowd other oestrous ewes in their attempt to gain attention of the ram. It is difficult to determine with certainty in many cases how the initial sexual contact between the oestrous ewe and ram is made. In some instances it appears that the ram makes the initial contact while actively seeking ewes in oestrus. However, in many instances the ewe seeks out the ram (Inkster, 1957).

Ewe lambs reach puberty at an age of 6 to 16 months according to breed, climate, season of birth, and level of nutrition. Behavioural manifestations of the first oestrus are weak and incomplete; the females are not attracted to males and allow copulation reluctantly. At the second oestrus, the females are attracted to the male after some sexual approaches by the male. It is not known whether such behavioural development is due to physiological factors (ovarian cycle) or to learning.

If the sexes have been separated, the oestrous ewe frequently urinates when the ram is introduced. Females may occasionally mount each other, but such behaviour is much less frequent and less intense than in cattle. Under farm conditions, the ewe is mated 0 to 18 times during an oestrous period with an average of 6·3; the frequency varies with the age of the ewe (Table 28).

Table 28. Sexual Behaviour of the Ewe. (Data from Hafez, 1952; Lambourne, 1956; Hulet *et al.*, 1962c.)

Pattern	Mean Values		Breed
	Lambs	Adults	
Duration of behavioural oestrus (hr)	23	36	Suffolk
No. of times mated	1·9	4·3	Romney Marsh
No. of rams mating each ewe	1·3	2·6	
Interval between 1st and last matings (hr)	3·3	15·0	
No. of copulations during oestrus	—	6·3	Columbia

PLATE VI

a. Investigatory behaviour and curiosity are often manifested by lambs when they hear unusual sounds in neighbouring pen.

b. Mother-young relationship. During periods of rest or rumination, very young lambs remain close to their mothers.

c. Sleeping posture. The animal is in complete relaxation and ignores or is ignorant of exteroceptive stimuli such as camera flash. Note the closed eyes and the reclining head. (Photograph by Munro, 1957.)

PLATE VII

a. Investigatory behaviour in sheep. This pattern is often followed by a typical pattern of agonistic behaviour: the sheep "bunch" and then take flight, usually led by an older animal. (Photograph by permission of Roscoe B. Jackson Mem. Lab.)

b. Allelomimetic behaviour in goats: the flock in flight. (Photograph by permission of Roscoe B. Jackson Mem. Lab.)

c. Agonistic behaviour in male goats: note the typical behavioural pattern of rearing on the hind feet, then falling back to butt with a lateral thrust of the head. (Photograph by R. Mayo-Smith, by permission of McGraw-Hill Book Co.)

3. DURATION OF OESTRUS

In the ewe, the onset of oestrus may be related to the time of the day since at the equator, the number of onsets of oestrus from 6 a.m. to noon may be three times greater than the number from midnight to 6 a.m. During the Australian summer (during the long days), the higher proportion of onsets has been observed during the day (Kelly, 1937). The number of onsets of oestrus was highest from 7 a.m. to 6 p.m. and lower from 7 p.m. to 6 a.m. at the U.S. Sheep Experiment Station (Hulet et al., 1962c). Initial sexual activity of the ewe is only about half as great at night as during the daytime. This may be highly associated with disturbances incident to feeding, watering and other management activities.

The duration of oestrus is 20 to 30 hours although it may last from a few hours to 5 days (McKenzie & Terrill, 1937; Goot, 1949; Asdell, 1964). Genetic and environmental factors affect duration of oestrus as does age (Hafez, 1952); yearlings and adult ewes have a longer oestrus than do ewe lambs (Table 28). The first oestrus of any breeding season tends to be short. Pro-oestrus and met-oestrus are not usually observed in all individuals of the flock (Quinlan et al., 1941) because these stages of the cycle are short and not easily detected. The length of time from first tease by the ram to first mount is 2 to 4 hours. Data presented by McKenzie & Terrill (1937) suggest that copulation shortens oestrus. The length of the normal oestrous cycle in the ewe ranges from 16 to 18 days (Hafez, 1952).

In goats, the duration of oestrus is 40 hours which is longer than that in the ewe. The length of the oestrous cycle is generally given as 21 days, but it is highly variable (Robinson, 1959).

4. INDUCTION OF OESTRUS BY PRESENCE OF THE MALE

The introduction of rams to Merino ewes in late spring and early summer in Australia (November–December) leads to a synchronization of oestrus in a high proportion of ewes (Underwood et al., 1944; Thompson & Schinckel 1952; Schinckel, 1954a) (Fig. 48).

Riches & Watson (1954) have demonstrated the effect on the occurrence of oestrus of continuous association of ewes with rams throughout the year in Australia, as compared with the incidence of oestrus in a changing group of ewes. These ewes were changed at monthly intervals so that a new group of ewes from the same flock as the continuous group were introduced each month. The greatest influence of the ram is to increase the incidence of oestrus during the normally anoestrous season (Fig. 49). The primary immediate effect of introducing rams to Merino ewes during the transition from the "non-breeding" to the "breeding" season was to stimulate ovulation without oestrus in the majority of those ewes which had not already commenced cyclic breeding activity (Schinckel, 1954b).

It is believed that the stimulus provided by the introduction of the ram acts through the hypothalamus-hypophyseal portal system, leading to the

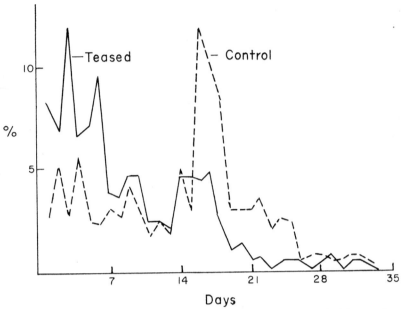

Fig. 48. Showing the daily incidence of first oestrus (per cent per day) in teased and control ewes in Australia during November–December. (From Schinckel, 1954a.)

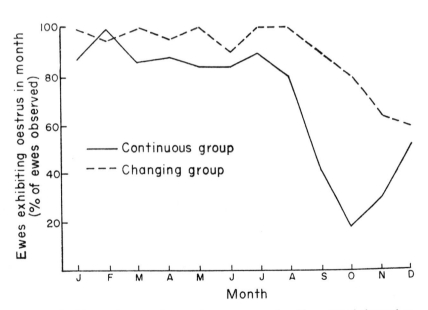

Fig. 49. Incidence of oestrus in ewes kept continuously with rams and changed at monthly intervals. Two years' data combined. (From Riches & Watson, 1954.)

production or release of sufficient gonadotrophins by the anterior pituitary gland to bring about maturation and ovulation of the more advanced follicles. Edgar & Bilkey (1963) suggest that the presence of the male may possibly represent a stress factor to the susceptible female, as the response is similar to that observed from trucking and rectal electrical stimulation. They suggest that the effect may be transmitted from the adrenal gland via the pituitary to the ovary.

Lamond (1962) observed an effect of time of ram introduction in relation to time of the last progesterone synchronization treatment on the incidence of oestrus or on the interval from the last injection until oestrus was exhibited. The nature of the response appeared to be related to the season of year. Later Lamond (1964) showed that during all seasons of the year in Australia, except the period March to May when the majority of Merino ewes showed oestrus, introduction of rams on the day after the final injection of a 15-day series of progesterone injections resulted in more ewes showing oestrus than in the group receiving similar treatments, but that had rams with them throughout the year.

These and other studies clearly indicate that there is some psychic stimulus to ovulation and subsequent oestrus associated with the introduction of the ram during those months of the year when not all ewes are cycling normally. This stimulus is not evident when the ram runs continuously with ewes.

5. HORMONAL AND NEURAL MECHANISMS

It is well established that the breeding season coincides with the short days of the year and that the photoperiodic control is mediated by hormonal and neurohumoral mechanisms via the anterior pituitary gland. Oestrous behaviour *per se* is normally under the control of steroid hormones, mainly oestrogens and progesterone. Robinson (1959) has reviewed the early attempts of hormonal control of oestrus and ovulation during the breeding season and non-breeding season. Numerous studies have been conducted since that time using orally active progestogens and vaginal implant sponges impregnated with progestogens. These synthetic progestogens have proved quite effective in synchronizing oestrus during the breeding season and as a progestogen priming agent before a gonadotrophin during the non-breeding season. Labban (1954) reported that the oral administration of L-thyrozine to ewes during the breeding season caused male-like mounting behaviour; such behaviour did not interefere with normal ovulation and pregnancy. Large dosages of oestrogen will also induce this same male-like mounting behaviour. Several studies have indicated that mating behaviour and pituitary gonadotrophic output in the ewe may be interrupted by lesions in the anterior hypothalamus. Bilateral electrolytic lesions were stereotaxically placed in the anterior hypothalamus, systematically covering the area ranging from the pre-optic region to the median eminence (Johnson et al., 1961). The effect of these lesions on the recurrence and regularity of oestrus was variable, indicating the possibility of different neural mechanisms. (See Chapter 7.)

C. Sexual Behaviour and Reproductive Efficiency

1. Social Dominance

When a particular combination of rams of varying ages are placed with a group of ewes during the breeding season (Hulet *et al.*, 1962b), a dominance rating or "peck order" is quickly established. The sexual activity of the subordinate rams in confined areas is always greatly inhibited. However when more than one ewe is in oestrus, especially when three rams are with a group of ewes, the dominant ram often finds it impossible to completely curtail the sexual activity of the subordinate ram or rams. This partial loss of mating control by the "boss" ram in breeding groups containing three rams and numerous ewes in oestrus is also observed in a study by Lindsay & Robinson (1961a, b).

Mature rams dominate or "boss" yearling rams. The degree of their dominance over yearling rams (as measured by the depression in the number of teases, mounts and matings) is greater than the dominance of certain yearling rams over other yearlings (Table 29). However, the degree of dominance of mature rams over other mature rams is less than it is for yearlings over other yearlings. The dominant ram performs a large majority of the services and may prevent other rams from mating a ewe, even when temporarily unable to do so himself. Some aggressive rams average 12 to 15 copulations per day over fairly long periods, whereas less aggressive (less dominant) rams in the same pastures average only 2 to 5 matings per day. The significance of dominance rating on number of teases, mounts and matings per day and matings per mount is shown in Table 29. When groups of rams are to be used together in a mass mating system, uniform age groups and adequate space will give the most satisfactory mating performance. Thus it is more important to semen test dominant rams. The genetic merit of dominant rams should also be given careful consideration because of their potentially greater quantitative contribution in progeny.

2. Intensity of Sexual Behaviour

Sexually inhibited rams may cause a delay in the lambing season. Semen characteristics of such rams do not differ from that of normal rams. However, semen traits of the normal rams are more highly correlated with reproductive efficiency than those of inhibited rams.

The proportion of ewes showing "weak" oestrus also affects lambing percentage of the flock. Improved management during the mating season increases chances of ewes with short or "weak" oestrous periods being mated by the ram. Frequently, it is necessary to try several rams before such ewes are mated. In artificial insemination centres, the ease of semen collection is modified by the reaction of the male to thermal and pressure stimuli in the artificial vagina (see also Chapter 9). High temperature (65°C) or high pressure (80 to 140 mm Hg) disturbs the ejaculatory mechanisms. The results are absence of ejaculation, "energetic" ejaculation and/or post-coital ejaculation (Rodin, 1940). These disturbances are usually associated with ejaculates with poor-quality semen.

Table 29. The Effects of Dominance Rating[1] and Various Combinations of Ages and Numbers of Rams on Physical Competition and Sexual Behaviour. (Data from Hulet et al., 1962b.)

Ram Combinations[2]	Age	Dominance Rating[1]	Unweighted average per day			
			Times Bunted	No. Mounts	No. Matings	No. Matings per Mount
Single		D_0	—	24·1	5·5	0·23
Y, Y	Y	D_1	0·0	26·0	5·4	0·21
	Y	D_2	1·1	16·2	1·1	0·07
Y, M	Y	D_1	0·0	47·9	5·0	0·10
	M	D_2	4·4	3·1	0·2	0·06
M, M	M	D_1	0·1	29·8	7·0	0·23
	M	D_2	5·2	13·0	3·8	0·29
Y, M, Y	M	D_1	0·8	24·1	2·8	0·12
	Y	D_2	2·8	4·8	0·2	0·04
	Y	D_3	2·5	0·5	0·0	0·00
M, M, Y	M	D_1	0·5	33·1	7·4	0·22
	M	D_2	8·4	30·8	3·4	0·11
	Y	D_3	12·1	22·2	1·8	0·08

[1] D_0 = Only ram in pen; D_1 = Dominant over all other rams in the pen; D_2 = Subordinate to D_1 but dominant over D_3 when the pen contained three rams; D_3 = Subordinate to all rams in the pen containing three rams.
[2] Y = Yearling rams; M = Mature rams.

Wiggins *et al.* (1953) measured the ejaculation times for three successive ejaculations and the number of ejaculates produced in a 30-minute period (Table 27). There was a negative correlation between ejaculation times and lambing percentage. Lambing percentage and number of ejaculates in a 30-minute period were positively correlated. It is possible that the increased lambing percentage was due to increasing the conception rate of the portion of the females that exhibit weak or short oestrus; the latter ewes are easily detected and bred by rams with vigorous sex drive. The number of ewes in heat or number of different stimulus animals and individual ram differences appear to be the most important factors contributing to variation in the number of copulations in single sire pens (Table 30). The time of day at the

Table 30. Number of Copulations per Hour for Rams with High and Low Intensity of Sexual Behaviour as Affected by the Number of Ewes in Oestrus. (From Hulet *et al.*, 1962a.)

No. of ewes in oestrus	Average no. matings/hour	
	High Ram	*Low Ram*
1	0·69	0·09
2	1·05	0·19
3	1·21	0·20

initial introduction of the ram to the ewes has an effect on the frequency of copulations. Initial introduction has a rather dramatic effect for about 8 hours (Fig. 50), after which the number of copulations per hour declines to a lower and somewhat constant level. Light or darkness, independent of time of day, appear to have little influence on sexual activity. A rather high incidence of copulations occurs at midnight.

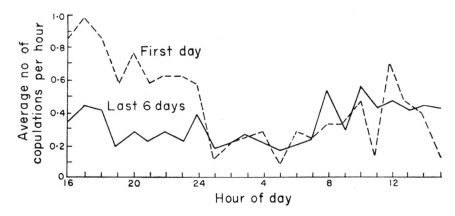

Fig. 50. Average number of copulations per ram by hour for day 1 versus days 2 to 7.

The effect of the number of different stimulus animals on the number of copulations per hour agrees with the findings of Almquist & Hale (1956) in dairy bulls. Hulet (1966) measured the frequency of copulation of several rams under experimental conditions of continuous observation. The ram with the highest intensity of sexual behaviour copulated 48 times the first day (Table 31). Lambourne (1956) states that individual rams have been known to copulate 25 times in 9 hours. High lambing percentages can be expected when temperature conditions are satisfactory, even when rams are producing an average of 26 ejaculates per day.

Table 31. Effect of Ram Differences and Day of Breeding on the Number of Copulations per Day in Pens Containing Numerous Ewes in Oestrus. (From Hulet, 1966.)

Day of Breeding	Ram Number		
	I	2	3
I	48	33	35
2	30	15	26
3*	25	12	14
Average	34	20	25

[1] There were fewer ewes in oestrus on day 3 than on days 1 and 2.

Rams with high initial fertility are capable of satisfactory breeding performance, when placed with as many as 63 progesterone-synchronized ewes per group for a 37-day breeding period during the cool autumn months (Hulet, 1966). It is probable that warm or hot temperatures typical of the summer season in many areas would have a depressing effect on breeding capacity.

One may wonder how it is that a ram distributes his copulations so uniformly over a large number of ewes. Part of this may be explained by the preference rams appear to have for ewes newly in oestrus and recently unmated (Pepelko & Clegg, 1964). Rams easily distinguish one individual ewe from another and deliberately choose, in most instances, to mate ewes which they have not mated before or have mated less frequently. Also rams mate ewes more frequently during the first half of oestrus than during the last half (Hulet *et al.*, 1962c).

IV. SOCIAL ORGANIZATION IN SHEEP AND GOAT FLOCKS

A social relationship is defined as regular and consistent behaviour between two individuals, usually members of the same species. Once a social

11*

relationship is established, certain response patterns of the two animals become mutually dependent, in that the behaviours of the one function as cues for the responses of the other. A primary and most important relationship is that which develops between the female and the newborn young. The maternal-filial interaction has been of particular interest to investigators because of its strength and rapidity of formation.

A. Mother-Young Relationship

Maternal sheep and goats are placentophagic (as are most other mammals; see Lehrman, 1961). During and immediately following birth, ewes and does chew and lick the afterbirth, consume parts of it, and groom the neonate (Collias, 1956; Smith, 1965). As noted in a preceding section, the maternal stimulation facilitates the development of successful suckling orientations in the offspring. Suckling by the neonate, in turn, likely has a stimulating influence upon the ewe, in that it reduces tension in the udder and provides the occasion for further licking of the lamb. The response patterns of the mother of offspring rapidly become mutually dependent. In the normal course of events, the initial interactions between the mother and the neonate occur under circumstances that are highly favourable for the development of a social bond.

Maternal-attachment to the young. Removal of the offspring from the maternal animal leads, under some conditions, to the later rejection of the lamb or kid. In observations of a small number of sheep (four ewes), Collias found that the maternal animal is likely to reject her own offspring, if it is removed at birth and kept away from her for $4\frac{1}{2}$ hours. The two lambs separated for a shorter period (4–5 minutes) were accepted at once upon return to the mother. Evidence for a very rapid social attachment formation has also been found in goats and their offspring. Kids that have been removed from the dam immediately following parturition and maintained in isolation for a period as brief as one hour are later rejected (Klopfer et al., 1964). If, however, the doe is permitted to nuzzle her kid for a few minutes prior to separation, the chances for subsequent acceptance will be greatly enhanced. This acceptance occurs even if the doe is partially anosmic during the initial interaction (Klopfer & Gamble, 1966).

Maternal rejection, however, is not an inevitable consequence of early separation of the newborn lamb from its mother. Strong maternal-young attachments form in sheep even when *post-partum* interaction is delayed for a period up to 8 hours after birth (Smith et al., 1966). In these studies, the ewes were unable to lick the lambs or consume birth fluids because the lambs were born on disposable polyethylene sheets. The lambs and the sheets were removed immediately. Though the period of separation ranged up to 8 hours, only one instance of maternal rejection was observed. Even this single instance of rejection was reversed after the lamb and ewe had been together for 4 hours.

Ewes and does can be induced to accept unfamiliar young and the offspring of alien species. Ewes accept both unfamiliar lambs and kids under conditions of forced confinement (Hersher et al., 1963a, b). Because ewes ordinarily will butt away unfamiliar young, it was necessary to physically restrain them in stanchions. Following a period of forced cohabitation, the ewe stopped reacting aggressively toward the alien young, and permitted the young to approach and suckle. The interval required for acceptance ranged from a few minutes to 10 days. The procedure of confining lambs individually with a non-maternal ewe in order to insure adoption is commonly practised by commercial breeders.

In summary, the available evidence indicates that both sheep and goats rapidly form an attachment to their neonate. A short period of interaction at parturition seems sufficient for initial preference formation. There is strong evidence, moreover, that maternal-young attachments can develop at various intervals following birth. Whether such later formed social relationships are equivalent in all essential respects to those which develop immediately following birth remains to be experimentally demonstrated.

Attachment formation in the lamb and kid. Young sheep and goats, like their dams, form strong social bonds. Reports of cross-specific attachment formation in lambs indicate that the range of potential attachment-objects is broad, including human caretakers (Grabowski, 1940; Scott, 1945), goats (Hersher et al., 1963b), dogs (Cairns & Johnson, 1965), cattle (Bond et al., 1967), and inanimate objects (Cairns, 1966b).

Because the strongest bonds are usually formed with respect to the animal's mother or primary caretaker, it has been thought that food reinforcement was an essential component of attachment formation (Miller, 1951). Recent investigations show, however, that attachments occur even in the absence of food or even direct physical interaction (Cairns, 1966b). If 4- to 8-week-old lambs are separated from other sheep and are maintained in continuous confinement with a dog as the sole cohabitant, the lambs will demonstrate a strong preference for the dog with which they had been paired over alternative choices.

A noteworthy feature of an experimentally induced cross-specific preference is that it normally is stronger than the lamb's preference for other sheep. In one of the preference tests employed, young lambs were permitted to approach either their canine cohabitant or a yearling ewe in a U-maze apparatus. The data (Fig. 51) show clearly that lambs, following 9 weeks of cohabitation, will select the animal with which they have been reared, regardless of its species. Other forms of attachment behaviour with respect to the dog (e.g. following the cohabitant, becoming disturbed upon being separated from the cohabitant) were readily elicited in the experimental lambs. Further analysis of this phenomenon demonstrated that lambs develop a strong preference for any salient environmental event to which they have been continuously exposed (Cairns, 1966a).

Stability of attachment behaviour in lambs. Under normal conditions, the social attachments of sheep, particularly ewes, are highly stable. They tend to remain in the same locale and interact with the same animals, even when free to move elsewhere. The stability of the response can be disrupted, however, by forcibly removing the lamb from one setting and placing it in a different context. Following a period of readaptation, the lamb typically develops a preference for its contemporaneous cohabitants and environment.

Forrester & Hoffmann (1963) report, for instance, that an orphaned male Rocky Mountain Bighorn lamb which was raised from about one week of age in a domestic environment developed a strong attachment to its human caretakers. At 11 months of age, the lamb was returned to the wild. Within weeks after being freed, the lamb associated with and adopted the behaviour patterns of undomesticated Bighorn sheep, the lamb's attachment to humans

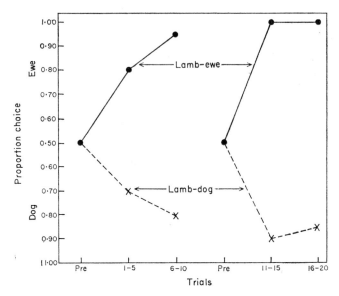

Fig. 51. U-maze performance of eight Dorset Horn lambs reared with another sheep (*lamb-ewe*) or with a mature dog (*lamb-dog*). A ewe was placed in one goal area, and a dog in the other, one of the choice animals being the lamb's cohabitant. In addition to the free-choice trials, each lamb was given six forced-choice trials (two to either side prior to the first trial, and one to either side prior to the eleventh trial). Each lamb demonstrated a strong preference for the animal with which it had been reared ($t = 7.83$, df = 6, $p < 0.001$). (Cairns & Johnson, 1965.)

apparently extinguished in that it ran away when approached by its former caretakers. In another demonstration of the plasticity of social and habitat preferences, Hunter & Davies (1963) observed that domestic lambs which have been weaned and separated from their dams for 6 months tended to remain with each other. These animals adopted a new home range that cut across the boundaries of existing ones.

Some of the conditions under which such reversals of social preferences occur have been experimentally manipulated (Cairns, 1966b; Cairns & Johnson, 1965). If a lamb is reared with an alien species, as noted above, it will develop a strong preference for the animal with which it has been cohabitated. Should the lamb be removed and placed with other sheep, however, its behaviour will typically become indistinguishable from normally reared animals. The reversal of preference occurs rapidly when cohabitation with other sheep is continuous, and is near complete after one week of co-habitation. Although the bulk of the evidence concerns the adjustment of lambs to a normal same-species environment, lambs will also adapt to an alien species after having been reared with other sheep. Lambs were reared from birth until 4 months of age with other sheep, then maintained in exclusive and continuous contact with a dog (Cairns, 1966b). By the seventh week of co-habitation, the lambs had demonstrated a significant shift in preference for their canine cohabitants relative to other sheep. In the case of ewes, there does not appear to be an age limit beyond which the capacity for the formation of new social attachments extinguishes. The formation of preferences for cross-specific animals has been demonstrated in both maternal and non-maternal ewes (Hersher et al., 1963b).

Aberrant behavioural patterns. The extent to which disruption of the mother-young relationship immediately following parturition prejudices the survival of the lamb or kid remains a controversial problem. Brief separation from the mother has been linked to several behavioural anomalies in the young animal, including increased susceptibility to stress (Liddell, 1956), heightened mortality following conditioning (Moore, 1960), and deficient depth perception (Lemmon & Patterson, 1964). Other studies by Smith et al. (1966), however, fail to yield any differences (in terms of weight losses or mortality rate) among lambs which were separated from the ewe at birth and those which were not. Similar negative findings have been reported by other investigators who have failed to observe differential mortality rates among animals reared in isolation and those reared with other sheep (e.g. Cairns, unpublished data).

Further studies of the disruptive consequences of early separation might well be concerned with the identification of the mechanisms by which such consequences are mediated. Hersher et al. (1958) suggest that some of the apparent effects of early experience in the kid can be attributed in part to the modified social organization of the herd. Maternal goats that were removed from their young for a brief period after birth tended to permit strange kids, as well as their own, to suckle. Such indiscriminative "mothering" behaviour by some does disrupted the normal maternal-young interactions among dams that had not been separated from their young. In the group of control animals, there were several instances of maternal rejection by the non-separated does, in that they would not permit even their own offspring to suckle.

B. Leader-follower Relationships

A simple test of leadership can be made by placing two sheep together in a small pen and attempting to separate them by repeatedly walking directly between them (Scott, 1945). One sheep usually takes the lead and is followed by the other, with separation occurring quite rarely. When the same test is given to goats, the animals separate more often (16 per cent instead of 2 per cent).

In sheep flocks, young lambs follow their mothers almost invariably, extending this behaviour to older animals. When older pairs are tested, there is more tendency for separation to take place, indicating that as sheep grow older they become more independent. In a naturally formed flock, the oldest ewes lead, followed immediately by their young lambs. Each is followed less closely by her descendants of previous years, female descendants being followed by their own lambs in turn. Thus the leader in such a flock is usually the oldest female with the largest number of descendants. In an all-male flock, the younger rams follow the older ones under most circumstances. This precise type of leadership is, of course, broken up in many commercial flocks where unrelated animals are brought together and where there is consequently no natural leader.

Feral domestic sheep keep in strict file and travel upon well-defined tracks when approaching or leaving a grazing area (Grubb & Jewell, 1966). Once a suitable area was reached, the group formed in sub-pockets. The lines of sheep were never led by lambs if older animals were present. On the other hand, no single animal stood out as the "leader" of the subgroups in the home range. Frequently if the file was halted by the leader, some other animal would move to the track ahead and continue the movement. The most clearly defined routes of travel were the tracks that the sheep followed in their early morning movement toward the grazing areas.

In a naturally formed goat flock the situation is somewhat similar, but leadership less definite. When alarmed the flock may form a thin line extending across in front of the disturbance, instead of a compact bunch following the single leader, and there is a much greater tendency for the flock to break up when pursued. This greater independence of goats is brought out in mixed flocks, where a goat is likely to make the first move away from the source of alarm and to be followed by the sheep. The "Judas goat" sometimes used at stockyards, makes use of these behavioural tendencies, combined with training the leading animal to go through a certain routine day after day. The same thing could be done with an older sheep.

C. Dominance-subordination Relationships

In the ordinary activity of grazing, sheep and goat flocks show little or no sign of dominance. With feed scattered uniformly over a wide area, there is no reason for competition of this sort. The same thing is true of grazing in the wild Rocky Mountain sheep flocks. However, Bighorn rams show competition for position when they bed down for the night. They also fight long and strenuously in the breeding season, the winner staying with the flock and mating with females then in oestrus. The same male may not be able to maintain its social position throughout the breeding season.

Neither domestic nor Bighorn sheep defend territorial boundaries (Buechner, 1960). Sheep do establish "monopolized zones", i.e. areas that a particular subgroup of sheep habitually occupies at a given time of day (Grubb & Jewell, 1966; Jewell, 1966). Unfamiliar ewes that tend to wander into a subgroup are subject to investigation by members of the group. The "investigation" includes sniffing of the anal-genital area and the flanks and, on rare occasions, will include butting. Under normal circumstances, there are few similar overt interactions among members of a ewe group. The resistance demonstrated by the subgroup towards intruders seems, in most instances, sufficient to maintain the coherence of the group and their space monopoly.

Although sheep in domestic flocks will compete for small amounts of food, the instrumental activity observed is limited to pushing and shoving towards the food rather than active butting. In general, the dominance order among sheep is less marked than among goats (Scott, 1945). The following information relates to studies of a naturally formed goat flock.

When a young kid is born it at first shows no fighting behaviour. Its relationship with its mother is peaceful and may remain so after the animal becomes an adult. Likewise, the relationship between twins or between kids born at about the same time is also peaceful. However, a kid which approaches an adult female other than the mother is almost sure to be butted away. Older males respond in the same way if kids attempt to compete with them. This establishes an important basis for the dominance hierarchy, namely that older animals are usually dominant over younger ones.

Table 32. Types of Dominance Relationships in a Flock of Goats, graded according to Degree of Control by One Animal. (Scott, 1948.)

Grade	Type	Occurrence in a Flock of 14 Goats; %	Ave. Attacks per Relationship; when Hungry
0	No dominance. Peaceful; no fighting, eat side by side.	24	1·8[1]
1	Unsettled dominance; both animals initiate attacks while eating.	8	6·7
2	Partial dominance; dominant animal attacks, subordinate shows only defensive fighting; both eat.	27	3·5
3	Partial dominance; dominant attacks; no fighting by subordinate, but competes for food.	35	3·3
4	Complete dominance; subordinate leaves immediately after threat or attacks.	7	1·1

[1] All these animals had previously eaten together peacefully and hence were graded 0; some of the pairs showed fighting when hungry.

We may now say a word about the nature of the dominance-subordination relationship. When two goats fight with each other, one usually wins and the other loses. As they meet over and over again the same result is repeated, so that one animal forms a strong habit of winning and the other of losing. Eventually the dominant animal merely has to threaten in order to make the subordinate one stay away.

This, however, is not the only possible outcome. If neither animal wins, the pair may form a habit of constant fighting whenever they meet. In another relationship the dominant animal may do all the aggressive fighting, but the subordinate butts back if attacked; or, the dominant animal may continually attack the subordinate one, but the latter will remain near and take the punishment. In the most complete type of dominance, the subordinate animal leaves immediately, as soon as threatened or attacked. Thus we have five grades of dominance, with the least fighting occurring in the lowest and highest grades (Table 32).

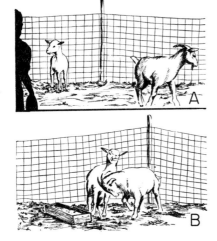

Fig. 52. Two adult female goats show no leader-follower relationship in a separation test when a handler is approaching (A), but the horned female is dominant when they compete over feed (B). (Redrawn from motion picture frames, Stewart & Scott, 1947.)

There are two important situations which lead to increased competition and the development of dominance in a goat flock. The first of these is crowding, and applies to either males or females. As Blauvelt (1956) has pointed out, each member of the flock attempts to preserve an empty living space around it. This corresponds to the phenomenon of minimum distance toleration in flocks of domestic birds. If the flock is crowded together or confined closely, the goats begin to butt each other. If the animals are now fed in a very small area, so that they have to come into contact in order to eat, fighting is further intensified (Fig. 52). This can be made into a test for the existence of a dominance order, either between single pairs or within larger groups.

A second cause of fighting applies to males alone. The presence of a female in oestrus attracts all the males, each of which attempts to drive away the others. In this situation the most dominant male attempts to herd all the females away from the rest, while the subordinate males continually attempt to approach the females and evade the dominant male. The subordinate males are much more persistent than they are in the first situation, the dominance order is less clear-cut, and the precise rank order may not be identical (Scott, 1948).

Once a dominance relationship between two goats is firmly established, it may remain stable over a period of years. However, the total organization of the flock changes as individuals are born, mature, and die (Ross & Berg, 1956).

D. Causes of Dominance in Goats

The most important factor determining dominance is the biological relationship between the two animals involved. Mothers control their offspring by calling and allowing them to suckle. They eat peaceably with them in close quarters and almost never attack them. The same peaceful relationship exists between twin kids born to the same mother.

On the other hand, goats not related in the above ways typically form a dominance order. In these dominance relationships the most important factor is relative age. When a young goat approaches an unrelated adult it is immediately butted and driven off. There is no opportunity to resist or fight back, and the young animal is soon trained to be subordinate. In a small flock of goats, older animals were dominant over younger in 74 out of 86 cases, including all relationships in which there was a difference in age of as much as six months (Scott, 1948; Ross & Scott, 1949).

In unrelated animals of the same age a dominance order is gradually set up, beginning with playful fighting in very young goats. In this case, the eventual dominance order is based on relative size and strength. The possession of horns gives a goat an important advantage over a hornless rival.

Males are usually larger and stronger than females and have larger horns. In the same flock described above, males were dominant over females in 21 out of 31 cases. However, a hornless male is usually no match for a horned female.

The most dominant animal is not necessarily the leader of the flock. Stewart & Scott (1947) found that leadership and dominance were associated in a proportion which could be accounted for purely by chance. In the relationships in which leadership is best developed (between mother and offspring), there is little or no dominance, while in those relationships where dominance is best developed (between older and younger unrelated animals), there is only a weak system of leadership. Consequently, the oldest females with the greatest numbers of descendants become the leaders of the flock, whereas the oldest and strongest males tend to become the most dominant members. It is possible that in a strictly male flock there might be some correlation between dominance and leadership, since both are related to age.

E. Effect of Animal Husbandry on Flock Organization

In the above pages we have described behaviour as it occurs in a naturally formed flock, this being defined as a flock formed by animals born in the same area and increased only by birth. This is the way in which a herd of wild sheep or goats is formed, and small flocks on farms are sometimes formed in a similar way. In modern practices of animal husbandry the number of males is always reduced, and the males and females brought together only during the breeding season. In herds of dairy goats, the odorous male must be kept separate from the females. With the use of frozen semen and artificial insemination, contact between males and females may be entirely eliminated. However, the female flock under such farm conditions is not an unnatural social unit, for in the wild species of sheep the males typically form separate flocks except during the breeding season. Likewise, most of the mating is normally done by a few males, as the result of conflicts within the male flock. We may conclude that farmers and shepherds throughout the ages have generally made use of the natural social behaviour of domestic animals and have not departed from it to any large extent.

The chief difference is that domestic flocks are usually somewhat disorganized, either by the introduction of strange adults, or by forcibly maintaining extremely large flocks, as in the case of sheep "bands". The result of introducing strange adult females into a flock of goats is to greatly increase the amount of fighting, which may result in injury or possibly improper nourishment. In sheep, one result of artificially large flocks is that its members are constantly being separated from their relatives. The small naturally formed flock is a silent one, contrasting greatly with the continual baaing in large commercial flocks. In short, the commercial flocks are comparatively disorganized but are protected from the natural consequences of disorganization by fencing and herding.

F. Communication

In keeping with the fact that sheep and goats are chiefly active in the daytime, members of the flock appear to maintain contact with each other largely through vision. As the flock grazes, each individual throws up its head and presumably responds to the position of other members. As might be expected if vision were a dominant source of information about the social environment, totally blind sheep are grossly abnormal in some behaviours. They cannot be approached without causing panic or injury, and they tend to run wildly into unseen obstacles (Arnold, 1966a). Apparently there is significant compensation from other sensory channels, however. A blind ewe was able to rear two normal lambs (Smith, 1965).

Sheep and goats are also responsive to the vocal events produced by other animals. If a dam and her lamb are separated, typically both animals "baa" or "bleat" until they are brought together again. Older sheep also call in this way if separated from the flock, even when they do not have young lambs (Grubb & Jewell, 1966). The vocalization ordinarily is accompanied by heightened movement, which ranges in degree from mild hyperactivity to extreme agitation. The correlation between vocalization and activity level in lambs is high and positive (Cairns, 1966b). If the lamb is maintained in isolation for 4 hours or longer, the vigour and intensity of the vocalization drop off sharply. Normal (that is, pre-separation) levels of activity and vocalization are reached after the third day of isolation. Apparently the young animal rapidly adapts to the stimuli that are present in the "isolation" compartment (Fig. 53).

Lambs and older sheep will "baa" or "bleat" under other conditions.

During water deprivation, the incidence of "baaing" is correlated positively with the number of hours since water was removed. The lamb will also vocalize, on occasion, if frightened or injured. The extent to which these vocalizations can be reliably differentiated from those observed during periods of maternal-young separation has yet to be empirically determined.

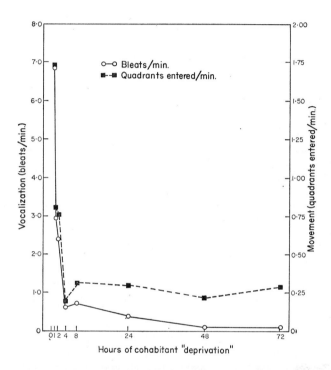

Fig. 53. Vocalization and activity level in 10 Hampshire Down lambs (4-5 months of age) as a function of the number of hours since being separated from its cohabitant. In the case of five lambs, the cohabitant was a ewe, and for the remaining lambs, the cohabitant was a mature dog. The results for the two groups of lambs were essentially identical. (Cairns, 1966b.)

At least two other types of vocalization have been distinguished. During the sexual season, rams sometimes produce a hoarse interrupted baaing sound as they approach the female. This vocalization is sometimes described as a "grumble". The "parturition call", a low "m-m-m", frequently is produced by a dam in the care of her newborn young.

Olfactory and gustatory events function as cues for both sheep and goats. Smell is apparently critical, for instance, in the identification of kids by their dams. If the maternal animal is made anosmic at parturition, she will later fail to discriminate between her own offspring and the young of other dams (Klopfer & Gamble, 1966).

V. LEARNING AND INTELLIGENCE

Sheep and goats have been used infrequently in laboratory investigations of learning and conditioning. It seems likely that such practical considerations as the expense of procuring and maintaining the animals, and their size, have served to limit the number of occasions that they have been studied in psychological laboratories. Their infrequent use in the analysis of learning phenomena cannot be attributed, however, to a deficient capacity for learning. On the contrary, the available reports indicate that both species are highly responsive to new environmental requirements and that they rapidly adapt to modifications in patterns of reward and punishment.

Classical conditioning. The earliest attempt to study systematically classical (Pavlovian) conditioning in sheep and goats was reported by Liddell *et al.* (1934). The procedure followed involved the presentation of an auditory stimulus (metronome operating at the rate of 120 beats a minute) 2–10 seconds before the animal was given a brief tetanizing shock of moderate intensity (UCS) to his upper forelimb. After the 8th to 9th recurrence of the sound-shock pairing, the beating of the metronome alone (CS) became sufficient to elicit defensive movements.

The parameters of conditioned reflex activity in response to painful stimulation have been more recently examined by Soviet investigators (Pavlik, 1958; Dedashev, 1959b). Using a technique similar to that of Liddell *et al.*, Pavlik found that mature sheep rapidly developed conditioned responses regardless of the nature of the conditioned stimulus. Either photic, auditory, or mechanical cutaneous stimuli were effective in the establishment of the CS–UCS association. Defensive conditioning was apparently obtained even in very young sheep (20 days old). With the younger animals, however, conditioning was less stable and required a great many more pairings to establish the effectiveness of the previously neutral stimulus. The 20-day-old animals also failed to discriminate between two levels of the auditory stimulus, when one level was paired with shock and the other level was not. This failure to differentiate between the CS and its alternative raises some question as to whether conditioning had occurred or whether the young animal was simply sensitized by the experimental procedures to respond to any discrete environmental stimulus.

Certain internal responses, including digestive activity in the rumen and reticulum, can be conditioned. To obtain measures of activity within the stomach, Dedashev (1959a) inserted balloons into the reticulum and rumen of fistulated sheep. One of two tones was then systematically paired with the delivery of a small amount of hay or oats. After approximately 100 pairings of the events, both the rumen and reticulum showed increased activity upon the occurrence of the tone associated with food delivery, and no change upon the occurrence of the tone not paired with food. Subsequent studies (Dedashev, 1959b) indicate that the digestive processes can be disrupted by the delivery of shock either internally or externally. Shock delivery to the skin or to the mucous membranes of the reticulum, rumen, or rectum inhibits activity in the various compartments of the stomach. Combination of an auditory stimulus with the internal or external pain stimulation leads to the establishment of conditioned internal disruption; the tone presented alone is sufficient to inhibit the digestive functions. The role that classical conditioning plays in the control of sexual activity has been discussed.

Instrumental learning. Liddell and his colleagues (1925, 1954) were also among the first to study instrumental learning phenomena in sheep. These training procedures differ from the classical Pavlovian methods in that the subject's responses are *instrumental* to the occurrence of a reward or to the avoidance or escape from punishment (Kimble, 1961). Young sheep and goats rapidly adapt to the available reward contingencies in a simple two-choice maze. In one of Liddell's typical experiments, lambs or kids learned to turn either to the right or left in order to reach a goal area where they were fed and placed with other members of their species. The correct turning response was quickly acquired (0–7 trials) in most instances, though Liddell reports that some animals simply remained in the runway. Young sheep and goats also learned single alternation problems, where the subject must learn to turn left on one trial, right on the next trial, left on the next, etc., in order to be rewarded.

Subsequent investigations confirm Liddell's findings that sheep rapidly learn the "correct" response in a modified two-choice maze, if the goal event is sufficiently attractive. The reward for correct performance need not involve the delivery of consumable food. Equally strong effects are obtained when the goal object is a "social" reinforcement event, i.e. the presence of another animal. As Fig. 51 indicates, lambs will acquire the appropriate turning responses in order to approach their cohabitant regardless of its species. The extent to which the lamb becomes disrupted upon being separated from his cohabitant is a sensitive index of the social reinforcement effectiveness of the cohabitant as a goal object. Those lambs that are highly disrupted upon separation when tested in the maze (a) obtain shorter response latencies (i.e. run faster to reach the goal area), and (b) select the "correct" goal area more frequently, than lambs which show little behavioural distress upon separation. These results support the empirical generalization that other animals, usually but not always members of the lamb's subgroup, play a significant role in the control of the lamb's instrumental activities.

Preference acquisition. Observations of the behaviour of sheep in both the laboratory and natural settings also underscore the role that experiential factors play in the development of preferences for particular environmental states or events. Evidence thus has been obtained which demonstrates that a wide range of preferences (e.g. food, habitat, range, species, filial) can be modified. The available evidence indicates that continuous or very frequent exposure to an event is sufficient to produce a change in the animal's acceptance of that event.

The exact process by which such preferences are acquired remains a matter of some debate. A primary issue concerns the problem of whether filial attachment acquisition in sheep and goats should be subsumed under the general concept of imprinting. On the one hand, the very rapid development of a social preference by the dam for her offspring at parturition suggests an "imprinting-like" process (Collias, 1956; Klopfer *et al.*, 1964). But not all characteristics of the filial relationship are so easily fitted to the imprinting construct. Current evidence indicates that attachment formation is not limited to the *post-partum* period (Hersher *et al.*, 1963a, b; Smith *et al.*, 1966), that social preferences can be reversed (Cairns, 1966b; Forrester & Hoffmann, 1963), and that early experience does not necessarily determine sexual preferences at maturity. Such outcomes are not consistent with the primary characteristics of imprinting outlined by Lorenz (1955). Even if the social preferences of sheep and goats could be classified as imprinting, it is not obvious that such an identification would advance the understanding of the essential learning processes involved. As Bateson (1966) has indicated, the available evidence does not support early claims that imprinting should be regarded as a special process and thus isolated from more general investigations of learning.

Alternative formulations have attempted to explain the role that conditioning plays in the formation of habitat and social preferences. According to one view, learned preferences reflect the extent to which a given behaviour sequence of the animal has become associatively linked to a particular event in the environment (Cairns, 1966a; Guthrie, 1952). Once a behaviour (e.g. grazing, suckling, drinking) has become conditioned to the presence of the event, the activity will be disrupted by its removal. Certain features of the event (e.g. salience, involvement in the activity) or the context (e.g. forced exposure to or interaction with the event) should facilitate the conditioning process. The parturition setting, as a special case, should be optimal for the conditioning of maternal behaviours to the presence of the offspring, in that the young animal would be directly involved in the support of maternal care-taking activities. If the dam can be induced to perform such activities with respect to an alien animal, at parturition or afterwards, that animal should acquire the capacity to support maternal behaviours (Hersher *et al.*, 1963a). Similarly, non-filial behaviours of the animal (e.g. bedding, grazing) are expected to become conditioned to the locations in which they have previously occurred.

VI. SLEEP

Since sheep usually spend about half of their daily time in a relatively inactive state the problem of whether or not true sleep exists is an interesting one. Ruminants do not enter the state of deep sleep shown by other domestic animals such as horses, dogs, and cats. Sheep and goats have their periods of "inactivity" scattered throughout the 24 hours but mainly during the night. They also show signs of torpidity during bouts of rumination which take place mostly at night.

It has been possible to correlate the electric activity of the brain with various stages of sleep by means of an electroencephalograph (EEG). Bell (1960) showed that goats do not sleep, but they do exhibit periods of somnolence when the EEG shows a hypersynchronous character similar to that recorded in other species in deep sleep or anaesthesia. "The animals usually lie down, the recumbency is followed by some loss of muscle tone so that the carriage of the head is lowered and the ears droop away sideways. The thresholds of sensory stimuli are also raised during these periods of languor of somnolence and rumination. The somnolent ruminant would appear to be cut off from its environment because of the rise in receptor threshold, but to a much lesser degree than other animals in whom a state of unconsciousness develops with sleep. The ruminant, moreover, during somnolence has its eyes open which is not normally an accompaniment of sleep" (Bell, *loc. cit.*).

However, Munro (1957) observed that 18-months-old sheep slept for short periods of up to 38 minutes' duration. The sheep assumed a relaxed position in which most parts of the body were maintained in unsupported position. The sleeping animal ignored and did not perceive exteroceptive stimuli such as odour of offered feed, sound of a ticking clock or sight of a camera flash (Plate VIc). Balch (1955) and Gordon (1955) thought that if ruminants do sleep it must be transient and polyphasic.

According to Balch (1955), sleep in the ruminant is linked directly to the digestive needs of the animal: rumination requires both time and consciousness, and the thorax must be maintained in an upright position for the proper functioning of the reticulo-rumen. Interesting support for this proposal has been reported by Morag (1967). Six ewes fed only water and finely ground grass were given hourly doses of oxytocin. The animals were soon observed to enter states of deep sleep (e.g. closed eyes, lying in an atypical position with head between forelegs and thorax inclined, non-responsive to noise). A marked change in behaviour was produced when hay *ad lib.* was added to their diet: the animals remained upright, kept their eyes open, and all signs of exhaustion vanished. Deep sleep was possible apparently only when the ewes' roughage intake did not require rumination.

VII. BEHAVIOURAL DIFFERENCES BETWEEN SHEEP AND GOATS

Most of the patterns of behaviour are quite similar in the two species, and under conditions of domestication sheep and goats readily mix in common flocks. However there are several distinct behaviour differences, the most basic of which is in ingestive behaviour—mainly the choice of food. Sheep prefer grass and succulent herbage while goats prefer twigs and leaves; such differences may be related to different structure of their upper lips. It has been suggested that goats need more food variety than sheep do and that this is the explanation of their chewing on the most unlikely substances (Gordon, 1961a); sheep can be maintained on the same monotonous diet for years. Another difference related to habitat selection is the greater tendency of goats to climb on rocks and other elevations when this is possible.

Differences in agonistic behaviour are related to the characteristic shapes of horns in the two species, but the characteristic patterns of behaviour appear even in hornless individuals. Sheep butt head on, but goats do this with a sideways hooking motion (Plate VII). When two goats fight they do not back off, but rear on their hind feet and turn their heads sideways so that they meet head to head, but with a lateral thrust. The lack of correspondence between these patterns of behaviour results in the fact that sheep and goats in the same flock rarely get into individual combats. Ordinary butting from a standing posture is quite similar in the two species (Scott, 1960; Blauvelt & Moore, 1960).

Sexual behaviour is quite similar, except that it seems to be exaggerated in male goats as compared to sheep. Male goats show much more tendency to "herd" females in oestrus than do rams. Goats and sheep mate readily with each other if kept in the same herd, but the cross-species matings are always sterile, since the embryos die early in development (Warwick & Berry, 1949).

A major difference occurs in the behaviour of the newborn young. Newborn lambs do not leave their mothers by more than a few feet and follow them constantly. On the other hand, the goat mothers leave their young and go off grazing at considerable distances, and young kids show the habit of "freezing", a type of behaviour also found in young antelope and deer.

A qualitative difference is found in the alarm signal. Sheep snort and stamp one forefoot when alarmed. Goats show this behaviour much more frequently, and the sound produced is more high pitched, sounding like a sneeze.

Although goats and sheep belong to different genera, their basic behaviour patterns remain relatively similar, indicating that behaviour traits are conservative in evolution. The differences between the two species are related to many things: presumed differences in the habitat of the wild parent species, as in feeding preferences; differences in morphology, as in agonistic behaviour and horn shape; and differences in social organization, as in the behaviour of the young. Originally closely related, wild sheep and goats evolved in two different directions: sheep towards highly co-ordinated flocks living in open pastures and feeding on grass, and goats towards more loosely organized flocks living in rocky and brushy areas. Much of the behaviour of their domesticated descendants may be understood in terms of these species-typical tendencies.

REFERENCES

ALEXANDER, G. & WILLIAMS, D. (1964). Maternal facilitation of sucking drive in newborn lambs. *Science*, **146**, 665–666.

ALEXANDER, G. & WILLIAMS, D. (1966a). Teat-seeking activity in lambs during the first hours of life. *Anim. Behav.*, **14**, 166–176.

ALEXANDER, G. & WILLIAMS, D. (1966b). Teat-seeking activity in newborn lambs: the effects of cold. *J. agric. Sci.*, **67**, 181–189.

ALMQUIST, J. O. & HALE, E. B. (1956). An approach to the measurement of sexual behavior and semen production of dairy bulls. *Plenary Pap. 3rd Intern. Congr. Anim. Reprod.* (Cambridge), pp. 50–59.

ANDERSSON, B. (1951). The effect and localization of electric stimulation of certain parts of the brain stem in sheep and goats. *Acta physiol. scand.*, **23**, 8–23.

ANDERSSON, B. & JEWELL, P. (1957). Studies on the thalamic relay for taste in the goat. *J. Physiol.*, **139**, 191–197.

ANDERSSON, B., KITCHELL, R. L. & PERSSON, N. (1959). A study of central regulation of rumination and reticulo-ruminal activity. *Acta physiol. scand.*, **46**, 319–338.

APPLEMAN, R. D. & DELOUCHE, J. C. (1958). Behavioural, physiological and biochemical responses of goats to temperature, 0° to 40°C. *J. Anim. Sci.*, **17**, 326–335.

ARNOLD, G. W. (1962). The influence of several factors in determining the grazing be-haviour of Border Leicester × Merino sheep. *J. Br. Grassld Soc.*, **17**, 41–51.

ARNOLD, G. W. (1964). Some principles in the investigation of selective grazing. *Proc. Aust. Soc. Anim. Prod.*, **5**, 258–271.

ARNOLD, G. W. (1966a). The special senses in grazing animals. I. Sight and dietary habits in sheep. *Aust. J. agric. Res.*, **17**, 521–529.

ARNOLD, G. W. (1966b). The special senses in grazing animals. II. Smell, taste, and touch and dietary habits in sheep. *Aust. J. agric. Res.*, **17**, 531–542.

ASDELL, S. A. (1964). *Patterns of Mammalian Reproduction*. New York: Comstock, p. 360.

BALCH, C. C. (1955). Sleep in ruminants. *Nature, Lond.*, **175**, 940–941.

BANKS, E. M. (1965). Some aspects of sexual behaviour in domestic sheep, *Ovis aries*. *Behaviour*, **23**, 249–279.

BARNICOAT, C. R., LOGAN, A. G. & GRANT, A. I. (1949). Milk-secretion studies with New Zealand Romney ewes. *J. agric. Sci.*, **39**, 44–55.

BATESON, P. P. G. (1966). The characteristics and context of imprinting. *Biol. Rev.*, **41**, 177–220.

BELL, F. R. (1959). Preference thresholds for taste discrimination in goats. *J. agric. Sci.*, **52**, 125–258.

BELL, F. R. (1960). The electroencephalogram of goats during somnolence and rumination. *Anim. Behav.*, **8**, 39–42.

BELL, F. R. (1961). Personal communication, Dept. Vet. Physiol., Royal Veter. Coll., London, England.

BELL, F. R. & LAWN, A. M. (1955). Localization of regions in the medulla oblongata of sheep associated with rumination. *J. Physiol.*, **128**, 577–592.

BELL, F. R. & LAWN, A. M. (1957). Patterns of rumination behaviour in housed goats. *J. Anim. Behav.*, **5**, 85–89.

BLAUVELT, H. (1955). Dynamics of the mother-newborn relationship in goats. In: *Group Processes*; Transactions of the First Conference. B. Schaffner (ed.). New York: May Foundation, pp. 221–258.

BLAUVELT, H. (1956). Neonate-mother relationships in goat and man. In: *Group Processes*; Transactions of the Second Conference. B. Schaffner (ed.). New York: May Foundation.

BLAUVELT, H. & MOORE, A. U. (1960). *Aggressive Behavior in the Social Organization of Goats and Sheep*. Film. Cornell Univ., Ithaca, N.Y.

BLAXTER, K. L., WAINMAN, F. W. & WILSON, R. S. (1961). The regulation of food intake by sheep. *Anim. Prod.*, **3**, 51–61.

BOND, J., CARLSON, G. E., JACKSON, C., Jr. & CURRY, W. A. (1967). Social cohesion of steers and sheep as a possible variable in grazing studies. *Agron. J.*, **59**, 481–482.

BOYD, J. M., DONEY, J. M., GUNN, R. G. & JEWELL, P. A. (1964). The Soay sheep of the island of Hirta, St Kilda. A study of a feral population. *Proc. zool. Soc. Lond.*, **142**, Part 1, 129–163.

BROWN, C. J. & CAVENESS, J. W. (1959). Preference for feed preparations by sheep. *J. Anim. Sci.*, **18**, 1158 (Abstract).

BROWN, T. H. (1959). Parasitism in the ewe and the lamb. *J. Br. Grassld Soc.*, **14**, 216–220.

BUECHNER, H. K. (1960). The Bighorn sheep in the United States. *Wildl. Monogr.*, No. 4, 1–174.

CAIRNS, R. B. (1966a). Attachment behavior of mammals. *Psychol. Rev.*, **73**, 409–426.

CAIRNS, R. B. (1966b). Development, maintenance, and extinction of social attachment behavior in sheep. *J. comp. physiol. Psychol.*, **62**, 298–306.

CAIRNS, R. B. & JOHNSON, D. L. (1965). The development of interspecies social prefer-ences. *Psychon. Sci.*, **2**, 337–338.

COLLIAS, N. E. (1956). The analysis of socialization in sheep and goats. *Ecology*, **37**, 228–239.

COOK, C. W., HARRIS, L. E. & STODDART, L. A. (1948). Measuring the nutritive content of a foraging sheep's diet under range condition. *J. Anim. Sci.*, **7**, 170–180.

CORY, V. L. (1927). Activities of livestock on the range. *Texas Agric. exp. Sta. Bull.*, No. 367.

COWLISHAW, S. J. (1961). Personal communication. Grassld Res. Inst., Hurley, Nr Maidenhead, Berks., England.

COWLISHAW, S. J. & ALDER, F. E. (1960). The grazing preferences of cattle and sheep. *J. agric. Sci.*, **54**, 257–267.

CRESSWELL, E. (1960). Ranging behaviour studies with Romney Marsh and Cheviot sheep in New Zealand. *Anim. Behav.*, **8**, 32–38.

CRESSWELL, E. & HARRIS, L. E. (1959). An improved rangemeter for sheep. *J. Anim. Sci.*, **18**, 1447–1451.

DEDASHEV, Ia. P. (1959a). Conditioned reflexes of motor activity in the reticulum and rumen of sheep. *Fiziol. Zh. SSSR* (Trans.), **45**, 104–108.

DEDASHEV, Ia. P. (1959b). Exteroceptive and interoceptive conditioned reflex effects on the motor activity of the reticulum and rumen in sheep. *Fiziol. Zh. SSSR* (Trans.), **45**, 95–99.

DENTON, D. A. & SABINE, J. R. (1961). The selective appetite for Na^+ shown by Na^+-deficient sheep. *J. Physiol.*, **157**, 97–116.

DUNCAN, D. L. (1953). The effects of vagotomy and splanchnotomy on gastric motility in the sheep. *J. Physiol.*, **114**, 156–169.

DUTT, R. H. & BUSH, L. F. (1955). The effect of low environmental temperature on initiation of the breeding season and fertility in sheep. *J. Anim. Sci.*, **14**, 885–896.

EDGAR, D. G. & BILKEY, D. A. (1963). The influence of the ram on the onset of the breeding season in ewes. *Proc. N.Z. Soc. Anim. Prod.*, **23**, 79–83.

ENGLAND, G. J. (1954). Observations on the grazing behaviour of different breeds of sheep at Pantyrhuad Farm, Carmarthenshire. *Br. J. Anim. Behav.*, **2**, 56–60.

EWBANK, R. (1964). Observations on the suckling habits of twin lambs. *Anim. Behav.*, **12**, 34–37.

EWBANK, R. (1967). Nursing and suckling behaviour amongst Clun Forest ewes and lambs. *Anim. Behav.*, **15**, 251–258.

EWBANK, R. & MASON, A. C. (1967). A note on the suckling behaviour of twin lambs reared as singles. *Anim. Prod.*, **9**, 417–420.

FORRESTER, D. J. & HOFFMANN, R. S. (1963). Growth and behavior of a captive Bighorn lamb. *J. Mammal.*, **44**, 116–118.

GOOT, H. (1949). Studies on some New Zealand Romney Marsh stud-flocks. III. Tupping season. *N.Z. Jl. Sci. Technol.* (Sect. A), **30**, 330–344.

GOOT, H. (1962). Training sheep for selective grazing. *Anim. Behav.*, **10**, 232.

GORDON, J. G. (1955). *Rumination in the Sheep*. Ph.D. thesis, Univ. Aberdeen, Scotland.

GORDON, J. G. (1958a). The relationship between fineness of grinding of food and rumination. *J. agric. Sci.*, **51**, 78–83.

GORDON, J. G. (1958b). The act of rumination. *J. agric. Sci.*, **50**, 34–42.

GORDON, J. G. (1961a). Personal communication. The Rowett Instit., Bucksburn, Aberdeen, Scotland.

GORDON, J. G. (1961b). The relationship between rumination and frequent feeding. *Anim. Behav.*, **9**, 16–19.

GORDON, J. G. & TRIBE, D. E. (1951). The self-selection of diet by pregnant ewes. *J. agric. Sci.*, **41**, 187–190.

GORDON, J. G., TRIBE, D. E. & GRAHAM, T. C. (1954). The feeding behaviour of phosphorus-deficient cattle and sheep. *Brit. J. Anim. Behav.*, **2**, 72–74.

GRABOWSKI, K. (1940). Pragung eines Jungschafs auf den Menschen. *Z. Tierpyschol.*, **4**, 326–329.

GRUBB, P. & JEWELL, P. A. (1966). Social grouping and home range in feral Soay sheep. *Symp. Zool. Soc. London*, No. 18, 179–201.

GUTHRIE, E. R. (1952). *The Psychology of Learning* (2nd edn.). New York: Harper.

HAFEZ, E. S. E. (1951). Mating behaviour in sheep. *Nature, Lond.*, **167**, 777–778.

HAFEZ, E. S. E. (1952). Studies on the breeding season and reproduction in the ewe. V. Mating behaviour and pregnancy diagnosis. *J. agric. Sci.*, **42**, 255–265.

HAFEZ, E. S. E. (1953). Ovarian activity in Egyptian (fat-tailed) sheep. *Cairo Fac. Agric. Bull.* No. 34.

HERSHER, L., MOORE, A. U. & RICHMOND, J. B. (1958). Effect of post-partum separation of mother and kid on maternal care in the domestic goat. *Science*, **128**, 1342–1343.

HERSHER, L., RICHMOND, J. B. & MOORE, A.   . (1963a). Maternal behavior in sheep and goats. In: *Maternal Behavior in Mammals*. H. L. Rheingold (ed.). New York: Wiley & Sons, Inc. Ch. 6, pp. 203–232.

HERSHER, L., RICHMOND, J. B. & MOORE, A. V. (1963b). Modifiability of the critical period for the development of maternal behavior in sheep and goats. *Behavior*, **20**, 311–320.

HUGHES, G. P. & REID, D. (1951). Studies on the behaviour of cattle and sheep in relation to the utilization of grass. *J. agric. Sci.*, **41**, 350–366.

HULET, C. V. (1966). Behavioral, social and psychological factors affecting mating time and breeding efficiency in sheep. *J. Anim. Sci.*, **25**, 5–20.

HULET, C. V., ERCANBRACK, S. K., PRICE, D. A., BLACKWELL, R. L. & WILSON, L. O. (1962a). Mating behavior of the ram in the one-sire pen. *J. Anim. Sci.*, **21**, 857–863.

HULET, C. V., ERCANBRACK, S. K., BLACKWELL, R. L., PRICE, D. A. & WILSON, L. O. (1962b). Mating behavior of the ram in the multi-sire pen. *J. Anim. Sci.*, **21**, 865–869.

HULET, C. V., BLACKWELL, R. L., ERCANBRACK, S. K., PRICE, D. A. & WILSON, L. O. (1962c). Mating behavior of the ewe. *J. Anim. Sci.*, **21**, 870–874.

HUNTER, R. F. (1960). Aims and methods in grazing-behaviour studies on hill pastures. *Proc. Int. Grassld Congr.* (Reading), paper 1B/6, pp. 21–24.

HUNTER, R. F. & DAVIES, G. E. (1963). The effect of method of rearing on the social behaviour of Scottish Blackface hoggets. *Anim. Prod.*, **5**, 183–194.

HUNTER, R. F. & MILNER, C. (1963). The behaviour of individual, related and groups of south country Cheviot hill sheep. *Anim. Behav.*, **11**, 507–513.

IGGO, A. (1951). Spontaneous reflexly elicited contractions of reticulum and rumen in decerebrate sheep. *J. Physiol., Lond.*, **115**, 74–75.

INKSTER, I. J. (1957). The mating behavior of sheep. *Sheepfarming Annual*. Massey Agric. Col., Palmerston North, New Zealand, p. 163.

JEWELL, P. A. (1966). Breeding season and recruitment in some British mammals confined on small islands. In: *Comparative Biology of Reproduction in Mammals*. I. W. Rowland (ed.). Symp. Zool. Soc. Lond. London: Academic Press. No. 15, pp. 89–116.

JOHNSON, W. H., BANKS, E. M. & NALBANDOV, A. V. (1961). Long term effects of hypothalamic lesions on reproduction in sheep. *Fedn. Proc.*, **20**, 186 (Abstract).

KECK, E. (1961). Personal communication, Dept. Poult. Husb., Penn. State Univ., University Park, Pa., U.S.A.

KELLY, R. B. (1937). Studies in fertility of sheep. *Bull. Coun. scient. ind. Res. Aust.*, No. 112.

KIMBLE, G. A. (1961). *Hilgard and Marquis' Conditioning and Learning*. New York: Appleton-Century-Crofts.

KLOPFER, P. H., ADAMS, D. K. & KLOPFER, M. S. (1964). Maternal imprinting in goats. *Proc. natn. Acad. Sci. U.S.A.*, **52**, 911–914.

KLOPFER, P. H. & GAMBLE, J. (1966). Maternal "imprinting" in goats: The role of the chemical senses. *Z. Tierpsychol.*, **23**, 588–592.

LABBAN, F. M. (1954). Effect of thyroxine on sexual behaviour in the ewe. *Nature, Lond.*, **174**, 993.

LAMBOURNE, L. J. (1956). *Mating Behaviour*. Proc. Ruakura Farmers' Conference Week, Hamilton, N.Z., *N.Z.D.A.*, pp. 16–20.

LAMOND, D. R. (1962). Anomalies in onset of oestrus after progesterone suppression of oestrous cycles in ewes, associated with introduction of rams. *Nature, Lond.*, **193** 85–86.

LAMOND, D. R. (1964). Seasonal changes in the occurrence of oestrus following progesterone suppression of ovarian function in the Merino ewe. *J. Reprod. Fert.*, **8**, 101–114.

LEHRMAN, D. S. (1961). Hormonal regulation of parental behavior in birds and infrahuman mammals. In: *Sex and Internal Secretions*. W. C. Young (ed.). Baltimore: Williams & Wilkins. 3rd edn. **II**, pp. 1268–1382.

LEMMON, W. B. & PATTERSON, G. H. (1964). Depth perception in sheep: Effects of interrupting the mother-neonate bond. *Science*, **145**, 835–836.

LIDDELL, H. S. (1925). The behavior of sheep and goats in learning a simple maze. *Am. J. Psychol.*, **36**, 544–552.

LIDDELL, H. S. (1954). Conditioning and Emotions. *Scient. Amer.*, **190**, 48–57.

LIDDELL, H. S. (1956). *Emotional Hazards in Animals and Man.* Springfield, Illinois: C. C. Thomas.

LIDDELL, H. S., JAMES, W. T. & ANDERSON, O. D. (1934). The comparative physiology of the conditioned motor reflex. Based on experiments with the pig, dog, sheep, goat and rabbit. *Comp. Psychol. Monogr.*, **11**, Ser. No. 51, p. 89.

LINDSAY, D. R. (1965). The importance of olfactory stimuli in the mating behaviour of the ram. *Anim. Behav.*, **13**, 75–78.

LINDSAY, D. R. & ROBINSON, T. J. (1961a). Studies on the efficiency of mating in the sheep. I. The effect of paddock size and number of rams. *J. agric. Sci.*, **57**, 137–140.

LINDSAY, D. R. & ROBINSON, T. J. (1961b). Studies on the efficiency of mating in the sheep. II. The effect of freedom of rams, paddock size, and age of ewes. *J. agric. Sci.*, **57**, 141–145.

LOFGREEN, G. P., MEYER, J. H. & HULL, J. L. (1957). Behavior pattern of sheep and cattle being fed pasture or silage. *J. Anim. Sci.*, **16**, 773–780.

LOHLE, K. (1954). Sexual excitement with special reference to the male goat, and its influence on semen quality. *Z. Tierzücht. ZüchtBiol.*, **62**, 356–366.

LORENZ, K. (1955). The companion in the bird's world. *Auk*, **54**, 245–273.

McKENZIE, F. F. & TERRILL, C. E. (1937). Oestrus, ovulation and related phenomenon in the ewe. *Mo. Agric. Expt. Sta., Res. Bull.*, 264.

MILLER, N. E. (1951). Learnable drives and rewards. In: *Handbook of Experimental Psychology.* S. S. Stevens (ed.). New York: John Wiley & Sons, pp. 435–472.

MOORE, A. U. (1960). Studies on the formation of the mother-neonate bond in sheep and goats. *Am. Psychol.*, **15**, 413.

MORAG, M. (1967). Influence of diet on behaviour pattern of sheep. *Nature, Lond.*, **213**, 110.

MUNRO, J. (1955). Observations on the suckling and grazing behaviour of lambs. *Agric. Prog.*, **30**, 129–131.

MUNRO, J. (1956). Observations of the suckling behaviour of young lambs. *Brit. J. Anim. Behav.*, **4**, 34–36.

MUNRO, J. (1957). Sleep in sheep. *Proc. Br. Soc. Anim. Prod.*, pp. 71–75.

MUNRO, J. & INKSON, R. H. E. (1957). The effects of different suckling frequencies on the quantity of milk consumed by young lambs. *J. agric. Sci.*, **49**, 169–170.

NELSON, E. A. (1958). Factors influencing semen quality and reproductive efficiency in rams. *Diss. Abstr.*, **18**, 1192 (Abstract 865).

NICHOLS, J. E. (1944). The behaviour of sheep browsing during drought in Western Australia. *Proc. Br. Soc. Anim. Prod.*, pp. 66–73.

PAVLIK, L. G. (1958). Features of conditioned reflex activity in sheep. *Fiziol. Zh. SSSR* (Trans.), **44**, 45–49.

PEIRCE, A. W. (1959). Studies of salt tolerance of sheep. II. The tolerance of sheep for mixtures of sodium chloride and magnesium chloride in the drinking water. *Aust. J. agric. Res.*, **10**, 725–735.

PEPELKO, W. E. & CLEGG, M. T. (1964). Factors affecting recovery of sex drive in the sexually exhausted male sheep, *Ovis aries. Fedn. Proc.*, **23**, 362 (*Fedn. Am. Socs. exp. Biol.* Abstract).

PEPELKO, W. E. & CLEGG, M. T. (1965). Influence of season of the year upon patterns of sexual behavior in male sheep. *J. Anim. Sci.*, **24**, 633–637.

PFAFFMAN, C. (1941). Gustatory afferent impulses. *J. cell. comp. Physiol.*, **17**, 243–258.

QUINLAN, J., STEYN, H. P. & DE VOS, D. (1941). Sex-physiology of sheep. Studies on the nature of the onset on oestrus in ewes following a period of sexual inactivity. *Onderstepoort J. vet. Sci.*, **16**, 243–262.

RADFORD, H. M. (1961). Photoperiodism and sexual activity in Merino ewes. I. The effect of continuous light on the development of sexual activity. *Aust. J. agric. Res.*, **12**, 139–146.

REED, C. A. (1959). Animal domestication in the Prehistoric Near East. *Science*, **130**, 1629–1640.

RICHES, J. H. & WATSON, R. H. (1954). The influence of the introduction of rams on the incidence of oestrus in Merino ewes. *Aust. J. agric. Res.*, **5**, 141–147.

ROBINSON, T. J. (1959). The estrous cycle of the ewe and doe. In: *Reproduction in Domestic*

Animals. H. H. Cole & P. T. Cupps (eds.). New York: Academic Press. **1**, Ch 9, p. 291.

RODIN, I. I. (1940). The influence of thermal and mechanical stimuli on ejaculation into the artificial vagina in rams. *Trud. Lab. inskusst. Osemen. Životn.* (Mosk.), **1**, 46–53.

ROSS, S. & BERG, J. (1956). Stability of food dominance relationship in a flock of goats. *J. Mammal.*, **37**, 129–131.

ROSS, S. & SCOTT, J. P. (1949). Relationship between dominance and control of movement in goats. *J. comp. physiol. Psychol.*, **42**, 75–80.

SCHINCKEL, P. C. (1954a). The effect of the ram on the incidence of oestrus in ewes. *Aust. vet. J.*, **30**, 189–195.

SCHINCKEL, P. C. (1954b). The effect of the presence of the ram on the ovarian activity of the ewe. *Aust. J. agric. Res.*, **5**, 456–469.

SCHNEIDER, B. H., SONI, B. K. & HAM, W. E. (1953). Digestibility and consumption of pasture forage by grazing sheep. *J. Anim. Sci.*, **12**, 722–730.

SCOTT, J. P. (1945). Social behavior, organization and leadership in a small flock of domestic sheep. *Comp. Psychol. Monogr.*, **18**, Ser. No. 96, p. 29.

SCOTT, J. P. (1948). Dominance and the frustration-aggression hypothesis. *Physiol. Zool.*, **21**, 31–39.

SCOTT, J. P. (1960). Comparative social psychology. Ch. 9 in: *Principles of Comparative Psychology*, R. H. Waters, D. A. Rethlingshafer & W. E. Caldwell (eds.). New York: McGraw-Hill.

SHUKLA, D. D. & BHATTACHARYA, A. P. (1953). Seasonal variation in "reaction time" and semen quality of goats. *Indian J. vet. Sci.*, **22**, 179–190.

SIMPSON, H. C., RICE, E. C., STEELE, D. G. & DUTT, R. H. (1959). Summer semen characteristics of rams having free access to a plastic air-conditioned room. *J. Anim. Sci.*, **18**, 1157 (Abstract).

SMITH, F. V. (1965). Instinct and learning in the attachment of lamb and ewe. *Anim. Behav.*, **13**, 84–86.

SMITH, F. V., VAN-TOLLER, C. & BOYES, T. (1966). The "critical period" in the attachment of lambs and ewes. *Anim. Behav.*, **14**, 120–125.

SOJETADO, R. M. (1952). The growth and habits of lambs. *Philipp. Agric.*, **35**, 572–578.

SOKOLOVA, L. M. (1940). A study of conditioned sexual reflexes in rams. *Trud. Lab. inskusst. Osemen. Životn.* (Mosk.), **1**, 23–25.

STEWART, J. C. & SCOTT, J. P. (1947). Lack of correlation between leadership and dominance relationships in a herd of goats. *J. comp. physiol. Psychol.*, **40**, 255–264.

SYKES, J. F. (1955). *Animals and Fowl and Water*. U.S. Dept. Agric. Yearbook, Washington, D.C.

SYMINGTON, R. B. (1961). Studies on the adaptability of three breeds to a tropical environment modified to altitude. V. The annual fluctuation in breeding ability of rams maintained on Rhodesian highveld. *J. agric. Sci.*, **56**, 165–172.

TAYLOR, E. L. (1954). Grazing behaviour and helminthic disease. *Br. J. Anim. Behav.*, **11**, 61–62.

THOMPSON, D. S. & SCHINCKEL, P. G. (1952). Incidence of oestrus in ewes. *Emp. J. exp. Agric.*, **20**, 77.

TOMME, M. F. & ODINEC, R. N. (1940). Energy expenditure of rams during mating. *Sov. Zootekh.*, No. 7, 61–69 (*Anim. Breed. Abstr.*, **9**, 326).

TRIBE, D. E. (1949a). The importance of the sense of smell to grazing sheep. *J. agric. Sci.*, **39**, 309–312.

TRIBE, D. E. (1949b). Some seasonal observations on the behaviour of sheep. *Emp. J. exp. Agric.*, **17**, 105–115.

TRIBE, D. E. (1950). Influence of pregnancy and social facilitation on the behaviour of grazing sheep. *Nature, Lond.*, **166**, 74.

TRIBE, D. E. (1955). The behaviour of grazing animals. In: *Recent Progress in the Physiology of Farm Animals*. J. Hammond (ed.). London: Butterworths. Vol. 2, p. 585.

TRIBE, D. E. & TRIBE, E. M. (1949). North Ronaldshay sheep. *Scott. Agric.*, **29**, 105–108.

UNDERWOOD, E. J., SHIER, F. L. & DAVENPORT, N. (1944). Studies in sheep husbandry in Western Australia. The breeding season of Merino crossbred and British breed ewes in the Agricultural Districts. *J. Dep. Agric. West. Aust.*, **21**, 135.

WALLACE, L. R. (1948). The growth of lambs before and after birth in relation to the level of nutrition. *J. agric. Sci.*, **38**, 93–153, 243–302, 367–401.

WARWICK, B. L. & BERRY, R. O. (1949). Inter-generic and intra-specific embryo transfers in sheep and goats. *J. Hered.*, **40**, 297–306.

WEIR, W. C. & TORRELL, D. T. (1959). Selective grazing by sheep as shown by a comparison of the chemical composition of range and pasture forage obtained by hand clipping and that collected by esophageal-fistulated sheep. *J. Anim. Sci.*, **18**, 641–649.

WIGGINS, E. L., TERRILL, C. E. & EMIK, L. O. (1953). Relationships between libido, semen characteristics and fertility in range rams. *J. Anim. Sci.*, **12**, 684–696.

YEATES, N. T. M. (1954). Daylight Changes. In: *Recent Progress in the Physiology of Farm Animals*. J. Hammond (ed.). London: Butterworths, Vol. 1, ch. 8, p. 363.

Chapter 11

The Behaviour of Swine

E. S. E. Hafez and J. P. Signoret[1]

The fossil record of the pig is restricted to the Old World. *Propalcaeo-choerus*, a fairly small, four-toed animal with well-developed canine teeth, represents one of the earliest pig-like artiodactyls which inhabited Europe during Oligocene period. By the late Cenozoic period body size had increased, the canines had become large outcurving tusks, and, although all four toes were retained, only the middle two bore the weight (Colbert, 1955). The pig was a forest-dwelling animal from the beginning of its history. In some parts of the world it has been domesticated for at least 7,000 years (Ensminger, 1960). The European breeds of domestic swine were derived from the local wild pig, *Sus scrofa* (Mellen, 1952). Herds were ranging in pastures and forests, and kept indoors only for fattening. The breeds in the Far East were derived from another wild pig, *Sus vittatus*, a smaller animal with shorter legs and a higher reproduction ability (Mohr, 1960; Zeuner, 1963). The two types interbred readily (Gray, 1954). The modern breeds of pig evolved from different crossings between the two original types. They form a rich diversity of genetic material and more than 200 breeds have been catalogued. Certain wild types of pig-like animals, such as the African bush pig (*Potamochoerus larvatus*) and the wart hog (*Phacochoerus aethiopecus*), have never been domesticated (Asdell, 1946).

Little is known of the behavioural changes imposed by domestication. Feral swine live in herds of 5 to 8, usually under the leadership of a senior boar. The aggressive disposition of the wild boar equips him well for both territoriality and herd protection, but the expression of such behaviour is not well documented. Under domestication the pig has been modified from a pugnacious, free-ranging, foraging beast to a more docile animal which is handled readily in large groups under conditions of confined rearing. The behavioural plasticity of swine is emphasized by the rapidity with which wild pigs have adapted to restricted laboratory conditions (Dettmers, 1959). Swine are raised in many different environments ranging from small, unsheltered, muddy styes to modern insulated buildings, from large grass or legume pastures, to a semi-wild state ranging freely in wooded areas.

For generations the common names pig, or hog, have been synonymous with gluttony, hoarding, dirt, and unwholesome habits. It is probable that this concept of the pig has restricted interest in the behaviour patterns of the species. Most of our knowledge of the behaviour of swine has been gained

[1] During the preparation of this manuscript we have drawn very freely on the first edition authored by E. S. E. Hafez, L. J. Sumption and J. S. Jakway.

from data collected incidentally to research on nutrition, breeding, physiology, and management. Observations of some behaviour patterns have been made simply because most pigs are raised in restricted quarters.

I. ACTIVITY

A. Climatic Adaptation

Piglets are unable to regulate their body temperature until 2 to 3 days post partum. Within a few minutes after birth they begin to huddle together to conserve body heat (Mount, 1960). The inclination towards community heating by huddling persists to adulthood. Pigs as heavy as 200 lb may lie on top of one another in sub-zero temperatures when shelter is limited (Fig. 54).

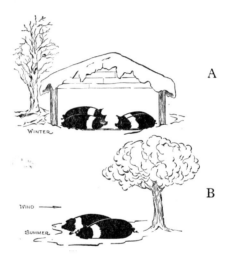

Fig. 54. Climatic adaptation. A. *Community heating*: during cold weather pigs huddle together while on their bellies to conserve body heat. B. During hot weather, pigs lie down on their side with their snouts facing the wind.

Adaptive behaviour under hot, dry conditions has been mentioned previously (see Chapter 4). Evaporative cooling is of major importance for the adaptation of pigs to hot environments (Heitman & Hughes, 1949). If they have access to soil, they root up the ground and lie on the cooler subsoil. If a water pool is present, they root around in it and wallow to cool themselves (Plate VIII). The use of wallows or sprinkling devices will lower body temperature and thereby maintain comfort and performance during hot seasons (Bray & Singletary, 1948).

B. The Use of the Territory

If pigs are not provided with shelters for nesting, they establish a definite area for this purpose. Territorial behaviour may be exhibited in shelter seeking. Newcomers to an established group are usually segregated in a separate part of a large shelter or forced to use a separate shelter. When large numbers of pigs are brought together in stockyards, they bed down in groups according to the farms from which they originated (Self, 1961). Littermates also tend to nest together. Laboratory-reared pigs are inept in shelter seeking. It is necessary to shut them in their house for several days before allowing them free range. Otherwise they may lie in the open, even when it is cold or raining.

Eliminative behaviour does not take place anywhere in the territory: swine are probably the cleanest and most orderly of the farm animals when circumstances permit the expression of their normal patterns of elimination. They keep their sleeping area as clean and dry as possible, moving away from the nest to urinate and defaecate. When confined to a relatively small pen, they will use one corner for this purpose. If housed in shelters on pasture, they usually move out of the shelters for elimination. This tendency is encouraged by placing feed and water at some distance from the house.

When a shelter is not occupied to capacity, pigs may use a corner of the house for elimination. If this habit develops when they are small, some pigs may be forced to bed down in the soiled area as they grow larger and occupy more floor space. The natural eliminative pattern is disturbed when pigs are raised in crowded, indoor pens with less than 9 sq. ft of floor space per animal (Rippel, 1960). The degree to which pigs will control elimination is illustrated by the behaviour of pregnant females. Sows are frequently confined in maternity pens just before parturition and during nursing. It is common to move them to a feeding floor several times daily to be fed and watered. In a few days, most of their excreta will be deposited on the feeding floor, rather than in the maternity pens.

The significance of "maternal training" is underscored by the behaviour of a large number of pigs reared without their dams. The early pattern of elimination was seemingly random; the nesting area, and consequently the pigs, were continually dirty. This tendency persisted in certain of the females that were raised to adulthood. "Normal" eliminative behaviour was quickly re-established in a majority of the first generation of naturally reared pigs. However, litters reared by "dirty" sows tended to behave similarly to the laboratory-reared pigs. The ability of swine to adapt to a wide range of environmental conditions is reflected in eliminative behaviour. Under the severe stress of high temperature, pigs urinate and attempt to wallow in their urine, thus augmenting evaporative cooling (Heitman & Hughes, 1949).

Micturition occurs during the preliminary phases of sexual behaviour in both sexes. In the wart hog, micturition seems to have a role in territorial marking (Frädrich, 1965).

Eliminative behaviour and management. Gates can be used in a permanent structure to restrict the size of the nesting area so that the animals are forced to eliminate elsewhere. Some buildings are constructed with a dunging alley in which the animals are trained to excrete. Similarly, on pastures one can provide small temporary shelters to which additions are made to provide only sufficient space for nesting. Patterns of elimination are influenced by the location of food and water. Generally, much of the excreta is deposited near the water source.

C. Activity Patterns

Swine are not well adapted to running. In the wild pig, trotting is used for long range displacements and galloping for short periods during emergencies.

The activity pattern of adult swine is largely diurnal. In temperate climates, pigs will feed and move around at night during the hottest weather. During cooler months they are quiet at night and active during daylight hours (Decker, 1961). In tropical or subtropical environments they are active at night throughout the year (Arcelay, 1961). When pigs are kept indoors and fed on concentrates, they may rest up to 80 per cent of the time (Lips, 1965; Dinusson, 1965); on pastures, they spend most of the day foraging, rooting, and walking (Puhac & Pribicevic, 1954). The nocturnal pattern of activity in the feral pig seems to be a consequence of hunting.

D. Investigatory Behaviour

The insatiable curiosity of the pig (Plate VIII) deserves careful study. When a person approaches a herd an alarm call, or "woof", is given and the animals may scatter in apparent fright. If the intruder stands still or sits down, the pigs invariably return to investigate by smelling, rooting, and nibbling. They are particularly apt to bite rubber boots or root at gates and doors that could allow them access to an unexplored area. When large groups of pigs are raised in close confinement, on concrete, they have little area to explore and a limited distance to travel for food and shelter. Rooting, nibbling and chewing is prevented. In such a situation, destructive activities like tail biting may occur. The frequent distribution of fresh straw reduces the occurrence of this abnormal behaviour (Van Putten, 1967). A play object, such as a chain or rubber hose, in each pen will also occupy the group's attention and may minimize tail biting.

In moving about their pen, individually reared pigs are influenced by the activity of animals in adjacent pens. Altmann (1941a) described the development of rigid patterns of spontaneous activity when pigs were kept in adjacent individual runways. They limited their walking to a few pathways. The drives for food and shelter were the major determining factors in the orientation of their paths. However, the presence of pigs in neighbouring pens caused them to deviate from the shortest route.

E. Temperament and Emotionality

Breed differences in temperament are well known. Simpson (1907) contrasted the activity of the "dull plethoric" Duroc with the "vitalistic"

PLATE VIII

a. Investigatory behaviour: Four-week-old pigs are sniffing and nibbling at an unfamiliar object, a piece of sod thrown into the pen.

b. Ingestive behaviour: Hampshire piglets, as yet unweaned, learning to take solid food along with their mother.

c. Rooting and wallowing: As a means of temperature regulation, the animals are keeping moist on a hot day.

PLATE IX. SEXUAL BEHAVIOUR

a. Preliminary sexual behaviour: The sow stands quietly while the boar sniffs and nuzzles at her sides and flanks.

b. Mount: The male has urinated and then mounted: the penis has not yet been unsheathed.

c. Abnormal sexual behaviour: Head mount between two yearling Large White boars kept in groups of males from weaning.

Tamworth. Hodgson (1935) found genetic segregation in temperament among a series of highly inbred lines. One line became aggressive, while sows of a second line were placid to the point of utter indifference towards their litters. There was a high percentage of neonatal mortality in both lines. Sows of the former line were so nervous that piglets were trampled to death, whereas in the latter line sows were so lethargic that they did not respond to the squeals when they accidentally lay on their young. Susceptibility to alarm may be heritable (Dawson & Revens, 1946). Considerable individual variation occurred among 42 sows in the time taken to return to a feed trough after being frightened away by an electric sparking device. The within-litter variation in response was not significant. Susceptibility to alarm, so measured, was not associated with the tendency of sows to crush their young by careless nesting behaviour. Strain, sex, season, and individual variations exist in the "open field behaviour" of pigs (Beilharz & Cox, 1967b).

Marcuse & Moore (1944), Moore & Marcuse (1945) showed that both increased and decreased physical restriction may produce disordered behaviour. The effective direction of change depends upon the animal. One sow would come out of her pen when the experimenter whistled, but violently resisted restraint. Although it was necessary to force a second sow from her pen, she stood quietly in restraint. Swine on a high dietary plane are less emotional and learn more readily than those on an inadequate feeding régime (Audley & Klopfer, 1953). Stress reactions may occur when pigs are transported or penned together for the first time (Cena, 1965). Some breeds, such as the Belgian "Pietrain", are much affected by transport stress. How temperament may affect learning ability is brought out in the next section.

F. Sensory Capacities and Communication

Olfactory and auditory cues play a role in the feeding behaviour of feral pigs (Bobak, 1957). The light wave-lengths to which pigs are maximally sensitive are slightly lower than those for humans. Within the range of 465 to 680 millimicrons, pigs can distinguish wave-length differences as small as 20 millimicrons (Seymore & Klopfer, 1960). If strong responses to non-visual cues for food are developed early, later on vision will not be used by the animal in this context (Klopfer, 1961b). Wesley (1955) found that 3 of 16 pigs distinguished between a black and a white card for a food reward if they were stopped for 5 seconds about 2 feet from the stimulus cards. Animals allowed immediate access to the cards and those halted about 4 feet from the cards did not learn. The three successful animals were quiet during arrest and orientated towards the stimulus. The unsuccessful animals were restless during this period.

Audition. Hearing is well developed in feral pigs. Since the ears are relatively short and immovable, localization of sounds is made by movements of the head.

Marcuse & Moore (1946) trained pigs to lift a lid for food in response to the higher of two tones, 480 and 1,000 cycles, respectively. The upper tone was then reduced in 3- to 4-cycle increments. Until the tones differed by 54 cycles, the animals' general activity was lower during the positive signal and the lid was raised appropriately. Until the tones differed by 24 cycles, the vocal response to the positive signal was a series of short grunts as the lid was raised, whereas during the negative signal it was a high-pitched squeal. There was no discrimination when the tones differed by only 22 cycles.

Acoustical signals seem to be of major importance in the organization of social behaviour. Grauvogl (1958) has described more than 20 different

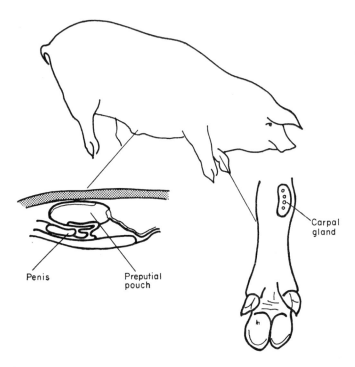

Fig. 55. The specialized scent or musk glands of the pig. Carpal glands occur in both sexes. Preputial glands occur only in the boar and are responsible for his odour.

sounds emitted during resting, social, agonistic, feeding, play, maternal and sexual behaviour.

Smell. The sense of smell is well developed, both in feral and domestic pigs. Rooting for buried food is accurately oriented by olfactory cues. In the Perigord Region of France, pigs are trained to search for truffles (subterranean mushrooms). Smell is also effective in discriminative learning. Although sows did not learn in 500 trials to discriminate among cards differing in form, colour, or size (Klopfer & Wesley, 1954), they learned in 10 to 20 trials to respond correctly to one of 3 neutral gray cards from odours they deposited during the trial by touching a card with their snouts. The odours were still perceptible to the animals several hours after the trial, and after the stimulus card had been washed.

The pig's characteristic odour is produced by skin gland secretions. The carpal glands are well differentiated on the front legs (Fig. 55). In the male, secretions of the preputial pouch, which gives pork its "boar" odour (Dutt et al., 1959), are involved in sexual behaviour.

G. Learning Ability

Pigs learn easily to open the door of a shelter and to manipulate mechanical devices to reach food and water. They are readily trained to push a panel, lift a lid, or follow an experimenter for food. They have even been used for hunting (Zeuner, 1963). Pavlov (1934) reported that pigs cannot be used for conditioning experiments. However, Yerkes & Coburn (1915) showed that pigs were able to learn the correct response by spatial discrimination when faced with five choices. This ability was demonstrated in four different problem solving situations. The marked influence of the first in an established series of habits (primacy) in the subsequent location of a food reward was shown by Myers (1916). Conditioned responses are easily obtained to sound, colour, or olfactory stimuli, but the response of a pig to a learning situation depends largely on the sensory modality used in the discriminatory process. Pigs develop conditioned responses to sound much earlier than to light (Koniukhova, 1955). Klopfer (1961a) has placed spatial discrimination well before visual discrimination when food is used as a reward. Visual discrimination for the avoidance of an aversive stimulus occurs, however, with only one to two trials. Pigs learned more rapidly to avoid an electric shock (by responding to an auditory stimulus) when one long series of trials was divided into sets which were presented over a period of 4 days (Karas et al., 1961). It is highly probable that the hierarchy of sensory capacities is specific for each situation.

II. INGESTIVE BEHAVIOUR

A. Rooting Behaviour

Ingestive behaviour appears to be associated with the general curiosity of the pig, its tendency to forage, and its strong impulse for rooting. When provided access to the soil, pigs usually root vigorously, and spend much time turning over the surface soil (Plate VIII). There are genetic differences in the tendency to root. For example, of the three Minnesota breeds, the Minnesota No. 2 pigs root most extensively (Fausch, 1961). It is often necessary to insert a ring in the margin of the snout to discourage rooting and the destruction of pasture forage plants.

The olfactory sense is acute and the pig is able to search out subsurface food. If a potential delicacy is detected, a pig will root to a considerable depth to uncover it. Pigs will root up large areas if they discover earthworms in their pasture. Rooting is high in the initial response hierarchy and is easily learned through simple reinforcement. The incidence of rooting can be increased or decreased by reward. Floor-fed animals which walk on their food and then root to loosen it show considerably more rooting on concrete than do trough-fed animals (Klopfer, 1961b).

B. Feeding Behaviour

Swine are omnivorous and their diet may include a wide variety of foods. Feral pigs eat plants, tubers, roots, seeds, grasses, buds, and leaves. They also consume earthworms, caterpillars, and slugs; snakes and frogs; eggs and young birds; rodents; and any ill or dead animal (Bubenik, 1959; Porzig, 1967).

Domestic pigs on pasture spend 6 to 7 hours a day searching for food and eating (Puhac & Pribicevic, 1954). However, the eating time is often less than 10 minutes a day when they are hand-fed with concentrates. With self-feeders, the feeding-time is prolonged (Dinusson, 1965; Lips, 1965).

Qualitative regulation of diet. The presence and relative amounts of certain dietary constituents markedly affect food intake. Baby pigs prefer rations to which sugar or saccharin has been added (Nelson et al., 1953; Aldinger et al., 1959). They also prefer water which contains saccharin (Kare et al., 1961). Salt (Grummer & Bohstedt, 1947) and quinine (Kare et al., 1961) reduce feed intake. The addition of 5 per cent lard to the ration increases the feed consumption of baby pigs (Combs & Wallace, 1959).

The palatability of various protein sources and cereals has been studied by Salmon-Legagneur & Fevrier (1959), Salmon-Legagneur & Aumaitre (1961) and Aumaitre & Salmon-Legagneur (1964). Yeast and fish meal additives increase feed acceptance, whereas meat meal depresses it. Soya cake was the only oil cake which increased feed consumption. Wheat was preferred to rye, barley, oats, and corn.

The amount of cellulose in the diet also influences feeding behaviour: hulling increases the palatability of cereals, and large amounts of alfalfa meal depress consumption.

Pigs eat balanced amounts of corn and protein supplement when the two are offered in separate parts of a self-feeder (Ellis et al., 1936; Lassiter et al., 1955; Hutchinson et al., 1956).

Mixed rations can be presented in a variety of physical forms from fine meals to compressed pellets of different sizes. Pigs sort out and eat palatable parts of meal, particularly if the meal contains particles of different sizes (Gill, 1956). They cannot, however, sort out the less palatable ingredients of pellet feeds. None the less, baby pigs consume six times more pellets than meal of the same composition (Salmon-Legagneur & Fevrier, 1955).

An appraisal of feed preference can be confounded by the arrangement of the experimental equipment. Time lapse cinematography studies reveal that pigs eat most frequently from the feeder closest to the source of water, regardless of its specific contents (Facto et al., 1959).

Quantitative regulation of food intake. The feeding patterns of both weanling and adult pigs are influenced by systems of management. Immediately before the usual feeding time, hand-fed pigs move to the feeding equipment or to where the herdsman enters the area. They scuffle, squeal, and crowd as they compete for food, even when the feeding space and the food supply are adequate. The temporal patterns of ingestion are more casual among self-fed pigs, more nearly geared to individual appetites. Pigs may be stimulated to eat by the movement of others from the nesting area to the feeder. The sight of general activity around the feeders, the sounds created by the manipulation of feeding equipment and electric lights, and occasional vocalization are some of the apparent stimuli.

Group size and interaction of the sexes affect growth and feed consumption (Jonsson, 1955). Among pre-puberal animals fed in groups of four (2 gilts and 2 barrows), barrows grow more rapidly than gilts; the reverse is true among individually fed animals. Daily gains do not differ significantly under the two systems of feeding, but the within litter variance is 4 times larger under group feeding than individual feeding, suggesting influences of dominance order. Pigs also grow more rapidly when there is competition for space at a self-feeder. Rippel (1960) found that the optimum number of feeder cups was one for each five pigs. At a ratio of 1:7, growth rates were satisfactory for lighter weight pigs, but were sharply reduced after pigs reached 130 lb. There was simply insufficient time for each pig to eat enough feed to sustain maximum growth.

As early as 1883, Shelton reported that feed intake increases when pigs are kept outdoors during the winter. In a detailed study of environmental conditions, Heitman & Hughes (1949) showed that the feed consumption of adult pigs decreased as the temperature increased from 40° to 100°F. The weight of the pig has a marked influence on this relationship since for small pigs an increase in temperature from 40° to approximately 55°F may actually increase feed consumption rather than decrease it. Humidity did not alter feeding behaviour appreciably until it rose above 90 per cent. In a cold environment the thermogenesis uses energy from feed that would be converted to meat at higher environmental temperature. Yet, litters reared outdoors are healthier than those reared indoors (McLagan & Thomson, 1950). Feed requirement does not increase during the winter, if nest areas are provided with electrically heated floors (Barber et al., 1955a). Appetite levels can be modified by genetic selection (Fowler & Ensminger, 1960). The Duroc, for example, has a greater appetite and grows more rapidly than the Poland China, although fewer ingested nutrients are converted to muscle (Gregory & Dickerson, 1952).

Feeding equipment and management. To reduce unfavourable competition during feeding, troughs can be subdivided with crossbars or stalls and scattered over a large area. The design of equipment for self-feeding is designed around the rooting habits of the pig. Self-feeders hold a supply of feed in a series of feeding cups, usually protected from the weather by hinged metal lids. When a pig is 3 to 5 weeks old, it learns to root up the lids for feed. The generally recommended floor space requirements for maximum growth to 200 lb is 14 to 20 sq. ft per pig; however, pigs on 7 to 8 sq. ft gain as well as those with 15 to 16 sq. ft (Rippel, 1960; Nofziger, 1961).

C. Drinking Behaviour

Patterns of drinking. Self-fed pigs alternate between eating and drinking until satisfied. When hand-fed they eat until all the feed is gone and drink water afterwards. Young pigs can learn to manipulate mechanical self-watering devices before 2 months of age. A pig will raise a protective lid with its snout, root against a pressure plate, or bite on a pressure valve inside a short length of rubber hose to obtain water.

Water intake. Water consumption changes with age, weight, environmental temperature (Table 33), humidity, and stage of reproduction. As temperature increases from 50°F to 100°F, water consumption increases markedly in pigs weighing 75 lb or more (Heitman & Hughes, 1949). A water:dry feed ratio of 3:1 is recommended, particularly for young, rapidly growing pigs (Braude, 1954).

Table 33. Water Consumption of Swine (lb/animal per hr) as Affected by Environmental Temperature and Live Weight.

Temperature (°F)	Live Weight (lb)		Pregnant Sows
	75–125	275–380	
50	0·20	0·50	0·95
60	0·25	0·50	0·85
70	0·30	0·65	0·80
80	0·30	0·85	0·95
90	0·35	0·65	0·90
100	0·60	0·85	0·80

III. SEXUAL BEHAVIOUR

Under farm conditions there are two general mating systems. With pasture mating, the boars and females are kept together and the nature of courtship and the frequency of mating are unrestricted by man. With hand mating, males and females are penned separately and a boar is brought to the oestrous female. It is common for the herdsman to provide the boar with considerable assistance during mating.

A. Sequence of sexual behaviour

The courtship behaviour lasts only a short time when a boar is placed in a small pen with an oestrous female. The male mounts and ejaculates quickly. The latency of ejaculation is generally less than 3 minutes. However, when females are living freely in a herd, sexual behaviour begins long before the first mating and lasts several days. In this section, we shall be concerned with the detailed description of this behaviour.

1. PRELIMINARY PHASE

Behavioural changes occur in the female several days before the onset of oestrus; she becomes nervous and moves about at the slightest disturbance, while her di-oestrous pen-mates continue to rest. The switching on of electric lights or sounds coming from gates or doors are sufficient to elicit her attention. Pro-oestrous females in a pasture will leave the herd to follow a tractor, or almost any large moving object. When kept indoors, females will use any opportunity to move out of the pen. When pro-oestrous females are tested with boars at regular intervals, they often leave their sleeping quarters to await the arrival of the male, as the regular time of testing approaches. Movements in an open field are more frequent (Signoret, 1968c). Spontaneous activity increases to about twice the normal level during oestrus (Altmann, 1941b) (Fig. 56). Di-oestrous females behave like males: they show

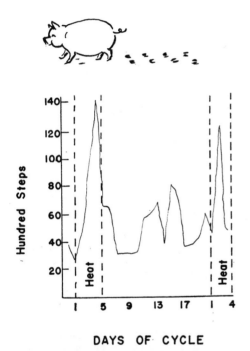

Fig. 56. Activity of sows during oestrus. (Adapted from Altmann, 1941b.)

frequent genital control, nose the flanks of their pen-mates, and attempt to mount. At this time and during oestrus, the sow utters characteristic sounds. The most common vocal response is a soft rhythmic grunt (Grauvogl, 1958).

Meeting of sexual partners. Detailed observations of sexual behaviour in herds during testing of a group of females with a boar indicate that the female has an active role: the boar detects heat largely because of the behaviour of the pro-oestrous or oestrus female. When placed in a group of sows, the male controls each female successively, sometimes pursuing an anoestrous one fiercely. In such a group, the oestrous sow will immediately go to the boar, sniff the anal and preputial regions, and remain close to him. In fact, some females actively pursue males on the day prior to oestrus (Jakway & Sumption, 1959). T-maze experiments indicate that the male attracts the oestrous and pro-oestrous female (Signoret, 1967b, 1968a). The sow shows a definite orientation towards the boar 1 day before oestrus.

12*

This reaction becomes more intense during oestrus and lasts for two days after oestrus (Fig. 57; Table 34). On the other hand, the male shows little preference in a similar choice experiment between an oestrous and an anoestrous sow. These results confirm the herd observations on the critical role played by the female in the meeting of sexual partners.

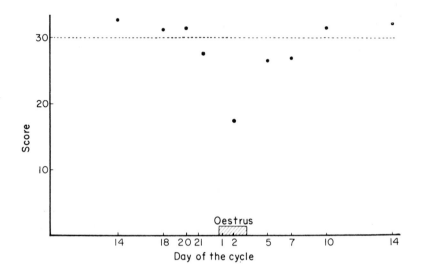

Fig. 57. The reactions of a sow to a boar in a T-maze at various times during the oestrous cycle. A score below 30 indicates that the female is attracted by the boar: this is particularly noticeable during oestrus, but also occurs on the day before and several days after oestrus.

Table 34. Experimental Study of the Attraction of the Female by the Male. Females were placed in a T-maze for 5 minutes and given a choice between a Male, a Female, or an Isolated Stimulus. The Time Spent near each Stimulus was Recorded

Female Subject	Stimulus	Time Spent near the Stimulus (*Seconds*)
Oestrous gilt	Boar	243
	Gilt	34
Anoestrous gilt	Boar	69
	Gilt	67
Oestrous female	Boar castrated as an adult	165
	Odour of preputial secretion at 40°C	43
	Broadcasting of boar "courting song"	59

2. Precopulatory Patterns

Both male and female may exhibit some preliminary courtship behaviour before mating. The sequence of events, which occur in precopulatory behaviour patterns, is presented schematically in Fig. 58.

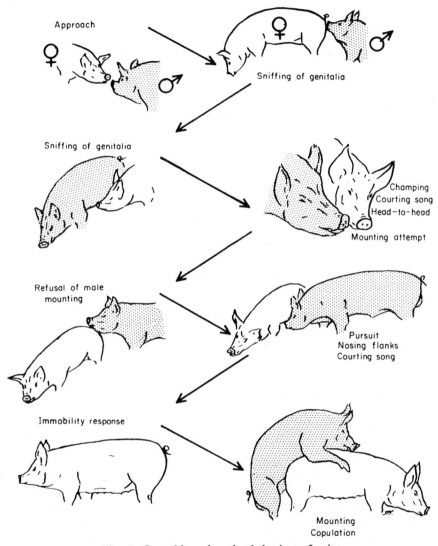

Fig. 58. Courtship and mating behaviour of swine.

When a male approaches an oestrous female which runs from him, he will follow her persistently and attempt to herd her to a standstill. Should the opportunity afford itself, he will nose her sides, flanks, and vulva during the pursuit. During this activity, and much of what follows, the boar emits "mating song" (chant de cour), a regular series of soft, guttural grunts, about 6 to 8 per second at an intensity of 85 to 95 decibels at a distance of 1 metre; he pauses every 15 to 20 sounds to catch his breath (Signoret *et al.*, 1961). A male rarely uses the "mating song" when he approaches a stationary oestrous female. He lightly presses his nose against her head, shoulder, or flank, and as he reaches her genito-anal region, his nuzzling generally becomes more vigorous. When he sniffs her genitalia, he often pokes his snout between her hind legs and, with a sudden jerk, lifts her hindquarters into the air. During this phase of courtship, the boar grunts continuously, grinds his teeth, moves his jaws from side to side, and foams at the mouth. A rhythmic emission of urine is characteristic of the boar during sexual arousal and occurs during the preliminary phases of sexual behaviour. The female shows a special interest in the male and nuzzles his scrotum, flanks, and genitals. She bites his ears lightly and the animals often remain head-to-head. The female also sometimes tries to mount the male. Mutual mounting attempts or genital sniffing leads to the inverted (reversed) parallel position frequently described in other ungulates. The female urinates more frequently during this phase of sexual behaviour (Quinto, 1957). This attracts the male, who sniffs at the sow's urine, but exhibits no characteristic behaviour pattern such as the "flehmen" of other ungulates. The response of the sow which is willing to mate is the standing reaction or mating stance. This is finally adopted by the female and ends the preliminary phase of sexual behaviour by allowing the boar to mount and copulate. The sow becomes stationary, arches her back, cocks her ears, and is rigid in every limb (Plate IX). In this stance, she is moved only with great effort, and cannot be herded in the ordinary way.

All elements of the courtship pattern are not displayed in all circumstances; considerable variation occurs, for example, in the frequency and duration of pursuit and nuzzling. During extremely hot weather, sex drive declines, and boars cease to mate about mid-morning and do not resume such activity until dusk.

Certain variations of frequency and intensity of behaviour are attributable to experimental manipulation although, as yet, there are no data on free-ranging bisexual herds. When males and females were penned separately except for daily test periods, courtship behaviour was affected by the testing procedure (Jakway & Sumption, 1959).

The duration of the chase depended not only on the level of the boar's behaviour, but also on the female's receptivity. Should more than one boar pursue a single female, seemingly unwitting co-operation in herding often shortened the chase for one. Courtship time was reduced when each of the same males was placed in a small enclosure with a single oestrous female.

Females with the lowest level of receptivity elicited the most prolonged pursuit before mating. Their shrill squeals tended to draw the attention of the boars away from other oestrous females. The relative sizes of the male and female also affect sexual receptivity because small oestrous females may be unable to support large males. An oestrous female may also resist when the dewclaws of clumsy boars have caused painful injuries to her back or flanks (Hancock, 1961). In addition, receptivity may be reduced if the boar becomes aggressive toward the female or uses his tusks playfully during courtship. An unknown environment may similarly have an adverse effect. Boars have refused to mate fully receptive females in a strange pen, but mounted readily when returned to their own area.

Mate preference. Some sows seem to be more attractive to boars than others, and some females are exceptionally receptive to specific boars (Hodgson, 1935). The expression of mating preference by one Yorkshire (white) gilt in a group of 20 for a boar of another colour was noticed by Jakway & Sumption (1959). A female may also occasionally display a decided aversion to a particular boar. This may be associated with the order, frequency, and/or intensity of the elements in his mating pattern. Different feeds apparently produce different odours in pigs; olfactory cues may be involved in mate preference, since boars fed on garbage may be rejected by oestrous sows fed on grains.

3. MOUNTING

The boar mounts a female in the mating stance immediately. Jakway & Sumption (1959) note that ejaculation began within 200 seconds in 54 out of 70 tests. In our own records of 202 natural matings (Signoret, 1968b), the average mount latency was 43 seconds; during semen collection with a dummy, 3,260 ejaculates were obtained in an average of 72 seconds. Only 17·4 per cent of fully receptive females are not mounted immediately when presented to a boar (Schenk, 1967).

Some boars mount and dismount a female repeatedly before coitus; other boars mount once and copulate. During mounting, the boar swings his trunk upwards to a nearly vertical position and rests his forefeet on the female (Plate IX). If he faces her, he usually dismounts without proceeding further, although occasionally a boar will thrust his pelvis against the female's side or head. Usually, if he approaches from the rear, he next takes a few quick steps forward, rests his trunk along her back, and grasps her firmly between his forelegs. Should he not dismount, he begins a rhythmic series of pelvic thrusts, champing his jaws and foaming frothy saliva at the mouth. The frequency and duration of mounting for each mating depends on the receptivity of the female, her ability to support the weight of the male, and the dominance interactions among boars. During multiple sire mating, it is not unusual to see a female being mated by one male, while one or two other boars are mounting her head or sides.

Although boars occasionally attempt posterior mounts with di-oestrous females, mounts elsewhere on the unreceptive sow are virtually non-existent (Jakway & Sumption, 1959). The rate of posterior mounting with thrusts increases rapidly on the first day of oestrus to reach a maximum on the second day. The frequency of mounting of the forequarters follows a similar pattern. The mounting of the forequarters is more common in young boars than in those with sexual experience (Self, 1961).

4. Intromission and Ejaculation

The mounting boar thrusts until the tip of the partially exposed penis penetrates either the vaginal or rectal orifice. Only then is the penis fully unsheathed. The penis is of fibro-elastic type, spirally shaped with a pre-scrotal sigmoid flexure, and does not enlarge during erection (Sisson & Grossman, 1953).

When intromission is vaginal, the boar seldom withdraws or dismounts and ejaculation occurs. A series of pelvic thrusts may occur after a rectal intromission, but the penis is most often withdrawn without ejaculation. Abortive ejaculation may take place if the sow refuses intromission or if the boar fails to penetrate either orifice. During ejaculation, the boar's haunches are clenched together and pressed forward, and a muscular wavelike movement of the perineum is visible as anal winking. One of the testes is retracted so that a visible contraction occurs on one side of the scrotum. Two or three separate waves of semen may be ejaculated before the boar dismounts and slides limply from the back of the female. The penis is retracted into the sheath within several seconds.

Among farm animals, the boar has the longest ejaculation time. Copulation is performed within 3 to 20 minutes (Jakway & Sumption, 1959; Self, 1961) with an average of 4 to 5 minutes (Burger, 1952; Bascos, 1953; Signoret, 1968b). The female generally remains completely immobile until the boar dismounts. The female sometimes (18 per cent of observed cases) moves at the end of copulation, but this seldom disturbs ejaculation (Schenk, 1967). After mating, the sow remains by the boar and often licks the flocculent discharge which accumulates on his penis.

When Burger (1952) allowed each of the two boars free access to a female during a complete oestrous period, they mated 7 and 11 times respectively; the interval between matings ranged from 0·2 to 15·4 hours (Fig. 59). In similar circumstances, other boars mated 4 to 8 times (Quinto, 1957) on an average of 6·6 times (Bascos, 1953). When a group of boars was allowed access to a group of females each morning until 30 minutes after the cessation of sexual activity, each of three boars ejaculated 8 times during a 2·25 hour test, in which 9 oestrous females were available (Jakway & Sumption, 1959). For some boars, the number of ejaculations per test depended on the number of available oestrous females; for other males, the two factors were unrelated. Ejaculation frequency was inversely related to the per cent of rectal ejaculations.

Presentation of a new oestrous sow was the most successful stimulus in eliciting repeated ejaculation, as described in cattle by Hale and Almquist (1960). During 40 minute observations, sexual activity was not renewed after arousal or physical activity if the boar remained with the same female. The boar ejaculated immediately when a new female was presented. An attempt was made to find out how often this response could be elicited by presenting a new sow to boars immediately after each ejaculation, or after 30 minutes, whichever occurred first. Up to five successive ejaculations took place; the latency of ejaculation, but not the duration, was prolonged with successive ejaculations.

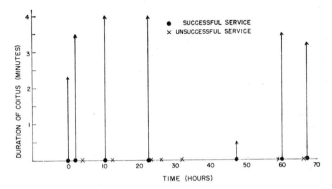

Fig. 59. Frequency and duration of copulation of a Large White boar when allowed free access to a single sow during one oestrous period. (Data from Burger, 1952.)

5. Competition Among Males

Competition during mating affects individual boars differently. Some boars will mate less frequently when tested in pairs than when tested alone, whereas others do not. Competition is seldom directly aggressive. If the female is approached by a second boar during courtship, one of the pair may lunge toward the competitor giving a short threatening grunt, but a fight rarely ensues. Competition more often takes the form of harassment than of a serious challenge. If a female is being mated in the presence of a second boar, the latter disturbs copulation by nuzzling and nibbling at the face and sides of the female until she becomes skittish and moves about. If the challenging male ranks high in the dominance order, he may mount her head or side, and slide around towards her rump, pushing until the copulating male releases his hold and dismounts.

B. Sexual Function in the Male

1. Puberty

The boar generally reaches puberty before 220 days of age (Wiggins *et al.*, 1951; Hauser *et al.*, 1952; Lagerlöf & Carlquist, 1958). The age at physiological maturity (i.e. presence of spermatozoa capable of fertilization) may differ from the time when the essential elements of mating behaviour appear. Spermatozoa are present in the testis as early as 110 to 125 days of age (Andrews & Warwick, 1949; Hauser *et al.*, 1952), but there is some delay before gametes are capable of fertilizing ova (Wiggins *et al.*, 1951). A co-ordinated behavioural response to oestrous females is seldom evident at less than 130 days of age (Table 35). Some boars mount and mate before spermatozoa are present in the ejaculate. The first ejaculation generally occurs when the male is between 5 and 8 months old (Niwa *et al.*, 1959). However, testis weight and sperm output increase until the age of 1 year.

Table 35. Sexual Functions in the Boar

Patterns	Values
Age at first mating	5–8 months
Mounting latency	30–70 seconds
Duration of ejaculation	5 minutes
Semen volume[1]	300 ml
Gel volume[1]	40 ml
Number of sperm in the ejaculate[1]	1×10^{11}

[1] Highly variable according to how often semen is collected. The figures given here are for one collection per week.

2. STIMULI RELEASING THE MALE SEXUAL RESPONSE

All the patterns of courtship behaviour may be released in the boar by a di-oestrous female, another male, or a variety of inanimate objects ranging from dummy sows to oversimplified devices for semen collection looking more like a table than an artificial female (Ito *et al.*, 1948; Aamdal *et al.*, 1958; Paredis, 1961). According to Büchlmann's (1950) hypothesis for cattle, the main stimulus which releases mounting behaviour in the boar is a visual one: an immobile structure looking more or less like a conspecific animal. This mechanism seems to have an innate basis: young boars reared in groups of males will readily mount a simplified dummy at the first opportunity. If ejaculation is achieved, i.e. if semen is collected in an artificial vagina, the response will be retained. However, visual stimuli from the shape of the body, head, and ears are increasing the value of dummies (Rothe, 1963). A young pig tethered under the dummy has the same effect (Smidt, 1965). The male readily identifies the oestrous female in the herd, but when tested in a T-maze with an oestrous and di-oestrous sow, he shows no preference. The behavioural response of the female, namely the standing reaction, seems to have major importance in evoking male sexual reflexes (Signoret, 1968b).

3. ABNORMALITIES IN SEXUAL BEHAVIOUR

The mounting reactions of the male, released by a simple visual stimulus, may explain the abnormal mounting (Plate IX) and homosexuality which are frequently observed in swine.

Boars reared in pairs or all-male groups often form stable homosexual relationships. Such relationships persist for many months, although both members repeatedly copulate with females (Jakway & Sumption, 1959). The identity of the active and passive partner is maintained. The behaviour pattern of the aggressor differs not at all from the pattern which occurs in normal heterosexual copulation and may include rectal intromission and ejaculation while the passive male stands quietly. If such a pairing is well established, homosexual coitus occurs even in the presence of oestrous females. It is probable, at least for the aggressor, that a pairing is reinforced by the rank of its members in the male social hierarchy. Although it has not been documented, the possibility of promiscuous homosexuality should not be excluded.

The so-called bizarre sexual objects may take their origin in the same mechanism: if the sex-drive is sufficiently high, an oversimplified schema may release the sexual response. Some boars for example, will mount a watering device, a straw pack, etc.

4. Ejaculation and Semen Criteria

The stimuli which cause ejaculation appear to be essentially tactile. The fixation of the spiral shaped penis in the cervix seems to have major importance. This is obtained in an artificial vagina with increasing pressure (Melrose & O'Hagan, 1959; Niwa et al., 1959). Thermal stimuli are unnecessary: ejaculation occurs without warming the artificial vagina (McKenzie, 1931; Ito et al., 1948). Boar semen is charactersized by its large volume and relatively low concentration and high total number of spermatozoa (Table 35). The presence of a gel-like material should also be noted. Semen is ejaculated in different fractions: a preliminary one; one rich in sperm; and a third one with the largest volume, but a low concentration of sperm. Gel-like material is emitted throughout ejaculation. Sperm emission through a fistula in the vas deferens occurs only during collection of the rich fraction (Wierzbowski & Wierzchos, 1968).

5. Genetic Factors

The inheritance of patterns and intensity of sexual behaviour is evident when clear-cut differences between breeds are demonstrated. Hauser et al. (1952) reported differences in sex drive among several breeds and their crosses. In a series of group studies, Wiggan (1961) found that Hampshire boars mounted and ejaculated twice as frequently as the Minnesota No. 1 or the Yorkshire. Critical studies employing quantitative methods are needed to determine the genetic mechanisms involved in the expression of sexual behaviour.

The broader differences between breeds are brought out clearly only when experiential and environmental factors remain constant, and individul variation is consequently minimal. Even when both environment and experience are carefully controlled, variation within any one breed is considerable.

C. Sexual Functions in Females

1. Puberty and Cycle

Puberty. Most gilts reach puberty at 6 to 8 months of age; breeds differ in this respect and heterosis is expressed by earlier maturity of some breed crosses (Foote et al., 1956). Gilts may be either investigative or antagonistic towards boars before puberty, but become sexually interested in them during their first pro-oestrus. None the less, some prepubertal gilts exhibit such sexual behaviour patterns as mounting and restlessness, although they never allow a boar to mount (Schmidt & Breitschneider, 1954; Ito *et al.*, 1960).

The length of the normal oestrous cycle is 21 days, with some variation according to breed (reviewed by Boda, 1959; Signoret, 1967a). Once oestrous cycles begin, they continue to occur regularly, unless the sow becomes pregnant. "Abnormal" cycles longer than 30 days represent only 5 per cent of the 1800 cycles observed by Signoret (1967a). They increase to 10 to 20 per cent among sows which were not successfully mated or inseminated during the summer, which suggests that they have a physiological origin, such as embryonic death (Corteel *et al.*, 1964).

During pregnancy, females elicit limited sexual responses; they may nuzzle and mount oestrous females. Pregnant females occasionally exhibit one or more "false heats" at about 20-day intervals after their first mating and yet subsequently produce a litter from that mating.

There is no evidence of seasonal changes in the female sexual cycles of domestic breeds (Burger, 1952; Signoret, 1967a). Feral swine reportedly have a breeding season in November–December, although this may begin as early as July and continue until February. No detailed study of this surprising difference between domestic and the feral swine has apparently been made.

Fragmentary evidence suggests that a sudden alteration in management, such as feeding régime or movement of females from one pasture or one farm to another, may cause a change in the oestrous cycle (Dziuk, 1960). Thirty per cent of prepubertal gilts exhibit their first oestrus on the 4th, 5th and 6th days after transport (du Mesnil du Buisson & Signoret, 1962). Similar upsets occur among heifers (Hafez, 1961).

Oestrus. Burger (1952) found no diurnal pattern in the onset of heat. Oestrus normally lasts 40 to 80 hours (Signoret, 1967a), its length varying with breed, reproductive cycle, and season (cf. Table 36). The duration of

Table 36. Variations in the Duration of Oestrus in Swine

Cause of Variation	Duration (Hours)
Puberty (Signoret, 1967a)	
First oestrus	47
Other oestrous periods	56
Genotype (Burger, 1952)	
Large White	50
Large Black	68
Season (Signoret 1967a)	
Winter	53
Spring	55
Summer	59
Autumn	57

PLATE X. FARROWING AND SUCKLING BEHAVIOUR

a. (left) *Farrowing behaviour:* The umbilical cord is still attached; the placenta has not been expelled; the search for a teat has already begun.

b. Nosing phase: Competition between two piglets for the same nipple.

c. Quiet phase: The young are completely immobile with their ears drawn back.

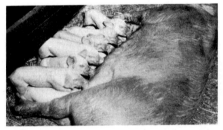

d. True suckling: Full milk flow has begun, and piglets are intent on nursing as fast as possible.

e. Suckling order and teat preference: Older piglets (white) are fostered to a nursing Hampshire sow with younger pigs that have not established a suckling order. The older foster pigs appropriate the preferred pectoral nipples.

PLATE XI. AGONISTIC BEHAVIOUR

a. Threat: An adult Miniature boar threatens a rival. Note that the animal produces foaming saliva (champing) and scratches at the straw.

b. Circling: The boars circle shoulder to shoulder, exerting continual side pressure.

c. (above) *Shoulder-to-shoulder contact:* Each of two adult Large White sows tries to bite the opponent's flank or shoulder.

d. (right) *Shoulder-to-shoulder contact:* Shoulder-to-shoulder pressure forces one of two adult Large White sows into a sitting position. Her opponent takes this opportunity to nip at her ear.

oestrus is not apparently affected by mating or insemination according to Burger (1952) and Signoret (1967a), although other authors report a shorter heat after mating (Lebedev & Pitkjanen, 1951; Grauvogl, 1958).

Endocrine balance of female sexual behaviour. In the ovariectomized sow, one oestrogen injection induces normal sexual behaviour. The response is unaltered after cyclic treatment for one year. The duration of oestrus increases logarithmically with the dose of oestrogen (Fig. 60), but the delay

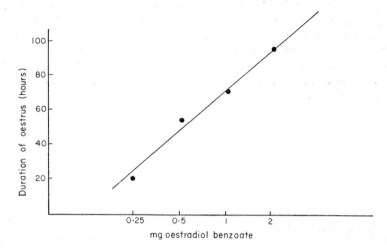

Fig. 60. The effects of oestradiol benzoate injections on ovariectomized sows. The duration of oestrus is proportional to the logarithm of the dose of hormone.

between injection and the onset of oestrous behaviour is unchanged (about 70 hours). A dose of 0·6 to 0·7 mg of oestradiol benzoate induces an oestrous period of normal length in gilts weighing approximately 120 kg. Individual and seasonal variations also occur in the duration of heat. The synergistic action of progesterone and oestrogen reported in the ewe by Robinson (1954–1956) does not exist in the sow, although sexual behaviour can be inhibited by progesterone injected after oestrogen (Signoret, 1968c).

2. Signs of Oestrus

Restlessness, capricious appetite, mounting of other females, and frequent attempts to urinate, particularly in the presence of a boar, are characteristic behavioural signs of oestrus. Spontaneous activity reaches about twice the normal level during late oestrus (Altmann, 1941b) (Fig. 56).

The vulva begins to enlarge 2 to 8 days before the onset of heat. It becomes red and a mucous discharge appears. Changes in basal body temperature and the vagina cannot be directly correlated with oestrous behaviour (Altmann, 1941b; Sanders *et al.*, 1964). Arborization of cervical mucus (Betteridge & Raeside, 1962) and changes in vaginal pH (Schilling & Rostel, 1964) are reportedly good physiological indicators of oestrus.

3. Searching for the Male

A tendency to go to a male in a T-maze appears in oestrogen-injected, spayed females, and the attraction of the male increases with the dose of hormone (Signoret, 1967b). The stimuli involved in this attraction have not been determined. The adult castrated boar is not as effective as a normal male, but the isolated odour and courting song do not influence the sow (Table 34).

The response persists in gilts reared in isolation from birth: they do not react to boar odour or sounds when spayed and injected with oestrogen. When they are presented to a boar after hormone treatment, they go to him immediately. On the other hand, if females are spayed, but not injected with hormone, they react identically towards a male or a female at first contact. The attraction for the male is apparently endogenous and not acquired by social experience.

Table 37. The Effect of a Variety of Stimuli on the Receptivity of the Gilt.

Conditions under Which Pressure Was Exerted on the Female's Back	Stimuli Presented	Per Cent Positive Responses[1]
Females alone	None	48
Females separated from boars by a wooden wall	Olfactory and auditory	90
Females separated from boars by wire fence	Olfactory, auditory, and visual	97
Females with boars	All	100
Broadcasting of boar courting song	Auditory	71
Broadcasting of an artificial signal	Auditory	61
Odour of preputial secretion at 40°C	Olfactory	81

[1] The assumption of a mating stance when a herdsman pressed on the back of the female.

4. Immobility Reaction

When an oestrous female is touched even lightly on the back by a boar or by a handler, she assumes a stationary position (the mating stance or immobility response). Every oestrous female assumes this stance more or less quickly when presented to a boar; however, only 48 per cent of the oestrous gilts tested by the herdsman do so in the absence of a male. The frequency of response increases from 40 per cent before the 10th hour of oestrus to 60 per cent between the 24th and 36th hours.

Gilts reared in isolation from birth react in the same way that normal females do when tested after ovariectomy and hormonal treatment: some become immobile immediately, whereas others do so only when presented to a boar. Schenk (1967) was unable to demonstrate any difference between gilts and sows.

Signoret (1961) has determined the stimuli which initiate the immobility response. Olfactory and auditory stimuli appear to be most important; 90 per cent of the oestrous gilts tested responded in the presence of these stimuli only (Table 37). The addition of visual and tactile stimuli increased the number responding by 7 and 3 per cent, respectively. Oestrous gilts do not respond to hand pressure.

When a recording of the mating chant was broadcast during the test, 50 to 60 per cent of the females assumed the mating stance. The number of positive responses decreased to 9 per cent when the rhythm of grunts was reduced by one-half. Twenty-six per cent of the gilts responded positively to an entirely artificial courting song. Hence, the rhythm of the boar's "chant de cour" appears to be an important cue.

The odour of a pen from which a boar has been recently removed also elicits the immobility response in 62 per cent of oestrous gilts. The odour of preputial secretion has the same effect. The fluid can be easily collected from the preputial pouch, but must be warmed to body temperature: it is not effective below $20°$ to $25°C$ (Signoret, 1965). 5α-androst-16-ene-3-one isolated by Patterson (1968) is apparently responsible for "boar" odour and for this olfactory stimulation (Melrose, personal communication).

V. MATERNAL BEHAVIOUR

Females become increasingly docile during pregnancy and reduction of physical activity is influenced by the increased body weight associated with advancing pregnancy. When pastured, many pregnant females graze more actively than non-pregnant contemporaries, and during the last fourth of pregnancy exhibit maternal tendencies. They show signs of excitement and often bite walls and fences seeking an outlet. In confinement they tend to cover any available holes, and in doing so will pack their water containers with straw. The pitch of the voice is lower and when such sows are either startled or crowded into small areas, their attitude becomes more aggressive, suggesting maternal protection. Although such signs appear clear-cut to a qualified herdsman, none is really characteristic, since typical behaviour patterns do not appear before nest-building.

In this section we shall be concerned with nest-building, farrowing, care-giving, nursing (maternal activity), and suckling (action of the young).

A. Nest-building

The female selects a site and begins to prepare a nest for her litter one to three days before farrowing. If shelters are not available, she establishes a definite area for nest building, and attempts to keep it clean and dry (see Eliminative Behaviour). The sow carries straw or pasture grass for the nest considerable distances. If the site of the nest is on soil, she roots and scratches the dirt and vegetation into a pile, then periodically roots out the centre for a place to lie down. In a concrete pen, her nesting tendencies are frustrated by the surroundings, but she still selects a nest site. Ordinarily, a female is less distressed if some bedding material is provided to augment even limited nest-building. From the time she begins to gather nest material until the time she farrows, the sow builds and rebuilds her nest many times, resisting human efforts to relocate the site. Some females build elaborate nests, whereas others only gather some straw. The nests of multiparous sows are more elaborate than those of gilts (Grauvogl, 1958).

Nest-building follows the same pattern among feral swine: large amounts of straw, hay, and other dry material are gathered and a nest is fashioned from this. The nest is sometimes so large that the female is completely hidden in it (Verlinden, 1953; Gundlach, 1967).

Within a few hours of farrowing, the vocalization of the sow changes radically. The undisturbed sow begins to utter gentle, deep-throated grunts, interspersed with short, shrill, whining sounds. If disturbed, most females defend their nesting site, just as they would defend newborn litters, with threatening movements of the body and open mouth. The vocal response is a loud, low-pitched, barking grunt. There is considerable individual variation in the expression of these patterns: some females appear to be nervous, whereas others remain placid. The nervousness of impending parturition may be manifested as panting grunts interspersed with loud squealing. While restricted in a farrowing stall, multiparous pregnant sows are ordinarily more placid than are nulliparous pregnant gilts. Females frustrated may continually fight the stall, rooting, biting, and shoving, often to the point of prostration. Females are less excited before and during parturition when they are unrestricted in a large rectangular pen (e.g. 8 ft × 10 ft) or a smaller round one (about 6 ft diameter) (Peo, 1960). However, piglet mortality by crushing is considerably reduced in farrowing stalls, where the movements of the sow are greatly restricted (Robertson et al., 1966).

B. Farrowing Behaviour

The gestation period of swine is 113 to 116 days (reviewed by Clegg, 1959). Farrowing appears to follow a diurnal pattern: parturition is rare in the morning and early afternoon, more frequent in late afternoon, and most frequent at night (60 per cent) (Grauvogl, 1958; Friend et al., 1962; Aumaitre, personal communication). The rate of farrowing in the first half of the day also increases from 35 per cent in winter and later autumn to 50 per cent in summer (du Mesnil du Buisson, personal communication) (Fig. 61).

The lips of the vulva generally swell 1 to 7 days before parturition. An increasing size of the mammary glands is apparent for several days. However, the best sign of approaching parturition is the presence of a milky secretion in the udder, which indicates that farrowing will occur within 24 hours (Jones, 1966) (Table 38).

As farrowing approaches, the sow rolls over more frequently or gets up and down repeatedly. She grunts intermittently and champs her teeth. Respiration gradually increases and the skin becomes hot and dry. During farrowing, the recumbent sow may occasionally stretch forward and kick backward with the upper hind leg or turn over Each motion forces fluid out of the vulva until the foetus is expelled (Plate XI). Before each foetus emerges, the efforts of the sow to expel it become noticeable: her body trembles and respiration stops.

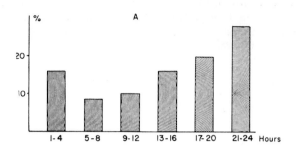

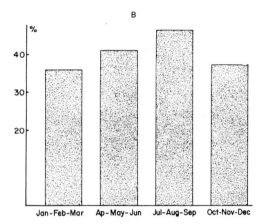

Fig. 61. Variations in farrowing rates: (A) Percentage of total number of farrowing throughout the day. Farrowing occurs more frequently during the afternoon and night than in the morning.
(B) Percentage of farrowing during the first half of the day. Farrowing rates also vary with the season of the year.

Table 38. Farrowing Data for Swine. (Adapted from Grauvogl, 1958; Friend, *et al.*, 1962; Jones, 1966.)

Pattern or Data	Value
Vulvar swelling	4 days before parturition
Milky secretion in the udder	24 hours before parturition
Duration of farrowing	4 hours (25 minutes to 8 hours)
Interval between each piglet	15 minutes (1 to 230)
Total piglet weight	13 kg
Placental weight	3 kg
Fluids weight	1·4 kg
Breech births	30 per cent

Sows usually farrow while lying on one side, but in a few cases they may deliver part of their litter while lying on their belly or standing up. If a sow is restless, she may stand up after farrowing each piglet; the reaction may be stimulated by either the release of pressure in the reproductive tract, or the faint squealing of the newborn. Farrowing lasts an average of 3 hours for gilts and 4 hours for sows (Table 38); it tends to be shorter in pastures than in confinement.

Some nervous sows are cannibalistic during or immediately after farrowing. If piglets are removed from such a sow as they are born and returned when she has quieted after farrowing, she will usually express normal protective behaviour rather than pathological aggression. Nervousness also leads to higher neonatal mortality in another way, since when the sow continually changes position she is more apt to lie on the young and crush them. Breeds differ in the degree of nervousness expressed during farrowing. The Palouse breed, for example, is more excitable than the Hampshire.

Each neonate is covered by transparent foetal membranes. Stronger piglets are freed from the drying membranes by their repeated struggles to get up, but weaker ones may not escape. The umbilical cord is usually intact when the piglet is born (Plate X); however, it is ordinarily detached within 15 minutes as the neonate moves about. Unless the expelled placenta is removed from the area immediately after farrowing, the sow will often eat all or a part of it (Hafez & Mauer, 1960).

The farrowing patterns of domestic and feral swine are apparently the same (Hediger, 1954).

C. Care-giving

The sow seldom licks her newborn and usually pays little attention to them until the last one is delivered. She roots the piglets along her belly toward the udder or pushes them back with a front foot. When standing she roots the piglets together in a nest away from where she wants to lie down. These movements are accompanied by low, gentle humming grunts. After 4 to 10 days, a sow begins to lead her piglets away from the nest area. She coaxes them with vocalizations similar to the food call she utters before nursing.

The protective behaviour of the sow is highly developed. During the first week after parturition, the more nervous sows are ready for defence after unusual sounds of any kind, especially a squeal from any piglet. Even the most docile sow can be quickly aroused by a threat to her young. The sow's first response is a series of short but loud grunts combined with a nasal whistling which serves as a warning call to her piglets. This reaction may be followed immediately by the typical aggressive barking grunt. If the sow is threatened when in close quarters, the two responses may be inseparable. After the warning call, the piglets crouch motionless and the sow either attacks the intruder or tries to lead it away from the piglets by dashing away herself. The piglets may rise after the initial warning and scatter, only to crouch again when the sow gives a new warning. The sow may threaten to bite; whether the threat is carried out depends on the reaction of the intruder, the aggressive tendencies of the sow, and the extent to which her defensive attitudes have been stimulated. Aggressive maternal tendencies can be enhanced by exteroceptive stimuli, such as deliberate teasing with a squealing pig, careless handling of piglets, and unusual disturbances in the nesting area.

The care-giving behaviour of the domestic boar has not been investigated, probably because males are usually separated from the sow herd. However, a feral boar is very tolerant of young, allowing them to mount his flanks while he is lying down (Frädrich, 1965).

D. Nursing

Milk is let down in response to suckling, massage of the udder, or injection of posterior pituitary hormone. Suckling excites a nervous reflex which augments secretion of hormones from the sow's posterior pituitary. Fright or adrenaline injections inhibit milk let-down. Breed differences in lactation

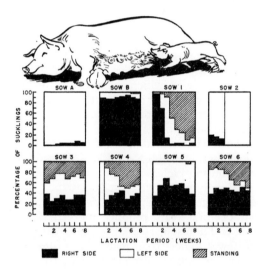

Fig. 62. Nursing positions of sows during an 8-week lactation period. (Adapted from Gill & Thomson, 1956.)

performance have been reported (Smith, 1952; Braude, 1954; Allen *et al.*, 1959; Hartman & Pond, 1960). Caution is necessary when interpreting data on the milk yield of sows because the frequency of milking or suckling and litter size influence milk production. Frequent suckling, for example, increases it: sows nursed at 1-hour intervals produced about one-third more milk than those nursed at 2-hour intervals (Barber *et al.*, 1955b).

Nursing position. The sow selects a definite place in the pen where she lies down for nursing. She is induced to assume the appropriate position when the piglets first massage the udder. Sows usually spend an equal amount of time on the right and left sides while nursing (Höpler, 1943) (Fig. 62). Sows do not turn over during a single nursing period. Some sows always lie on one particular side and others favour the standing position while nursing. Feral and phacochoerus pigs frequently nurse in a standing position (Frädrich, 1965).

E. Suckling

The young may reach the udder within a few seconds after birth, while the umbilical cord is still attached (Plate X). They express the suckling response as soon as their noses make contact with other objects. Piglets which are milk-fed and reared in the laboratory must be cared for individually during the first week. If two or more are joined together at birth, they will suckle each other on any available soft part. These piglets frequently perish because they suckle each other and do not learn to drink the milk furnished in shallow pans.

Orientation of piglets on the udder. During the first days after birth, piglets fight for individual teats and in time assume a definite order on the udder. Unemployed teats regress within a few days.

The fixation to an individual teat and the establishment of the social hierarchy occur simultaneously (Van Loen & Molenaar, 1967). The most anterior pectoral and most posterior inguinal teats are appropriated earliest and by the end of the second week order is also established for the middle teats. Regardless of the side on which the sow lies for nursing, the piglets maintain the same suckling order on the udder. Perhaps recognition is facilitated visually by perception of body and udder conformation, since more errors in locating the appropriate teat are made at the centre of the udder than at either end (Donald, 1937). The way the hair lies on the sow's belly may also orient the piglets on the udder (McBride, 1963). Odour and recognition of neighbouring piglets may also play a role. However, when a sow's udder is coated with mud or a scented substance, the suckling order of the young does not change (Donald, 1937). A number of investigators have demonstrated that pectoral teats produce more milk than inguinal ones (Donald, 1937; Gill & Thomson, 1956).

When there are fewer piglets than teats, the posterior teats remain unemployed. With some exceptions (England *et al.*, 1961), piglets which are heavier at birth tend to appropriate the anterior teats (Allen & Lasley, 1958; McBride *et al.*, 1965). Even when litters are experimentally homogenized with respect to body weight, the correlation between birth weight, teat order, and subsequent growth persists (Lodge & Pratt, 1963).

Piglets which change their preference for a particular nipple during the early part of the suckling period attempt to move anteriorly. Occasionally, one piglet is left to suckle the two rearmost teats when all others are occupied. Coincidentally, pectoral teats are in a safer region than those near the dam's hind legs: piglets suckling anteriorly are less apt to be trampled if the sow should rise.

Suckling order is generally established more rapidly and is better maintained in large, vigorous litters than in small or weak ones (Höpler, 1943). The errors made by some piglets in locating their "own" teat are due to factors summarized by Donald (1937) as follows: (a) the milk supply for large litters is inadequate, so that the piglets are often hungry, particularly after 3 weeks of age. In this condition the young are less inclined to insist on having the correct nipple; (b) the litter is so large in relation to the length of the udder that it is very difficult for a late-comer to reach the correct teat; (c) in small litters, there are more opportunities for a piglet to appropriate spare nipples; (d) the number of nipples may not be the same on both sides of the sow, in which case the piglets have difficulty arranging themselves when the sow changes from one side to the other; (e) some sows let down their milk before all the young have found their proper places. This encourages piglets to be less discriminating.

Phases of suckling. Three phases of suckling have been described by Barber *et al.* (1955b) and are discussed below.

Nosing phase. As soon as the sow takes the nursing position, the young nose the udder vigorously (Plate X) for 55 to 140 seconds. The duration of nosing increases steadily from birth to weaning. Barber *et al.* (1955b) suggest that the secretion of neurohypophysial hormones in response to the suckling stimulus is delayed as lactation advances.

Quiet phase. The nosing phase ends abruptly. The young become quieter and draw back their ears, but do not get any measurable amount of milk (Plate X); this phase lasts 16 to 23 seconds. Possibly the vigorous udder nosing results in the passive withdrawal of a small quantity of milk which cannot be detected by the routine technique of weighing before and after suckling (Cowie *et al.*, 1951). Perhaps, as in rabbits (Cross, 1954), a small milk let-down is caused by mechanical stimulation of the mammary glands, whereas oxytocin is not released until a few seconds later.

True suckling. The piglets are quiet during this phase (Plate X). Milk let-down lasts for only 13 to 37 seconds, with an average of 19 (Niwa *et al.*, 1951; Barber *et al.*, 1955b). The longer suckling periods reported by some investigators may include the quiet phase and true suckling. Whittlestone (1953) finds little variation in the duration of milk-flow among Berkshire sows, using standard doses of oxytocin.

Gill & Thomson (1956) describe a final phase of udder massage by the piglets when no milk flows and the massage rhythm is slower than that of the nosing phase.

During early lactation, nursing ends when the piglets fall asleep at the udder. Later, the sow often brings nursing to a halt by turning over on her udder or walking away. Sows with tender teats do not allow the piglets to massage the udder after milk let-down ceases. During the first 4 to 5 weeks of lactation, once the piglets have satisfied their hunger they fall asleep cuddling to the sow for warmth. Older piglets will first suckle the sow and then eat supplementary feed before resting (Allen *et al.*, 1959).

Piglets suckle at about 1-hour intervals (Berge & Indrebo, 1953; Braude, 1954; Barber *et al.*, 1955b; Hafez, 1959). Intervals between suckling are shorter during the day than at night; the number per day declines 7 days after farrowing (Fig. 63). However, individual variations in frequency occur at every stage of lactation. The nursing-suckling behaviour of swine is summarized in Table 39; values represent the results of several studies using different breeds, management, and experimental techniques. The suckling rhythm of feral swine in zoological gardens is identical to that of domestic breeds, but in the wild nursing occurs less frequently (Frädrich, 1965).

Stimuli eliciting suckling. Suckling may be initiated by the sow or the young. It is common for the sow to issue as a food call a series of quiet grunts, arousing the piglets and bring them to the udder. Some piglets may become hungry and begin to suckle independently; in this case the sow still grunts, inducing the other resting piglets to arise.

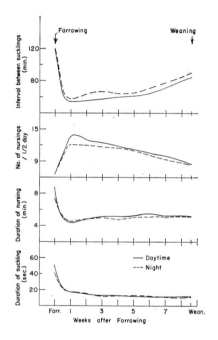

Fig. 63. Nursing behaviour of Middle White sows (5 litters) during an entire lactation period. (Data from Niwa *et al.*, 1951.)

The vocalization and activity of one litter can have a manifold effect on the behaviour of other litters housed in the same piggery. Most of the litters will become active within minutes after the first litter is aroused. An electric shock to the sow or sudden unusual sounds, such as slamming a door or gate, may also initiate suckling. Sleeping piglets may be aroused and induced to suckle when a herdsman simulates the smacking sound of suckling (Hartman & Pond, 1960). Kristjansson (1960) suggests that it might be possible to increase suckling frequency with appropriate sounds given at regular intervals.

Foster piglets. The domestic sow is able to recognize her own litter by olfactory cues just as feral pigs do (Steinbacher, 1954). Swine are much more tolerant of foreign young than any other domestic mammal. In the wild, two females sometimes have their litters in the same nest. The mothers do not behave differently toward their own piglets and the foster ones, and the young suckle both sows (Steinbacher, 1954; Frädrich, 1965). Fostering is also rather easy among domestic pigs. A small litter can be fostered to a sow, which also has a small litter of similar age, by placing the new piglets in the same nest area with the sow's progeny. Fostering is most successful when it is initiated within the first 48 hours after birth: not only does the sow recognize her own litter by olfactory cues, but her ability to do so increases

Table 39. Nursing and Suckling Behaviour in Swine. (Adapted from Wells et al., 1939; Niwa et al., 1951; Smith, 1952; Berge & Indrebo, 1953; Braude, 1954; Barber et al., 1955b; Hafez & Mauer, 1960.)

Suckling Pattern	Values[1]
Time taken by sow to lie down after piglets enter the pen (sec)	11–17
Suckling frequency (periods/day)	18–28
Interval between sucklings (min)	51–63
	up to 160 (rare)
Duration of nursing (min)	4–8
Duration of suckling (sec)	
Nosing phase	55–140
Quiet phase	16–23
True suckling	13–37
Milk consumption of piglet/suckling period (g)	24–28
Milk consumption of piglet/day (g)	546–676
Milk consumption of piglet/lactation period (kg)	30–37

[1] These represent average values for different breeds under different environmental conditions using different techniques.

with the age of the litter. It may consequently be necessary to smear the skins of older piglets with the sow's excreta to thwart their identification as strangers. Putting the sow's own piglets in a basket with the foster piglets for 2 to 3 hours may also be effective. Although an occasional sow will refuse to accept foster piglets under any circumstances, it is usually easy to foster piglets when the sow is being handled in a farrowing stall. If the foster piglets avoid walking in front of the sow they may remain undetected until they are an integral part of the litter. The newcomers will attempt to take up the suckling position they had on their own mother (Plate X), and the new suckling order of the combined piglets will depend upon their relative weights, aggressive tendencies, and previous suckling order.

The litters of several sows may be combined for community rearing. Such a technique gives best results when the piglets are combined at 7 to 10 days of age and with groups of not more than 4 to 5 sows (Novak & Pytel, 1959; Krjutcenko, 1962; Kazanceva, 1964). The best results are obtained when the sows are kept together and only isolated for a few days while farrowing (Holub, 1959).

VI. SOCIAL BEHAVIOUR

Olfaction appears to play a major role in individual recognition and social relations among swine. Naso-nasal or naso-ano-genital contact frequently occurs between animals (Table 40). The urine is frequently smelled and is apparently used for territorial marking in Phacochoerus (Frädrich, 1965).

Table 40. Type and Frequency of Olfactory Contact Between Pigs of Different Ages and Sexes. The Number of ($+$) or ($-$) Indicates the Relative Frequency of Each Type of Contact. (Adapted from Frädrich, 1965.)

Contact Between	Naso-Nasal Contact	Naso-Ano-Genital Contact
Adult males	$++$	$--$
Juveniles	$++$	$-$
Females and males	$++$	$+$
Adult females	$++$	$++$
Males and females	$++$	$+++$
Female and young	$+++$	$++$

Tactile stimuli are also very important in the social behaviour of swine, which are "contact type" animals, according to Hediger (1954). Resting pigs try to have the maximum possible bodily contact with other animals, even in hot environment.

Social grooming rarely occurs among pigs (Fig. 64).

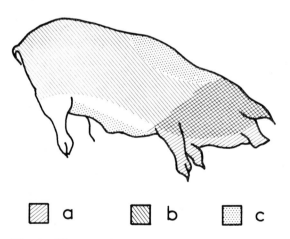

Fig. 64. The grooming behaviour of the pig: (*a*) The area scratched by the hind legs; (*b*) the area scratched on vertical objects; and (*c*) areas groomed by other animals (social grooming).

A. Social Structure

The social structure of swine populations has been modified by commercial breeding. Feral pigs live in small herds of 4 to 20 individuals, including 1 to 4 female with their yearlings and the young of the year. Adult males join these herds during the reproduction season, but range separately during the rest of the year. The social structure is not rigid, but changes frequently; for example, herds with up to 80 animals sometimes form (Snethlage, 1934; Heck, 1950).

B. Agonistic Behaviour

Young-young interactions. Agonistic behaviour appears during the first 2 days of life. Piglets are born with sharp temporary canine teeth which they use to bite or slash their opponents as they fight during suckling. Once the suckling order is established, the level of aggression declines. Play patterns include some agonistic behaviour, but seldom involve prolonged vicious attacks on litter-mates. The temporary canines are later replaced by permanent teeth, which in boars develop into long formidable tusks if left untrimmed.

Adult interactions. When animals are penned together for long periods, they seldom show agonistic behaviour, except for sudden pushing or rooting of a pen-mate. They may quickly turn their heads and bite, or snap their jaws without bodily contact. This generally occurs during feeding.

When separately-reared animals are assembled, some fighting ensues, especially among more dominant animals. The agonistic behaviour of adult boars and sows is similar. None the less, the jaw clicking and saliva production (champing) so characteristic of fighting boars is seldom seen in barrows or gilts unless they are treated with testosterone propionate (Slebodnick & Klopfer, 1953). A sow will try to bite, whereas the boar strafes his opponent with his tusks.

When strange boars are first penned together, they smell one another and begin to circle as if "sizing up" their foe (Plate XI). They frequently strut shoulder to shoulder with their hair bristled, ears cocked, and heads raised in an alert, threatening attitude. Some boars paw the ground, throwing loose soil in the air (Fig. 65). In a serious encounter, the opponents utter deep-throated, barking grunts, begin to grind their teeth and snap their jaws (champing), and produce large amounts of foaming saliva which falls in clumps on the ground. Champing continues throughout the fight. After a minute or two, the opponents assume the typical shoulder-to-shoulder fighting position and apply side pressure (Plate XI). They circle continually always shoving and watching for an advantage. As the tension mounts, each boar repeatedly thrusts his head and neck sideways and up, with his jaws open and teeth bared. These slashes are the most damaging blows the boar can deliver. If the boars have tusks, their opponent's shoulders are soon severely lacerated. Experienced boars often manoeuvre the point of contact forward to their neck region, so that they may swing their bodies away from the opponent and decrease the target area.

Fighting boars await the opportunity to cease shoulder contact and to nip at a front leg, neck, or ear. They also charge the side of an opponent with mouth open to bite. Dominant experienced boars often resolve a conflict with 2 or 3 quick attacks.

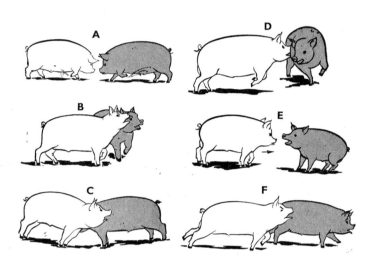

Fig. 65. Agonistic behaviour in boars. A. Initial encounter. B. Strutting. C. Shoulder-to-shoulder contact and slashing. D. Perpendicular biting attack. E. Submission (or "defeat posture"). F. Pursuit of loser.

Fighting may continue for 30 to 60 minutes. The dominant boar is seldom satisfied until he has pursued the loser; he will continue to bite and slash the retreating boar. The loser backs away in haste with his mouth open and head held high, squealing loudly; if pursued, he will turn and run. A single glance or a short warning grunt is sufficient to induce a defeat response once the dominance order is established. Occasionally, a dominance order is not established and the boars may then accept one another as equals.

The intensity of a breed's general activity is not a good indication of its fighting ability. The Yorkshire is alert, curious, and has a strong aggressive vocal response that may bluff some opponents. The Duroc, on the other hand, is sluggish and inattentive, but it is a more tenacious fighter. Although all three of the Minnesota breeds are alert, the Minnesota No. 1 is the quietest and most tractable. However, the No. 1 is a more vicious fighter than either the No. 2 or No. 3 (Fausch, 1961).

The expression of aggression after weaning depends upon management practices. Usually, pigs are reared to market weight in larger than litter groups. With each addition of new animals to a group, fighting ensues and newcomers are attacked. Fighting is frequently less intense if two groups are joined in an environment which is new to both, presumably because their attention is divided between fighting and investigation of the new surroundings. Boars which are placed in adjacent pens run along their common fence, champing and even attempting to charge and bite. Tranquillizers have been used successfully to diminish fighting when strange adult boars are brought together (Fausch, 1961).

C. Dominance Order

Gilts and barrows. Several authors report that a stable social hierarchy exists among swine (Reebs, 1960; Rasmussen *et al.*, 1961; McBride *et al.*, 1964; Beilharz & Cox, 1967a). Barrows are generally more dominant than gilts.

A pig's social rank between weaning and adulthood appears to be influenced by its suckling rank. The teat order gives dominant animals access to the most productive teats: their higher weight gain is an advantage for future competition (McBride *et al.*, 1964). According to McBride *et al.* (1965), 13 per cent of the variation in weight gain is reflected in the social hierarchy. Similarly, Jonsson (1959) found that dominant barrows gain more weight than gilts when reared in groups, but when fed separately, the reverse is observed.

Males. When strange boars are penned together, fighting generally continues until social rank is established. Each strange boar added to the group must encounter all its new pen-mates. Even if boars are separated for only a day or two, fighting occurs when they reaggregate. Wiggan (1961) studied the development of dominance order among eight crossbred boars which had been reared as two groups of 4 after weaning. When they reached sexual maturity, paired encounters were staged between boars of the two groups in the presence of an oestrous female. Both boars initially nuzzled the female, but ignored her completely during the peak of the fight. Encounters between boars with high ranks in their home pens were always classified as undecided after 40 minutes: on no occasion was a dominant boar conquered. When low ranking boars were paired, the dominance order was established within 10 to 20 minutes. The boar which first attacked its opponent was always the victor. Dominant boars were more conservative than the less dominant boars in the exchange of slashes, but in all cases the boar receiving the greater number of slashes was ultimately the loser. During the encounters, one boar often tried to postpone the settling of dominance relationships in favour of mating, but the opponent always prevented copulation.

When 11 yearling boars were housed together, the dominant males continued to test their social rank by brief encounters and threats, especially if females were present (Jakway & Sumption, 1959). Although a general descending order was established, there was no continuous and precise ranking from one extreme to the other. When this group of boars was later subdivided and then recombined, the same social order was quickly re-established. Most of the new encounters were brief and harmless.

VII. SCOPE OF FUTURE RESEARCH

The pig can be used for the observation of many well-defined patterns of individual and group behaviour. Because its behaviour can be modified, this species invokes an exciting prospect for future research concerning alteration of stimuli, environment, and experience. Recent technical developments make it possible to evaluate the role of experience in the emergence of behavioural patterns. Piglets can now be delivered by hysterectomy and reared successfully in the laboratory away from their mothers (Young *et al.*, 1955). Such animals may be kept indefinitely in complete isolation to provide new possibilities for studying imprinting, learning, and sexual and maternal behaviour. For example, observations of the behaviour of piglets reared without maternal contact suggest that imprinting occurs in this group. After the pigs are removed from isolation boxes in the laboratory and placed in a large heated and lighted building, their attention is immediately directed toward their caretaker. Critical studies under controlled conditions will be required to distinguish between imprinting and food-seeking behaviour. If auditory isolation is also maintained, studies of vocalization could be most instructive.

Breed differences afford a glimpse of the possibilities for studying the genetics of behaviour. Such work would require only limited changes in conventional methods of handling. The size of the animal imposes certain limitations if adults are used. If studies require precise timing, for example, the difficulties in manoeuvring such animals are obvious. New strains of genetically small swine (Dettmers, 1959) may help overcome the problem of size. Early puberty and large litters make swine the most suitable large animal for genetic studies.

Since some types of feral pigs interbreed with domestic swine, the natural patterns of behaviour and their evolution during domestication may be studied. The influence of genetic selection, taming, and management practices may also be observed.

The animal husbandman is currently paying more attention to the influence of behaviour on the expression of important quantitative traits in nutrition, physiology, engineering, and breeding research (McBride, 1959). As a result, the behaviourist will find it easier to enlist co-operation in the development and joint use of facilities for swine research.

REFERENCES

AAMDAL, J., HOGSET, I., SVEBERG, O. & KOPPANG, N. (1958). A new type of artificial vagina and a new collection technique for boar semen. *J. Am. vet. med. Ass.*, **132**, 101–104.

ALDINGER, S. M., SPEER, V. C., HAYS, V. W. & CATRON, D. V. (1959). Effect of saccharin on consumption of starter rations by baby pigs. *J. Anim. Sci.*, **18**, 1350–1355.

ALLEN, A. D. & LASLEY, J. F. (1958). The location of mammae on the sow as correlated with the performance of pigs. *J. Anim. Sci.*, **17**, 1223.

ALLEN, A. D., LASLEY, J. F. & TRIBBLE, L. F. (1959). Milk production and related performance factors in sows. *Mo. agric. exp. Stn tech. Bull.*, No. 712.

ALTMANN, M. (1941a). A study of patterns of activity and neighborly relations in swine. *J. comp. Psychol.*, **31**, 473–479.

ALTMANN, M. (1941b). Interrelations of the sex cycle and the behavior of the sow. *J. comp. Psychol.*, **31**, 481–498.

ANDREWS, F. N. & WARWICK, E. J. (1949). Comparative testicular development of some inbred, purebred, crossbred and linecross boars. *J. Anim. Sci.*, **8**, 603 (Abstract).

ARCELAY, C. P. (1961). Personal communication. Dept. of Anim. Husb., Penn. State Univ., University Park, Pa., U.S.A.

ASDELL, S. A. (1946). *Patterns of Mammalian Reproduction*, Ithaca, N.Y.: Comstock.

AUDLEY, P. & KLOPFER, F. D. (1953). Dietary influences on the acquisition of a plate pressing response in pigs. Paper read at meeting of Western Psychological Association, Seattle, Washington, U.S.A.

AUMAITRE A. Personal communication. Station de recherches sur l'élevage du porc, C.N.R.Z. 78 Jouy en Josas, France.

AUMAITRE, A. & SALMON-LEGAGNEUR, E. (1964). Les préférences alimentaires du porcelet. VII. Appétibilité des tourteaux. *Annls Zootech.*, **13**, 237–240.

BARBER, R. S., BRAUDE, R. & MITCHELL, K. G. (1955a). The value of electrically warmed floors for fattening pigs. *J. agric. Sci.*, **46**, 31–36.

BARBER, R. S., BRAUDE, R. & MITCHELL, K. G. (1955b). Studies on milk production of Large White pigs. *J. agric. Sci.*, **46**, 97–118.

BASCOS, C. D. (1953). Breeding habits of Berkjala swine. *Philipp. Agric.*, **37**, 28–35.

BEILHARZ, R. G. & COX, D. F. (1967a). Social dominance in swine. *Anim. Behav.*, **15**, 117–122.

BEILHARZ, R. G. & COX, D. F. (1967b). Genetic analysis of open field behaviour in swine. *J. Anim. Sci.*, **26**, 988–990.

BERGE, S. & INDREBO, T. (1953). Milk production by sows. *Meld. Norg. LandbrHøisk.*, **33**, 389–424.

BETTERIDGE, K. J. & RAESIDE, J. I. (1962). Investigations of cervical mucus as an indicator of ovarian activity in pigs. *J. Reprod. Fert.*, **3**, 410–421.

BOBAK, A. W. (1957). *Das Schwarzwild*. Neuman Verlag, Radebeul.

BODA, J. M. (1959). The estrous cycle of the sow. In: *Reproduction in Domestic Animals*. H. H. Cole & P. T. Cupps (eds.). Vol. I, Ch. 10, p. 335. New York, N.Y.: Academic Press.

BRAUDE, R. (1954). Pig nutrition. In: *Progress in the Physiology of Farm Animals*. J. Hammond (ed.). Vol. I, Ch. 2, pp. 40–105. London: Butterworths Scientific Publ.

BRAY, C. E. & SINGLETARY, C. B. (1948). Effect of hog wallows on gains of fattening swine. *J. Anim. Sci.*, **7**, 521–522 (Abstract).

BUBENIK, A. (1959). *Grundlagen der Wildernährung*.

BÜCHLMANN, E. (1950). Das sexuelle Verhalten der Rindes. *Wien. tierärztl. Mschr.*, **37**, 153–156 and 225–230.

BURGER, J. F. (1952). Sex physiology of pigs. *Onderstepoort J. vet. Res.*, **2** (suppl.), 1–218.

CENA, M. (1965). Verhalten der Schwein beim Transport. *Int. Koll. K. Marx Univ., Leipzig*, 5–14.

CLEGG, M. T. (1959). Factors affecting gestation period and parturition. In: *Reproduction in Domestic Animals*. H. H. Cole & P. T. Cupps (eds.). Vol. I, Ch. 15, p. 509. New York N.Y.: Academic Press.

COLBERT, E. H. (1955). *Evolution of the Vertebrates*. New York, N.Y.: John Wiley.

COMBS, G. E. & WALLACE, H. D. (1959). Palatable creep feeds for pigs. *Univ. Fla. Agric. Exp. Stn Bull.*, 610.

CORTEEL, J. M., SIGNORET, J. P. & DU MESNIL DU BUISSON, F. (1964). Variations saisonnières de la reproduction de la truie et facteurs favorisant l'anoestrus temporaire. *V⁰ Cong. Reprod. Anim. (Trento)*, Vol. III, 536–540.

COWIE, A. T., FOLLEY, S. J., CROSS, B. A., HARRIS, G. W., JACOBSOHN, D. & RICHARDSON, K. S. (1951). Terminology for use in lactational physiology. *Nature, Lond.*, **168**, 421.

CROSS, B. A. (1954). Milk ejection resulting from mechanical stimulation of mammary epithelium in the rabbit. *Nature, Lond.*, **173**, 450–451.

DAWSON, W. M. & REVENS, R. L. (1946). Varying susceptibility in pigs to alarm. *J. comp. Psychol.*, **39**, 297–305.

DECKER, E. (1961). Personal communication. Rd.1. Emlenton, Pa., U.S.A.

DETTMERS, A. (1959). Über die Zucht des Miniaturschweines. *Z. Tierzücht. ZuchtBiol.*, **73**, 311–315.

DINUSSON, W. E. (1965). A day in the life of a pig. *Feedstuffs, Lond.*, **37**, 5.

DONALD, H. P. (1937). The milk composition and growth of suckling pigs. *Emp. J. exp. Agric.*, **5**, 349–360. Suckling and suckling preference in pigs. *Ibid.*, **5**, 361–368.

DUTT, R. H., SIMPSON, E. C., CHRISTIAN, J. C. & BARNHART, C. E. (1959). Identification of preputial glands as the site of production of sexual odor in the boar. *J. Anim. Sci.*, **18**, 1557 (Abstract).

DZIUK, P. J. (1960). Personal communication. Dept. Anim. Sci., Univ. of Illinois, Urbana, Ill., U.S.A.

ELLIS, N. R., CRAFT, W. A. & ZELLER, J. H. (1936). A preliminary study of variations in free choice intake of the components of a standard ration and breeding as a possible factor of the occurrence of lameness in pigs. *Rec. Prod. Am. Soc. Anim. Prod.*, 77–81.

ENGLAND, D. C., BERTUN, P. L., CHAPMAN, V. M. & MILLE, J. C. (1961). Nursing position of baby pigs in relation to birth weight. *J. Anim. Sci.*, 20, 682.

ENSMINGER, M. E. (1960). *Animal Science.* Danville, Illinois: Interstate.

FACTO, L. A., DIAZ, F., HAYS, V. W. & CATRON, D. V. (1959). Time lapse cinematography as a method of studying animal behavior and ration preferences. *J. Anim. Sci.*, 18, 1488 (Abstract).

FAUSCH, H. D. (1961). Personal communication. California Polytech. Coll., Pomona, California, U.S.A.

FOOTE, W. C., WALDORF, D. F., CHAPMAN, A. B., SELF, H. L., GRUMMER, R. H. & CASIDA, L. E. (1956). Age at puberty of gilts produced by different systems of mating. *J. Anim. Sci.*, 15, 959–969.

FOWLER, S. H. & ENSMINGER, M. E. (1960). Interaction between genotype and plane of nutrition in selection for rate of gain in swine. *J. Anim. Sci.*, 19, 434–449.

FRÄDRICH, H. (1965). Zur Biologie und Ethologie des Warzenschweines (Phacochoerus Aethiopicus Pallas) unter Berücksichtigung des Verhaltens anderer Suiden. *Z. Tierpsychol.* 22, 328–393.

FRIEND, D. W., CUNNINGHAM, H. M. & NICHOLSON, J. W. G. (1962). The duration of farrowing in relation to reproduction performance of Yorkshire sows. *Can. J. comp. Med.*, 26, 127–182.

GILL, J. C. & THOMSON, W. (1956). Observations on the behaviour of suckling pigs. *Br. J. Anim. Behav.*, 4, 46–51.

GRAUVOGL, A. (1958). Über das Verhalten der Hausschweinen mit besonderer Berüchsichtigung der Fortpflanzungsverhaltens. *Vet. Med. Diss.*, Berlin.

GRAY, A. P. (1954). *Mammalian Hybrids.* Tech. Comm. No. 10. Commonwealth Bureau of Anim. Breeding & Genet., Slough, Bucks., England.

GREGORY, K. E. & DICKERSON, G. E. (1952). Influence of heterosis and plane of nutrition on rate and economy of gains, digestion and carcass composition of pigs. *Mo. agric. exp. Stn Res. Bull.*, No. 493.

GRUMMER, R. H. & BOHSTEDT, G. (1947). Can salt poisoning kill pigs? *Univ. Wis. Agric. Exp. Stn Bull.*, 472.

GUNDLACH, H. (1967). *Sus Scrofa: Nestbau Verhalten.* Encyclopoedia cinematographica, Film E 1254. Inst. Wiss. Film Univ. Göttingen.

HAFEZ, E. S. E. (1959). Nursing-suckling interactions in the domestic pig. *Anat. Rec.*, 134, 576 (Abstract).

HAFEZ, E. S. E. (1961). Unpublished data on superovulatory responses of cattle after shipment. Dept. Anim. Sci., Wash. State Univ., Pullman, Washington, U.S.A.

HAFEZ, E. S. E. & MAUER, R. E. (1960). Unpublished data on sexual and maternal behavior of Palouse Pigs. Dept. Anim. Sci., Wash. State Univ., Pullman, Washington, U.S.A.

HALE, E. B. & ALMQUIST, J. O. (1960). Relation of sexual behavior to germ cell output in farm animals. *J. Dairy Sci.*, 43 (suppl.), 145–169.

HANCOCK, J. L. (1961). Personal communication, Anim. Breeding Res. Org., Agric. Res. Coun., Edinburgh, Scotland.

HARTMAN, D. A. & POND, W. G. (1960). Design and use of a milking machine for sows. *J. Anim. Sci.*, 19, 780–785.

HAUSER, E. R., DICKERSON, G. E. & MAYER, D. T. (1952). Reproductive development and performance of inbred and crossbred boars. *Mo. agric. expt. Stn Res. Bull.*, No. 503.

HECK, L. (1950). *Das Schwarzwild: Lebensbild der Wildschweins.* Bay. Landwirt. Verlag München.

HEDIGER, H. (1954). *Skizzen zu einer Tierpsychologie im Zoo und in Zirkus.* Büchergilde Gütemberg, Zürich.

HEITMAN, H., JR. & HUGHES, E. H. (1949). The effects of air temperature and relative humidity on the physiological well being of swine. *J. Anim. Sci.*, 8, 171–181.

HODGSON, R. E. (1935). An eight generation experiment in inbreeding swine. *J. Hered.*, 26, 209–217.

HOLUB, A. (1959). Vilv Chladu na Teplotu Kuze u Selat. *Sb. vys. Sk. zeměd. les. Fac. Brne*, **27**, 291–302.

HÖPLER, E. v. (1943). Trinken Ferkel tatsächlich immer am gleichen Strich ? *Z. Schweinez.*, **50**, 147–148.

HUTCHINSON, H. D., JENSEN, A. H., TERRILL, S. W. & BECKER, D. E. (1956). Comparison of free choice and complete rations on pasture and dry lot. *Univ. of Illinois Mimeo.*, A.S. **435.**

ITO, S., KUDO, A. & NIWA, T. (1960). Studies on the normal oestrus in swine with special reference to proper time for service. *Bull. nat. Inst. agric. Sci. Chiba-shi, Japan*, 27–28.

ITO, S., NIWA, T. & KUDO, A. (1948). Studies on the artificial insemination in swine. I. On the method of collection of semen and the condition of ejaculation. *Zoot. Exp. Stn Res. Bull.*, **55**, 1.

JAKWAY, J. S. & SUMPTION, L. J. (1959). Unpublished data on the behavior of swine. Dept. Anim. Husb., Univ. of Nebr., Lincoln, Nebraska, U.S.A.

JONES, J. E. T. (1966). Observations on the parturition in the sow. *Br. vet. J.*, **122**, 420–426 and 471–478.

JONSSON, P. (1955). Fortsatte statistiske undesgelser over grisenes daglige tilvaekst samt foderforbruget pr. kg tilvaekst. *Tidsskr. Landøkon.*, **11–12**, 405–429.

JONSSON, P. (1959). Investigations on group feeding and on the interaction between genotype and environment in pigs. *Acta agric. scand.*, **9**, 204–228.

KARAS, G., McKINNEY, A. C., COX, D. F. & WILLHAM, R. L. (1961). Avoidance learning in offspring of irradiated and non-irradiated swine. Paper read at meeting of the Psychonomic Society. New York, N.Y., U.S.A.

KARE, M. R., POND, W. C. & CAMPELL, J. (1961). Individual variation in taste reactions of pigs. *J. Anim. Behav.* (In press).

KAZANCEVA, A. P. (1964). Gruppovoe soderzanie podsosnych matok S Porosjatami. *Svinovodstvo*, **18**, 12–14.

KLOPFER, F. D. (1961a). Early experience and discrimination learning in swine. Paper read at American Institute of Biological Sciences, August, 1961, West Lafayette Indiana, U.S.A.

KLOPFER, F. D. (1961b). Personal communication. Dept. of Psychology, Wash. State Univ., Pullman, Washington, U.S.A.

KLOPFER, F. D. & WESLEY, F. (1954). Observation of discrimination learning in swine. Paper read at Oregon and Washington Psychological Association, May, 1954, Pullman, Washington, U.S.A.

KONIUKHOVA, V. A. (1955). Ob uslovnykh refleksakh u svinei. (Conditioned reflexes in pigs.) *Fiziol. Zh. SSSR*, **41**, 326–333; *Referat. Zh., Biol.* (1956). No. 20248. (Translation); *Biol. Abstracts.* (1957), **31**, 26882.

KRISTJANSSON, F. K. (1960). Personal communication. Central Expt. Farm, Ottawa, Canada.

KRJUTCENKO, E. A. (1962). Kak my peresly ka gruppovoe soderzanie podsosnych matok. *Svinovodstvo*, **16**, 18–19.

LAGERLÖF, N. & CARLQUIST, H. (1958). A study of the semen of boars of the Yorkshire breed between the ages of 5–9 months. *VIII Nordiska Veterinärmötet, Helsingfors.* Section D, Report No. 1.

LASSITER, J. W., TERRIL, S. W., BECKER, D. E. & NORTON, H. W. (1955). Protein level for pigs as studied by growth and self selection. *J. Anim. Sci.*, **14**, 482–491.

LEBEDEV, M. M. & PITKIANEN, I. G. (1951). Increasing fertility in pigs. *Sovetsk. Zool.*, **6**, 34–43.

LIPS, C. (1965). Untersuchungen zum Verhalten von Mastschweinen. *Agric. Diss. Iena.*

LODGE, G. A. & PRATT, P. D. (1963). Birth weight and the subsequent growth of suckled pigs. *Anim. Prod.*, **5**, 225.

McBRIDE, G. (1959). The influence of social behaviour on experimental design in animal husbandry. *Anim. Prod.*, **1**, 81–84.

McBRIDE, G. (1963). The teat order and communication in young pigs. *Anim. Behav.*, **11**, 53–56.

McBRIDE, G., JAMES, J. W. & HODGENS, N. (1964). Social behaviour of domestic animals. IV. Growing pigs. *Anim. Prod.*, **6**, 129–140.

McBRIDE, G., JAMES, J. W. & WYETH, G. S. F. (1965). Social behaviour of domestic animals. VII. Variations in weaning weight in pigs. *Anim. Prod.*, **7**, 67–74.

McKENZIE, F. F. (1931). A method for the collection of boar semen. *J. Am. vet. med. Ass.*, **31**, 244–246.

McLAGAN, J. R. & THOMSON, W. (1950). Effective temperature as a measure of environmental conditions for pigs. *J. agric. Sci.*, **40**, 367–374.

MARCUSE, F. L. & MOORE, A. U. (1944). Tantrum behavior in the pig. *J. comp. physiol Psychol.*, **37**, 235–241.

MARCUSE, F. L. & MOORE, A. U. (1946). Motor criteria of discrimination. *J. comp. physiol. Psychol.*, **39**, 25–27.

MELLEN, I. M. (1952). *The Natural History of the Pig*. New York, N.Y.: Exposition Press.

MELROSE, D. R. (1968). Personal communication. P.I.D.A. London (Great Britain).

MELROSE, D. R. & O'HAGAN, C. (1959). Some observations on the collection of boar semen and its use for artificial insemination. *Coll. Reprod. Insem. Artif. Porc. Ann. Zootech.* **8** (Suppl.), 69–79.

DU MESNIL DU BUISSON, F. (1968). Personal communication. Laboratoire de physiologie de la Reproduction, 37, Nouzilly, France.

DU MESNIL DU BUISSON, F. & SIGNORET, J. P. (1962). Influences de facteurs externes sur le declenchement de la puberte chez la Truie. *Annls Zootech.*, **11**, 53–59.

MOHR, E. (1960). *Wilde Schweine*. Die neue Brehmbücherei, A. Ziemsen Verlag, Wittember Lütherstadt.

MOORE, A. U. & MARCUSE, F. L. (1945). Salivary, cardiac and motor indices of conditioning in two sows. *J. comp. physiol. Psychol.*, **38**, 1–16.

MOUNT, L. E. (1960). The influence of huddling and body size on the metabolic rate of the young pig. *J. agric. Sci.*, **55**, 101–105.

MYERS, G. C. (1916). The importance of primacy in the learning of a pig. *J. Anim. Behav.*, **6**, 64–69.

NELSON, L. F., HAZEL, N. & CATRON, D. V. (1953). Baby pigs have a sweet tooth. *Feedstuffs, Lond.*, **25**, 16–17.

NIWA, T., ITO, S., YOKOYAMA, H. & OTSUKA, M. (1951). Studies on milk secretion in the sow. I. On nursing habits, milk yield, milk composition, etc. *Bull. natn Inst. agric. Sci. Tokyo* (Ser. G.), **1**, 135–150.

NIWA, T., MIZUHO, A. & SOEJIMA, A. (1959). Studies on the artificial insemination in swine. IV. Devices of a new type of artificial vagina and improvement of dummy and semen injector for swine. *Bull. natn. Inst. agric. Sci., Tokyo* (Ser. G), **18**, 45.

NOFZIGER, J. C. (1961). *Effects of some Environmental Stresses on Swine*. Ph.D. Dissertation. Wash. State Univ., Pullman, Washington, U.S.A.

NOVAK, U. & PYTEL, J. (1959). Neu Methode der Viehhaltung. *Int. Z. Landwirtsch., Sofia-Berlin*, **5**, 130–141.

PAREDIS, F. (1961). Onderzoekingen over vuchtbaarheidt en kunstmatige inseminatie by het varken. *Meded. VeeartsSch. Rijksuniv. Gent*, **5**, 63–133.

PATTERSON, R. L. S. (1968). *Investigations on boar taint*. European Association for Animal Production, Commission on Pig production, Dublin.

PAVLOV, I. P. (1934). Cited in Liddel, Manuscript of Pavlovian seminar, 25 April 1934.

PEO, E. R., JR. (1960). Round stall for farrowing. *Nebr. exp. Stn Q.*, **7**, 12.

PORZIG, E. (1967). Verhaltensforschung bei Schweinen. *FortschrBer. Landw.*, **5**, 76 pp.

PUHAC, I. & PRIBICEVIC, S. (1954). Prilog posnavanju ponaskanja i pasknich navika svinga. *Acta vet., Beogr.*, **6**, 65–69.

QUINTO, M. G. (1957). The breeding habits of Berkshire swine. *Philipp. Agric.*, **41**, 319–326.

RASMUSSEN, O. G., BANKS, E. M., BERRY, T. H. & BECKER, D. E. (1961). Social dominance in swine. *J. Anim. Sci.*, **20**, 982.

REEBS, H. (1960). *Das Verhalten des Schweines bei der Futteraufnahme*. Vet. Diss. F. Univ. Berlin.

RIPPEL, R. H., JR. (1960). *The effect of varying amounts of Floor Space and Feeder Space on the Growing-finishing Pig*. M.Sc. Thesis, South. Ill. Univ., Carbondale, Illinois, U.S.A.

ROBERTSON, J. B., LAIRD, R., HALL, J. K. S., FORSYTH, R. J., THOMSON, J. M. & WALKER-LOVE, J. (1966). A comparison of two indoor farrowing systems for sows. *Anim. Prod.*, **8**, 171–178.

ROBINSON, T. J. (1954). Relationships of oestrogen and progesterone in oestrous behaviour of the ewe. *Nature, Lond.*, **173**, 878.

ROBINSON, T. J. (1956). The necessity for progesterone with oestrogen for the induction of recurrent oestrus in the ovariectomized ewe. *Endocrinology*, **55**, 403–408.

ROTHE, K. (1963). Die kunstliche Besamung beim Schwein. *Arch. exp. VetMed.*, **17**, 957–1018.

SALMON-LEGAGNEUR, E. & AUMAITRE, A. (1961). Les préférences alimentaires du porcelet. VI. Appétibilité des farines animales. *Annls Zootech.*, **10**, 313–319.

SALMON-LEGAGNEUR, E. & FEVRIER, R. (1955). Les préférences alimentaires du porcelet. I. Influence du mode de présentation des aliments: granulés ou farine. *Annls Zootech.*, **4**, 215–218.

SALMON-LEGAGNEUR, E. & FEVRIER, R. (1959). Les préférences alimentaires du porcelet. III. Appétance de quelques céréales. *Annls Zootech.*, **8**, 139–146.

SANDERS, D. P., HEIDENREICH, C. J. & JONES, H. W. (1964). Body temperature and the oestrous cycle in swine. *Am. J. vet. Res.*, **25**, 851–853.

SCHENK, P. M. (1967). An investigation into the oestrus symptoms and behaviour of sows. *Z. Tierzücht. ZüchtBiol.*, **83**, 86–110.

SCHILLING, E. & ROSTEL, W. (1964). Methodische Untersuchungen zur Brunstfestellung beim Schwein. *Dt. tierärztl. Wschr.*, **71**, 429–436.

SCHMIDT, K. & BRETSCHNEIDER, W. (1954). Über die äusseren Ablauf der Sexualzyklus bei der Sau. *Tierzucht*, **8**, 119.

SELF, H. L. (1961). Personal communication. Iowa Agric. Exp. Sta., Ames, Iowa, U.S.A.

SEYMORE, R. & KLOPFER, F. D. (1960). Personal communication. Dept. Psychology, Wash. State Univ., Pullman, Washington, U.S.A.

SIGNORET, J. P. (1965). Untersuchungen der geschlechtlichen Verhaltensweise der Sau. Fragen der Verhaltensforschung bei Rind und Schwein. *Int. Koll. K. Marx Univ., Leipzig*, 15–21.

SIGNORET, J. P. (1967a). Durée du cycle oestrien et de l'oestrus chez la truie, action du benzoate d'oestradiol chez la femelle ovariectomisée. *Annls Biol. anim. Biochim. Biophys.*, **7**, 407–421.

SIGNORET, J. P. (1967b). Attraction de la femelle en oestrus par la mâle chez les porcins. *Rev. Comp. Anim.*, **4**, 10–22.

SIGNORET, J. P. (1968a). Attraction de la femelle en oestrus par le mâle chez les porcins. *VI° Cong. Int. Reprod. Anim. (Paris)* (to be published).

SIGNORET, J. P. (1968b). *Verhalten bei Schweinen in Verhalten landwirtschaftlicher.* Nutztiere Deutsche Land wirtschaftsverlag, Berlin (In press).

SIGNORET, J. P. (1968c). Unpublished data on reproductive behaviour in swine. Laboratoire de Physiologie de la Reproduction, 37, Nouzilly, France.

SIGNORET, J. P. & DU MESNIL DU BUISSON, F. (1961). Etude du comportement de la truie en oestrus. *IV° Cong. Int. Reprod. Anim. (La Haye)*, 171–175.

SIMPSON, Q. I. (1907). Rejuvenation by hybridization. *Am. Breeder's Ass.*, **3**, 76–81.

SISSON, S. & GROSSMAN, J. (1953). *The Anatomy of the Domestic Animals.* Philadelphia, Pa.: Saunders.

SLEBODNICK, E. B. & KLOPFER, F. D. (1953). Situational and hormonal influences on social dominance in gilts. Paper read at meeting of Western Psychological Association, Seattle, Washington, U.S.A.

SMIDT, D. (1965). *Die Schweinebesamung.* Verlag M. H. Schaper, Hannover.

SMITH, D. M. (1952). Milk production in the sow. *Proc. N.Z. Soc. Anim. Prod.*, **12**, 102–114.

SNETHLAGE, K. (1934). *Das Schwarzwild.* P. Parey Verlag, Berlin.

STEINBACHER, D. (1954). Zur Biologie der europäischen Wildschweines (*sus scrofa* Linne 1758). *Säugetierkdl. Mittl.*, **2**, 216.

VANLOEN, A. & MOLENAAR, B. A. J. (1967). A behavioural study in pigs. Methodology in measuring the evolution of the teat order. *Tijdschr. Diergeneesk.*, **92**, 297–307.

VANPUTTEN, G. (1967). Tail-biting in pigs. *Tijdschr. Diergeneesk.*, **92**, 705–712.

VERLINDEN, C. (1953). *Notes sur l'histoire naturelle et la chasse du sanglier*. Liège.

WESLEY, F. (1955). *Visual Discrimination Learning in Swine*. M.Sc. Thesis. Dept. of Psychology, Wash. State Univ., Pullman, Washington, U.S.A.

WHITTLESTONE, W. G. (1953). The milk-ejection response of the sow to standard doses of oxytocic hormone. *J. Dairy Res.*, **20**, 13–15.

WIERZBOWSKI, S. & WIERZCHOS, E. (1968). Analysis of the copulatory pattern in the boar. *VI° Cong. Reprod. Anim. (Paris)* (to be published).

WIGGAN, L. S. (1961). *Some Influences of Genotype and Environment on the Mating Behaviour of Swine*. M.Sc. Thesis., Univ. Nebr., Lincoln, Nebraska, U.S.A.

WIGGINS, E. L., WARNICK, A. C., GRUMMER, R. H., CASIDA, L. E. & CHAPMAN, A. B. (1951). Variation in puberty phenomena in inbred boars. *J. Anim. Sci.*, **10**, 494–504.

YERKES, R. M. & COBURN, C. A. (1915). A study of the pig *Sus scrofa* by the multiple choice method. *J. Anim. Behav.*, **5**, 185–225.

YOUNG, G. A., UNDERDAHL, N. R. & HINZ, R. W. (1955). Procurement of baby pigs by hysterectomy. *Am. J. vet. Res.*, **16**, 123–131.

ZEUNER, F. E. (1963). In: *A History of Domesticated Animals*. London: Hutchinson, pp. 256–271.

Chapter 12

The Behaviour of Horses

E. S. E. Hafez, M. Williams and S. Wierzbowski

The wild ancestor of the horse was the four-toed *Hyracotherium* of the Eocene. Small and short-limbed, probably with padded feet, they were well adapted to their forest habitat. Their teeth were low crowned, and their jaws short, suggestive of a diet of succulent titbits gathered from low woodland shrubs. During the Eocene, Miocene, and Pliocene the first digit was lost, while the second and fourth became reduced to vestigial splint bones. The remaining third metapodial became stronger and elongated as the only hoof bearing digit. During the Oligocene, as the prairies developed, the limbs became longer and horses evolved from forest browsers to fleet-footed inhabitants of the grasslands. The teeth became high crowned, the enamel pattern of the premolars more complex, increasing the grinding surfaces, and the jaws longer, as the diet became predominantly tough prairie grasses. There is evidence from the very early Pleistocene, when the modern genus *Equus* arose, that the horse was a herd animal for many fossil specimens from this period are found in close association.

The horse is a native of North America. There were migrations to the Old World in Early Eocene times and the horse diversified in Asia, Europe, and Africa. By the Late Pleistocene *Equus* was extinct in North America (Simpson, 1951). The present varieties of domestic horses (*Equus caballus*), draught, light, and pony, probably originated from different wild stocks such as the Asian, European, Tarpan[1], and Przewalsky's[2] horse (Ensminger, 1956). These breeds have been developed mainly for draught as in the Shire breed, for speed as in the Arab and English Thoroughbred, or for pleasure such as the Shetland ponies. The breeds vary with regard to body weight, compactness, conformation, and length of limbs; such anatomical differences are partly associated with differences in tameness, temperament, and other behaviours.

In general, there have been few scientific studies of horse behaviour except certain specialized studies; much of the literature has been anecdotal. With any species which has been closely associated with human beings, the untrained human observer is likely to be anthropomorphic.

I. INGESTIVE AND ELIMINATIVE BEHAVIOUR

A. Ingestive Behaviour

The small wild horses of Asia and the so-called wild, or feral, horse of the American western plains still live entirely on pasture, and when captured,

[1] Tarpan is a small dun-coloured wild species, formerly abundant in southern Russia and central Asia.

[2] The only surviving species of wild horses, discovered in Mongolia by the Soviet explorer, Przewalsky.

they do not readily change their dietary habits to include concentrated feeds. The domestic breeds of horses have acquired the habit of eating concentrates and herbage simultaneously or in successive seasons.

During grazing, horses cover a large area: they seldom take more than one or two mouthfuls before moving a step. Two animals may graze well apart if pasture area is unlimited, but close up at the appearance of a hostile herd within their pasture (Altmann, 1951).

The upper lip is sensitive, strong, and mobile and during grazing it pushes the herbage between the incisor teeth, which then cut the grass. Loose herbage is collected by the lips and tongue. The patterns and levels of grazing behaviour of horses have not been closely studied but in general grazing behaviour varies primarily with the climate, breed, and condition of pasture. A pasture which has been grazed by horses evidences patches of long grass where they defaecate but do not graze, and overgrazed parts where they graze but do not defaecate (Taylor, 1954). A grazing horse may sometimes cease grazing and walk a considerable distance to defaecate in a rough part of the field. Thoroughbred horses seem to show this pattern more than other breeds.

The suckling foal starts to nibble herbage and eat concentrates at the age of 5 to 10 days. If feed is available continuously the feeding times of adults are irregular. Colin (1871) reported that it requires a horse $1\frac{1}{4}$ hours for the mastication of 4 lb of dry hay out of which 60 to 65 boluses are made; the rate of mastication is 70 to 80 strokes per minute.

During the fox hunting season, overtired horses often refuse to eat when returned to their stables. Horses may bark trees or hedges or chew wood from gates even when the herbage is abundant, probably as a result of some nutritional deficiencies.

B. Eliminative Behaviour

Unlike cattle, horses interrupt any of their activities in order to urinate or defaecate; i.e. horses do not urinate or defaecate while eating or walking. During urination, the stallion attains a characteristic stance: the hindlegs are spread apart and extended backwards (no squatting as in sheep), the back is curved in a concave manner and urination takes place while the penis is still in sheath. After urination, the stallion turns to *smell the urine* and walks away, flexing the penile muscles and lashing the tail. The mare shows a less marked pattern than the stallion: the hindlegs are spread apart during urination, and the muscles of the vulva are flexed after urination. Before and after defaecation, the stallion (but not the mare) smells the spot of defaecation. After defaecation, the anus muscles contract and the tail is slightly lashed, probably to relieve the tension of the anus muscles. In general the mare with a foal exhibits more elaborate patterns during elimination than a mare without a foal, presumably to avoid contamination of the skin of the udder.

Hammond (1944) described the patterns by which horses deposit their excreta in the field as follows: "In a field grazed by mares exclusively the rough grass patches where the faeces fell gradually extended from year to year because the mare went up to the patch and faced it as they defaecated and so it grew from year to year. This alone, quite apart from parasites, might cause a pasture to become horse-sick in a few years. The stallions, however, behaved differently—when the stallions came up to the rough patch, they backed down on it, and so the size of the patch remained constant. Castration caused the behaviour of the stallion in this respect to change to that of the mare."

Vigorous adult horses defaecate 5 to 12 times per day, weaker ones more often; the daily number of urinations varies from 7 to 11 (Dukes, 1955). These values vary with the climate, type of feed, breed of animal, and amount of physical work. Although adult horses reject the faeces of their own kind, young foals tend to eat a considerable amount of fresh faeces of adult horses. Such behaviour ensures the acquisition of the proper bacterial flora for the foals' intestines.

II. ACTIVITY

A. Investigatory Behaviour

The horse appears to use all sensory modalities in investigatory activity. The newborn foal shows great curiosity and little fear approaching and smelling everything it sees while the mare spends much of its time shepherding the young foal away from danger. As the foal grows and learns to accept certain visual patterns as representing harmless phenomena, it begins to show fear of new objects, circling around these first at a distance and then approaching to smell them. Only after new objects have been investigated in this manner are they accepted.

The role of gustatory stimuli in investigatory behaviour is still uncertain. Young horses are often seen picking up unfamiliar objects in the mouth and teeth to chew them and spit them out again; this may be due to tactile or gustatory factors.

Recognition of home range. Even in unfenced areas horses move around only within their own home range, keeping to set routes at certain times of the day (see also section on Social Life). Recognition of home range seems to depend largely on olfactory stimuli, probably the smell of excreta since young horses when ridden away for the first time stop and smell piles of faeces on the road. The Vaqueros of Mexico claim that a well-trained horse can smell a lost cow up to half a mile away and follow the track of a cow or wild animal as a hunting dog does.

The tendency to return to home range is strong, and "*homing*" can be achieved from great distances. In some experiments carried out to determine the factors involved in homing, it was noted that there was a strong tendency for unguided horses in strange surroundings to walk up-wind. When the wind was blowing *from* the homeward direction homing was plainly evident, but when the experiments were carried out up-wind of the stables, the horses walked more willingly away from home than towards it, although the direction of their choice seemed uncertain. Visual cues did not seem to play an important part. The main dependence seemed to be on wind-borne cues but the sensory modality of the cues themselves could not be established (Williams, 1956). If olfactory cues are involved one would expect a sort of trial-and-error searching until very close to home.

B. Grooming

Unlike cattle, horses do not groom each other but they may nibble one another along the withers, or stand head-to-tail in summer, flicking the flies off each other's faces. The "wither-nibbling" pattern is usually a sign of recognition and is frequently noted when two animals meet after an interval.

In the spring, the horse sheds its winter coat which is replaced by short silky hair for the summer. The coat along the top of the body contains a large amount of dust and salt to protect the skin from the weather. During the winter much of this is accumulated by rolling in the mud. During the hot seasons, the horse scratches the dust out of its neck and back by rubbing along trees or fences and may develop the habit of persistently rubbing the tail on different objects as a result of irritation by external or internal parasites. If tail-rubbing is indulged in for a long time, it may become an established habit after the cause has been removed (Ensminger, 1956).

The readiness with which horses allow themselves to be groomed by man varies with the animal and with the part of the body that is being touched; the thin-skinned "Thoroughbreds" usually show more resentment than other thick-skinned breeds to grooming with a stiff brush. When not accustomed to handling, horses resent being touched around the head and ears and under the belly but not along the neck, withers, and back. Some horses retain such resentment permanently; this is not necessarily because of rough handling. The legs and feet are the last parts of the body which are allowed to be touched by man. Zebras seem to bear the same resentment and may require general anaesthesia for paring their feet; after prolonged domestication the rest of the body can be handled without difficulty.

C. Resting

During the day it is rare to see all members of a herd lying down together. One is almost always on the look-out; even if it appears to be asleep, it reacts to the slightest sound or movement. In lying down, the horse brings all its legs under its body and it bends its knees and hocks; the chest makes contact with the ground before the hindquarters. When actually lying down a horse either rests on one side of its chest with two legs (a fore and a hind) underneath, or else it lies on its side with legs stretched out. A cow can rest vertically on the ventral ridge of the sternum, but this attitude is impossible in a horse owing to the sharpness of the edge of that bone. In getting up, the horse stretches out both its forelegs in front, pressing its hindquarters upwards by fixing its hoofs firmly on the ground; the fore part of the body is the first to rise (Marshall & Halnan, 1946).

D. Sleeping

The fact that ruminant herbivores do not sleep or sleep very little is in marked contrast to the horse which may sleep profoundly 7 hours out of the 24 (Steinhart, 1937), mostly during the hottest period of the day. Most horses sleep while standing and lie down in the sun to expose the body to warmth. Intervals of sleep are short and irregular, depending on the degree of hunger and climatic conditions. Careful records of nocturnal behaviour of horses in the stable have not shown regular patterns of resting or sleeping. The duration of sleep at different temperatures depends to some extent on the type and breed (Barmincev, 1951). White and grey horses tend to be more alert and suspicious, especially at night, than darker animals; possibly as a result of a colour-linked hereditary tendency since the light colour renders the individual more conspicuous to its enemies than the dark colour.

E. Temperament

The term "temperament" is an inexact concept; yet it is still used in the Soviet literature (see Chapter 9). The term temperament in relation to a horse denotes the general consistency with which the animal behaves; its tendency to perform certain actions in certain situations. Temperament depends on physiological and genetical factors as well as experience and learning, i.e. temperament is the outcome of an interaction between heredity and the life history of the animal. Thus different breeds have different conformations and different temperaments. Keeler (1947) studied the inheritance of temperament in relation to coat colour. He claimed that white, palomino, and dapple grey horses tend to be more tractable than other colours, while duns, chestnuts, and sorrels are less so. On the other hand, Estes (1948) and Sloan (1949) could not find a correlation between temperament and coat colour. It is possible that differences in the sensory modalities are related to coat colours and so account for possible differences in the temperament.

Soviet investigators claim that the sire transmits its temperament to the offspring more frequently than does the dam, and that the "energetic" temperament is more readily transmitted than the "inert" (Manakov & Volkov, 1951; Ugrjumov, 1952; Pruski, 1958).

III. SOCIAL BEHAVIOUR

A. Social Life

The newborn foal shows no appreciation of social order but the dam usually tries to prevent the foal from joining other groups until it is old enough to run away should it be attacked. Mares vary in the extent to which they let foals other than their own approach them.

Feral horses are gregarious by nature and tend to wander in herds. Imanishi (1953) studied the social life of some 70 feral horses (of which only 3 were males) in Toimisaki, Japan. The animals spend the summers on open pastures. Winters are spent in their territories in the mountain forests where they live either alone or in groups of 2 to 6.

Horses show preference to certain individual herd-mates and show indifference or distaste to others. In large herds of horses, pairs of horses are often recognized grazing or resting together; they are usually of a similar social rank. Two horses may show agonistic behaviour on one occasion but react very differently on another.

Tactile and olfactory stimuli are used for social recognition. When two horses meet for the first time, four stages of introduction can be clearly recognized: (*a*) the horses circle round one another a short distance apart, (*b*) they touch nostrils, (*c*) each horse investigates the other's tail and body with the tip of its nose, (*d*) if mutual tolerance is decided upon they nibble one another along the crest of the neck.

B. Social Dominance

Grey horses, which tend to be cautious, often give warning of an approaching stranger and so seem to be rounding up the herd for flight, but this does not signify that they are dominant in other situations.

Social relations of horses were studied by placing a pail of oats within a group of horses (Grzimek, 1949) or by joining two horses and recording their dominance-submission interactions in terms of the intensity of kicks, bites, head bumps, or passive reactions (Montgomery, 1957). The latter procedure is a modification of that used in social studies in chickens (see Chapter 16). The percentage of total interactions in which an individual horse was involved when tested, is of special interest. Montgomery (*loc. cit.*) described the existence of a triangular relationship; Grzimek (*loc. cit.*) did not find triangular relationships among any of his combinations of animals. The two authors indicated that old or heavy horses as well as stallions tend to have a high dominance order; the relative importance of age, weight, and sex is not clearly understood.

In feral horses, the social dominance exists from the neighbourhood interactions: a group of 4 to 8 horses living in one territory are dominant over a smaller group (Imanishi, 1950).

C. Leadership

Although there are no clearly documented reports, it is believed that the leader is usually a stallion. In a herd consisting entirely of mares, one of the mares may take over leadership (Altmann, 1951; Williams, 1956). It is not clear whether the stallion *leads* a herd of mares or whether he rounds them up and *drives* them.

D. Communication

Vocal-auditory signals. These signals consist of patterns of sound, produced by motions of the respiratory and upper alimentary tract. A *snort* of the horse is a danger signal to the whole herd; the sound carries over great distances and is better developed in wild horses than in domesticated ones (Molison, 1959). The familiar *neigh* is a distress call: if one horse is separated from its companions and placed either alone or in the company of strangers, it signals its distress by a neigh. The *nicker* is a sign of relief, midway between the *whinny* and the soundless wrinkling of the nostrils seen in close communication or at courtship. There may be also innumerable other signals recognized by horses but unknown to us; one single snort may mean something quite different from a double one. Differences in the structure of sound-producing and sound-detecting mechanisms have an important bearing on intra-species communication including the variety thereof involved when human behaviourists study the sounds of other animals (Hockett, 1960).

Visual signals. Horses may communicate many of their feelings and intentions to one another by slight muscular movements of the feet, ears, tail, and skin around the nose and lips. Ancestral forms, armed with large eyeteeth, such as *Miohippus acutidens*, may have displayed these weapons by drawing up the corners of the mouth (Trumler, 1959). If the ears are placed at the back of the head, in the domestic forms, this means "beware", if the tail is lashed as well it means that a kick will follow. Raising the head and waving the upper lip in the air indicates nausea but may have another significance too. A foot lifted at one angle may have a different meaning from a foot lifted in another way. The signals may be picked up either visually or tactically. The horse's ability to perceive and interpret slight muscular movements is very acute as shown in the well-known case of *Clever Hans*, the horse which was believed to be able to count but was found to be responding to unconscious movements of its trainer's head (Katz, 1927).

If a group of horses are moving together, only the leader keeps its head and eyes focused on the distance while the others tend to follow in single file, each with its nose to the tail of the animal in front, apparently ready to receive and interpret any signals.

E. Agonistic Behaviour

Under wild conditions, natural enemies are avoided by flight or fight, according to the situation. Under farm and range conditions, the tendency to flee or fight depends on whether the horse is dominant or subordinate; fighting occurs among animals with similar social rank. The teeth and hind feet are the organs used for fighting. In wild Zebras (*Equus quagga*) Backhaus (1960) analysed and described different patterns of fighting: neck-fighting, circle-fighting, biting, and fighting by blows.

It is possible that in the more solitary ancestors of recent *Equidae* the sexes fought before mating (Trumler, 1959). Later on, when the mares formed herds, all that remained of this fighting was the "driving" of the mare by the stallion found in asses and other species.

The herd leader shows agonistic behaviour towards any other animal which tries to usurp its position. In isolation, aggressiveness may be due to excitement, fear, or habit. The tendency to aggressiveness diminishes as the confidence of the horse is increased, but the tendency increases if the horse is further frightened by punishment for its aggressive acts. The view that a tendency towards agonistic behaviour can be recognized in morphological characteristics such as the size of the ears, the look in the eyes, the breadth of the head, the prominence of the poll, or the colour of the coat is now generally discarded.

F. Play

Play may well be designed to increase locomotive skills and to test out social dominance. Young horses may be seen kicking and biting at the doors and sides of the stable as a form of play. Some horses develop the habit of stamping at their tails or kicking the sides of the stable for hours, apparently to hear the noise they make.

IV. SEXUAL BEHAVIOUR

A. Sexual Behaviour in Males

I. ANATOMICAL CONSIDERATIONS

As we mentioned in previous chapters, sexual behaviour of the male is determined to a large extent by the anatomical structures of the penis. The stallion has a typical vascular-muscular penis (comparable to the penis of humans) with no sigmoid flexure and the efficient function of the penis as an organ of intromission depends on the power of erection as a result of sexual excitement. Thus courtship[1] is important for copulation and successful horse breeding. The size, shape, and length of penis varies greatly between the flaccid and erect state (Figure 66) and the retractor penis

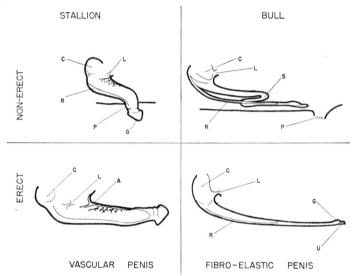

Fig. 66. Diagram of the anatomy of a vascular-muscular type penis (stallion) and a fibro-elastic type penis (bull) in the non-erect and erect positions. The anatomy of the penis determines, to a great extent, the ejaculatory reflexes of the species. A, dorsal artery. C, cavernous muscle. G, glans. L, suspensory ligament of penis. P, prepuce. R, retractor penile muscle. S, sigmoid flexure. U, urethral orifice. (From Hafez, 1960.)

muscle is relatively undeveloped and adherent to the ventral surface of the penis. Erection and protrusion of the penis are effected by gradually increasing tumescence of the erectile vascular tissue in the corpus cavernosum penis (Walton, 1955). The motor erectile mechanisms include the

[1] Commonly called foreplay or teasing.

synergism of two processes: (*a*) the expansion of the helicine arteries and the contraction of corresponding venules (under nervous control), permitting additional blood to enter the erectile tissues and slowing its exit, and (*b*) compression of the dorsal vein of the penis by the ischio-cavernosus muscle, thus limiting total flow from the penis (Julian & Tyler, 1959).

2. Patterns of Male Behaviour

The stallion exhibits three distinct phases: courtship—erection and mounting—and intromission and ejaculation.

Courtship. The sight of the mare elicits characteristic neighing and signs of "excitability" by the stallion. It smells the external genitalia and groin of the mare and extends its neck with an upcurled upper lip. Smelling is accompanied by "pinching": the stallion grasps with his teeth the folds of skin near the mare's croup (rump) and holds it more or less tightly.

Erection and mounting. Erection of the penis takes place gradually while the stallion is some distance from the mare. The penis at first protrudes so that half of its length is covered by the internal folding of the prepuce, which becomes smoother with progressing erection. At full erection, the free part of the penis is 12 to 20 inches long at the dorsal side. In vigorous stallions, the penis may reach full erection when they anticipate mating and even before they see the mare. Young stallions with no sexual experience often mount the mare at first laterally or from the front, and then move their hind legs behind the croup of the mare.

Wierzbowski (1959) measured the latency of erection and the latency of mounting in young and adult stallions (Table 41). Although the latency of

Table 41. Sexual Behaviour in Stallions. (Adapted from Wierzbowski, 1958, 1959; Nishikawa, 1959; Bielanski, 1960.)

Behaviour Component		Young Stallions	Adult Stallions (9 years)
Latency of erection[1] (sec)		163	119
Latency of mounting[2] (sec)		206	101
Latency of ejaculation[3] (sec)	Natural mating	11	13
	Artificial vagina	—	16
No. mounts per ejaculation	Natural mating	5·7	1·4
	Artificial vagina	—	2·2
Maximum no. of ejaculates	in 24 hr	—	11
	in 2½ hr	—	9

[1] Interval when stallion first sees mare until full erection.
[2] Interval when stallion first sees mare until first mounting.
[3] Interval from intromission to first emission of semen.

erection was not different in the two age groups, the latency of mounting was much shorter in adult stallions than in young ones, indicating the effect of sexual experience. The number of mounts per copulation ranges from 1 to 4 mounts with an average of 1·4 mounts; at each mount the stallion performs several pelvic oscillations, then dismounts (Wierzbowski, 1958). There are individual differences in the number of mounts per copulation. Year to year differences may be due to variations in the stimuli acting upon the tactile receptors of the penis as well as the exteroceptive environment during copulation. In general, the number of mounts per copulation is a characteristic of the individual.

Intromission and ejaculation. Intromission takes place after several pelvic oscillations which stimulate the engorgement of penis with blood. At full erection and after intromission, the glans penis forms the shape of a basin. Before ejaculation commences the penis is kept quite still adhering tightly to the vaginal wall for some seconds, then semen is ejaculated with great pressure entering the uterus directly (Walton, 1960). On an average the first ejaculation of semen occurs after some 13 seconds after intromission. The cessation of the pelvic oscillations is an external sign of the beginning of ejaculation. Rhythmical shrinking of the urethral muscles passes to the tail muscles, causing characteristic up and down movements of the tail. Most stallions, upon ejaculation, exhibit fast respiration, drooping of the head, and relaxation of the whole body. The stallion then dismounts the mare with a flaccid penis which is soon withdrawn into the prepuce.

3. Intensity of Sexual Behaviour

Erection and mounting may be observed at the age of 6 to 8 months but proper copulatory patterns are manifested at the age of 10 to 12 months. Sex desire is manifested throughout the year; Nishikawa (1961) collected semen from stallions throughout the year using ovariectomized mares injected with oestrogens. Sex drive was higher in the spring than in the autumn or winter, in agreement with the breeding season of mares. Young stallions returning from the race tracks may show seasonal variation in the expression of sex drive, presumably because of an inhibition of sexual behaviour (Parshutin & Rumjanceva, Jr., 1953; Bielanski et al., 1957). It is believed that intense sex drive is associated with low running performance of trotter stallions. With advancing age, the stallion loses its reproductive capacity before the disappearance of sex drive.

In general, the maximum number of ejaculates obtained from a given stallion using an artificial vagina is much lower than that in other species with a fibro-elastic penis.

4. Stimuli Releasing Sexual Behaviour

Erection in the stallion takes place rather slowly and depends normally upon the continued reception of erotic stimuli derived from courtship which seems to be important for the sequence of events leading to vascular engorgement of the penis. In contrast to the stallion, the bull exhibits little or no

PLATE XII. SEXUAL RESPONSES OF A STALLION TO A DUMMY

a. Visual and olfactory cues are not important to elicit erection, mounting and ejaculation; stallion blindfolded and with nose mask.

b. Stallion mounting the dummy.

(Photographs by Plewinski and Wierzbowski.)

PLATE XIII

a. Suckling behaviour.

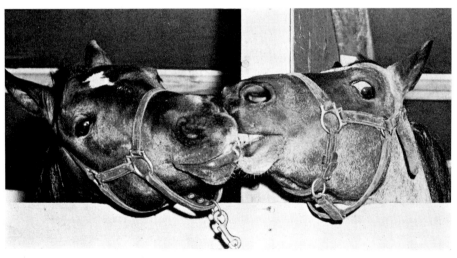

b. Fighting instinct and "self-right".

(Photograph *a* by Sumption; photograph *b* by Hafez.)

courtship before ejaculation; the contractile tissue is small and the penis enlarges little on erection.

It has been shown in previous chapters that sexual behaviour in the male is elicited via sensory modalities (visual, auditory, tactile, and olfactory); the relative importance of these modalities varies among species.

Visual stimuli. Erection and mounting may be also elicited by the presence of certain sexual objects other than mare, i.e. cow or a dummy. When adult stallions are blindfolded they react to a dummy (in 38 per cent of the cases) and semen can be collected by an artificial vagina (Plate XII). In blindfolded stallions, the mounting reflex is provoked by the stallion's touching the mare with his shoulder. After mounting, the stallion leans with his sternum upon the sacral region of the mare, pressing the mare with stiffly straightened forelegs and embracing her flanks. The stallion grasps and bites the mane of the mare with his teeth, keeping his hind legs close to that of the mare.

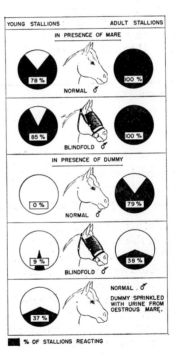

Fig. 67. Effect of sexual experience on copulatory responses of stallions under normal and experimental conditions. Young stallions did not react sexually to the dummy. The percentage of adult stallions which showed sexual responses to the dummy was lower in blindfolded stallions than in normal. In young stallions, sexual response was increased towards the dummy when sprinkled with urine from an oestrous mare. (Adapted from Wierzbowski, 1959.)

Visual stimuli are of some importance for erection and mounting. However, modification of sexual responses to visual stimuli are accomplished by training and experience (Fig. 67).

Olfactory stimuli. It is interesting to note that olfactory stimulation of the stallion is derived from different parts of the mare's body: the external genitalia, muzzle, and particularly the groin and the neighbouring parts, also the urine of oestrous mares (Berliner, 1959). Young stallions usually show poor or no response to a dummy; yet sexual behaviour is elicited when the

dummy is sprinkled with urine from an oestrous mare (Fig. 67). Wierzbowski (1959) inhibited olfactory stimuli of adult stallions using nose masks
applied with trichloroethylene and he found no inhibition in sexual behaviour.
When both olfactory and visual stimuli were inhibited, the stallion manifested erection and mounting even when a cow was used as a sexual stimulus
(Table 42).

Table 42. Erection and Mounting Responses in 3 Stallions when a Cow was used
as a Sexual Stimulus for Mounting. (Adapted from Wierzbowski, 1959.)

Condition of Stallion	No. of Trials	No. of Erection Responses	No. of Mounting Responses
Normal	4	0	0
Visual stimuli inhibited[1]	4	1	0
Olfactory stimuli inhibited[2]	3	0	0
Visual[1] and olfactory[2] stimuli inhibited	7	7	5

[1] Blindfolded.
[2] A nose mask applied with CCl_2:$CHCl$ was used.

It would appear that sexual experience and conditioned reflexes are involved in the expression of sexual response as affected by olfactory cues;
i.e. adult males can show sexual responses even in the absence of olfactory
cues while young males depend on these cues.

Tactile stimuli. Intromission and ejaculation are elicited by stimuli acting
upon the tactile receptors on the surface of the penis. Unlike the bull, the
penis of the stallion is not so sensitive to temperature and ejaculation is
initiated by the pressure and friction exerted on the penis by the vagina and
fornix (Walton, 1960).

Semen is usually collected in an artificial vagina; electro-ejaculation and
the massage of ejaculatory ducts are not as successful in the stallion as in the
bull, probably a result of neuro-anatomical differences. The ease with which
the semen is collected in an artificial vagina is about the same with heavy
and intermediate type stallions (Nishikawa, 1959). The number of mounts
per ejaculation as well as the latency of ejaculation are higher during semen
collection than during normal copulation (Table 41). The latency of ejaculation increases with the number of mounts required for ejaculation; this is
true for both the use of artificial vagina and for normal copulation.

Sexual experience and conditioned reflexes. Sexual experience accelerates
the speed of copulatory responses since young stallions copulating for the
first time exhibit certain awkwardness and require more mounts per ejaculation than older ones. The latency of erection and of mounting are shorter

in adult stallions than in young stallions copulating for the first time (Table 41).

Wierzbowski (1959) used a dummy and blindfolded stallions to perceive the essential differences between inexperienced and experienced stallions (Fig. 67). Young stallions did not show any sexual excitement in the presence of a dummy, while in most adult stallions normal sexual responses were elicited. This suggests the importance of associations developed through "stimulus generalization". The importance of conditioned reflexes was shown in castrated adult stallions which maintained normal sex desire for some 516 days and erection for 618 days following castration (Nishikawa, 1954).

5. ABNORMAL MALE BEHAVIOUR

Abnormalities in copulatory patterns are quite common in easily excited or sexually exhausted stallions. Failure to ejaculate may occur in spite of complete erection and energetic intromission. In Belgian stallions, the abnormality is manifested in the first breeding season or at the peak of any of the successive seasons (Vandeplassche, 1955). In some cases ejaculation is inhibited when semen is collected by an artificial vagina but no inhibition occurs with natural mating (Bielanski, 1960). Such inhibition is temporary and is mainly due to faulty application of the artificial vagina.

Another anomaly is the incomplete intromission and lack of pelvic oscillations after intromission. This irregularity, which is partly hereditary, may appear in young stallions at the onset of their sexual life and may persist during following years.

Excessive biting of the mare during copulation may be associated with disturbances in the copulatory mechanisms when the stallion performs the usual pelvic oscillations without ejaculation or with incomplete erection. Excessive biting of the mare may be caused by inhibition. When tests on mares were made using stallions throughout the year over a long time, sexual behaviour of the stallion became abnormal (Nishikawa, 1961). When brought close to the mares, their eyes became glaring, their ears dropped backwards, and they started to bite the mare excessively. Such abnormal excitement tapers out gradually when the tests were suspended for a considerable length of time.

Masturbation either complete or incomplete (without ejaculation) is common. In complete masturbation the stallion rhythmically rubs his rigid erected penis against his hypogastrium followed by rapid lowering of the croup, several forward movements of the pelvis, and abortive ejaculation. Temporary lack of sexual expression is also common in young stallions and in older stallions isolated in box stall without exercise.

B. Sexual Behaviour in Females

The information in the literature pertaining to the nature and length of the breeding season in the mare is conflicting and inconsistent. The main confusion seems to arise from a misinterpretation of the terms "polyoestrus" and "breeding season" (reviewed by Berliner, 1959). Mares can be classified into 3 categories: (*a*) monoestrus: the wild breeds of horses manifest several oestrous cycles during a restricted breeding season coinciding with the longest days of the year; the foals are born during a restricted period of the year; (*b*) transitory polyoestrus: some domestic breeds and some individual mares manifest sexual receptivity throughout the year, but ovulation accompanies oestrus only during the *breeding season,* the foals are born during a limited foaling season; (*c*) typical (true) polyoestrus: some domestic breeds and some individual mares exhibit sexual periodicity accompanied by ovulation throughout the year and foals are born throughout the year. So, in general, mares may show sexual receptivity all the year round but they do not necessarily breed all the year round.

The duration of pregnancy varies from 322 to 344 days according to the breed and the season of foaling (Clegg, 1959; Hammond, 1960). The interval to post-partum oestrus (foal heat) ranges from 2 to 40 days with an average of 11 days (Andrews & McKenzie, 1941; Cummings, 1942; Berliner, 1959); such an interval can be prolonged by lactation. It was suggested that copulation stimulates milk-ejection since in lactating mares, milk often flows freely during copulation. Van Rensburg & Van Heerden (1953) reported that oestrus and ovulation may occur during the 8th week of pregnancy.

1. Patterns of Oestrous Behaviour

During dioestrus, the mare is not receptive to the stallion and shows defensive reactions when the stallion attempts to mount. The degree of non-receptivity varies from mare to mare; some become aggressive, some are disinterested, while others are mildly interested. The onset of oestrus is more gradual in the mare than in other domestic animals. With approaching oestrus, the mare allows the stallion to smell and bite her, she urinates frequently in small quantities and discharges mucus from the vulva. These are associated with gross changes in the external genitalia including erected clitoris, elongated vulva, and swollen, partly everted labia (these symptoms are commonly called *showing*). The readiness of the mare to copulate is manifested by spreading the hind legs, lifting the tail sidewise, lowering the pelvis, shrinking the labia, and exposing the clitoris.

2. Intensity of Oestrus

The intensity of oestrous behaviour varies throughout the cycle and between individual mares. In this respect, Andrews & McKenzie (1941) presented a graphic form for the intensity of behavioural responses graded in 8 phases ranging from *very receptive* to *very actively resistant.* The intensity of oestrus reaches a maximum just before ovulation (Fig. 68) which often occurs at night (Studiencow, 1953). Whatever the duration of oestrus, ovulation bears a closer relation to its cessation than to its onset. In general the intensity of oestrous behaviour in the mare is much less than that in female asses. Following ovulation, sexual receptivity decreases until oestrus ceases 1 to 2 days later. It is generally believed that oestrous behaviour is intensified if the mares are pastured with a stallion which has undergone an operation for retroversion of the penis (surgery by which the penis points backwards during erection).

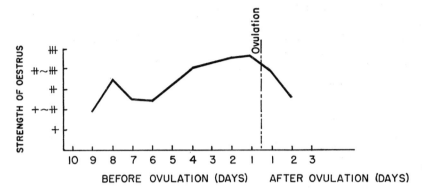

Fig. 68. Level of oestrus in relation to ovulation time in Korean, Chinese, and Japanese mares. (Adapted from Nishikawa, 1959.)

3. DURATION OF OESTRUS

The duration of oestrus in the mare is very variable and ranges from 5 to 9 days (Asdell, 1946). Oestrus is shorter in light mares than in draught mares (Fig. 69). Breed differences in the duration of oestrus in different localities (Table 43) are attributable to genetical and environmental factors,

Table 43. Duration of Oestrous Behaviour in the Mare as affected by Genetic and Environmental Factors. (Data from the literature.)

Breed or Type	Duration of Oestrus (days)		Locality	Author
	Range	Mean		
Draught	4–21	7	U.S.A.	Aitken (1926)
	1–14	5	U.S.A.	Andrews & McKenzie (1941)
Korean and other	4–14	9	Japan	Satoh & Hoshi (1932)
	1–15	6	Japan	Nishikawa (1959)
Light	1–37	6	U.S.A.	Andrews & McKenzie (1941)
Thoroughbreds	3–15	7	England	Hammond (1934)
	4–7	5	U.S.S.R.	Krat (1933)
Welsh and Shetland	3–30	7	England	Hammond (1938)

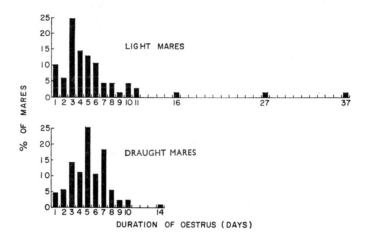

Fig. 69. Frequency distribution showing the duration of oestrus in light and in draught mares. The average duration in light mares was shorter than that in draught mares (5·5 and 5·2 days respectively). (From Andrews & McKenzie, 1941.)

and to differences in methods used for testing oestrus, and for statistical analysis.

4. ATYPICAL OESTRUS BEHAVIOUR

"Split" oestrus and prolonged oestrus are common in mares. In split oestrus, the manifestation of oestrus ceases for a short period followed by recurrence of sexual receptivity during what is evidently one full oestrous period (Fig. 70). Prolonged oestrus may last from 10 to 40 days. Skurgin (1939) and Buiko-Rogalevic & Skatkin (1946) have stated that prolonged oestrus usually occurs in mares which have failed to conceive or have aborted, and those used for heavy draught. Mintscheff & Prachoff (1960) reported that a dietary régime consisting of 1 day's complete starvation (followed by controlled level of feeding) caused a reduction in duration of oestrus from 12·2 days to 6·9 days.

V. SCHOOLING AND TRAINING

A. Sensory Capacities and Perception

Vision. The young foal is very quick to learn the boundaries and constituents of its home territory as shown by its reluctance to leave this and its fear-reactions to new objects after the first 24 hours. Quick visual learning may, however, continue well into adult life (Williams, 1960), and this life-long ability to absorb new visual patterns may account for the lack of species misidentification ("imprinting") seen in bottle-fed and hand-reared foals.

In its natural habitat, the adult horse keeps a sharp look-out for its natural enemies, even while grazing. Although horses, like other herd animals,

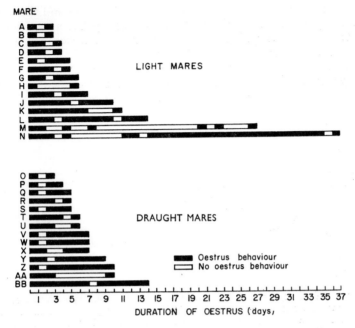

Fig. 70. The occurrence and duration of "split" oestrus in draught and light mares.

Split oestrus was characterized by an initial heat period followed by a non-receptive period of 1 or 2 days followed by a subsequent return of oestrus. Graafian follicles were present in the initial heat period and continued to develop during the non-receptive interval. (From Andrews & McKenzie, 1941.)

are very active during cool parts of the day, the eye of the horse is adapted in several ways for darkness vision as in nocturnal animals. Also, its eyes are set on the sides of the head so that each eye receives a very wide but largely different scene (Fig. 71). When the images of the two eyes are superimposed, the result is not one small three-dimensional field but a flat panorama. The lens of the horse's eye is non-elastic, but the retina is arranged on a slope or ramp, the bottom part being nearer the lens than the top (Rochon-Duvigneaud 1943; Detwiller, 1943). In order to focus on objects at different distances, the horse has to raise or lower its head so that the image is brought on to that part of the retina at the correct distance to achieve a sharp image. This arrangement has the advantage for animals in the wild state that while the head is down grazing, both near objects on the ground and distant ones on the horizon are in focus simultaneously.

A French army horse, in North Africa, was able to distinguish its master in a red uniform from other men in blue ones at a distance of ⅓ mile (Walls, 1942); however, the discrimination may have been based on brightness or intensity rather than hue since colour vision has not been established in horses by scientific methods.

Other senses. Olfaction is well developed in the horse and the epiglottis plays an important role by forming with the soft palate a partition which prevents the breathing of air through the mouth (Dukes, 1955). During feeding, and at other times when the mouth is open, air flows through the nasal passages. This efficient mechanism for air passage is advantageous for olfactory activity. Horses hear over a great range in frequencies and many of them can pick up sounds too slight for us to perceive.

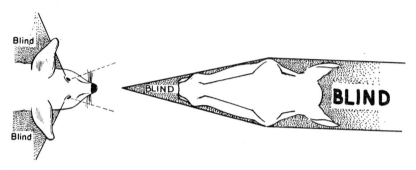

Fig. 71. The shape of the horse head and the position of his eyes allows vision of a wide panorama and a very extensive field of view. The predatory animals on the other hand have the eyes well forward and concern themselves less with pursuit by other animals. (Adapted from Prince, 1956.)

B. Intelligence

There are many difficulties in testing a horse's basic intelligence because of lack of problems suited to the horse's normal environment and physical constitutions. Because some horses have been known to use their teeth or fore-feet on occasions, and because they have done so to open gates or doors, it does not follow that such acts demonstrate intelligence. There is always the diffi-culty of knowing whether a horse which fails to see its way around a detour is obeying its intellect or its training.

It is true that a horse loose in the field (such as *X* in Fig. 72) when

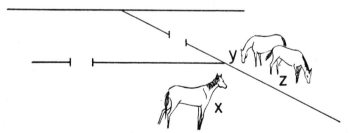

Fig. 72. A type of detour problem which horses frequently encoun-ter. *X*, which has inadvertently become separated from its companions *Y* and *Z*, is unlikely to see the way back to them through the open gates. Unless it encounters the gaps in the course of its random gallop-ing, it will probably try to rejoin *Y* and *Z* by jumping the intervening hedge at the nearest point to them. (From Williams, 1956.)

s eparated from its companions (Y and Z), may show almost unbelievable stupidity in finding the way back to them. Even if gates adjoining X's field to that occupied by the others are wide open, so that by retracing its steps a little way X could reunite with its friends without difficulty, it will very seldom realize this simple solution to its predicament. It may charge up and down the fence closest to the others, pushing against it, stumbling into it, until perhaps in desperation, horse X will start galloping wildly in all directions and finally hit upon the open gate by accident.

The Wonder Horse "Clever Hans". About 1900, an extraordinarily perceptive horse in Germany, named "Hans" was able to perform very intelligently after a 2-year training period (Katz, 1927). "The horse would stand in front of its trainer and paw the ground with its hoof. If it was asked the sum of $2+2$, it would paw the ground 4 times. Hans could not only do addition but also multiplication and division; it could spell out words and sentences by pawing the ground the appropriate number of times for each letter of the alphabet. Remembering that R is the 18th letter of the alphabet in the midst of spelling a complicated sentence is a pretty difficult task" (Scott, 1958). A committee of zoologists and psychologists studied the case and found that someone had to be present who knew the answer to the problem posed. It was concluded that Hans was watching this person, who, unknown to himself would relax his head and neck muscles when the correct number of taps had been made, so giving Hans the signal to stop.

C. Training

Skinner (1951) has described some techniques used to train animals, but most of these would be difficult to apply to the horse in practical situations. It is difficult to predict from the way a horse behaves in one situation just how it will behave in another, or to assume from its reaction to one stimulus that it will react to similar ones in the same way if it is presented with them in different situations. On the whole, horses show little generalization of either perception or learning. What they have learned in one setting is not carried over to others, so that they have to be taught each act in the context in which it is required. They do not appear to generalize even from one side of their bodies to the other. A horse which is led or handled from the left (near) side will not allow the same things to be done on the right (off) side until it has been accustomed to these separately.

1. PRINCIPLES OF TRAINING HORSES

In describing his own techniques and methods, each trainer shows his idiosyncrasies. All trainers, however, have agreed on certain basic principles.

(a) The young animal must never be overworked, but should be allowed to relax as soon as it has grasped a new idea. Much of the skill in training lies in knowing when to stop.

(b) The animal should never be given an opportunity to exercise an undesired reaction to a signal. In this way the training of horses for practical purposes differs from the trial-and-error training of laboratory animals designed to study the factors influencing their behaviour. The latter are given a choice of right or wrong response, and learning of the desired response is encouraged by reward. When training horses, every effort is made to ensure a correct response and prevent an incorrect one. Thus when the circus horse is being taught to recognize its name, it is held on the end of a lunge-rein; each time its name is called it is pulled towards the trainer.

(c) Signals must be easily distinguished by the animal and applied in a uniform manner. Thus, if the signal is a vocal one (e.g. a word) care must be taken to give the same word in the same tone of voice on each occasion. With increased training, sensory discrimination tends to be improved. The riding horse taught to respond to touch signals on its mouth and flanks becomes sensitive to very slight changes of pressure (Wynmalen, 1953). This may account for a highly schooled horse's anxiety under a tense and frightened

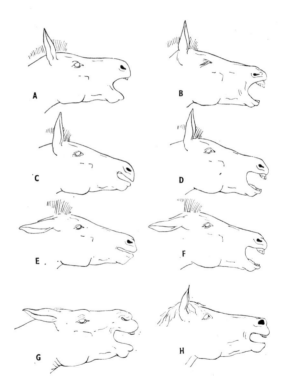

Fig. 73. Communication and facial expression in the zebra *Equus* sp. A. Begging for feed, a simple mouth opening. B. Yawning. C, D. Greeting expression in two successive stages; note the erect ears. E, F. Threatening expression in two successive stages, note the ears drawn backwards. G. Another threatening position, note the corner of the mouth and up-curled lips. H. Acoustical greeting with corners of the mouth drawn up (horse). (Adapted and redrawn from Trumler, 1959.)

rider. The circus horse trained to respond to a visual stimulus (i.e. the angle of the trainer's whip) is able to pick up very slight changes of direction. One horse trained to lie down every time its trainer bent down and touched his feet was able to pick up this signal even when its trainer was walking among members of the audience outside the ring. The discriminatory power of the horse "Clever Hans" in picking up slight movements of the head which its trainer was unconsciously making has already been referred to.

There is some doubt about the sensory mechanism by which signals are most readily learned. The manner in which it is most convenient to give signals depends on the action it is intended the horse should do.

(d) The trainer should understand the characteristic idiosyncrasies of the individual horses as well as the meaning of different actions and facial expressions of the horse during and after the training period. The significance of different facial expressions have been described (Trumler, 1959) for zebras (Fig. 73).

2. Techinques of Training

The different uses to which horses are to be put involve variations or additions to the above principles.

Driving. As well as teaching it to obey signals of command the horse must be taught to overcome its innate fear of being followed. A fully trained driving horse not only shows no fear of its cart, but always allows room for it when turning through gates or trotting along roads. This is achieved by habituation and the use of blinkers to prevent the horse from seeing objects behind it.

Riding. Different degrees of sensitivity to the riding aids are required by different types of riders. The high school or Dressage rider is dissatisfied by a horse which cannot distinguish and respond to very slight pressure changes. On the other hand, a horse suitable for a novice must be prepared to disregard many types of unintended pressures from the rider. The actual aids to which different movements are trained by different riders are matters of personal choice.

Jumping. Few horses jump regularly until they are taught to. The unschooled horse can be kept in an enclosure by fences 2 feet high unless the drive to escape is very strong. The ability developed by the schooled horse depends on its physique, temperament, and confidence acquired during training. While boldness and the style of a horse can often be recognized by an experienced rider in the early stages of schooling, the manner in which it reaches its potential and the mistakes it makes before doing so may vary with individuals. Owing to their lack of binocular vision, horses seem to have difficulty in judging the correct take-off distances for fences in the early stages. Thus unless it is told when to jump a fence, a young horse tends to get too close to it before taking off. Possibly for the same reason, horses prefer to jump vertical than horizontal obstacles and normally avoid ditches.

Racing. Training for racing consists mainly in building up the horse's physique and developing speed and stamina. At the same time, the horse must be taught what is expected of it, i.e. not to exert itself unduly until its rider gives the command. Experienced horses differ greatly in the amount they either depend on their riders during a race or use their own judgement (Hislop & Skeeping, 1951; Francis, 1957). Much of a horse's success in racing depends on the confidence acquired during the first races. The memory of horses for different courses can often be recognized by jockeys who sense a horse's anxiety as it approaches a fence where it had a previous fall, or feel it begin to quicken its stride at the point where an effort was made in a previous winning race. The memory of the horse for different hunting trips can also be recognized.

Circus. The training of circus horses involves special techniques based on three fundamental principles: (*a*) repetition, (*b*) insistence on obedience; the horse is gently pushed or pulled into the required position each time the signal is given, and (*c*) rewarding by feeding or petting each time the required response is given. Each horse is first trained to recognize its name, and then to perform certain acts in response to signals. No act is ever permitted to be performed without its appropriate signal, for anticipation would disrupt a performance in which many animals have to work in a team. After a horse has learned its routine, much time has to be spent in accustoming it to the distractions (loudspeakers, crowd, flags, lights, etc.) under which it has to perform. Only healthy, contented horses make good performers, and efforts should be made to keep circus horses fit and to prevent boredom once training has been completed.

VI. ATYPICAL BEHAVIOUR (VICES)

Atypical behaviours in the horse (generally known as vices) are largely produced by confinement and loneliness (Levy, 1944, 1954). Little has been recorded about atypical behaviours in the natural state, but a great deal has been reported about vices developed in the stable and during training. Most vices are caused by improper management and faulty handling but the importance attached to heredity and constitution varies among authors.

Once an abnormal pattern has been established, it is difficult to eradicate it and even if the animal can be prevented from exercising it for a short time, the habit tends to recur during stress. Methods of inhibiting abnormal behaviour are based on rewards and removing the causes of discomfort during what is considered by the horse as unpleasant. The most common atypical patterns are discussed below.

Neurosis. Neurosis may take various forms. There may be signs of physical stress, such as sweating, shivering, rolling of the eyes, and excessive salivation; or neurosis may be associated with loss of appetite, restlessness, and tendency to attack other herd-mates. There may be defects of perception: the animal fails to see or hear in the normal manner and tends to bump into obstacles. Neurotic horses may develop certain habits, such as licking the lips, swaying on the feet, stamping the ground, or pacing up and down and around in circles.

Rearing. In rearing, a horse brings its hind legs some distance under its body, and at the same time throws up its head, and all the legs are then used to raise the body upwards. Rearing is used to evade forward action and may be employed as a resistance to the rider. A rearing horse may fall over backwards on a rider. The confirmed rearer can be felt to balance itself and prepare for this act. The unschooled horse may stand on its hind legs for a few seconds in the course of its antics but seldom remains poised for long. A tendency to choose this form of evasion in preference to any other may be due to heredity or faulty management. Habitual rearing is usually a sign of fear so that increasing the animal's confidence should be aimed at rather than further frightening. Hitting a horse over the head for rearing is seldom successful and only increases its anxiety.

Bucking. Bucking takes several forms but in all cases the horse stands on its forelegs and kicks up with the hind legs. Bucking before or during a gallop usually is due merely to excitement. Bucking with the head between the forelegs to dismount the rider is an evasion and originates in the same way as rearing. "Fly-bucking" is usually developed only in the highly schooled horse in response to perplexity; the head is kept high and the horse kicks out either to the side or behind. Bucking can be treated by regular exercise with consistent rewards and punishments to reinforce the desired reaction to signals and to increase the horse's confidence. Bucking horses are in demand in rodeos; thus bucking may not be considered as typical abnormal behaviour.

Bolting. The horse does not stop when ordered to do so despite all signs or signals. The first time a horse bolts is usually due to fear, but the tendency to bolt rapidly becomes established as a habit. The use of more severe bits is often recommended but is seldom effective for very long unless reinforced by other treatment; a satisfactory method is to re-school the horse to vocal signals. Bolting is a term also used to describe eating feed too fast, but this is not discussed here.

Jibbing. The horse refuses to go forward, usually when asked to leave the stable or the company of others, or may rear and refuse to enter confined stables. Confidence must be increased by rewards and firmness of the rider, and a counter-habit must be established. As soon as the horse has been per-suaded to move on after jibbing, it should be taken back to the spot where it stopped and made to leave it again; the procedure should be repeated within a short time. Very often a lead from another horse (especially into closed spaces) helps to overcome initial reluctance; as soon as it has been persuaded to follow the lead the horse should be made to repeat the manoeuvre by itself.

Wind-sucking. The animal stretches its head forwards, holds any fixed object with its teeth and sucks in gulps of air until the stomach is full of wind; the victim is usually in poor condition. The exact causes of wind-sucking are not fully understood. The tendency to wind-suck may be due to nutritional deficiencies, illness of the mother, or poor pasture. The pattern first appears in the foal at weaning, and may originate as a substitute for suckling. In Japan, 1 per cent of the horses are wind-suckers; Hosoda (1950) considered wind-sucking as a genetic factor of a recessive inheritance.

Weaving. Weaving is common in horses confined to stalls: a series of rhythmic side to side oscillations of head and neck with forelegs usually spread. The pattern may be analogous to head-shaking in chickens. Weaving movements are especially prominent before feeding, urination, and defaeca-tion (Holzapfel, 1938) and as a result of external stimuli such as noise, restless movement of a horse in the next stall, approach of the stableman, or sudden opening of a window (Levy, 1944). The explanation of weaving in terms of movement restraint or frustration is supported by the following considerations: (*a*) weaving occurs in riding and buggy horses, never in work horses regularly employed; (*b*) removal from stall to a wide box stall greatly eliminates weaving; (*c*) weaving is ameliorated by increasing the range of movement (Levy, *loc. cit.*). Summerhays (1952) recommended hanging bricks outside the door on to which the horse knocks its head if it moves.

Kicking. When a horse kicks, the head is lowered and the croup is raised, and the hind legs are thrust suddenly and forcibly backwards. Kicking at others may be either aggressive or self-defensive. Kicking at walls is usually a sign of boredom; such kicking is discouraged if soft padding is placed in all accessible parts of the stable.

Biting. Sensitive or thin-skinned horses often bite when they are being girthed up or groomed under the stomach. Grooming with a soft brush minimizes the desire to bite and re-education is usually effective, but if it fails the horse can be fitted with a muzzle.

Shying. Shying is a tendency to turn around and run away from certain objects—usually unfamiliar scenes such as a lake, traffic, or unusual crowd. A horse which once starts shying at an object or at a particular place on a road may continue to do so for no apparent reason. Shying may be caused also by defective vision or hearing, nervousness, or playfulness.

Other abnormal behaviour includes rolling, stumbling, toe-dragging, over-bending, rug-stripping, bed-eating, and resistance to be mounted by rider, bridled, cantered, or boxed (Summerhays, 1959).

REFERENCES

AITKEN, W. A. (1926). The estrous cycle of the mare. *Vet. Practitioner's Bull.*, Iowa State College, **8**, No. 1, 178–187.

ALTMANN, M. (1951). The study of behavior in a horse-mule group. *Sociometry*, **14**, 351–366.

ANDREWS, F. N. & McKENZIE, F. F. (1941). Estrus, ovulation, and related phenomena in the mare. *Mo. Agric. Exp. Stn Res. Bull.*, No. 329.

ASDELL, S. A. (1946). *Patterns of Mammalian Reproduction.* New York, N.Y.: Comstock.

BACKHAUS, VON D. (1960). Über das Kamferhalten beim Steppen-Zebra (*Equus quagga*, H. Smith, 1841). *Z. Tierpsychol.*, **17**, 345–350.

BARMINCEV, N., JR. (1951). Behavior characteristics of horses under taboon management. *Konevodstvo*, (**10**) 16–22.

BERLINER, V. R. (1959). The estrous cycle of the mare. In: *Reproduction in Domestic Animals.* H. H. Cole & P. T. Cupps (eds.). Ch. 8, Vol. **1**, p. 267. New York, N.Y.: Academic Press.

BIELANSKI, W. (1960). *Reproduction in Horses.* 1. *Stallions.* Krakow, Poland: Publ. No. 116, Instit. Zootech. & Agric. Coll.

BIELANSKI, W., WIERZBOWSKI, S. & ZAKRZEWSKA, G. (1957). The results of large-scale investigations on the semen and sexual reflexes of stallions. *Zesz. Nauk. W.S.R. Krakow, Z.*, **1**, 97–114.

BUIKO-ROGALEVIC, A. N. & SKATKIN, P. N. (1946). The sexual cycle in draft mares. *Konevodstvo*, (**3**) 12–15.

CLEGG, M. T. (1959). Factors affecting gestation length and parturition. In: *Reproduction in Domestic Animals.* H. H. Cole & P. T. Cupps (eds.). Vol. I, Ch. 15, p. 509. New York, N.Y.: Academic Press.

COLIN, G. C. (1871). *Traité de Physiologie Comparée des Animaux*. 2nd edn Paris. (Cited by H. H. Dukes, 1955.)

CUMMINGS, N. J. (1942). A study of estrus and ovulation in the mare. *J. Anim. Sci.*, **1**, 308–313.

DETWILLER, S. R. (1943). *Vertebrate Photoreceptors*. New York, N.Y.: Macmillan.

DUKES, H. H. (1955). *The Physiology of Domestic Animals*. 7th edn Ithaca, N.Y.: Comstock.

ENSMINGER, M. E. (1956). *Horses and Horsemanship*. Danville, Illinois: Interstate.

ESTES, B. W. (1948). Lack of correlation between coat color and temperament. *J. Hered.*, **39**, 84 (Abstract).

FRANCIS, D. (1957). *The Sport of Queens*. London: Michael Joseph Ltd.

GRZIMEK, G. (1949). Rangordnungsversuche mit Pferden. *Z. Tierpsychol.*, **6**, 455–464.

HAFEZ, E. S. E. (1960). Analysis of ejaculatory reflexes and sex drive in the bull. *Cornell Vet.*, **50**, 384–411.

HAMMOND, J. (1934). Some new progress made in the science relative to reproduction in horses. Special Report. Proc. 16th Cong. Agric., Budapest, Sec. VI.

HAMMOND, J. (1938). Recent scientific research on horse breeding problems. *York. agric. Soc. J.*, **95**, 11–25.

HAMMOND, J. (1944). In a discussion on "The importance of prosecuting the study of animal behaviour and suggestions for its development as a field of veterinary medicine," by J. T. Edwards, *Proc. Br. Soc. Anim. Prod.*, p. 103.

HAMMOND, J. (1960). *Farm Animals, Their Breeding, Growth and Inheritance*. London: E. Arnold & Co.

HISLOP, J. & SKEEPING, J. (1951). *Steeplechasing*. London: Hutchinson Co.

HOCKETT, C. F. (1960). Logical considerations in the study of animal communication. In: *Animal Sounds and Communication*. W. E. Lanyon & W. N. Tavolga (eds.). P. 392, Publ. No. 7, Washington, D.C. Amer. Inst. Biol. Sci.

HOLZAPFEL, M. (1938). Über Bewegungssterotypien bei gehaltenen Saugern. *Z. Tierpsychol.*, **2**, 46–72.

HOSODA, T. (1950). On the heritability of susceptibility to wind-sucking in horses. *Jap. J. zootech. Sci.*, **21**, 25–28.

IMANISHI, K. (1950). Social life of semi-wild horses in Toimisaki. III. Summary for the three surveys undertaken in 1948–1949. *Physiology & Ecology* (Japan), **4**, 29–42.

IMANISHI, K. (1953). Social life of semi-wild horses in Toimisaki. II. Horses in their winter quarters. *The Annual of Anim. Psychol.* (Japan), **3**, 11–32.

JULIAN, L. M. & TYLER, W. S. (1959). Anatomy of the male reproductive organs. In: *Reproduction in Domestic Animals*. H. H. Cole & P. T. Cupps (eds.). Ch. 2, Vol. 1, p. 29. New York, N.Y.: Academic Press.

KATZ, D. (1927). *Animals and Men*. London: Longmans Green.

KEELER, C. E. (1947). Coat-color, physique and temperament. *J. Hered.*, **38**, 271–277.

KRAT, A. V. (1933). Refinement of tests of mating in the Thoroughbred and its derivatives. *Konevodstvo* (6), 16–20.

LEVY, D. M. (1944). On the problem of movement restraint. Tics, stereotyped movements, hyperactivity. *Amer. J. Orthopsychiat.*, **14**, 644–671.

LEVY, D. M. (1954). *The Relation of Animal Psychology to Psychiatry in Medicine and Science*. New York, N.Y.: International Universities Press.

MANAKOV, I. D. & VOLKOV, D. A. (1951). The relation between types of nervous activity, constitution and performance. *Konevodstvo* (6), 16–20.

MARSHALL, F. H. A. & HALNAN, E. T. (1946). *Physiology of Farm Animals*. Cambridge: Cambridge University Press.

MINTSCHEFF, P. & PRACHOFF, R. (1960). Versuche zur Erhöhung der Befruchtungsfahigkeit der Stute durch Hungerdiät wahrend der Brunst. *Zuchthyg. Fort-PflStör. Besam. Haustiere*, **4**, 40–48.

MOLISON, S. (1959). *Stable Studies*. London: Hollis & Carter.

MONTGOMERY, G. G. (1957). Some aspects of the sociality of the domestic horse. *Trans. Kans. Acad. Sci.*, **60**, 419–424.

NISHIKAWA, Y. (1954). Strength of sexual desire and properties of ejaculate of horse after castration. *Bull. Nat. Inst. Agric. Sci.* (Japan), Ser. G, No. 8, 161–167.

NISHIKAWA, Y. (1959). *Studies on Reproduction in Horses*. Tokyo: Japan Racing Assoc.

NISHIKAWA, Y. (1961). Personal communication, Dept. Anim. Husb., Kyoto Univ. Kitashirakawa, Kyoto City, Japan.

PARSHUTIN, G. V. & RUMJANCEVA, E., JR. (1953). The occurrence of sexual reflexes in young stallions. *Konevodstvo*, 23 (7), 12–17.

PRINCE, J. H. (1956). *Comparative Anatomy of the Eye*. Springfield, Illinois: Charles C. Thomas.

PRUSKI, W. (1958). Inheritance of speed from the sire. *Z. Tierz. Biol.*, **71**, 69–90.

ROCHON-DUVIGNEAUD, A. (1943). *The Eyes and Vision of Vertebrates. Les Yeux et la Vision des Vertèbres*. Paris: Mason & Co.

SATOH, S. & HOSHI, S. (1932). A study in reproduction in mares. *J. Jap. Soc. vet. Sci.*, **11**, 257–268.

SCOTT, J. P. (1958). *Animal Behavior*. Chicago: University Press.

SIMPSON, G. G. (1951). *Horses; The story of the Horse Family in the Modern World and through Sixty Million Years of History*, New York, N.Y.: Oxford Univ. Press.

SKINNER, B. F. (1951). How to teach animals. *Scient. Amer.*, 185 (6), 26–29.

SKURGIN, A. V. (1939). Prolonged œstrus in the mare. *Voprosy Plodov. Rabotosp. Losad. Moscow: Seljhozgiz:* 29–35.

SLOAN, F. (1949). Color and temperament in mares. *J. Hered.*, **40**, 12 (Abstract).

STEINHART, P. (1937). Der Schlaf des Pferdes. *Z. Veterinark*, 49; 145–157; 193–232.

STUDIENCOW, A. (1953). Veterinary Obstetrics and Gynæcology. "*Wieterinarnoje akuszerstwo i ginekologija.*" Moscow: Gos. Izd. Sielsk. Lit.

SUMMERHAYS, R. S. (1952). *Encyclopædia for Horsemen*. London: Frederick Warne & Co.

SUMMERHAYS, R. S. (1959). *The Problem Horse*. London: J. A. Allen.

TAYLOR, E. L. (1954). Grazing behaviour and helminthic disease. *Br. J. anim. Behav.*, **2**, 61–62.

TRUMLER, E. (1959). Das "Rossigkeitsgesicht" und ähnliches Ausdrucksverhalten bei Einhufern. *Z. Tierpsychol.*, **16**, 478–488.

UGRJUMOV, B. A. (1952). The effect of the temperament of the sire on the transmission of his type. *Konevodstvo*, **22** (8), 37.

VANDEPLASSCHE, M. (1955). Ejaculation disturbance in the stallion. *Dtsch. tierärztl. Wschr.*, **62**, No. 43/44.

VAN RENSBURG, S. W. J. & VAN HEERDEN, J. S. (1953). Infertility in mares caused by ovarian dysfunction. *Onderstepoort J. vet. Res.*, **26**, 285–313.

WALLS, G. L. (1942). *The Vertebrate Eye*. Cranbrook Inst. Sci. Bull. No. 19. Bloomfield Hills, Michigan: Cranbrook Press.

WALTON, A. (1955). Sexual Behaviour. In: *Recent Progress in the Physiology of Farm Animals* J. Hammond (ed.). Vol. 2, Ch. 13, p. 603. London: Butterworths.

WALTON, A. (1960). Copulation and natural insemination. In: *Marshall's Physiology of Reproduction*. A. S. Parkes (ed.), 3rd edn, Ch. 8, Vol. I, Part 2, London: Longmans.

WIERZBOWSKI, S. (1958). Ejaculatory reflexes in stallions following natural stimulation and the use of the artificial vagina. *Zesz. Probl. Postep. Nauk Roln. P.A.N.*, No. 11, 153–156.

WIERZBOWSKI, S. (1959). The sexual reflexes of stallions. *Roczn. Nauk Roln.*, 73-B-4, 753–788.

WILLIAMS, M. (1956). *Horse Psychology*. London: Methuen & Co.

WILLIAMS, M. (1960). *Adventures Unbridled*. London: Methuen & Co.

WYNMALEN, H. (1953). *Dressage*. London: Museum Press.

Chapter 13

The Behaviour of Rabbits[1]

V. H. Denenberg, M. X. Zarrow and Sherman Ross

The domestic rabbit is derived from the European rabbit, *Oryctolagus cuniculus* Linnaeus, 1758. In many respects it resembles the cottontails of the North and South Americas, but differs from them in its capacity to dig its own burrows. The cottontails, by contrast, live either above ground or in the burrows of other animals. Although fertilization occurs when European rabbits and cottontails are mated by artificial insemination, embryonic development does not occur (Thompson, 1956). The same is true in the case of hare-rabbit crosses (Adams, 1957).

There is some palaeontological evidence that rabbits may perhaps have originated in North America, although the bones of extinct species of rabbits have been found in the Upper Pleiocene deposits of central France and Italy. The light and fragile nature of rabbit and hare bones has not, however, permitted the collection of adequate fossil evidence and any discussion of the classifications of the rabbit and its relatives is best confined to living forms. The antiquity of many of the rabbit bones found in caves and other deposits is suspect because of the rabbit's burrowing habit, and the bones are frequently found to be of more recent origin than other fauna from the same level.

The Order Lagomorpha includes one genus of hares, eight genera of rabbits, and one genus of pikas. Lagomorphs share with rodents the absence of canine teeth and the presence of characteristic, large, rootless, chisel-shaped incisors, separated from the pre-molars and molars by a space or diastema; they differ from them in possessing two pairs of upper incisors and in a more primitive (less reduced) arrangement of the pre-molars and molars. Rabbits differ from hares in having a posterior nasal opening that is narrower than, or no wider than, the narrowest part of the palatal bridge. Generally, they are smaller than hares, with less well developed hind limbs (Thompson & Worden, 1956).

I. INGESTIVE BEHAVIOUR

A. Patterns of Feeding

The digestive system of the rabbit shows many adaptations to its mode of life, including the specialized nature of the teeth, heavy bile production, and voluminous intestines with a caecum ending in an appendix (Blount, 1945).

---

[1] In preparing this chapter we have drawn freely from the chapter by Worden and Leahy in the first edition of this book.

Mastication movements ensure that the food is cut and ground, so that it is usually in a finely divided condition on entering the stomach. Domestic rabbits seldom protrude the tongue while eating, but do so after drinking to lick surplus water from the lips. When grass is offered, the rabbit raises its head and chews rapidly (Blount, *loc. cit.*). In the wild rabbit, the same method of dealing with long food may be observed, and in cropping short herbage the average number of jaw movements is 120 per minute (Thompson & Worden, 1956). Although the wild rabbit is a highly selective feeder, its range of foods is nevertheless wide.

Rabbits vary much in alertness, as detected from the movements of their ears. For much of the time, the ears lie backward, but, periodically, either ear is erected and directed forwards. When both ears face forward, some sight or sound has made the rabbit unusually alert. The animal will then cease to crop, raise its head, and look around. Edible herbage around the burrows is kept extremely short, although the feeding area of the rabbits may cover several acres.

Normal feeding is true cropping, usually on areas already heavily grazed. Each animal crops in a semicircle, moving its head from side to side. It may move forward in any direction and so describe a zig-zag path, but usually grazes a small area, hops or shuffles forward six to nine inches, and crops again, thus moving forward more or less in a straight line (Southern, 1939).

1. REINGESTION

The rabbit reingests its faeces rapidly. The animal sits and bends its head down between its legs, or sometimes around its flank, to procure the faecal pellets directly from the anus. These pellets are swallowed immediately.

The faecal pellets differ in physical consistency from the familiar, hard, round pellets that are voided onto the ground. They are lined with mucus, and much of the pellet is apparently made up of secretion from the caecum. Reingestion is sometimes called *pseudo-rumination*. Certain B vitamins (Kulwich et al., 1953) and nitrogen (Thacker & Brandt, 1955) are utilized better as a result of this process. Southern (1940) observed reingestion in the wild rabbit and Stephens (1952) in West Wales found that 22 out of 49 rabbits still suckling their does had well-formed faecal pellets in their stomach. In New Zealand, Watson (1954) similarly reported evidence of reingestion in rabbits under 3 weeks of age.

2. FEEDING AREAS

The wild rabbit has definite feeding areas. Southern (1939) studied a warren in England for 3 years and observed that these areas varied little, except in so far as newly formed burrows acted as new focal points. The areas were clearly marked by the effect of grazing on the vegetation. Some warrens observed by Southern and by Thompson & Worden (1956) had alternate feeding areas.

Southern (*loc. cit.*) computed that a warren of half an acre, maintaining a population of around 150 during spring and early summer, used a total area of two acres. Of this, about one acre was heavily grazed, while the rest was only used occasionally. Later in the summer, when there was more cover for lying out, the feeding areas were extended.

3. TIME OF FEEDING

Wild rabbits feed principally at dawn and dusk. They may, however, be seen grazing throughout the day in undisturbed places, and they also feed at night. Rabbits dislike grazing in heavy rain and retire to their burrows during heavy downpours. They may, however, feed particularly voraciously before a storm.

B. Stimuli Releasing Feeding

Sunshine stimulates non-grazing wild rabbits to feed and grazing rabbits to start basking or washing. Cold does not seem to stop rabbits from feeding, unless it is combined with wind, particularly wind from the north or east. Rabbits lie close to the ground with ears flattened back. They sometimes shelter from the wind and graze in the lee of nettles or in depressions in the ground. A strong wind with snow keeps rabbits underground, but they will graze when snow is falling if there is little wind.

The ecological implications of the forms of feeding practised by the wild rabbit are considerable. Thus, in chalk grassland the number of botanical species is considerably reduced by rabbit grazing (Tansley, 1949). This is not always the case on other types of soil, where the disturbance of the surface by rabbits encourages the growth of annual weeds. This subject is discussed at length by Thompson & Worden (1956).

II. SEXUAL BEHAVIOUR

The wild rabbit has a sharply defined main breeding season. In Great Britain, this extends from January to June, although sporadic breeding also occurs during other months of the year (Thompson & Worden, 1956). The breeding season of domestic rabbits is much less sharply defined, although mating occurs more often in spring and summer than in the autumn and winter. Hammond (1925) believed that environmental factors, such as temperature and nutrition, influenced the extent of breeding in laboratory colonies at different times of the year.

A. Sexual Behaviour in Males

1. COURTSHIP

This must be distinguished from aggressive chasing, such as that between adult does and young rabbits, in which a definite attack occurs if the quarry is caught. Courtship chasing is the early form of sexual activity observed most frequently. In the open field, the buck usually follows some 10 to 20 yards behind the doe. It is rare to see a continuous or fast chase in the field, for both animals lope along with frequent halts. The buck sometimes begins by rushing the doe, which spurts ahead to maintain her lead; after this the pace slows again.

The persistence of these courtship chases varies greatly. In some cases, the rabbits continue to weave around almost the entire field; in other cases, both animals appear to lose interest and begin to feed. Other chases are interspersed with long intervals of false feeding, during which both animals remain alert, and the buck will sometimes edge toward the doe until he is near enough to attempt a rush.

Courtship chases are often fast and furious, but are easily and quickly broken off if the doe goes to earth. Such chases are frequently accompanied by the other forms of behaviour described below.

Tail "flagging". In this commonly seen form of behaviour, the buck elevates his haunches, so that he walks with a stiff-legged gait, and lays his tail flat along his back, so that he displays its white underside or scut. Various accompanying movements are performed, including "false retreat", in which the buck walks stiffly away from the doe for some 6 yards, giving her a full view of his elevated tail. He will then return to her and repeat the performance, sometimes 3 or 4 times in succession. Another accompanying movement is the "parade", in which the buck circles the doe at about 2 yards' distance with tail elevated and rear quarters twisted toward her.

Tail flagging is almost entirely intersexual and used by the males. In many cases it is combined with *epuresis*. Although the display obviously provides a visual stimulus, it may also provide an olfactory stimulus from the inguinal glands.

Epuresis (enurination). The rabbit emits a jet of urine at the partner in the display. Enurination is performed in several ways. The buck may merely turn his hindquarters toward the doe and shoot out a jet of urine backwards, but some form of circling is often involved, frequently preceded by tail flagging. A very common method is for the buck to run past the doe about a yard away from her and twist his hindquarters. A less common method is downward enurination, in which the buck leaps over the doe and emits a jet of urine as he passes over her.

Enurination is sometimes aggressive. A doe may enurinate a buck which is pestering her, and an old buck may attack and drive off a younger one in this way. Enurination may also occur during acute danger and stimulation.

2. COPULATION

Attempts to copulate may be associated either with indefinite preliminaries or with none at all. A buck may suddenly approach a doe from behind and attempt to mount by placing his fore-paws on her flanks. More commonly, the buck rests his head on the flanks of the doe or nuzzles her hindquarters. An attempt to copulate is usually preceded by tail flagging and enurination. The reactions of females to such attempts are generally negative (Southern, 1948).

Actual mating is rarely witnessed in wild rabbits. Rowley & Mollison (1955) describe how the buck mounts from behind and grasps the female's body with a foreleg on each flank. The male's hindquarters are then vibrated energetically, presumably in a process of orientation, and this is followed by several plunges, during which the back is arched until the white of the scut becomes invisible to an observer in the rear. Coitus is terminated by the breaking of contact.

Blount (1945) describes mating in the domestic rabbit, which generally follows when a doe is placed in the hutch of a buck, provided that both animals are healthy. If the doe is fully receptive, she will at once lie in the correct position, raising her hindquarters to allow copulation. The buck mounts the doe and performs 8 to 12 rapid copulatory movements, endeavouring to effect intromission. If successful, there is forcible penetration by the erect penis and ejaculation occurs. A vigorous buck may attempt to serve the head of the doe, but speedily corrects his mistake. As a terminal feature the buck falls backwards or sideways, after emitting a cry, but soon rights himself and, if healthy and vigorous, will again perform the same mating act quickly—often within a minute. Should the doe be less receptive, she may refuse to raise her hindquarters and tail, in which case the buck usually licks the base of her ears or goes through other courtship manoeuvres. Slow-motion cinematography has revealed that the buck completes copulation by a thrust so vigorous that both his hind feet are off the ground. His forelegs are unable to grip the doe's chest sufficiently to compensate for this imbalance and he promptly falls backwards or sideways. If the buck is less vigorous, there is less loss of balance, and he may scramble off sideways without necessarily having ejaculated. The cry, regarded as an indication of pain, may arise from either the doe or the buck. In the doe, it probably arises from too forceful an entry into the vagina; while in the buck, the erect penis is sometimes painfully displaced as he falls away from the doe.

In domestic does, a vaginal plug of mucus, similar to that seen in many rodents, frequently forms a few minutes after coitus. However, no cervical seal forms during pregnancy (Blount, *loc. cit.*).

3. SEX DRIVE

The sequence of events in the male has been summarized as: exploration, smelling, jumping, chin rubbing, mounting, gripping with the teeth, pelvic oscillations, exploratory movement of erect penis, intromission, and orgasm with ejaculation (Walton, 1955). Hafez (1960), in experimental studies on sex drive, found that this sequence of events remains fairly constant in all individuals, but that the rate and persistency with which the pattern is performed vary considerably from one male to another and in the same male at different times. He recognized seven degrees of sex drive, ranging from aggression, through mild aggression, smelling of the skin for several seconds before mounting, great caution, prolonged smelling with occasional attempts to mount, and fear reaction to offensive reaction. The latency of ejaculation of fertile bucks varied from 5 to 300 (average 70·5) seconds and that of vasectomized bucks from 5 to 180 (average 19·8) seconds. In 70 per cent of the cases, intact bucks failed to complete the act compared with only 45 per cent among vasectomized animals.

Doggett (1956) collected semen samples daily for up to 13 months from male laboratory rabbits and found considerable variation, not only between individuals, but also in the same individual at different times. Maximal sperm counts occurred on the average every 3·16 days, the interval between peaks being longer in younger animals. Differences were not related to environmental effects, nor did the number of collections per day affect sperm counts. Sperm count was related to sperm motility and to the total volume of the ejaculate, but not to the amount of coagulum in the semen. In a later study, Doggett & Ett (1958) evaluated sex drive in male rabbits by three methods (reaction time in the presence of a dummy female, number of ejaculations achieved before exhaustion, and interest in a dummy female) and did not find significant correlations between these criteria.

Cordts (1953) recorded maximum heart rate and an intense electrical output from the brain in rabbits of both sexes during coitus. He also measured the various changes quantitatively (1955). Pulse rate rises more rapidly from sexual stimulus than from the stimulus of pain, and rises progressively from first contact to a peak which coincides with orgasm. Orgasm, in fact, is accompanied by maximum elevation of blood pressure and maximum heart rate. There are changes in the frequency amplitude of the waves detected electroencephalographically during sexual excitement. The amplitude declines even more than under the stimulus of pain, the values during orgasm being 270 to 320 millivolts, as compared with values of 330 to 550 millivolts for testing rabbits.

The rabbit does not sweat and the possible relationship to male fertility of continued high temperature and of modification of thyroid function has been studied by Oloufa et al. (1951). High environmental temperature (91°F) has an adverse effect upon all semen characters, while thyroxine improves the fertility of animals kept intermittently at high temperatures.

Krushinsky (1947) studied the behaviour of male laboratory rabbits following extirpation of the penis. Such animals showed every sign of sexual excitement and mounted the female, but did not attempt to serve her.

B. Sexual Behaviour in Females

1. OESTRUS

The occurrence of oestrus is less easily detected in the rabbit than it is in mammals with regular oestrus cycles. Wild rabbits are generally anoestrous outside the breeding season proper, although a varying number conceive. Non-oestrous does will accept the buck even during this time and indeed it appears that acceptance may occur throughout most of adult life (Brambell, 1948), although there appear to be many individual occasions when mating is refused (Southern, 1940).

Domestic, adult, non-pregnant, well-fed does vary in their receptivity. There are days and occasions when non-pregnant does will refuse to mate, and also times when an individual doe will refuse one buck and readily accept another (Blount, 1945). Moreover, some does will accept a buck once, but refuse him a second time. Blount records that a doe which is willing to be mated usually shows a congested vagina characterized by its purplish colour.

The width of the vaginal orifice is not important, since it varies with age and previous matings and pregnancies. The vagina is normally moist: a pale dry vagina almost always indicates non-receptivity.

Hamilton (1949) found evidence from vaginal smears of cyclic changes in the non-ovulating laboratory rabbit. There are cyclic peaks of relative abundance of distinct cell types at intervals of 4 to 6 days. Cudnovskii (1957) attempted to classify the sexual cycle on the basis of vulval changes, of which he recognizes five, and which he regards as parallel to changes in the uterine mucosa. Myers & Poole (1958) studied the sexual activity of wild rabbits which were confined in an enclosure for 17 weeks, and described an apparent cycle of sexual behaviour at intervals of 7 days or multiples of 7 days. This cycle continues during pregnancy and is especially evident on the 14th day of pregnancy, when many females mate again. A 6-day cycle of sexual behaviour has also been described in the domestic rabbit (Hughes & Meyers, 1966).

Brambell (1942) noted that males copulate frequently with non-oestrous females during the 3 months before the breeding season. While the onset of the breeding season is determined by the female, the end appears to be associated with a decline in male fecundity, for many young females born earlier in the season are sexually mature by this time.

Klein & Gagnière (1956) studied the effect of various hormonal preparations on oestrous behaviour in ovariectomized and intact female rabbits and concluded that complex neurohormonal mechanisms were involved. An interesting feature of their studies was the spontaneous appearance of what appeared to be complete oestrus in a certain number of ovariectomized animals.

2. OVULATION

Rabbits do not ovulate spontaneously, but only 10 to 11 hours after coitus or some other mechanical or hormonal stimulus (e.g. intravenous LH injection). Ovulation may be induced experimentally by electrical or manipulative treatment and may occur naturally through the orgasm induced by contact with other females. In this respect the rabbit resembles the ferret and some other mustelids, the cat, and the shrew. It is possible that this type of ovulation is itself more primitive than that which accompanies a regular oestrous or menstrual cycle, and that even in species which do ovulate spontaneously induced ovulation sometimes occurs as a result of pairing or of some strong emotional or physical stimulus (Thompson & Worden, 1956; Zarrow & Clark, 1968).

3. POST-PARTUM OESTRUS

Wild rabbits lactate during gestation (Brambell, 1942; Stephens, 1952; Thompson & Worden, 1956), although domestic rabbits do not (Hammond, 1925). Domestic and wild does will mate at the post-partum oestrus. When the young from the previous pregnancy are removed, the willingness to mate persists for at least 36 days after parturition. If suckling of the young is permitted, however, sexual desire wanes. Hammond (loc. cit.) recorded that

14*

100 per cent of rabbits will copulate on the first day after parturition, 71 per cent on the 4th day, only 42 per cent on the 8th day, and only 11 per cent on the 12th day. From this time onwards, all the does he studied refused to copulate until they became receptive again when the young began to feed themselves 50 to 60 days after parturition. Desire for coitus is therefore lost as long as the mammary glands of the domestic rabbit are actively functioning.

The lactating domestic rabbit seldom conceives when mated before the 8th day after parturition (Hammond, *loc. cit.*). If a small litter of only 1 to 2 young is left to the suckling doe, conception usually occurs, i.e. the percentage of fertile matings is inversely related to litter number. The only instances recorded by Hammond of does conceiving while suckling 4 or more young occurred during May and June, when nutritional conditions for breeding are optimal.

III. MATERNAL BEHAVIOUR

A. Normal Maternal Behaviour Pattern

Several days to a few hours before parturition, the doe begins to collect hay, straw, excelsior, or other similar material, which she carries to the nest site. In nature, this would be one of her burrows; in the laboratory, it is usually the nest box. She then plucks hair from her body and interweaves this with the hay or straw. The hair is normally fast, but becomes loose near the time of parturition. The centre of the nest is hollowed out and this is where the young are generally born. In an excellent nest, the top is roofed over with nesting material, which effectively helps them maintain their body temperature.

As the young are born, the mother usually eats the foetal placenta and membranes and severs the umbilical cord. The behaviour pattern is similar to that of the rat. See Neektgegoren (1963) for a detailed description of birth and maternal care in the rabbit.

The above description is of normal parturition. Abnormalities associated with parturition include failure to build a nest, birth of young outside the nest, and scattering or cannibalism of the young.

This section will update the material presented in Ross *et al.* (1963a) and Zarrow *et al.* (1962a, b). However, the reader is referred to those chapters for their greater detail and tabular material.

B. Behavioural and Genetic Studies

1. THE INITIAL WORK

In the early 1950s, Sawin and his associates (Sawin & Curran, 1952; Sawin & Crary, 1953) first posed the problem of systematic investigation of maternal behaviour in the rabbit. In their initial paper, Sawin and Curran discussed the more general problem of reproduction and stated that six major factors were involved: fertility, fecundity, milk production, maternal behaviour, growth rate, and viability. They presented data from five closely bred rabbit races showing that there were genetic differences among the races on all of these factors.

Their subsequent paper (Sawin & Crary, 1953) expanded upon the discussion of racial differences in maternal behaviour. They listed the following characters as part of the maternal behaviour complex: location of the nest (inside or outside the nest box), time of nest building (pre-partum, partum, or post-partum), time when the nest was lined with hair, quality of the nest (poor or good), quality of nest lining, scattering of young, cannibalism of young, maternal interest in the young, and aggressive protection of the young when approached by an attendant. Here, too, genetic factors were found to be significant.

2. SOME ASSOCIATIONS AMONG BEHAVIOURAL PATTERNS

In subsequent studies (Denenberg et al., 1958; Denenberg et al., 1959), further evidence was found for genetic influence upon the variables of time of nest building, post-parturient aggression toward the attendant, scattering, and cannibalism.

In these studies, certain correlations were obtained among some of the measures. Sawin & Crary found a strong association between the time the nest was built and the time that the nest was lined. Another association was between the quality of nest constructed and the quantity of hair used in preparing the nest. Denenberg et al. (1958) reported a weak positive relationship between nest quality and time of nest building, a strong positive relationship between nest quality and the percentage of live-born young suckled on the first day of life, and a weak positive relationship between the percentage suckled and the time of nest construction. Positive associations were later obtained between scattering, cannibalism, poor nest quality, and later onset of nest construction (Denenberg et al., 1959). These studies indicated that better nest quality, earlier time of nest building, greater percentage of live-born young suckled on the first day of life, lack of scattering, and lack of cannibalism were positively interrelated. These behaviours were tentatively classified as a "maternal care" complex. Other statistical analyses indicated that this complex appeared to be independent of a second group of factors involving interest in young and aggressive protection from the attendant.

3. NEST BUILDING BEHAVIOUR

The core event in maternal behaviour appears to be construction of the maternal nest. Initially, the doe builds a nest of straw, excelsior, hay, or other such material. We have called this the "straw nest" (Zarrow et al., 1961). At the end of pregnancy or pseudopregnancy the female will pluck hair from her body and use this to line the nest. This is called the "maternal nest". This nest is only built under conditions of pregnancy pseudopregnancy, or appropriate experimental manipulations involving the hormones of pregnancy.

Our initial question concerning nest building behaviour was to ask whether the quality of the nest improved as a function of successive litters (Ross et al., 1956). An analysis was made of the nest quality scores for 84 rabbits which had had at least four litters. Five different races were represented among the 84 animals. The quality of nest built by the mother improved in a linear fashion from the first through the third litter and then levelled off at the fourth litter. In addition, the usual genetic differences among races were obtained. The mechanism underlying the improvement in nest quality is not known. It could be a function of learning experiences, or there could be systematic endocrine changes taking place with successive parturitions which are expressed in nest construction.

Although nest building appeared to be centrally involved in maternal behaviour, this assumption had never been put to an experimental test. The functional significance of nest building was demonstrated by Zarrow et al. (1963), who deprived pregnant rabbits of nesting material or hair (by clipping their entire coat several days before parturition), both of these materials, or neither of these materials. Among control females, more than 87 per cent of the young survived until weaning. For those females which had to build a nest out of their own body hair, 92 per cent of the young were alive at weaning. However, when straw alone was provided (because the females had been shaved) only 39 per cent of the young survived. When neither straw nor hair was present, approximately 6 per cent of the young lived through weaning. Therefore, although both hair and straw contribute to the survival of the young, the mother's hair is the more important element, thus establishing the functional value of the maternal nest.

Because of the difficulties and dangers in rating scales of nest quality, it was thought advisable to attempt to obtain a more quantitative method of measuring nest construction. Eventually, a technique for the quantification of maternal nest building was developed (Denenberg et al., 1963a, b). A rack containing 48 small, frozen fruit juice cans was mounted on one side wall of the female's cage. Each can contained a "wad" of excelsior to be used as nesting material. The rabbit pulled these wads out of the cans and shredded several of them in constructing her nest. Analyses of two different strains of rabbit indicated that the number of wads pulled out and shredded reached a peak on the day that the pregnant animal built her maternal nest. Control females failed to show any systematic trend in shredding behaviour. The shredding score was found to be independent of strain differences, amount of time spent in the maternity cage prior to parturition, and whether the mother was primiparous or multiparous. These data were initially collected from rabbits of the ACEP strain and the III strain at the Jackson Laboratory. The findings were later confirmed and extended using the Dutch-belted rabbit at Purdue University.

The influence of maternal nutrition on maternal nest building and cannibalism was studied by Hafez et al. (1966). Maternal food intake had no effect on the quality of the maternal nest or the time when nest construction began. However, the percentage of viable young decreased in both overweight and underweight does. Cannibalism of the young was not affected by maternal food intake.

4. RETRIEVING AND SUCKLING

Although retrieving of young is quite common among mammals, it does not appear to occur among rabbits. In over 30 years of work with this animal, Sawin reports no evidence of retrieving, and a review of the naturalistic literature also fails to provide any convincing evidence that the rabbit retrieves its young (Ross et al., 1959). In order to investigate this question systematically, Ross et al. (loc. cit.) gave nine parturient domestic rabbits opportunities to retrieve their young under a variety of conditions. Under no circumstances was retrieving observed: the attention of the mother toward her young ranged from lack of concern to solicitous licking.

The lack of this behaviour can perhaps be better understood when one considers the daily nursing pattern between mother and young. There were suggestions from the naturalistic literature (Seton, 1909; Cahalane, 1947; Naaktgegoren et al., 1963) that rabbits and hares only nurse their young once a day. In an experimental investigation of this phenomenon, Zarrow et al. (1965a) allowed one group of rabbits to feed their young ab lib., while a second group was permitted to feed their young twice a day, and a third group was only allowed to nurse their young once daily. Body weight data were obtained between birth and 30 days of age. There were no differences among the three weight curves and, indeed, the three curves overlapped considerably. Therefore, nursing once a day is sufficient to maintain adequate body weight and growth in the rabbit.

For the once-a-day group, the amount of time the doe spent nursing the young ranged from 2·7 to 4·5 minutes, with a mean of 3·4 minutes. Nursing always ended when the mother moved away from the young. In the twice-a-day situation, observations confirmed that the mother would only nurse the young once a day. When the young attempted to suckle a second time, the mother would prevent this either by running away or by lying flat so that the teats were not accessible. It would therefore appear that the restriction of suckling to once a day is due to the mother and not the young.

Rongstad & Tester (personal communication) have established that the wild snowshoe hare visits her nest only once a day, for 4 to 12 minutes, returning at approximately the same time each evening. This was determined by trapping a pregnant doe and, after birth, placing radio transmitters on her and her young and then releasing them. The distance of the female from her nest ranged from 200 to 1,200 feet, except while nursing. The radio-tracking system has also been used to study activity in both the cottontail rabbit and showshoe hare (Mech et al., 1966).

These several observations seem to fit into a meaningful, adaptive pattern. The mother places the young in a nest which is completely covered and thereby well hidden from predators. It is constructed so that the young would have difficulty in escaping. She only visits the nest once a day, and then only for a period of four or five minutes while she nurses. This is sufficient for the young to grow and develop at an adequate rate. Under these conditions, there is no necessity for the mother to engage in retrieving behaviour, since that is a characteristic common to females who remain near their young.

5. SOME METHODOLOGICAL ISSUES

Genetic differences in maternal behaviour among different strains of rabbits have been clearly established. However, the use of different genetic stocks reared in a laboratory raises two critical methodological issues. One is whether the behaviours which have been studied are stable with time or whether marked changes have occurred due to genetic drift, selection, or environmental pressures. The second methodological point concerns the generality of findings obtained in a laboratory setting. To answer the first question, analyses were made of the behaviour of 941 does which had successful pregnancies between January 1953 and June 1959 (Ross *et al.*, 1961). Analyses were made of the percentage of animals which failed to build nests and the time of construction for those rabbits which did build nests. Neither of these measures showed any significant change over the six-and-a-half years which were investigated. The findings both extend and confirm the work of Sawin & Crary (1953), who obtained similar data on nest building.

To investigate the question of the generality of findings obtained in a laboratory setting, 18 pregnant rabbits of different genetic backgrounds were placed in a semi-natural field situation and their behaviour was observed as they approached parturition and during early rearing of the young (Ross *et al.*, 1963). The results were compared with findings obtained from does reared in the laboratory. The same qualitative and quantitative results were obtained in the field as in the laboratory. In addition, burrowing behaviour occurred in the field. Although burrowing could not occur in the laboratory, the occurrence of this event in the field in no way changed the characteristics of the maternal behaviour pattern which had been previously observed in the laboratory. Naaktgegoren (1963) also reported that he was able with domestic rabbits to confirm descriptions in the literature concerning the wild rabbit.

6. EFFECTS OF CHANGING THE HOUSING ENVIRONMENT

One approach to the experimental analysis of maternal behaviour is to isolate those events which will disrupt this behaviour pattern. In one study (Denenberg *et al.*, 1963b), we investigated the effects of a marked change in housing environment subsequent to mating. Several studies using rodents (e.g. Parkes & Bruce, 1961; Calhoun, 1962; Eleftheriou *et al.*, 1962) have shown that manipulation of dimensions of the external environment significantly affects reproductive physiology and maternal behaviour. We have found a similar phenomenon with the rabbit. Subsequent to mating, the females were placed in a metal and wire standard colony cage or a large wooden experimental test cage. The colony cage was more comparable than the experimental cage to the quarters in which the rabbits had been reared until mating. The maternal behaviour of the females in the experimental cages was markedly inferior to that of control females. A small percentage built maternal nests; they also scattered and cannibalized their young. Finally, 87·3 per cent of the young born to control mothers survived through weaning, whereas only 25·9 per cent of the young born in the experimental cages survived. The studies suggest that placing the pregnant female in a markedly different environment after mating is enough to interfere with normal maternal behaviour.

C. Endocrine Factors

1. Effects upon Maternal Nest Building

In examining endocrine factors underlying maternal behaviour, the major end point chosen has been the construction of the maternal nest because of the demonstrated significance of this event. The major generality stemming from several studies is that the construction of the maternal nest by a normal female rabbit is caused by a change in the ratio of the two hormones, progesterone and oestrogen. Progesterone is dominant to oestrogen during normal pregnancy; however, toward the end of pregnancy, this ratio shifts to one in which there is oestrogen dominance. It is this shift which brings about nest construction. The data supporting this generalization will now be briefly summarized.

Maternal nest building will occur at the end of pregnancy and pseudopregnancy. In both instances, there is a progesterone dominance which shifts to an oestrogen dominance. Another way to shift the ratio of progesterone to oestrogen is by means of foetectomy (complete removal of the conceptus mass). If this is done late enough in pregnancy, maternal nest building will ensue subsequent to the operation. Rather than removing the conceptus mass directly, one can ovariectomize the female. This will lead to abortion and will be followed by the construction of the maternal nest (Zarrow et al., 1961, 1963).

The above data established an association between the shift in the progesterone-oestrogen ratio and the occurrence of nest building. Once these associations had been established, Zarrow et al. (1963) then carried out a number of experiments which demonstrated that this association was of a causal nature. Females were castrated and subjected to various regimens of progesterone, oestradiol, and prolactin. Certain regimens were found which would cause the female to build a maternal nest. Prolactin was found not to be a necessary hormone in these treatments. The regimen which required the minimal time to induce nest building was as follows: 5 μg oestradiol on days 1 to 18 and 4 mg progesterone on days 2 to 15. With this regimen, 10 out of 10 rabbits built a maternal nest. If the oestradiol and progesterone treatments were terminated at the same time, no nest was built.

Given the hypothesis that the cause of nest building is a shift to oestrogen dominance in the pregnant animal, it follows that nest building should be blocked if progesterone dominance is maintained. Therefore, Zarrow et al. (1963) injected 12 pregnant rabbits with 4 mg progesterone starting on day 29 of pregnancy. Nest building was inhibited in all but one animal. Using the same reasoning as above, it should be possible to induce early nest building in a pregnant animal by injecting exogenous oestrogen in order to shift the ratio. Zarrow et al. (1963) injected 10 pregnant rabbits with 10 μg oestradiol benzoate on days 20, 21, and 22 of pregnancy. Nine of these animals built a maternal nest between days 22 and 24 of gestation, while normal control rabbits did not build a maternal nest until days 30–32 of gestation.

Minimum pregnancy length for nest building. A relevant question concerns the length of time that the ovarian steroids must act upon the pregnant animal before maternal nest building can occur. To answer this, Zarrow *et al.* (1962a, b) castrated Dutch-Belted rabbits on days 13, 14, 15, 16, and 17 of pregnancy and observed whether or not maternal nest building occurred subsequent to abortion. No animal castrated on day 13 built a nest, whereas all those castrated on day 17 did build the maternal nest. The function relating gestation age to probability of nest building was an ogive. Thus, the hormones of pregnancy must act for 16 to 17 days in this strain in order to establish the appropriate conditioning for nest building to occur.

Strain differences in nest building. Much of the research relating endocrine factors to maternal nest building had been done with the Dutch-Belted rabbit. In order to establish the generality of the findings Zarrow *et al.* (1965b) repeated some of their basic experiments using Grey Chinchilla rabbits. The latter are long-haired, whereas Dutch-Belted rabbits are short-haired; in addition, the Chinchilla weighs approximately twice as much as the Dutch-Belted animal. The same qualitative results were obtained with both strains of rabbits, but major quantitative differences occurred between them: (1) castration on day 20 of pregnancy resulted in 100 per cent nest building in the Dutch-Belted rabbit, whereas only 80 per cent nest building occurred in Chinchilla rabbits castrated as late as day 25; (2) maternal nest building occurred in all Dutch-Belted rabbits receiving 10-μg injections of oestradiol benzoate on days 20 to 22 of pregnancy, whereas Chinchillas required 40-μg injections; (3) castrated Dutch-Belted does treated for 18 days with oestrogen and progesterone built nests, whereas Chinchilla rabbits required 29 days of hormone treatment at double the dosages. Racial differences in maternal nest building may then be quantitative in nature.

Although oestrogen and progesterone were sufficient to induce maternal nest building, this does not eliminate participation of the hormone prolactin. Since all animals used in these studies had intact pituitaries, and since oestradiol causes sythesis and release of prolactin by the pituitary gland, this hormone may be involved in the induction of nest building behaviour.

2. Effects upon Hair Loosening

The unique feature of the maternal nest is the hair which does pluck from their bodies and use as lining maternal. Early observations indicated that the hair becomes relatively loose near the time of parturition, but the phenomenon had never been actually measured until Sawin *et al.* (1960) described a procedure for measuring the degree of hair loosening during gestation. These investigations developed a standardized combing technique and weighed the amount of hair obtained to the nearest milligram. Pregnant animals showed a marked increase in combed hair weight between 5 days prepartum and the day of parturition, whereas control females showed no such increase during an equivalent period of time. Furthermore, the increase in hair weight was also seen in a pseudopregnant rabbit at the termination of pseudopregnancy.

Once a method for measuring hair loosening existed, it was possible to determine if this event was under endocrine control. Such an analysis was conducted by Farooq *et al.* (1963). The results were similar to those obtained in studies of the endocrine factors involved in maternal nest building: (1) hair loosening increased at the end of pregnancy and pseudopregnancy; (2) hair loosening was significantly depressed if parturition was delayed by progesterone injections; (3) castrated females showed a marked increase in combed hair weight after treatment with progesterone, oestrogen, and pro-lactin. However, other experiments indicated that nest building and hair loosening could be separated experimentally. Hair loosening failed to occur in pregnant animals with 2 mg progesterone; yet, 7 out of 8 animals still built a maternal nest. Maternal nest building occurred if castrated females were treated with oestrogen and progesterone for 18 or 31 days; yet, there was no increase in combed hair weight. Finally, maternal nest building also occurred when does were treated with oestrogen, progesterone, and pro-lactin for 18 days; again, there was no change in combed hair weight. Such data indicate that maternal nest building and hair loosening occur inde-pendently of each other and suggest that they may be controlled by separate mechanisms.

D. Summary

In examining the variety of events involved in maternal behaviour, positive associations were obtained between nest quality, time of nest building, percentage of live-born young suckled on the first day, lack of scattering of the young, and lack of cannibalism. A second set of associations was found between the mother's interest in young and her aggressive pro-tection of the young from an attendant.

The functional significance of the maternal nest (i.e. the nest containing hair from the mother's body) was demonstrated by showing a marked reduction in the percentage of young surviving when the mother was unable to use her hair in constructing the nest.

Quantitative techniques for measuring nest construction and the degree of hair loosening during parturition have been developed.

Retrieving behaviour does not appear to be part of the normal repertoire of the doe, and suckling of the young is limited to approximately 5 minutes per day. This pattern of behavioural events appears to have adaptive signi-ficance.

Analysis of the behaviour of pregnant animals placed into a semi-natural setting showed behaviour patterns similar to those of laboratory animals, suggesting that the results obtained in the laboratory may be generalized to the field situation.

It is possible to disrupt the maternal care pattern by placing the pregnant female in a markedly different environment subsequent to mating.

Analyses of the endocrine factors involved in maternal nest building indicate that the shift of the progesterone-oestrogen ratio from progesterone dominance to oestrogen dominance triggers off maternal nest building. The ratio of these two hormones also affects hair loosening. However, hair loosening and maternal nest building can occur independently of each other, which suggests that they may be governed by separate mechanisms.

IV. SOCIAL BEHAVIOUR

A. The Warren System

The characteristic wild rabbit community inhabits a warren (from the French, *garenne*). Since rabbits excavate mainly with their fore-paws and throw much soil backwards with their powerful hind feet, entrances without a mound of soil must have been opened from the inside. Such entrances often descend vertically, are small and concealed by vegetation, and are known as bolt-holes.

Only those burrows used for breeding contain bedding materials. Elsewhere in the warrens, the rabbits rest on bare earth—frequently in chambers opening from the main passages. The ramifications of a warren are such that digging out an area that has been colonized for some years is an arduous process. They may run to a depth of more than 9 feet. Phillips (1953) found, only a short while after the beginning of her experiment, that a rabbit hole was present 54 feet from the nearest burrow. Myers & Parker (1965) give a detailed description of warrens of wild rabbits.

B. Social Structure

The social structure of the wild rabbit colony during the breeding season is apparently related to the territorial conservation of the does and the dominance of certain bucks. During the summer, young males are driven out by the older bucks, and only the most vigorous of those that remain until the next year are able to survive amid the older animals. Southern (1940) found that remarkably few bucks took an active part in the life of the warren, at least during daylight hours. No buck was found to become a leader during his first season, and only one before it was 1 year old. Most dominant males were old bucks weighing 1,600 g or more.

Polygamy is therefore probable. Only once did Southern (*loc. cit.*) see a female approached on two separate occasions by two separate bucks. In all other instances, the bucks chased away other males who approached the does they were attending. Southern believed that the does formed the more stable element in the social organization. They often drove bucks away, although it was not known whether such action was directed towards bucks other than the dominant ones of the particular area of the warren. Older does were generally hostile toward young rabbits, especially young does, and this was thought to be a factor reducing the swollen summer populations to levels permitting the operation of the characteristic type of social organization.

Myers & Poole (1963) reported that the increased population density of wild rabbits kept in confinement resulted in some reduction in fecundity, but much less than expected. They concluded that the rabbit possesses no innate behavioural or physiological mechanisms capable of regulating numbers at levels below the food ceiling. Myers (1966) reviewed the effects of population density on sociality and health in mammals.

C. Aggressive Behaviour

Southern (1940) found that aggressive behaviour in the wild rabbit colony was not as clearly defined as sexual behaviour. He nonetheless distinguished two types of aggressive behaviour: aggressive chasing and leaping and bouncing.

Aggressive chasing. Aggressive chasing occurs primarily within the warren enclosure and is the result of trespass into the sphere of an adult doe. Adult does are particularly aggressive towards younger rabbits at the end of the breeding season and will not tolerate them within several yards of their burrows. The encounter is usually brief and the trespasser retreats rapidly.

Bucks are not nearly as aggressive as does, but will also respond when another buck approaches too near their mate.

Leaping and bouncing. It is somewhat difficult to distinguish the aggressive leaping said to occur between two bucks in combat from the leaping that involves enurination and may be intersexual. Southern (*loc. cit.*) observed two instances of aggressive leaping. In one case, both animals indulged in false feeding, at the end of which both turned simultaneously and rushed at each other. This behaviour was repeated for 2 to 3 minutes. Another form of behaviour, perhaps akin to this, involves two rabbits that circle each other with haunches elevated for a while. They face each other, and then leap into the air in turn, each landing on the same spot from which it sprang. In addition to this, single animals are often seen to leap in the air. This is probably related to play activity and is commonly indulged in by young rabbits.

Southern found that the types of sexual behaviour described above increased during spring and summer, with an expected peak of activity during March and April and a second peak in August. The latter peak occurred because the young of the early litters suffered considerable mortality and scattering, and the young that were born in May contributed most to the summer population of the warren. They were therefore just reaching sexual maturity during late July and August.

Southern showed that the development of sexual behaviour in young bucks was inhibited by the dominance of the older ones. He found, however, that the behaviour of young females, while never so positive, was much more clearly affected by enurination. This took the form of much leaping about and frisking, which in turn made the buck even more excited, although no case of copulatory behaviour was seen in consequence.

V. OTHER BEHAVIOUR

A. Latrines

The wild rabbit uses latrines or earth closets, which are special patches of ground near to the burrow—often old molehills. They are usually close to a run, are circular in shape and a foot or more in diameter. The herbage within the area is brown and withered and has a collection of dry faecal pellets at its centre. A single latrine is often related to a particular burrow.

Tame rabbits tend to void their droppings at one particular site, such as the corner of the cage. According to Blount (1945), they may be trained quite readily to use a special hutch lavatory unit. The rabbit is first coaxed to the unit by spreading a small quantity of soiled bedding on it.

The use of pellet counts for estimating the density of wild rabbit populations is fraught with difficulties, not the least of which is the varying degree of decay in the field. However, it would appear that a successful adaptation of the method has been made in New Zealand (Taylor & Williams, 1956).

B. Locomotor Activity

Rabbits may travel considerable distances in search of food or water and it has been stated that in the more arid parts of Australia they may travel as much as four miles daily from their feeding grounds to a waterhole. In temperate regions, however, their feeding movements are usually much more restricted. Territorial conservatism is marked among adult does, which tend to occupy particular areas, usually patches of bare earth. This conservatism is combined with fiercely aggressive behaviour toward other females, young rabbits, and occasionally bucks. Adult males also appear to have their territories in which they are dominant.

The maximum speed attained by rabbits under observation is said to approach 24 m.p.h. (Neithammer, 1937), despite an unspecialized forearm modified neither for digging nor running. The rabbit's gait is a hopping one, and when it is moving slowly the forefeet are placed in front of the hind feet. As speed increases, and the suspension becomes longer, the hind feet tend to come further forward and to land in front of the forefeet. Individuals vary considerably in their capacity to climb (Thompson & Worden, 1956). Rabbits are also capable of swimming.

C. Resistance to Thirst

Anghi (1954) has advanced the view that one may classify the constitutional strength of rabbits by measuring resistance to thirst. On this basis, he reports that Chinchillas are superior to Angoras, and that both of these strains are superior to Vienna Blues.

REFERENCES

ADAMS, C. E. (1957). An attempt to cross the domestic rabbit and hare. *Nature, Lond.,* **180**, 853.

ANGHI, G. (1954). Constitutional strength of domestic rabbits. *Hung. Agric. & Rev.,* **3** 20–21. (*Anim. Breed. Abstr.,* **23**, 275.)

BLOUNT, W. P. (1945). *Rabbits' Ailments.* Idle, Bradford: Fur & Feather.

BRAMBELL, F. W. R. (1942). Intra-uterine mortality of the wild rabbit, *Oryctolagus cuniculus* (L.). *Proc. roy. Soc.* (B), **130**, 462–479.

BRAMBELL, F. W. R. (1948). Prenatal mortality in mammals. *Biol. Rev.,* **23**, 370–405.

CAHALANE, V. H. (1947). *Mammals of North America.* New York: Macmillan.

CALHOUN, J. B. (1962). A "behaviour sink". In: *Roots of Behavior*. E. L. Bliss (ed.). New York: Harper.

CORDTS, H. (1953). Zur Kenntnis der Sexuellen Erregung bei Haustieren. *Z. Tierzücht*, **61**, 305–352.

CORDTS, H. (1955). Elektrokardiographische und Elektrencephalographische Messungen während der Begattung von Haustieren. *Z. Tierzücht*, **64**, 389–399.

CUDNOVSKII, L. A. (1957). A method of determining sexual activity in rabbits. *Kara Kulevodstvo i Zverovodstvo*, **10**, 37–38. (*Anim. Breed. Abstr.*, **25**, 1420.)

DENENBERG, V. H., HUFF. R. L., ROSS, S., SAWIN, P. B. & ZARROW, M. X. (1963a). Maternal behaviour in the rabbit: the quantification of nest building. *Anim. Behav.*, **11**, 494–499.

DENENBERG, V. H., SAWIN, P. B., FROMMER, G. P. & ROSS, S. (1958). Genetic, physiological and behavioural background of reproduction in the rabbit. IV: An analysis of maternal behaviour at successive parturitions. *Behaviour*, **13**, 131–142.

DENENBERG, V. H., PETROPOLUS, S. F., SAWIN, P. B. & ROSS, S. (1959). Genetic, physiological and behavioural background of reproduction in the rabbit. VI: Maternal behaviour with reference to scattered and cannibalized newborn and mortality. *Behaviour*, **15**, 71–76.

DENENBERG, V. G., ZARROW, M. X., KALBERER, W. D., FAROOQ, A., ROSS, S. & SAWIN, P. B. (1963b). Maternal behaviour in the rabbit: effects of environmental variation. *Nature, Lond.*, **197**, 161–162.

DOGGETT, V. C. (1956). Periodicity in the fecundity of male rabbits. *Am. J. Physiol.*, **187**, 445–450.

DOGGETT, V. C. & ETT, J. C. (1958). Libido and the relationship to periodicity in the fecundity of male rabbits. *Fedn. Proc.*, **17**, 36.

ELEFTHERIOU, B. E., BRONSON, F. H. & ZARROW, M. X. (1962). Interaction of olfactory and other environmental stimuli on implantation in the deermouse. *Science, N.Y.*, **137**, 764.

FAROOQ, A., DENENBERG, V. H., ROSS, S., SAWIN, P. B. & ZARROW, M. X. (1963). Maternal behavior in the rabbit: endocrine factors involved in hair loosening. *Am. J. Physiol.*, **204**, 271–274.

HAFEZ, E. S. E. (1960). Sex drive in rabbits. *Southwest Vet.*, **14**, 46–49.

HAFEZ, E. S. E., LINDSAY, D. R. & MOUSTAFA, L. A. (1966). Some maternal factors affecting nest building in the domestic rabbit. *Z. Tierpsychol.*, **6**, 691–700.

HAMILTON, C. (1949). Some indications of cyclic reproductive activity in the non-ovulating rabbit. *Am. J. Physiol.*, **159**, 573.

HAMMOND, J. (1925). *Reproduction in the Rabbit*. Edinburgh: Oliver & Boyd.

HUGHES, R. L. & MYERS, K. (1966). Behavioural cycles during pseudopregnancy in confined populations of domestic rabbits and their relation to the histology of the female reproductive tract. *Aust. J. Zool.*, **14**, 178–183.

KLEIN, M. & GAGNIÈRE, E. (1956). Sur les regulations hormonales du comportement sexuel chez la lapine. *C.R. Soc. Biol.* (Paris), **150**, 1440–1442.

KULWICH, R., STRUGLIA, L. & PEARSON, P. B. (1953). The effect of coprophagy on the excretion of B vitamins by the rabbit. *J. Nutrit.*, **49**, 639–643.

KRUSHINSKY, L. V. (1947). The role played by peripheral impulses in the sexual form of behaviour of mates. *C.R. (Dokl) Acad. Sci. URSS NS*, **55**, 461–463.

MECH, L. D., HEEZEN, K. L. & SINIFF, D. B. (1966). Onset and cessation of activity in cottontail rabbits and snowshoe hares in relation to sunset and sunrise. *Anim. Behav.*, **14**, 410–413.

MYERS, K. (1966) The effects of density on sociality and health in mammals. *Proc. ecol. Soc. Aust.*, **1**, 40–64.

MYERS, K. & PARKER, B. S. (1965). A study of the biology of the wild rabbit in climatically different regions in eastern Australia. I. Patterns of distribution. *C.S.I.R.O. Wildl. Res.*, **10**, 1–32.

MYERS, K. & POOLE, W. E. (1963). A study of the biology of the wild rabbit, *Oryctolagus cuniculus* (L.), in confined populations. V. Population dynamics. *C.S.I.R.O. Wildl. Res.*, **8**, 166–203.

MYERS, K. & POOLE, W. E. (1958). Sexual behaviour cycles in the wild rabbit. *C.S.I.R.O. Wildl. Res.*, 3, 144–145.

NEEKTGEBOREN, C. (1963). *Untersuchungen über die Geburt die Säugetiere.* Doctoral thesis, University of Amsterdam.

NEITHAMMER, G. (1937). Ergebnisse von Markierungsversuchen an Wildkaninchen. *Z. Morphol. Okol. Tiere*, 33, 297–312.

OLOUFA, M. M., BOGART, R. & McKENZIE, F. F. (1951). Effect of environmental temperature and the thyroid gland on fertility in the male rabbit. *Fertil. & Steril.*, 2, 223–228.

PARKES, A. S. & BRUCE, H. M. (1961). Olfactory stimuli in mammalian reproduction. *Science, N.Y.*, 134, 1049–1054.

PHILLIPS, W. M. (1953). The effect of rabbit grazing on a reseeded pasture. *J. Br. Grassl. Soc.*, 8, 169–181.

ROSS, S., DENENBERG, V. H., SAWIN, P. B. & MYERS, P. (1956). Changes in nest building behaviour in multiparous rabbits. *Br. J. Anim. Behav.*, 4, 69–74.

ROSS, S., DENENBERG, V. H., FROMMER, G. P. & SAWIN, P. B. (1959). Genetic, physiological and behavioral background of reproduction in the rabbit. V: Nonretrieving of neonates. *J. Mammal.*, 40, 91–96.

ROSS, S., SAWIN, P. B., DENENBERG, V. H. & ZARROW, M. X. (1961). Maternal behavior in the rabbit: yearly and seasonal variation in nest building. *Behaviour*, 18, 154–160.

ROSS, S., SAWIN, P. B., ZARROW, M. X. & DENENBERG, V. H. (1963a). In: *Maternal Behavior in Mammals.* H. L. Rheingold (ed.). New York: Wiley.

ROSS, S., ZARROW, M. X., SAWIN, P. B., DENENBERG, V. H. & BLUMENFIELD, M. (1963b). Maternal behaviour in the rabbit under semi-natural conditions. *Anim. Behav.*, 11, 283–285.

ROWLEY, I. & MOLLISON, B. C. (1955). Copulation in the wild rabbit, *Oryctolagus cuniculus. Behaviour*, 8, 81–84.

SAWIN, P. B. & CRARY, D. (1953). Genetic and physiological background of reproduction in the rabbit. II: Some racial differences in the pattern of maternal behaviour. *Behaviour*, 6, 128–145.

SAWIN, P. B. & CURRAN, R. H. (1952). Genetic and physiological background of reproduction in the rabbit. I. The problem and its biological significance. *J. exp. Zool.*, 5, 165–202.

SAWIN, P. B., DENENBERG, V. H., ROSS, S., HAFTER, E. & ZARROW, M. X. (1960). Maternal behavior in the rabbit. Hair loosening during gestation. *Am. J. Physiol.*, 198, 1099–1102.

SETON, E. T. (1909). *Lives of Game Animals.* Garden City, N.Y.: Doubleday.

SOUTHERN, H. N. (1939). Coprophagy in the wild rabbit. *Nature, Lond.*, 145, 262.

SOUTHERN, H. N. (1940). The ecology and population dynamics of the wild rabbit, *Oryctolagus cuniculus. Ann. appl. Biol.*, 27, 509–526.

SOUTHERN, H. N. (1948). Sexual and aggressive behaviour in the wild rabbit. *Behaviour*, 1, 173–194.

STEPHENS, M. N. (1952). Seasonal observations on the wild rabbit, *Oryctolagus cuniculus* (L) in West Wales. *Proc. zool. Soc. Lond.*, 122, 417–434.

TANSLEY, A. G. (1949). *The British Islands and their Vegetation.* Cambridge: University Press.

TAYLOR, R. H. & WILLIAMS, R. M. (1956). The use of pellet counts for estimating the density of populations of the wild rabbit, *Oryctolagus cuniculus* (L). *N.Z. Jl. Sci. Technol.* (B), 38, 236–256.

THACKER, E. J. & BRANDT, C. S. (1955). Coprophagy in the rabbit. *J. Nutrit.*, 55, 375–385.

THOMPSON, H. V. (1956). Myxomatosis: A summary. *Agriculture*, 63, 51–57.

THOMPSON, H. V. & WORDEN, A. N. (1956). *The Rabbit.* London: Collins (New Naturalist Monograph No. 13).

WALTON, A. (1955). Sexual Behaviour. In: *Progress in the Physiology of Farm Animals.* J. Hammond (ed.). Vol. 2, pp. 603–616. London: Butterworths.

WATSON, J. S. (1954). Breeding season of the wild rabbit in New Zealand. *Nature, Lond.*, 174, 608.

ZARROW, M. X., SAWIN, P. B., ROSS, S., DENENBERG, V. H., CRARY, D., WILSON, E. D. & FAROOQ, A. (1961). Maternal behaviour in the rabbit: evidence for an endocrine basis of maternal-nest building and additional data on maternal-nest building in the Dutch-belted race. *J. Reprod. Fert.*, **2**, 152–162.

ZARROW, M. X., FAROOQ, A. & DENENBERG, V. H. (1962a). Maternal behavior in the rabbit: critical period for nest building following castration during pregnancy. *Proc. Soc. exp. Biol Med. (N.Y.)*, **8**, 537–538.

ZARROW, M. X., SAWIN, P. B., ROSS, S. & DENENBERG, V. H. (1962b). Maternal behavior and its endocrine basis in the rabbit. In: *Roots of Behavior*. E. L. Bliss (ed.). New York: Harper.

ZARROW, M. X., FAROOQ, A., DENENBERG, V. H., SAWIN, P. B. & ROSS, S. (1963). Maternal behaviour in the rabbit: endocrine control of maternal-nest building. *J. Reprod. Fert.*, **6**, 375–383.

ZARROW, M. X., DENENBERG, V. H. & ANDERSON, C. O. (1965a). Rabbit: frequency of suckling in the pup. *Science, N.Y.*, **150**, 1835–1836.

ZARROW, M. X., DENENBERG, V. H. & KALBERER, W. D. (1952b). Strain differences in the endocrine basis of maternal nest-building in the rabbit. *J. Reprod. Fert.*, **10**, 397–401.

ZARROW, M. X. & CLARK, J. H. (1968). Copulation following vaginal stimulation in a spontaneous ovulator and its implications. *J. Endocr.*, **40**, 343–352.

Chapter 14

The Behaviour of Dogs[1]

J. L. Fuller and M. W. Fox

Verified evidence of domesticated dogs goes back at least to the 7th millennium B.C. in the Near East (Reed, 1959). In this area the dog may share with the goat the distinction of being the earliest domesticated animal. The kitchen middens of Denmark and Sweden have yielded skeletons of domesticated dogs dated at 6000 to 8000 B.C. (Dahr, 1936). Although any reconstruction of the actual process of domestication must be speculative the origin may well have involved a commensal relationship between scavenging wolves and a primitive tribe of hunters. The wolves fed on refuse heaps; man captured the puppies in time of food scarcity (Scott, 1954). Authorities are not in complete agreement on the wild ancestor of the dog. The wolf, *Canis lupus*, appears to be the most likely ancestor, but the introduction through hybridization of genes from the golden jackal, *C. aureus*, has been suggested (cf. Reed, 1959). Dogs are widespread and wherever remnants of Mesolithic culture still exist, for example the Caribou Eskimo, the dog has been found to be the only domestic animal.

The dog offers a greater number of varieties than any other species of domestic animal. One hundred and fifteen breeds are recognized by the American Kennel Club and hundreds of other true breeding types are scattered throughout the world, the probable total being estimated as 800.

Great differences occur in size and conformation as well as in behaviour. A Saint Bernard may weigh 100 times more than a Chihuahua; a 25-lb Whippet is twice as tall as a 25-lb Dachshund. Terriers, which are fighting breeds, are usually so aggressive towards each other that even young puppies cannot be safely kennelled together; on the other hand, as a result of selection, groups of Beagles are generally peaceful (Fuller, 1953; Scott, 1958b). Pointers, Beagles, and Greyhounds may be equally eager hunters; yet, they show widely divergent patterns of hunting behaviour. Basenjis have an annual seasonal mating cycle in contrast to the semi-annual non-seasonal cycles of other breeds.

Despite this variability, most and probably all breeds may be successfully interbred. Although matings of such strongly contrasted breeds as the Chihuahua and the Great Dane have not been reported, there is no known reason why such matings would not be fertile. Turning to the wild *Canidae*,

[1] Miss Edna Dubuis was a co-author of this chapter in the first edition and the present authors acknowledge her contribution to the unrevised sections.

fertile hybrids have been produced by matings of dogs with the wolf (*C. lupus*), coyote (*C. latrans*), jackal (*C. aureus*), and dingo (*C. dingo*) (Gray, 1954).

For classification purposes, the American Kennel Club has divided recognized breeds into six main groups according to their primary utility. It is interesting to note that the divisions are primarily based upon behavioural characters rather than phylogeny.

The *Sporting Group* includes bird-hunting specialists—the various Spaniels which flush or start their game; the pointing breeds which set their game; and the Retrievers which deliver the game to hand.

The *Hound Group* lists breeds which for the most part have been developed for hunting game other than birds. This group comprises the trail hounds which hunt by scent (e.g. Foxhounds, Bloodhounds) and the sight hounds which depend upon keen vision and speed in pursuing their game (e.g. Greyhounds, Salukis).

The third group, *Working Dogs*, consists of breeds best fitted and most consistently used for work not of a sporting nature. It includes the breeds built for sled, pack, and other draught purposes; the herding, driving, and guarding dogs; and those adapted for army, police, and patrol work.

Under *Terriers* are included breeds originally selected for the ability to hunt and attack various types of game, particularly rats, foxes, badgers, weasels, and similar vermin which are apt to go to earth when pursued. Large or small, rough- or smooth-coated, Terriers are noted for persistence and readiness to attack.

The *Toy Group* includes the Chihuahua, the smallest pure breed in the world; also various diminutive Spaniels and Terriers which seldom weigh more than 10 lb when mature.

The sixth group, designated *Non-Sporting*, contains breeds which have been preserved primarily for their companionable attributes and tractability. Some were initially developed for particular sporting or working purposes but are now largely unemployed in these capacities (e.g. Poodles, Bulldogs).

A number of publications, dealing with domestication (Zeuner, 1963), breeding, genetics and inherited abnormalities (Burns & Fraser, 1966; Fox, 1966c) and behavioural abnormalities (Fox, 1965; Fox, 1968; Brunner, 1968), provide valuable information on aspects of canine behaviour which will be only briefly alluded to in this chapter. Scott & Fuller's (1965) studies on the genetics and social behaviour of the dog present a detailed analysis of behavioural differences within 5 breeds of dog (and hybrids) together with a comprehensive description on the development of behaviour and the process of socialization.

I. BEHAVIOURAL DEVELOPMENT

Although development is a continuous process it is convenient to consider a life history as a series of phases which are separated by important episodes. The early life history of the dog has been divided into stages by Scott (1958c). He recognizes a neonatal stage during which the immature puppy remains in its nest (weeks 1 and 2), a transition period marked by rapid sensory and motor maturation (week 3), a period of socialization which terminates at about the time of natural weaning (weeks 4–10), and a juvenile period (to sexual maturity). These stages will be considered in some detail, followed by brief remarks on the remainder of the life span.

A. Neonatal and Transitional Periods

The pup is frequently delivered within its foetal membranes. The bitch licks these off and ingests them and also severs the umbilical cord with the carnassial teeth. Her attention is not focused exclusively on the pup at this time, for she also licks her vulva and any foetal fluids that are on the ground. As the pup is stimulated by the tongue of the bitch during removal of the foetal membranes, it responds violently by squirming from side to side, and eventually vocalizes when spontaneous respiration is established and excess foetal fluids escape from the upper respiratory tract. Prior to initial stimulation by the bitch, the pup lies with little or no spontaneous movement, within the intact foetal membranes. Tactile stimulation by the mother subsequently evokes a "mass response" and the pup squirms, rights itself, crawls (rooting reflex) (James, 1952; Welker, 1959), and after a short period of irregular gasping respiration, commences to breathe regularly and to vocalize (mewing). The severed umbilical cord continues to exude blood for some time, and this area is licked intensively by the mother. Such contact reflexively stimulates respiration (Fox, 1964a). Other parts of the body are licked also. When the face and dorsum of the head are stimulated, the pup crawls forward. Mere contact with the leg of the mother will stimulate rooting and eventually the pup locates the mammary region. Individual nipple preference, as reported for the cat (see Chapter 15) has not been observed in the dog. During this immediate postnatal period, several patterns can be recognized: rooting reflex, negative geotaxis, positive thermotaxis, auriculonasocephalic reflex, positive thigmotaxis, righting reflex, reflex stimulation of respiration by umbilical-anogenital tactile stimulation, Galant's reflex, labial and sucking reflexes (Fox, 1964a). These reflex components (together with olfactory-mediated responses (Fox, 1967a)) are evoked by the mother and the "nest environment" or "situation structure" (i.e. warmth of nest, inverted U position of the mother, etc.). When the pup is removed from the nest and is placed on a cold surface, distress vocalization, pivoting and rooting head movements are seen; this behaviour ceases upon contact with a warm, soft object (Fredericson, 1952; Fredericson et al., 1956).

Through the first 2 to 3 weeks, the bitch stimulates urination by licking and also ingests faeces from the pups. By 3 to $3\frac{1}{2}$ weeks, the pups are able to stand and to follow the mother. They show the rooting response, and reach up to the teats of the bitch while she is standing up, especially seeking the generally more productive and more pendulous inguinal teats. Pups at this age remain quite still while the bitch is licking the ano-genital region.

B. Socialization Period

Occasionally, the bitch will regurgitate food for the litter from 4 to 5 weeks onwards. This is not uncommon in the domesticated dog, and is elicited partly spontaneously and partly by the pups licking the face and mouth of the bitch.

Prior to 4 to 5 weeks of age, the pups sleep in a "pile" when the mother is out of the nest, undoubtedly a heat-conservation mechanism. When two or three pups are placed together on a cold surface, they "crawl around" and over each other (contactual circling). After 3 weeks of age, thermoregulation is well developed, and the pups tend to sleep in a row instead of a heap, unless the whelping quarters are excessively cold.

From $3\frac{1}{2}$ to 4 weeks of age, the pups begin to interact playfully with each other; occasional chewing on the ears, or licking the face of one pup by another, elicits avoidance and distress vocalization, if the stimulation is too intense, or mutual chewing, licking and mouthing. From this age the pups learn, possibly through play, how much pain they can inflict upon each other as a result of chewing and biting. These early play responses become more variable as locomotor and perceptual abilities improve. Playful fighting, scruff holding and "prey-killing" head-shaking movements, appear between 4 and 5 weeks together with pouncing, snapping and aggressive vocalizations, such as growling and snarling with the teeth bared. Co-ordinate attacks upon subordinate animals may occur from about the ninth week and can lead to serious injury or death if not controlled (Fuller, 1953). Submissive postures, care and play soliciting gestures appear also, including lateral recumbency, hindleg raising and whining (as distinct from "distress") vocalization, licking and foreleg raising.

The facial "expressiveness" of the pup at 5 weeks contrasts with the mask-like appearance of 3-week-old pups. This is due especially to the development of expressive ear movements, elongation of the muzzle and possibly to improved functioning of the muscles controlling elevation of the lips and display of the teeth. Also the repertoire of vocal patterns improves during this period from 3 to 5 weeks. As early as 3 to 4 weeks of age, pups show the "rooting approach" to the inguinal region of their littermates, who remain quite still during such contacts. This approach may be related to the earlier response associated with rooting and reaching towards the inguinal region of the mother, and the immobility of the pup evoking this response may be derived from its responses when the mother is stimulating and cleaning the ano-genital region. This inguinal approach persists as a highly stereotyped and ritualized pattern (Plate XV).

At 4 to 5 weeks of age, pups frequently carry small objects in their mouths, and engage in tugs-of-war. Soon afterward a defensive-protective agonistic pattern emerges in which the pup vigorously guards (prey-guarding?) a particular object or morsel of food. Several pups may follow one littermate who is carrying something in its mouth; these are the first signs of co-ordinated group activity. A sudden disturbing noise will frequently cause the entire litter to withdraw rapidly; again, this is a co-ordinated group or allelomimetic response. At this age, pups begin to defaecate and urinate in one particular area of their living quarters, usually some distance from the nest.

By 6 weeks of age, most of the species-specific behaviour patterns are present, notably the facial-lingual licking or greeting response, inguinal approach and ano-genital investigation (Plate XIV). The bite is seen less as a chewing-licking, jaw biting response, but as a more specific grab-and-hold of the scruff of the opponent's neck. Although the mature male hindleg-raising urination pattern does not develop until around puberty, fragments of sexual behaviour, such as mounting, clasping and pelvic thrusts are seen during play, as early as 6 weeks, predominantly in male pups (Plate XIV).

C. Period of Socialization

The 4th to the 6th weeks are characterized by rapid maturation of motor patterns. Skills are acquired in walking, climbing, and manipulation with jaws and forelimbs. Visual orientation to distant objects becomes more precise and, possibly as a consequence, the puppy begins to follow moving objects such as its mother, a human being, or a rolling ball.

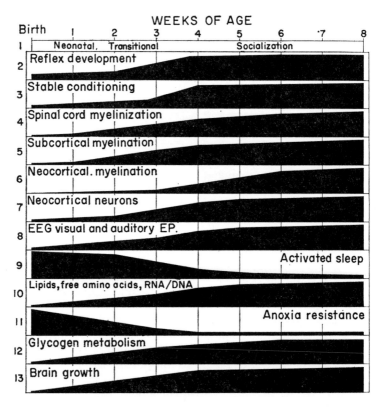

Fig. 74. Neuro-ontogeny of the dog. Attainment of relatively mature function or structure of various components of CNS. Most rapid development between 2 and 5 weeks. (1) Scott & Marston, 1950; (2) Fox, 1964a; (3) see Fig. 75; (4) Fox *et al.*, 1967; (5) and (6) Fox, 1967a; (7) Fox *et al.*, 1966; (8) Fox, 1967b and 1968a; (9) Fox & Stanton, 1967; (10) Fox, 1967a; (11) Mandel *et al.*, 1962; (12) Dravid *et al.*, 1965; (13) Himwich & Fazekas, 1941; (14) Fox, 1964b, 1966b.

CENTRAL NERVOUS SYSTEM

Development of the CNS. Underlying the gradual development of behaviour are changes in the structure and function of the Central Nervous System (CNS), which tend to parallel the development of overt behaviour (Scott & Marston, 1950). One can distinguish a neonatal period, characterized by adaptation to the extra-uterine environment and "low level" organization of behaviour, followed by a transitional period, during which time adult-like

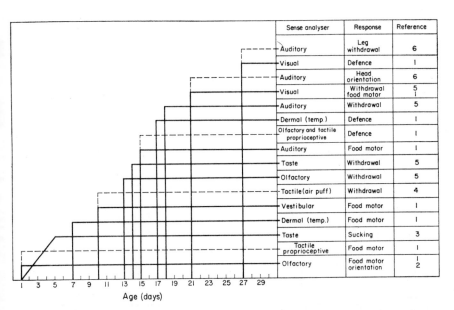

Sense analyser	Response	Reference
Auditory	Leg withdrawal	6
Visual	Defence	1
Auditory	Head orientation	6
Visual	Withdrawal food motor	5 1
Auditory	Withdrawal	5
Dermal (temp.)	Defence	1
Olfactory and tactile proprioceptive	Defence	1
Auditory	Food motor	1
Taste	Withdrawal	5
Olfactory	Withdrawal	5
Tactile (air puff)	Withdrawal	4
Vestibular	Food motor	1
Dermal (temp.)	Food motor	1
Taste	Sucking	3
Tactile proprioceptive	Food motor	1
Olfactory	Food motor orientation	1 2

Age (days)

Fig. 75. Age of first appearance of positive conditioned responses in the dog (time of stable conditioning not included). Data from: (1) Volokhov, 1959; (2) Fox, 1967a; (3) Stanley *et al.*, 1963; (4) Cornwall & Fuller, 1961; (5) Fuller *et al.*, 1950; (6) James & Cannon, 1952.

patterns emerge and neonatal behaviour patterns decline. The socialization period is represented by increasing perceptual and motor abilities and a more complete interaction of the organism with its environment and social milieu. Various parameters of CNS development, such as myelinization of the cerebrum and spinal cord, ontogeny of visual and auditory evoked potentials, ontogeny of conditioned reflexes, sleep and wakefulness, etc., are summarized in Figs. 74 and 75. Most parameters of CNS development undergo gradual development during the neonatal period and a more rapid maturation during the transitional period so that by the onset of the period of socialization, most physiological (perceptual and motor) functions are relatively mature. Thus a close correlation between behavioural development and CNS maturation is demonstrated; the occurrence of relatively mature perceptual and motor abilities at the onset of the socialization period has been described as a period of integration (Fox, 1966d). However, development at this stage is by no means completed. Although the retina is mature by 6 weeks of age (Parry, 1953a) and around this age pups show avoidance (Clark, 1961; Fox, 1964a) of the visual cliff (developed by Gibson & Walk, 1960), more complex visual-motor tasks, such as visual discrimination, develop more gradually over the

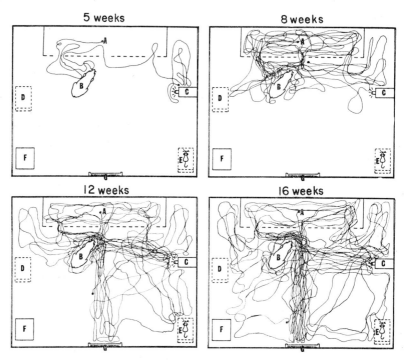

Fig. 76. Development of exploratory behaviour in the dog. With increasing age, pup becomes more exploratory, especially to mirror (G). Other stimuli include a toy pup (B), flashing light (C), empty cage (D), rat in cage (E) and click-in-box (F). (Records from activity in 8 ft × 8 ft arena during 15 minutes' observation.)

first 4 months of life (oddity discrimination using modified W.G.T.A. Apparatus, Fox *et al.*, 1967). Myelinization of association areas in the neo-cortex is extremely protracted in the dog (Fox, 1967a), and this may be correlated with the gradual improvement of learning abilities during the juvenile period. For example, delayed response performance gradually improves during the first 3 months of life (Fox & Spencer, 1967) and explora-tory behaviour in a novel environment increases with increasing age, there being a marked preference for more complex stimuli as the organism matures (Fox & Spencer, 1969, Fig. 76).

Socialization and filial attachments. Freedman *et al* (1961) demonstrated that pups lacking contact with human beings during the first 3 months of life were almost impossible to socialize or domesticate. They concluded that the socialization period, which extends from approximately the first to the third months of life, is an optimum time for the dog to develop social attachments (to its own or to an alien species). There is much practical evidence to support these findings; pups raised in kennels with little human contact during the period of socialization generally make inferior house pets (Fox, 1965; Brunner, 1968) and their resistance to subsequent attempts at socializing them may be increased by familial traits such as timidity (Fox, 1965; Krushinski, 1962). Such delayed socialization may also severely interfere with subsequent training for specialized tasks such as guide dogs for the blind (Pfaffenberger & Scott, 1959).

The question, what brings the socialization (or imprinting) period to an end, has received much discussion. The gradual onset of a "fear" or anxiety period following an earlier period of "approachfulness" may be a normal ontogenetic sequence, being accompanied by improved perceptual abilities, greater qualitative and quantitative input from the environment and a greater ability of the organism to perceive sudden changes in its environment. The initial "approach" period ensures attachment to conspecifics; the later emerging avoidance period ensures some protection against potential danger as the organism becomes more independent and explores beyond the nest. These ontogenetic sequences in approach/withdrawal have been discussed in detail by Schneirla (1965) and their significance in relation to socialization in the dog emphasized by Scott (1962) and Fox & Stelzner (1966b). Traumatic experiences, especially during the early part of the avoidance period, tend to consolidate the emotional bond established earlier between the organism and its conspecifics or caretakers (Fox & Stelzner, 1966b).

Another question arises; namely, do the early social experiences and attachments persist throughout life? Woolpy & Ginsburg (1967) find that wolf cubs socialized with human beings and subsequently given little human contact, "regress" to the wild state, while adults, when once socialized, remain attached and also generalize their attachments from one or two caretakers to people in general. Similarly, many dog breeders have reported to one of us (Fox) that they will keep a litter of pups in the house and give them plenty of attention, in order to socialize them, during the first 12 weeks or so of life. Then they are put back into the breeding (rearing) kennels, if not sold, and by 6 months of age, many of them are fearful when handled and can only be taken out on a leash with difficulty. In this example, we are dealing with the problem of restricted rearing (i.e. environmental impoverishment, and possibly a mass "fear" response on emergence from the confines of the kennel) complicated by individual and breed differences in timidity and other inherited capacities which, as a result of genotype-environment interaction during development, produced variable effects in those subjects raised in kennels compared to those from the same litter which were sold earlier as pets and develop normally. The possibility of a lack of human contact after the critical period of socialization in causing a loss of attachment to man cannot be ruled out, but contributory factors, especially an impoverished environment and lack of manipulatory and exploratory freedom should be considered (Fox, 1966a). Genetic factors, in relation to the effects of isolation rearing, emergence from isolation and ability to recover have been discussed by Fuller & Clark (1966), Fuller (1967) and Krushinski (1962). Dykman *et al.* (1966) have selectively bred two strains of German Pointers, one of which is "normal" and the other extremely shy and withdrawn. The findings of Dykman *et al.*, in that genetic factors not only influence learning abilities and trainability, but may restrict adaptation to the normal environment and give rise to severe behavioural disturbances, lead to the conclusion that such over-fearful animals develop under a more or less self-imposed environmental impoverishment, as they withdraw from any novel stimulus. Experiments in which dogs have been raised in isolation show that emergence from isolation is characterized by excessive arousal (abnormal EEG patterns and short-latency visual evoked potentials), together with biochemical changes and bizarre behaviour and responses to novel and painful stimuli (Melzack & Scott, 1957; Thompson *et al.*, 1956; Thompson & Heron, 1954; Melzack & Burns, 1965; Fuller & Clark, 1966; Fox & Stelzner, 1966a; Fox, 1967c; Agrawal *et al.*, 1967). This behavioural state, in which selective attentiveness is impaired, finds close parallels with the model proposed by Hebb (1955) and Hutt & Hutt (1965) who postulate a state of chronic arousal in cases of human infantile autism (see also Fox, 1966a, 1967c). In close agreement with these interpretations, Fuller (1967) considers the post-emergence behavioural deficits a result of competing emotional responses rather than from a failure of behavioural organization during isolation or atrophy of established patterns as postulated by Lessac (1965) and earlier by Hebb (1949). The effects of environmental enrichment, or giving the organism excessive stimulation early in life, have been studied in the dog by Fox & Stelzner (1966a); the handled or excessively stimulated subjects, receiving 1 hour daily stimulation of various modalities from birth to 5 weeks of age, showed a more rapid maturation of EEG, superior performance in a simple problem-solving situation and were more easily handled but were more aggressive and exploratory than non-handled controls. Adrenal epinephrine was significantly raised in the handled subjects and heart rate showed a significant increase from 2 weeks of age onwards, i.e. tachycardia. These data find close agreement with earlier studies in rodents.

Rearing animals in a quantitatively or qualitatively selectively-controlled environment (e.g. with contact only with an alien species or with scheduled contact with specific stimuli) provides a means to determine to what extent certain behaviour patterns are "species specific" or are acquired. Also, depending upon the duration of exposure to the alien species, the ability of the organism to "recover" when placed with conspecifics, and its later social and sexual preferences as a result of early rearing experiences, can be evaluated. Fox & Stelzner (1967) have evaluated the effects of raising dogs with varying increments of human contact and varying decrements of contact with their own species and Cairns & Werboff (1967) have studied the effects of raising pups with visual or physical contact with rabbits.

Raising pups from $3\frac{1}{2}$ weeks of age until 16 weeks with a litter of kittens of approximately the same age (1 pup per litter of kittens) revealed some interesting phenomena when such groups were tested with dog-raised littermates and with kittens having had no prior exposure to pups (Fox, unpublished data). The most dramatic finding was that cat-raised dogs responded minimally to their own mirror reflection (see Fig. 77), from which we may conclude that species-identity and self-recognition are a function of early exposure learning to a given species or social majority. Cat-raised dogs were passive and submissive in the presence of their own species, and preferred to maintain contact with their foster-kittens. Kittens raised with one pup generalized from this experience with one animal and played vigorously

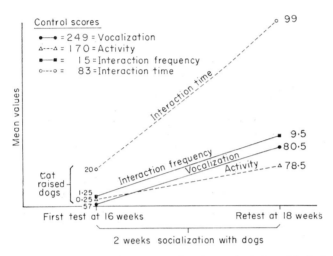

Fig. 77. Averaged results in control and cat-raised dogs. Marked recovery especially interaction time response to mirror in cat-raised dogs 2 weeks after forced-socialization with dogs.

with dog-raised dogs while kittens that had not previously been exposed to pups, with few exceptions, avoided any form of playful contact. It would be of interest to determine the effects of prolonged socialization with an alien species on later sexual preferences. Fox & Stelzner (1967) found that when differentially raised dogs were placed in a group for the first time, individuals segregated themselves and showed a preference for conspecifics having had a similar rearing history. Sackett *et al.* (1967) have reported comparable findings in the social and sexual preferences of monkeys raised with varying increments of conspecific and human contact. A good deal of evidence in house-pets suggests that dogs will develop close affectional ties with other species (both cats and human beings) and direct their sexual responses towards these species. Generally such reactions arise from restricted opportunities in that the dog receives little socialization with its own species, but the selective-deprivation of the domestic environment may result in some form of sexual imprinting (Fox, 1965; Scott, 1961; Brunner, 1968).

It is beyond the scope of this chapter to discuss further aspects of the ontogeny of behaviour and experiential variables in the development of social relationships in the dog. Future research in this species should be directed towards a more vigorous analysis of genetics as well as experimental/environmental factors, as exemplified by the studies of Scott & Fuller (1965).

15+

The social position of the domesticated dog and the various socio-emotional relationships between pet and owner present unique opportunities for the application of animal behaviour studies in veterinary medicine (Fox, 1968b). Also, on the basis of experimental findings to date, some practical applications for optimal breeding, rearing and training of pets and for the recognition, treatment and prevention of behavioural abnormalities can be formulated (Burns & Fraser, 1966; Fox, 1965, 1968b).

D. Remainder of Life Span

The period of socialization, as herein defined, ends with complete weaning, which takes place ordinarily from the 8th to 10th week. The interval from weaning to puberty is characterized by growth, increased strength, and greater activity. Except for sexual behaviour, there is no real boundary between juvenility and early adulthood, and even here adult patterns of courtship and copulation are grafted on to a series of incomplete responses which have been established for many months. Physical growth continues into the 2nd year and increased size and strength have behavioural consequences, particularly in the area of dominance relationships between males.

Old age may begin at about 7 years for breeds such as Great Danes, or as late as 12 years for some of the toy breeds (Comfort, 1960). Behaviourally, senescence has been a neglected area of study, a situation which may improve.

II. SENSORY AND LOCOMOTOR CAPACITIES

As in any species, the behaviour of the dog has limits set by sensory and motor capacities. Locomotor capacities have been of less interest to behaviourists, but have engaged the attention of physiologists studying work capacity and of trainers preparing dogs for useful work.

In the early part of the twentieth century considerable attention was given to the determination of sensory capacities of small mammals. Because of its availability the dog was frequently used in such investigations. Many of the early studies were rather crude, some because the physical nature of the stimulus was not specified, others because the behavioural task used as an indicator was poorly chosen. Some of the problems used to test discrimination were psychologically complex and performance failure was improperly considered to reflect sensory deficits. Possibly the most sensitive method for the determination of sensory thresholds is the conditioned reflex method of Pavlov, although the operant conditioning method of Skinner has great potentials for such applications.

It is beyond the scope of this chapter to discuss the behavioural differences between breeds and between individuals. These differences invariably stem from inherited factors (Scott & Fuller, 1965; Burns & Fraser, 1966) which broadly influence the organism's capacity to respond and its response threshold. Behaviour is the product of both environmental and genetic influences, and is not inherited as such. The Pavlovian school has developed conditioning procedures to evaluate the nervous typology of individual dogs, and

a. Facial-lingual approach.

b. Inguinal approach.

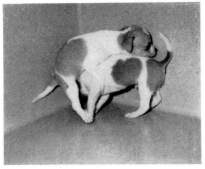

c. Fragmentary sexual activities during play; disoriented mount with pelvic thrusts.

d. Inguinal approach.

e. Fragmentary sexual activities during play with pelvic thrusts.

f. Lateral recumbency, genital exposure and urination in submissive pup when handled.

PLATE XV. ONTOGENY OF INGUINAL APPROACH IN THE DOG

a. Food-oriented inguinal approach in 6-week-old pups.

b. Inguinal approach in pup during play.

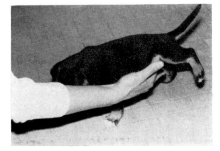

c. Passivity during inguinal approach by mother (who is cleaning ano-genital area).

d. Passivity following inguinal contact by handler.

e. Highly ritualized inguinal approach in adult to a model.

several types can be classified in relation to conditionability, ability to inhibit or to switch conditional responses to different signals. The Pavlovian types represent extremes; in a non-selected population one would expect most dogs to be somewhere between the two extremes of excitation and inhibition. By breeding for temperament, man has come close to creating the pure type, and indeed certain breeds do show a predominance of one nervous typology over another. Kurtsin (1968) and Peters (1966) have reviewed several aspects of conditioning in dogs. They generally agree that the individual nervous typology (or higher nervous activity) which is principally genetically determined, influences the organism's reactivity to various stimuli (i.e. behaviour), its learning abilities and even resistance or susceptibility to disease or psychological trauma (Kurtsin, 1968; Krushinski, 1962). The Pavlovian school also recognizes the fact that restricted or unrestricted rearing conditions (i.e. experiential or environmental factors) can greatly modify the innate nervous typology, which in turn may increase or decrease the effects of such experiential influences (Krushinski, 1962).

A. Olfaction

The dog has long been noted for its olfactory acuity and its ability to discriminate between complex mixtures of odours. A well trained tracking dog can follow the trail of a specific individual when it is criss-crossed by other trails, being distracted only when the confusing trail is made by an identical twin of the original track layer (Kalmus, 1955). Other feats which have been well documented include the ability of a trained dog to select from a group of wooden billets the particular piece which had been handled by its keeper. Only 2 seconds of hand contact were necessary for identification and the presence of pieces handled by other persons did not interfere with correct choice (Löhner, 1926). Dogs can readily be trained to select objects which have been in contact with a particular individual (Kalmus, 1955; Whitney, 1947b).

The dog's accomplishments in tracking and in locating objects or individuals imply two different kinds of ability. First, the dog must have remarkable olfactory acuity, the ability to detect extremely small concentrations of odorous material. Secondly, the dog must be able to distinguish between the odours of specific individuals, even though the chemistry of human perspiration is fairly similar from person to person. The problem of acuity has received more attention. A rather primitive method of measuring olfactory ability is in terms of the layers of paper through which a dog can detect the odour of meat (Krushinskii & Fless, 1959). These investigators used this method to determine the effect of injections of amphetamine on olfaction. In tests on 23 dogs under control conditions the mean number of layers of paper permitting detection was 7·4; after injection of amphetamine the same dogs could detect meat through 15·2 layers of paper. Other investigations of acuity of smell have employed the discrimination technique. For example, two glass containers are placed before panels concealing food dishes: one contains a solution of the test substance, the other contains solvent only. The dog is trained to push open the door associated with the

positive stimulus and receive a food reward. Other methods involve retrieving of small pails in which test solutions can be placed; food rewards are
given for correct retrieval only. Using this method, dogs easily recognized
organic salts (lactic, acetic, formic, and caprylic acid at a dilution of
1:1,000,000; sulphuric acid was recognized at a dilution of 1:10 million)
(Buytendijk, 1936). A more accurate olfacto-meter which mixes known
amounts of air and odorant has been used to determine thresholds of detecting fatty acids (Table 44). Neuhaus (1953) claimed that the olfactory sensi-

Table 44. Odour Thresholds for Some Fatty Acids in a Trained Dog. (From
Neuhaus, 1953.)

Substance (Fatty Acids)	Number of Carbon-atoms/ molecules	Threshold Concentration	
		In molecules/cm³	In g/cm³
Acetic	2	5×10^5	$5 \cdot 0 \times 10^{-17}$
Propionic	3	$2 \cdot 5 \times 10^5$	$2 \cdot 9 \times 10^{-17}$
Butyric	4	9×10^3	$1 \cdot 3 \times 10^{-18}$
Valerianic	5	$3 \cdot 5 \times 10^4$	$5 \cdot 9 \times 10^{-18}$
Capronic	6	$4 \cdot 0 \times 10^4$	$7 \cdot 7 \times 10^{-18}$
Caprylic	8	$4 \cdot 5 \times 10^4$	$1 \cdot 1 \times 10^{-17}$

tivity of the dog is from 1 to 100 million times greater than that of man, but
less extreme differences between man and dog have been found by other
investigators (Moulton *et al.*, 1960). Using a discrimination method with
odorous substances in solution, these workers found thresholds about 100
times lower than that of man. These authors suggested that the individual
olfactory cells of the two species may be fairly similar in sensitivity, and
attributed the greater acuity of the dog to a larger olfactory area which permits pooling of responses from many cells. The olfactory region in man has
an area of some 500 mm², containing 2×10^7 cells (Ehrensvärd, 1942),
while that of a large dog has an area of 7,000 mm², containing $2 \cdot 8 \times 10^8$
cells (Wieland, 1938).

In summary, though different methods have yielded contradictory absolute values for the threshold of acuity, it is most reasonable to believe that
the superiority of the dog over man is a function of the greater area of olfactory epithelium rather than of sensitivity of individual elements.

B. Hearing

Audition is well developed in the dog. Below frequencies of 250 cycles/
second man and dog have substantially equal auditory acuity; above this
frequency the dog's threshold becomes progressively lower compared with
that of man (Lipman & Grassi, 1942). The dog's upper frequency limit for
audition is higher than that of man, a fact utilized in the high-pitched "silent
dog whistle" of commerce. The statement is frequently made that the dog's

upper limit is between 70,000 and 100,000 cycles per second, but this value is probably too high, as it is apparently based upon experiments with a Galton whistle, an instrument which produces impure tones (Warden *et al.*, 1936). Using the conditioned reflex method and an audio-oscillator for stimulus control, a limit of 26,000 cycles per second has been found (Andréev, 1925). Dogs have also been credited with a sense of absolute pitch, but this again overstates the evidence. Accurate discrimination on the conditioned reflex situation has been obtained only between sounds differing in pitch by about one tone on the musical scale. Laboratory studies may not show the real limits of a dog's ability to discriminate auditory stimuli. It is a common observation that dogs apparently recognize the sounds of particular automobiles, or footsteps of particular individuals. Such complex stimuli are, of course, difficult to specify in an experiment and it is also true that the free-ranging dog may be utilizing non-auditory cues.

A considerable amount of attention has been paid to the dog's ability to understand spoken words. Much of the literature is anecdotal and most of it suffers from the methodological flaw that tests of understanding have been carried out under conditions which permitted the dog to see its trainer. The dog is adept at utilizing subtle cues as a guide to responding and care is needed to eliminate the possibility of the "Clever Hans" error (Scott, 1958a, p. 149). Under ordinary conditions of dog training, vocal commands are accompanied by gestures and it is difficult to separate the effects of various stimulus components in the control of the animal's behaviour. Remarkable control can be achieved by patient training, but only rigorously controlled tests will separate out the influence of words by themselves, divested of variations in emphasis and other "contaminations". Thus, the noted German Shepherd "Fellow" did less well in demonstrating understanding of language when tested by psychologists standing behind a screen than when tested by his handler in full visual contact. However, even under the more rigorous conditions, "Fellow" could select and retrieve objects such as gloves, packages, and balls when directed by words alone (Warden *et al.*, 1936).

C. Vision

The extent to which dogs utilize vision in their daily life varies with breed. So-called sight hounds pursue their prey in open country predominantly by means of vision, while trail hounds rely much more on olfaction. Dogs can apparently discriminate between light intensities almost as well as man (Stone, 1921), but they are much inferior in discrimination of form and pattern. Accuracy of form discrimination is greater when a dog is free to move about than when it is restrained and presented with stationary patterns (Buytendijk, 1936). Even though pattern discrimination is poor, dogs attend to strange visual stimuli and often show strong fearful responses to sudden movements or variations in familiar forms (Mahut, 1958). Furthermore, in a relatively barren environment dogs investigate minute bits of straw or flecks of light on the floor. Guide dogs for the blind can be trained to avoid overhead obstacles under which they could pass, but which might injure their charge. Herding dogs can respond to hand signals at a distance of a mile. Thus, vision is adequate for a wide range of behaviour which does not require fine discrimination of static forms.

D. Locomotor Capacities

As enormous differences in physical type occur, wide divergence in locomotor capacities may be expected. In general, the dog is an agile, sure-footed animal, exhibiting great speed, strength, and endurance. For limited distances, the coursing hounds of the Greyhound family are among the faster mammals. Greyhounds of average racing quality run at a rate of 37 miles/ hour (Davis, 1949); the smaller Whippet, weighing about 25 lb, can maintain 35 miles/hour on a 200-yard course. Afghans and Salukis, bred for hunting in rocky, mountainous country, are particularly adept at hurdling, dodging, and turning while maintaining speed.

Racing Greyhounds cover about 18 feet in a running stride; German Shepherds may clear more than 20 feet in a single broad jump. Dogs in training for special army, police, or guard work are often required to scale a sheer wooden wall 8 to 10 and even 12 feet high. In wall scaling, climbing ability is also demonstrated, as the dog must scramble upward until he can grasp the top of the wall with his front feet, pull himself over, and then climb down a ladder on the other side. Dogs do not have retractile claws or prehensile paws and cannot climb trees as do cats, bears, or raccoons. Nevertheless, they are not devoid of grasping prowess; hunting dogs may go up trees which have limbs low enough for them to reach by jumping from the ground; police dogs and guard dogs may be trained to climb regulation length ladders or to stop a running man by seizing his leg, wrapping the forepaws and legs around it, and holding him securely. Some breeds (e.g. Basenjis) show considerable skill in climbing and manipulating objects with their paws; others (e.g. Cocker Spaniels) show neither ability nor inclination in these capacities, but are adept at using their jaws for pulling and carrying (Scott, 1954). Terriers seem particularly inclined to use their feet for digging, perhaps as a result of selection. All dogs can swim, but some breeds such as Newfoundlands and the Retrievers have been particularly bred to withstand cold and endure rough water.

Great individual differences are found in activity; an apartment pet may walk less than 1 mile per day. In contrast, a sheep dog may cover close to 100 miles a day for nearly 6 consecutive weeks (Robertson, 1957); a bird dog may travel 75 to 100 miles a day while hunting. Teams of racing sled dogs average 12 to 14 miles/hour on courses of up to 30 miles. One team of 6 dogs was driven 522 miles in 118 hours (Dawson, 1937). Beagles on a treadmill were able to run for 75 minutes at 5 miles/hour on grades of zero to 15°, but became exhausted rapidly when required to run 7·2 miles/hour on a 15° incline (Jetter et al., n.d.). These short-legged dogs panted heavily after their exercise periods and remained quiet for an hour or two following cessation of running.

A pack dog can carry a burden equalling about half its body weight, the size of pack depending upon the distance it must be carried and general travelling conditions. In the wilds of British Columbia, where sleds cannot be used, the dog is regularly employed as a carrier. Tahltan Indian dogs, no larger than small Collies and usually in very poor condition, carry at least 20 lb each on long trips; two larger dogs packed 25 and 36 lb each over rough terrain for a distance of 15 miles (Stanwell-Fletcher, 1946). Working 8 hours a day at 40° to 60° below zero sled dogs can pull double their own weight at a rate of 2 to 3 miles/hour (Davis, 1949). On a long trip a freight team of 6 or 7 dogs, pulling about 200 lb each plus sleds and men, can average 20 miles a day.

III. INGESTIVE BEHAVIOUR

The wild *Canidae* must adjust their behaviour to alternate feast and famine and this background still retains some influence on the domestic dog. Given access to abundant palatable food, most dogs will gorge themselves. In the authors' colony there is a great variation in the ability of individuals to adjust food intake to energy requirements. When fed *ad libitum,* some animals become extremely overweight, while others remain at a uniform level for many years. Dogs have survived two months and more without eating, and deprivation for a week is readily sustained without serious effects. They are, however, less resistant to stress when body weight falls more than 15 per cent below normal. In this section we shall consider the development of ingestive behaviour, the food and water requirements, and factors which regulate ingestion. Emphasis will be placed upon behavioural rather than physiological regulation of metabolic balance.

In passing we should mention a type of hoarding behaviour which is somewhat related to ingestion. Many dogs bury bones, others pick up food and carry it from one place to another. Such behaviour may be a residue of the behaviour patterns of adults bringing food back to their puppies. Under domesticated conditions it has no real value, nor is much known regarding factors leading to its sporadic expression. It is doubtful that bone-burying actually represents storage of food for future use.

A. Ingestive Behaviour in the Puppy

The orientating responses by which a puppy locates its mother were described in the section on the neonatal stage. When a puppy has located its mother it pushes against her side, searching for the breast, the hind legs churning to maintain position while the head bobs and swings back and forth. When a nipple is found the puppy makes several attempts to nurse before it grasps the nipple properly and produces the necessary suction. Considerable differences are seen, some puppies appearing to be much more proficient in this respect than others. Suckling is accompanied by rhythmic movements of the forelegs, which usually continue until the puppy has become satiated.

One of the first complex patterns of behaviour exhibited by the puppy is the suckling response, an essential component of ingestive behaviour, which is also manifested apart from the hunger drive. All puppies show some degree of non-nutritional sucking. Feeding under natural conditions, puppies commonly remain fastened on a nipple even though their stomachs are well filled. They often fall asleep with a nipple in the mouth but resume suckling activity at once if attempts are made to dislodge them. Vigorous suckling on the mother for 15 to 20 minutes has been observed in food-satiated puppies (James, 1957).

A number of researches have been concerned with the role of learning in suckling behaviour and the extent to which suckling continues when it is nonproductive of nutriment (Ross et al., 1957). A fair summary of the somewhat contradictory evidence is that the strength of suckling in puppies is increased by food reinforcement, though some innate oral drive may be presumed to initiate the behaviour. Non-nutritive sucking on the bodies of littermates is occasionally intense enough to be a problem. Sucking of this type appears to be facilitated by hunger but may persist for weeks in some well-fed individuals, presumably because the oral contacts themselves are reinforcing.

Stanley & Bacon (1963) have found that stomach loading puppies with milk decreased the amount of sucking on the mother, in contrast to the earlier findings of James (1957). Stanley and Bacon found that stomach loading depressed sucking in pups either deprived of food over an extended period, or non-deprived.

The necessary frequency of feeding is a matter of some dispute. Many puppies will thrive with a minimum of care and handling, adjusting to a schedule of 3 feedings per day of concentrated food (Sheffy, 1957). These methods have not been universally proved, and feeding at 4-hour intervals is often recommended for the first week of life (Fox, 1966a).

B. Ingestive Behaviour in the Adult

Adult domestic dogs are generally fed once a day, though feeding twice daily is recommended for lactating bitches as it is difficult for them to meet their requirements in a single feeding. The regulation of calorie intake under such conditions is generally in the hands of the handler, who can be guided by estimates of needs (Mayer, 1953). In general, small dogs require more calories per pound than larger breeds and free-ranging animals expend some 25 to 35 per cent more energy than restricted ones (James & McCay, 1950). If given excessive palatable food dogs will become obese, a status all too common in household pets. In many kennels dry food is kept continuously before the dogs, a procedure which reduces labour costs. For the majority of animals this is satisfactory, but some overeat and others need additional attention in order that they eat enough.

Palatability. A great deal of research on palatability has been conducted by the manufacturers of dog food. Results of such studies tend to be trade secrets, but the factors which make a diet palatable are readily observed. Odours of meat and fish extracts, moisture, and warmth all contribute to palatability. Preference for such stimuli appears to be innately determined. However, we have had to train dogs reared on dry food to accept ground meat used as a vehicle for drug administration. Once the initial aversion to meat, probably based on strangeness, has been overcome, it is preferred to the dry food. Sick and pregnant animals have depressed appetites and are much more sensitive to palatability factors. This fact is important in practical dog management.

Bulk. The same amount of calories may be contained in a large or small volume, depending upon the amount of non-nutritive material in the diet. A number of studies have shown that the dog regulates its intake largely according to calorie content and that dilution with inert fillers has only a temporary influence on total intake (Adolph, 1947; Mayer, 1953). Other investigators have reported a tendency of the dog to maintain a constant volume of ingestion for some time after inert material was added to a balanced diet (Janowitz & Grossman, 1949). The dog apparently learns a relationship between stomach distension and a feeling of well-being which leads to regulation of calorie intake in terms of the bulk of food ingested. A sudden change in the nature of the food, however, can disrupt the regulatory system.

Social factors. Puppies eat 14 to 50 per cent more food when fed in groups than when fed independently (Ross & Ross, 1949a). In another variation of this experiment, puppies were fed to "satiation" while alone. When hungry littermates were introduced the "satiated" puppy resumed feeding and ingested 30 to 200 per cent additional food (Ross & Ross, 1949b). From this it might be inferred that group-reared puppies would gain weight more rapidly than isolated animals, but experiments do not confirm this inference. Presumably the effects of social facilitation are temporary.

Actually, group-fed puppies tend to centre their dominance relationships about the food pan and the dominant animal frequently takes so much of the food that the subordinate is inadequately nourished (James, 1949). Inequality of intake in competitive feeding is particularly marked when the animals are fed once a day from a single pan; it is less when individual pans are provided, or when animals have continuous access to dry food. Social factors affect less obvious features of ingestive behaviour as well. The mere presence of a dominant animal serves to inhibit conditioned salivary secretion (James, 1936).

C. Drinking Behaviour

Body water is a reservoir which is being continuously depleted by evaporation from the lungs and by elimination of urine and faeces, and periodically replenished by drinking and eating. Dogs weighing an average of 18·6 kg ingested 1,100 ml of water per day, 540 by drinking and 560 in food; the average number of draughts per day was 9, with an average volume of 60 ml (Adolph, 1943). Variation was extreme, however, and one draught of 1,390 ml was recorded.

When dogs are deprived of water and then permitted to drink, they will, within a few minutes, take in almost precisely the amount of deficit incurred. Since the dog stops drinking long before the water can have been absorbed into the tissues, it is clear that stimuli arising from the drinking act itself play some role in regulation. Dogs with oesophageal fistulae drink an amount equal to their water deficit and then stop, even though the ingested water

15*

never reaches the stomach. Hence, the pharyngeal receptors probably play some role in stopping the act of drinking. As in the case of food intake, there is a close relationship between physiological systems operating internally and regulation operating through behaviour.

IV. ELIMINATIVE BEHAVIOUR

In *Canidae*, defaecation and urination serve not only as acts of elimination but as important means of communication with other animals. Like wolves, jackals, and other wild members of this family, the domestic dog gives and receives information from the scent posts distributed throughout the locality he frequents. Male dogs are particularly inclined to investigate these areas and mark them with their own urine or faeces, but females, especially when they are in oestrus, also show intense interest. Thus the social consequences of eliminative behaviour are given special consideration as we discuss its development, its manifestations in adults of both sexes, and the endocrine control of variations between the sexes. Kleiman (1966 & 1967) has discussed and compared certain aspects of social behaviour and scent marking in *Canidae*; it is unfortunate that there are not more studies of a comparative nature in this group of mammals.

A. Development

Puppies between birth and about $2\frac{1}{2}$ weeks of age do not generally defaecate or urinate until they are stimulated by the mother's vigorous licking of their anogenital regions. Elimination is easily evoked in bottle-reared puppies by stroking with a piece of warm, moist cotton. Such stimulation is not absolutely necessary, but without it urine and faeces may be retained to the point of apparent discomfort.

At about 18 days the elimination response to external contact becomes more difficult to elicit. At this time the puppy is able to walk and will deposit its urine and faeces in a corner of the whelping box, away from the nest proper. The mother continues for some time longer to clean the puppy after elimination, but she no longer initiates the act.

At 3 weeks of age most puppies leave the nest box entirely to defaecate and urinate, gradually moving to a far corner of the pen. Usually the sleeping area is not soiled if it is kept clean by the kennel attendant and if free access is provided to an outside site. Also, groups of puppies tend to restrict defaecation to a particular section of the kennel which is regularly selected for the deposition of faeces by all members of the group (Ross, 1950). Such behaviour is highly adaptive for a "lair-dwelling" species.

Very young puppies of both sexes squat when urinating, but there are subtle postural differences between males and females. The female simply squats deeply and the infantile pattern normally persists throughout life. Male puppies spread their hind legs and bend them slightly in a less deep squat. As puberty approaches, body torsion appears during micturition and one hind leg is lifted slightly. Eventually the male is able to elevate one leg and rotate its body so as to direct the stream of urine horizontally.

B. Adult Patterns

The adult male commonly lifts one hind leg and directs urine against an object, thus marking it with his scent. Transition from the infantile micturition posture occurs gradually; some males begin to assume the adult pattern at 5 or 6 months of age, but most do not consistently display leg-elevation until they are 8 or 9 months old. However, puppies treated with testosterone from the 3rd day of life may lift a hind leg during micturition as early as the 39th day (Martins & Valle, 1946, 1948). After leg-elevation has been adopted some dogs continue to urinate in the straddle position occasionally, and a few, even though they may be experienced sexually, seldom employ leg-elevation. During illness, males regress into the infantile pattern of squatting.

Accompanying leg elevation is a tendency to expel only a little urine at a time, thus making it possible for the male to mark many different areas or to mark one area repeatedly. Although the strongest stimulus for micturition is the fresh urine of another dog, the behaviour may be released in response to other stimuli. Like wild *Canidae*, the domestic dog responds to the scent of oil, tobacco, and excreta of other animals or birds (Heimburger, 1959). Occasionally dogs attempt to roll on these substances, but more often urinate on or near them.

Micturition is also released when the dog is under emotional stress. Puppies frequently dribble involuntarily when they are excited or are displaying a submissive attitude, and adult dogs may expel urine when fighting or resisting restraint.

C. Marking Behaviour

Ordinarily, male dogs who are allowed to roam in a village check the scent posts in their neighbourhoods at least once a day. The route covered usually encompasses several miles, and 2 or 3 hours may be spent carefully "reading" each stop along the way. Small groups of dogs meet at various points on the route and continue together. Scent post checking may be interrupted while individuals pursue other interests, but the average adult male does not return home until the usual course has been completed.

D. Hormonal Control

Sex difference in posture during urination appears to be dependent upon the action of sex hormones on the developing nervous system. Males castrated in the first weeks of life may never abandon the infantile micturition posture, though we have seen elevation of the leg in one old castrate male. The injection of androgens into early castrates induces leg-lifting within a few days (Berg, 1944). However, later castration does not abolish the typical male pattern nor prevent its appearance (Martins & Valle, 1946). Ordinarily, the injection of androgens into spayed bitches induces male micturition behaviour only when the treatment is begun during the first few days of life (Martins & Valle, 1948).

The results of the experiments referred to above are of general interest with respect to hormonal control of behaviour. The appearance of the male patterns is not based upon imitation, but appears to be the result of activation of neural control systems by androgens. Presumably the potentiality for leg-torsion during micturition is present in both sexes, but the determination of development in the male direction must take place early in life; hence, females can be stimulated towards leg-lifting only if endocrine treatment starts early. A degree of residual lability is shown, however, by exceptional instances of leg-elevation in elderly bitches and castrated males.

V. SEXUAL BEHAVIOUR

Individual differences in the onset of puberty, variations in experience and in heredity have important influences upon sexual behaviour. Any description of typical behaviour will, therefore, not necessarily apply to all members of the species. In this section, we have chosen to describe primarily sexual behaviour as shown in Beagles under kennel conditions. This is followed by discussion of factors modifying sexual behaviour. Fuller (1967) has described a method for measuring sexual activity in the dog and Beach & LeBoeuf (1967) have studied sexual behaviour in dogs with particular emphasis on mate-preferences in the bitch.

A. Development

Male puppies as young as 5 weeks of age may show distinct sexual responses. A littermate of either sex is chosen as a partner and mounted; if the partner remains relatively passive, prolonged and strenuous pelvic thrusts may follow. Older puppies stimulated by play often clasp and "mount" children or other animals; if the response is not punished or prevented, the puppy may repeat the behaviour persistently.

Most males are capable of mating when they are 10 to 12 months old. Beagle males of 6 to 8 months will often mate readily at their first opportunity, but males of some other breeds may not do so until they are 2 or even 3 years old. Since the gonads are functional at an earlier age it seems that psychological factors play a large role in the development of appropriate sexual responses.

Beagle females usually reach puberty between 9 and 12 months of age and continue to come into heat about every 6 months. Irregularities in the intervals between oestrous periods are prevalent, both between individuals and in the same individual at different ages. Long quiescent periods of 9 to 12 months occur more often than short intervals of less than 6 months. Apparently either decreasing or increasing length of day can affect oestrus. Andersen & Wooten (1959) found in their colony of 400 Beagles a greater frequency of oestrus in early winter and late spring. However, in most domestic breeds oestrus may occur in any month (Engle, 1946) and seasonal peaks are insignificant.

The bitch puppy typically does not display juvenile sexual responses to the same degree as the male. When a female puppy is the object of the juvenile male's mounting activity she is relatively passive. The beginning of active courtship behaviour coincides with the first oestrous period.

B. Patterns

Courtship

Under kennel management bitches in heat are generally brought to the home pen of the male selected for breeding. In free-ranging animals the situation is somewhat different. When a male on his daily course of checking scent posts encounters the urine of a female in oestrus he urinates on the same area and, if the female is well advanced in her sexual cycle, he remains in the area where her odour is strong or follows her trail. Where many male dogs run free it is common to see aggregations gathered about a residence where a female is in heat.

The male's normal precoital responses to receptive females range from comparative indifference to elaborate courtship. Initial advances include nosing about the neck and ears and wagging the tail. The female may stand quietly to accept these attentions, but frequently initiates spurts of running which the male follows closely. We have called this pattern "running together". Both males and females show playful approaches. Characteristically they face each other with forelegs lowered, hind quarters elevated, and suddenly bounce upward to a normal position. The male often interrupts his courtship to urinate, but he remains intently orientated to the female. Sometimes he crowds against her or places his feet on her back.

As the intensity of courtship increases the male continues to investigate the head and body of the female, but gradually devotes more time to licking her vulva. A receptive female responds to this stimulation by elevating her rump and turning her tail to one side. When she is ready to accept copulation she stands quietly in this posture and allows the male to mount. The male grips her thighs with his forelegs, thus making it difficult for her to crouch or pull away. Some males also grip the neck of the female with their teeth. Experienced males mount from the rear and are well orientated to effect intromission. Novice males may mount from the side or in front.

Behaviour of the female during courtship is closely correlated with her stage of oestrus. During pro-oestrus a bitch may run and play with males and investigate their genital regions, but at this stage she is not ready to mate and will resist sexual advances on the part of the male. Such resistance ranges from crouching or repeatedly turning her hindquarters away from him to vicious attack. As pro-oestrus wanes, running together and sexual posturing increase and the female stands more quietly for male investigation. Ordinarily, acceptance of the male occurs towards the end of oestrus, corresponding to the time of ovulation (Whitney, 1947a). Older bitches tend to become receptive earlier in the cycle than maidens, and mate more readily over a longer period of days. Occasionally a bitch mounts the male and executes a few pelvic thrusts. This behaviour is seen most frequently when the male is inexperienced compared with the female. The reversal of roles is temporary, however, and does not interfere with later successful copulation by the male.

Copulation

The receptive female stands quietly as the male mounts, with her tail turned to one side. The male executes rhythmic pelvic thrusts which accelerate greatly as the penis enters the vagina. Following insertion the posterior portion of the penis (*bulbus glandis*) swells rapidly within the vagina and is held by the constricted vaginal muscles, resulting in the locking or tying which is characteristic of *Canidae*. Often intromission is not accomplished at the first mounting and the courtship sequence may be repeated over and over again. Each time the male dismounts the female typically becomes active, investigating the male's genitals, bumping roughly against his shoulder or chest, and repeatedly placing herself in coital position. If her "invitations" are ignored, she may further stimulate the male by putting her forefeet on his back or rump. This usually causes him immediately to resume mounting activity.

Experienced studs placed with receptive females may achieve intromission in less than a minute. In such circumstances, the preliminaries of body investigation, running and playing together, are compressed. After a brief period of licking the vulva the male mounts and inserts his penis into the vaginal orifice. If the female is not fully receptive and particularly if the male is sexually inexperienced, courtship goes on intermittently for hours without a mating.

The Tie (lock)

Following successful intromission and locking, the male usually remains mounted for a minute or so and then dismounts by dropping his fore quarters to the ground at one side of the female. He turns away from the bitch and places himself in a more comfortable position by throwing one hind leg over her back, thus permitting the animals to stand together. The attachment of the penis to the body permits the pair to stand facing in opposite directions during copulation. Ejaculation begins at about the same time as locking and continues until nearly the end of the tie.

The locked animals may stand quietly or, more frequently, move their feet occasionally, strain against one another, or walk slowly about. It is not unusual for a female to try to break away from the male immediately upon becoming tied and she may succeed in dragging the male or throwing him down. She may also throw herself in the process, and, whirling about on the ground, may exert considerable tension on the penis. This activity does not appear, however, to interfere with ejaculation, nor does it result in apparent injury to the male.

A tie usually lasts 10 to 30 minutes. As the *bulbus glandis* deflates the pair pulls apart and the male licks his genital region for a few minutes. Immediately following he shows no sexual interest, but recovery is rapid and there are records of 5 copulations of one male within one day.

C. Factors influencing Sexual Behaviour

Physiological Factors

Sexual responses of both males and females are mediated primarily through scent and touch receptors. Males are attracted to the urine of females in heat more than to other samples of urine (Beach & Gilmore, 1949). Neither vision nor hearing play important roles; the breeding performance of males and females totally blind for years continues normally (Parry, 1953a, b). A Collie born blind and deaf bred normally upon reaching puberty and raised a litter without difficulty (Fuller & DuBuis, 1961).

Prepubertal castration of males greatly reduces sexual interest, though the writers have observed an early castrate which courted a bitch in heat, mounted, and executed pelvic thrusts in a manner indistinguishable from an intact animal. Interestingly, previous tests over a period of 3 years with receptive females had elicited no sexual responses. We have no explanation for the late appearance of sexual patterns. Castration of experienced males has less drastic effects; full capacity for coitus and orgasm persist in some males for at least several months (Beach, 1950).

As in other species, sexual behaviour of females is more intimately correlated with the oestrous cycle, and spayed bitches are sexually non-receptive. False heat, which resembles the pro-oestrous stage of the normal cycle, is widespread in females of all breeds and all ages, particularly maidens approaching puberty (Whitney, 1947a; Andersen & Wooten, 1959). While males are attracted, the female seldom allows a mating to take place. Normal oestrus generally follows in 2 to 3 months.

Breed Differences

There are no exhaustive scientific studies of genetic variations in sexual behaviour. Nevertheless, a number of dog breeders and scientists using dogs in the laboratory have observed variations in sexual behaviour. Individual differences in onset of puberty are influenced by heredity, but they are not rigidly correlated with body size or conformation. For example, both Beagles and German Shepherds tend to reach puberty at an early age compared with Greyhounds. In some breeds the male is usually capable of mating before the female, but in others (Chows; Salukis) not until several years later than the female. Salukis are particularly likely to be "difficult breeders", responding to receptive bitches in a timid or an aggressive manner (Stockard et al., 1941). Stockard also commented that it was difficult to hybridize Salukis as they did not seem to recognize other breeds as suitable objects for sexual attention. Basenjis are unusual among domestic breeds in having one seasonal breeding period per year. In the northern hemisphere, this occurs in early autumn. However, the period can be shifted by artificially modifying the length of day (Fuller, 1956). Both types of oestrous cycle occur in hybrids between this breed and Cocker Spaniels (Scott et al., 1960).

Dominance Relationships

One indicator of social dominance among males is preferential access to females in heat. Because bitches are often spayed, intact males usually outnumber intact females in village populations. Large numbers of males frequently congregate near the residence of a bitch in oestrus. Fighting appears to be common under such conditions, though it is perhaps more the outcome of increased general excitation than direct competition for sexual favours. The role of dominance in securing access to females is shown by the observations of James (1951). In a mixed kennel of Wirehaired Fox Terriers and Beagles all the puppies born were sired by the dominant

Terriers. Similar results would undoubtedly occur in free-ranging dogs if no restraints were placed upon breeding.

The most important influence of dominance upon sexual behaviour is exerted through the male-female relationship. Successful coitus requires that the female permits mounting from the rear, and that she stands quietly during intromission. Some females, although sexually excited, continue to face their partners and to behave as though courtship were a testing ground for dominance. Established female dominance over the male probably explains occasional sexual incompatibility between littermates. Usually, however, there are no psychological barriers to the very closest forms of inbreeding.

Environment

Males in particular are highly sensitive to unfamiliar surroundings and a strange environment is definitely inhibitory to sexual responses. Breeding is facilitated when the female is brought to the male's quarters as compared with the reverse arrangement. If matings are to be consummated in a special enclosure (for observational purposes, perhaps) the male should be accustomed to the place in advance. One successful copulation in a strange enclosure dispels the original inhibitions associated with the area (Beach, 1950).

The presence of a human observer may be considered as an environmental factor affecting mating. Some males are inhibited under observation but mate successfully when left alone. Other males are aided by the presence of a familiar person, particularly if the bitch is over-active. It is a common practice of breeders to restrain an active bitch or hold up one which crouches when mounted. Such measures facilitate copulation only if the male has been trained to accept human attention.

VI. MATERNAL BEHAVIOUR
A. Parturition

Useful reviews of maternal behaviour and discussion of some environmental factors which may disrupt the norm have been presented by Bleicher (1962); Rheingold (1963); Fox (1966c) and Freak (1968).

A pregnant bitch nearing the end of the gestation period spends much time resting and sleeping, particularly if she is carrying a large litter. The onset of parturition is marked by increasing restlessness, frequent changes of position, and usually retirement to a dark, quiet place. House pets may enter closets or crawl under sofas, while dogs in an outside kennel may burrow under it and hollow out a nest. As uterine contractions begin, the female strains lightly every 10 minutes or so, urinates often, and rearranges her bed. Some females sit with their heads somewhat elevated, shivering occasionally, but remaining relatively calm. Others whine and become increasingly distraught, tearing up the bedding material and pushing it into a heap, then digging the floor and scratching the walls of the whelping box. House pets may rip rugs or corners of the linoleum or blankets on a bed. As imminent parturition requires greater effort the female strains heavily, usually remaining in a sitting position with her body pressed against a corner of the whelping box. Occasionally she will assume the urinating posture while straining.

During contraction intervals, she pants freely and may investigate and lick her vulva. Few bitches, even in the sensitive Toy breeds, cry out during the birth process, regardless of difficult or prolonged delivery.

Fecundity varies both within and among breeds and may range from a single puppy to 23, the largest number ever reported. Length of parturition, therefore, depends upon litter size and such complications as structural abnormalities or uterine inertia which are known to be more prevalent in some breeds than in others (Burns & Fraser, 1966). Whelping of an average litter of 5 puppies takes about 3 hours and larger litters may require 10 hours or more.

During birth the puppies are presented at irregular intervals, occasionally within 5 to 10 minutes but more frequently spaced 20 to 60 minutes apart. When the first puppy is born, the female attends to it at once, breaking the amniotic sac and removing it by vigorous licking. With her jaws she then detaches the afterbirth if present; more often it is not passed until later. Within a few minutes the female chews through the umbilical cord 2 or 3 inches from the puppy's body and consumes it with all other waste. She attempts to clean her own hindquarters, but spends more time cleaning and drying the puppy and directing it to her side. When this has been accomplished, she curls herself protectively around the puppy and rests until the next birth begins.

Most bitches, even at their first parturition, carry out efficiently the sequence of acts described above. Occasionally parts of the pattern are missing and puppies are whelped outside the nest or neglected after birth. Prior familiarization with the whelping area when it differs from the usual living quarters reduces the probability of such disruptions of maternal care. The presence of a familiar person may facilitate the establishment of patterns of care in a bitch prone to emotional disturbance.

B. Nursing Care

During the first few days following birth of the litter the dam remains in the nest, leaving only for feeding or elimination or for brief responses to human handlers. Aggressive females may also leave the nest to drive away animal intruders; some threaten or attack unfamiliar humans. Frequently young bitches must be forced to leave their puppies to attend to bodily functions and permit cleaning the whelping quarters. Gradually the time spent away from the puppies is increased until at 2 weeks the female remains out of the nest for 2 or 3 hours at a time.

While in the nest the dam customarily lies stretched out on her side with the puppies resting on or near her body. Most females remain in this position while nursing, although some assume a sitting position. The dam responds to disturbances among the puppies by investigating and nosing those who cry. Frequently she grooms them and stimulates urination and defaecation by licking their perineal regions.

A bitch often persists in keeping her puppies in the spot where she whelped and will retrieve them repeatedly if they are moved. Some females will retrieve a puppy which wanders from the group and whines, but most do not make this response. The lost puppy may be attended regularly, but not returned to the group. Older puppies which fall out of the whelping box or into water dishes are seldom retrieved even though they may be shrieking for help. New-born puppies that fail to thrive are usually rejected by the dam, who may be responding to their lower temperature or lack of movement. Such pups may simply be shoved out of the nest, but if they cry incessantly the dam may kill them or bury them outside the nest.

About 4 weeks after whelping some bitches begin regurgitating their food, which is quickly consumed by the puppies (Martins, 1949). The response is especially apt to be elicited if the female has been removed from her brood for several hours and is returned soon after eating. Usually the dam's milk supply gradually decreases after 5 weeks of lactation. As the growing puppies exert heavier demands upon her, the dam restricts nursing to a few short periods daily and may growl at the puppies if they attempt to suckle more often. At the same time, some females compete with their puppies over food, even to the point of punishing them severely when they approach the food dish. Bitches receiving adequate nourishment, however, usually allow their puppies to eat freely.

C. False Pregnancy

A rather common phenomenon in the bitch is false pregnancy. Beginning a month or so after ovulation there is a slight abdominal enlargement and a more readily noticeable enlargement of the mammae. The teats are erect and in maidens are two or three times their former size. These signs may persist for a month and then regress, often with no significant behavioural changes taking place at any time. In some females, mammary development continues until lactation results; in others, the most distinct symptom is the turgidity of the abdomen, which may increase in size until the bitch appears to be carrying a large litter and there seems no doubt of her impending parturition. Near the estimated whelping time the abdomen suddenly reduces in size and the bitch no longer appears pregnant.

A female with more extreme symptoms of pregnancy is apt to show the behavioural changes characteristic of the approach of parturition. Generally she is restless; she rearranges her bed frequently, whines, pants, and retires to a quiet recess. House pets may tear up any available material for nest building, and in severe cases may appear frenzied, whining or crying almost constantly, carrying toys or other inanimate objects about, or curling around them as though they were a litter of puppies. In this situation even obedient house dogs may fail to respond normally to their owners. These symptoms may persist for a week or longer if not relieved, and are apt to recur with each successive heat in susceptible females. There are definite breed tendencies towards false pregnancy; maiden Beagles, for instance, seldom show symptoms of pregnancy following oestrus (Andersen & Wooten, 1959).

VII. SOCIAL BEHAVIOUR

When given the opportunity, dogs are highly social animals. Their association with one another is by no means limited to sexual relationships and the care of young. In villages where dogs are allowed to run freely, small troops of dogs can frequently be observed making the rounds of the neighbourhood.

The legendary devotion of the dog to its master appears to be a modification of this tendency of dogs to act together. Such co-ordinated action has been called allelomimetic behaviour (Scott, 1958a). In other words, to the domestic dog, a human being becomes a member of his pack.

Not all social relations between dogs, however, are of the co-operative variety. Fighting, defence of territory, and dominance-submission relationships represent varying degrees of competition. The term agonistic behaviour has been given to such types of activity. In this section, we shall discuss allelomimetic and agonistic behaviour.

A. Allelomimetic Behaviour

Although group action is highly characteristic of dogs, relatively few studies of allelomimetic behaviour have been reported. The four groups of dogs studied by King (1954) spent from 43 per cent to 85 per cent of their time within 50 feet of each other. The lowest value was found in a group of Cocker Spaniel males. The cohesiveness of each group depended upon previous acquaintance among its members; strangers tended to be isolated.

Attraction of dogs to one another is modifiable by early experience. Puppies reared in isolation to the age of 16 weeks make few body contacts when first placed with one another (Fuller, 1963). Contacts increase in number as opportunity for social interaction is provided, but delaying experience appears to reduce indefinitely the amount of close association.

The motivation underlying allelomimetic behaviour has not been clearly defined. It has obvious adaptive value in pack hunting, and appears early in development as the following response given to the mother or to a human being. In these early stages, no reward other than the act itself has been clearly demonstrated.

B. Agonistic Behaviour

Social Hierarchies

We have already described the development of fighting in the young puppy. It will be recalled that "play fighting" occurs as early as 5 or 6 weeks, and in some breeds vicious fighting may occur by 9 or 10 weeks. Ordinarily, such competition leads to the establishment of a stable social hierarchy which, once established, may remain constant for a long period. Diagrams of social hierarchies involved show that, in general, relationships are transitive; that is, if A dominates B, and B dominates C, A also dominates C. This is not always true, however, as is shown by the social relationship diagrams in Fig. 78. These diagrams are based upon observations of two litters in a feeding situation (James, 1939). In Litter 1, one animal is dominant over all the others; two are approximately equal and continue to compete for food. In a second litter, two animals share dominance and one animal is dominated by all the other members of the group. The absence of arrows between 1514 and 1515 indicates that this pair shows mutual toleration and no agonistic behaviour.

Weight and sex are important factors in determining dominance and their

effects are somewhat dependent upon each other. In male pairs the heavier
dog has considerable advantage, but in female pairs weight appears less
decisive (Scott, 1958b, p. 75). In Wirehaired Fox Terriers and Basenjis
males are usually dominant over females, while in Cocker Spaniels and
Beagles, sex appears to have less influence on dominance (Pawlowski & Scott,
1956). In general, Basenjis and Wirehaired Fox Terriers typify breeds which

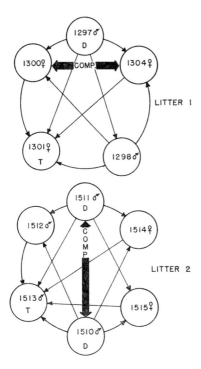

Fig. 78. Dominance relationships in two
litters of dogs. The letter D in circles indicates the dominant individuals. Double-headed
arrows (thick line) connecting individuals denote competition (comp.) without clear dominance; single-headed arrows point to the
subordinate member of the pair; absence of
connecting arrows indicates mutual tolerance.
(After James, 1939.)

form rather rigid dominance hierarchies; Beagles and Cocker Spaniels typify
breeds which are less competitive. These differences influence the acceptance
of new individuals into established groups. Even when space is provided
Basenji groups tend to be closed, while Cocker Spaniel groups accept new
members without great disturbance (King, 1954).

Many of the factors which influence relationships between dogs apply also
to the association between man and the dog. Owners who acquire dogs as
young puppies and have handled them regularly acquire dominance over
their charges. This appears to be particularly true if the dog is lifted by his
neck skin and is punished when necessary by vigorous shaking while he is
held in this fashion. Even large and powerful breeds can thus be managed by
relatively small humans who have acquired dominance. The subordination
reaction may not transfer to other human beings, however. Thus, if it is
desired that an animal be friendly with many people it should be handled by
a number of persons while it is young.

Patterns of Agonistic Behaviour

Dominance-submission relationships within litters become established after a series of tests of strength. However, their maintenance and often the establishment of temporary relationships between previously unacquainted dogs are not always dependent upon overt fighting. Posture and vocalization are effective responses with which a pair may compete for dominance. An erect and rigid posture with raised tail, accompanied by low growls, serves to intimidate an opponent. In some breeds conspicuous pilo-erection along the dorsal midline is an external indicator of arousal of the sympathetic nervous system. These patterns of behaviour may occur in both members of a pair who encounter each other in the course of their daily reconnaissance. The issue may be resolved by a sort of truce in which each individual cautiously investigates the other, particularly sniffing at the anogenital region, and then withdraws. Alternatively, this display may be preliminary to a battle which is terminated by submission, flight, or death of one participant, or by mutual exhaustion. Still another outcome is the immediate adoption of a submissive attitude by one individual; the tail is lowered or drawn tightly between the legs, the animal crouches or rolls over on its back. Vivid descriptions of these sequences, which are characteristic of wolves as well as domestic dogs, have been given by Lorenz (1952).

Factors influencing Agonistic Behaviour

Possibly the strongest incentive for fighting is defence of territory. A dog which will attack invaders viciously on his home ground may respond with friendly investigation when he meets a strange animal away from his territory. Quite possibly the increased aggressiveness of dogs which are kept tied up is attributable to the formation of an extremely strong habit of territorial defence.

Although avoidance behaviour is often considered as the opposite of fighting, the two may be quite closely related. Some dogs, known as fear biters, avoid human beings but when finally caught or cornered attack viciously. Such animals have occasionally appeared in our colonies without any ascertainable proof of either hereditary or environmental aetiology. Littermates reared in the same fashion have been tractable and friendly animals. Puppies which run wild for the first 3 or 4 months of their lives are extremely fearful and aggressive when captured. Some of our puppies reared in isolation cages until 4 months of age have attacked the experimenter when they were first handled. These agonistic responses are not permanent, however, particularly if the animals are kept for a time in small cages and handled frequently. As a matter of fact, *complete* isolation from 4 to 16 weeks appears to decrease agonistic behaviour in later life (Fuller, 1963). Such animals are less competitive and less vigorous in their contacts with humans. As we gain knowledge of the effects of schedules of experience upon agonistic behaviour it seems possible that rather precise control over temperament can be obtained by modifying rearing procedures.

Agonistic behaviour is also affected by changes in the social environment in later life. Any break in routine may lead at times to aggression directed at other dogs or people (Worden, 1959). Prevention, preferably when the first anticipatory threatening responses are seen, is much simpler than curing the condition after it has appeared.

For certain types of guard duty and hunting large animals, dogs may be especially trained for aggressiveness. The specialized procedures needed for this are beyond the scope of this chapter, but selective breeding is essential for success. Agonistic behaviour is definitely heritable and aggressiveness is considered as an essential character in some breeds (Scott, 1958b). A considerable amount of intra-breed variation still exists, however, and must be considered by prospective breeders.

C. Communication

Dogs communicate with each other through a variety of vocalizations, body postures, and the system of marking scent-posts with urine or faeces. The intense interest displayed by both males and females in the scent fields of the dog's environment has been discussed in the section on eliminative behaviour.

Postural Communication

The carriage of head and body and the manner in which one dog approaches another are significant factors affecting social behaviour. For example, a dog can quickly interpret the threatening attitude of another dog bent upon immediate attack even though the aggressor may be 50 ft away. The aggressor's intentions are revealed by his demeanour which is characterized by a stealthy approach, crouched body posture, lowered head, erect ears, and intense orientation towards the prospective victim. In contrast, a dog that bounds forward with head level with the body, ears folded back, and tail loosely carried is at once recognized as a potentially friendly animal. Various other postures and the responses they elicit are discussed elsewhere in this chapter.

The role of tail-wagging, commonly observed when dogs meet, is not completely understood. Possibly it serves to distribute characteristic body odours which serve for recognition, but it also appears to function as a visual cue by which a dog signals its peaceful intentions. In many dogs tail-wagging is evoked by a wide variety of stimuli and seems to be merely a sign of generalized excitement. Selection for animals which are readily trained to become socially dependent upon man, and who thus remain permanently in a juvenile status with respect to man, has undoubtedly been responsible for the high frequency of tail-wagging in some modern breeds.

Vocalization

Vocalization varies considerably both among and within breeds. Chows seldom whine, bark, or howl; Basenjis do not bark, but often whine and occasionally howl or utter chortling noises and loud cries characteristic of the breed. Some strains of Cocker Spaniels bark excessively, while others are relatively quiet. Basenji-Cocker Spaniel hybrids bark, but less frequently than their Spaniel ancestors.

Newborn puppies have a surprisingly strong distress call which is emitted during pain or intense emotional disturbance. Puppies accidentally lain upon by the mother utter piercing yelps. The mother responds at once by getting up and nosing or licking the puppies. In a preliminary study of early learning, puppies were placed on a small table and allowed to move about freely for the duration of a 2-minute trial. Puppies 1 to 2 weeks old often backed partly off the table and clung to the edge, shrieking as long as their precarious positions were maintained. In such emergencies several shrill cries are emitted with each respiration. Though the mother may become highly upset by such a loud and prolonged distress signal she is more apt to run around ineffectively than she is to attempt rescue.

In less stressful situations puppies may whine or whimper and while being groomed by the mother they sometimes make soft, chirping noises. Very young puppies do not usually cry when hungry if they are warm and otherwise comfortable and are not disturbed. Three-day-old puppies deprived of food as long as 23 hours remained quiescent when kept warm (Welker, 1959). Puppies separated from the mother for more than 4 hours may emit spasmodic fretful cries of low intensity if the retention of urine and faeces causes discomfort. Draughts of cold air or a marked reduction in the temperature of the sleeping surface evokes crying almost at once (Fredericson et al., 1956). The "cold" cry is at first a thin wail of protest ending on a quavering note as the puppy shivers and presses close to its littermates. If the cold persists or is intense, vocalization increases until the cries are loud yelps accompanying each respiration. The "hot" cry of neonates is a distinctly different sound. Placed on a heating pad which is too hot, puppies react by crawling about and uttering short, fretful cries. As the high temperature and active efforts of the puppy to escape increase respiration and induce panting, the cries are rapidly emitted and have a flat sound. If the heat is not reduced the puppies become highly distressed and their vocalizations more shrill in character.

Adult dogs also have the "hot" and "cold" cries and a distress call similar to that of puppies. In addition they employ a variety of sound patterns ranging from the low, throaty maternal whine to the sharp, threatening barks of the watchdog. Some of the calls common in wild Canidae are commonly heard in dogs, particularly the prolonged howl of loneliness and the baying of the hunt. These evoke response from other dogs even at a considerable distance and may result in lengthy communication between dogs who have no other contacts. Like wolves, Malamutes and Huskies often howl in the evening and at break of day. While males of certain breeds tend to howl when restrained in the vicinity of a female in oestrus, there are no vocalizations universally associated with courtship or copulation in the dog. Salzinger & Waller (1962) have demonstrated experimentally that vocalization in the dog can be subjected to operant control.

VIII. TRAINING

Dogs are trained for so many tasks that a series of manuals would be necessary to treat this subject in detail. The economic value of the dog is based upon its capability to acquire specialized responses which serve the interests of man. Training a dog does not produce new patterns of behaviour, but establishes control over the appearance of the patterns. We have chosen to illustrate psychological principles which run through all forms of training by examples from the laboratory. Among the better books on practical training are those of Pfaffenberger (1947), Saunders (1952), and the editorial board of *Sports Illustrated* (1960).

Dogs are probably unique among animals in the extent to which learning can be motivated by intangible rewards such as praise from a handler. Because of the importance of social motivation in training much interest exists in the conditions for its development. A critical period for socialization between the 4th and 12th week has been postulated by Pfaffenberger and Scott (1959). German Shepherd puppies, intended as guide dogs for the blind, which were not taken into homes before 13 weeks of age were much less successful in training at a year of age. Factors other than social motivation may also have been involved, but it is certain that early socialization facilitates later training. Food rewards are, of course, useful adjuncts to training, but they cannot be scheduled as flexibly as words or body contact.

Stanley & Elliot (1962), Bacon & Stanley (1963), Stanley (1965) and Stanley *et al.* (1965) have convincingly demonstrated the motivational or reward value of a human being (who may remain silent and passive) to a dog; indeed, when deprivation levels of reinforcement by contact with a passive person are varied, the asymptotes of the pups' instrumental response speed curves were affected analogously to those of rats under food or water deprivation (Bacon & Stanley, 1963). Thus the presence of a passive person alone is of great reward value to a dog, although food, voice and petting have reward properties also.

A. Training Procedures

Training is carried out by a method of rewarding successively closer and closer approximations to the desired pattern. For example, we start training dogs on discrimination problems by inducing them to eat from food pans in the discrimination apparatus. When this is achieved, a slide is placed over the dish so as to cover about one-fourth of it. The dog soon pushes this aside to reach the food which it can still see. Gradually the slide is advanced until it completely covers the food and must be pushed back to permit eating. In all of this training the dog has access to two dishes covered with identical slides and which always contain food. In discrimination training proper, the two slides are identified by prominent differences (one black, the other white, for example) and one of these stimuli always designates a slide which has been locked in place so that it cannot be moved. Dogs can be trained to select the stimulus associated with food with almost 100 per cent accuracy and many complex variations can be added to the basic problem.

Note that this method of training involves working "backwards". In the complete sequence, eating from a dish is the last step and this is where the training starts. Gradually increasing the area of the dish covered by the slide during training reverses the process which the dog employs in pushing the slide to get at the food. All of these motor acts are acquired before training in discrimination is attempted, but in the final stages discrimination precedes them. Backward training of this sort is also employed in training dogs for tracking (Whitney, 1947b).

For some tasks training is more conveniently carried out by a "forward" procedure, employing the principle of rewarding successive approximations to the desired pattern. Suppose that one wishes to train a dog to cross a narrow plank which it initially avoids. A wider plank may be substituted and reward made contingent upon crossing it. As this task is mastered a narrower plank is employed, and the process is repeated until a limit is reached. Each step involves trial and error learning in which the trainer selectively rewards the responses which fit into his prearranged plan, and the criteria for reward are made successively more stringent.

B. Avoidance Training

Aversive behaviour can be elicited in various ways, by stimuli which have been associated with pain, by strange objects or sudden movements, by lack of secure support, and by restraint of freedom of movement. Wide individual variations are found, partly ascribable to heredity (Thorne, 1944; Mahut, 1958) and partly to experience. One of the major problems of dog training is the achievement of a satisfactory balance between approach and avoidance patterns. Any dog should avoid motor cars, but not panic in heavy traffic; a Pointer should be attracted to game birds, but should not flush the covey. It may be desired to train a dog to accept food only in one situation and refuse it elsewhere. It is usually easy to establish avoidance behaviour in dogs, but not always at the desired balance point. Over-severe punishment for proscribed acts may produce generalized avoidance behaviour which interferes with positive training. Insufficient aversive training produces an animal which cannot be controlled.

Avoidance training is most efficiently accomplished by pairing a mildly aversive stimulus closely with the response to be trained against. A loudly shouted "no" is mildly aversive at short range and can be made much more so by simultaneously shouting the word while striking the puppy firmly with a rolled-up newspaper. Some trainers never strike a dog, but prefer to shake it vigorously while holding the loose skin of the neck (Worden, 1959). This mode of punishment is believed to be effective because it is derived from the species pattern of establishing social dominance.

Once established, avoidance behaviour is extinguished with difficulty (Solomon & Wynne, 1953). However, persistent effort will usually reduce unwanted timidity. A dog which fears people improves if it is approached quietly and if contacts are invited rather than forced. Confinement in a small area which renders escape impossible has proved to be advantageous in training shy dogs. Consistent application of such techniques permits the formation of man-dog relationships for some time after the ordinary period of socialization has ended. However, the association appears to be less stable than one established during the first 3 or 4 months.

IX. SPECIAL BEHAVIOUR

From the many special tasks for which dogs have been trained we have selected tracking behaviour and sheep herding as examples for discussion. The versatility of dogs is in part a function of selection of genetic types adapted for particular ends, and in part attributable to a dog's capacity to discriminate numerous postural and vocal cues emitted by its trainer.

A. Tracking Behaviour

Perhaps the dog's most extraordinary ability is the tracking of animals and man. The combination of olfactory sensitivity and discrimination with investigatory habits leads more or less naturally to the location of objects or individuals by tracking. The highly skilled tracker acquires his ability, however, through learning.

Despite the interest in tracking behaviour critical studies are few in number. Published reports tend to emphasize successful tracking and to neglect cases where the dog failed. Nevertheless, a number of facts are well established. Because human track-layers are more predictable than animals, most scientific work has been concerned with the tracking of human beings. Many breeds have been employed as trackers, the German Shepherd and Bloodhound being most widely acclaimed in this area. The nature of tracking behaviour can be understood best through a consideration of methods used in training. The objective of such training is to produce a dog which can follow the trail of a particular individual or animal and not be diverted. Naturally, olfactory sensitivity is involved, particularly if an old track must be followed. The critical point is training the animal to discriminate odours of different human beings (Whitney, 1947b). Whitney, who has observed many Bloodhounds, states that animals of this breed require very little training in following a trail, but must be schooled to differentiate one trail from all others. His preferred method is to train animals to discriminate between people and then to match individuals with articles of clothing or objects which have been handled. Once the dog has been trained for such matching, he can be taken into the field and will follow a track which resembles the odour of the objects presented to him. The procedure parallels the "backwards" learning described in Section VIII.

In actual practice, of course, discriminated cues may not be the most prominent feature of a trail. Studies on police dogs (Most, 1928) have shown that odorous cues for tracking may result from compression of the earth and crushed vegetation, as well as from particles emanating from the track layer's clothing, shoes, and body. Obviously, the first two classes of odours would not be specific to individuals and could not form a basis for discrimination between trails. The fact that dogs often do select the correct trail among several is good evidence that the third class of odours is important, at least to a trained animal. Most (1928) pointed out that dogs were fallible as trackers and that they sometimes did change from one track to another. Hence, tracking from the scene of the crime should not, in his opinion, be considered as conclusive evidence of guilt. Many ingenious methods, such as overhead rails, mechanical trail-making devices and the like, were employed in his researches and excellent English summaries may be found in Buytendijk (1936) and Warden *et al.* (1936).

The ability of the dog to distinguish trails and to match pieces of clothing or other objects with the appropriate human being has been used in an original way to demonstrate the chemical similarity of identical twins (Kalmus, 1955). This study will be presented in some detail, since it illustrates some of the methods employed. The technique for discrimination is shown in Fig. 79. In studies of this sort, trained animals showed almost

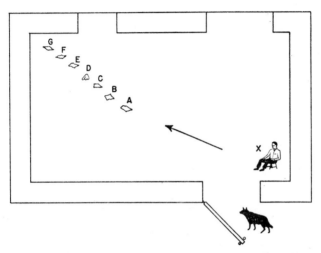

Fig. 79. A row of handkerchiefs (A to G) laid out for retrieving. The person whose body odour is sought sits in the chair at X; the dog is led in through the door, made to sniff at the person's hand, and directed as shown by the arrow. (After Kalmus, 1955.)

perfect performance in selecting a handkerchief handled by an indicated person from a series which included handkerchiefs handled by other persons. This discrimination broke down, however, when handkerchiefs were introduced which had been handled by an identical twin of the person whose body odour was being sought. On the other hand, one trained dog was able on several occasions to discriminate between identical twins in a tracking experiment laid out according to Fig. 80. The solid single line is the track of the person sought; the dotted line with loops represents the decoy tracks laid down by another person. At point "P" the track-layer placed a pole in the ground and dropped a handkerchief. The trained animal on fresh trails readily discriminated between the trails of the person dropping the object and the decoy. In fact, correct selection was usually possible even when both track-layers were identical twins. The improved discrimination between twins in the tracking experiment was attributed to the fact that at point "P" both stimuli were present simultaneously and could be compared directly; in the handkerchief selection experiment each object was sniffed separately.

The literature on tracking contains many observations on conditions which favour accurate trailing. Temperature, humidity, wind, and the absence of competing odours are all important factors. It would be beyond the scope of this account to describe them all in detail. Depending upon conditions, dogs may track closely or down wind from the path of the track

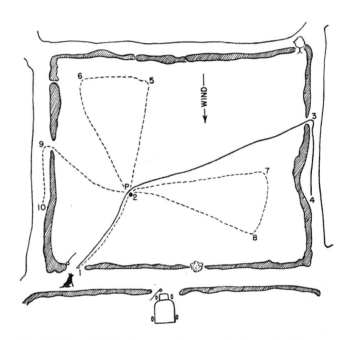

Fig. 80. The solid line (1, 2, 3, 4) represents the trail of the person to be sought; an object was dropped at 2. The dotted loops (1, 2, 5, 6, 7, 8, 9, 10) represent the path of a second individual. The dog is brought on to the course at 1; at 2 he is given the problem of matching the odour of the dropped article to the one track out of six made by its owner. This arrangement, originated by Budgett, is from Kalmus (1955).

layer. Frequently the dog runs in loops, crossing and recrossing the trail. Possibly this is a behavioural means of avoiding olfactory adaptation. By swinging back and forth across the trail, or following it in a general direction, the dog takes advantage of the contrast in odour between the trail and adjacent territory. Recent experiments have indicated that trained dogs preinjected with 10 to 20 mg of amphetamine follow trails more closely and more accurately (Krushinskii & Fless, 1959). The results were interpreted as an enhancement of olfactory acuity, but it is equally possible that heightened motivation is responsible for the improved performance.

B. Herding Behaviour

A well-trained, experienced sheep dog works quickly, quietly, and efficiently. Since sheep naturally gather together in compact groups and run away when disturbed (see Chapter 10) the dog's task appears to be fairly simple. However, a more complete analysis indicates that the requirements for a good herding dog are specific and complex.

Sheep dogs at an early age stalk and herd anything that moves. Young Working or Border Collies orientate towards insects, baby chicks, even ripples on a pool, and stare intently at the object of interest, a behaviour pattern known as "showing-eye" (Burns & Fraser, 1966). Later, a successful sheep dog shows-eye to control the sheep flock and refrains from rushing in and attacking even when greatly stimulated. Physically, the dog must be of moderate size, fast, and capable of long periods of exertion.

These physical and psychological characters are heritable, since they respond to selection (Kelly, 1947). In the development of a superior strain of Border Collies, dogs were selected on the basis of their reactions towards sheep at the age of 10 weeks before they had received training of any sort. Choice of puppies for training and breeding was based upon the extent to which they approached the established ideal for adults.

The principles of training to herd are the same as those employed in training dogs for any specialized task. Working Collies are usually employed with sheep, although they can be trained to work cattle, hogs, or barnyard fowl. The 10 or 12 basic commands that are taught enable the shepherd to convey his wishes and maintain control. Every effort is made to prevent the dog from ever giving an undesired response to a signal, while correct responses are consistently encouraged. One to two years' experience in working sheep under varied conditions is necessary to develop the dog's skills to the highest degree. Audible signals are gradually replaced by arm semaphore to prevent flock disturbance and to allow the shepherd to direct the dog from greater distances. Sheep dogs are trained to check regularly with the herder by looking back, and if necessary will jump on a boulder to extend their vision.

Because sheep are easily frightened, a sheep dog must approach them slowly and work carefully. Large spreads of range-bred sheep require more aggressive handling than small, barnyard flocks; nevertheless, stampeding must be prevented in all sheep handling, and control must be executed largely by showing-eye. Although sheep resist being separated and become panic stricken when cornered or restrained a dog must be able to shed (separate out) individuals from the flock, hold them, or drive them into a pen.

In North American Sheepdog Trials a dog is required to demonstrate this ability while working with range-bred sheep unused to confinement and rigid control. A group of 20 sheep is run between the shepherd and the dog; the dog is directed to stop and turn back 5 marked sheep within a shedding distance of 50 yards. When all 5 have been shed off, the dog is directed to pen them or drive them up a ramp and into a truck without assistance from the shepherd. In these trials a dog is also required to go away from the shepherd about 800 yards, gather an unseen lot of 10 sheep, and drive them on a straight line as fast as possible through a gate obstacle in the centre of the field. The dog is then directed to gather another unseen lot at the opposite side of the field, drive them over the same course, and unite them with the first lot. Only one trial is permitted in negotiating the gate obstacle with either lot. In a similar test the dog must go forward 200 yards, gather 5 sheep, drive them to a pen, then turn and drive them back over a 700-yard course, through two gate obstacles at opposite sides of the field, return to the shepherd, and pen the sheep. A course of this sort can be completed by a good dog in about 6 minutes.

The value of the dog as a domestic animal depends primarily upon its adaptability for training. In addition to tracking and herding, dogs find use in such diverse functions as guard duty in department stores, guides for the blind, and companions. This functional versatility is based upon selective reinforcement of desired behaviour patterns supplemented by genetic selection to produce physique and temperament adapted to specialized purposes. Principles of learning, of social behaviour, and of heredity have been simultaneously applied to convert the wanderer and scavenger of the wilds to its present status as "man's best friend".

REFERENCES

ADOLPH, E. F. (1943). *Physiological Regulation*. Lancaster: Cattell.

ADOLPH, E. F. (1947). Urges to eat and drink in rats. *Am. J. Physiol.*, **151**, 110–125.

AGRAWAL, H. C. & FOX, M. W. (1967). Neurochemical and behavioral effects of isolation rearing in the dog. *Life Sci.*, **6**, 71–78.

ANDERSEN, A. C. & WOOTEN, E. (1959). Estrus cycle of the dog. In: *Reproduction in Domestic Animals*. H. H. Cole & P. T. Cupps (eds.). New York, N.Y.: Academic Press.

ANDREEV, A. A. (1925). The resonance theory of Helmholtz in the light of new observations upon the function of the peripheral and of the acoustic analyzer in the dog. Pavlov Jubilee Volume, Leningrad, pp. 339–363. Cited from Warden *et al.*, 1936. Comparative Psychology, Vol. III. *Vertebrates*. New York, N.Y.: Ronald Press.

BACON, W. E. & STANLEY, W. C. (1963). Effect of deprivation level in puppies on performance maintained by a passive person reinforcer. *J. comp. physiol. Psychol.*, **56**, 783–785.

BEACH, F. A. (1950). Sexual Behavior in Animals and Man. *The Harvey Lectures*, Series 43. Springfield, Ill.: C. C. Thomas, pp. 254–280.

BEACH, F. A. & GILMORE, R. W. (1949). Response of male dogs to urine from females in heat. *J. Mammal.*, **30**, 391–392.

BEACH, F. A. & LEBOEUF, B. (1967). Coital behaviour in dogs I. Preferential mating in the bitch. *Anim. Behav.*, **15**, 546–558.

BLEICHER, N. (1962). Behavior of the bitch during parturition. *J.A.V.M.A.*, **140**, 1076–1082.

BERG, I. A. (1944). Development of Behavior: The micturition pattern in the dog. *J. exp. Psychol.*, **34**, 343–368.

BRUNNER, F. (1968). Application of behavioural studies in small animal practice. In: *Abnormal Behaviour in Animals*. M. W. Fox (ed.), Philadelphia: W. B. Saunders & Co.

BURNS, M. & FRASER, M. N. (1966). *Genetics of the Dog*. London: Oliver & Boyd.

BUYTENDIJK, F. J. J. (1936). *The Mind of the Dog*. Boston: Houghton-Mifflin.

CAIRNS, R. & WERBOFF, J. (1967). Behavior development in the dog: an interspecific analysis. *Science*, **158**, 1070–1072.

CLARK, F. H. (1961). Unpublished data. Roscoe B. Jackson Mem. Lab., Bar Harbor, Maine, U.S.A.

COMFORT, A. (1960). Longevity and mortality in dogs of four breeds. *J. Gerontol.*, **15**, 126–129.

CORNWALL, A. C. & FULLER, J. L. (1961). Conditioned responses in young puppies. *J. comp. physiol. Psychol.*, **54**, 13–15.

DAHR, E. (1936). Studien über Hunde aus Primitiven Steinzeitkulturen in Nordeuropa. *Acta Univ. Lund. N.S.*, **32**, 1–63.

DAVIS, H. P. (ed.) (1949). *Modern Dog Encyclopedia.* New York, N.Y.: Stackpole & Heck.

DAWSON, W. M. (1937). Heredity in the dog. In: *Yearbook of Agriculture*, U.S. Dept. of Agriculture. Washington: Government Printing Office.

DRAVID, A. R., HIMWICH, W. A. & DAVIS, J. M. (1965). Some free amino acids in dog brain development. *J. Neurochem.*, **12**, 901–906.

DYKMAN, R. A., MURPHREE, O. D. & ACKERMAN, P. T. (1966). Litter patterns in the offspring of nervous and stable dogs. II. Autonomic & motor conditioning. *J. nerv. ment. Dis.*, **141**, 419–431.

EHRENSVÄRD, G. (1942). Über die Primärvorgänge bei Chemoreptorenbeeinflussung. *Acta physiol. scand.*, **3**, Suppl. 9, 1–151. Cited by Neuhaus, 1953.

ENGLE, E. T. (1946). No seasonal breeding cycle in dogs. *J. Mammal.*, **27**, 79–81.

FOX, M. W. (1964a). The ontogeny of behaviour and neurologic responses in the dog. *Anim. Behav.*, **12**, 301–310.

FOX, M. W. (1964b). The postnatal growth of the canine brain and correlated anatomical and behavioral changes during neuro-ontogenesis. *Growth*, **28**, 135–141.

FOX, M. W. (1965). *Canine Behavior.* Springfield, Ill.: C. C. Thomas.

FOX, M. W. (1966a). A syndrome in the dog resembling human infantile autism. *J.A.V.M.A.*, **148**, 1387–1390.

FOX, M. W. (1966b). Brain to body relationships in ontogeny of canine brain. *Experientia*, **22**, 111–112.

FOX, M. W. (1966c). *Canine Pediatrics.* Springfield, Ill.: C. C. Thomas.

FOX, M. W. (1966d). Neuro-behavioral ontogeny. A synthesis of ethological and neuro-physiological concepts. *Brain Research*, **2**, 3–20.

FOX, M. W. (1967a). Postnatal neuro-ontogeny and behaviour of the dog. Unpublished doctoral dissertation, University of London.

FOX, M. W. (1967b). Postnatal development of the EEG of the dog. *J. small Anim. Pract.*, Parts I, II and III, **8**, 71–111.

FOX, M. W. (1967c). The effects of short-term social and sensory isolation upon behavior, EEG and averaged evoked potentials in puppies. *Physiol. Behav.*, **2**, 145–151.

FOX, M. W. (1968a). Neuronal development and ontogeny of evoked potentials in auditory and visual cortex of the dog. *EEG & Clin. Neurophysiol*, **24**, 213–226.

FOX, M. W. (1968b). Socialization, handling and isolation. In: *Abnormal Behaviour in Animals.* M. W. Fox (ed.). Philadelphia: W. B. Saunders & Co.

FOX, M. W., INMAN, O. & HIMWICH, W. A. (1966). The postnatal development of neocortical neurons in the dog. *J. comp. Neur.*, **127**, 199–206.

FOX, M. W., INMAN, O. R. & HIMWICH, W. A. (1967). The postnatal development of the spinal cord of the dog. *J. comp. Neur.*, **130**, 233–240.

FOX, M. W. & SPENCER, J. (1967). Development of the delayed response in the dog. *Anim. Behav.*, **15**, 162–168.

FOX, M. W. & SPENCER, J. (1969). Development of exploratory behaviour in the dog. *Develop. Psychobiol.* (in press).

FOX, M. W., SPENCER J. & BRAIN, C. (1967). Development of black-white discrimination and reversal learning in the dog. Unpublished data, Galesburg State, Research Hospital, Galesburg, Ill.

FOX, M. W. & STANTON, G. (1967). The maintenance of sleep and wakefulness: an ontogenetic study in the dog. *J. small Anim. Pract.*, **8**, 605–611.

FOX, & M. W. STELZNER, D. (1966a). Behavioural effects of differential early experience in the dog. *Anim. Behav.*, **14**, 273–281.

FOX, M. W. & STELZNER, D. (1966b). Approach/withdrawal variables in the development of social behaviour in the dog. *Anim. Behav.*, **14**, 362–366.

FOX, M. W. & STELZNER, D. (1967). The effects of early experience on the development of inter and intraspecies social relationships in the dog. *Anim. Behav.*, **15**, 377–386.

FREAK, M. J. (1968). Abnormal behaviour in the bitch during parturition. In: *Abnormal Behaviour in Animals.* M. W. Fox (ed.). Philadelphia: W. B. Saunders & Co.

FREDERICSON, E. (1952). Perceptual homeostasis and distress in vocalization in puppies. *J. Personality*, **20**, 472–477.

FREDERICSON, E., GURNEY, N. & DuBUIS, E. M. (1956). The relationship between environmental temperature and behavior in neonatal puppies. *J. comp. physiol. Psychol.*, **49**, 278–280.

FREEDMAN, D. G., KING, J. A. & ELLIOT, O. (1961). Critical period in the social development of dogs. *Science*, **133**, 1016–1017.

FULLER, J. L. (1953). Cross-sectional and longitudinal studies of adjustive behavior in the dog. *Ann. N.Y. Acad. Sci.*, **56**, 214–224.

FULLER, J. L. (1956). Photoperiodic control of estrus in the basenji. *J. Hered.*, **47**, 179–180.

FULLER, J. L. (1963). Effects of experiential deprivation upon behavior in animals. *Proc. III Int. Psychiatric Congress*, Montreal, **3**, 223–227, Univ. Toronto Press.

FULLER, J. L. (1967). Experimental deprivation and later behavior. *Science*, **158**, 1645–1652.

FULLER, J. L. & DuBUIS, E. M. (1961). Unpublished data. Roscoe B. Jackson Mem. Lab., Bar Harbor, Maine, U.S.A.

FULLER, J. L., EASLER, C. & BANKS, E. (1950). Formation of conditioned avoidance responses in young puppies. *Am. J. Physiol.*, **160**, 462–466.

FULLER, J. L. & CLARK, L. D. (1966). Effects of rearing with specific stimuli upon postisolation behavior in dogs. *J. comp. physiol. Psychol.*, **61** (2), 258–263.

FULLER, J. L. & CLARK, L. D. (1966). Genetic and treatment factors modifying the postisolation syndrome in dogs. *J. comp. physiol. Psychol.*, **61** (2), 251–257.

GIBSON, E. J. & WALK, R. D. (1960). The "Visual Cliff." *Scient. Am.*, **202** (4), 64–71.

GRAY, A. P. (1954). *Mammalian Hybrids*. Commonwealth Agric. Bureaux. (Cited in Worden, 1959.)

HEBB, D. O. (1949). *Organization of Behavior*. New York: John Wiley & Sons, Inc.

HEBB, D. O. (1955). Drives and the C.N.S. (conceptual nervous system) *Psychol. Rev.*, **62**, 243–254.

HEIMBURGER, VON N. (1959). Das Mar kierungsverhalten einiger Caniden. *Z. Tierpsychol.*, **16**, 104–113.

HIMWICH, H. E. & FAZEKAS, J. F. (1941). Comparative studies of the metabolism of the brain of infant and adult dogs, *Am. J. Physiol.*, **132**, 454.

HUTT, C. & HUTT, S. J. (1965). Effects of environmental complexity on stereotyped behaviors of children. *Anim. Behav.*, **12**, 1–4.

JAMES, W. T. (1936). The effect of the presence of a second individual on the conditioned salivary response in dogs of different constitutional types. *J. genet. Psychol.*, **49**, 437–449.

JAMES, W. T. (1939). Further experiments in social behavior among dogs. *J. genet. Psychol.*, **54**, 151–164.

JAMES, W. T. (1949). Dominant and submissive behavior in puppies as indicated by food intake. *J. genet. Psychol.*, **75**, 33–43.

JAMES, W. T. (1951). Social organization among dogs of different temperaments, terriers and beagles, reared together. *J. comp. physiol. Psychol.*, **44**, 71–77.

JAMES, W. T. (1952). Observations of behavior of newborn puppies: method of measurement and types of behavior involved. *J. genet. Psychol.*, **80**, 65–73.

JAMES, W. T. (1957). The effect of satiation on the sucking response in puppies. *J. comp. physiol. Psychol.*, **50**, 375–378.

JAMES, W. T. & CANNON, D. (1952). Conditioned avoiding responses in puppies. *Am. J. Physiol.*, **168**, 251–253.

JAMES, W. T. & McCAY, C. M. (1950). A study of food intake, activity and digestive efficiency in different type dogs. *Am. J. vet. Res.*, **11**, 412–413.

JANOWITZ, H. D. & GROSSMAN, M. I. (1949). Effect of variations in nutritive density on intake of food of dogs and rats. *Am. J. Physiol.*, **158**, 184–193.

JETTER, W. W., LINDSLEY, O. R. & WOHLWILL, F. J. (Undated). The effects of X-irradiation on physical exercise and behavior in the dog. Report on Contract AT(30-1) 1201, U.S. Atomic Energy Commission and Boston University.

KALMUS, H. (1955). The discrimination by the nose of the dog of individual human odours and in particular of the odours of twins. *Br. J. Anim. Behav.*, **3**, 25–31.

KELLY, R. B. (1947). *Sheepdogs. Their breeding, maintenance and training.* Sydney & London: Angus & Robertson. Cited from Burns (1952).

KING, J. A. (1954). Closed social groups among domestic dogs. *Proc. Am. philos. Soc.,* **98,** 327–336.

KLEIMAN, D. (1966). Scent marking in the canidae. *Symp. zool. Soc. Lond.,* **18,** 167–177.

KLEIMAN, D. (1967). Some aspects of Social Behavior in the Canidae. *Am. Zool.,* **7,** 365–372.

KRUSHINSKI, I. V. (1962). *Animal Behavior.* New York: Consultant Bureau.

KRUSHINSKII, L. V. & FLESS, D. A. (1959). Strengthening of olfaction in police dogs. *Pavlov J. High. Nerv. Act.,* **9,** 266–272.

KURTSIN, I. T. (1968). Physiological mechanisms of behaviour disturbances and cortico-visceral interrelations in animals. In: *Abnormal Behaviour in Animals.* M. W. Fox (ed.). Philadelphia: W. B. Saunders & Co.

LESSAC, M. S. (1966). The effects of early isolation and restriction on the later behavior of beagle puppies. Thesis, University of Pennsylvania, 1965. *Diss. Abstr.,* **26** (a), 556a.

LIPMAN, E. A. & GRASSI, J. R. (1942). Comparative auditory sensitivity of man and dog. *Am. J. Psychol.,* **55,** 84–89.

LÖHNER, L. (1926). Untersuchungen über die Geruchsphysiologische Leistungsfähigkeit von Polizeihund. *Pflügers Arch.,* **212,** 84–94.

LORENZ, K. (1952). *King Solomon's Ring.* New York, N.Y.: Crowell.

MAHUT, H. (1958). Breed differences in the dog's emotional behaviour. *Canad. J. Psychol.,* **12,** 35–44.

MANDEL, P., REIN, H., HARTH-EDEL S. & MANDEL, R. (1962). Distribution and metabolism of ribonucleic acid in vertebrate CNS, D. Richter (ed.). *Comparative Neurochemistry,* Pergamon Press, New York, 149–163.

MARTINS, T. (1949). Disgorging of food to the puppies by the lactating dog. *Physiol. Zool.,* **22,** 169–172.

MARTINS, T. & VALLE, J. R. (1946). A atitude do cão na micção e os hormônios sexuais. *Mem. Instit. Oswaldo Cruz.,* Brazil, **44,** 343–360.

MARTINS, T. & VALLE, J. R. (1948). Hormonal regulation of the micturition behavior of the dog. *J. comp. physiol. Psychol.,* **41,** 301–311.

MAYER, J. (1953). *Caloric Requirements and Obesity in Dogs.* New York, N.Y.: Gaines Veterinary Symposium.

MELZACK, R. & BURNS, S. K. (1965). Neurophysiological effects of early sensory restriction. *Exper. Neurol.,* **13,** 163–175.

MELZACK, R. & SCOTT, T. H. (1957). The effects of early experience on the response to pain. *J. comp. physiol. Psychol.,* **50,** 155–161.

MOST, K. (1928). Neue Versuche über Spurfähigkeit. Das Problem der Spurenreinheit auf der menschlichen Spur in Lichte der zumal mit der Farhtenbahn erzielten Versuchsergebnisse. *Hund,* 1928, 31–35.

MOULTON, D. G., ASHTON, E. H. & EAYRS, J. T. (1960). Studies in olfactory acuity. 4. Relative detectability of n-aliphatic acids by the dog. *Anim. Behav.,* **8,** 117–128.

NEUHAUS, W. (1953). Über die Riechschärfe des Hundes für Fettsäuren. *Z. vergl. Physiol.,* **35,** 527–552.

PARRY, H. B. (1953a). Degenerations of the dog retina. I. Structure and development of the retina of the normal dog. *Br. J. Ophthal.,* **37,** 385–404.

PARRY, H. B. (1953b). Degenerations of the dog retina. II. Generalized progressive atrophy of hereditary origin. *Br. J. Ophthal.,* **37,** 487–502.

PAWLOWSKI, A. A. & SCOTT, J. P. (1956). Hereditary differences in the development of dominance in litters of puppies. *J. comp. physiol. Psychol.,* **49,** 353–358.

PETERS, J. E. (1966). Typology of dogs by the conditional reflex method. *Conditional Reflex,* **1,** 235–250.

PFAFFENBERGER, C. J. (1947). *Training your Spaniel.* New York, N.Y.: Putnam.

PFAFFENBERGER, C. J. & SCOTT, J. P. (1959). The relationship between delayed socialization and trainability in guide dogs. *J. genet. Psychol.,* **95,** 145–155.

REED, C. A. (1959). Animal domestication in the prehistoric Near East. *Science,* **130,** 1629–1639.

RHEINGOLD, H. L. (1963). (ed.). *Maternal Behavior in Mammals.* New York: John Wiley & Sons, pp. 196–202.

16+

ROBERTSON, R. B. (1957). *Of Sheep and Men.* New York, N.Y.: Knopf.

ROSS, S. (1950). Some observations on the lair-dwelling behaviour of dogs. *Behaviour,* **2,** 144–162.

ROSS, S., FISHER, A. E. & KING, D. (1957). Sucking behavior: A review of the literature. *J. genet. Psychol.,* **91,** 63–81.

ROSS, S. & ROSS, J. (1949a). Social facilitation of feeding behavior in dogs. I. Group and solitary feeding. *J. genet. Psychol.,* **74,** 97–108.

ROSS, S. & ROSS, J. (1949b). II. Feeding after satiation. *J. genet. Psychol.,* **74,** 293–304.

SACKETT, G., GRIFFIN, G. A., JOSLYN, C., DANFORTH, W. & RUPPENTHAL, G. (1967). Mother-infant and adult female choice behaviour in Rhesus monkeys after various rearing experiences. *J. comp. physiol. Psychol.,* **63,** 376.

SALZINGER, K. & WALLER, M. B. (1962). The operant control of vocalization in the dog. *J. exp. Analysis Behav.,* **5,** 383–389.

SAUNDERS, B. (1952). *Training You to Train Your Dog.* Rev. edn, Garden City, N.Y.: Doubleday.

SCHNEIRLA, T. C. (1965). Aspects of stimulation and organization in approach/withdrawal processes underlying vertebrate behavioral development. Chapter 1 in *Advances in the Study of Animal Behavior.* D. S. Lehrman, R. A. Hinde and E. Shaw (eds.). Vol. 1. New York: Academic Press.

SCOTT, J. P. (1954). The effects of selection and domestication upon the behavior of the dog. *J. Nat. Cancer Inst.,* **15,** 739–758.

SCOTT, J. P. (1958a). *Animal Behavior.* Chicago: Univ. Chicago Press.

SCOTT, J. P. (1958b). *Aggression.* Chicago: Univ. Chicago Press.

SCOTT, J. P. (1958c). Critical periods in the development of social behavior in puppies. *Psychosom. Med.,* **20,** 42–54.

SCOTT, J. P. (1961). Animal sexuality in *Encyclopedia of Sexual Behavior.* A. E. Ellis & A. Abarbenel (eds.). New York: Hawthorn.

SCOTT, J. P. (1962). Critical periods in behavioral development. *Science,* **138,** 949–958.

SCOTT, J. P. & FULLER, J. L. (1965). *Genetics and Social Behavior of the Dog.* Chicago: Univ. Chicago Press.

SCOTT, J. P., FULLER, J. L. & KING, J. A. (1960). The inheritance of annual breeding cycles in hybrid Basenji-Cocker Spaniel dogs. *J. Hered.,* **50,** 255–261.

SCOTT, J. P. & MARSTON, M. V. (1950). Social facilitation and allelomimetic behaviour in dogs. II. The effects of unfamiliarity. *Behaviour,* **2,** 135–143.

SHEFFY, B. D. (1957). Hand rearing of puppies. New York, N.Y.: *Gaines Vet. Symposium.*

SOLOMON, R. L. & WYNNE, L. C. (1953). Traumatic avoidance learning: acquisition in normal dogs. *Psychol. Monogr.,* **67** (4), 1–19.

SPORTS ILLUSTRATED (Eds.) (1960). *Sports Illustrated Book of Dog Training.* Philadelphia: Lippincott.

STANLEY, W. C. (1965). The passive person as a reinforcer in isolated beagle puppies. *Psychonomic Science,* **2,** 21–22.

STANLEY, W. C. & BACON, W. E. (1963). Suppression of sucking behavior in non-deprived puppies. *Psychol. Rep.,* **13,** 175–178.

STANLEY, W. C., CORNWALL, A. C., POGGIANI, C. & TRATTNER, A. (1963). Conditioning in the neonate puppy. *J. comp. physiol. Psychol.,* **56,** 211–214.

STANLEY, W. C. & ELLIOT, O. (1962). Differential human handling as reinforcing events and as treatments influencing later social behavior in Basenji puppies. *Psychol. Rep.,* **10,** 775–788.

STANLEY, W. C., MORRIS, D. D. & TRATTNER, A. (1965). Conditioning with a passive person reinforcer and extinction in shetland sheep dog puppies. *Psychonomic Science,* **2** (1), 19–20.

STANWELL-FLETCHER, T. C. (1946). *Driftwood Valley.* Boston: Little Brown.

STOCKARD, C. R., ANDERSON, O. D. & JAMES, W. T. (1941). *Genetic and Endocrinic Basis for Differences in Form and Behavior.* Philadelphia: Wistar Inst. Press.

STONE, C. P. (1921). Notes on light discrimination in the dog. *J. comp. Psychol.,* **1,** 413–431.

THOMPSON, W. R. & HERON, W. (1954). Exploratory behavior in normal and restricted dogs. *J. comp. physiol. Psychol.,* **47,** 77–82.

THOMPSON, R., MELZACK, R. & SCOTT, T. H. (1956). "Whirling Behavior" in dogs as related to early experience. *Science*, **123**, 939.

THORNE, F. C. (1944). The inheritance of shyness in dogs. *J. genet. Psychol.*, **65**, 275–279.

VOLOKHOV, A. A. (1959). Comparative-physiological investigation of conditioned and unconditional reflexes during ontogeny. *J. Higher Nerv. Activity*, **9**, 49–60.

WARDEN, C. J., JENKINS, T. N. & WARNER, L. H. (1936). Comparative Psychology. Vol. III. *Vertebrates*. New York, N.Y.: Ronald Press.

WELKER, W. I. (1959). Factors influencing aggregation of neonatal puppies. *J. comp. physiol. Psychol.*, **52**, 376–380.

WHITNEY, L. F. (1947a). *How to Breed Dogs*. Rev. edn, New York, N.Y.: Orange Judd.

WHITNEY, L. F. (1947b). *Bloodhounds and how to Train Them*. New York, N.Y.: Orange Judd.

WIELAND, G. (1938). Über die Grösse des Riechfeldes beim Hunde. *Vet.-med. Diss.* Berlin. Cited by Neuhaus (1953).

WOOLPY, J. H. & GINSBURG, B. E. (1967). Wolf socialization: a study of temperament in a wild social species. *Am. Zool.*, **7**, 357–364.

WORDEN, A. N. (1959). Abnormal behaviour in the dog and cat. *Vet. Rec.*, **71**, 966–981.

ZEUNER, F. E. (1963). *The History of Domesticated Animals*. London: Hutchins.

Chapter 15

The Behaviour of Cats

A. Kling, J. K. Kovach and T. J. Tucker

The domestic cat (*Felis catus*) is one of the most frequently used subjects for studies of animal behaviour. Our aim is to present a brief, coherent picture of the behaviour of the cat with an emphasis on neural and hormonal mechanisms. Our principal guide in deciding what to include and what to omit of the material available prior to 1962 was a former review of the behaviour of the domestic cat by Rosenblatt & Schneirla (1962), and we wish to acknowledge our indebtedness to them.

In order to appraise those behaviour characteristics that set the cat apart from other domesticated animals, it should be useful to take a brief look at the evolutionary history of the cat. The *Felidae*, or felids, include all cats, large and small, such as the lion, leopards, cougars, lynx, and various small wild cats. They, together with all modern carnivores, originated some 50,000,000 years ago with the carnivore family known as *Miacidae*. The latter were small animals about the size of an ordinary house cat and they compare well with the modern Old World civets. The felids, by comparison, are characterized by a unique evolutionary history. Already at the beginning of Oligocene they were well-established predatory animals. In contrast to the generalized carnivore characteristics of other predatory animals, the felids were highly specialized at the earliest times in their evolutionary history. The skeletal structure, skull, and dentition of the felids are very narrowly specialized for silent ambush, sudden dash, grabbing, stabbing and shearing. These specialized characteristics were acquired by a sudden leap at the very beginning of the cats' evolutionary history and have been retained ever since (Colbert, 1958).

The primary selective factor in the domestication of the cat appears to be the highly specialized capacity for lone hunt. Cats were used in the times of the earliest recorded history for keeping farms and granaries free from rodents, and the cat continues to serve man as an important agent in the control of rodent populations. However, the domestication of the cat was never carried to the extent where selective breeding for specific morphological and behavioural capacities would result in an animal completely dependent on man and unable to survive in its original wild habitat. Thus, the student of feline behaviour and neurophysiology deals with a highly specialized, phylogenetically old, and relatively wild organism having a limited capacity for social organization.

I. INGESTIVE BEHAVIOUR

The ingestive behaviour of the cat can be divided into three developmental stages: (a) neonatal stage lasting approximately 5 weeks during which the ingestive behaviour is suckling; (b) a transitional stage lasting from the 5th week until the end of the 2nd month between which the initial response to solid food appears and weaning is completed; (c) predatory feeding which begins to develop in the 3rd month of life, and only gradually reaches a high level of effectiveness (Wilson & Weston, 1947; Leyhausen, 1956a, 1956b).

A. Suckling

Immediately after birth the mother cleans the newborn kitten of birth membranes. The kitten rights itself, begins to crawl forward with paddle-like movements of the forelegs and simultaneous pushing movements with both hind legs. A characteristic side to side movement of the head brings the mouth and nose of the kitten into contact with objects in the environment. Gentle tactile stimulation of the facial areas and body surface results in a slow turning of the head and total body with alternate movements of the front legs towards the source of stimulation. This capacity to orient toward the source of stimulation enables the kitten to find the abdomen of the mother.

The initial contact of the crawling kitten with the mother is usually at her legs while she is lying in an outstretched nursing position. The kitten follows the legs and climbs upwards onto the mother's body and promptly nuzzles in her fur. Nuzzling continues until one of many contacts with a protruding nipple elicits first the head withdrawal and then nipple grasping and suckling (Tilney & Casamajor, 1924; Langworthy, 1929; Windle, 1930; Stavsky, 1932; Prechtl, 1952; Rosenblatt et al., 1959). Thus the feeding behaviour of kittens includes a searching component and a specific response to the nipple.

While the searching and suckling pattern occurs immediately after birth, the kitten usually does not feed until 1 or 2 hours later. During the first few days the suckling response becomes restricted to a specific single nipple, but this declines as the kitten matures.

The development of suckling and the mother-kitten interaction during development has been traced by Rosenblatt et al. (1959). They found three major stages with respect to the role of the female and the young in the initiation of feeding. In the 1st stage the female initiates feeding by approaching the kittens in the home area, but in the 2nd stage the feeding is initiated by a reciprocal interaction of the female and young. As the 2nd stage proceeds, the female may remain at rest at a distance from the home locality, and the kittens walk up to her, nuzzle, and attach to a nipple. In the 3rd stage suckling is initiated mainly by the kittens. They approach the female, which wanders around and increasingly avoids the kittens. Through persistent following, the kittens force her to lie down or to remain at rest while they nuzzle and attach to her nipples.

An analysis of the contribution of each developmental period to the development of suckling behaviour was done utilizing isolation experiments in which the normal mother-kitten interaction was interrupted for periods of different duration (Rosenblatt et al., 1961). The isolated kittens were placed in specially constructed chambers fitted at one end with an artificial feeder

from which a special milk formula could be withdrawn through a small
rubber nipple. Within 2 or 3 days, the kittens learned to locate the nipple
independently on these feeders. After a set period in the brooder without
contact with the mother, each kitten was returned to its mother and its
suckling and related social behaviour was observed. Results showed that
continued contact with the female was essential for the development of
suckling pattern characteristic of the kittens at each age period. Isolation from

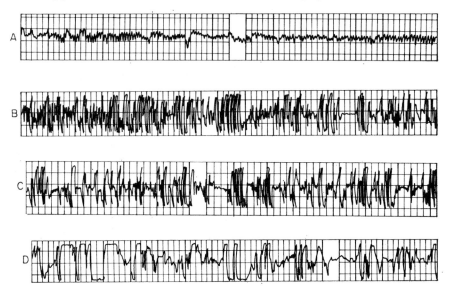

Fig. 81. Recordings of the initial sucking response at birth (line A), and at the 4th (line
B), 10th (line C), and 20th (line D) days of age. (From Kovach & Kling, 1967.)

birth to 7 days of age resulted in inadequate orientation to the home area of
the cage. Once contact was established, they also had difficulty in locating
the nipples. Suckling in these kittens was delayed an average of 3 hours. A
more extended period of isolation (from the 6th to the 23rd day) resulted in
delays of nearly 20 hours after introduction. Kittens isolated from birth
to the 44th day and from the 23rd to the 44th day in all but one instance
failed to suckle from the female. Kittens isolated from the 34th to the 49th
day were delayed in suckling by an average of nearly 2 days. Once contact
was established, however, nipple grasping was rapid and suckling was
efficiently performed (Rosenblatt *et al.*, 1959).

In our studies on the mechanisms of suckling behaviour in the kitten
(Kovach & Kling, 1967), suckling movements and oral manipulation of an
artificial nipple were recorded on a dynograph. The insertion of the nipple
into the mouth shortly after birth elicited an immediate, regular, low-
amplitude suckling pattern (Fig. 81, line A). At the 4th day of age (Fig. 81,
line B) there was a considerable increase in amplitude and decrease in the
regularity of initially elicited movements. The pattern, however, was pre-

dominantly suckling. At 10 days of age (Fig. 81, line C) the pattern included some irregular biting movements and attempts to expel the nipple from the mouth. The flow of milk into the mouth did not elicit the immediate suckling seen at 10 days of age, and it took hours of forced practice with the nipple before the appearance of the first regular suckling pattern. Kittens beyond 40 days of age learned to obtain milk from the feeding bottle by chewing on the nipple and did not learn to suckle. These data, together with our direct observations, indicate that the establishment of a regular

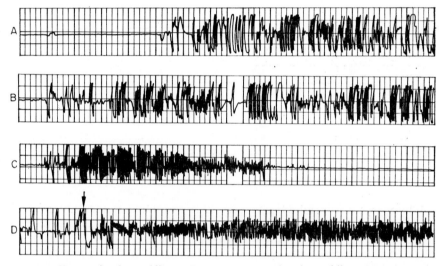

Fig. 82. Changes in the suckling pattern on consecutive days while training to nurse from the bottle. Line A shows the initial response to the nipple on the 1st day of training. Lines B and C represent the recordings taken at the 1st daily forced trials on the 2nd and 3rd days. The arrow on line D indicates the point of the 1st independent approach to the nipple, and the record that follows shows the 1st independent nursing response on the 4th day of training.

suckling pattern is sudden, rather than gradual, and that the regular suckling pattern appears before the kitten learns to approach the bottle independently.

The complex behaviour associated with suckling on the mother cannot be attributed to neonatal reflexes alone. The early association of learning with suckling behaviour is shown by the improvement in the suckling pattern as a result of experience with an artificial nipple (Fig. 82) and by the kittens' learned response of locating and suckling from the artificial nipple.

We also undertook to study the effects of complete suckling deprivation on the elicitability of suckling behaviour (Kovach & Kling, 1967). A special formula of milk was injected into the kittens' stomach through a polyethylene tube 4 times daily. None of the kittens was given opportunity for nutritive suckling until a pre-set time limit for their return to the mother.

The length of isolation for complete suckling deprivation, and the number of kittens at each deprivation period are indicated in Table 45.

Table 45. The Effects of Complete Suckling Deprivation and Social Isolation on the Kittens' Ability to Initiate Suckling on the Mother.[1] (Adapted from Kovach & Kling, 1967.)

No. of kittens tested	Age of return to mother (days)	Time for initiation of suckling after return to mother (hr)
I	6	$\frac{1}{2}$
2	9	$\frac{1}{2}$
I	16	5
2	19	3
I	23	No suckling on the mother
2	24	,,
I	35	,,

[1] All kittens were separated from the mother within 4 hours after birth.

This procedure resulted in a very high mortality of the subjects and 69 per cent of the individually isolated and force-fed kittens died before reaching the set age for return to the mother. Table 45 shows that the complete suckling deprivation and social isolation did not interfere with the kittens' ability to initiate suckling if the deprivation lasted from birth to 19 days of age. Beyond 22 days of deprivation, however, the kittens were not able to find the nipple or initiate suckling during a 96-hour experience with the mother.

The kittens raised by force feeding in groups from birth to the 30th day of age thrived well during the deprivation period, none died, and they all initiated suckling on the mother upon return to her. The suckling deprivation of these kittens, however, was not complete. Shortly after the beginning of the deprivation period they started to suckle on each other's anal and genital regions. Since the stimulation of the anal and genital areas in the young kittens produces elimination, it is possible that this mutual suckling was directed toward these regions because of the reinforcing value of the liquid obtained. It appears that the continuous mutual suckling practice seen in these kittens was sufficient to initiate the development of suckling on the mother.

These data seem to imply that the initiation and establishment of persistent suckling behaviour depends upon the kittens' reflex response to the mother's nipple. The disappearance of this reflex response at 23 days of age renders the kittens, which did not have previous suckling experience, unable to initiate suckling on the mother. While the initiation of suckling behaviour of the kitten seems to depend on the presence of innately given reflexes, the maintenance of this behaviour depends on practice and learning.

B. Weaning

The initial response to solid food appears at about the end of the 5th week in the kitten. Farm-raised kittens begin to eat food provided by the mother who has hunted, killed prey and brought it back to the nest (Wilson & Weston, 1947). Weaning is a gradual process and it is usually completed by the end of the 2nd month. Suckling, however, may continue for many months—even for as long as a year in kittens which remain with the mother (Leyhausen, 1956a). Normally, after the 5th week of age, a new tendency of following the mother on her hunting trips appears, which ultimately leads to the development of predatory habits. This is also the time when intensive play behaviour can be seen. All the characteristic stereotyped movements associated with stalking, attacking, jumping, pawing, and biting the prey in adult animals can be seen at this time. These movements, however, appear in irregular and incomplete order and out of the normal context of predatory behaviour. The kittens obtain experience catching and eating prey through the tendency to accompany the female on hunting trips (Wilson & Weston, 1947). Thus, the full development of predatory activities is dependent upon social experience obtained in the context of play with the littermates, and more significantly while observing the predatory behaviour of the mother.

There are several studies showing that kittens raised with rats and mice develop a playful reaction towards these rodents rather than showing the predatory activity seen in normally raised cats (Kuo, 1930, 1938; Tsai, 1963). Those kittens which were raised from birth in a situation where they could not have seen the mother killing a prey were much less predatory than the kittens raised with females that killed a rat or a mouse within sight of her kittens. The relationship between killing, feeding and hunger was also studied by Kuo (1930). He showed that kittens raised on a vegetarian diet killed as many animals as those raised on a non-vegetarian diet, but tended to eat the dead animals less frequently. Hunger also did not seem to exert special effects on killing.

C. Adult Predatory Feeding

The predatory activity of the cat must be tempered by the specific factors of domestication. Domestic cats are not allowed to control their own population density though they may be able to establish a limited ecological balance between themselves and available rodents upon which they prey. Consequently, neither the territorial patterns of hunting activities nor the specific behaviour patterns associated with hunting can be considered as representing the characteristic behaviour of a lone hunter in the wild.

Predatory behaviour is initiated by the visual and auditory stimuli of the moving prey. If it is not too large, the cat approaches it with a characteristic "stalking run" (Leyhausen, 1956a), which is a smooth quiet run with the body close to the ground. When close to the prey, the cat pauses, hides under

16*

cover, moves slowly, and, when it is within an appropriate distance, pounces quickly forward and grabs the prey with outstretched claws and teeth. The bite which wounds or kills the prey is almost invariably aimed at the neck. The prey is carried to a secluded place, and the cat consumes it. In laboratory and house cats, often only a partial execution of this sequence is seen. The stalking and grabbing of the prey in the upper neck area is very much the same as described in feral cats (Leyhausen, 1956a), but the killing is often postponed for long periods. We have seen cats catching and releasing a prey repeatedly, going through the movements and behaviour patterns of stalking, pouncing, grabbing, and shaking the prey for long periods without actually killing it. In such instances the prey may collapse from exhaustion and the ingestive behaviour may be initiated by the lack of motion in the unwounded but unconscious rodent. The question as to what extent such behaviour can be related to motivational needs for hunting, and the possible reinforcing value of such prolongation of hunting activities deserves serious investigation.

D. Neural Mechanisms of Ingestive Behaviour

1. Suckling

Since olfaction is likely to be the sense modality involved in the mediation of the suckling reflex, we studied the effects of destruction of the olfactory bulbs on suckling from the mother and on the learned habit of suckling from the bottle (Kovach & Kling, 1967).

None of the kittens with bilateral olfactory lesions was able to initiate suckling on the mother during the post-operative test. As can be seen in Table 46, the length of pre-operative suckling experience with the mother

Table 46. The Effects of the Removal of Olfactory Bulbs on the Kittens' Ability to Suck. (Adapted from Kovach & Kling, 1967.)

No. of kittens tested	Age at operation (days)	Extirpated structure	Time required to initiate sucking from mother in post-operative test	Average length of time for learning to suck from bottle after post-operative test with mother
2	2	Olf. bulbs biltaterally	No sucking from mother in 96 hours	Not tested
2	4	,,	,,	145 hr (1 only)
2	5	,,	,,	Not tested
2	16	,,	,,	,,
3	18	,,	,,	104 hr (2 only)
1	20	,,	,,	Not tested
1	26	,,	,,	,,
4	7	Frontal areas and caudate nucleus	20 minutes	Not tested
4	8	Sham	80 minutes	,,

did not alter this phenomenon. The subjects which were left with the mother and whose force feeding was discontinued died of starvation.

However, we were successful in training operated subjects to suckle independently from a bottle. Thus, it appears that the removal of the olfactory bulbs renders the kittens unable to locate and respond to the mother's nipple, while sham operates and subjects with lesions of other areas (see Table 46) were able to initiate suckling within a short period after the beginning of the post-operative test. Kittens trained to suckle independently from the artificial nipple on a brooder and operated on after the acquisition of the behaviour retained the learned habit of suckling from a bottle (Table 47).

Table 47. Effects of Removal of the Olfactory Bulbs at 2 to 4 Weeks of Age on the Retention of the Learned Habit of Suckling from the Bottle. (Adapted from Kovach & Kling, 1967.)

Age of kitten at beginning of training on bottle (days)	Age at first independent response to bottle (days)	Extirpated structures	Length of time before suckling on bottle in post-operative test in hours	Response to mother after post-operative test on bottle
7	11	Olf. bulbs bilaterally	13	No sucking
11	14	,,	8	,,
22	24	,,	22	,,
7	11	Olf. bulbs unilaterally	26	Sucking in 3 hr.

These data indicate that while behaviour of suckling on the mother develops immediately after birth, the kittens need a long learning experience before they are able to suckle on the bottle independently. Furthermore, the destruction of the olfactory bulbs renders the kittens unable to initiate suckling on the mother, but it does not interfere with the learned response of suckling on the artificial nipple, nor does it interfere with the kittens' ability to learn to suckle from the bottle.

A variety of lesions to neocortical or paleocortical areas did not interfere with suckling from the mother (Kling, 1965). However, lesions of lateral hypothalamus in the neonatal kitten had the same anorexic effect as in adults (see below). In the kitten, then, direct olfactory, hypothalamic connections would appear to be the essential reflex pathway for the initiation of suckling behaviour.

2. EATING AND DRINKING

Much of the evidence on the neural regulation of eating and drinking points toward diencephalic mechanisms, which, modified by connections

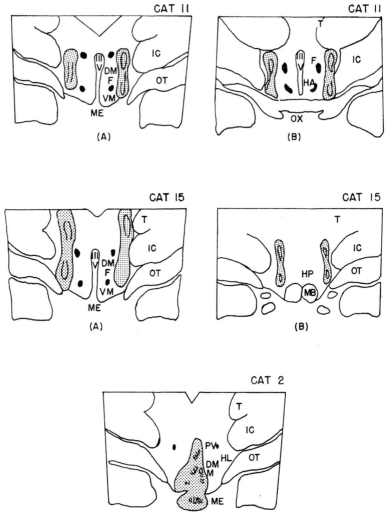

Fig. 83. Drawing of sections through diencephalon of cat 11. Lesions have destroyed the lateral parts of middle (*a*) and anterior hypothalamus going up to suprachiasmal region (*b*), producing aphagia. Lettering for this and later figures: DM, nucleus hypothalamicus dorso-medialis; F, fornix; HA, nucleus hypothalamicus posterior; IC, internal capsule; MB, mammillary body; ME, median eminence; OT, optic tract; OX, optic chiasma; PV, nucleus paraventricularis; T, Thalamis; VM, nucleus hypothalamicus ventro-medialis; III V, third ventricle. Sections through diencephalon of cat 15. Lesions have destroyed the lateral parts of middle (*a*) and posterior hypothalamus going up to mammillary region (*b*), producing aphagia. In cat 2: Lesions destroying the medial hypothalamus in the region of median eminence, producing hyperphagia. Lateral hypothalamus is spared.

with limbic and midbrain structures, control hunger and satiety as well as the motor reflexes necessary for co-ordinated feeding behaviour.

Anand et al. (1955) found that, in the cat, bilateral lesions of the lateral hypothalamus result in total aphagia. Conversely, electrical stimulation of the medial hypothalamus results in total aphagia. Conversely, electrical stimulation of the lateral hypothalamus elicits eating behaviour (Anand & Dua, 1955).

Lesions of the ventral medial nucleus of the hypothalamus (Wheatly, 1944; Anand et al., 1955) result in excessive eating and lead to obesity in the cat. After a degree of obesity occurs, food intake seems to level off.

Grossman (1960) was able to elicit eating by adrenergic stimulation of the lateral hypothalamus but not from cholenergic stimulation. The latter form of stimulation caused drinking instead.

Levels of glucose in the blood have been shown to affect the electrical activity of both medial and lateral hypothalamus. Anand (1961) found that hyperglycaemia increased the electrical activity of the ventral medial nucleus with a corresponding reduction in the activity of the lateral hypothalamus, while hypoglycaemia resulted in the reverse condition.

Other areas of the brain shown to have a profound effect on feeding behaviour in the cat include the amygdaloid nuclei. Complete bilateral ablation in the adult cat results in a profound anorexia for 1 to 2 weeks after operation. Following recovery, the cat appears to have normal food intake but will mouth and lick inedible objects and show hyperoral exploratory behaviour (Schreiner & Kling, 1953).

Lesions restricted to the lateral group of the amygdaloid nuclei have been reported to result in hyperphagia and obesity (Morgane & Kosman, 1957; Green et al., 1957; Kling et al., 1960), while damage to the anterior portion of the amygdala has been reported to produce temporary aphagia as does complete ablation.

As with direct chemical stimulation of the hypothalamus, Grossman (1964) found that cholinergic stimulation of the amygdala resulted in drinking, while adrenergic stimulation caused eating.

Another neural region which may grossly affect food intake in the cat is the posterior orbital gyrus (which lies just anterior to the anterior amygdala). Injury to this structure has been demonstrated to decrease food intake (Anand et al., 1958).

The effects of lesions of limbic and other structures on eating may be age dependent. Kling (1965) showed that neonatal kittens with amygdala lesions did not show the temporary anorexia characteristic of the adult, nor was there any adverse effect on growth for the 1st year of life. Lesions of olfactory cortex or hippocampus in the kitten tended to result in increased weight gain, while kittens sustaining large cortical lesions showed deficient growth (Fig. 84).

Thus, the integrity of the medial and lateral diencephalic nuclei is necessary from birth on for adequate regulation of feeding behaviour. With maturation, limbic and neocortical connections with the hypothalamus become increasingly important along with the behavioural requirements for independent food seeking or predation.

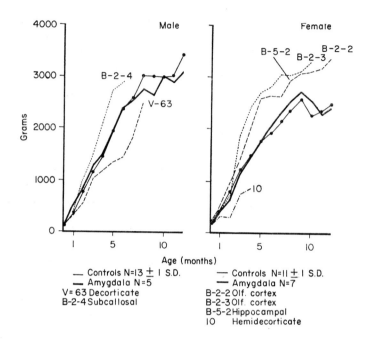

Fig. 84. Growth curves, from birth to one year, for kittens sustaining brain lesions in the neonatal period. (Kling, 1965.)

II. SEXUAL BEHAVIOUR

The cat is a seasonally polyoestrous animal with several oestrous cycles in the female, and a stable sexual responsiveness in the male throughout the year. In the northern latitudes, the peaks of oestrus in the female occur from mid-January to early March and during May and June. In more southern latitudes, the first seasonal peak occurs earlier (Eckstein & Zuckerman, 1956; Scott & Lloyd-Jacob, 1959). The frequency of mating is the highest during these periods, but it is not completely restricted to them. Sexual behaviour in experienced males can be elicited at all seasons of the year. However, desensitization of the glans penis reveals a latent sexual cycle in male cats which corresponds in time with the female cycle (Aronson & Cooper, 1966).

A. Development of Sexual Behaviour

In the laboratory, males are found to show the 1st complete copulation at about 11 months of age. Some elements of mating behaviour, such as mounting and neck grip, can be seen as early as 4 months of age, but such kittens cannot achieve intromission (Rosenblatt & Aronson, 1958). In the laboratory, the female shows the first behavioural signs of oestrous cycle at 6 to 8 months of age (Rosenblatt & Schneirla, 1962). Other reports indicate the age of puberty between 15 to 18 months in free-ranging female cats (Eckstein & Zuckerman, 1956). Once the oestrous cycle appears, receptivity lasts from 4 to 6 days if the female is mated, and from 6 to 10 days if the female is not mated. Parallel with the variations in the length of oestrous cycle, there is a corresponding adjustment in the length of the anoestrous phase. Subsequent oestrous cycles appear at intervals of 2 to 3 weeks (Scott, 1955; Scott & Lloyd-Jacob, 1955). Ovulation does not occur unless coitus has taken place during the oestrous period (Greulich, 1934; Windle, 1939; Eckstein & Zuckerman, 1956).

Before initiating sexual behaviour, the male investigates the area, sprays an odorous substance from his anal glands and urinates on objects in the area. He may also rub his face against previously unsprayed objects. The female cat also investigates the environment, and occasionally sprays urine on objects (Leyhausen, 1965). Her sexual behaviour, however, appears to be less dependent on familiarity with the mating area (Rosenblatt & Schnierla, 1962).

Dominance ranking is very indefinite and unstable in the cat (Leyhausen, 1965). Free-ranging male cats meeting for the first time usually fight fiercely, but in subsequent encounters will rarely engage in actual fighting—even in the presence of an oestrous female. Females may remain faithful to inferior males from one heat period to the next for years without any serious interference from dominant males. Thus, the social system of the cat, with regard to sexual behaviour, appears to be designed to insure an equal chance of reproduction to all strong, healthy males, rather than favouring a single dominant individual (Leyhausen, 1965).

B. Patterns of Sexual Behaviour

When exposed to an oestrous female, the experienced male orients toward her, sniffs the genital area, utters a "sex call" not heard at other times, and grips the skin on her neck as she assumes a crouched position. The courtship pattern of the female is more elaborate. She rubs the floor with her chin and neck, rolls over repeatedly, makes a low throaty vocalization, and crouches with occasional treading movements. Up to the point of neck gripping by the male, there is no definite sequence in the occurrence of these behavioural patterns. The male may initiate his courtship with vocalization or may omit it. The female may respond by crouching and treading early in the mating sequence and then revert to chin rubbing and rolling over, or she may only crouch in response to the neck grip by the male. Occasionally the male releases his neck grip without mounting, but more commonly he assumes a position astride the female, executes a series of rapid pelvic thrusts, and maintains the neck grip until the copulation is completed. When intromission is effected, the female emits a copulatory cry, the male stops the pelvic thrusts but maintains full penetration for several seconds, after which he releases the neck grip and dismounts. After the male dismounts, the female may pull forward and turn on the male, hissing and pawing at him (Beach et al., 1956; Rosenblatt & Aronson, 1958). After a variable delay, new mounting and copulation follows.

Experienced pairs may mate ten times in an hour, with a progressive increase in the refractory period after each copulation. In her post-copulatory reaction, the female licks her paws and genital areas and rolls on the ground or rubs against nearby objects. The male sits near the female, licking the fur and genital areas (Bard, 1940; Beach *et al.*, 1956; Leyhausen, 1956a; Rosenblatt & Aronson, 1958).

The reciprocal behaviour patterns of courtship, copulation and post-copulatory activity of the cat are shown in Table 48 (adapted from Rosenblatt & Aronson, 1958).

Table 48. Synchronization of Male and Female Courtship, Copulation, and Post-copulatory Reactions. (Adapted from Rosenblatt & Aronson, 1958.)

Time	Male	Female
	Courtship	
	Orientates towards trails	crouches
	sniffs genitalia	rubs nose and mouth
10 sec to 5 min	circles	rolls
	mating call	vocalizes
	runs towards	treads
	Copulation	
	takes neck grip	crouches
	mounts with front legs	
	mounts with hind legs	
	rubs with forepaws	treads
1 to 3 min	steps	
	arches back	
	pelvic thrusts	swings tail to one side
	penile erection	bends hind end
	pelvic lunge	vaginal dilatation
5 to 10 sec	intromission	copulatory cry
	ejaculation	pulls forward
	penis withdrawn	turns on male
	Post-copulatory Reaction	
	licks penis and forepaws	licks genitalia
Less than 1 min	sits near female	rolls
		rubs nose and mouth
		licks paws
		watches male
About 5 min		paws male

C. Hormonal Mechanisms

1. THE MALE

Puberty in the male cat typically appears at about 8 months of age. While some androgen exists from birth on, a gradual rise occurs until shortly before puberty when it increases at a greater rate. Sufficient androgen is available at approximately $3\frac{1}{2}$ months of age to stimulate the growth of spines on the penis.

Castration prior to the age of puberty prevents the appearance of normal mating behaviour, but administration of exogenous testosterone can induce normal mating behaviour (Rosenblatt & Aronson, 1958). If oestrogen instead of testosterone is used, it will cause female-type behaviour, such as treading and tail deviation. Males treated with oestrogen will also permit mounting (Green et al., 1957). The effects of post-pubertal castration on sexual behaviour in the male cat was studied extensively by Rosenblatt & Aronson (1958). They found a variable decline in sexual behaviour following castration. Some showed only a decrease in frequency and longer latency to copulate, but retained the ability for intromission; others showed a rather rapid decline of all sexual behaviour.

Rosenblatt & Aronson (1958) found that castrated males with prior sexual experience showed a greater retention of sexual behaviour than those males who had no prior sexual experience. The sexually experienced males started at a higher level and remained so for a $3\frac{1}{2}$-month test period. Cooper (1960) similarly showed that pre- and post-puberty castrated males with a history of sexual behaviour who were given androgen over a period of 3 months performed at higher levels than castrates without prior sexual experience.

2. THE FEMALE

While the female has been described as having breeding seasons in the northern hemisphere, i.e. January to March and May to June, mating may take place throughout the year. The period of sexual inactivity which ordinarily lasts from September to January can be artificially altered by increasing the illumination under artificial situations. The onset of puberty or the first oestrus occurs from 6 to 8 months of age, and from then on the female has regular oestrous cycles lasting from 2 to 3 weeks. The period of receptivity is approximately 4 to 6 days but if unmated may last to 10 days. Ovulation occurs with coitus or with artificial vaginal stimulation, which has been shown to result in alterations in the electrical activity of the anterior hypothalamus (Porter et al., 1957). Ovulation can also be induced by direct stimulation of the hypothalamus. While oestrous behaviour is dependent on the presence of oestrogen, it is the direct effect of oestrogen on the brain which is important in sexual receptivity and not its effect on peripheral organs (Harris et al., 1958). While testosterone also has the capacity to induce oestrous behaviour, it is not associated with ovulation or endometrial growth of the uterus (Green et al., 1957).

D. Atypical Sexual Behaviour

Normal intact males may occasionally assume a feminine crouch with tail deviation when mounted by another male but do not show more complete forms of oestrous behaviour unless treated with oestrogens (Green et al., 1957). It is rare to have females assume a male posture even when treated with testosterone. Male-male mountings frequently occur in the pubertal stage of development as well as when males are deprived of the presence of a female. However, it is rare for such mountings to lead to ejaculation (Aronson, 1949). It is even a rarer occurrence for females to display mutual mountings. It is not uncommon to observe male house cats mounting inanimate objects, especially furry toys, in the absence of a female. Occasionally, laboratory cats will masturbate with their forepaws, and Rosenblatt & Schneirla (1962) report behaviour resembling masturbation in females treated with oestrogen. This includes rubbing the perianal region against the floor and vocalizing, or licking the genital region. In most instances these behaviours can be attributed to either environmental deprivation, or, as will be discussed later, the effects of certain brain lesions.

E. Neural Regulation of Sexual Behaviour

The degree to which sexual behaviour may be influenced by the central nervous system is dependent on a variety of internal and external factors. Thus, similar brain injury may have different effects, depending on the sex of the subject, degree of maturation, past experience, levels of circulating gonadal hormones, and the environment where the behaviour takes place. Since neocortical mechanisms for copulation do not appear as influential in the cat as in higher species, the following discussion will concentrate on subcortical and paleocortical regions of the brain.

Reflex control of motor patterns for copulation in the female and male exists in the lumbar region of the spinal cord. Maes (1939) demonstrated that female cats with transections of the spinal cord above the lumbar region will assume a coital posture when stimulated either manually or by a mounting male. Both erection and ejaculation can be elicited from the male cat with spinal cord transection using appropriate manual or electrical stimulation of afferent pathways to the lumbar cord.

The oestrous reflexes of the female cat have been shown to be sustained in spite of complete decortication. It is suggested that the posterior hypothalamus and more caudal brain-stem structures are responsible for the reflexes essential for copulation in the female (Bard, 1940).

Small lesions located in the supraoptic region of the hypothalamus were effective in producing anoestrus and elimination of all sexual responses in the female cat (Sawyer & Robinson, 1956). These effects were not reversible by oestrogen treatment. Lesions placed further posteriorly in the ventralis medialis or in the premammillary region gave the same behavioural result but the effect was reversible with oestrogen. In addition to the lack of sexual behaviour, lesions of the posterior hypothalamus resulted in ovarian atrophy, presumably because of the lack of gonadotrophic activity by the pituitary. These results indicate that two separate areas in the hypothalamus regulate sexual behaviour by two different mechanisms. The anterior portion regulates the behavioural mechanisms of sexual behaviour, while the posterior hypothalamus controls the gonadotrophic secretion necessary to sustain adequate levels of gonadal hormones for sexual behaviour (Fig. 85). Further demonstration that the anterior hypothalamus is important in the

PLATE XVI

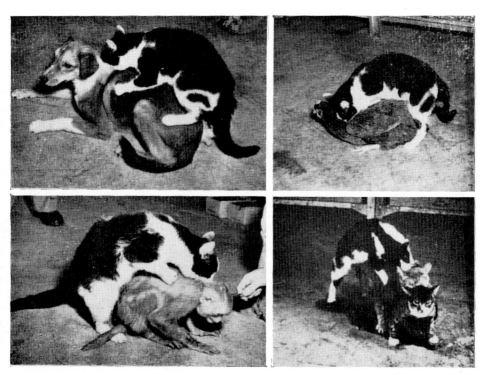

Illustrations of various phases of sexual activity of male amygdalectomized cats. (From Schreiner & Kling, 1953.)

regulation of sexual behaviour in the female was found by Porter *et al.* (1957) who demonstrated EEG changes in the anterior and lateral hypothalamus after artificial vaginal stimulation. This effect could only be produced in oestrous or oestrogen-primed females, but not in those that were anoestrus. Harris *et al.* (1958), using direct hormonal stimulation on localized areas of the hypothalamus by implanting minute amounts of solid

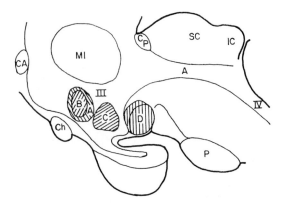

Fig. 85. Site of lesions in the female cat which interfere with reproductive behaviour. A and B are anterior hypothalamic lesions which produce permanent anoestrus, which is not reversible by administration of oestrogen. C is a lesion of the ventromedial hypothalamus, which induced ovarian atrophy. However, mating behaviour could be evoked by administration of oestrogen. D are mammillary lesions with the same effect as in lesion illustrated in C. In E the reconstructions of the lesions are all projected on a midsagittal map; CA, anterior commissure; Ch, optic chiasm; CP, posterior commissure; MI, massa intermedia; P, pons; IC, inferior colliculi; SC, superior colliculi; III, third ventricle; IV, fourth ventricle. (From C. H. Sawyer, *Handbook of Physiology*, 1960, p. 1232.)

hormones, found that ovariectomized females could display mating behaviour if the implants were made in the anterior hypothalamus. That the oestrous behaviour was not a result of generalized hormonal activity was demonstrated by the lack of effect on the uterus.

Much less information is available on the diencephalic regulation of sexual behaviour in the male cat. However, as in the female, lesions of the midtubural region eliminate sexual behaviour. In contrast to the female, however, cortical lesions reduce the ability of the male to perform the complex motor acts necessary for copulation—the larger the cortical lesion the less effective the male (Beach *et al.*, 1956).

Lesions of the central nervous system can also cause an increase in sexual behaviour (Fig. 85). Bilateral ablation of the amygdaloid nuclei in male adult cats results in an increase in frequency of indiscriminate copulatory attempts on receptive or non-receptive females, males, other species, or inanimate objects. Tandem copulations have also been observed among males. [XVI]. Females with similar ablations tend to display exaggerated rolling, squirming, and vocalization similar to normal females in oestrus (Schreiner & Kling, 1953).

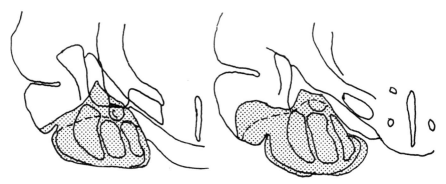

Fig. 86. Serial sections through brain of cat sustaining bilateral ablation of amygdaloid nuclei. Shaded area indicates lesion. (From Schreiner & Kling, 1953.)

Green *et al.* (1957) found that a similar hypersexual syndrome could be produced by lesions of the pyriform cortex alone. They described behaviour which included copulation with other species and anaesthetized animals, and engaging in sexual behaviour beyond their own territory. Female kittens sustaining lesions of the amygdaloid nuclei show evidence of precocious puberty (Kling, 1965).

A number of hypotheses have been suggested to explain the increase in sexual behaviour following lesions of the amygdaloid nuclei. One such hypothesis would include the inability of the animal to recognize the inappropriateness of the situation, territory, etc. Thus it would mount indiscriminately without regard to species or sex. This would be consistent with other aspects of the syndrome, which include indiscriminate mouthing of objects and tameness, or lack of fear, towards objects which previously would have called for defensive behaviour. Other possibilities include an increase in the sensitivity of the hypothalamus to circulating hormones and an increase of gonadotrophic secretion following these lesions.

III. MATERNAL BEHAVIOUR

A. Parturition

Pregnancy in the cat lasts from 63 to 65 days. Several days before parturition, the female may seek out a dark undisturbed corner which will serve as a place for delivery and later as the area for raising the young. The physiological events of parturition can be divided into four phases: (1) the contraction phase; (2) an emergence phase; (3) the delivery phase; and (4) the placental phase. This sequence of events is fairly constant but their respective durations are highly variable. Each of these physiological phases is marked by characteristic behaviour patterns of the female (Cooper, 1944; Failla et al., 1951).

When parturition is initiated, the female frequently changes her posture, and her movements appear to be specifically related to facilitating the expulsion of the foetus. Squatting and crouching predominate but there is also floor scratching, lordosis, and bracing the body against objects in the immediate environment. Usually the amniotic sac breaks prior to delivery of the foetus; or the unbroken sac, protruding from the vulva, may be torn by the female's licking or by the movements of the foetus. Occasionally, the sac is expelled completely without being broken and the female breaks the sac on the ground. In each case, the female licks the spilled amniotic fluid, first licking it from her body, her genital area, the floor, and finally from the kitten. Multiparous cats, in contrast to primiparous cats, direct their attention more promptly to the newborn. At this stage the abdominal contractions and the related behaviour patterns of squatting and crouching are absent. As the birth membranes are licked and pulled off by the mother, the kitten starts breathing and the female's activities are soon interrupted by the appearance of new contractions which lead to the expulsion of the placenta. The delivery of the placenta also elicits excessive licking. Ultimately, the female eats the placenta and severs the umbilical cord close to the body of the kitten (Failla et al., 1951).

B. Maternal Care

The activities of the female during parturition are directed toward expulsion of the foetus, freeing it of the birth membranes, and initiating respiration. Following completion of the litter's delivery, the female begins to retrieve those kittens that crawl away from the immediate vicinity of the home area. She then presents the abdominal regions to the kittens, and directs the kittens to the mammillary region by enclosing them in a small area encircled by her legs and body. She also licks the genital areas of the kittens. During the first 2 weeks this genital stimulation by the mother is necessary for initiating elimination in the kittens (Rosenblatt & Schneirla, 1962).

Nursing is initiated by the female, who arouses them by vigorous licking and presents them the mammillary region. The kittens orient to the mammillary area, nuzzle in her fur, locate a nipple and attach to the nipple for suckling. Toward the middle of the 3rd week, the kittens occasionally leave the home region, and then crawl or walk unsteadily across short distances to reach the female. From this time on until the 6th week, the kittens progressively extend the range of their movement and gradually become adept at attaching to the nipple while the female is sitting or standing. From the 5th week on, the maternal behaviour declines and the kittens begin to take food from other sources. There is increasing tendency at this stage for the female to leave the kittens and avoid them. In the early period of maternal behaviour, each kitten suckles nearly 25 per cent of total time, and the nursing activity of the female may take up nearly 70 per cent of her time, since kittens often suckle individually and in succession. After the 5th week, nursing declines to less than 20 per cent of the female's time. From birth until the end of the 3rd week, the kittens usually aggregate in one part of the cage in the home region, where maternal activities and nursing take place. Beyond the 3rd week, while the individual interaction and play activities grow among the kittens, the attachment to one locality weakens, as does the tie with the mother. At this time the kittens begin intensive play activities directed towards littermates and towards the female. From the 4th week on, the female increasingly withdraws from the play activities by avoiding the kittens. The apparent annoyance with the play and related avoidance of the kittens may contribute greatly to the acceleration of the weaning process at this time (Rosenblatt et al., 1961).

C. Early Somatic and Behavioural Maturation

The maturation of the maternally reared cat in the laboratory follows a relatively consistent sequence. Under laboratory conditions the following schedules of maturation for 40 kittens were obtained.

Table 49. Behavioural Development in Days of Age

Walk	Eye Opening	Climbing	Grooming	Play (complete)	Aggression (complete)
22	12	28	23	35	52

1. Effects of Maternal Deprivation

The maturation of kittens deprived of maternal experience and reared separately in incubators did not reveal major differences in development except in the time of walking which was accelerated to 17 days. Kittens reared under this condition tend to be more active and vocal, as if searching for a source of nutrition and warmth. Grooming behaviour was accelerated in the deprived group to 15 days, but was difficult to evaluate since the kittens tended to perform suckling movements on their own bodies very soon after birth. The time appearance of play and aggressive behaviour was no different than that found in normally reared kittens. Somatic growth of the deprived group was also similar to maternally reared kittens up to 60 days of age (Fig. 87).

Seitz (1959) studied the later behaviour of kittens separated from their mothers at the ages of 2, 6, and 12 weeks. He found that the kittens which were separated at 2 weeks of age, as adults, showed the most randomly active, but least goal directive activity. They were the most anxious in novel situations, were disorganized in efforts to get food, and were the most aggressive but least successful in competitive feeding. They were also the slowest to learn a simple feeding routine, and showed a chronic, asthma-like respiratory syndrome following a feeding frustration test. Cats separated at later ages displayed little deviant behaviour.

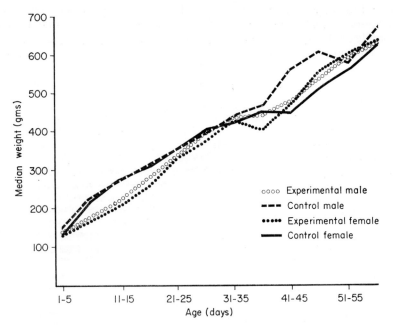

Fig. 87. Comparison of weight gain for maternally reared (control) and maternally deprived (experimental) kittens.

2. Effects of Brain Injury and Stimulation

The somatic growth for cats sustaining early amygdala lesions was no different than normal littermate controls up to 8 months of age (Fig. 84). It is interesting that three of four cats with damage to olfactory bulb and para-olfactory cortex showed weight gains above the control groups, while two cats with large cortical lesions showed growth which was below both control and other operate cats.

The onset of walking, eye opening, and play behaviour was somewhat accelerated in the amygdala-lesioned kittens (Kling, 1965), while aggressive behaviour was somewhat delayed.

In an attempt to determine whether affective responses could be elicited by subcortical electrical stimulation prior to their natural appearance, stimulation via implanted electrodes in the amygdaloid nucleus and hypothalamus was carried out in a series of 30 kittens varying in age from 4 hours to 50 days (Kling & Coustan, 1964). Alerting and somato-motor responses were consistently elicited from both areas at all ages (Fig. 88). Beginning in the 2nd week of life, most autonomic responses could be evoked from the hypothalamus before they could be evoked from the amygdala. These included salivation, urination, and defaecation. Affective defence reactions were obtained from the hypothalamic locations at 12 days, and similar responses from the amygdala by 3 weeks of age. Flight and attack were the last to appear (Fig. 88).

The functional maturation of the hypothalamus for autonomic and behavioural responses precedes that of the amygdala. In addition, a variety of complex behavioural responses seems to be elicited prior to their natural appearance in the normally developing kitten.

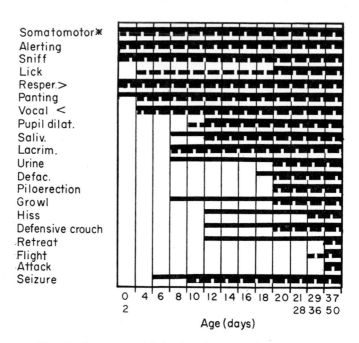

Fig. 88. Summary of behavioural response in kittens to electrical stimulation of hypothalamus and amygdala from birth to 50 days of age.

* Somato-motor responses include: circling, head turning, rolling over, extension of leg with clonic movements, claw extension, twitching or more sustained contractions of eyelid and ear, and mouth and jaw movements. (From Kling & Coustan, 1964.)

D. Neural Regulation of Affective Behaviour

Lesions of the neocortex which spare paleocortical tissue result in a raised threshold for noxious stimulation, and these animals typically appear quite placid. If, however, paleocortical tissue is removed, the classical decorticate preparation results in the syndrome known as "sham rage" (Bard, 1938). Decerebrate preparations which included ablation of the cortex plus the caudal half of the diencephalon resulted in rage responses which were more circumscribed, less integrated, and required a higher threshold for stimulation. In general, as decerebration is continued caudally, the more fragmentary the rage responses become. The term "sham rage" was introduced by Cannon and Britton in 1925. It was based upon the fact that the behaviour occurred in the absence of conscious experience or awareness and that the animal was merely responding to an immediate tactile stimulus. It has also been suggested that the picture of decorticate "sham rage" is a release of sensory inhibitory mechanisms, so that the animal is hyperalgesic and its responses to noxious stimulation are greatly exaggerated.

Other aspects of the syndrome, such as the autonomic responses of pilo-erection, pupillary dilation, defaecation and urination, which occur at the same time as the emotional discharge, have been suggested as evidence for the rage being an exaggeration of responses to sensory stimulation. Whatever the essential changes are due to, it must be considered that these animals have little in the way of control of sensory motor mechanisms and, in addition, lack the normal inhibitory action of the cerebral cortex on lower brain-stem structures. More localized brain lesions which have been reported to produce lowering of thresholds to noxious stimulation and rage behaviour include the septal region, orbital frontal cortex and hippocampal fornix system (Speigal et al., 1940).

Rothfield & Harmon (1954) found that the effect of the latter lesion resulted only when the neocortex had also been removed. Kling et al. (1960) reported that lesions of the posterior septum fornix, and anter-commissure did not result in heightened irritability, as did Bond et al. (1957). Large lesions of this area, in fact, produced a preparation which was stuporous.

In the cat, lesions of the ventral medial nuclus of the hypothalamus caused a dramatic change in behaviour as regards irritability and rage (Wheatly, 1944). De Molina and Hunsberger (1959) contend that hypothalamic mech-anisms are less critical than midbrain mechanisms in the elicitation of rage, since midbrain lesions will abolish electrically induced affective responses obtained from stimulating more forebrain mechanisms. Lesions in the posterior portion of the hypothalamus have been reported by Ranson (1939), Ingram et al. (1936) and Masserman (1941) to produce a loss in emotional responsiveness. However, Kelly et al. (1946) reported that cats in which the hypothalamus was destroyed at the level of the mammillary bodies exhibited clear, irritative responses such as growling, crying and spitting, when pro-voked by noxious stimulation. It appears that integrity of the midbrain is essential for the elicitation of rage responses, and while lesions anterior to this area may result in rage behaviour, it is possible that this is the result of their influence on lower midbrain structures.

Perhaps one of the most striking sets of behavioural changes occurring after central nervous system lesions is that which occurs following ablation of the amygdaloid nuclei (Schreiner & Kling, 1953). In the adult male cat the syndrome is characterized by a period immediately following operation of stupor and anorexia lasting from 1 to 2 weeks. During this time the cats are typically unresponsive to surrounding stimuli, sit in their cages in a sphinx-like posture, and require gastric lavage to maintain nutrition. Following recovery from this initial post-operative effect, there is an increase in cage activity with frequent pacing, rubbing against the sides, and purring. The cats are solicitous of petting and stroking, rub against observers, and appear as though there were an enhancement of pleasurable responses. They are less responsive to noxious stimulation and are inattentive to threatening stimuli in their environment, such as dogs or other animals, which would normally arouse a defensive posture. With sufficient stimulation, they can be aroused to hiss, spit and strike. Also, there is an increase in oral exploratory activity, licking and sniffing. The cats may pick up and mouth faeces or other objects ordinarily ignored. Their physical appearance is generally sleek and well-muscled. After a period of 2 weeks to a month, the male cats typically become hypersexual.

The behaviour of the female is not as clear. In fact, we have had several female operates who became quite vicious after operation. The behaviour of those females was similar to that reported by Bard & Mountcastle (1948) and later by Wood (1958) for lesions restricted to the basal and central nuclei of the amygdala. Wood (1958) suggests that partial lesions, restricted to the basal lateral group, are responsible for the increased aggressiveness.

The behavioural syndrome resulting from ablation of the amygdaloid nuclei seems to be dependent on the integrity of the hypothalamus (Kling & Hutt, 1958). De Molina & Hunsberger (1962) found that unilateral coagulation in the midbrain ipsilaterally suppressed the affective pattern evoked from the amygdala, and bilateral coagulation of the midbrain field suppressed the pattern evoked from hypothalamic stimulation. They suggested that the system governing defensive behaviour is organized at three levels of progressively increasing importance, the most important being in the midbrain.

IV. LEARNING

A. Classical Conditioning

Various appetitive and defensive reactions of the domestic cat can be conditioned by classical (Pavlovian) procedures, and as such do not differ markedly from classical conditioned response learning in other higher mammals. However, contemporary usage of classical conditioning techniques with cats most frequently revolves about investigations of acquired fear reactions or the use of such reactions as a means of investigating neurophysiological correlates of the learning process itself.

Conditioned fear reactions are rapidly established in the cat by pairing any one of a variety of afferent stimuli with shock. For example, in studying the behaviour effects of limbic system lesions in cats (McCleary, 1961), animals were initially trained to approach a feeding compartment and eat from a metal food dish while standing on a metal floor section. After short-latency food approaching had been established, following several days of training, the cats were suddenly shocked as they began eating from the food dish. After two shock trials, normal cats no longer entered the feeding compartment when food was again made available on successive test trials. Clearly, the cats had acquired a distinct fear of the feeding compartment and/ or the food approach response itself.

Grossly observable, conditioned fear responses of the cat, such as that described above, may be contrasted with conditioned brainwave activity in so far as this technique provides a means of studying those electrophysiological events underlying the learning process itself. For example, there is the work of Galambos & Sheatz (1962) on the effects of a classical conditioning procedure on changes in auditory evoked potential amplitude in the cat brain. Following a habituation procedure, in which the CS (click) alone was repeatedly presented, there occurred a diminution in the average EP amplitude. At this point, a classical conditioning procedure was initiated, in which a puff of air (UCS) was delivered to the cat immediately after the presentation of the click stimulus. After several paired presentations of CS and UCS, it was found that the previously habituated EP to click was now greatly enhanced. Subsequent unpaired presentations of click alone led to a progresive diminution in the EP amplitude (a conditioning procedure known as extinction), and reinstating the CS-UCS presentations resulted in a progressive reaugmentation of the click EP (reconditioning).

B. Instrumental Conditioning

This type of learning is currently discussed under a variety of titles, including problem solving, principle learning, selective learning, and discrimination learning. Regardless of the particular nomenclature used, the major operational distinction between this form of conditioning and the classical form of conditioning involves a consideration of the relationship between the organism's response and the reward (incentive; UCS) used in establishing the learned response. In contrast to the classical conditioning paradigm, there is present in the instrumental paradigm a fundamental response-reward contingency, such that the behaviour of the learner is "instrumental" in altering the nature of the incentive.

The vast majority of learning studies using domestic cats as subjects are of the instrumental conditioning type. Among these, studies of sensory discrimination learning and problem solving predominate. In the paragraphs that follow, brief examples of some of the more representative learning situations and associated types of apparatus used in them will be given.

1. Sensory Discrimination Learning

The visual and auditory capacities of the cat have been the sensory modalities most intensively studied. Using stimuli presented to the cat either successively (i.e. only one of the stimuli to be discriminated is presented on each training trial) or simultaneously (i.e. two or more stimuli are presented together on each trial), a variety of qualitative and quantitative stimulus dimensions for each of these modalities has been examined. For example, within the visual modality, cats have been trained to make discriminations based upon form or shape (Winans, 1967), brightness and flux (Schilder, 1966), and acuity, flicker and movement (Smith, 1937). Within the auditory modality, cats have been trained to make discriminations relating to frequency (Goldberg & Neff, 1961), intensity (Rosenzweig, 1946), onset (Meyer & Wooley, 1956), sound localization (Neff, 1961), tonal pattern (Diamond & Neff, 1957) and duration (Scharlock et al., 1965).

The usual discrimination paradigm employed in these studies requires the cat to learn to respond selectively to only one of the several stimuli presented to it.

Audition. Auditory discriminations are necessarily presented successively, rather than simultaneously, since the former mode of presentation obviates such interference effects as overtones, harmonics, and intensity summation effects, which would occur if two or more auditory stimuli were presented at the same time. Successive discriminations of this type are most conveniently conducted using a double-grille box apparatus and shock avoidance as the incentive. As an example in the testing of frequency discrimination, the cat initially learns that on any given trial the onset of a 1,000 c.p.s. tone is always followed by foot-shock, whereas an 800 c.p.s. tone, on a subsequent trial or following the 1st tone within the same trial, is not. In response to the shock, the animal learns to escape into the adjoining compartment of the double-grille box. After the classically conditioned fear response to the 1,000 c.p.s. tone is established, the cat eventually learns to avoid the shock (instrumental learning) by crossing over into the adjoining compartment as soon as the 1,000 c.p.s. tone is presented but *before* the shock actually comes on. When the cat is able to make 90 per cent or more such avoidance responses in a daily 20-trial test session, it is concluded that the cat has learned this particular frequency discrimination. The testing of such discriminations as tonal intensity, onset, duration and pattern may be conducted using similar procedures.

The testing of the cat's ability to localize spatially a sound source is typically conducted in an open-field maze. The hungry animal is placed in a large enclosure, within which are located two or more feeding stations. Located in proximity to each of these stations is a small audio speaker which can selectively deliver an auditory signal (either pure tone or multifrequency signals) to be used as a cue to the correct location of a food reward.

Vision. In contrast to the procedures used in testing auditory discrimination, the testing of visual discrimination is usually conducted using a simultaneous method of cue presentation. The fact that the cat can attend selectively first to one visual stimulus and then to the other, without significant interactive interference, makes a more efficient comparison and contrast of the several stimuli to be discriminated than is possible with a successive type of stimulus presentation. This efficiency is often reflected in a more rapid acquisition of the discriminatory response under the one condition than the other (Kimble, 1961).

Although shock has been used as the unconditioned stimulus (Tucker & Kling, 1966), most visual discrimination studies in the cat have used hunger and its partial reduction consequent to a "correct" response as the major drive-state, or motive, for learning. For discriminations of form, colour, brightness and flux, the usual testing apparatus consists of a two-alley "maze". Of the two stimuli to be discriminated, one is located at the end of one of the alleys and the other is placed at the end of the other alley. These stimuli consist typically of stationary cues, either painted on a food-compartment door or projected onto the door by light projectors located behind the door. The placement of the stimuli in the two alleys is randomized over successive trials in order to prevent fixed positional responding. In such situations, the cat is required to make an approach response in one or the other of the alleys and, depending upon the correctness of the choice, secure a small food reward located behind the stimulus door. As with other varieties of discrimination learning, choice of one of the stimuli is consistently rewarded, while choice of the opposing stimulus does not lead to any reward and may actually result in a mild punishing shock in some cases.

2. PROBLEM SOLVING

In the discrimination paradigms described above, the major cues to learning consist of rather obvious sensory discriminanda which serve to guide the cat in making its response. In contrast, problem solving situations often require the cat to base its response predominantly upon some implicit principle or "rule" which may or may not be signalled by the presence of some discriminable sensory feature of the situation. If sensory discriminanda are present, they are not necessarily restricted to a single modality and do not usually provide the only cues to a correct response. Even with sensory cues present, many problem-solving situations are so constructed that, in order to make a correct response, the animal must "remember" either the response given on a previous trial or, correspondingly, a specific type of preceding stimulus display. This apparent inter- or intra-trial involvement of a memory function, or the capacity to make what may be termed "symbolic" responses (in the sense of responding to a stimulus which, at the time of actual response, is no longer present), clearly differentiate the problem-solving or principle-learning situation from the sensory discrimination paradigms previously described.

One of the earliest attempts to investigate problem-solving behaviour in animals systematically was provided by Edward L. Thorndike (1898). Using a specially constructed "latch box", hungry cats were required to learn how to manipulate the locking mechanism on the door of the box so as to escape and secure a visible food reward placed outside the box.

Somewhat later, the escape behaviour of cats in a similar situation was studied by Guthrie (1930) and Guthrie & Horton (1946). Rather than focusing upon the element of gradual response acquisition, these investigators emphasized the marked stereotypy of the escape response itself. Although the mode of escape differed from one cat to another, the precise sequence of escape movements for an individual animal were highly consistent and unvarying over hundreds of practice trials.

Avoidance conditioning is a widely-used learning paradigm with cats. Both the active and passive varieties of this paradigm have been previously described. However, it should be pointed out that, in addition to the element of classically conditioned fear, the problem-solving aspect in both varieties requires the animal to learn how to deal with the acquired fear. In the active avoidance paradigm, the cat learns to reduce the aroused fear by changing its physical location within the apparatus, while in passive avoidance, it learns to maintain its physical locus rigidly so as to avoid shock-pain.

The element of "symbolic" responding in problem solving is nowhere more evident than in the delayed response learning paradigm. Indeed, this type of learning, with its apparent dependence upon a memory factor, has been characterized as involving what might be termed "higher mental function". As studied in the cat (Lawicka & Konorski, 1961; Warren et al.,

1962), the learning situation is so constructed that, in order to secure a food reward, the animal must respond to a previously experienced, but currently unavailable, stimulus event. In the usual situation, the cat is allowed to attend visually to the placement of a concealed food reward under or behind one of several identical-looking stimulus locations. Following the concealment of the reward, there occurs an interposed period of time during which the cat is prevented from responding to the location. Visualization of the "correct" food location may or may not be permitted during the delay period. Finally, the animal is allowed to approach and secure the food reward only after the specified period of time has elapsed.

Delayed alternation, conditional discrimination and learning set tasks are examples of problem-solving situations in which the cat must learn a specific principle of responding in order to gain the food reward. In the delayed alternation task (Stewart & Warren, 1957), the animal must learn that food rewards can only be secured if a particular pattern of responding is adhered to. For example, in single (delayed) alternation, the response pattern in a two-stimulus, two-position situation might be L, R, L, R, L, etc. in unbroken sequence, with delay periods of pre-set duration interposed between successive choices. Double alternation problems (L, L, R, R, L, L, etc.), with or without delay, are almost beyond the cat's capacity to learn.

Although conditional discriminations and learning set tasks are basically built upon a sensory discrimination learning paradigm, the inclusion of additional features involving the learning of a particular principle of responding sets them apart from the usual sensory discrimination learning tasks. For example, in the learning of a triangle vs. circle form discrimination, the "correct" stimulus to be selected may be contingent upon the size, as well as the shape, of the two stimuli as they are presented to the cat (Warren et al., 1962). If the two forms (either stereometric or planometric) are large, the triangle may be the rewarded stimulus, whereas if the two forms are relatively small, the circle may be the rewarded stimulus. Thus, in addition to discriminating the shape differences between triangles and circles, the animal must abstract from the situation the principle that size, in conjunction with form, is a fundamental cue to correct response.

The learning set paradigm, first described by Harlow in connection with primate learning (1949), is also built upon a sensory discrimination learning paradigm, but, as with the conditional discrimination problem, is designed to assess the cognitive (or principle learning) capacities of the learner.

As an example of a discrimination learning set paradigm, cats may be trained to discriminate which one of three simultaneously presented shapes is distinctly different from the other two shapes—such a stimulus display might include two crosses and a circle. The fact that cats are able to solve this type of "oddity discrimination" (Warren, 1960), as well as many other problems of the same type, permits an assessment of what has been termed a "learning set" in the cat. If such a "set" (or approach to learning) for oddity discrimination problems does occur, one can trace its development in all succeeding problems of the same type by inspecting the percentage of correct responses made by the cat after the first trial choice on each of these succeeding problems. Typically, there is a progressive improvement in the animal's performance from the second trial forward as each new problem of the same type is presented. For example, a comparison of performance in trials 2 to 6 in literally hundreds of oddity discrimination problems of this type shows that, although correct responding in the 1st trial of each problem remains at chance level, the percentage of correct responding in the next five trials steadily increases. It is as though the cat had learned to maximize the information gained in the first trial of any such problem, regardless of the particular form of discriminanda used, and apply that information to all succeeding trials of both the same problem and different problems as well. The abstraction of such a "learning set", which overrides any particular oddity form discrimination problem, constitutes the cognitive learning principle inherent in such tasks.

REFERENCES

ANAND, B. K. (1961). Nervous regulation of food intake. *Physiol. Rev.*, **41**, 677–708.

ANAND, B. K. & DUA, S. (1955). Feeding responses induced by electrical stimulation of hypothalamus in cat. *Int. J. med. Res.*, **43**, 113–122.

ANAND, B. K., DUA, S. & CHINA, G. S. (1958). Higher nervous control over food intake. *Int. J. med. Res.*, **46**, 277–287.

ANAND, B. K., DUA, S. & SCHOENBERG, K. (1955). Hypothalamic control of food intake in cats and monkeys. *J. Physiol., Lond.*, **127**, 143–152.

ARONSON, K. R. (1949). Behavior resembling spontaneous emission in the domestic cat. *J. comp. physiol. Psychol.*, **42**, 226–227.

ARONSON, L. & COOPER, M. (1966). Seasonal changes in mating behavior in cats after desensitization of glans penis. *Science, N.Y.*, **152** (3719), 226–230.

BARD, P. (1938). Central nervous mechanisms for emotional behavior patterns in animals. *Res. Publs Ass. Res. nerv. ment. Dis.*, **19**, 190–218.

BARD, P. (1940). The hypothalamus and sexual behavior. In: The hypothalamus and central levels of autonomic function. *Res. Publs. Ass. Res. nerv. ment. Dis.*, **20**, 551–594.

BARD, P. & MOUNTCASTLE, V. B. (1948). Some forebrain mechanisms involved in the expression of rage with special reference to suppression of angry behavior. In: *The Frontal Lobes*. J. F. Fulton (ed.). Baltimore: Williams and Wilkins, pp. 362–404.

BEACH, F. A., ZITRIN, A. & JAYNES, J. (1956). Neural mediation of mating in male cats. I. Effects of unilateral and bilateral removal of the neocortex. *J. comp. physiol. Psychol.*, **49**, 321–327.

BOND, D. D., RANDT, C. T., BIDDER, H. T. & RAWLAND, V. (1957). Posterior septal, formical and anterior thalamic lesions in the cat. *A.M.A. Archs Neurol. Psychiat.*, **78**, 143–162.

CANNON, W. B. & BRITTON, S. W. (1925). Studies on the conditions of activity in endocrine glands. XV. Pseudoaffective medulli adrenal secretion. *Am. J. Physiol.*, **72**, 283–294.

COLBERT, E. H. (1958). Morphology and behavior. In: *Behavior and Evolution*. A. Roe and G. C. Simpson (eds.). Yale University Press, pp. 24–27.

COOPER, J. B. (1944). A description of parturition in the domestic cat. *J. comp. Psychol.*, **37**, 71–79.

COOPER, K. K. (1960). The significance of past sexual experience in the reappearance of sexual behavior in castrated male cats treated with testosterone propionate. M.Sc. Thesis, New York Univ., New York.

DIAMOND, I. T. & NEFF, W. D. (1957). Ablation of temporal cortex and discrimination of auditory patterns. *J. Neurophysiol.*, **20**, 300–315.

ECKSTEIN, P. & ZUCKERMAN, S. (1956). The oestrous cycle in the mammalia. In: *Marshall's Physiology of Reproduction*. Vol. I, Part I. 3rd edn. A. S. Parkes (ed.). London: Longmans, Green & Co., pp. 226–396.

FAILLA, M. L., TOBACH, E. & FRANK A. (1951). A study of parturition in the domestic cat. *Anat. Rec.*, **III**, Abstract No. 90.

GALAMBOS, R. & SHEATZ, G. C. (1962). An electroencephalographic study of classical conditioning. *Am. J. Physiol.*, **203**, 173–184.

GOLDBERG, J. M. & NEFF, W. D. (1961). Frequency discrimination after bilateral ablation of cortical auditory areas. *J. Neurophysiol.*, **24**, 119–128.

GREEN, J. D., CLEMENTE, C. D. & DE GROOT, J. (1957). Rhinencephalic lesions and behavior in cats. *J. comp. Neurol.*, **108**, 505–545.

GREULICH, W. W. (1934). Artificially induced ovulation in the cat (*Felis domestica*). *Anat. Rec.*, **58**, 217–224.

GROSSMAN, S. P. (1960). Eating and drinking elicited by direct adrenergic or cholinergic stimulation of the hypothalamus. *Science, N.Y.*, **132**, 301–302.

GROSSMAN, S. P. (1964). Behavioral effects of chemical stimulation of the ventral amygdala. *J. comp. physiol. Psychol.*, **57**, 29–36.

GUTHRIE, E. R. (1930). Conditioning as a principle of learning. *Psychol. Rev.*, **37**, 412–428.

GUTHRIE, E. R. & HORTON, G. P. (1946). *Cats in a Puzzle Box.* New York: Rinehart.

HARLOW, H. F. (1949). The formation of learning sets. *Psychol. Rev.*, **56**, 51–65.

HARRIS, G. W., MICHAEL, R. P. & SCOTT, P. P. (1958). Neurological site of action of stilboestrol in eliciting sexual behavior. In: *Neurological Basis of Behavior.* Ciba Found. Symp. London: Churchill, pp. 236–254.

INGRAM, W. R., BARRIS, R. W. & RANSON, S. W. (1936). Cotatepsy: an experimental study. *Archs Neurol. Psychiat., Chicago*, **35**, 1175–1197.

KELLY, A. H., BEATON, L. E. & MAGOUN, H. W. (1946). A midbrain mechanism for facio vocal activity. *J. Neurophysiol.*, **9**, 181–189.

KIMBLE, G. A. (1961). *Hilgard and Marquis' Conditioning and Learning.* New York: Appleton-Century Crofts, Inc.

KLING, A. (1965). Behavioral and somatic development following lesions of the amygdala in cat. *J. Psychiat., Res.*, **3**, 263–273.

KLING, A. & COUSTAN, D. (1964). Somato-motor autonomic and affective responses to subcortical stimulation in the infant cat. *Fedn. Proc.*, **23**.

KLING, A. & HUTT, P. (1958). Effects of hypothalamic lesions on the amygdala syndrome in cat. *Archs Neurol. Psychiat., Chicago*, **75**, 511–517.

KLING, A., ORBACH, J., SCHWARTZ, N. & TOWNE, J. (1960). Injury to the limbic system and associated structures in cats. *A.M.A. Archs Psychiat.* **3**, 391–430.

KOVACH, J. K. & KLING, A. (1967). Mechanisms of neonate sucking behaviour in the kitten. *Anim. Behav.*, **15**, 91–101.

KUO, Z. Y. (1930). The genesis of the cat's behavior toward the rat. *J. comp. Psychol.*, **25**, 1–8.

KUO, Z. Y. (1938). Further study on the behavior of the cat toward the rat. *J. comp. Psychol.*, **25**, 1–8.

LANGWORTHY, O. R. (1929). A correlated study of the development of reflex activity in foetal and young kittens and the myelinization of tracts in the nervous system. *Contr. Embryol.*, **20**, No. 114, 127–171.

LAWICKA, W. & KONORSKI, J. (1961). The effects of prefrontal lobectomies on the delayed responses in cats. *Acta Biol. exp., Vars.*, **21**, 141–156.

LEYHAUSEN, P. (1956a). *Verhaltensstudien an Katzen.* Berlin: P. Perey, Verlag.

LEYHAUSEN, P. (1956b). Das Verhalten der Katzen (*Felidae*). *Handb. Zool.*, **10**, 1–34.

LEYHAUSEN, P. (1965). Communal organization of solitary mammals. *Symp. zool. Soc. Lond.*, **14**, 259–263.

MAES, J. P. (1939). Neural mechanisms of sexual behaviour in the female cat. *Nature, Lond.*, **144**, 598–599.

MASSERMAN, J. H. (1941). Is the hypothalamus a center of emotion? *Psychosom. Med.*, **3**, 3–25.

McCLEARY, R. A. (1961). Response specificity in the behavioral effects of limbic system lesions in the cat. *J. comp. physiol. Psychol.*, **54**, 605–613.

MEYER, D. R. & WOOLEY, C. N. (1956). Effects of localized cortical destruction upon auditory discriminative conditioning in the cat. *J. Neurophysiol.*, **19**, 500–512.

DE MOLINA, F. A. & HUNSBERGER, R. W. (1959). Central representation of affective reactions in forebrain and brain stem: electrical stimulation of amygdala, stria terminalis and adjacent structures. *J. Physiol., Lond.*, **145**, 251–265.

DE MOLINA, F. A. & HUNSBERGER, R. W. (1962). Organization of subcortical system governing defence and flight reactions in the cat. *J. Physiol., Lond.,* **160,** 200–213.

MORGANE, P. V. & KOSMAN, A. V. (1957). Alterations in feline behaviour following bilateral amygdalectomy. *Nature, Lond.,* **180,** 598–600.

NEFF, W. D. (1961). Discriminatory capacity of different divisions of the auditory system. In: *Brain and Behavior.* Vol. I. M. A. D. Brazier (ed.). Washington, D.C.: Am. Inst. Biol. Sci., pp. 205–262.

PORTER, R. W., CAVANAUGH, E. B., CRITCHLOW, E. B. & SAWYER, C. H. (1957). Localized changes in electrical activity of the hypothalamus in estrous cats following vaginal stimulation. *Am. J. Physiol.,* **189,** 145–151.

PRECHTL, H. F. R. (1952). Amgeborenen Bewungsweisen junger Katzen. *Experientia,* **8,** 220–221.

RANSON, S. W. (1939). Somnolence produced by hypothalamic lesions in the monkey. *A.M.A. Archs Neurol. Psychiat.,* **41,** 1–23.

ROSENBLATT, J. S. & ARONSON, L. R. (1958). The decline of sexual behaviour in male cats after castration with special reference to the role of prior sexual experience. *Behaviour,* **12,** 285–338.

ROSENBLATT, J. S. & SCHNEIRLA, T. C. (1962). The behavior of cats. In: *The Behavior of Domestic Animals.* E. S. E. Hafez (ed.). Baltimore: The Williams and Wilkins Co., pp. 453–488.

ROSENBLATT, J. S., TURKEWITZ, G. & SCHNEIRLA, T. C. (1959). Early socialization in the domestic cat as based on feeding and other relationships between female and young. In: *Determinants of Infant Behavior.* Foss (ed.). New York: Wiley, pp. 51–74.

ROSENZWEIG, M. (1946). Discrimination of auditory intensities in the cat. *Am. J. Psychol.,* **59,** 127–136.

ROTHFIELD, L. & HARMON, P. (1954). On the relation of the hippocampalfornix systems to the control of rage responses in cats. *J. comp. Neurol.,* **101,** 265–282.

SAWYER, C. H. & ROBINSON, B. A. (1956). Separate hypothalamic areas controlling pituitary gonadotropic function and mating behavior in female cats and rabbits. *J. clin. Endocr. Metab.,* **16,** 914.

SCHARLOCK, D. P., NEFF, W. D. & STROMINGER, N. L. (1965). Discrimination of tone duration after bilateral ablations of cortical auditory areas. *J. Neurophysiol.,* **28,** 673–681.

SCHILDER, P. (1966). Loss of a brightness discrimination in the cat following removal of the striate area. *J. Neurophysiol.,* **29,** 888–897.

SCHREINER, L. & KLING, A. (1953). Behavioral changes following rhinencephalic injury in cat. *J. Neurophysiol.,* **16,** 643–659.

SCOTT, P. P. (1955). The domestic cat as a laboratory animal for the study of reproduction. *J. Physiol., Lond.,* **130,** 47P–48P.

SCOTT, P. P. & LLOYD-JACOB, M. A. (1955). Some interesting features in the reproductive cycle of the cat. *Stud. Fert.,* **7,** 123–129.

SCOTT, P. P. & LLOYD-JACOB, M. A. (1959). Reduction in the anoestrus period of laboratory cats by increased illumination. *Nature, Lond.,* **184,** 2022.

SEITZ, P. F. D. (1959). Infantile experience and adult behavior in animal subjects. *Psychosom. Med.,* **21,** 353–378.

SMITH, K. W. (1937). The relation between visual acuity and the optic projection centers of the brain. *Science, N.Y.,* **86,** 564–565.

SPIEGAL, E. A., MILLER, H. R. & OPPENHEIMER, M. J. (1940). Forebrain and rage reactions. *J. Neurophysiol.,* **3,** 538–548.

STAVSKY, W. H. (1932). The geotropic conduct of young kittens. *J. genet. Psychol.,* **6,** 441–446.

STEWART, C. W. & WARREN, J. M. (1957). The behavior of cats on the double alternation problem. *J. comp. physiol. Psychol.,* **50,** 26–28.

THORNDIKE, E. L. (1898). Animal intelligence. An experimental study of the associative processes in animals. *Psychol. Monogr.,* **2,** No. 8.

TILNEY, F. & CASAMAJOR, L. (1924). Myolenogeny as applied to the study of behaviour. *Archs Neurol. Psychiat., Chicago.,* **12,** 1–66.

TSAI, L. S. (1963). Peace and cooperation among "Natural Enemies": educating a rat-killing cat to cooperate with a hooded rat. *Acta psychol. Taiwan.,* **5,** 1–5.

TUCKER, T. & KLING, A. (1966). Differential effects of early vs. late brain damage on visual duration discrimination in cat. *Fedn. Proc.*, **25**, 207.

WARREN, J. M. (1960). Oddity learning set in a cat. *J. comp. physiol. Psychol.*, **53**, 433–434.

WARREN, J. M., WARREN, H. & AKERT, K. (1962). Orbito frontal cortical lesions and learning in cats. *J. comp. Neurol.*, **118**, 17–41.

WHEATLY, M. D. (1944). The hypothalamus and affective behavior in cats. *Archs Neurol. Psychiat., Chicago*, **52**, 296–316.

WILSON, C. & WESTON, E. (1946). *The Cats of Wildcat Hill*. New York, N.Y.: Duell, Sloan & Pearce.

WINANS, S. C. (1967). Visual form discriminations after removal of the visual cortex in cats. *Science, N.Y.*, **158**, 944–946.

WINDLE, W. F. (1930). Normal behavioral reactions of kittens correlated with postnatal development of nerve-fibre density in the spinal grey matter. *J. comp. Neurol.*, **50**, 479–503.

WINDLE, W. F. (1939). Induction of mating and ovulation in the cat with pregnancy urine and serum extracts. *Endocrinology*, **25**, 365–371.

WOOD, C. D. (1958). Behavioral changes following discrete lesions of temporal lobe structures. *Neurology, Minneap.*, **8**, 215–220.

Part Four

Behaviour of Birds

Chapter 16

The Behaviour of Chickens

A. M. Guhl and Gloria J. Fischer

The origin of the common domestic fowl is lost in antiquity. Four species have been recognized (Beebe, 1926). The Red Jungle Fowl, *Gallus gallus* Linnæus, sometimes referred to as *G. ferrugineus*, is widely distributed from north central and eastern India through Siam, Cochin China, Malay peninsula to Sumatra; and is considered by Beebe to be the only species from which the domestic fowl arose. A grey species, *G. sonneratii* Temminick, is found in western, central, and southern India; a red and yellow *G. lafayetii* Lesson, inhabits Ceylon; and *G. varius* Shaw, called either the green or black species, is found in Java and adjacent islands. There is evidence of a polyphyletic origin, and Hutt (1949) suggests that those accepting the monophyletic theory should stick to *G. gallus*, whereas the familiar *G. domesticus* is probably permissible under the assumption that all domestic fowls were produced by hybridization from several wild species and have become distinct from any of them. The numerous breeds are classified according to place of origin, such as Asiatic, Mediterranean, English, and American breeds. At the time when poultry breeding was dominated by fanciers for exhibitions (Brown, 1929; A.S.P., 1953), colours, feather patterns, comb characteristics, and range of body weights were of utmost concern. The attention then shifted to inbred lines and is now directed towards crosses and the so-called hybrids.

Among domestic animals chickens are unique as to the conditions under which they are maintained, and this is reflected in the type of behaviour which has received the most attention. Unlike other domestic species, chickens under modern management are kept indoors in compact groups or isolated in laying cages. Incubation and broodiness (parental behaviour) are performed by mechanical devices, and genetic selection has been exercised against these traits.

Chickens differ from other domestic animals in several other respects. Feeding behaviour is simplified by means of the ever-present feed trough containing a single balanced ration, mash or pellets. Scratching in the litter is a displacement activity. Unlike the dog or cat, elimination is a simple process including the passage of urine through the cloaca. Wet droppings may be voided without regard to place or other concurrent activity, such as feeding. Furthermore, the relations between chickens and man are not as close as with dairy cattle, horses, or pets. As a result behaviour traits have not received as much attention during recent years as they deserve, and this chapter will emphasize behaviour concepts and their interrelationships.

I. SENSORY CAPACITIES

Although there have been a number of studies on sensory and perceptual processes, precise information on the range and sensitivity of the sense organs is still lacking. Progress in certain aspects of behavioural research has been correspondingly limited. What is known about avian sensory physiology can be found in Sturkie (1965). Vision and hearing are the most acute senses in birds. The relative importance of vision is reflected in the gross morphology of the brain (Portman & Stingelin, 1961). Optic lobes are conspicuous, whereas olfactory lobes vary in size, with those of the chicken moderate among all birds compared. Vision and hearing play a major role in the life of free-living fowl, and the less acute chemical senses apparently do not delimit the wide variety of food that they consume. Red Jungle fowl are wary at the approach of man and predators and give alarm calls to which the flock responds. These birds are omnivorous, accepting a great variety of seeds, fruits, and insects (Collias & Collias, 1967). Similar observations were made on domestic fowl that have been feral for many generations on an island off the north-east coast of Australia (McBride, 1967).

Vision

The chick's eye responds to light as early as the 17th day of incubation (Kuo, 1967), consistent with rod and cone differentiation (Peters *et al.*, 1958). The Purkinje shift is present (e.g. Armington & Frederick, 1956), with the range and maxima of spectral sensitivity similar to man's except for a slight displacement away from the blue end. Such a displacement is probably related to the presence of only red, orange and yellow oil droplets in chick cones. Their many more cones, as compared to man (Pumphrey, 1948), probably relates to the greater importance of colour vision in chick behaviour. Unfortunately, neither following nor pecking preference studies have yielded consistent findings with respect to colour preferences, but there may be relatively greater following of colours in the red-orange area (e.g. Gray, 1961) as compared to pecking preference for colours in the orange-yellow area (e.g. Gunther & Jones, 1962).

The structure of the eyeball of a bird differs principally from that of a mammal in that it has a pecten, which projects into the vitreous humour near the optic nerve. Various functions have been ascribed to the pecten, one of which is an aid in the perception of moving objects (Pumphrey, 1948, 1961a; Kare, 1965). In addition to the upper and lower eyelids which close in sleep, the chicken has a nictitating membrane for blinking. There is little eye movement, but the position of the eyes in the orbits gives the fowl a visual field of approximately 300°, with a possible binocular field of about 26° (Benner, 1938). Distance vision is normal or slightly far sighted, and acuity is very good (Kare, 1965).

Behavioural data indicate innate perception of depth (Walk & Gibson, 1961), though use of shadow is an important acquired cue (Hess, 1950). Behavioural data also support perception of differences in size, shape, pattern complexity, and colour. Newly hatched chicks, for example, prefer to peck at round over angular objects and solid over flat ones (Fantz, 1957; 1958). Also, like young rats, they prefer more certain (less random) patterns (Karmel, 1966).

Hearing

Except for the lack of a pinna or ear lobe, the ears and semicircular canals of the fowl are similar to those of mammals. Although the cochlea resembles that of man, it is straight in birds and that of *Gallus* is shorter than in many birds. Unlike most mammals, a sense organ, the *lagena*, is found at the apex of the cochlea and beyond the end of the basilar membrane. This structure is a carry-over from reptiles and is believed to be responsive to low frequencies (Pumphrey, 1948, 1961b). The frequency range of hearing in the domestic fowl may be narrower than in higher mammals. Von Bekesy (1960) found various portions of the chick's basilar membrane to vibrate to sound frequencies between 100 and 2,800 c.p.s. Though this method yields a reliable lower limit, an upper limit cannot be determined, because vibrations are too small to be measured (von Bekesy, 1960). Thus behavioural studies are needed to estimate the upper limit of the chick's frequency range. Unfortunately vocalizations do not contribute to our knowledge of this limit, because vocalizations fall within the same relatively narrow range as the one determined by von Bekesy. For example, natural sounds that baby chicks follow, e.g. maternal broody calls, do not extend much below 250 c.p.s.; and sounds that attract hens, e.g. chick distress calls, do not extend above 3,000 c.p.s. (Collias & Joos, 1953). Behavioural data are available on sound preferences. Baby chicks are attracted by maternal calls and repetitive tapping (Collias & Joos, 1953). More specifically, they approach and follow repetitive pure tones (emanating from a surrogate model) within a rather narrow range of about 350 to 700 c.p.s. (Fischer, 1969a). Chicks follow within a wider range of sound intensity. In fact the range is comparable to what is tolerable to man, i.e. just audible through extremely loud (Fischer & Gilman, 1967).

Taste

Taste buds in chickens vary from about 8 in the day-old chick to 24 in the 3-month-old cockerel and are found on the base of the tongue and floor of the pharynx. They resemble those of mammals, but taste recognition differs fundamentally (Lindenmaier & Kare, 1959). Unlike mammals, for example, chicks are indifferent to the common sugars (Kare & Medway, 1959) and evidence a wider range of tolerance for acidity and alkalinity (Fuerst & Kare, 1962). They do, however, discriminate among carbohydrates and are sensitive to salt. Only up to about a 0·9 per cent solution will be accepted, with chicks dying of thirst before drinking a toxic 2 per cent salt solution (Pick & Kare, 1962). In general, sensitivity to flavours of additives is greater when in water than in feed (Kare & Pick, 1960), suggesting aqueous solutions as more appropriate for investigations on taste. Because the domestic fowl will drink water down to freezing temperatures, but will reject it if above the body temperature (Kare, 1965), this factor is also a variable for consideration. Even with aqueous solutions to choose from, chicks may not correct for nutritive deficiencies. Calcium-deprived chicks, for example, increase their preference for calcium carbonate; but consumption of the corrective nutrient appears to be learned and related to the palatability of the nutrient offered (Wood-Gush & Kare, 1966).

Sense of Smell

Precise information on this sense is lacking. Birds do have an olfactory epithelium but the sense of smell is considered to be poorly developed. The literature on olfaction is largely anecdotal, and most of the positive evidence has been questioned. When electric cautery destroyed the olfactory bulbs in chicks, and they were then exposed to various compounds odorous to man, no change in preference behaviour was noted. Normal chicks failed to show a preference between waterers when one contained odorous material (Kare, 1965).

Tactile Sensitivity

Although the manipulation of grain and grit with the beak and tongue indicate that food preferences may be guided in this manner, the tactile sense has not been investigated. There is some evidence of chick preference for smooth over rough texture in their living quarters (Taylor et al., 1967). Observations on preening suggest a tactile sense associated with the feathers, because foreign matter adhering to them is readily located, and picking and preening are evoked. Also, particles of sand and litter are worked between and beneath the feathers when dusting and are shaken off on arising. A tactile sense may initiate and guide this behaviour. Contact with eggs, especially by an incubating hen, may involve not only the sense of touch, but also of temperature, and merits consideration in studies of incubation. During brooding, the chicks provide contactual stimuli in addition to auditory and visual signals. Tactile stimulation is almost certainly involved in the development and maintenance of broodiness. Hens that had no physical contact with chicks, or had contact when anaesthetized in the breast region, did not become broody, and hens that had contact with chicks and developed broodiness, lost it when contact was discontinued (Maier, 1963).

Temperature Sensitivity

Corpuscles of Merkel are reported in the skin, buccal cavity, and tongue of birds (Portman, 1961). Ability to detect radiant heat is evident in "sunbathing" and preening in the sun. Infra-red and photo-flood lamps evoke the same response, and the latter can be a problem when photographing behaviour for motion pictures because the ensuing preening terminates other types of behaviour. Cold reduces activity, as indicated by the huddling of chicks and adults and by the reduced following in imprinting that occurs if chicks are first adapted to a normal room temperature (Salzen & Tomlin, 1963). Subjective or experienced cold, on the other hand, seems to stimulate activity. For example, chicks that had been adapted to a warm temperature showed increased following when imprinted at colder temperatures (Fischer, 1968a). Similarly, chicks were found to run faster to heat in an end box from the colder of two start boxes (Zolman & Martin, 1967).

Other Senses

The interrelation between proprioceptors in the skeletal musculature of the neck, body, wings, legs, and tail and certain reflexes involved in righting and flight were determined in the chicken by Kleitman & Koppanyi (1926).

Some of these reflexes are described for the turkey in Chapter 17. Although impulses from the semicircular canals activate reflexes of the head, as shown by their abolition with bilateral labyrinthectomy, the intact bird can maintain a normal head position only with the aid of normal visual impulses.

II. INGESTIVE BEHAVIOUR

Wood-Gush (1955) summarized several studies on food preferences, but these used natural, rather than ground balanced rations, and appetite, hunger and learning were inadequately controlled. More is known about factors influencing feeding. For example, eating is directly facilitated by social stimulation and indirectly by sounds associated with social stimulation and/or feeding, such as pecking-like tap sounds (Tolman, 1963). Unlearned visual preferences in feeding behaviour may lead to a learned drinking response (Hunt & Smith, 1967). The unlearned visual pecking preferences are clearly modifiable through early experience. Preference for circular forms and for orange-yellow hues, for example, were modified by early experience pecking at other forms or colours (Fantz, 1957; Capretta, 1961).

Feeding rights. In groups of chickens precedence at the feed trough is associated with rank in the peck-order, and the importance of this factor decreases with the progressive integration of the flock (Guhl & Allee, 1944). Hens at lowest levels might be at the point of starvation if inadequately assimilated into the group, and therefore nutritionists usually isolate birds to eliminate this factor and to promote a more uniform intake.

Social facilitation. One or more birds feeding will stimulate others to join. A dominant individual, after having fed, may return when its inferiors begin active feeding, thereby increasing its consumption and reducing that of those in lowest rank, which avoid it. Social facilitation also occurs among birds in adjoining wire cages.

Feeder space. Availability of feed is more important than the amount of feed present (Grosslight et al., 1966). Dominance relations as well as degree of hunger influence the number of birds feeding at a given time. Poultrymen vary somewhat in their recommendations, but about 5 linear inches per bird in a pen is considered adequate for laying hens. For chicks the ratio increases in relation to growth.

Form of feed. Balanced rations may be fed as mash, crumbles, or pellets. Well-fed birds may scatter mash with their beaks as they search for some choice morsel, and in doing so, feed may be wasted and the actual intake may be out of balance. Both of these problems are solved by offering the mixture as crumbles or the larger pellets. However, hens can fill their crops readily with pellets and then may have too much "leisure" and as a result engage in activities which create social problems, e.g. feather pulling. Comparative studies of behaviour as related to mash or pellet feeding are suggested.

III. COMMUNICATION

Communication between animals involves the giving by one individual of some sign or signal that on being perceived by another, influences its behaviour (Frings & Frings, 1964; Marler, 1967). In chickens these stimuli are essentially visual and auditory (see Sense Organs and Perception). Among the non-vocal signs only the most obvious postures and movements have received attention, e.g. raised hackles denote aggressive intentions whereas a low crouch is a submissive response, and the waltz plus wing-flutter is a displacement reaction (see Sexual Patterns). More subtle and definitive signs are common interactions in a well-integrated flock. Experimental analysis of these and other possible signals in social behaviour is needed, and application of an information theoretical model has been generally advocated (e.g. Marler, 1961; Sebeok, 1965). The most specific approaches to experimental analysis and its associated problems have been outlined in a notable paper on communicative functions of animal signals (Marler, 1967).

Vocal Communication

The syrinx in domestic fowl is less developed than in passerine birds, and vocalizations are correspondingly less complex. In domestic fowl, in fact, vocalizations are all of the simple call note variety (Lanyon, 1960). Schjelderup-Ebbe (1913) and Collias & Joos (1953) discuss a number of these sounds, and the method of spectrographic analysis has contributed importantly to their description (e.g. Collias & Joos, 1953). Plates XVII and XVIII summarize the various calls for which spectrographic data exist. The seven categories of calls that chicks make include four basic ones shown in the upper part of Plate XVII. These four represent a sequence that involves progressively louder, longer and higher pitched sounds, suggesting a single controlling system (Andrew, 1963). As for their significance, twitters (pleasure calls) seem to be a response to moderately intense sensory stimulation. Peeps (distress calls) seem rather to be a response to withdrawal of normal stimulation (e.g. sudden isolation or removal to a cold environ). Trills are a startle-like response to sudden, intense stimulation; shrieks, a response to traumatically intense stimulation, such as electric shock or capture (Andrew, 1963).

The ontogeny of vocalization in domestic fowl has received little experimental attention. Simner (1966) found some support for the conditioning of vocal activity in embryos to tactual components of the heartbeat. It seems likely too, that at least some adult calls develop from chick calls, e.g. the maternal cluck from the chick twitter (see Plate XVII). Andrew (1963) suggested that modifications of chick calls into adult calls may be mediated by testosterone. Other adult calls may not develop from chick calls. One example, the crow, develops from hormonal stimulation (e.g. Hamilton, 1938; Andrew, 1963). Because it develops from hormonal stimulation, the crow is shown in Plate XVIII as a sexual call, but it also has territorial and warning significance (Collias & Joos, 1953). A crow may function too as a means of recognition, since a given cock crows much the same crow, whereas crows vary greatly between cocks (Marler et al., 1962). Crow duration differences between breeds have been found to be highly heritable (P. B. Siegel et al., 1965).

Vocalization as an Index of Emotion

Vocalizations could be more widely used in behavioural research. Hess (1959) demonstrated a positive relation between age and incidence of distress calls, suggesting use of the latter as a measure of fear. Similarly, distress calls might inversely measure imprinting strength, i.e. incidence of distress might vary with absence from the imprinted-to stimulus. Pleasure calls were shown to accompany self-stimulation of the medial forebrain, suggesting a so-called pleasure centre in aves, as well as in mammal (Andrew, 1967). Electrical stimulation of the brain-stem produced crowing in cocks (von Holst & von St. Paul, 1960). It would be of interest to know whether or not cocks would self-stimulate this area or avoid it. Both pleasure and distress calls are initially unconditioned, but have been shown to be under operant control as well, e.g. Bantam chicks learned to emit calls in order to obtain food (Lane, 1960). These speculations and findings point to a general relation between chick calls (if not calls more generally) and emotionality. They also suggest methods by which calls might measure behaviours having emotional components, e.g. avoidance, imprinting, and other forms of early learning.

IV. LEARNING AND RETENTION
A. Learning

Assessment of learning ability in chickens has not been nearly so extensive or systematic as in mammalian forms such as the rat, monkey or man. Two trends are apparent in the research that has been done. One is comparative assessment of chick ability relative to other avian species and to mammalian forms. The measurement most often used has been one of the most promising for comparative purposes, that is, the rate of improvement in solving successive reversals of a spatial or visual discrimination. A second trend is assessment of developmental factors affecting learning. Age has been related mostly to the ease of learning various types of avoidance problems. Avoidance learning is especially appropriate for developmental studies, since the ontogeny of avoidance behaviour seems largely maturational (Schaller & Emlen, 1962).

Discrimination reversal learning. Young and adult chickens improve on successive reversals of both spatial (e.g. Fischer, 1966b; Warren *et al.*, 1960) and visual discriminations (e.g. Bacon *et al.*, 1962; Gossette, 1967). Asymptotic performance in such studies averages about 70 per cent choice accuracy. This does not compare unfavourably with rat performance, but does compare unfavourably with the one-trial reversal learning (90 per cent accuracy) that has been demonstrated in monkeys (see Plotnik & Tallarico, 1966). Rate of improvement may be a more sensitive measure of species differences than accuracy at asymptote. Initial performance on successive spatial reversals, for example, suggested avian differences (e.g. chick inferiority to magpie and parrot), whereas asymptotic performance was comparable for the several species tested (Gossette *et al.*, 1966).

When chicks were overtrained either on a spatial discrimination, or a spatial discrimination with an additional cue of brightness, reversal learning was retarded (Brookshire *et al.*, 1961; Warren *et al.*, 1960). In contrast, overtraining facilitated a spatial plus brightness discrimination in rats. Mackintosh (1965) demonstrated that chick stereotypy contributed to this difference, but unequal magnitude of the reward given chicks and rats also may have been a factor.

Other forms of learning. Chickens are inferior to cats in escape conditioning (Krieckhaus & Wagman, 1967). Still, chicks do readily learn both active (escape) (James & Binks, 1963; Gray *et al.*, 1967) and passive conditioned avoidance responses (e.g. Fischer & Campbell, 1964). Passive avoidance (e.g. inhibition of an approach response to avoid shock) is more difficult to learn than active avoidance. This may account for the slowness with which chickens solve detour problems (see Ball & Warren, 1960; Scholes, 1965).

Chicks can learn to alternate turning responses in a temporal maze (Gunther & Jones, 1961), which is a fairly difficult task for lower mammals. Chicks also may be capable of latent learning (Vitulli & Tallarico, 1967), although learning by observing others learn has not been demonstrated (Sexton & Fitch, 1967).

To summarize the results on learning in chickens: the few comparative studies that have been done sometimes show learning as good as that found for lower mammals; sometimes not. Where chickens appear to be inferior to lower mammals, greater response stereotypy may be a contributing factor, with multiple cues necessary to overcome it. Measures apparently sensitive to phyletic differences, and inter-avian differences as well, are rate of improvement and asymptotic accuracy on successive discrimination reversals.

Developmental factors. Escape conditioning can be demonstrated on the first day of life, though it is much easier on the second day (Gray *et al.*, 1967). Passive avoidance conditioning, however, is difficult to demonstrate prior to the third or fourth day of life (Peters & Isaacson, 1963; Fischer & Campbell, 1964). This finding probably reflects developing avoidance responses, especially inhibitory control, rather than developing ability to learn. For example, avoidance responses can be observed on the first day of life, but they do not peck until the second or third day (Phillips & Siegel, 1966; Schaller & Emlen, 1962). Also, younger chicks, though unable to learn to inhibit an approach response to avoid shock, nevertheless distress on approaching the shock area (Peters & Isaacson, 1963; Fischer & Campbell, 1964).

Warren *et al.* (1960) found improvement with age up to 2 months in spatial reversal learning. Solution of detour problems (Ball & Warren, 1960) and maze learning (Vitulli & Tallarico, 1967), however, improved only up to 2 to 3 weeks of age. Such findings suggest improvement in learning with age up to 2 weeks to 2 months, depending on task difficulty. Yet the findings of one study contradict this conclusion. Instead of improvement on detour problems up to 2 to 3 weeks of age (Ball & Warren, 1960), Scholes & Wheaton (1966) found ability to solve a detour problem decreased sharply at about 2 weeks of age. The basis for such a discrepancy in results is not apparent.

It seems reasonable to conclude that chick learning ability, especially when involving inhibitory or avoidance behaviour, is influenced by developmental factors. Effects appear to decrease with age up to 2 weeks to 2 months, depending upon the nature of the task or, perhaps, upon task difficulty.

B. Retention

Difficulty with short-term memory is suggested by the finding that hens were unable to associate a discriminative cue for non-reward with a previously rewarded stimulus when the cues were separated by a 10-seconds interval. This compares unfavourably with mammals (e.g. dogs and baboons) and even a marsupial (rabbit) (Voronin, 1962). Hens' long-term memory also may be inferior to that of lower mammals. After only 2 weeks in isolation, for example, hens were no longer able to recognize members of their flock (Guhl, 1953). After 6 days, on the other hand, chicks still remembered the solution of a detour problem (Scholes, 1965).

More research on retention is needed, and it is especially important that such work replicate findings in rat and man that have related the memory process to time (e.g. Lee-Teng & Sherman, 1966) and to underlying molecular changes (e.g. Byrne, 1968).

V. IMPRINTING

Imprinting refers to the formation of primary social bonds (e.g. maternal and sexual) during so-called "critical periods" of development (Lorenz, 1935). Two processes may be involved. One is the activation of an approach response, which may be a function of arousal (i.e. CNS stimulation). The second is a selective attachment of the approach response to the stimulus that elicited it, which appears to be a function of learning (Fischer, 1966a). Tinbergen (1948) cites as an example of such learning how a chick initially follows any hen that crosses its path, but after a time, follows only its mother. The strength of the bond, once formed, is indicated by the intimacy of the attachment, its extreme resistance to extinction and replacement by another stimulus (Steven, 1955), and by effects the attachment has on the direction of subsequent social (e.g. sexual) behaviour (see Early Experience).

A. The Critical Period

Normally the critical period for maternal imprinting begins soon after hatching and lasts for about 3 days (Spalding, 1873; Jaynes, 1957). Both lower, and especially upper, limits may be extended, however. A lower limit is imposed by the minimum sensorimotor development necessary for an organism to respond to an imprinting stimulus. For example, chicks approach and follow a moving object just as soon as they are able to locomote (1 to 4 hours from hatching (Hess, 1959)). Yet imprinting may occur sooner, possibly even prior to hatching, since sensory systems are responsive as early as the 17th day of incubation to light and even earlier to sound (Kuo, 1967). In fact, prenatal auditory imprinting has been reported (Grier et al., 1967), but the finding needs to be confirmed.

The upper limit of the critical period is longer than the 2 to 3 days suggested by early work. Jaynes (1957), for example, found little imprinting in 3-day-old chicks, but they had been reared communally and probably imprinted to each other (see Guiton, 1958; Polt & Hess, 1964). If kept isolated, on the other hand, chicks imprint well on their 3rd day (Guiton, 1958) and imprint, though to a lesser degree, as late as their 7th day (James, 1960). So the critical period may be as long as 10 days to 2 weeks, that is, 3 to 4 days pre-hatch to 7 to 10 days post-hatch. The upper limit can be extended slightly by social facilitation, for example, imprinting an older chick with other chicks that are already imprinted (Ramsey & Hess, 1954). A more general method, perhaps, is to reduce fear. This is suggested by extension of the critical period through the use of tranquillizing drugs (Hess, 1960) and through habituation of fear and avoidance responses (Steven, 1955).

B. Arousal and Learning Processes

Psychologists, especially in America, have emphasized the role of arousal in imprinting, and thus an approach response (e.g. following) to measure it. In contrast, ethologists have emphasized that imprinting could be assumed only when learned preference for the imprinting stimulus could be demonstrated. The psychologists' assumption that following measures imprinting is not altogether wrong. If animals do not follow, for example, they subsequently show indifference to the imprinting stimulus. The opposite assumption, however, is not warranted. Animals that follow, even strongly, are by no means necessarily imprinted. In fact, a substantial number of strong followers may subsequently follow a strange model as much as they follow the model they were imprinted to (Fischer, 1967). Therefore, following can be regarded as a sufficient criterion only of the approach component or arousal process in imprinting. One implication is that the two processes, arousal and learning, need to be assessed independently, since factors that influence one may not influence the other. Thus a factor known to affect learning, such as distribution of practice, might be expected to affect the learning process in imprinting, but not necessarily the arousal process (see Fischer, 1966c).

Arousal and Following

Arousal stimuli. Pitz & Ross (1961) first viewed imprinting as a function of arousal, that is, the total amount of stimulation impinging on the organism. They clearly substantiated stronger following in Vantress chicks when a "clapper" was sounded each time the chick was within 6 inches of a moving model. Even painful stimulation can increase following. When intense shock was administered to Vantress chicks through a wire extending from the imprinting model to electrodes attached to the chick's wings, a resulting increase in activity was found to be directed toward the model (Kovach & Hess, 1963). Increased environmental stimulus (e.g. cold) intensity may also function as an arousal stimulus. Chicks whose ambient and imprinting room temperatures were warmer-to-colder, respectively, followed much more during imprinting than chicks did whose ambient and imprinting room temperatures were the same or colder-to-warmer (Fischer, 1968a).

In addition to the indicated external sources of arousal there are internal sources that appear to enhance following. In day-old chicks magnitude of following has been related to level of activity, or reactivity to stimulation, which in turn may be related to level of arousal (Fischer, 1967). Several CNS stimulant drugs were also found to increase following in day-old chicks (Kovach, 1964). In older chicks, on the other hand, these stimulants depressed following. It was suggested that the more mature organism has a higher level of arousal (see also Tolman, 1963), and additional CNS stimulation becomes excessive. Regardless, both external and internal arousal seems to enhance following.

Stimulus intensity. Fischer & Gilman (1967) found the magnitude of auditory stimulus intensity to relate to following in an approximately normal manner, with greater following to normal intensities than to either lower or higher ones. The relation then, between magnitude of eliciting stimulus intensity and following, may follow the inverted U function that has been found to relate drive and behaviour. Such a relation would be consistent with a broadly based developmental theory of approach/withdrawal processes (Schneirla, 1959, 1965). This theory views stimulus intensity as having arousal and orienting properties in the newborn. Low intensities up to an optimal level arousal parasympathetic (A) processes, and the neonatal responses are orientation and approach. High intensities on the other hand, arouse sympathetic (W) processes, and responses are disturbance and avoidance. As applied to imprinting, A/W theory predicts increased following with increased intensity up to an optimum, but withdrawal at high intensities. The Fischer & Gilman (1967) results, cited previously, were supportive of A/W theory, but there was no negative slope to the curve of following, as would be expected from predicted withdrawal responses at very high intensities. It was suggested, therefore, that active avoidance may be a function of sudden onset, rather than high stimulus intensity *per se.*

Learning and Imprinting-Stimulus Preference

Problems of measurement. Two criteria for learning in imprinting are: preference for the imprinting stimulus, and demonstration that preference was acquired via the imprinting experience. The first criterion has been measured almost exclusively by stimulus preference tests, where the imprinting stimulus and another stimulus are presented either simultaneously (e.g. Polt & Hess, 1964) or successively (e.g. Moltz *et al.*, 1960). Preference is measured then by choice of stimulus or by time in proximity to the respective stimuli. Unfortunately, the validity of both these methods can be questioned (Klopfer, 1965; Fischer, 1967). The simultaneous preference test, for example, has been found to yield variable results, depending upon whether tests were made with stationary or moving models (Klopfer, 1965). Further, this method is unlikely to be sensitive to stimulus generalization and thus may overestimate preference for the imprinting stimulus. The successive method probably does not have this problem, since it produces substantial generalization (Fischer, 1966c). However, a major problem with the successive method is that the second stimulus presented is followed more than it would have been followed if it had been presented first (Fischer *et al.*, 1965; unpublished data from Fischer & Gilman, 1967). This latter finding suggests a sensitization-like effect that would produce too conservative a preference measure, when order of presentation was imprinting-stimulus first. Clearly, these traditional methods of measuring preference for the imprinting stimulus, subsequent to the imprinting experience, need to be modified and compared directly in order to assess their relative adequacy. This means that inferences about the learning process in imprinting, from results of studies that have used stimulus preference tests, must be made with caution.

The second criterion for learning in imprinting is that preference for the imprinting stimulus must have been acquired via the imprinting experience. This citerion has been measured by one of two control methods. Traditionally, results of stimulus preference tests on an imprinted group have been compared with results of such tests on a control group that had not been imprinted. This control method is almost certainly invalid, because Ss that have not been imprinted show little if any following on subsequent preference tests. An alternative method, introduced in research on the primacy effect in imprinting (Hess, 1959), has the important feature of eliciting following in all Ss. One-half of them are imprinted to stimulus A; one-half to stimulus B, then all are given a stimulus preference test with both A and B present.

The "primacy" method is excellent, but its adequacy clearly depends on the use of equally effective stimuli A and B, as well as on an adequate stimulus preference test. The latter assumption has been questioned, but the need for more knowledge about effective stimuli is an even greater problem. One reason is that so few studies have included parametric analysis of imprinting stimuli. Thus far, these have been restricted to colour (Schaeffer & Hess, 1959; Gray, 1961; Smith & Bird, 1964) and sound intensity and frequency (Fischer & Gilman, 1967; Fischer, 1969a). These studies suggest that a colour combination such as red and orange, repetitive taps at about 5 through 25 dB of signal above ambient noise and repetitive pure tones between about 350 and 700 cps would be equally effective stimuli. This is paucious knowledge indeed.

In an important paper Klopfer (1967) has shown that equally effective stimuli may still not qualify as imprinting stimuli. Whereas ducklings imprinted to red or yellow decoys followed appropriately on simultaneous re-test, Ss imprinted to horizontal or vertically striped decoys followed either stimulus on re-test. One obvious possibility is that equally effective stimuli may be equally effective because they are not discriminated. Even if discriminated, they still may not be useful as imprinting stimuli if there is an unlearned or previously learned preference for one of them. For example, chicks followed repetitive taps or maternal cackles equally well, but preferred taps on successive-preference test, regardless of which sound they had been imprinted to (Fischer, unpublished data).

The learning process. Critical research on the learning process in imprinting cannot be done until measurement problems are overcome. Still, data from studies, especially those using the primacy method combined with a test of stimulus preference, are relevant, and some tentative inferences may be warranted. One is an apparent overestimation of the importance of a single brief experience in learning to recognize the imprinting stimulus. This is suggested by the finding that the number of chicks following an imprinting stimulus more than they follow a different stimulus increased with successive days of additional imprinting sessions. After the first day's 20-minute session, for example, just a bare majority of chicks preferred the imprinting stimulus, whereas by day 2 and thereafter, all the chicks preferred it, by increasingly greater margins (Jaynes, 1958). Similarly, chicks given a brief imprinting experience did not show sexual preference for the imprinting stimulus at maturity (Guiton, 1961). When reared with, and thus repeatedly exposed to, an imprinting stimulus, on the other hand, turkeys preferred the imprinting stimulus to a biologically appropriate releaser (Schein & Hale, 1959).

Learning to recognize the imprinting stimulus from a brief early experience was one of the characteristics of imprinting thought to distinguish it from "ordinary" associative learning (Lorenz, 1935; Hess, 1964). It does not, and other supposedly distinguishing characteristics have also been challenged (e.g. Fischer, 1966c). Some investigators (e.g. Hess, 1964; Schneirla, 1965) still regard imprinting as different, however, and Schneirla emphasizes the fact that imprinting occurs only during a critical period in infancy, whereas learning more generally can occur at any time in the life of the organism. But if one accepts the distinction between arousal and learning components of imprinting, then it is really only the arousal component that is so restricted. Learning to discriminate an imprinting stimulus from another stimulus by differential reinforcement could undoubtedly occur then, and at any other time. In any case, most investigators (e.g. Hinde, 1962) presently view imprinting as an example of early learning.

VI. DEVELOPMENT OF SOCIAL BEHAVIOUR IN CHICKS

The development of behaviour involves the maturation of the nervous and endocrine systems and indirectly of other systems. Conditioning and learning, or experience, are continuous processes. Scott (1963) has outlined steps in the development of behaviour that depend upon pre-existing levels of organization from the genetic pool, through prenatal environment, postnatal environment, to the biotic environment. Kuo (1967) takes a different view and develops the concepts of behavioural gradients and behavioural potentials as the major postulates of an epigenetic approach to behaviour. Kruijt (1964) in a study of the development of behaviour in Junglefowl concludes that there are no great differences between domesticated and wild *Gallus gallus*.

A. Socialization

The calls described in an earlier section (see Communication) play an important role in the early socialization of the chick, especially in relation to the hen. Distress calls by the chick stimulate clucking, which in turn causes the chick to emit pleasure notes. These social relationships originate before hatching when the egg is pipped, and continue for several days post-hatch. In response to the mother hen's warning note the chicks remain silent. Positive responses are given to the hen's purring sound when she settles down to brood, and to sources of warmth and to moving objects. Thus the broody hen offers the chick a complex of attractive stimuli. The repetition of these exposures, as well as food guidance and protection, strengthen the family bond; but during the first 10 days the chick's responsiveness to these stimuli begins to wane (Collias, 1950b). It has been shown experimentally that some behaviours, such as feeding (e.g. Tolman, 1964), are enhanced by the presence of other chicks. This phenomenon has been termed social facilitation. Furthermore, chicks become imprinted or conditioned to the mother hen and to one another, and as a result, cease to follow unfamiliar objects. This occurs before avoidance behaviour is sufficiently well developed to interefere with socialization (Guiton, 1959).

B. Developmental Sequence of Behaviour

There is much variability among individuals in the age at which certain behaviour patterns appear. In a given set of eggs in an incubator there may be many hours difference in time of hatching, and some chicks require much more time than others to emerge free from the shell after pipping. Nevertheless there is a general sequence in which behaviour patterns appear, such as escape or fear reactions, frolicking, sparring, aggressive pecking, avoidance, and fighting.

Phillips & Siegel (1966) found that responses to loud noises increased to a peak at about 2–3 days post-hatching and then dropped to a fairly high plateau. Their results are discussed in respect to Salzen's (1962) hypothesis that fear is present from hatching or before, and that the increase is a product of experience rather than of maturation. There is the question of whether the concepts of "escape" and "fear" are unitary drive systems or whether withdrawal and fear are independent systems with mutual inhibitory relations (Hogan, 1965). However, as noted previously, chicks subjected to an electric grid do not appear to be capable of learning a passive avoidance response prior to 3 days of age. Maturational changes are involved that may be related to the development of inhibitory control (Fischer & Campbell, 1964).

Frolicking may be viewed as an incipient agonistic interaction because it leads to sparring. When chicks spar, the stance and subsequent movements resemble fighting but no pecks are actually delivered. With the addition of pecking and a refined foot action suggesting the use of spurs which are still undeveloped, true fighting emerges. Such interactions are repeated without any decision as to dominance. Not until avoidance or submissive behaviour appear are peck-rights established. This stage is reached at about 6 weeks of age between some individuals, and not until 10 weeks or later for others. Capons begin to establish peck-rights much later than controls and require more time to form a peck order. Males show aggressive behaviour earlier than females, and there is a disproportionate pecking frequency between the sexes in a group containing males and females. During weeks 10 to 15, the peck order divides gradually into two unisexual social organizations (Guhl, 1958). The development of various behaviour patterns from hatching to 70 days of age were observed in more detail by Dawson & Siegel (1967). Apparently domestication has not affected the sequence in which they appear, because descriptive observations of Junglefowl by Kruijt (1964) are similar to the above.

C. Gonadal Hormones and Chick Behaviour

The question now arises as to whether the emergence of pecking, fighting, and submission are limited to certain levels of hormone concentrations secreted from the developing gonads, or whether the timing is related to the

PLATE XVII

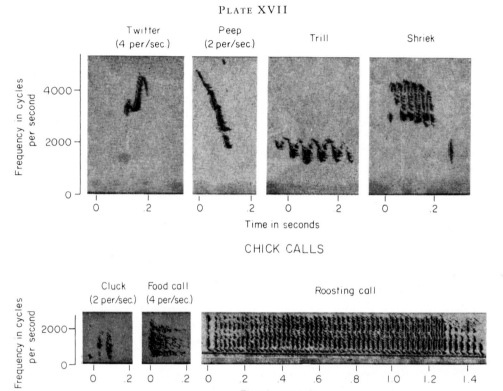

Acoustic analysis (spectrograms) of common calls by young chicks and broody hens. (Adapted from Collias & Joos, 1953.)

PLATE XVIII

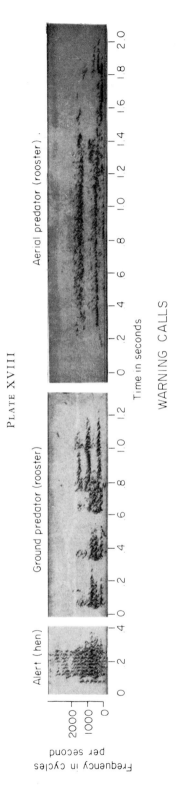

Alert (hen)   Ground predator (rooster)   Aerial predator (rooster)

WARNING CALLS

Time in seconds

Frequency in cycles per second

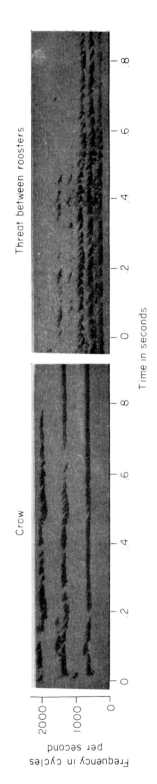

Crow   Threat between roosters

SEXUAL CALLS

Time in seconds

Frequency in cycles per second

Acoustic analysis (spectrograms) of common calls by adult chickens. (Adapted from Collias & Joos, 1953.)

development of neural mechanisms. Chicks of both sexes treated with either androgen or oestrogen did form a social order somewhat earlier than controls. However, the androgen-treated birds were aggressive, whereas those receiving oestrogen were highly submissive. The mean differences between treated and normal birds in the age at which peck orders formed was less than one week. This suggested that neural mechanisms were not developed, but the group-reared chicks had been adjusting to each other, a form of social inertia which interfered with the display of agonistic behaviour as it developed (Guhl, 1958).

It may be postulated that the various behaviour patterns influenced by gonadal hormones are present at hatching although poorly, if at all, organized, and that the sequence in development reflects progressively higher thresholds of response. Normally the production of sex hormones develops gradually and thereby lowers the response thresholds successively. Thus the behaviour patterns can be evoked, or organized, precociously by treatment with appropriate exogenous hormones and/or by the presentation of appropriate and intense stimuli. Experience may facilitate or inhibit the initial appearance of such behaviour. (See Hormones and Sexual Behaviour for developmental comparison.) Andrew (1963) investigated the neurophysiological interrelationships in several vocalizations by chicks treated with testosterone, and his discussion presents some concepts for the above hypothesis.

VII. AGONISTIC BEHAVIOUR IN ADULT CHICKENS

A. Behaviour Patterns and the Peck Order

Agonistic behaviour includes attack, escape, avoiding, and submissive behaviour. These patterns of activity vary in intensity and can be recognized by differences in posture and movement. Attacks include fighting, pecking and threatening. The negative reactions in social interaction may be differentiated as follows. In submission a bird bows or crouches and thus yields to its social superior. Avoidance often involves locomotion, and the individual tends to keep away from a superior, that is, withdraws and avoids social contact. In well-integrated flocks at low intensities of social interaction avoidance may be indicated by merely moving the head away from a flockmate. Escape is marked by flight and is the most intense negative reaction. In males crouching is absent, and escape is more common.

The peck order can be ascertained by recording agonistic behaviour as to actor and reactor in each possible combination in a flock (see Chapter 5). Since males are more aggressive than females and are passive in intersexual domination, a mixed flock has two peck orders, each unisexual. It is unusual for a hen to attack a cock. Chickens may not recognize a peck order, *per se*, since tests in a discrimination cage show that discriminations are not related to ranks in the peck order.

B. Estimating Levels of Aggressiveness

Although a peck order suggests variability in levels of aggressiveness, it does not necessarily follow that ranks represent an accurate gradient for this trait. When a number of unacquainted birds are assembled, the ensuing agonistic encounters are random, often with the most aggressive birds initiating a series of attacks. Fatigue, and other factors, may cause aggressive birds to lose encounters that they would be expected to win. Within flocks of birds reared together there are individual differences in rates of development that yield an advantage to those maturing at a higher rate.

A technique for estimating levels of aggressiveness was devised by Allee et al. (1939) and Collias (1943), known as the initial paired contest method. It is based on the tendency of unacquainted chickens to establish dominance-subordinate relationships soon after they meet. When a number of birds are tested, the relative aggressiveness is indicated by the number of contests won. There are several procedures. In the caged contest method all birds to be paired are isolated in cages for 2 or 3 weeks to extinguish previous dominance relations prior to a round-robin of contests (Guhl, 1953; 1958). Another procedure is to stage contests between individuals of different flocks. If large numbers of birds are involved, a random sampling in the inter-pen method may be sufficient (Siegel, 1960). It is also possible to use panels or teams against which many birds may be paired (McBride, 1958).

There have been tests to determine the coefficients of correlation between ranks in the ontogenetic peck order and ranks in the number of initial paired contests won; also between ranks by paired encounters and subsequent peck order ranks after assembly. Resulting correlations have varied from non-significant to highly significant (Guhl, 1953). Subsequent experiments have considered both rankings, but with the initial paired contest method considered the more reliable.

Results of paired contests may be used experimentally to correlate various performances to relative aggressiveness, or for selective breeding to determine whether agonistic behaviour has any genetic basis.

C. Selection for Agonistic Behaviour

The extent to which characteristics of behaviour are inherited or learned have been questioned for some time (Chapter 3). The concept of imprinting is one in which genetics and conditioning or learning appear to merge. A heritable basis for the following response is probable. Though estimates of heritability were small from within-breed analyses (Fischer, 1969b; Smith & Templeton, 1966), heritability of following was substantial when a between-breeds analysis was used (Graves & Siegel, 1968).

In attempts to determine the inheritance of aggressiveness the impossibility of completely separating innate and conditioned behaviour must be recognized in devising tests for measuring aggressiveness. The domestication of chickens is historically closely related to cock-fighting (Wood-Gush, 1959) which suggests that early selection was for aggressiveness and related traits. In observing fighting, Fennell (1945) found that Games were shiftier, faster, and not as clumsy as domestic cocks; also that varieties of Games showed differences in their methods of attack. Breed differences in ability to win initial encounters were reported by Potter (1949). Inter-breed and inter-strain comparisons in aggressiveness were made between hens of two breeds and their crosses (Guhl, 1961c). Initial encounters were conducted between the breeds, crosses, and breeds and crosses. Ten New Hampshire hens won 216 out of 340 contests (63 per cent); 10 White Rock hens won 90 out of 324 (28 per cent); 8 WR × NH crosses won 22 out of 160 (14 per cent); and 16 NH × WR crosses won 216 out of 304 (71 per cent). Another series of contests was run between 18 caged hens of each of two strains of White Leghorns. Of 324 contests only 24 failed to establish dominance, and strain "H" won 289 and strain "G" only 11.

The evidence of breed differences in aggressiveness suggested a programme for selective breeding within breeds. Selection for mateships were based on the number of initial paired contests won or lost, that is, high and low lines of aggressiveness. Significant differences were found from the F_2 to the F_4 generations (Guhl et al., 1960). With larger flocks Komai et al. (1959) found that strain and breed differences were repeatable.

Observations of a large number of initial paired encounters suggested that aggressiveness and submissiveness may be independent factors, and that selection might involve the relative levels of each factor in a given bird. Therefore, it may be more accurate to assume that paired contests measured ability to establish social dominance, rather than aggressiveness *per se*. Selection was continued with each of two breeds and large intra-strain differences were found (Craig et al., 1965).

D. Hormonal and Neural Mechanisms

Androgen. Both psychological and physiological factors play a role in the attainment of social status. Some physiological factors are indicated by comb size. Cocks are more aggressive than hens and have larger combs, whereas capons and poulards have very small combs and have lost most of their pugnacity (Domm, 1939). The application of androgen directly to the comb's surface induces comb growth (Koch, 1937). Female birds normally secrete an androgenic substance (Juhn et al., 1932; Witschi & Miller, 1938). These early observations suggest that male hormone induces aggressive behaviour. Following these indications Allee et al. (1939) injected testosterone propionate into low-ranking hens, with resultant upward social mobility in the hierarchy, even to top rank for some hens. The data also suggested a positive relation between comb size (within strain) and ability to win initial encounters. The data of Guhl et al. (1960) appear to be in general agreement, but Hale's (1954) experiment showed that individual differences in comb size were not related to androgen level. Individual differences within breed in the effects of treatment may be influenced by experience, such as social inertia (Guhl, 1968), neural or genetic differences (Hale, 1954), or anti-androgens (Dorfman, 1960, 1962).

Oestrogen. Although oestrogen is more closely related to female sex behaviour, it may also influence agonistic behaviour. When Allee and Collias (1940) injected oestradiol into hens, there was some indication of lowered aggressiveness. However, when chicks of both sexes were treated with oestrogen, they were definitely more submissive than their controls and showed more instances of submission or avoidance than of domination

(Guhl, 1958). In another experiment hens receiving oestrogen not only maintained their social ranks, but also their ability to win initial paired contests (Guhl, 1968). The hormone appeared to induce a relaxed, rather than alert stance which, as a social signal, tended to induce a mild attack. Revolts against treated hens were typically initiated by controls, whereas most revolts against androgen-treated birds were usually initiated by andogen-treated hens. The effects of oestrogen on agonistic behaviour need further investigation.

Adrenal hormones. The injection of adrenaline into normal hens failed to alter social position or aggressiveness as measured by paired encounters (Allee & Collias, 1938). Adrenal hormones function under social stress. Males rotated into organized groups have heavier adrenals than controls (H. S. Seigel & Siegel, 1961). The general adaptation syndrome in chickens has been reviewed by von Faber (1964).

Neural mechanisms. Differences in agonistic behaviour between individuals, strains, and breeds suggest variation in neural mechanisms, especially when levels of pertinent hormones are considered. Regions of the brain involved may be elucidated by direct stimulation or by hormonal implants. Breed differences in effects or exogenous androgen on initial paired encounters were found by Allee *et al.* (1955). New Hampshire capons won 92 per cent of encounters with Barred Plymouth Rock capons when all were given equal doses of androgen (Hale, 1954). With capons from "high" and "low" aggressiveness strains treated equally, the "high" group maintained its greater amount of agonistic behaviour (Ortman, 1964). Agonistic behaviour was elicited by direct brain stimulation (von Holst & von St. Paul, 1960; Phillips, 1965). The implantation of an androgen ventral to the paleostriatum augmented aggressive behaviour (Barfield, 1965) and could be activated independently of a copulatory area (Barfield, 1964).

VIII. PSYCHOBIOLOGICAL FACTORS

A. Recognition of Individuals

The existence of a peck order is evidence that birds recognize one another, otherwise pecking would be promiscuous, and unidirectional pecking would not occur. Close observation of hens suggests that recognition is based on features of the head, because a subordinate bird may not avoid its superior until the latter's head is raised and becomes visible. In fact, if hens' combs are not visible or have been removed, the hens are attacked as strangers, even by inferiors (Guhl, 1953; Schjedelderup-Ebbe, 1922). Also, dubbed fowl attack each other more often than undubbed fowl do (James, 1873; Marks *et al.*, 1960; Siegel & Hurst, 1962).

A study of visual cues used in the recognition of pen-mates altered the contour and/or colour of various parts of the body and head. Contour was changed by either adding feathers (white or coloured) or by denuding certain areas which simulated changes occurring during moulting. Dummy combs of red flannel were sewed to some combs, or shapes were changed with the use of transparent adhesive tape and some colour changes on feathers were made with alcohol solutions of various dyes. Dominance relations were recorded, and the experiments were made after peck orders appeared to be stable. Because strangers are usually attacked, aggression towards a modified bird by its inferior was taken as evidence of a loss of recognition (Guhl & Ortman, 1953). It was concluded that disguised features of the head and neck are more effective in producing a loss of recognition than are those made on the trunk. Intense colours on white birds are more effective than shades or tints. No single feature is the sole means of recognition. There is some evidence that a large comb offers some psychological advantage in initial inter-pen contests, probably due to resemblance to a dominant bird in the home flock. The stance or deportment of an altered bird is a factor in evoking either aggression or submission in pen-mates.

B. Breed Recognition

Potter (1949) presented evidence that previous experience with a member of another breed may influence the outcome of an encounter with a stranger of that breed. Hens may distinguish between breeds which are dissimilar, but are unable to distinguish between individuals of such breeds (Douglis, 1948; Potter & Allee, 1953). However, hens in intermingled flocks of as many as 48 hens seemed to respond to members of other breeds on the basis of individual rather than breed recognition (Tindell & Craig, 1960).

The hypothesis that the type of agonistic behaviour between breeds in mixed flocks is based on breed recognition rather than individual recognition was tested by Hale (1957). A single experience with one individual of another breed was sufficient to determine subsequent response to other members of that breed in initial encounters. Hale suggested that the factor of breed recognition may mask breed differences in levels of aggressiveness when small flocks are used for comparative purposes. In small (less than 10 hens) multibreed flocks the members of one breed held dominance over the other breeds. Apparently flock size is important because Tindell & Craig 1960), with large flocks, did not find breed recognition an important factor. The question of early experience has not been considered experimentally as yet, i.e. would results differ from the above in small flocks if the chicks of different breeds were reared together? Studies of unlearned auditory preference suggest that the role of at least a brief experience may be limited. Chicks whose first sound exposure was to the maternal call of another species, subsequently preferred the maternal call of their own species without any prior exposure to it (Gottlieb, 1965). Care must be taken in interpreting preference for own species' visual and auditory characteristics as species or breed recognition, however. As cited previously, chicks whose first sound exposure was either to repetitive taps or recorded maternal cackles, subsequently preferred repetitive tapping! (Fischer, unpublished data).

C. Memory of Other Individuals

As noted previously (see Retention), separation of two or three weeks results in failure to recognize former pen-mates (Schjelderup-Ebbe, 1935; Guhl, 1953), and James (1873) noted that separation of Games increases fighting when the birds are brought together again. It is reasonable to assume that there is a limit to the number of individuals that one bird can remember, although there is evidence of a peck order in a flock of 96 pullets (Guhl, 1953). Because frequent meetings between any pair of individuals in a flock reinforce memory, the number of birds in the flocks and the size of the enclosure influence the extent to which pair reactions (peck rights) are maintained. Maier (1964) has shown that physical contact is essential in maintaining dominance-submission relationships. In penned flocks of about 10 hens the peck orders may remain unchanged for months. With 40 or more in the flock some revolts may occur, and with an expansive outdoor range reversals of dominance might be expected.

Related to recognition of individuals is the question of whether chickens react differentially to flock members on the basis of social rank. Guhl (1942) tested the members of two flocks in a discrimination cage. When the discriminations were compared with respective ranks in the peck orders, it was found that neither cocks nor hens, as groups, discriminated the social position of individuals of either sex. Reactions in the cage were influenced by individuality and by various inter-individual experiences in the home pen. This lack of recognition of social levels partially explains other observations. Revolts, when they do occur or when attempted, may involve birds at any two levels, not necessarily contiguous. Discriminations between individual cocks and hens result in non-random or "preferential" mating.

D. Social Inertia and Peck Order Stability

When unacquainted chickens are assembled, there is considerable strife until all dominance relationships have been established. In time, depending on various factors, especially the number in the flock, agonistic behaviour becomes less frequent and less intense. Siegel and Hurst (1962) observed flock integration in four flocks of 85 females each and found a gradual decrease in agonistic behaviour, followed by a slight increase before strife decreased again to the same low level. This indicates that each bird learns and accepts its dominance-subordinate relationships, and when the inter-individual habits within the social order become habituated so as to resist change, then social inertia is operative and promotes the stability of the peck order (Guhl, 1968). Another attribute of an integrated social order is the toleration for the presence of others in limited space, that is, small social distances. In such flocks slight increases in pecking may arise from incidences of competition or excitement. Also, habituation of pecking or threatening responses by dominant members may approach extinction, thereby inducing revolts. In these cases, repeatedly reinforced contacts will again produce social inertia and stability.

Experiments have shown that learning is involved in stability; that social inertia can maintain a bird's rank even when its potential aggressiveness is relatively low, and when its aggressiveness is increased by treatment with androgen. Other experiments show that the peck order can be altered. The peck order that developed among cockerels reared together was completely reversed as follows. After isolation for one month the male in lowest rank was placed into a pen for one week. When others were added, one at a time, in reverse order of their previous ranking, the former inferior male became the dominant bird (Guhl, 1961b). When a dominant male received electric shocks whenever it attacked a subordinate bird, he avoided his inferiors (Radlow et al., 1958). Similar treatment modified the dominance relationships among hens (Smith & Hale, 1959), with the hens at the top and bottom ranks requiring the least training. The new relations were stable for several weeks.

Still another method of altering social ranking is to treat low-ranking birds with androgen which makes them more aggressive. This method is not always successful (Allee et al., 1939, 1955; Williams & McGibbon, 1956). Some of the factors involved were tested with several flocks (Guhl, 1968). Treatment with androgen increased social tension due to increased aggressiveness. This increased the interactions and thereby reinforced the dominance relationships. Although treated hens staged revolts, only a few of these resulted in a reversal of dominance because social inertia acted as feedback and suppressed attacks on social superiors.

E. Social Stress and Physiological Reactions

Chickens are subjected to a variety of stresses, for example, extremes of heat and cold, deficiencies in feed and water, diseases, and competition with flockmates. Adjustment to some stressors may be made by behavioural responses such as movement to more optimal conditions; other reactions involve the endocrine glands in a complex of processes known as the General Adaptation Syndrome (Selye, 1947). If stress persists, either or both of these means of adjustment may break down. Known relationships between various stressors and the syndrome have been reviewed by von Faber (1964) for chickens.

Some social tension can be observed in most flocks, especially in large ones and in groups of cockerels. Birds low in social status may remain on perches or in out-of-the-way places, and have less opportunity to feed (Guhl, 1953). Combining birds from two different groups introduces social stress that is reflected in reduced egg production (Morgan & Bonzer, 1959). The individuals of one of these groups were caged previously and the others penned; the formerly caged birds appeared to be under greater stress during the period of peck order establishment.

Crowding may introduce social tension. Egg production decreases with density but not with flock size (30, 60, and 100 birds), according to Fox and Clayton (1960). A decline in egg production was also found by H. S. Siegel (1959a), and it is interesting to note that it was clutch length, rather than number of birds laying, that decreased. Although increased population density caused adrenal cortical hypertrophy, histological observation indicated that the stressor was probably well within the bird's adaptive ability (H. S. Siegel, 1959b).

Chickens may show various degrees of moulting during any time of the year, when subjected to some stresses such as substantial changes in feed, vaccination and debeaking. Apparently moulting in hens can be stress-triggered and mediated by way of the pituitary and adrenal glands (Perek &

Eckstein, 1959). The seasonal moult is attributed to a decrease in ovarian activity, together with the deterioration of the feather follicle maintaining old feathers, but not by the activation of thyroid function (Tanabe & Toshimatsu, 1962).

IX. SEXUAL BEHAVIOUR

Since chickens are polygamous the social aspect of their reproductive behaviour may be more complex than in monogamous species. Sexual behaviour in chickens is usually referred to as mating behaviour, whereas with monogamous species the term "mating" has a different connotation, i.e. the formation of mateships or pair-bonds. Coitus in chickens is preceded by various behaviour patterns known as displays or courting, which synchronize sexual activities of the males and females.

A. Sexual Patterns

A number of behaviour patterns are associated with sexual behaviour in chickens (Skard, 1937; Davis & Domm, 1943; Guhl *et al.*, 1945; Wood-Gush, 1954, 1956; Fisher & Hale, 1957; Williams & McGibbon, 1955, 1957). Some are definite components of the stimulus-response sequence which terminate in coitus; others occur there but also appear in agonistic behaviour and still others are post-copulatory reactions. Those patterns that function in the initiation, progression, and culmination of the stimulus-response sequence (Fig. 89) are most significant. For certain broader studies of sex-limited behaviour, for example, crowing by the cock, and social dominance relations between the sexes would be pertinent.

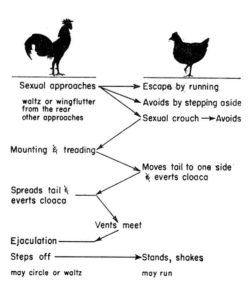

Fig. 89. A simplified signal-response sequence in the mating behaviour of chickens.

The cock typically takes the initiative in sexual behaviour moving among the hens as though testing each for sexual receptivity. If the male and females are well acquainted the *rear approach* with head extended towards or over the female is the most common (the hackle may be raised) (Guhl, 1961a). This approach varies in intensity and may be accompanied by the sex call. The male may place one *foot on the back* of a hen, *grasp the comb* or *hackle*, or he may drive the female. The *waltz* (also called wing-flutter or circling) is common when the cock and hens are strangers. In this conspicuous behaviour the male drops and flutters the wing on the side towards the female, takes several quick steps in an arc around the hen and kicks outward with the other leg. This is a displacement activity and, like hackle raising, has an element of aggression. (Cocks may waltz to cocks and raising hackle occurs during fighting in both sexes.) Such behaviour may assure the dominance of the male over the female and thereby facilitate mating (Guhl, 1949). If the hens are unresponsive the cock may attract them by *titbitting* (also called food-call, sham eating, feeding courts). All of these behaviour patterns collectively have been called *courting*.

A hen may respond negatively, positively, or be indifferent to courting. As a negative reaction she may step aside, walk or run away, or struggle if captured. Such *escape behaviour* may be accompanied by vocalization varying in intensity from faint squawks to loud shrieks. A positive reaction, *crouching*, usually occurs with head low and wings spread. The level of the male's sexual arousal is increased if the head and tail of the hen are prone rather than erect (Carbaugh *et al.*, 1962). The sexual crouch is a strong stimulus for the cock to *mount* and *tread*, especially when he approaches from the rear. The cock stands on the outstretched wings, grasps the comb or hackle, and moves his feet up and down in a treading motion. Subsequently he rears up, spreads his tail while the hen moves her tail to one side, and each *everts* the cloaca as the vents meet. The male usually steps off in a forward direction and the hen ruffles her feathers as she gets to her feet. She may run in a circle and the cock may execute a waltz.

There are marked differences in the rates at which several cocks tread individual females. This non-random mating has been called "preferential" mating and may be explained, in part, by some of the experiments discussed at the end of this chapter. Although matings may occur at any time of the day, they are most frequent during late afternoons (Heuser, 1916; Philips, 1919; Upp, 1928; Skard, 1937; Parker *et al.*, 1940).

Early experience. Chicks in the down do not usually display any sexual behaviour. The question arises as to whether the behaviour pattern is present and awaits further development of the endocrine system to reduce the response threshold. Does early experience play a role in the evocation of sexual behaviour in response to an appropriate stimulus under high excitation even under low hormonal levels? Bambridge (1962) found that chicks imprinted during the critical period and treated daily with androgen from 5 to 20 days of age, trod the imprinted object when tested on days 19 and 20. Andrew (1964) obtained mounting behaviour at 6 days with imprinted chicks, but imprinted chicks treated with androgen mounted at 5 days and were more consistent performers. Varied movements of the imprinting object resulted in more varied and intense behaviour. In another experiment (Andrew, 1966) chicks excited by repeated thrusts of the hand as an imprinting object tended to leap on the hand. In addition, when the hand was suddenly lowered for ease of mounting, chicks as young as 2 days of age displayed the full juvenile copulatory sequence. Experience with a moving object, hand or chick, during the first week was not essential for juvenile copulation, but it did increase the incidence, as did treatment with testosterone.

A sensitive period for sexual behaviour was found by Siegel and Siegel (1964) as cockerels approached sexual maturity. Males were reared with females and placed either into all-male flocks or into isolation at 57, 70, and 84 days of age and tested from 217 to 231 days. The earlier the separation from females, the less the sexual performance of caged males.

An experiment of broader scope (Salzen, 1965) suggests that unfamiliar objects evoke fear which masks sexual behaviour, but in the absence of fear, sexual behaviour will be elicited according to hormonal condition and depending on the object providing the appropriate stimulation.

B. Hormones and Sexual Behaviour

Castration is followed by a marked reduction or cessation of sexual behaviour in capons (Goodale, 1913). Bilateral ovariectomy produces similar results, and like the capon, the poulard is neutral in behaviour (Domm, 1927). When the left ovary is removed and the right rudiment is left intact, there develops either a testis-like organ or an ovotestis. Such poulards resemble the normal cock in appearance and behaviour depending on the extent of the hypertrophy in the rudiment. Intersexes are obtained by injecting oestrogen into eggs before the fourth day of incubation (Domm & Davis, 1941). Males from these eggs develop varying degrees of femininity.

The relation of hormones to behaviour has been reviewed by Beach (1948), Collias (1950a), and Guhl (1961b), but some experiments should be mentioned which were concerned with the influence of androgen and oestrogen on the behaviour of gonadless males and females, and on related experiments which demonstrate the most marked relationship of sex hormones to behaviour.

Davis and Domm (1943) injected androgen into capons and poulards, and oestrogen into capons, poulards, and cocks. Capons treated with testosterone propionate crowded, trod, and became aggressive. Both bilaterally and sinistrally ovariectomized poulards treated with an androgen crowed, waltzed, became aggressive but never copulated. Oestrogen-treated poulards became more timid and crouched for a normal cock, whereas similarly treated capons waltzed, copulated but did not crow or crouch. Two normal males receiving oestrogen ceased to be aggressive, stopped crowing, and copulated in a listless manner. Oestrogen induced copulation in a capon and in a poulard.

Collias (1950a) injected sex hormones into male chicks and noted an increase in the frequency of crowing, pecking, attempts to mate, and waltzes with an increase in dosage. Androgen was more effective in inducing male behaviour in male chicks than oestrogen, whereas testosterone propionate was far more effective in male chicks than in female chicks in inducing typical male behaviour. Cockerels treated with androgen at 35 days of age and tested at 6 months showed no significant differences from normals in sexual behaviour (Wood-Gush, 1963). Gonadal hormones may be used to locate specific brain areas that mediate sexual behaviour. Androgen implanted into the preoptic region of the capon's brain promotes copulatory behaviour independently of aggressive and courting behaviour (Barfield, 1964).

C. Measurement of Sexual Activity

Since there are a number of behavioural patterns included in sexual behaviour, the question arises as to which should be tabulated to obtain a reliable estimate of sex drive in either sex. The frequency of copulations alone may be an inadequate measure. Other significant measures of sexual behaviour include the sexual approaches by the cock, the responses of the hens, and the co-operative behaviours involved in treading (Fig. 89).

Since the sequence is completed in a few moments, recordings should be made in code. For example, A-c-T:25 may indicate that the cock approached; the hen crouched; the treading followed; the 25 indicates the wing badge number (or colour symbol) of the hen. If types of approaches are important, suitable symbols may be used in place of (A); and if more than one male is present, the whole sequence can be preceded by the wing badge number of the male. Other elaborations can be made as needed.

Variability in sexual behaviour may have genetic, physiologic, and psychologic backgrounds. The latter as effects of social status are most readily determined, and the results can point to controls needed for genetic selection.

D. Dominance and Sexual Behaviour

When reproductive cycles of several wild birds were compared (Guhl, 1960), a general agreement was found between the progressive recrudescence of the gonads and the sequence in which certain behaviour patterns developed. A relation between gonadal hormones and behaviour was indicated. There was a suggestion that the several behaviour patterns had a series of threshold levels which required progressively higher concentrations of certain hormones. It was pointed out that social organization formed, or changed, before sexual behaviour was evoked, and this suggested that some form of social organization may be essential for successful reproduction.

Heterosexual dominance and mating. Since cocks dominate the hens and have a peck order separate from that among the females, there is a suggestion that subordinate status of hens facilitates sexual synchronization. This assumption was tested with flocks of pullets and capons which established heterosexual peck orders (Guhl, 1949). When capons at mid-levels of the social orders were treated with oestrogen the sex drive was restored without introducing aggressiveness. The treated capons were definitely more successful in courting and treading pullets which were their inferiors than those superior to them, because dominant pullets repulsed the capons when courted.

Social dominance among cocks and mating. When several cocks were tested singly and successively with the same flock of hens, differences in the frequency of sexual behaviour were not related to ranks in the peck order (Guhl *et al.*, 1945; Guhl, 1951; Wood-Gush, 1957). However, when several males were later placed with the hens and permitted to remain there, the rate of sexual activity was in accord with social levels of the cocks (Guhl *et al.*, 1945; Guhl & Warren, 1946; Lill & Wood-Gush, 1965; Lill, 1966). Since the frequency at which males interfere with each other's matings is also related to dominance, it might be expected that the number of chicks sired by each cockerel would be disproportionate (Table 50). Since the cockerels are listed in the table in the order of dominance, it will be seen that the top-ranking male did not court or attempt to tread as often as his immediate inferior; nevertheless the most dominant male was most successful in attaining coitus; and he also fertilized a larger number of eggs and sired the most chicks. The lowest-ranking male was the least successful.

Table 50. Summary of the Data showing the Relation of the Number of Chicks sired by Cockerels to Social Dominance. (From Guhl & Warren, 1946.)

Peck order of Males	Courtings	Total Treadings	Completed Matings	Eggs Fertilized	Viable Chicks
Flock of 36 Barred Plymouth Rock pullets					
Dominant white	710	175	112	267	221
Recessive white	2,184	244	54	129	120
Rose comb	71	8	0	0	0
Total	2,965	427	166	396	341

Social dominance among hens and mating. Since the passive dominance of the male over the female facilitates mating and the submissive crouch is one of the behaviour patterns in the sexual signal-response sequence, the assumption may be made that the readiness with which the hen submits to the cock also facilitates mating. Conversely, the habit of dominating possessed by hens high in social rank should raise the threshold for the crouching response. Support for this concept is given by Schjelderup-Ebbe's (1935) observation that high social status among hens interferes with mating. Furthermore, Guhl *et al.* (1945) found negative correlations between rank of hens and (*a*) rate of mating, (*b*) rate of being courted, and (*c*) rate of giving the sexual crouch.

Among hens those at the top of the peck order are not in the habit of submitting, whereas those low in social status submit freely to their numerous superiors. These qualitative differences are reflected in their sexual responses to a cock (Guhl *et al.*, 1945), but it must be shown that the variations are not traits which might be correlated with others that influence the attainment of either high or low rank. The intensity of domination, or of submission, can be altered by subflocking, thus reducing the number of individuals each may dominate, or to which each must submit. When the top, middle, and bottom thirds of a peck order are separated, the middle and bottom thirds crouch less frequently, while hens in the top third crouch more often than when in the larger flock (Guhl, 1950). The conclusion that psychological factors associated with intensities of domination among hens may affect sexual receptivity was substantiated by Wood-Gush (1958). We may conclude that in breeding flocks containing several males, the variability in aggressiveness and sexual behaviour tends to remain normal since the greater success of dominant males is offset by the greater receptivity of subordinate females.

Aberrant sexual behaviour. In flocks of cockerels, whether penned or on range, unisexual "mating" is a common observation. Certain males are driven and trodden repeatedly to such an extent that some young males are killed or mutilated. In a flock of 65 cockerels it was found that "pseudo-mating" occurred most frequently between males ranking high in the social scale and males of low rank which were pursued and trodden (Guhl, 1949), indicating that dominance relations play a role in this aberrant behaviour. No indications of sexual receptivity were noted.

Heuser (1916) observed a hen which attempted 35 times to mate with 13 other hens. These "pseudo-matings" occurred at time of production and non-production, and in the presence or absence of males. Similar observations have been made by Guhl *et al.* (1945). Dominance relations facilitate "pseudo-matings" among hens (Guhl, 1948), and none of the hens showed any other indications of maleness, such as crowing or waltzing. Such aberrant behaviour is rare in birds of dimorphic species, but is not uncommon in such birds as pigeons which lack plumage differentiation in the sexes. The conditions under which this abnormal behaviour originates is still obscure, but an assumption might be made. Highly receptive hens may give a deep crouch when pecked by a superior hen, and if this occurs frequently, it may eventually evoke a male-like response in certain hens.

E. Sexual Satiation and Compensatory Sex Behaviour

When a sexually active cock was placed singly and daily for short periods into a pen of hens, there was a decrease in the frequencies of waltzes, crouches, and copulations indicating sexual satiation (Collias, 1950a; Guhl, 1951). On the other hand, Carpenter (1933) noted that a female pigeon paired with a castrated male showed more provocative behaviour than with a normal male, i.e. compensatory sexual behaviour.

Evidence of satiation and compensatory sexual behaviour was obtained when males were caged within the pen of hens and released singly and successively for short daily periods (Guhl, 1953). The data were tabulated as to the first, second, etc., cocks in the order of release. Since the order of individuals was rotated the data tend to eliminate individual differences and also differences in reactions of hens to individual males (Fig. 90).

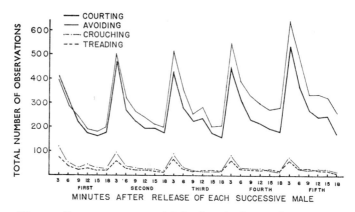

Fig. 90. Compensatory sexual behaviour is shown when 5 rotated males are introduced daily, singly, and consecutively for 18 minutes each, into a pen of hens. Each cock shows satiation by a temporal decrease in courting. As the hens crouch less often (avoid more) for each successive male the succeeding cocks compensate for lowered receptivity in the hens by increased courting, with the consequence that treadings tend to give similar results with each cock. (From Guhl, 1953.)

The figure shows that (*a*) all activities are most frequent during the first 3 minutes after the introduction of a male, decline, and tend to level off. (*b*) Satiation in each of the males is indicated by the progressive decline in courting during 18 minutes. (*c*) The peaks of crouching by the hens decrease with each successive male, and as the hens become satiated their frequency of escape behaviour (avoidance) increases. (*d*) Each succeeding male compensates for the lowered receptivity by initially courting at a higher rate, and thus (*e*) the level of peaks and troughs for treading are similar for each of the 5 males.

In mating flocks under the usual procedures, sex drive would be dissipated more gradually. Nevertheless satiation and compensatory behaviour would be expected to occur. If cocks were relatively inactive the hens may be expected to crouch more often, whereas an unusually active cock might cause the hens to avoid him. In laying flocks devoid of males, the hens crouch readily when threatened by a superior hen or by a caretaker. When cockerels are reared together and apart from pullets, the intensification of sex drive is a problem, because the most aggressive males tread their inferiors and often kill them.

F. Selection for Mating Ability

Variability in the sexuality of several cocks was found when tested with the same flock of hens. Differences included their rates of courting, treading, and ability to solicit crouching (Guhl, 1951, 1953). Males high in one category of display were not necessarily high in the others. After reviewing a number of reports concerned with sexuality of cocks, Guhl (1961a) devised an experiment to examine several methods of testing. The frequencies of courting, as well as the kinds of sexual approaches, and treading depend on the relative acquaintance between the sexes. Wood-Gush (1956) noted that strangeness of the male tended to increase the evasiveness of hens. Subsequently Justice et al. (1962) developed a technique for measuring the sexual effectiveness of a large number of males in each of two breeds. A number of minimum requirements and recommendations were presented.

Individual differences suggest a possible genetic basis (see also Selection for Agonistic Behaviour), and raise the question of whether or not aggressiveness and fertilizing capacities were correlated. Wood-Gush (1957) found no correlation between sexual activity and aggressiveness nor with rank in the peck order of cockerels. McDaniel and Craig (1959) found significant associations between social aggressiveness scores, sexual effectiveness scores and crouches elicited from females. Correlations of relatively low magnitude, although significant or nearly so, were found between sexual behaviour and semen traits.

Significant differences were found by Wood-Gush and Osborne (1956) in the mating frequencies of males belonging to six different sire families which offered an opportunity for selective breeding. Subsequently Wood-Gush (1960) selected for high and low sex drive through three generations. Differences in mating frequencies were found between the six sire families, and incidences of incomplete matings were higher in the high score males. Siegel (1959) also found differences among sire families and then (Siegel, 1965) conducted bidirectional selection through six generations. He obtained heritabilities of 0.18 ± 0.05 and 0.31 ± 0.11 in upward and downward directions for completed matings. Some correlated responses were also observed.

X. PARENTAL BEHAVIOUR

Parental activities are limited to the hen and since broodiness decreases egg production the interest of the poultryman in this behaviour has been one of elimination. Poultry geneticists soon discovered that it was heritable (Goodale et al., 1920) and now many common strains and crosses are devoid of broodiness. The modern incubator has replaced the hen and as a result the information on parental behaviour is sparse.

Among poultrymen the words "broodiness" and "parental" behaviour are apt to be used as synonyms, although "broodiness" often refers to maternal calls and care of young, especially the covering of chicks by the wings as well as incubation. However, parental behaviour begins with building of a nest, egg laying or nesting, and includes incubation, brooding, and feeding of the chicks.

For the behaviourist the study of parental behaviour in chickens offers opportunities to develop and test theoretical concepts. This domestic bird should be useful in attaining a better understanding of the physiological mechanisms and the stimuli which play a role in the shift of behaviour patterns between laying, incubation, brooding, and the return to egg production.

18+

A. Hormonal Mechanisms

The endocrine background of parental behaviour in birds has been investigated for some time by a number of experimenters and indications are that generalizations between species cannot be made with confidence (Lehrman, 1959, 1961; Eisner, 1960). In some of the reports on chickens it is not clear whether reference to broodiness also includes incubation. The indications are that prolactin induces incubation and broodiness in the hen (Sturkie, 1965; Eisner, 1960). Non-broody breeds like the Leghorn require 4 to 5 times as much prolactin to become broody than do the Cornish (Riddle & Bates, 1939). Chicken pituitaries have been assayed by their injection into pigeons to determine potencies by the effects on the crop sac. This method has shown that glands of broody chickens contain more prolactin than those of non-broody birds (Burrows & Byerly, 1936; Byerly & Burrows, 1936).

There are interactions between external stimuli and the production of pituitary and gonadal hormones which govern the sequence of behaviour patterns in several species of birds (Lehrman, *loc. cit.*; Eisner, 1960). Incubation and brooding may have some common aspects, such as separation from the flock and certain postures, but differ greatly in pattern and level of activity. It may well be that endocrine conditions may differ between these two behaviour patterns with less potency of prolactin during broodiness (Saeki & Tanabe, 1955). There are some suggestions of endocrine (prolactin) control of incubation with a shift towards control of broodiness by certain external stimuli discussed below. Some indirect evidence suggests that one function of prolactin, by way of an anti-gonadal effect, may be the suppression of aggressive and sexual behaviour (Nalbandov *et al.*, 1945). The same function has been ascribed to alcohol which, in conjunction with physical contact with chicks, induced maternal behaviour in cocks (Kovach, 1967). Consistent with this view is the finding that aggressiveness, as measured by social rank, appears to suppress or delay the onset of broodiness (Ramsay, 1953).

B. Stimuli

Some birds, for example, pigeons, will continue to lay for some time if eggs are removed. If eggs do not hatch, or dummy eggs are given, the incubation period is extended beyond the normal duration. The various stimuli which influence the onset and duration of incubation are still to be established. Eggs given to a hen in semi-darkness and at 85°F evoke incubation (Saeki & Tanabe, 1955); prolactin potency rises after nesting behaviour is induced and lasts until the chicks hatch.

Visual and auditory stimuli are relevant to the shift from incubation to brooding. However, Maier (1961, 1963) has shown that physical contact between the hen and chicks is a requisite for both the development and maintenance of broodiness. Chicks will induce broodiness whether the hen has incubated or not (Burrows & Byerly, 1936; Collias, 1946; Ramsay, 1953; Saeki & Tanabe, 1955). They must be in the down, and if replaced by baby chicks every 3 or 4 weeks the broody period can be extended. Semi-darkness, high temperature, and humidity are contributing factors.

REFERENCES

ALLEE, W. C. & COLLIAS, N. E. (1938). Effects of injections of epinephrine on the social order in small flocks of hens. *Anat. Rec.* (Suppl)., **72**, 119.

ALLEE, W. C. & COLLIAS, N. E. (1940). The influence of estradiol on the social organization of flocks of hens. *Endocrinology*, **27**, 87–94.

ALLEE, W. C., COLLIAS, N. E. & LUTHERMAN, C. Z. (1939). Modification of the social order in flocks of hens by the injection of testosterone propionate. *Physiol. Zool*, **14**, 412–440.

ALLEE, W. C., FOREMAN, D. & BANKS, E. M. (1955). Effects of an androgen on dominance and subordinance in six common breeds of *Gallus gallus*. *Physiol. Zool.*, **28**, 89–115.

American Standard of Perfection (1953). American Poultry Association, Davenport, Iowa.

ANDREW, R. J. (1963). Effect of testosterone on the behavior of the domestic chicken. *J. comp. physiol. Psychol.*, **56**, 933–940.

ANDREW, R. J. (1964). The development of adult responses from responses given during imprinting by the domestic chick. *Anim. Behav.*, **12**, 542–548.

ANDREW, R. J. (1966). Precocious adult behavior in the young chick. *Anim. Behav.* **14**, 485–500.

ANDREW, R. J. (1967). Intracranial self-stimulation in the chick. *Nature, Lond.*, **213**, 847–848.

ARMINGTON, J. C. & FREDERICK, C. T. (1956). Electroretinal demonstration of a Purkinje shift in the chicken eye. *Am. J. Psychol.*, **186**, 258–262.

BACON, H. R., WARREN, J. M. & SCHEIN, M. W. (1962). Non-spatial reversal learning in chickens. *Anim. Behav.*, **10**, 239–243.

BALL, G. G. & WARREN, J. M. (1960). Maturation of *Umweg* learning in White Leghorn chicks. *J. comp. physiol. Psychol.*, **53**, 273–274.

BAMBRIDGE, R. (1962). Early experience and sexual behavior in the domestic chicken. *Science, N.Y.*, **136**, 259–260.

BARFIELD, R. J. (1964). Induction of copulatory behavior by intracranial placement of androgen in capons. *Am. Zool.*, **4**, 301.

BARFIELD, R. J. (1965). Induction of aggressive and courtship behavior by intracranial implants of androgen in capons. *Am. Zool.*, **5**, 203.

BEACH, F. A. (1948). *Hormones and Behavior*. New York: Hoeber.

BEEBE, W. (1926). *Pheasants: Their Lives and Homes*. New York: Doubleday Page.

BENNER, J. (1938). Untersuchungen über die Raumwahrnehmung dier Hühner. *Z. wiss. Zool.*, **151**, 382–444.

BROOKSHIRE, K. H., WARREN, J. M. & BALL, G. G. (1961). Reversal and transfer learning following overtraining in the rat and chicken. *J. comp. physiol. Psychol.*, **54**, 98–102.

BROWN, E. (1929). *Poultry Breeding and Production*. Vols. I and II. London: Ernest Benn Ltd.

BURROWS, W. H. & BYERLY, T. C. (1936). Studies of prolactin in the fowl pituitary. I. Broody hens compared with laying hens and males. *Proc. Soc. exp. Biol. N.Y.*, **34**, 841–844.

BYERLY, T. C. & BURROWS, W. H. (1936). Studies of prolactin in the fowl pituitary. II. Effects of genetic constitution on prolactin content. *Proc. Soc. exp. Biol. N.Y.*, **34**, 844–846.

BYRNE, W. L. (ed.) (1968). *Molecular Approaches to Learning and Memory*. New York: Academic Press. (In press.)

CAPRETTA, P. J. (1961). An experimental modification of food preference in chickens. *J. comp. physiol. Psychol.*, **54**, 238–242.

CARBAUGH, R. T., SCHEIN, M. W. & HALE, E. B. (1962). Effects of morphological variations of chicken models on sexual responses of cocks. *Anim. Behav.*, **10**, 235–238.

CARPENTER, C. R. (1933). Psychological studies of social behavior in Aves. I. The effects of complete and incomplete gonadectomy on the primary sexual activity of the male pigeon. II. The effects of complete and incomplete gonadectomy on secondary sexual activity with histological studies. *J. comp. Psychol.*, **16**, 25–98.

COLLIAS, N. E. (1943). Statistical analysis of factors which make for success in initial encounters between hens. *Am. Nat.*, **77**, 519–538.

COLLIAS, N. E. (1946). Some experiments on broody behavior in fowl and pigeon. *Anat. Rec.*, **96**, 572 (Abstract).

COLLIAS, N. E. (1950a). Hormones and behavior with special reference to birds and the mechanisms of hormone action. In: *A Symposium on Steroid Hormones*, E. S. Gordon (ed.). Madison: University of Wisconsin Press, pp. 277–329.

COLLIAS, N. E. (1950b). Some basic psychological and neural mechanisms of behavior in chicks. *Anat. Rec.*, **108**, 552 (Abstract).

COLLIAS, N. E. & COLLIAS, E. C. (1967). A field of study of the Red Junglefowl in North-Central India. *Condor*, **69**, 360–386.

COLLIAS, N. E. & JOOS, M. (1953). The spectographic analysis of signals of the domestic fowl. *Behaviour*, **5**, 176–188.

CRAIG, J. V., ORTMAN, L. L. & GUHL, A. M. (1965). Genetic selection for social dominance ability in chickens. *Anim. Behav.*, **13**, 114–131.

DAVIS, D. E. & DOMM, L. V. (1943). The influence of hormones on the sexual behavior of domestic fowl. In: *Essays in Biology*. Berkeley: University of California Press.

DAWSON, J. S. & SIEGEL, P. B. (1967). Behavior patterns of chickens to ten weeks of age. *Poult. Sci.*, **46**, 615–622.

DOMM, L. V. (1927). New experiments on ovariotomy and the problem of sex inversion in the fowl. *J. exp. Zool.*, **48**, 31–173.

DOMM, L. V. (1939). Modification in sex and secondary sex characters in birds. In: *Sex and Internal Secretion*, 3rd edn, E. Allen, C. H. Danforth & E. Doisy (eds.). Baltimore: Williams & Wilkins, pp. 227–327.

DOMM, L. V. & DAVIS, D. E. (1941). Sexual behavior in intersexual domestic fowl. *Proc. Soc. exp. Biol. N.Y.*, **48**, 665–667.

DORFMAN, R. I. (1960). A test for anti-androgens. *Acta Endocrinol.*, **33**, 308–316.

DORFMAN, R. I. (1962). A subcutaneous assay for anti-androgens in the chick. *Acta Endocrinol.*, **41**, 268–273.

DOUGLIS, M. B. (1948). Social factors influencing the hierarchies of small flocks of the domestic hen: Interactions between resident and part-time members of organized flocks. *Physiol. Zool.*, **21**, 147–182.

EISNER, E. (1960). The relationship of hormones to the reproductive behaviour of birds, referring especially to parental behaviour: A review. *Anim. Behav.*, **8**, 155–179.

FANTZ, R. L. (1957). Form preference in newly hatched chicks. *J. comp. physiol. Psychol.*, **50**, 422–430.

FANTZ, R. L. (1958). Depth discrimination in dark-hatched chicks. *Percept. mot. Skills*, **8**, 47–50.

FENNELL, R. A. (1945). The relation between heredity, sexual activity and training to dominance-subordination in Game cocks. *Am. Nat.*, **79**, 142–151.

FISCHER, G. J. (1966a). Auditory stimuli in imprinting. *J. comp. physiol. Psychol.*, **61**, 271–273.

FISCHER, G. J. (1966b). Discrimination and successive reversal learning in chicks that fail to imprint and ones that imprint strongly. *Percept. mot. Skills*, **23**, 579–584.

FISCHER, G. J. (1966c). Distribution of practice effects on imprinting. *Psychon. Sci.*, **5**, 197–198.

FISCHER, G. J. (1967). Comparisons between chicks that fail to imprint and ones that imprint strongly. *Behaviour*, **24**, 262–264.

FISCHER, G. J. (1968). Temperature effects on following during imprinting. Paper presented at meetings of the Midwestern Psychological Association in Chicago, Ill.

FISCHER, G. J. (1969a). Sound stimulus frequency and following during imprinting. Paper presented at meetings of the Midwestern Psychological Association in Chicago, Ill.

FISCHER, G. J. (1969b). Heritability in the following response of white Leghorns. *J. genet. Psychol.* (In press.)

FISCHER, G. J. & CAMPBELL, G. L. (1964). The development of passive avoidance conditioning in Leghorn chicks. *Anim. Behav.*, **12**, 268–269.

FISCHER, G. J., CAMPBELL, G. L. & DAVIS, W. M. (1965). Effects of ECS on retention of imprinting. *J. comp. physiol. Psychol.*, **59**, 455–457.

FISCHER, G. J. & GILMAN, S. C. (1967). Following during imprinting as a function of auditory stimulus intensity. Paper presented at meetings of the Midwestern Psychological Association in Chicago, Ill.

FISHER, A. E. & HALE, E. B. (1957). Stimulus determinants of sexual and aggressive behavior in male domestic fowl. *Behaviour*, 10, 309–323.

FOX, T. W. & CLAYTON, J. T. (1960). Population size and density as related to laying house performance. *Poult. Sci.*, 39, 896–899.

FRINGS, H. & FRINGS, M. (1964). *Animal Communication*. New York: Blaisdell.

FUERST, W. F., Jr. & KARE, M. R. (1962). The influence of pH on fluid tolerance and preferences. *Poult. Sci.*, 41, 71–77.

GOODALE, H. D. (1913). Castration in relation to the secondary sexual characters in Brown Leghorns. *Am. Nat.*, 34, 127–148.

GOODALE, H. D., SANBORN, R. & WHITE, D. (1920). Broodiness in domestic fowl. *Mass. Agric. exp. Sta. Bull.*, 199, 93–116.

GOSSETTE, R. L. (1967). Successive discrimination reversal (SDR) performances of four avian species on a brightness discrimination task. *Psychon. Sci.*, 8, 17–18.

GOSSETTE, R. L., GOSSETTE, M. F. & RIDDELL, W. (1966). Comparisons of successive discrimination reversal performances among closely and remotely related avian species. *Anim. Behav.*, 14, 560–564.

GOTTLIEB, G. (1965). Prenatal auditory sensitivity in chickens and ducks. *Science*, 147, 1596–1598.

GRAVES, H. B. & SIEGEL, P. B. (1968). Chick's response to an imprinting stimulus: Heterosis and evolution. *Science, N.Y.*, 160, 329–330.

GRAY, P. H. (1961). The releasers of imprinting: Differential reactions to color as a function of maturation. *J. comp. physiol. Psychol.*, 54, 597–601.

GRAY, P. H., YATES, A. T., VANDIVER, D. R. & KIRWAN, K. (1967). A reverse Kamin effect in the escape and avoidance conditioning of newly hatched chicks. *Psychon. Sci.*, 9, 507–508.

GRIER, J. B., COUNTER, A. S. & SHEARER, W. M. (1967). Prenatal auditory imprinting in chickens. *Science, N.Y.*, 155, 1692–1693.

GROSSLIGHT, J. H., SCHEIN, M. W., ROSS, S. & LYERLY, S. B. (1966). Perceptual factor: Quantity of food available and consummatory behavior in chickens. *Psychon. Sci.*, 4, 97–98.

GUHL, A. M. (1942). Social discrimination in small flocks of the common domestic fowl. *J. comp. Psychol.*, 34, 127–148.

GUHL, A. M. (1948). Unisexual mating in a flock of White Leghorn hens. *Trans. Kans. Acad. Sci.*, 51, 107–111.

GUHL, A. M. (1949). Heterosexual dominance and mating behavior in chickens. *Behaviour*, 2, 106–120.

GUHL, A. M. (1950). Social dominance and receptivity in the domestic fowl. *Physiol. Zool.*, 23, 361–366.

GUHL, A. M. (1951). Measurable differences in mating behavior of cocks. *Poult. Sci.*, 30, 687–693.

GUHL, A. M. (1953). Social behaviour of the domestic fowl. *Tech. Bull. Kans. Agric. exp. Sta.*, No. 73, 3–48.

GUHL, A. M. (1958). The development of social organization in the domestic chick. *Anim. Behav.*, 6, 92–111.

GUHL, A. M. (1960). Psycho-physiological factors and social behavior related to sexual behavior in birds. *Trans. Kans. Acad. Sci.*, 63, 85–95.

GUHL, A. M. (1961a). The effect of acquaintance between the sexes on sexual behavior in White Leghorns. *Poult. Sci.*, 40, 10–21.

GUHL, A. M. (1961b). Gonadal hormones and social behavior in infra-human vertebrates. In: *Allen's Sex and Internal Secretions*, 4th edn, W. C. Young (ed.). Baltimore: Williams & Wilkins, pp. 1240–1267.

GUHL, A. M. (1961c). Unpublished data. Department of Zoology, Kansas State University, Manhattan, Kansas, U.S.A.

GUHL, A. M. (1968). Social inertia and social stability in chickens. *Anim. Behav.*, 16, 219–232.

GUHL, A. M. & ALLEE, W. C. (1944). Some measurable effects of social organization in flocks of hens. *Physiol. Zool.*, 17, 320–347.

GUHL, A. M., COLLIAS, N. E. & ALLEE, W. C. (1945). Mating behavior and the social hierarchy in small flocks of White Leghorns. *Physiol. Zool.*, 18, 365–390.

GUHL, A. M., CRAIG, J. V. & MUELLER, C. D. (1960). Selective breeding for aggressiveness in chickens. *Poult. Sci.*, **39**, 970–980.

GUHL, A. M. & ORTMAN, L. L. (1953). Visual patterns in the recognition of individuals among chickens. *Condor*, **55**, 287–298.

GUHL, A. M. & WARREN, D. C. (1946). Number of offspring sired by cockerels related to social dominance in chickens. *Poult. Sci.*, **25**, 460–472.

GUITON, P. (1958). The effect of isolation on the following response of Brown Leghorn chicks. *Proc. R. Soc. Edinb.*, **27**, 9–14.

GUITON, P. (1959). Socialization and imprinting in Brown Leghorn chicks. *Anim. Behav.*, **7**, 26–34.

GUITON, P. (1961). The influence of imprinting on the agonistic and courtship responses of the Brown Leghorn chick, *Anim. Behav.*, **9**, 167–177.

GUNTHER, W. C. & JONES, R. K. (1961). Effect of nonoptimally high incubation temperatures on T-maze learning in the chick. *Proc. Ind. Acad. Sci.*, **71**, 327–333.

GUNTHER, W. C. & JONES, R. K. (1962). Effect of nonoptimally high incubation temperature on frequency of pecking and on color preferences in the chick. *Proc. Ind. Acad. Sci.*, **72**, 290–299.

HALE, E. B. (1954). Androgen levels and breed differences in the fighting behavior of cocks. *Bull. Ecol. Soc. Am.*, **35**, 71 (Abstract).

HALE, E. B. (1957). Breed recognition in the social interactions of domestic fowl. *Behaviour*, **10**, 240–254.

HAMILTON, J. B. (1938). Precocious masculine behavior following administration of synthetic male hormone substances. *Endocrinology*, **23**, 53–57.

HESS, E. H. (1950). Development of chick's responses to light and shade cues of depth. *J. comp. physiol. Psychol.*, **43**, 112–122.

HESS, E. H. (1959). Two conditions limiting critical age for imprinting. *J. comp. physiol. Psychol.*, **52**, 515–518.

HESS, E. H. (1960). Effects of drugs on imprinting behavior. In L. M. Uhr & J. G. Miller (eds.). *Drugs and Behavior*. New York: Wiley, ch. 16.

HESS, E. H. (1964). Imprinting in birds. *Science, N.Y.*, **146**, 1128–1139.

HEUSER, G. F. (1916). *A Study of the Mating Behavior of the Domestic Fowl*. Thesis, Master of Agric., Cornell University, Ithaca, N.Y.

HINDE, R. A. (1962). Some comments on the nature of the imprinting problem. *Behaviour*, **10**, 181.

HOGAN, J. A. (1965). An experimental study of conflict and fear: an analysis of behavior of young chicks towards a mealworm. Part I. The behavior of chicks which do not eat the mealworm. *Behaviour*, **25**, 45–97.

HUNT, G. L., Jr. & SMITH, W. J. (1967). Pecking and initial drinking responses in young domestic fowl. *J. comp. physiol. Psychol.*, **64**, 230–236.

HUTT, F. B. (1949). *Genetics of the Fowl*. New York: McGraw-Hill.

JAMES, E. (1873). *The Game Cock*. New York: E. James.

JAMES, H. (1960). Imprinting with visual flicker: Evidence for a critical period. *Canad. J. Psychol.*, **14**, 13–20.

JAMES, H. & BINKS, C. (1963). Escape and avoidance learning in newly hatched domestic chicks. *Science, N.Y.*, **139**, 1293–1294.

JAYNES, J. (1957). Imprinting: The interaction of learned and innate behavior: II. The critical period. *J. comp. physiol. Psychol.*, **50**, 6–10.

JAYNES, J. (1958). Imprinting: The interaction of learned and innate behavior: IV. Generalization and emergent discrimination. *J. comp. physiol. Psychol.*, **51**, 238–242.

JUHN, M., GUSTAFSON, R. G. & GALLAGHER, T. F. (1932). The factor of age with reference to reactivity to sex hormones in fowl. *J. exp. Zool.*, **64**, 133–176.

JUSTICE, W. P., McDANIEL, G. R. & CRAIG, J. V. (1962). Techniques for measuring sexual effectiveness in male chickens. *Poult. Sci.*, **41**, 732–739.

KARE, M. R. (1965). The special senses. In: *Avian Physiology*. P. D. Sturkie (ed.). Ithaca: Cornell Press.

KARE, M. R. & MEDWAY, W. (1959). Discrimination between carbohydrates by the fowl. *Poult. Sci.*, **38**, 1119–1127.

KARE, M. R. & PICK, H. L., Jr. (1960). The influence of the sense of taste on feed and fluid consumption. *Poult. Sci.*, **39**, 697–706.

KARMEL, B. Z. (1966). Randomness, complexity, and visual preference behavior in the hooded rat and domestic chick. *J. comp. physiol. Psychol.*, **61** 487–489.

KLEITMAN, N. & KOPPANYI, T. (1926). Body-righting in the fowl (*Gallus domesticus*). *Am. J. Physiol.*, **78**, 110–126.

KLOPFER, P. H. (1965). Imprinting: A reassessment. *Science, N.Y.*, **147**, 302–303.

KLOPFER, P. H. (1967). Stimulus preferences and imprinting. *Science, N.Y.*, **156**, 1394–1396.

KOCH. F. C. (1937). The male sex hormone. *Physiol. Rev.*, **17**, 152–238.

KOMAI, T., CRAIG, J. V. & WEARDEN, S. (1959). Heritability and repeatability of social aggressiveness in the domestic chicken. *Poult. Sci.*, **38**, 356–359.

KOVACH, J. K. (1964). Effects of autonomic drugs on imprinting. *J. comp. physiol. Psychol.*, **57**, 183–187.

KOVACH, J. K. (1967). Maternal behavior in the domestic cock under the influence of alcohol. *Science, N.Y.*, **156**, 835–837.

KOVACH, J. K. & HESS, E. H. (1963). Imprinting: Effects of painful stimulation upon the following response. *J. comp. physiol. Psychol.*, **56**, 461–464.

KRIECKHAUS, E. E. & WAGMAN, W. J. (1967). Acquisition of the two-way avoidance response in a chicken compared to rat and cat. *Psychon. Sci.*, **8**, 273–274.

KRUIJT, J. P. (1964). Ontogeny of social behaviour in Burmese Red Junglefowl (*Gallus gallus spadiceus*). *Behaviour*, Suppl. XII, 1–201.

KUO, Z. (1967). *The Dynamics of Behavior Development*. New York: Random House.

LANE, H. (1960). Operant control of vocalizing in the chicken. *J. exp. Analysis Behav.*, 171–177.

LANYON, W. E. (1960). The ontogeny of vocalization in birds. In: *Animal Sounds and Communication*. W. E. Lanyon & W. N. Tavolga (eds.). Washington, D.C.: Amer. Inst. Biol. Sci., 321–347.

LEE-TENG, E. & SHERMAN, S. M. (1966). Memory consolidation of one-trial learning in chicks. *Proc. natn. Acad. Sci. U.S.A.*, **56**, 926–931.

LEHRMAN, D. S. (1959). Hormonal responses to external stimuli in birds. *Ibis*, **101**, 478–496.

LEHRMAN, D. S. (1961). Hormonal factors in parental behavior in birds and infra-human mammals. In: *Allen's Sex and Internal Secretions*, 4th edn, W. C. Young (ed.). Baltimore: Williams & Wilkins, Ch. 21.

LILL, A. (1966). Some observations on social organization and non-random mating in captive Burmese Red Junglefowl (*Gallus gallus spadiceus*). *Behaviour*, **26**, 228–242.

LILL, A. & WOOD-GUSH, D. G. M. (1965). Potential ethological isolating mechanisms and assortive mating in the domestic fowl. *Behaviour*, **25**, 16–44.

LINDENMAIER, P. & KARE, M. R. (1959). The taste end-organs of the chickens. *Poult. Sci.*, **38**, 545–550.

LORENZ, K. (1935). Der Kumpan in der umwalt des vogels. *J. Orthinol.*, **83**, 137–213.

MACKINTOSH, N. J. (1965). Overtraining, reversal and extinction in rats and chicks. *J. comp. physiol. Psychol.*, **59**, 31–36.

MAIER, R. A. (1961). Personal communication. Department of Psychology, Kansas State University, Manhattan, Kansas, U.S.A.

MAIER, R. A. (1963). Maternal behavior in the domestic hen: the role of physical contact. *J. comp. physiol. Psychol.*, **56**, 357–361.

MAIER, R. A. (1964). The role of the dominance-submission ritual in social recognition of hens. *Anim. Behav.*, **12**, 59.

MARLER, P. (1961). The logical analysis of animal communication. *J. theoret. Biol.*, **1**, 295–317.

MARLER, P. (1967). Animal communication signals. *Science, N.Y.*, **157**, 769–774.

MARLER, P., KRIETH, M. & WILLIS, E. (1962). An analysis of testosterone induced crowing in young domestic cockerels. *Anim. Behav.*, **10**, 48–54.

MARKS, H. L., SIEGEL, P. B. & KRAMER, C. Y. (1960). Effect of comb and wattle removal on the social organization of mixed flocks of chickens. *Anim. Behav.*, **8**, 192–196.

McBRIDE, G. (1958). The measurement of aggressiveness in the domestic hen. *Anim. Behav.*, **6**, 87–91.

McBRIDE, G. (1967). Personal communication.

McDaniel, G. R. & Craig, J. V. (1959). Behavior traits, semen measurements and fertility of White Leghorn males. *Poult. Sci.*, **38**, 1005–1014.

Moltz, H., Rosenblum, L. & Stettner, L. J. (1960). Some parameters of imprinting effectiveness. *J. comp. physiol. Psychol.*, **53**, 297–301.

Morgan, W. C. & Bonzer, B. J. (1959). Stresses associated with moving cage layers to floor pens. *Poult. Sci.*, **38**, 603–606.

Nalbandov, A. V., Hochhauser, M. & Dugas, M. (1945). A study of the effect of prolactin on broodiness and on cock testes. *Endocrinology*, **36**, 251–258.

Ortman, L. L. (1964). Developmental and physiological differences produced by selection for agonistic behavior in chickens. Ph.D. Dissertation, Kansas State University.

Parker, J. E., McKenzie, F. F. & Kempster, H. L. (1940). Observations on the sexual behavior of New Hampshire males. *Poult. Sci.*, **19**, 191–197.

Perek, M. & Eckstein, B. (1959). The adrenal ascorbic acid content of molting hens and the effect of ACTH on the adrenal ascorbic acid content of laying hens. *Poult. Sci.*, **38**, 996–999.

Peters, J. J. & Isaacson, R. J. (1963). Acquisition of active and passive responses in two breeds of chickens. *J. comp. physiol. Psychol.*, **56**, 793–796.

Peters, J. J., Vonderahe, A. R. & Powers, T. H. (1958). Electrical studies of functional development of the eye and optic lobes in the chick embryo. *J. exp. Zool.*, **139**, 459–468.

Philips, A. G. (1919). Preferential mating in fowls. *Poult. Husb. J.*, **5**, 28–32.

Phillips, R. E. (1965). Agonistic behavior elicited by brain stimulation. *Bull. Ecol. Soc. Am.*, **46**, 117.

Phillips, R. E. & Siegel, P. B. (1966). Development of fear in chicks of two closely related genetic lines. *Anim. Behav.*, **14**, 84–88.

Pick, H. L., Jr. & Kare, M. R. (1962). The effect of artificial cues on the measurement of taste preference in the chickens. *J. comp. physiol. Psychol.*, **55**, 342–345.

Pitz, G. F. & Ross, R. B. (1961). Imprinting as a function of arousal. *J. comp. physiol. Psychol.*, **54**, 602–604.

Plotnik, R. J. & Tallarico, R. B. (1966). Object-quality learning-set formation in the young chicken. *Psychon. Sci.*, **5**, 195–196.

Polt, J. M. & Hess, E. H. (1964). Following and imprinting: Effects of light and social experience. *Science, N.Y.*, **143**, 1185–1187.

Portman, A. (1961). Sensory organs: Skin, taste and olfaction. In: *Biology and Comparative Physiology of Birds*. A. J. Marshall (ed.), Vol. II, 37–48. New York: Academic Press.

Portman, A. & Stingelin, W. (1961). The central nervous system. In: *Biology and Comparative Physiology of Birds*. A. J. Marshall (ed.). Vol. II, 1–36. New York: Academic Press.

Potter, J. H. (1949). Dominance relations between different breeds of domestic hens. *Physiol. Zool.*, **22**, 261–280.

Potter, J. H. & Allee, W. C. (1953). Some effects of experience with breeds of *Gallus gallus* on behavior of hens toward strange individuals. *Physiol. Zool.*, **26**, 147–161.

Pumphrey, R. J. (1948). The sense organs of birds. Smithsonian Report for 1968, 305–330.

Pumphrey, R. J. (1961a). Sensory organs: Vision. In: *Biology and Comparative Physiology of Birds*. A. J. Marshall (ed.). Vol. II, 55–68. New York: Academic Press.

Pumphrey, R. J. (1961b). Sensory organs: Hearing. In: *Biology and Comparative Physiology of Birds*. A. J. Marshall (ed.). Vol. II, 69–86. New York: Academic Press.

Radlow, R., Hale, E. B. & Smith, W. I. (1958). Note on the role of conditioning in the modification of social dominance. *Psychol. Reps.*, **4**, 579–581.

Ramsey, A. O. (1953). Variations in the development of broodiness in fowl. *Behaviour*, **5**, 51–57.

Ramsey, A. O. & Hess, E. H. (1954). A laboratory approach to the study of imprinting. *Wilson Bull.*, **66**, 196–206.

Riddle, O. & Bates, R. W. (1939). The preparation, assay, and actions of lacto-genetic hormones. In: *Sex and Internal Secretions*, 3rd edn. E. Allen, C. H. Danforth, & E. Doisy (eds.). Baltimore: Williams & Wilkins, pp. 1088–1117.

SAEKI, Y. & TANABE, Y. (1955). Changes in prolactin content of fowl pituitary during broody periods and some experiments on the induction of broodiness. *Poult. Sci.*, **34**, 909–919.

SALZEN, E. A. (1962). Imprinting and fear. *Symp. Zool. Soc. Lond.*, **8**, 199–217.

SALZEN, E. A. (1965). The interaction of experience, stimulus characteristics and exogenous androgen in the behaviour of domestic chicks. *Behaviour*, **26**, 286–322.

SALZEN, E. A. & TOMLIN, F. J. (1963). The effect of cold on the following response of domestic fowl. *Anim. Behav.*, **11**, 62–65.

SCHAEFER, H. H. & HESS, E. H. (1959). Color preferences in imprinting objects. *Z. Tierpsychol.*, **16**, 161–172.

SCHALLER, G. B. & EMLEN, J. R., Jr. (1962). The ontogeny of avoidance behaviour in some precocial birds. *Anim. Behav.*, **10**, 370–380.

SCHEIN, M. W. & HALE, E. B. (1959). The effect of early social experience on male sexual behavior of androgen injected turkeys. *Anim. Behav.*, **7**, 189–200.

SCHJELDERUP-EBBE, T. (1913). Hönsenes stemme. Bidrag til hönsenes psykologi. *Naturen*, **37**, 262–276.

SCHJELDERUP-EBBE, T. (1922). Beiträge zur Social-psycholgie des Haushuhns. *Z. Psychol.*, **88**, 225–252.

SCHJELDERUP-EBBE, T. (1935). Social behavior in birds. In: *Murchison's Handbook of Social Psychology.* C. Murchison (ed.). Worcester, Mass.: Clark University Press.

SCHNEIRLA, T. C. (1959). An evolutionary and developmental theory of biphasic processes underlying approach and withdrawal. In: *Nebraska symposium on motivation.* M. R. Jones (ed.). Lincoln: University of Nebraska Press, pp. 1–41.

SCHNEIRLA, T. C. (1965). Aspects of stimulation and organization in approach/withdrawal processes underlying vertebrate behavioral development. In: *Advances in the Study of Behavior.* D. L. Lehrman, R. Hinde & E. Shaw (eds.). New York: Academic Press, Ch. 1.

SCHOLES, N. W. (1965). Detour learning and development in the domestic chick. *J. comp. physiol. Psychol.*, **60**, 114–116.

SCHOLES, N. W. & WHEATON, L. G. (1966). Critical period for detour learning in developing chicks. *Life Sciences*, **5**, 1859–1865.

SCOTT, J. P. (1963). Principles of ontogeny of behavior patterns. *Proc. XVI Internat. Congr. Zool.*, Washington, D.C., 363–366.

SEBEOK, T. K. (1965). Animal communication. *Science, N.Y.*, **147**, 1006–1014.

SELYE, H. (1947). The general-adaptation syndrome and the diseases of adaptation. *Textbook of Endocrinology*, Acta Endocrinologica, Montreal: Montreal University, 837–866.

SEXTON, O. J. & FITCH, J. (1967). A test of Klopfer's empathic learning hypothesis. *Psychon. Sci.*, **7**, 181–182.

SIEGEL, H. S. (1959a). Egg production characteristics and adrenal function in White Leghorns confined at different floor space levels. *Poult. Sci.*, **38**, 893–898.

SIEGEL, H. S. (1959b). The relation between crowding and weight of adrenal glands in chickens. *Ecology*, **40**, 495–498.

SIEGEL, P. B. (1959). Evidence of a genetic basis for aggressiveness and sex drive in the White Plymouth Rock cock. *Poult. Sci.*, **38**, 115–118.

SIEGEL, P. B. (1960). A method for evaluating aggressiveness in chickens. *Poult. Sci.*, **39**, 1046–1048.

SIEGEL, P. B. (1965). Genetics of behavior: selection for mating ability in chickens. *Genetics*, **52**, 1269–1277.

SIEGEL, P. B. & HURST, D. C. (1962). Social interactions among females in dubbed and undubbed flocks. *Poult. Sci.*, **41**, 141–145.

SIEGEL, P. B., PHILLIPS, R. E. & FOLSOM, E. F. (1965). Genetic variation in the crow of adult chickens. *Behaviour*, **24**, 229–235.

SIEGEL, H. S. & SIEGEL, P. B. (1961). The relationship of social competition with endocrine weights and activity in male chickens. *Anim. Behav.*, **9**, 151–158.

SIEGEL, P. B. & SIEGEL, H. S. (1964). Rearing methods and subsequent sexual behaviour of male chickens. *Anim. Behav.*, **12**, 270–271.

SIMNER, M. L. (1966). Relationship between cardiac rate and vocal activity in newly hatched chicks. *J. comp. physiol. Psychol.*, **61**, 496–498.

18*

SKARD, A. G. (1937). Studies in the psychology of needs: Observations and experiments of the sexual needs of hens. *Acta Psychol.*, **2**, 175–232.

SMITH, F. V. & BIRD, M. W. (1964). The sustained approach of the domestic chick to coloured stimuli. *Anim. Behav.*, **12**, 60–63.

SMITH, F. V. & TEMPLETON, W. B. (1966). Genetic aspects of the response of the domestic chick to visual stimuli. *Anim. Behav.*, **14**, 291–295.

SMITH, W. & HALE, E. B. (1959). Modification of social rank in the domestic fowl. *J. comp. physiol. Psychol.*, **52**, 373–375.

SPALDING, D. A. (1873). Instinct: With original observations on young animals. *MacMillan's Magazine*, **27**, 282–293. Reprinted in *Br. J. Anim. Behav.*, **2**, 2–11 (1954).

STEVEN, D. M. (1955). Transference of "imprinting" in a wild gosling. *Br. J. Anim. Behav.*, **3**, 14–15.

STURKIE, P. D. (1965). *Avian Physiology*. Ithaca, N.Y.: Comstock.

TABER, E., DAVIS, D. E. & DOMM, L. V. (1943). Effect of sex hormones on the erythrocyte number in the blood of the domestic fowl. *Am. J. Physiol.*, **138**, 479–487.

TANABE, Y. & TOSHIMATSU, K. (1962). Thyroxine secretion rates of molting and laying hens and general discussion of the hormonal induction of molting in hens. *Bull. Nat. Inst. Agric. Sci.* (Japan), Series G, **21**, 49–49.

TAYLOR, A., SLUCKIN, W., HEWITT, R. & GUITON, P. (1967). The formation of attachments by domestic chicks to two textures. *Anim. Behav.*, **15**, 514–519.

TINBERGEN, N. (1948). Social releasers and the experimental method required for their study. *Wilson Bull.*, **60**, 6–51.

TINDELL, D. & CRAIG, J. V. (1960). Genetic variation in social aggressiveness and competition effects between sire families in small flocks of chickens. *Poult. Sci.*, **39**, 1318–1320.

TOLMAN, C. W. (1963). A possible relationship between the imprinting critical period and arousal. *Psychol. Rec.*, **13**, 181–185.

TOLMAN, C. W. (1964). Social facilitation of feeding behaviour in the domestic chick. *Anim. Behav.*, **12**, 245–251.

UPP, C. W. (1928). Preferential mating of fowls. *Poult. Sci.*, **7**, 225–232.

VITULLI, W. F. & TALLARICO, R. B. (1967). Visual, motor and age factors in maze learning in chicks. *J. gen. Psychol.*, **77**, 111–120.

VON BEKESY, G. (1960). *Experiments in Hearing*. New York: McGraw-Hill.

VON FABER, H. (1964). Stress and general adaptation syndrome in poultry. *World's Poult. Sci.*, **20**, 175–182.

VON HOLST, E. & VON ST. PAUL, U. (1960). Vom Wirkungsgefüge der Triebe. *Naturwissenschaften*, **47**, 409–422.

VORONIN, L. G. (1962). Some results of comparative-physiological investigations of higher nervous activity. *Psychol. Rev.*, **59**, 161–195.

WALK, R. D. & GIBSON, E. J. (1961). A comparative and analytical study of visual depth perception. *Psychol. Monogr.*, **75** (15, Whole No. 287).

WARREN, J. M., BROOKSHIRE, K. H., BALL, G. G. & REYNOLDS, D. V. (1960). Reversal learning by White Leghorn chicks. *J. comp. physiol. Psychol.*, **53**, 371–375.

WILLIAMS, C. & McGIBBON, W. H. (1955). Courtship behavior of the male domestic fowl, *Gallus domesticus. Poult. Sci.*, **34**, 1172–1173.

WILLIAMS, C. & McGIBBON, W. H. (1956). An analysis of the peck-order of the female domestic fowl, *Gallus domesticus. Poult. Sci.*, **35**, 969–976.

WILLIAMS, C. & McGIBBON, W. H. (1957). The relationship of the various mating behavior activities of the male domestic fowl. *Poult. Sci.*, **36**, 30–33.

WITSCHE, E. & MILLER, R. A. (1938). Ambisexuality in the female starling. *J. exp. Zool.*, **79**, 475–487.

WOOD-GUSH, D. G. M. (1954). The courtship of the Brown Leghorn cock. *Br. J. Anim. Behav.*, **2**, 95–102.

WOOD-GUSH, D. G. M. (1955). The behaviour of the domestic chicken: A review of the literature. *Br. J. Anim. Behav.*, **3**, 81–110.

WOOD-GUSH, D. G. M. (1956). The agonistic and courtship behaviour of the Brown Leghorn cock. *Br. J. Anim. Behav.*, **4**, 133–142.

WOOD-GUSH, D. G. M. (1957). Aggression and sexual activity in the Brown Leghorn cock. *Br. J. Anim. Behav.*, **5**, 1–6.

WOOD-GUSH, D. G. M. (1958). Fecundity and sexual receptivity in the Brown Leghorn female. *Poult. Sci.*, **37**, 30–33.

WOOD-GUSH, D. G. M. (1959). A history of the domestic chicken from antiquity to the 19th century. *Poult. Sci.*, **38**, 321–326.

WOOD-GUSH, D. G. M. (1960). A study of sex drive of two strains of cockerels through three generations. *Anim. Behav.*, **8**, 43–53.

WOOD-GUSH, D. G. M. (1963). The relationship between hormonally-induced sexual behaviour in male chicks and their adult sexual behaviour. *Anim. Behav.*, **11**, 400–402.

WOOD-GUSH, D. G. M. & KARE, M. R. (1966). The behaviour of calcium-deficient chickens. *Br. Poult. Sci.*, **7**, 285–290.

WOOD-GUSH, D. G. M. & OSBORNE, R. (1956). A study of differences in the sex drive of cockerels. *Br. J. Anim. Behav.*, **4**, 102–110.

ZOLMAN, J. F. & MARTIN, R. C. (1967). Instrumental aversive conditioning in newly hatched domestic chicks. *Psychon. Sci.*, **8**, 183–184.

Chapter 17

The Behaviour of Turkeys

E. B. Hale, W. M. Schleidt and M. W. Schein

Although wild turkeys (*Meleagris gallopavo*) ranged over North America from southern Mexico to the north-eastern United States, domestication occurred only in the seed-planting culture of the Aztec Indians of Mexico. Another wild species native to Mexico, the ocellated turkey (*Agriocharis ocellata*), is present on the Yucatan Peninsula of Central America but has not been domesticated. Wild turkeys may have been exterminated around centres of advanced Indian culture prior to the Spanish conquest (Leopold, 1948). Domestic turkeys were present in southern Mexico at the time of the Spanish conquest, but wild turkeys in Mexico were restricted to the northern provinces in historical times. Further information on the history of domestication and on the management of the wild turkey is contained in two recently published extensive studies (Schorger, 1966; Hewitt, 1967).

Meleagris gallopavo includes five subspecies of wild turkeys and all domestic varieties. Recognized subspecies are the Mexican turkey (*gallopavo*) of central Mexico, Merriam's turkey (*merriami*) in the mountain areas of south-western United States and northern Mexico, the Rio Grande turkey (*intermedia*) of Texas and north-eastern Mexico, the Florida turkey (*osceola*), and the Eastern turkey (*silvestris*) ranging from Texas and Nebraska east to the Atlantic coast and north-eastern United States. A proposed subspecies (*onusta*) is found in Mexico on the western slopes of the Sierra Madre mountains.

Domesticated turkeys from Mexico were introduced into Europe through Spain and several European varieties were developed. The first domestic turkeys in the United States came from Europe rather than from the indigenous *M. g. silvestris*. Crosses with the native turkey were made in New England in the early 19th century and a local strain developed in this manner was the forerunner of the Narragansett and Bronze varieties (Marsden & Martin, 1955). According to the American Standard of Perfection, there is properly only one breed of domestic turkey with seven recognized varieties and several non-standard varieties (Marsden & Martin, 1955). Divergent trends in the behaviour of wild and domestic turkeys have been considered in Chapter 2.

A few individual birds have been produced by hybridizing turkeys with guinea fowl, pea fowl (Schorger, 1966), chickens (Olsen, 1960) and ring-neck pheasants (*Phasianus colchicus*), as well as with the ocellated turkey (Asmundson & Lorenz, 1957). A study of the behaviour of these unusual hybrids may offer an exciting opportunity in behavioural genetics.

I. SENSORY AND LOCOMOTOR CAPACITIES RELATED TO DISPLAY

A picture of the unique behaviour of turkeys requires a careful analysis of the *fixed action patterns* (species-characteristic movements) and unique stimulus situations effective in eliciting behaviour. Although chickens and turkeys may show only minor differences in gross sensorimotor equipment, the precise movements during locomotion, courtship, or fighting, the vocalizations and alarm calls, and the significance of these for other birds, are very different between the two species. Similar divergences may be observed in even more closely related species (Hinde, 1961). Only a brief resumé of the basic sensory capacities will be noted and in the absence of specific information on turkeys, inferences are made on the basis of the general capacities of birds (Sturkie, 1954; Portmann, 1961; Pumphrey, 1961). Only those motor patterns having general significance in a variety of behavioural displays will be considered.

A. Sensory Capacities

The highly developed visual and auditory capacities of turkeys emphasize their crucial role in social behaviour, communication, and responses to predators. As diurnal gallinaceous birds, turkeys are presumed to have visual characteristics similar to chickens, *viz.* colour vision, rapid accommodation, high visual acuity, a visual field of about 300 degrees, and a limited binocular field. As the eyes fit tightly in the orbits, there is minimum eye movement and moving objects are followed by moving the head and neck. Hearing is well developed with frequency discrimination and the ability to determine the direction of sound comparable with that of man (Schwartzkopff, 1950). However, the band of frequencies to which the ear is sensitive is generally more restricted in birds than in mammals. In turkeys the band of sensitivity includes frequencies from 200 to 6,000 cycles and may be slightly broader (Schleidt, 1955).

The skin of birds is rich in sensory receptors but the functional characteristics have not been clarified (Portmann, 1961). Tactile receptors may be presumed to be highly developed in turkeys since sexual responses in the female are readily elicited by tactile stimulation over extensive areas of the body (see section on Sexual Behaviour). The well-authenticated ability of birds to detect distant explosions has been demonstrated to be based on the excitation of sensory organs (corpuscles of Herbst) in the legs receiving vibrations through the ground (Schwartzkopff, 1949). These structures are particularly sensitive when the bird is in a resting posture on branches and may serve to detect the approach of predators.

Inferences from studies of taste in chickens (Chapter 16) suggest a well-developed gustatory sensitivity in turkeys. However, wide species differences in behavioural acceptance and rejection of various chemical compounds indicate the need for specific studies with turkeys to determine their actual gustatory capabilities (Kare & Halpern, 1961). Evidence for true olfactory functions in birds is contradictory (Portmann, 1961) and no positive evidence has been obtained for turkeys. For the present, we must assume that olfaction plays a negligible role in the behaviour of turkeys.

B. Locomotor Capacities and Display Patterns

The beak serves as the primary structure for manipulating and grasping objects and for exploration of the immediate environment. Limited grasping functions are also served by the feet, notably in roosting and male mating behaviour. Turkeys are typical of birds in specialization of the fore limbs for flight. However, the high sternum and poorly developed pectoral muscles of even the wild *Meleagris* is indicative of inferior capabilities on the wing relative to most gallinaceous birds and even in comparison with the ocellated turkey (Leopold, 1948). Apart from these structural specializations, postural reflexes, intention movements, feather postures, and pigmentation of the carunculated areas of the head and neck provide the basis for specific display patterns.

1. Postural Reflexes and Intention Movements

Displacement of one part of the body is followed by postural changes in other parts so that the body as a whole strikes a new attitude. Such attitudinal reflexes are most readily elicited by altering the position of the head and are termed *static reflexes* since the animal is not in motion. *Statokinetic reflexes* are evoked when the body is suddenly subjected to movement through space. Several static and statokinetic reflexes observed in chickens (Kleitman & Koppanyi, 1926) and ducks (Koppanyi & Kleitman, 1927) are also typical of turkeys. Static reflexes are evoked by proprioceptive impulses from the skeletal muscles and vestibular organs, and are in part influenced by visual cues maintaining the visual field (Dukes, 1955).

Static postural reflexes and intention movements. Most behavioural acts begin with a movement towards or away from some object or animal and preparatory to making a movement, the head may be raised, lowered, withdrawn, or extended. These movements are accompanied by postural adjustments evoked by static reflexes. Postural adjustments of this type are suggested to constitute the intention movements described by Daanje (1950). Body attitudes produced by tilting a bird closely parallel postures observed during the first and second phase of the jump (take-off leap) by a bird (Fig. 91). In turn, the components of the take-off leap play a prominent part in many display postures and Daanje (1950) has suggested the male sexual display of turkeys is derived from the first phase of the jump. Other behavioural displays in turkeys probably represent ritualization of one of the two phases of the take-off leap and some have acquired functions as social signals (Fig. 91). However, in other related gallinaceous birds we find displays of analogous function but of very different appearance as far as the tilt of the body axis is concerned. Therefore, only a thorough study of homologous displays in related species can give us decisive evidence for the origin of a particular behaviour pattern.

Statokinetic reflexes. These reflexes are evoked during movement and are elicited by angular or linear accelerations, both positive and negative, acting upon receptors in the semicircular canals of the inner ear (Dukes, 1955). Sudden lowering of a supported turkey evokes slight flapping of the wings with the legs extended and toes spread (*landing reflex*). If a supported turkey is rotated in a wide arc, the head and tail flex ipsilaterally to a side determined by the position of the body with respect to the direction of motion. For example, a bird held so that it travels anteriorly around a large circle characteristically flexes its head and tail to the side *away* from the centre of the

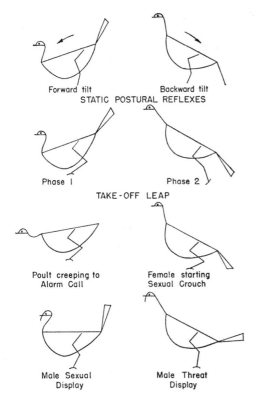

Forward tilt   Backward tilt
STATIC POSTURAL REFLEXES

Phase 1   Phase 2
TAKE-OFF LEAP

Poult creeping to
Alarm Call

Female starting
Sexual Crouch

Male Sexual
Display

Male Threat
Display

Fig. 91. Diagrammatic illustration of the possible origin of displays in turkeys. Movements constituting the two phases of the take-off leap are derived from static postural reflexes. Displays representing ritualization of different phases of the leap appear under the respective phase.

circle; if travel is reversed, the head and tail shift to the side nearest the centre of the circle (Fig. 92). Displays involving ipsilateral flexion of the head and tail are probably derived from statokinetic reflexes.

Physiologically related displays—feather posture and chromatophores. Physiological changes related to conflict and thwarting may produce related changes in feather posture or chromatophore action. These physiologically related changes may form the basis of feather and colour displays and evolve into ritualized social signals (Morris, 1956). Feathers may be relaxed, sleeked, fluffed, or ruffled depending upon the physiological state of the bird. In turkeys the sleeked feathers of the threat display and the ruffled feathers of a broody hen or courting male serve as examples.

The naked carunculated areas of the head and neck of turkeys are richly pigmented and chromatophore changes during display provide a rapidly changing spectrum of colours from brilliant red to blue, purple, and white. When the bird is disturbed all colours tend to pale; during threat or early stages of a fight the naked areas take on a brilliant red hue, and during prolonged courtship white becomes predominant. The red colour of the neck is produced by dilated blood vessels. Astaxanthin and other carotenoid pigments contribute to the red colour of caruncles covering the head. Blue areas have heavy deposits of melanin in the deeper epidermal layers (Sawyers, 1966). For a review of the literature on colour change, see Schorger (1966).

Turkeys also possess a pendant *snood* just above the beak which is usually retracted but becomes elongated during various displays (Plate XIX). If a bird is disturbed the snood is retracted and if sleeping or tranquillized with drugs it is flaccid and elongated. During sexual and fighting displays the snood is elongated and turgid. A spongy tissue on the breast of the male shows similar but less conspicuous changes and is collapsed by various tranquillizing drugs (Hale, 1969).

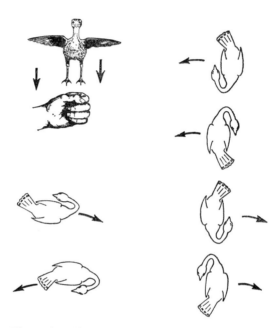

Fig. 92 Statokinetic reflexes evoked in turkeys during movement through space illustrate shifts in flexion of head and tail with changes in direction of rotation. A landing reflex is elicited at the upper left by removing support.

Displacement activities typical of conflict situations in many birds are not prominent in turkeys. This may imply that such movements have become ritualized and are now recognized components of specific behaviour patterns. Only a rare wiping of the head across the back (eye wiping) stands out as a possible displacement activity in turkeys.

II. INGESTIVE BEHAVIOUR

A. Feeding and Drinking Behaviour

Wild turkeys are not specialized in feeding habits and eat plant materials including berries, acorns, seeds, leaves, and tender grasses as well as insects. Consequently, formulated rations or supplemented pasture provide adequate and readily acceptable feed for domestic turkeys. The special diets of hard-boiled eggs, etc., once recommended have been replaced by low-fibre, high-energy rations with abundant mineral and vitamin supplement (Marsden & Martin, 1955; Schorger, 1966; Hewitt, 1967).

1. Pattern of Feeding and Drinking

Turkeys, like chickens, possess the basic avian pattern of feeding without specialized adaptations. Swallowing is accomplished without the necessity of raising the head although some adults may lift the head frequently while gulping large quantities of mash. Drinking is accomplished differently from feeding, with the turkey dipping its beak into the water to the level of the nares and making several rapid partial closures of the beak. It then raises its head, extends the beak upward and repeats several rapid closures of the beak. The latter pattern suggests some degree of active swallowing in addition to flow by gravity to the oesophagus.

Scratching behaviour is an important component of foraging for food and gallinaceous birds scratch with one foot while standing on the other. Wild turkeys extend the foot forward and scratch backward and outward in a sweeping motion to form an inverted V pattern on the ground. When feeding on mast (acorns) covered with leaves they may dig up several acres (Latham, 1956). In domestic turkeys scratching behaviour is rarely observed on range or in litter provided in pens and seems to depend upon the rearing conditions. Scattering grain in the litter readily induces vigorous scratching behaviour in domestic chickens but is generally ineffective with domestic turkeys. The apparent reduction of ground-scratching in domestic turkeys either suggests that a highly specific texture is required to elicit scratching in turkeys or that there has been a marked reduction of this behaviour in turkeys under domestication.

Stimuli. Newly hatched poults peck indiscriminately at bright spots and small objects which contrast with the background. These responses soon lead the poults to peck at feed and at images in water. In this manner the poults quickly learn the location of feed and water if these are placed in readily available areas. It is sometimes recommended that beaks be dipped in water as the poults are placed under the hover. However, in a well-lighted pen with clear accessible water, poults soon drink without any special care. The early indiscriminate pecking changes to selective eating and drinking over a period of a few days. Social facilitation of this behaviour is provided by other poults or by the hen as she pecks the ground and calls. If novel objects are presented, poults continue to engage in curiosity pecking.

2. Regulation of Feed Intake

Growing turkeys readily regulate feed intake in accordance with the energy demands for maintaining normal growth. Birds receiving pelleted rations more than doubled the amount of feed eaten as the caloric content of the ration was reduced (Table 51). Only when the lowest energy ration (20 per cent fibre) was fed in the form of mash did the turkeys fail to maintain normal growth (Dymsza et al., 1957). However, since these turkeys ate over 3 times as long as birds receiving the same ration in pellet form (Fig. 93) yet were able to consume only three-quarters as much feed, failure to maintain growth must be ascribed to difficulty in eating large quantities of the bulky feed.

Changes in energy requirements with seasonal changes in temperature bring about similar adjustments in feed intake. In addition to a general increase in total intake during colder months and a decrease in warmer months (Almquist, 1953; also Chapter 4), turkeys may select grains on the basis of their energy content. Turkeys hatched at different times of the year

Table 51. Ability of Turkeys to adjust Energy Intake to meet Growth Requirements and the Relation of Physical Characteristics of Pelleted Rations to Feather Picking. (Adapted from Dymsza et al., 1957; Hale, 1962.)

Ration		Feed Intake per Bird	Body Wt per Bird[1]	Feather Picking[1]	
Fibre	Cal/lb			Av Size of Denuded Area	Birds Denuded
%		lb	lb	Sq inch	%
5	848	18·9	10·8	1·8	95
10	639	24·2	11·2	1·1	84
15	433	29·4	11·0	0·6	61
20	226	40·3	10·7	0·1	10

[1] At 16 weeks of age.

and provided mash, oats, wheat, and corn ad lib., showed different preferences when compared at 4 months of age. During the summer, oats constituted 42 per cent of the total intake and the high energy grains only 18 per cent. Birds reaching the same age during the cold winter months reversed their preferences with oats constituting only 25 per cent and corn and wheat 48 per cent of the total intake (Margolf, 1961).

As turkeys grow older, the amount of feed eaten each week increases from 0·2 lb during the 1st week to almost 8 lb at 30 weeks of age. However, the intake each week per pound of body weight decreases from 0·7 lb during the 1st week to a relatively constant 0·3 lb by 20 weeks of age (Scott, 1960). Total water intake per bird each week increases from 0·08 U.S. gallons during the 1st week to more than one gallon by 15 weeks of age and then remains fairly stable (Morehouse, 1949).

3. BEHAVIOURAL CHANGES RELATED TO PHYSICAL CHARACTERISTICS OF RATION

Shifts in the general activity of birds and in feather picking or cannibalism occur with changes in the fibre content of the ration and with the form (mash or pellets) in which it is fed. More than twice as much time is required to eat the same ration in mash than in pelleted form and birds eating mash spend far more time wiping the beak through feathers of other birds in a cleaning movement (Fig. 93). The time required to eat a bulky ration (20 per cent fibre) in mash form is so great that the birds have no time to rest (roosting) and are either eating or moving to another spot at the feeder.

PLATE XIX

a. Threat display in male turkeys is characterized by the high head, sleeked feathers, and a trilling vocalization as the birds face each other in a lateral stance. The high head also serves as a social signal eliciting attack. (Photograph from Pennsylvania Agricultural Experimental Station.)

b. Females giving the sexual crouch to man may be used to determine the responses elicited by tactile stimulation of various areas of the body. The hen pictured is elevating her tail and exposing the cloaca in response to tactile stimulation of the erogenous area along the sides of the body. (Photograph from Pennsylvania Agricultural Experimental Station.)

PLATE XX

(*Above*) The size of the denuded area on a hen's neck is an excellent indication of her social status. The hen on the left, with all her feathers intact, is at the top of the peck order in her flock, and the hen on the right, with the completely denuded neck, at the bottom. (Photograph from Pennsylvania Agricultural Experimental Station.)

(*Centre*) Abnormal behaviour may develop in inbred lines or in certain crosses. The almost completely denuded hen pictured is from a cross showing extreme feather picking even though neither parent strain exhibits this abnormality. (Photograph from Pennsylvania Agricultural Experimental Station.)

(*Below*) Partial eversion of the oviduct by a turkey female in response to tactile stimulation. Sexual receptivity is terminated by complete eversion of the oviduct in an orgasmic-like response. (Photograph from Pennsylvania Agricultural Experimental Station.)

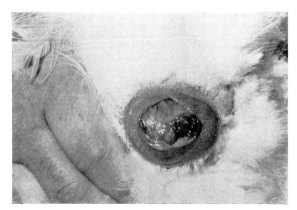

Activity patterns of birds receiving rations with 10 per cent and 15 per cent fibre differ little from those receiving 5 per cent.

Picking feathers from the backs of other birds is an undesirable form of behaviour and may predispose the flock to cannibalism. Pelleting a ration induces a marked increase in both the number of birds with denuded backs and in the size of the denuded areas. This tendency is counteracted by an increase in the fibre content of the feed (Table 51). In general, the low-fibre, high-energy, pelleted rations providing most efficient growth and production favour the development of this undesirable behaviour.

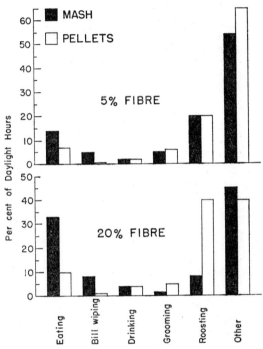

Fig. 93. Ingestive and general activities of turkeys at 12 weeks of age as influenced by the form (mash or pellet) and fibre content of the ration (Hale, 1969).

III. SEXUAL BEHAVIOUR

Turkeys under domestication have continued to be seasonal breeders but show sufficient individual variation to suggest that a more prolonged reproductive period could be established through selective breeding if man desires. Increasing day length brings domestic turkeys to sexual maturity in March (northern United States) in contrast with the later sexual maturity in May of wild turkeys in the same latitude. Sexual maturity may be advanced several months by subjecting the birds to 14 hours of artificial light starting in December (Margolf *et al.*, 1947) or to higher environmental temperatures (Burrows & Kosin, 1953). Both male and female turkeys achieve a condition of potential sexual maturity at approximately seven months of age. In males spermatogenesis is attained at the end of the normal growth period, after which the reproductive system regresses towards a quiescent state in the absence of artificial lighting. Females do not show complete ovarian development unless they attain the proper age at a time of the year when sufficient daylight is present to stimulate reproductive maturation (Margolf *et al.*,

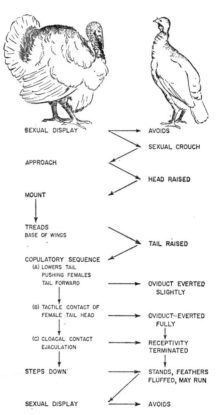

Fig. 94. Schematic illustration of the chain reaction type of mating behaviour in turkeys indicating interactions between male (*left*) and female (*right*). (Adapted in part from Hale, 1955b.)

1947). Gallinaceous birds are somewhat exceptional in that ovarian development in the female is independent of stimulation provided by the male (Lehrman, 1959). Turkey females actively courted and mated by males do not show any advance in egg laying over those isolated from males (Hale, 1969).

Mating behaviour in both wild and domestic turkeys is characterized by promiscuous mating without the formation of pair bonds before or after copulation. The females are either present with the male or are attracted to special mating areas. In *M. g. intermedia* males may alternate between more than one breeding area, with dominant males chasing subordinate males away and several dominant males remaining in the same area (Evans, 1961). Non-receptive females avoid the males while receptive ones approach and respond with a sexual crouch. Tendencies to attack or flee from the female during courtship are relatively insignificant in both wild and domestic males.

A. Synchronization of Male and Female Patterns

Mating behaviour in turkeys tends to follow a chain reaction type pattern. The behaviour of one sex partner elicits a specific response from the other and that response in turn elicits a further response from the first partner (Fig. 94). The sequence of events may be divided into the courtship and copulatory phases (Schein & Hale, 1958, 1960; Schleidt & Schleidt, 1962a).

1. COURTSHIP

Courtship display in the male turkey (usually referred to as "strutting") is characterized by slow and restricted movements and elaborate feather display. The tail is fanned vertically; feathers on the back, breast, and flank are erected; the wings are lowered with the primary feathers spread; the crop is inflated. All combine to create an impression of larger apparent size by an enhancement in all dimensions. The neck forms a tight S shape bringing the head in close to the body. The snood becomes elongated and turgid and the naked carunculated areas of the head and neck change from bright red to a predominantly white colour with blue contrasting cheeks after prolonged courtship.

While courting, the male slowly paces back and forth and at intervals takes a series of three to five steps, drags the primaries on the ground, and accompanies the sequence with a low vocalization described by Audubon as a pulmonic puff. This sequence is a good example of a "fixed action pattern" and we suggest for it the term "strut" (Fig. 95). There is usually some shift in direction before the next sequence of rapid steps. The frequency of struts tends to be low at the start of a courtship sequence and increases to a maximum after a few minutes. Males show considerable variation, ranging from 1 to over 10 per minute. During the display, the male orients the frontal aspect of his tail towards the female courted.

Non-receptive females avoid the courting males but females stimulated almost to the point of crouching may show signs of *semi-receptivity*. These semi-receptive females avoid the males less vigorously and may stand in a posture suggesting drowsiness or assume a posture similar to that of a female starting the sexual crouch (Fig. 91) but with the legs not flexed and with the back feathers fluffed. In this position they may stand or walk about among the courting males, sometimes for up to half an hour (Schleidt & Schleidt, 1961). Semi-receptive females may respond to the male vocalization during a strut by giving a reflex-like partial crouch.

Sexually receptive females enter the sexual crouch by assuming the posture illustrated in Fig. 91. The head is held high, the tail flexed ventrally and the legs are gradually flexed as the wings are drooped and fluttered. In the complete crouch, the head is usually held close to the body. As a courting male approaches, the female raises her head and then extends it to the maximum height as the male mounts.

Fig. 95. "Strut" of the male Turkey. *Left*: Sound spectrogram; note the wide frequency burst at the beginning, followed by a low frequency drumming sound (40 to 60 c.p.s.), which is very probably produced by the inflated airsacs. *Right*: Position of the turkey, about 0·1 seconds after the introductory burst. (Schleidt, 1964a.)

2. COPULATORY SEQUENCES

In response to a crouching female the male moves slowly to her, mounts deliberately, balances, and orients towards the female's head. After standing on the female's back for a period varying from a few seconds to several minutes with the hen's head frequently in contact with his breast, the male moves to grip with his feet the wing base of the female in preparation for making copulatory attempts. The female responds by raising her tail and exposing the vent. As the male lowers his tail, the female everts the oviduct slightly and then as the male comes into contact with the base of the female's tail she everts the oviduct fully. Normally the male brings the copulatory papillæ into contact with the everted oviduct and freezes momentarily as he ejaculates while the female stretches the head straight forward. After the male steps down, the female gets up, fluffs her feathers, and may run in an arc and vocalize in a typical after-response. Once the female everts her oviduct fully her sexual receptivity is terminated; she attempts to escape even if the male fails to inseminate her and in spite of continued copulatory attempts by the male. If the male dismounts at a stage prior to complete eversion of the female's oviduct, she remains in the sexual crouch.

Several deviations from the normal pattern may be seen in various strains. Some males may tread the ground in front of the female and press their breasts against the female's head without attempting to mount. Some males may repeat mountings several times before making copulatory attempts; in these instances the female remains receptive. If the male terminates the female's receptivity during copulatory attempts by eliciting eversion of the oviduct, he continues to make repeated attempts until the female escapes. Males show marked variation in their ability to complete attempted matings with some completing almost all attempts and others as few as 12 per cent (Hale, 1955b; Smyth & Leighton, 1953). If a female crouches against a barrier, making it difficult for the male to mount, he may peck lightly at her head or tramp her head with one foot. This usually disturbs the hen momentarily and she may stand and recrouch. If a male struts for long periods without mounting, a female may on rare occasions get up and attack him by pecking at his head.

B. Quantitative Aspects of Female Sexual Behaviour

In contrast to mammals, female turkeys do not exhibit oestrous periods within the breeding season. As noted in the previous section, extreme eversion of the oviduct terminates female sexual receptivity and the female does not crouch again until after an interval of time characteristic of individual females and highly variable from bird to bird (Table 52). An analysis of the

Table 52. Periods of Sexual Non-receptivity of Turkey Hens following Complete and Incomplete Matings. (Adapted from Hale, 1955b; Hale, 1969.)

Female	No. of Matings	Days Non-Receptive After	
		Complete Matings	Incomplete Matings
A	60	2	3
B	38	4	5
C	28	6	4
D	21	8	6
E	17	11	9
F	12	12	13
G	8	18	19
Av. of 36 females	767	5·4	5·2

quantitative aspects of female behaviour requires consideration of the stimuli terminating sexual receptivity in addition to those factors influencing the frequency of mating.

1. STIMULI TERMINATING FEMALE RECEPTIVITY

Although male courtship provides visual stimuli arousing sexual receptivity in the female, during the copulatory sequence tactile stimuli play the predominant role in eliciting female behaviour up to and including the termination of receptivity in an orgasmic-like response. Inferences as to the effective stimuli may frequently be made from observations of behaviour, but such judgments may be erroneous and therefore should be subjected to independent testing. Fortunately, some female turkeys give the sexual crouch to humans and provide excellent material for the precise analysis required (Plate XIX).

A step by step analysis of female behaviour indicates the importance of a series of highly specific stimuli in eliciting the sequence of female responses. The specific responses and typical effective stimuli are: sexual crouch—presence of the male; elevation of the head—approach of the male; elevation of head to maximum height—pressure on the back of the female; elevation of tail—light tactile stimulation of the ventral aspect of the wing base or the sides of body normally covered by the wings; partial eversion of the oviduct —pressing tail forward; complete eversion of oviduct—tactile stimulation of the base of the tail or area to the side of the cloaca (Schein & Hale, 1958; Hale, 1959b).

Sexual receptivity is terminated by complete eversion of the oviduct irrespective of the eliciting stimulus and whether or not copulation has been achieved (Plate XX). Similarly, the refractory period is just as long if termination of receptivity is achieved without copulation (*incomplete mating*) as when copulation occurs (*complete mating*) (Table 52). Curiously enough, palpation of the oviduct has no effect on terminating sexual receptivity.

Individual differences in sensitivity to tactile stimulation. As suggested above, the turkey female has a large erogenous area starting at the wing base and extending backwards along the sides of the body to include all the area normally covered by the wings. The vent, the area below the vent, and the exposed surfaces of the body are not a part of the sensitive zone. Most hens respond by elevating the tail, and perhaps everting the oviduct slightly, if any part of the erogenous area is stimulated. However, in a few highly sensitive females stimulation of any part of the erogenous zone elicits full eversion of the oviduct and terminates sexual receptivity. On the other hand, some females have such high thresholds to tactile stimulation that combined tactile stimulation of the erogenous zone, the bare skin of the tail head, and the area about the vent are required to induce complete eversion of the oviduct and even then termination of receptivity does not always follow. These differences in thresholds of sensitivity reflect individual or strain differences in sensitivity rather than differences in motivational levels. No differences are observed in the frequency of mating of females in the extreme categories of sensitivity (Hale, 1969).

2. FACTORS INFLUENCING MATING FREQUENCY

The onset of mating in turkey females follows closely the initiation of ovarian development. Mating begins within two weeks from the start of exposure to 14 hours of light daily and reaches a peak in frequency after 4 to 5 weeks. Thereafter there is a brief decline in frequency of mating until at about 7 weeks each female levels off at a relatively constant individual frequency (Fig. 96). Egg laying starts approximately 4 weeks after initial exposure to light and the increase in rate of lay is accompanied by the just noted decline in frequency of mating. That physiological changes, rather than a form of sexual satiation, are the basis for the decline in mating frequency is indicated by the observation that females permitted to mate during the pre-laying period, mate at the same frequency after the onset of laying as do females without prior mating experience (Fig. 96).

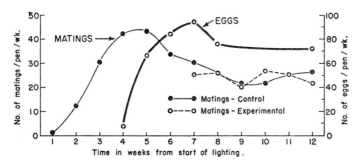

Fig. 96. Relation between mating behaviour and onset of egg production in turkeys (20 birds per pen). Experimental females were not permitted to mate until after they were in full production (Hale, 1969).

Broodiness. Mating in females ceases as the ovaries regress during broody periods. Incubation is maintained by tactile stimulation from the eggs. If broodiness is interrupted, as by removal of the eggs or by destruction of the nest, the hen starts to lay after a short period and again, as at the start of the breeding season, a peak in mating frequency occurs just prior to the onset of laying. Mating in certain females with exceptionally low sex drive may be restricted to these periods of rapid ovarian growth just before laying commences. Some inhibitory effect related to ovulation seems implicated in the reduced mating frequency associated with laying.

Genetic factors. Wide individual differences exist in frequency of mating with some hens commonly mating 10 times as often as others (Table 52). A high correlation between frequency of mating during the pre-laying period with subsequent mating during the remainder of the breeding season, makes it possible to predict mating frequencies of individual females on the basis of limited observations early in the breeding season (Smyth & Leighton, 1953; Hale, 1955b). The contribution of genetic factors to these individual differences is suggested by family differences in mating frequency within a strain (Hale, 1953b) and the demonstrated response under selective breeding for high and low levels of mating frequency (Fig. 97; Smyth, 1955; Harper, 1961). After nine generations of similar selection, a seven-fold difference in mating frequency between high and low lines has been attained (Smyth, 1961). Offspring of crosses between the lines were intermediate in mating frequency. Calculated estimates of heritability are so high as to suggest that mating frequencies are subject to limited environmental modification (Smyth, 1955).

Hormone levels. The observed individual differences in frequency of mating do not appear to be related to differences in levels of oestrogenic hormones.

Oestrogen implants produce negligible increments in mating of females previously mating at low frequencies (Hale, 1969). If mating behaviour is induced by diethylstilboestrol prior to normal stimulation from increasing daylight, females from low- and high-frequency mating lines mate at frequencies characteristic of the line (Smyth, 1961).

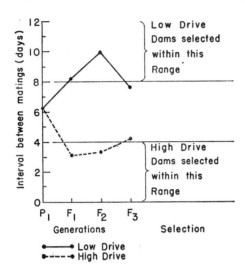

Fig. 97. Response to selection for frequent and infrequent mating in female turkeys. High and low lines were established by selecting as breeders those females mating more frequently than every 4 days or less frequently than every 8 days. Sires were not selected (Hale, 1969).

Correlated characteristics. Females with the highest mating frequency tend to be lower in social rank, smaller in body size, and lay fewer eggs than hens mating infrequently (Hale, 1953a; Harper, 1961). A similar negative correlation between social rank and mating occurs in chickens. Although mating frequency of chickens can be modified experimentally by changing the frequency at which females are dominated (Guhl, 1950), similar manipulation of turkeys fails to produce corresponding changes in mating (Hale, 1969).

3. BISEXUAL BEHAVIOUR AND RESPONSES TO MAN

The potential of turkey females for male as well as female sexual behaviour is indicated by the strutting behaviour induced in young female poults by administration of androgen (Table 54). Although treated females show male sexual behaviour, sensitivity to male hormone is well below that of males. Only 26 per cent of androgen-treated female poults showed male display with less than 15 per cent attempting copulation compared with 85 per cent and 46 per cent of the males, respectively (Schein & Hale, 1959). Secondary male sexual characteristics, such as gobbling and development of the snood, also develop in response to androgens but to a lesser degree in females than males (Table 54; Herrick, 1951).

Adult females otherwise showing normal reproductive characteristics may exhibit male sexual behaviour in addition to normal female behaviour. Such individuals may crouch sexually and, after being mated by a male, get up, court other females with the struts, mount crouching hens, and in some instances carry copulatory movements to the point of providing adequate stimulation to terminate the receptivity of the other hen. Similar behaviour has been observed in female Sage Grouse (Scott, 1942) and is of special interest because of the similarities in sexual behaviour of Sage Grouse and turkeys (Simon, 1940).

Females imprinted to humans at hatching would be expected to sexually crouch to man (Chapter 6). More surprising is the tendency of a relatively large number of females not imprinted to man to give the sexual crouch to man as well as to male turkeys. Females of several varieties were separated from males before reaching sexual maturity and their response to man determined by entering each pen for a standard time at the peak of sexual receptivity during the pre-laying period (Table 53). At least one

Table 53. Variety Differences in Percentage of Normally Reared Females giving Sexual Crouch to Man (Hale, 1969).

Variety or Cross	No. of Hens	% Crouching to Man
Bronze-Narragansett	48	0
Bronze	48	2
White-Bronze	48	4
Spotted-Bronze	48	4
Black	18	6
Grey-Bronze	48	10[1]
Buff-Black	48	12
Bronze-Black	48	20[1]
Grey	17	24[1]
Black Wing Bronze	17	70

[1] Although these females crouched as the observer entered the pen, they were likely to get up as he approached.

female in the Black Wing Bronze variety demonstrated a preference for the observer over a male turkey. With both a courting male and the observer in the pen, she consistently got up as the male approached and recrouched as the observer approached. In contrast, females of those varieties standing as the observer approached continued to crouch as a male in display came near.

C. Male Sexual Behaviour

1. STIMULI AROUSING AND DIRECTING MALE BEHAVIOUR

While strutting is observed to occur in many situations, even when the male is in a room by itself, there is no question that courtship and mounting motor patterns are readily triggered by visual stimuli provided by the female. However, the specific characteristics of the female eliciting sexual behaviour can be determined only by using models or dummies (Chapter 8). Sexual behaviour of the male is fully aroused and copulatory attempts elicited by a detached female head in an upright position, while a body alone elicts display but no mounting behaviour (Schein & Hale, 1957, 1958). Although a female head is the most effective stimulus arousing male behaviour, turkey males respond to male heads as well as to a series of carved models ranging from an abstract tear-drop, approximately the same size as a female head, to models with marked likeness to a female head (Schoettle & Schein, 1959).

Successful copulation with the female demands that the male be properly oriented in space. The essential role of the head in directing orientation can be tested by placing a detached head at various positions with respect to the body axis. In all instances the male orients after mounting by facing towards the head irrespective of the axis of the dummy body (Schein & Hale, 1957, 1958). Orientation towards the head may serve the additional function of maintaining visual stimulation or in bringing the male's breast into contact with the female's head to provide tactile stimulation as well. If several hens are in the sexual crouch and near each other at the same time, a male may mount one and orient to the head of another. The result is disoriented behaviour with the male attempting to mate the hen sideways or backwards. This situation frequently occurs during the period when many females are highly receptive just prior to the onset of egg laying.

Imprinting. Although young turkeys reared away from other turkeys develop normal patterns of sexual behaviour, the effective stimuli may be markedly modified by imprinting to biologically inappropriate objects (Räber, 1948; Schein & Hale, 1959). Adult males reared with other turkeys after being imprinted to man immediately after hatching, court humans in preference to turkeys. Although these males will mate with females if humans are not present, they continue to respond preferentially to man (Schein, 1960, 1963). Males of the Black Wing Bronze variety, in which many females crouch to man, show enhanced courtship display in the presence of humans and react much as imprinted birds except that their primary preference is for turkey females. This response seems to have resulted from selection of those males strutting most vigorously in the presence of the breeder. The behaviour of such males is in sharp contrast with that of most domestic turkeys which typically avoid man and cease strutting as humans approach them.

2. QUANTITATIVE ASPECTS OF MALE SEXUAL BEHAVIOUR

The level of sexual behaviour anticipated in a mature male at a given time depends upon several factors including his basic genetic potential, transitory response potential (responsiveness decreases with repeated copulations), and the appropriateness of the stimulus situation. Males with high genetic potential show copulatory behaviour under optimal conditions (Condition A, Fig. 98) but merely display without making copulatory attempts if response potential is low following a series of copulations or if the stimulus is deficient (Conditions B and C, Fig. 98). Quantitative comparisons of sexual activity are valid only if the birds compared are equated for response potential and stimulus conditions (Schein & Hale, 1959). Aspects of the situation evoking fear must also be eliminated since fear completely inhibits male sexual behaviour.

Hormone levels. Maximum expression of male sexual behaviour is dependent upon the normal androgen levels provided by functional gonads. Gross changes in androgen levels of adult males with the season of the year are accompanied by corresponding fluctuations in sexual behaviour. The essential role of androgens in male behaviour is illustrated by the induced precocious sexual display of poults treated with androgens (Table 54; Chapter 8). Although some untreated poults may show struts as early as one day

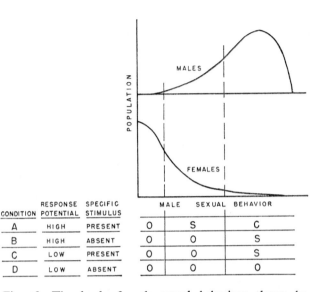

CONDITION	RESPONSE POTENTIAL	SPECIFIC STIMULUS	MALE	SEXUAL	BEHAVIOR
A	HIGH	PRESENT	O	S	C
B	HIGH	ABSENT	O	O	S
C	LOW	PRESENT	O	O	S
D	LOW	ABSENT	O	O	O

Fig. 98. The level of male sexual behaviour shown by androgen-treated male and female turkeys varies from none (O), to courtship strutting (S) and copulatory attempts (C) depending upon a bird's basic genetic potential (indicated by graph with that of females less than that of males), momentary responsiveness, and the adequacy of the stimulus presented (Schein & Hale, 1959).

after hatching and gobbling at an age of 4 weeks, all treated males exhibited these behaviours by 5 weeks of age and several made copulatory attempts before 3 weeks (Schein & Hale, 1959; Schleidt, 1968b). Males with underdeveloped gonads would be expected to be deficient in sexual behaviour but even males with fully functional gonads may vary markedly in levels of sexual behaviour.

Genetic factors. Mature, sexually rested males may copulate with as few as one or as many as 10 females in a 30-minute test period and males showing high sexual activity early in the breeding season tend to remain more active than others throughout the season (Adams, 1959; Hale, 1969). There is strong presumptive evidence that these wide individual differences are related to genetic factors but adequate documentation is not available. Males from heavy, broad-breasted varieties tend to exhibit greater difficulties in mating than varieties not selected for extremes in body type. Defects in level of sexual behaviour seem to be indicated as well as distorted anatomical dimensions, since some individual birds of extreme weight mate without difficulty.

Table 54. Behavioural and Morphological Development at 40 Days of Age in Turkey Poults Treated from Hatching with Androgens. (Hale & Hanford, 1962.)

Treatment	No. of Birds	Snood Length	Age 1st Strut	% Strutting	Age 1st Gobble	% Gobbling
		mm.	days		days	
Injected ♂	23	37	4	100	13	100
Injected ♀	24	27	13	54	15	12
Control (♂ & ♀)	33	18	17	21	29	3

Experience and degree of moult. Patterns of male sexual behaviour develop normally in the absence of early social experience and levels of sexual activity are not modified by varying degrees of early experience. Androgen-treated poults reared in groups or in isolation show similar patterns and levels of sexual activity (Schein & Hale, 1959). Although individual males differ markedly in their ability to complete attempted matings (mating efficiency), this ability does not improve with practice and experienced and inexperienced males differ little in this characteristic (Table 55). As the breeding season progresses a gradual decline in the percentage of fertile eggs typically occurs but this decline is independent of mating activity (Hale, 1955a). Males brought to early sexual maturity by pre-lighting moult earlier in the year than males exposed to normal daylight (Olsen & Marsden, 1952). However, the mating activity of males showing extensive moult does not differ from that of males showing negligible moult even when previous sexual experience is equated (Table 55). Strutting and gobbling are suppressed during the annual moult in summer (Fig. 99). However, injection of testosterone can restore strutting and gobbling activity to its full height (Schleidt, 1968b).

Table 55. Sexual Behaviour of Male Turkeys with Varying Degrees of Experience and Severity of Moult. (Hale, 1969.)

Measure	Pre-lighted		Daylight
	Experienced	Inexperienced	Inexperienced
No. of males	14	13	14
Degree of moult	Extensive	Extensive	Negligible
Mating efficiency (%)	72	66	70
Av. copulations per male	11·3	10·0	12·6
Av. time per copulation (min)	9·0	8·4	9·6

Social rank and mating activity. Although males of different social rank are equally active in mounting females, most of the actual matings are completed by the high-ranking males (Hale, 1953a; Adams, 1959). This difference in

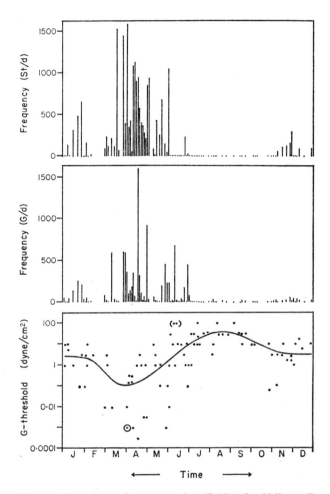

Fig. 99. Frequency of strut per day (St/d), of gobbling calls per day (G/d), and the gobbling-response threshold at various parts of the annual cycle; the solid line in the threshold diagram resembles a smoothed moving average curve. These data are obtained from individually isolated turkeys, kept for 48 hours in a soundproof room. (Schleidt, 1968a.)

total number of copulations achieved is directly related to the fact that the dominant males interfere with mating attempts by the low-status males (Table 56). Subordinate males commonly dismount as soon as a dominant male approaches or following a single peck by the dominant male. Subordinate males rarely interfere with dominant males; when they do, fighting generally occurs indicating that established dominance relations are unstable, since fights frequently follow interference in these instances. Low-ranking males tested in individual matings are as effective as the more dominant ones, thus there is no apparent relation between social rank and actual sexual capacity of the males (Hale, 1953a). In contrast with chickens, subordinate male turkeys continue to attempt mating despite repeated interference by dominant

Table 56. Social Rank and Total Mating Activity of Male Turkeys. (Adapted from Hale, 1953a.)

Social Rank	Total No. Mountings	Total No. Matings	Percentage of Mountings followed by		
			Mating	*Dismount*	*Inter-ference*
1	310	171	53	39	8
2	335	114	34	19	47
3	334	73	22	10	68

males. Obviously, males with underdeveloped gonads would be low in both sexual and fighting behaviour since both of these behaviours are mediated by androgens. Apart from this exception, levels of sexual and fighting behaviour probably vary independently.

D. Behavioural Differences between Chickens and Turkeys

In chickens movements during courtship and mating are more rapid and the feather display of the male is less elaborate than in turkeys. Although male chickens typically force matings with hens, this behaviour is not seen in turkeys and the male at most presses his breast gently against a female or

Table 57. Relative Sexual Stimulus Value of Head and Body in Turkeys and Chickens. (Adapted from Carbaugh *et al.*, 1962.)

Species	Structure	Effectiveness as Stimulus for [1]	
		Arousal	*Complete Pattern*
Turkey	Head	+ + + +	+ + + +
	Body	+ + +	+
Chicken	Head	+ + +	+
	Body	+ + +	+ + +

[1] Maximum value equals 4 +.

extends his wing over her back. Turkey females sometimes approach or follow the male before crouching and if he moves away they get up and crouch near him again. Female chickens show this behaviour less and frequently do not approach the cock before crouching. Although males of both species move away if another male attempts to mount, cockerels succeed in treading each other while male turkeys do not. This difference is related to the forced matings typical of chickens.

Although the female's head is an important cue in both species for achieving proper orientation in space during copulation, the sexual arousal value of different parts of the female vary. Whereas a female head is the primary stimulus eliciting complete mating patterns in turkeys, in chickens neither the head nor the body alone provide maximum stimulation (Table 57).

IV. PARENTAL BEHAVIOUR

Parental behaviour is typically absent in both wild and domestic male turkeys and all aspects of parental care are assumed by the female. Nevertheless, domestic males have occasionally been observed to attach themselves to hens with young and to show the complete set of maternal behaviour patterns (Schleidt & Schleidt, 1961). Male turkeys can be forced to broodiness by first being made drowsy (e.g. by an ample dose of brandy) and then being put on a nest with eggs. After recovery from the hangover broodiness is established. This method was used extensively by farmers in Europe before incubators were available (Schleidt, 1968b).

A. Nesting Behaviour

In wild turkeys the cohesiveness of the winter flocks disintegrates in early spring; the females spread out over a wide area and are mated by males moving through (Watts, 1967). Nesting sites are selected under low concealing vegetation at the base of a large tree or even in more open areas (Leopold, 1944; Ligon, 1946; Latham, 1956). Good fertility of eggs laid over a period of 15 to 20 days is ensured since the average duration of fertility following a single mating is approximately 5 to 6 weeks (Hale, 1955a). If the hen leaves the nest she covers the eggs with leaves by moving the head alternately to each side, picking up leaves in her beak, and depositing them over the eggs.

Domestic hens tend to pace slowly about the edge of the pen or enclosure and give a characteristic low vocalization (a type of yelp) for a short period just before laying an egg. If adequate artificial nests are not available, the hen may dig shallow depressions in a corner or along the edge of the enclosure and lay eggs in them. Although some hens readily adapt to laying in nests placed in the pen, others must be trained to use them. If hens exhibiting the just described pre-nesting behaviour are placed on the artificial nests a few times, they readily return to the nests subsequently and discontinue laying on the floor. Occasionally, certain hens may fail to show the pre-nesting behaviour or be so wary of man that the behaviour is difficult to observe. In these instances the oviduct can be palpated and all hens with oviductal eggs

19+

can be placed on trap nests to ensure laying in the nest a short time after placement. A variety laying only about 25 per cent of the eggs in nests was trained by this procedure to lay 90 per cent of the eggs in nests after less than two weeks of training (Fig. 100). Laying in the nests persisted thereafter even though no further training was provided.

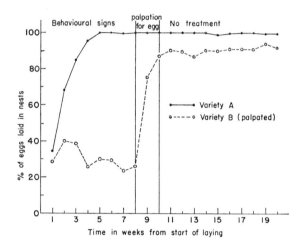

Fig. 100. Variety differences in nesting behaviour. Females of variety A are readily trained to the nests if placed in them while showing pre-nesting behaviour. In variety B pre-nesting behaviour is reduced and hens must be placed on nests when an egg is present in the oviduct (determined by palpation) if they are to be trained successfully (Hale, 1969).

B. Incubation

The termination of laying and the start of incubation are closely related but the exact relationship and regulating factors are not known (Eisner, 1960). During the broody period the hen's feathers tend to be fluffed and the hen shows variable degrees of defence behaviour including ruffling of her feathers and hissing at any intruder.

Decreased levels of broody behaviour are desirable under modern conditions of turkey production and selective breeding against broodiness is very effective (McCartney, 1956). The incidence of broodiness in two strains, one selected for non-broodiness and one unselected, was compared by housing them together to provide equivalent environmental conditions. In the unselected strains 67 per cent of the birds became broody as compared with only 9 per cent of the females in the non-broody strain (Hale, 1969). That the difference between the two strains was a difference in sensitivity to external stimuli rather than an absence of broody characteristics in the non-broody strain was determined by letting eggs accumulate on the nests.

Previously non-broody hens with eggs in the nest stopped laying and started incubating eggs within 7 weeks (Fig. 101). Eggs were removed from the nests of control birds soon after laying.

Fig. 101. Effect of presence of eggs in the nest on inhibiting egg laying in a "non-broody" strain of turkeys. All experimental hens were incubating eggs at the end of 7 weeks while none of the control birds showed evidence of incubation (Hale, 1969).

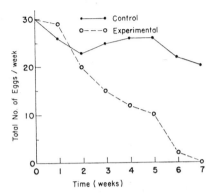

Attempts to interrupt broodiness also indicate the effectiveness of nests and eggs in inducing or prolonging broody behaviour. Although androgens and oestrogens have been used to interrupt broodiness (Kosin, 1948; Blakely et al., 1951), placing birds on a wire floor away from nesting materials is even more effective (Haller & Cherms, 1959). Of the various hormone treatments, progesterone seems most effective in terminating broody behaviour (van Tienhoven, 1958; Jeannoutot & Adams, 1961). Most surprising is the observation that sound (recorded jet plane noise played at 110 to 135 decibels) is more effective than either progesterone or the wire floor treatments (Jeannoutot & Adams, 1961).

C. Parent-young Interactions

Turkeys hatch approximately 27 days after eggs are set with pipping of the shell the first outward sign of activity. Pipping is followed by a quiescent period of 12 to 24 hours with a renewed attack on the shell and hatching taking place on the average about 28 hours after the initial pip (Abbott & Craig, 1960). Within about 8 hours after hatching the young poults are dry and start to move about slowly. A crucial role in hatching is played by a *hatching muscle* which appears during the last half of incubation, reaches its maximum development at the time of hatching, and gradually regresses over a period of days after hatching (Fisher, 1958).

Basic patterns of parent-young interactions in turkeys are grossly similar to those in chickens. Young turkeys imprint to the mother as previously indicated in Chapter 6 and soon are able to discriminate her individual calls (Ramsay, 1951). However, deafened poults follow the hen using visual characteristics (Schleidt et al., 1960). The peeping of the young poult serves both to attract the attentions of the mother and actually to protect the poults from attacks by the hen. Hens previously deafened as poults incubated eggs like normal hens but killed their own young immediately after hatching, behaving as if they could not differentiate between poults and small predators (Schleidt *et al.*, 1960). Furthermore, normal hens presented with silent dummy poults attacked them after a short interval while dummy poults with speakers emitting recorded peeps were not attacked (Schleidt & Schleidt, 1961).

Turkey poults are precocious and soon after hatching leave the nest with the mother in search of food. At first movement is slow but by 48 hours the pace approximates to that of adults (Ligon, 1946). The poults crouch, rest, or sleep under the mother's wings or body as she hovers over them. Young wild turkeys spend each night on the ground under the mother until about 4 to 5 weeks of age (Latham, 1956). After that the mother roosts on the lower branches of trees and the young fly up to follow her.

The female defends the brood against intruders and may attack by hissing, running in a low crouch, and jumping at the intruder with extended claws and beating wings. As an intruder approaches, the hen may hover over the poults and remain motionless. If disturbed, the hen may then fly off in a wobbly flight suggesting injury-feigning and directing the intruder's attention away from the poults. The mother sounds various alarm calls (cluck, rattling, singing; Fig. 102) as strange objects are sighted on the ground or in the air. Responses of poults to these calls will be considered in the section on communication and response to predators.

V. FIGHTING BEHAVIOUR AND SOCIAL ORGANIZATION

Fighting typically occurs when two strange birds meet for the first time and the winner of the fight subsequently dominates the loser. Groups of young turkeys living together show only sporadic sparring until about 3 months old. Over the next two months there is a gradual increase in threatening and fighting until a peak is reached at about 5 months of age. By that time social hierarchies are well established and fighting is reduced. Males fight more vigorously than females but the pattern of fighting is the same for the two sexes except that the associated morphological structures are reduced in females.

A. Fighting Behaviour
1. PATTERNS OF FIGHTING

Turkeys show a variety of threat displays, ranging from "mild threat" (head raised, looking toward the opponent), to strutting (against other males) and postures which diverge from strutting in various respects. The only common feature of these latter postures is the extended snood and the enhanced coloration of the caruncuiated areas of the head and neck, with a brilliant red predominating during threat display (Hale & Schein, 1960; Schleidt & Schleidt, 1962b). Body posture during threat tends to be lateral to the opponent with the head held high and facing the other bird. The beak is downward, snood pendant, wings drooped and held slightly away from the body, feathers over the entire body are sleeked, and the tail feathers may be only partially fanned and held horizontally or at an angle between a perpendicular and horizontal plane.

Vocalization during the threat display is a distinctive trill of relatively high pitch and is emitted repeatedly as the birds face each other. The birds may circle slowly while maintaining a wide stance and leaning slightly away from each other. Thus they maintain a posture from which they may jump quickly towards or away from the other bird. While the two birds are threatening, one may throw its head back and move it laterally across its back in a wiping-like movement (eye-wiping). Although this pattern is relatively rare, it suggests a displacement activity.

Actual fighting begins as both birds jump simultaneously or one just slightly before the other. As the birds jump their feet are extended forwards, with the toes spread, and moved in a raking fashion while directed at the opponent's body. During the jumping phase of the fight, the domestic turkeys rarely injure one another since their spurs are poorly developed and the pattern of movement does not bring the spurs into contact with the opponent's body. If one bird lands a stroke on the other's back, the latter gives up. However, if neither bird submits after a few jumps there is a shift to a tugging battle but as few as one or more than 20 jumps may occur first. The head is darted forward to grasp the caruncles, snood, wattle, or upper or lower beak of the opponent. Necks may become entwined as the birds grasp each other simultaneously and tug and push in an attempt to force the opponent's head downwards. During this phase of the fight, the skin about the head may be injured and bleed. In prolonged fights one bird may interrupt tugging by pressing its head over the opponent's back or under the opponent's wing. After a brief rest the fight is resumed. If the tugging is interrupted for longer than a few seconds, e.g. because the birds stumbled over an obstacle, the jumping starts over again before the tugging is continued.

A fight is terminated by the sudden submission of one contestant with no prior indication of the impending event. After a bird submits it retracts its snood, lowers the head as the wings and tail assume normal positions, and attempts to hide his head under the opponent's breast, and then flees; occasionally the hiding is omitted. The winner may follow closely (snood extended, head high) and threaten or peck the defeated bird. Usually the threat display is quickly terminated and the winner may shift to courtship. In some pairings between strange birds only one threatens and the other submits; in other pairings neither bird threatens and one or both may avoid or go into sexual display. Occasionally a male may continue sexual display even though attacked vigorously by the other bird.

2. STIMULI RELEASING FIGHTING

Threat displays in turkeys are highly ritualized and provide signals eliciting attack. This is most clearly seen in the behaviour of androgen-injected poults showing precocious fighting. A detached dummy head presented at a height corresponding to the relative head height of a bird in threat display releases fighting while heads presented at lower heights elicit sexual responses (Hale, 1960). There is also some suggestion that the tendency of a submissive individual to turn away from an attacking bird, to lower the head, in a manner obscuring the carunculated areas of the neck, and to contract the area of naked skin, may serve as an appeasement posture. In wild turkeys, feathers along the back of the neck are erected in a manner tending further to obscure the pigmented areas (Schleidt & Schleidt, 1961).

3. QUANTITATIVE ASPECTS OF FIGHTING BEHAVIOUR

Hormone levels. Maximum levels of fighting are dependent upon high androgen levels and consequently male turkeys fight more vigorously than females and fights between males and females occur only during the first year in young males, when they have to fight their way through the hierarchy of old females. Precocious fighting in both male and female poults is facilitated by androgen treatment. Treated females fight each other more readily than do males and are more successful in winning fights than are treated males (Schein & Hale, 1959; Hale, 1960). The apparent difference in fighting between treated poults of the two sexes reflects the tendency of the male poults to persist in sexual courtship even when attacked by other poults. Individual differences in fighting are probably related to factors other than levels of male hormone.

Genetic factors. Males from different varieties show considerable differences in ability to win fights in paired contests with other males. A genetic basis for these differences is suggested since males from crosses between two varieties tend to be intermediate in their success in winning fights. Similarly, males from back-crosses to the least successful variety are less successful than the first generation crosses (Table 58). If the Black variety shown in

Table 58. Variety Differences in Percentage of Fights won in Paired Encounters and the Intermediate Success of Crosses[1] (Hale, 1969).

Variety	Black	F_1	Back-cross	Bronze
Black	—	63	75	75
F_1	37	—	55	61
Back-cross to Bronze	25	45	—	57
Bronze	25	39	43	—

[1] Values represent percentage of contests won by variety or cross listed at left.

Table 58 is assigned a score of 100, the relative scores of the F_1, back-cross, and Bronze are 87, 80, and 75 respectively. Males of a Grey variety were less successful than the Bronze and their relative score of 45 indicates the wide variation existing between varieties.

Females also differ in ability to win fights as is demonstrated by the stratification of peck orders when two strains of white females are penned together. Peck order measurements in several strains so housed revealed that all the females of one strain were high in the peck order and those of the other low (Hale, 1969). Since the birds were all of the same colour, the possibility of stratification based on colour recognition rather than differences in aggression was eliminated (Hale, 1957).

B. Social Organization and Group Integration

I. SOCIAL HIERARCHIES

Both male and female turkeys develop a typical peck-right type of rank order with each sex forming an independent hierarchy in heterosexual groupings. Although groups become well integrated with time and fighting is greatly reduced, flocks of turkey females tend to show more aggressiveness than similar flocks of chickens. Female hierarchies tend to be highly stable while male hierarchies may be more labile. Strutting is common in all-male groups and rank-order interactions may be so infrequent that it is difficult to determine the precise rank of all birds in the flock.

The social rank is established by threat or fight, and maintained through a special display of the higher ranking bird toward the lower ranking one: The red dewlap is displayed close to the eyes of the opponent, the head is moved upward from horizontal to about 45°, the hyoid horns bulge the throat, and a vocalization is emitted which is reminiscent of a pigeon's coo; preliminarily, we suggest as an unbiased term to characterize this display the letter "T". This T-display is given repeatedly, in sequence of a few up to a hundred or more, until the lower ranking bird moves away. Newcomers in a flock are often welcomed with T-display by the top-ranking bird; this leads frequently to a fight (Schleidt, 1968b).

If wild or domestic turkeys are kept on range, with ample space to avoid close proximity of other flockmates, the only means of enforcing the rank in the hierarchy are threat, T-display, jumping, and tugging. Aggressive pecking, an integral part of the tugging match, is oriented only toward the naked skin on the head and neck. Occasional pecks may be directed at the back of the neck. All other pecking or picking at the feathered parts of the body, as observed conditions of crowding, is unrelated to rank order.

Denudation of the neck as an index of social rank in females. The social environment of high- and low-ranking birds in a rank-order may differ drastically under conditions of close confinement in a limited space. Since rank-orders usually must be determined by the laborious procedure of observing which birds peck the others, in the past only limited information involving a small number of individuals could be obtained. However a rapid method is now available which permits an evaluation of the peck order status of a turkey female (but not males) in a few seconds. Dominant hens peck the backs of the neck of subordinate birds and the area of denudation of a hen's neck is a good indicator of the number of birds pecking that individual (Plate XX). The correlation between the size of the denuded area and social rank was found to be 0·65 in 10 flocks of 20 females (Hale, 1956). By using the denuded area as an index, it is possible to evaluate the rank of hundreds of females in a single day and studies of the genetics of social rank become feasible. The possibility that certain genes associated with feather pigments may also influence social rank has been suggested by application of this technique (Hale & Buss, 1960).

2. INTERACTION WITHIN INTEGRATED HIERARCHIES

Observation of the degree of pecking in well-organized social hierarchies suggest that flocks differ markedly in the amount of interaction between birds. This observation is confirmed by denudation measurements of flocks of hens from several varieties (Table 59). Some varieties consistently

Table 59. Variety Differences in Peck-order Interactions in Flocks of Female Turkeys as measured by Denudation of Neck. (Adapted from Hale & Buss, 1960.)

Variety	Total No. of Pens	Year Observed			Av. Score[1]
		1st	*2nd*	*3rd*	
White Holland	8	7·7	8·0	6·0	7·6
Grey	4	5·5	5·1	—	5·3
Black	4	5·1	4·4	—	4·7
Bronze	12	7·3	5·1	2·0	3·2
Narragansett	5	2·9	2·5	—	2·7
Black Wing Bronze	4	2·5	2·7	2·3	2·5

[1] Scores correspond closely to size of denuded area in square inches.

develop large denuded areas while other varieties, with minimum interaction within rank orders, develop very small denuded areas. In general, the denudation scores remain stable for a particular variety under variable environmental conditions and provide presumptive evidence indicating a genetic basis for the observed differences. Ability to win fights and the degree of interaction within peck orders appear to be independent characteristics since females showing minimum interaction as a group within their own flock may be very successful in winning fights over birds from other flocks.

VI. COMMUNICATION AND RESPONSE TO PREDATORS

Although the several displays noted in the foregoing sections function as signals in social communication, the wide ranging vocalizations of turkeys comprise the most important class of social signals. The general character of various calls will be noted here (Fig. 102) but a full comprehension of communication in turkeys must await further spectrographic analysis.

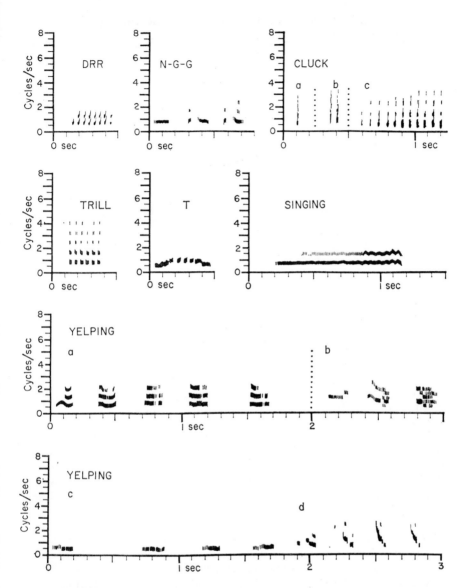

Fig. 102. Representative sound spectrograms of various vocalizations of turkeys. The predominant situations are as follows: DRR, relaxed; N–G–G, also two-syllabic (N–G) or four-syllabic (N–G–G–G), conspecific in close distance; *cluck,* (a) monosyllabic alert call, (b) irregular burst, and (c) rhythmical *rattling,* both in response to sudden disturbance, as an approaching predator; *trill,* during fighting; T, in showing dominance to a lower ranking bird; *singing* in response to slight disturbance, as a high-flying bird-of-prey; *yelping,* (a) in distress, (b) example of transient phase from distress-peep to distress-yelp by a 6-month-old female, (c) yelp of a broody hen, leading poults, at (d) breaking to higher pitch, under slight distress.

19*

A. Vocalization related to Social Interactions

Several vocalizations previously discussed in this category include the trill emitted during threat behaviour and the vocal component of the male strut. Hens give various yelps attracting the poults and the young turkeys give a variety of vocalizations including a distress peep, a trilling contentment call, and a high-pitched scream-like call when pecked. Another variation is heard as the poults quiet down before going to sleep.

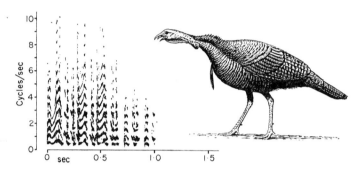

Fig. 103. Gobbling of the male Turkey. *Left*: Sound spectrogram; note the rhythmical structure, and the frequency modulated harmonics. *Right*: Position of the turkey, about o·1 seconds after gobbling had started; hyoid and beak are moved rhythmically. (Schleidt, 1964a.)

In older birds the peep is replaced by the yelp, a lower pitched call given frequently by adult females and less frequently by males. In addition to the rolling trill given during threat display, turkeys emit a somewhat similar call, at a lower intensity and in a shorter phrase, as another bird comes near (T). A soft trilling call ("Drr") is given when relaxed or browsing; if exceptionally desirable food is present, the Drr may be combined with the cluck commonly given to unusual or strange objects (Schleidt, 1968b).

The gobbling call of the male turkey (Fig. 103) very likely serves the interindividual recognition within a flock, especially over wider distances; it may also have part in the spacing of male subflocks and their roosts in the wild. If one turkey gobbles, usually most of the males in the vicinity are triggered to join in gobbling with surprisingly short latency (o·1 to 1 second). The releasing mechanism of the gobbling call appears to be rather unspecific, at least at the first glance. A male turkey at the height of the breeding season will respond with gobbling to a wide variety of acoustic stimuli: pure tones between 300 and 8,000 Hz mark the approximate frequency range; at 2,000 Hz, the pitch to which he is most sensitive, the sound pressure can be anywhere between 10^{-3} to 10^2 microbar (equivalent to a range of intensity of 100 decibel (Fig. 104). Amplitude modulation or frequency modulation of a carrier in the frequency range mentioned enhances the releasing value. White noise or band noise can serve as a stimulus too. The length of the stimulus tone may vary between o·1 second and 60 seconds; in a stimulus near the threshold, intensity can substitute for the duration, as stated in the reciprocity law. Stimuli of modalities other than acoustic do not elicit gobbling.

Under normal circumstances, i.e. in wild turkeys in their natural environment, the appropriate stimulus for the gobbling response appears to be the gobbling call of a conspecific male. Considering the wide variety of acoustic

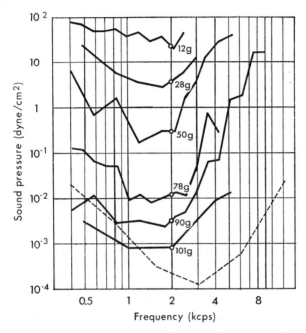

Fig. 104. Threshold of the gobbling response at various states of responsiveness. The g-values are based on an inverse decibel-scale at 2 kc.p.s., and are used as measure of responsiveness. The dotted line is the threshold of hearing of the bullfinch, for comparison. (Schleidt & Schleidt, 1958.)

stimuli effective in the laboratory one might wonder how safe this assumption is. However, field observations show that by far the highest percentage of gobbling calls in a single individual are elicited by the gobbling of another male, or appear to be spontaneous (i.e. without any noticeable external stimulus), and only a rather small proportion (up to 5 per cent) is elicited by acoustic stimuli other than gobbling, most frequently by other calls or conspecifics, or by "noise" from the environment such as songbird song or airplane sound. This indicates that the selectivity of the releasing mechanism is high enough to ensure a two-way communication between males, with the function of transmitting information on the identity and distance of the caller. These findings show that a certain selectivity of the gobbling response exists, but do not explain how it is achieved. The key to this problem is that the habituation of the gobbling response depends on the complexity of the stimulus. The turkey habituates relatively rapidly to pure tones, but slower to frequency modulated or amplitude modulated tones or to repetitive bursts or to signals with strong harmonics. Looking back to the sonogram of the gobbling call in Fig. 103, one finds all of these properties assembled there. To summarize: The selectivity of the gobbling response is achieved by the peculiar composition of the natural stimulus in a way that it circumvents the normal ability of the auditory system to habituate to repeated stimulation. The changes in frequency of occurrence of spontaneous gobbling as well as the changes of the gobbling threshold during the annual cycle were demonstrated earlier (Fig. 99). Finally, it must be mentioned that gobbling is not restricted to males only, but is also emitted by females in certain situations (Schleidt, 1955; Schleidt, 1964a, 1964b, 1965, 1968a, 1968b; Schleidt & Schleidt, 1958, 1961).

B. Response to Predators

For several days after hatching young turkeys do not respond directly to predators but react to various alarm calls sounded by the mother. As the poults grow older the escape responses to alarm calls disappear and direct responses to the predators become predominant.

1. Alarm Calls

Strange objects or animals on the ground elicit a highly segmented alert call or cluck which can be readily simulated by a tapping sound. Young turkeys respond to this call by dispersing slowly over a wide area in a creeping posture. On sighting an aerial object, the mother gives one of two related alarm calls depending upon the nearness of the object (Schleidt & Schleidt, 1961). When high-flying predators are detected the mother gives a "singing" note which elicits an attentive vigilance response in the young poults. A

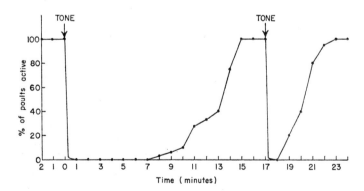

Fig. 105. Effect of a low-frequency tone of 5 seconds' duration on activity of 2-day-old turkey poults. Habituation of the response with repetition of the simulated alarm-call is illustrated by the shorter period of "freezing" behaviour following the second tone (Hale, 1969).

lower pitched call which resembles a sequence of clucks (CCC, rattling) is sounded by the mother as a predator comes close enough to indicate impending danger. Poults respond to this call by dashing violently to cover or to the corners of a pen, pile in clumps, and remain motionless for several minutes. The interval between notes in both calls is so small that the first call is readily simulated by any sustained high-frequency tone above 1,000 cycles per second and the latter call by a sustained tone below 1,000 cycles. Repetition of the low-frequency call produces a gradual habituation of the response with each response becoming less intense and the duration of freezing shorter following termination of the alarm call (Fig. 105). In response to all three of the above alarm calls, poults cease all other activities and become quiet in addition to giving the distinctive escape reactions. By the time the poults are approximately 8 weeks old, the distinctive escape reactions disappear and attentive vigilance is the only response given to any of the three alarm calls (Table 60).

Table 60. Responses of Domestic and Wild Turkey Poults to Auditory Signals. (Adapted from Hale, 1959a.)

Race	Age (weeks)	Segmented Alarm Call	Low Frequency Tone
Domestic	2	Creeping dispersal	Run, clump, and freeze
	4	Erect dispersal	Run, clump, and freeze
	8	Attention	Attention
Wild (*M. g. silvestris*)	2	Freeze	Creep and clump
	4	Erect dispersal	Run, clump, and freeze
	7	Partial dispersal	Run, clump, and freeze
Wild (*M. g. osceola*)	2	Erect dispersal	Run, clump, and freeze
	4	Erect dispersal	Run, clump, and freeze

2. VISUAL RESPONSES TO PREDATORS

About the same time that young poults cease to give escape responses to the alarm calls they begin to respond directly to various objects instead of reacting indirectly through communication from the mother. Turkeys are included among those species reported to respond differentially to hawk-like and goose-like models (Tinbergen, 1948; Chapter 6). Analysis of the responses of young poults at 11 weeks of age indicates that turkeys respond equally to both hawk- and goose-like models and show rapid habituation to both if not previously exposed to overhead objects (Schleidt, 1961a, 1961b). Nevertheless, protection against flying predators is provided since the number of hawk-like birds in natural populations is exceedingly small relative to the number of birds presenting a goose-like silhouette. Thus in natural habitats young turkeys rapidly become habituated to goose-like shapes but readily react to the infrequently observed hawk-like silhouettes.

Young poults give an especially violent reaction if an umbrella is opened above them suddenly. In contrast to the absence of vocalizations during responses to alarm calls, poults exposed to the umbrella exhibit a frenzy of vocalization, dart about, freeze momentarily, and dash again in all directions. Presumably these responses are similar to those elicited by the visual proximity of an attacking predator.

VII. MISCELLANEOUS BEHAVIOUR AND APPLICATIONS

A. Activity and Abnormal Behaviour

Following initial activity and feeding soon after daybreak, turkeys taper off to a period of relative inactivity near midday and engage in considerable feather preening and dusting. Dusting is carried out in a stereotyped fashion with the sitting bird extending its head and making raking movements in towards the body with its beak. The raking is continued until a fan-like semicircle has been raked and dirt piled against the body. Dust is then worked into the feathers by flapping the wings and kicking. The complete pattern may be carried out by birds on concrete floors in the absence of litter, and by young birds which have never had any experience with litter. As dusk approaches, turkeys become inactive and rest in clumps if roosts are not available. During the night, the inactive turkeys remain motionless if approached or exposed to a beam of light and can be caught or picked up with ease. Sleeping turkeys may place their heads under the wing but usually simply relax the neck and rest their head on the body with the snood flaccid and extended.

Frolicking is relatively rare in turkeys and is best seen when one bird picks up some desired object and is chased about by the others.

Abnormal behaviour and cannibalism. The most common type of abnormal behaviour developing in turkeys is some pattern of feather picking. Under confinement some varieties show no evidence of picking while others may show characteristic patterns of picking restricted to the tail feathers, to a small area at the base of the wings, or a fraying of the feathers at the base of the neck. In some highly inbred lines or certain crosses, feather picking may be so extreme that many members of a flock may be almost completely denuded (Plate XX). Breaking the skin during feather picking may produce bleeding and favour the development of cannibalism. Females of certain strains have a tendency to peck at and eventually injure the pendant snood of strutting males. If a bird is weak or cannot stand, the other birds peck it so actively on the head and back that severe damage and death are likely to follow if the bird is not removed.

B. Applications

Successful application of our basic knowledge of turkey behaviour to problems of management depends upon the insight and ingenuity of the caretaker. Handling of turkeys is greatly facilitated if advantage is taken of their strong herding tendencies. Catching birds is easiest if done at night but at other times holding one's hand over a bird's head and then catching it by the wings as it hesitates favours effective capture. Maintaining flocks as organized groups and avoiding mixing of different flocks can do much to reduce fighting. Broodiness can be controlled in part by frequent removal of eggs from the nest, and the laying of eggs on the floor reduced by training the birds to the nests. On a long-range basis, it is possible to select for non-broodiness, adequate nesting behaviour, and reduced pecking within organized flocks.

Females of some strains crouch so readily to man that they may become a nuisance if artificial insemination is used. As their sexual receptivity is not terminated, the hens may crouch on the floor in large numbers as the care-taker enters and make it difficult to move equipment through the pens for feeding or collecting eggs. Some breeders using artificial insemination also keep males in with the hens for the express purpose of keeping the hens off the floor. On the other hand, imprinting male turkeys to man would have the advantage of facilitating handling of the males during collection of semen for artificial insemination.

Low fertility related to behavioural defects can be overcome by selecting the sexually most active males. This can be done by observing the performance of males in live matings with hens during the peak of mating before laying begins or by testing their response to a dummy incorporating a female head. Males responding most readily to the dummy head are among those performing best in matings with live hens (Harper, 1961). Selective breeding of both males and females for improved sexual behaviour provides an effective long-range programme.

REFERENCES

ABBOTT, U. K. & CRAIG, R. M. (1960). Observations on hatching time in three avian species. *Poult. Sci.*, **39**, 827–830.

ADAMS, J. L. (1959). Observation of male turkey mating behavior in a commercial breeding flock. *Poult. Sci.*, **38**, 743–746.

ALMQUIST, H. J. (1953). *Feed Requirements of Turkeys as Related to Time of Year*. The Grange Co., Modesto, Calif. 9 pp.

ASMUNDSON, V. S. & LORENZ, F. W. (1957). Hybrids of ring-necked pheasants, turkeys, and domestic fowl. *Poult. Sci.*, **36**, 1323–1334.

BLAKELY, R. M., ANDERSON, R. W. & MACGREGOR, H. I. (1951). The estrogen interruptions of broodiness in turkeys. *Poult. Sci.*, **30**, 907 (Abstract).

BURROWS, W. T. & KOSIN, I. L. (1953). The effects of ambient temperature on production and fertilizing capacity of turkey spermatozoa. *Physiol. Zool.*, **26**, 131–146.

CARBAUGH, B. T., SCHEIN, M. W. & HALE, E. B. (1962). Effects of morphological variations of chicken models on sexual responses of cocks. *Anim. Behav.*, **10**, 235–238.

DAANJE, A. (1950). On locomotory movements in birds and the intention movements derived from them. *Behaviour*, **3**, 48–98.

DUKES, H. H. (1955). *The Physiology of Domestic Animals*. 7th ed. Ithaca, N.Y.: Comstock.

DYMSZA, H., BOUCHER, R. V. & MCCARTNEY, M. G. (1957). The influence of fiber content and physical form of the diet on the energy requirements of turkeys. *Poult. Sci.*, **36**, 914–917.

EISNER, E. (1960). The relationship of hormones to the reproductive behaviour of birds, referring especially to parental behaviour: a review. *Anim. Behav.*, **8**, 155–179.

EVANS, L. T. (1961). An interpretation of turkey courtship. *Am. Zool.*, **1**, 353 (Abstract).

FISHER, H. I. (1958). Problems in functional anatomy—the hatching muscle in the chick and diurnal variation in liver size in birds. *Anat. Rec.*, **132**, 438 (Abstract).

GUHL, A. M. (1950). Social dominance and receptivity in the domestic fowl. *Physiol. Zool.*, **23**, 361–366.

HALE, E. B. (1953a). Social factors in sexual behavior of turkeys. *Pa. Agric. exp. Sta. Prog. Rept.* No. 108, 4 pp.

HALE, E. B. (1953b). Family differences in mating activity and social rank in female turkeys. *Poult. Sci.*, **32**, 903 (Abstract).

HALE, E. B. (1955a). Duration of fertility and hatchability following natural matings in turkeys. *Poult. Sci.*, **34**, 228–233.

HALE, E. B. (1955b). Defects in sexual behavior as factors affecting fertility in turkeys. *Poult. Sci.*, **34**, 1059–1067.

HALE, E. B. (1956). Degree of denudation of neck as index of social rank in turkey females. *Poult. Sci.*, **35**, 1146 (Abstract).

HALE, E. B. (1957). Breed recognition in the social interactions of domestic fowl. *Behaviour*, **10**, 240–254.

HALE, E. B. (1959a). Distinctive responses of turkey poults to specific sounds. *Anat. Rec.*, **134**, 576 (Abstract).

HALE, E. B. (1959b). Stimuli terminating sexual receptivity in female turkeys. *Anat. Rec.*, **134**, 577 (Abstract).

HALE, E. B. (1960). Role of head height in releasing sexual versus fighting behavior in turkeys. *Anat. Rec.*, **138**, 354–355 (Abstract).

HALE, E. B. (1962). Unpublished data. Dept. Poult. Sci., Penn. State Univ., University Park, Pa.. U.S.A.

HALE, E. B. & BUSS, E. G. (1960). Pigment gene effects on peck-order behavior of female turkeys. *Anat. Rec.*, **137**, 362 (Abstract).

HALE, E. B. & HANFORD, P. V. (1962). Unpublished data. Dept. Poult. Husb., Penn. State Univ., University Park. Pa., U.S.A.

HALE, E. B. & SCHEIN, M. W. (1960). *Meleagris gallopavo (Meleagrididæ)—Fighting Behaviour Patterns.* Encyclopædia Cinematographica No. E 360, Institut für den Wissenschaftlichen Film. Göttingen.

HALLER, R. W. & CHERMS, F. L. (1959). A comparison of several treatments on terminating broodiness in broad breasted bronze turkeys. *Poult. Sci.*, **38**, 1211 (Abstract).

HARPER, J. A. (1961). Personal communication. Dept. Poult. Husb., Oregon State Univ., Corvallis, Ore., U.S.A.

HERRICK, E. H. (1951). The influence of androgens in a turkey female. *Poult. Sci.*, **30**, 758–759.

HEWITT, O. H. (ed.) (1967). *The Wild Turkey and its Management.* Washington, D.C.: The Wildlife Society.

HINDE, R. A. (1961). Behavior. In: Marshall, A. J. (ed.), *Biology and Comparative Physiology of Birds.* Vol. 2, pp. 373–411. New York, N.Y.: Academic Press.

JEANNOUTOT, D. W. & ADAMS, J. L. (1961). Progesterone versus treatment by high intensity sound as methods of controlling broodiness in broad breasted bronze turkeys. *Poult. Sci.*, **40**, 517–521.

KARE, M. R. & HALPERN, B. P. (eds.) (1961). *Physiological and Behavioral Aspects of Taste.* Chicago: University of Chicago Press.

KLEITMAN, N. & KOPPANYI, T. (1926). Body-righting in the fowl (*Gallus domesticus*). *Am. J. Physiol.*, **78**, 110–126.

KOPPANYI, T. & KLEITMAN, N. (1927). Body righting and related phenomena in the domestic duck (*Anas boscas*). *Am. J. Physiol.*, **82**, 672–685.

KOSIN, I. L. (1948). The use of testosterone propionate in controlling broodiness in turkeys. *Poult. Sci.*, **27**, 671 (Abstract).

LATHAM, R. M. (1956). *Complete Book of the Wild Turkey.* Harrisburg, Pa.: Stackpole.

LEHRMAN, D. S. (1959). Hormonal responses to external stimuli in birds. *Ibis*, **101**, 478–496.

LEOPOLD, A. S. (1944). The nature of heritable wildness in turkeys. *Condor*, **46**, 133–197.

LEOPOLD, A. S. (1948). The wild turkeys of Mexico. *Trans. 13th N. Amer. Wildlife Conference*, Washington, D.C., pp. 393–400.

LIGON, J. S. (1946). History and management of Merriam's wild turkey. *Univ. New Mexico Publ. Biol.*, No. 1, 84 pp.

MARGOLF, P. H. (1961). Personal communication. Dept. Poult. Sci., Penn. State Univ., University Park, Pa., U.S.A.

MARGOLF, P. H., HARPER, J. A. & CALLENBACH, E. W. (1947). Response of turkeys to artificial illumination. *Pa. Agric. exp. Sta. Bull.*, No. 486.

MARSDEN, S. J. & MARTIN, J. H. (1955). *Turkey Management*. Danville, Ill.: Interstate.

McCARTNEY, M. G. (1956). Reproductive performance in broody and non-broody turkeys. *Poult. Sci.*, **35**, 763–765.

MOREHOUSE, N. F. (1949). Water consumption in growing turkeys. *Poult. Sci.*, **28**, 152–153.

MORRIS, D. (1956). The feature postures of birds and the problem of the origin of social signals. *Behaviour*, **9**, 75–113.

OLSEN, M. W. (1960). Turkey-chicken hybrids. *J. Hered.*, **51**, 69–73.

OLSEN, M. W. & MARSDEN, S. J. (1952). Pre-seasonal molt in male turkeys. *Poult. Sci.*, **31**, 715–722.

PORTMANN, A. (1961). Sensory organs: skin, taste and olfaction. In: Marshall, A. J. (ed.), *Biology and Comparative Physiology of Birds*. Vol. 2, pp. 37–48. New York: Academic Press.

PUMPHREY, R. J. (1961). Sensory organs: vision; Sensory organs: hearing. In: Marshall, A. J. (ed.), *Biology and Comparative Physiology of Birds*. Vol. 2, pp. 55–86. New York: Academic Press.

RÄBER, H. (1948). Analyse des Balzverhaltens eines domestizierten Truthahns (*Meleagris*). *Behaviour*, **1**, 237–266.

RAMSAY, A. O. (1951). Familial recognition in domestic birds. *Auk*, **68**, 1–16.

SAWYERS, W. A. (1966). Unpublished Ph.D. thesis, Penn. State Univ., University Park, Pa., U.S.A.

SCHEIN, M. W. (1960). Modification of sexual stimuli by imprinting in turkeys. *Anat. Rec.*, **137**, 392 (Abstract).

SCHEIN, M. W. (1963). On the irreversibility of imprinting. *Z. Tierpsychol.*, **20**, 462–467.

SCHEIN, M. W. & HALE, E. B. (1957). The head as a stimulus for orientation and arousal of sexual behavior in male turkeys. *Anat. Rec.*, **128**, 617 (Abstract).

SCHEIN, M. W. & HALE, E. B. (1958). Stimuli releasing sexual behavior of domestic turkeys. 16 mm. film, *Psychological Cinema Register*, PCR-114K, Pennsylvania State Univ.

SCHEIN, M. W. & HALE, E. B. (1959). The effect of early social experience on male sexual behaviour of androgen injected turkeys. *Anim. Behav.*, **7**, 189–200.

SCHEIN, M. W. & HALE, E. B. (1960). *Meleagris gallopavo (Meleagrididae)—Sexual Behavior Patterns*. Encyclopaedia Cinematographica No. E 359, Institut für den Wissenschaftlichen Film, Göttingen.

SCHLEIDT, M. (1955). Untersuchungen über die Auslösung des Kollerns beim Truthahn (*Meleagris gallopavo*). *Z. Tierpsychol.*, **11**, 417–435.

SCHLEIDT, W. M. (1961a). Über die Auslösung der Flucht vor Raubvögeln bei Truthühnern. *Naturwissenschaften*, **48**, 141–142.

SCHLEIDT, W. M. (1961b). Reaktionen von Truthühnern auf fliegende Raubvögel und Versuche zur Analyse ihres AAM's. *Z. Tierpsychol.*, **18**, 534–560.

SCHLEIDT, W. M. (1964a). Über die Spontaneität von Erbkoordinationen. *Z. Tierpsychol.*, **21**, 235–256.

SCHLEIDT, W. M. (1964b). Über das Wirkundsgefüge von Balzbewegungen des Truthahnes. *Naturwissenschaften*, **51**, 445–446.

SCHLEIDT, W. M. (1965). Gaussian interval distribution in spontaneously occurring innate behaviour. *Nature, Lond.*, **206**, 1061–1062.

SCHLEIDT, W. M. (1968a). The annual cycle of courtship behavior in the male turkey. *J. comp. physiol. Psychol.* (in press).

SCHLEIDT, W. M. (1968b). Unpublished data. Dept. Zoology, Univ. of Maryland, College Park, Md., U.S.A.

SCHLEIDT, M. & SCHLEIDT, W. (1958). Kurven gleicher Lautstärke beim Truthahn (*Meleagris gallopavo*). *Naturwissenschaften*, **45**, 119–120.

SCHLEIDT, W. M. & SCHLEIDT, M. (1961). Personal communication. Max-Planck-Institut f. Verhaltensphysiol., Seewiesen, Germany.

SCHLEIDT, W. M. & SCHLEIDT, M. (1962a). *Meleagris gallopavo sylvestris (Meleagrididae)—Sexualverhalten.* Encyclopaedia Cinematographica No. E 486 T, Institut für den Wissenschaftlichen Film, Göttingen.

SCHLEIDT, W. M. & SCHLEIDT, M. (1962b). *Meleagris gallopavo sylvestris (Meleagrididae)—Kampfverhalten der Hähne.* Encyclopaedia Cinematographica No. E 487 T, Institut für den Wissenschaftlichen Film, Göttingen.

SCHLEIDT, W. M., SCHLEIDT, M. & MAGG, M. (1960). Störung der Mutter-Kind-Beziehung bei Truthühnern durch Gehörverlust. *Behaviour*, **16**, 254–260.

SCHOETTLE, H. E. T. & SCHEIN, M. W. (1959). Sexual reactions of male turkeys to deviations from a normal female head model. *Anat. Rec.*, **134**, 635 (Abstract).

SCHORGER, A. W. (1966). *The Wild Turkey, its History and Domestication.* Norman, Oklahoma: Univ. of Oklahoma Press.

SCHWARTZKOPFF, J. (1949). Über den Sitz und Leistung von Gehör- und Vibrationssinn bei Vögeln. *Z. vergl. Physiol.*, **31**, 529–608.

SCHWARTZKOPFF, J. (1950). Beitrag zum Problem des Richtingshörens bei Vögeln. *Z. vergl. Physiol.*, **32**, 319–327.

SCOTT, J. W. (1942). Mating behavior of the Sage Grouse. *Auk*, **59**, 477–498.

SCOTT, M. L. (1960). Feed consumption and growth standards. *Turkey World*, **35** (1), 48–49.

SIMON, J. R. (1940). Mating performance of the Sage Grouse. *Auk*, **57**, 467–471.

SMYTH, J. R., JR. (1955). Selection for differing levels of sexual receptivity in the female turkey. *Genetics*, **40**, 596 (Abstract).

SMYTH, J. R., JR. (1961). Personal communication. Dept. Poult. Sci., Univ. Mass., Amherst, Mass., U.S.A.

SMYTH, J. R., JR. & LEIGHTON, A. T., JR. (1953). A study of certain factors affecting fertility in the turkey. *Poult. Sci.*, **32**, 1004–1013.

STURKIE, P. D. (1954). *Avian Physiology.* Ithaca, N.Y.: Comstock.

TINBERGEN, N. (1948). Social releasers and the experimental method required for their study. *Wilson Bull.*, **60**, 6–52.

VAN TIENHOVEN, A. (1958). Effect of progesterone on broodiness and egg production of turkeys. *Poult. Sci.*, **37**, 428–433.

WATTS, R. (1967). Personal communication.

Chapter 18

The Behaviour of Ducks

F. McKinney

The ducks, geese and swans comprise the family Anatidae, one of the most familiar and well-studied groups of birds. The classic taxonomic work of Delacour & Mayr (1945) provided a sound classification of the group, but recently further refinement has come from the anatomical work of Woolfenden (1961) and the comparative behaviour studies of Johnsgard (1965). Three sub-families are now recognized by Johnsgard: (1) *Anseranatinae* (including a single species, the Magpie Goose); (2) *Anserinae* (Whistling Ducks, Swans, Geese and the Freckled Duck); (3) *Anatinae* (Sheldgeese, Shelducks, Steamer Ducks, Perching Ducks, Dabbling Ducks, Pochards, Sea Ducks and Stiff-tails). Only four species are known to have been domesticated: Greylag Goose (*Anser anser*), Swan Goose (*Anser cygnoides*), Mallard (*Anas platyrhynchos*) and Muscovy (*Cairina moschata*).

This chapter will deal almost entirely with wild or semi-domesticated Mallards since surprisingly few studies have been made on the behaviour of the truly domesticated forms and little attention has been paid to the Muscovy. Emphasis is placed on recent work and for background information the reader is referred to the important reviews of Anatid behaviour by Phillips (1922–26), Hochbaum (1944, 1955), Weller (1964) and Johnsgard (1965). The original version of this chapter (Collias, 1962) deals with several topics omitted here for lack of space.

Seven races of *Anas platyrhynchos* are generally recognized, but the nominate form (*A. p. platyrhynchos*) is the most widely distributed. It breeds in a wide belt across North America and Eurasia and has been introduced in New Zealand. Of the northern hemisphere Dabbling Ducks the Mallard is the hardiest when faced with cold temperatures. Some birds are year-round residents, wintering as far north as open water and food can be found. In most parts of the range, however, there is a southward migration in autumn and a return north in early spring. The breeding habitat is varied. Mallards will use many types of water areas, for example, marshes, lakes, rivers, and small ponds both open and wooded. Especially dense populations are found in the "pothole" country of the Dakotas and the prairie provinces of Canada. This is the region south of the boreal forest, now famous for grain-growing and ranching, where hundreds of small ponds dot the landscape.

For many years semi-domesticated stocks of Mallards have been produced on game-farms in North America and released to the wild to supplement the natural supply of birds for hunting. Addy (1964) estimates that as many as 40,000 propagated Mallards have been released each year in the Atlantic

Flyway alone since 1940 and similar releases are made in many parts of the United States. While many of these birds are shot locally, some undoubtedly survive to breed in the wild. The nature and magnitude of their contributions to natural populations are unknown. Further mixing of domesticated stocks and wild birds has occurred in city parks in many parts of the world. Most of these hatchery propagated Mallards are similar in appearance to wild birds but they tend to be heavier. Some game farms repeatedly introduce wild birds in an effort to "improve" the characteristics of their stock. Some game-farm Mallards are known to migrate with a similar pattern to wild birds (Brakhage, 1954). In short, it has become impossible to distinguish between "pure" wild Mallards and birds with characteristics produced through artificial selection in captivity. The degree to which wild stocks are contaminated in this way is likely to remain unknown for ever.

Wild Mallards were apparently kept in captivity for several centuries B.C., notably by the Egyptians and Greeks, and the Romans built large aviaries where ducks were fattened and encouraged to nest. According to Phillips (1923) and Delacour (1964), the earliest reference to distinctive domesticated forms comes in the twelfth century A.D. when St Hildegard distinguished between *aneta silvestris* and *aneta domestica*, but Delacour suggests that the Chinese may have domesticated them earlier.

According to Delacour (1956), "Mallards have probably been reared and domesticated earlier than any other birds", but as Phillips (1923) remarks, "the actual process of domestication is a rapid one and can easily be observed by anyone who cares to make the effort" (cf. also Darwin, 1887). After a few generations of inbreeding the wild-caught birds lose their natural wildness, they become more waddling in their gait, heavier and "coarser" in appearance, and lose much of their inclination to fly. With continued inbreeding patches of white feathers crop up.

The Muscovy duck was already domesticated in the New World at the time of arrival of the first Europeans (Phillips, 1922). When Columbus arrived in the West Indies his men found "ducks as large as geese" among the Indians, and the Spaniards early in the sixteenth century found the Muscovy domesticated by the South American Indians. The Muscovy was apparently imported to Europe in 1550.

Characteristics of domesticated forms of Mallard. Plumage variants of many types are found quite frequently in wild Mallard populations. These could result from mutations or from interbreeding with domestic stocks. Whole and partial albinism is not uncommon, males may lack the red breast and white neck ring, and melanistic birds often with white on the breast are well known. Such forms, as well as the major domesticated types, are illustrated in two colour plates by Peter Scott included in Delacour (1964).

Domestic Mallard types have been developed for three purposes: meat producers, egg layers, and ornamental birds. The most familiar of the large table birds is the Rouen, a giant form retaining Mallard-like plumage but weighing up to 8–9 lb. (Wild birds range from 2 to 3½ lb.) Another huge form is the white Aylesbury developed in England and reaching a similar weight to the Rouen. The Pekin, originally coming from China, is also white but has a more upright carriage than the Aylesbury. Among the best-known egg producers are the Indian Runner, a slim bird with an upright carriage, and the Khaki Campbell which can lay 340 or more eggs in a year. Several varieties appear to have been developed, at least in part, for their ornamental qualities, for example, the miniature Mallards referred to as Call Ducks, the English Magpie Ducks, and the Crested Duck which has a large downy tuft on top of the head.

I. DAILY ACTIVITIES

A. Feeding and Drinking

The natural diet of wild Mallards is likely quite variable, depending on local conditions. Seeds and fruits of water plants, aquatic crustaceans, insects, worms and molluscs undoubtedly are major items in most habitats. In late summer, autumn and winter, grain, rice and corn may become important foods, field-feeding traditions becoming rapidly established and often resulting in significant crop depredation problems (Bossenmaier & Marshall, 1958). In towns, wild Mallards may depend heavily on man for their food. In Helsinki, Finland, for example, a large overwintering population is apparently completely dependent on food provided by the citizens during the cold months (Raitasuo, 1964).

Several feeding methods are used to locate small food items which are not fixated individually. Most familiar is the *dabbling* of the bill along the water surface or in mud which achieves a straining of planktonic organisms and mud-dwelling invertebrates. By rapid opening and closing movements of the mandibles, water enters the mouth near the tip of the bill and is expelled through the lamellae on each side. Dabbling is most frequently performed as birds swim, but it also occurs as they wade through mud puddles or walk along shorelines.

Feeding from the bottom of ponds is achieved in three ways. In water a few inches deep, the head and neck are submerged as the bill explores the muddy bottom. In deeper water, the whole fore-part of the body is immersed, the tail sticking up in the air, giving rise to the familiar pattern of "*tipping-up*" or "*up-ending*". *Diving* to feed from the bottom is a common practice in ducklings. Adults have also been recorded diving to take acorns from the bottom (Heinroth, 1910).

Flying insects are taken by rapid snaps of the bill. Sometimes there is a short pursuit with sudden changes of direction, but such behaviour is more often seen in ducklings than in adults. On land, crawling insects may be snapped up.

Plant material is probably taken mostly in the process of dabbling, but Weidmann (1956) has drawn attention to a special method used to take seeds from plants. Typically a duck swimming along a shoreline reaches up to catch a weed in its bill. The seeds are stripped or shaken off as the stem is pulled down and released through the bill. The floating seeds are then ingested by dabbling. Grain and other kinds of dry food are much preferred if they can be eaten in or close to water when frequent dabbling alternates with ingestion of the food. For this reason flooded cornfields and ricefields are particularly favoured as rich feeding grounds.

When feeding in shallow water, where only the bird's head and neck need be immersed for the bottom to be reached, a vigorous *paddling* movement is commonly seen. The bird makes alternate thrusts with the feet similar to those used in raising the body off the surface for wing-flapping or shaking. The forepart of the body is lifted slightly out of the water and the bird's back rocks from side to side with the movements of the feet. Each bout of paddling lasts for a few seconds and is followed immediately by feeding with the head and neck submerged. Evidently the bottom mud is stirred up by these foot movements, disturbing invertebrates which are then eaten.

Drinking, as in most birds, involves a distinctive movement sequence. The bill is dipped in water and then lifted slightly above the horizontal. It is a common activity especially when a bird has been away from water for some time. In the case of an incubating duck, frequent drinking is especially characteristic at the beginning of periods off the nest.

B. Care of the Body Surface and Related Activities

All waterfowl spend a considerable amount of time each day caring for their plumage. A variety of shaking movements remove water from the feathers, several cleaning movements serve to remove sources of irritation caused by foreign bodies, an elaborate sequence is involved in distributing oil to the feathers from the uropygial gland above the tail and several specialized actions serve to wet the feathers during bouts of bathing. Apparently these activities are necessary to keep the feathers in good condition, of special importance in preserving both waterproofing and heat-regulatory properties. The comfort movements listed below are described in more detail in Mc-Kinney (1965b).

1. SHAKING MOVEMENTS

A general *body-shake* occurs very frequently in Dabbling Ducks such as the Mallard, after a bird comes out on land (Fig. 106). A similar *swimming-shake* occurs less often. Lateral *head-shakes* and rotary *head-flicks* occur as independent actions and also as components of many other more complicated movements. The *tail-wag* is completely lateral and the *foot-shake* involves a series of forward and back movements often performed before a foot is tucked away in the belly feathers prior to sleeping. The most conspicuous and familiar shaking movement is the *wing-flap*, in which the body is held erect as the wings are vigorously beaten a few times. Associated with oiling sequences following bathing, vigorous *wing-shakes* are seen and prolonged *wing-shuffling* and *tail-fanning* accompany the preening.

All these movements occur in rather well-defined situations and each appears to function in removing water or foreign bodies from various parts of the body surface.

Fig. 106. (a) Head-shake, (b) head-flick, (c) body-shake, (d) swimming-shake, (e) tail-wag, (f) wing-shake. (Drawings by Peter Scott from McKinney, 1965b.)

2. STRETCHING MOVEMENTS

Three stretching movements, presumably serving some physiological functions, occur in close association with activities obviously related to care of the body surface. The *wing-and-leg-stretch* and *both-wings-stretch* (Plate XXIa) are characteristic of resting birds and may occur both before and after a period of sleep. The *jaw-stretch* closely resembles yawning in mammals but does not appear to be accompanied by marked inhalation.

3. CLEANING MOVEMENTS

Scratching of the head is done with the inner edge of the middle toe which is specially modified as a semi-sharp edge (Plate XXIb). *Foot-pecking* is the typical response to a foreign body on the foot. *Bill-cleaning* is accomplished by inserting the bill in water and blowing out through the nares, but a much commoner activity, *bill-dipping*, no doubt removes feather fragments and dirt from the bill during preening sessions. *Shoulder-rubbing* appears to be a response to irritation around the eye, but very similar actions are also involved in distributing oil and wetting the feathers during bathing.

4. OILING PREENING

Bathing is normally followed by oiling, but oiling can occur at other times also. The bird comes out on land, shakes, wing-flaps and wipes the water off

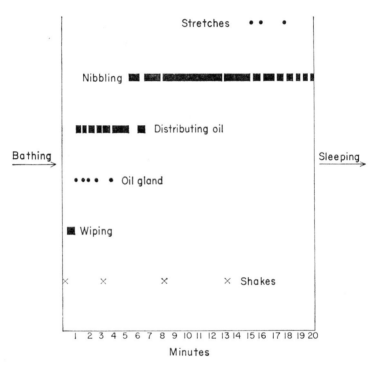

Fig. 107. The pattern of a long oiling sequence. After bathing, water is removed from the plumage by shaking and wiping movements. Then oil is taken from the oil-gland and distributed over the feathers. Prolonged nibbling decreases gradually in intensity, stretching movements are performed, and finally the bird stops preening and goes to sleep. (From McKinney, 1965b.)

its breast. After several head-shakes, the bird turns its head back to one side of the body and the tail tilts to the same side. The feathers overlying the oil-gland are erected. The bird nibbles at the tuft of feathers surmounting the gland and rubs its chin and head over the gland (Plate XXIc). Oil is then distributed over the plumage by rubbing movements. In the course of an oiling sequence, generally lasting less than 5 minutes, oil is taken from the gland about 4 or 5 times. Thereafter, the bird may continue treating its feathers for half an hour or longer, using nibbling movements. The course of a typical oiling session is diagrammed in Fig. 107. Mallards oil several times each day.

Oiling is believed to serve an important function in maintaining feather structure which is necessary for waterproofing (Elder, 1954; Fabricius, 1959; Rutschke, 1960). Elder removed the oil glands from Mallards and found that the plumage deteriorated. His experiments also suggested that the secretion is effective in maintaining the surface structure of the bill and legs. Now rejected is the earlier suggestion of Hou (1928) that the oil gland secretion contains provitamin D which is irradiated on the feathers to produce vitamin D and that this is then ingested during preening. Analyses of oil glands have revealed no significant amount of provitamin D (Koch & Koch, 1941; Rosenberg, 1942, 1953). Mallard ducklings have two sources of oil; they have functional oil glands which they use within the first day after hatching and they also receive oil from the female's breast (see 'Hatching Behaviour', p. 619 below). The value of the oil to ducklings is uncertain, since Elder found that ducklings with extirpated glands were normal in appearance and behaviour until the juvenal plumage was assumed.

5. NIBBLING PREENING

On many occasions throughout the day Mallards pause in their other activities to spend a few minutes nibbling their feathers (Plate XXId). Often these movements are coupled with dipping the bill in water. Presumably these activities play some role in removing sources of irritation on the skin and in keeping the feathers clean.

6. BATHING

Dabbling Ducks bathe several times each day, but prolonged sessions involving all the bathing movements do not result on every occasion. The commonest action is *head-dipping* by which water is thrown over the back as the bird swims or stands in shallow water (Fig. 108). At higher intensities these movements are accompanied by vigorous *wing-shuffling*. In thorough bathing sessions, *wing-thrashing* and *somersaulting* are incorporated and are often accompanied by *dashing and diving*. The latter performance, involving sudden diving, short flights and flapping over the surface, was considered a form of "play" by Heinroth (1910) and Lebret (1948) but it appears that these actions, closely resembling escape behaviour patterns, have become linked with bathing and presumably they help in wetting and washing the plumage.

C. Sleeping

Waterfowl seldom keep their eyes closed for more than a few moments at a time but they do have long periods (30 to 60 minutes or longer) when they are mainly concerned with resting and sleeping. Dabbling Ducks such as the Mallard normally come out on land to rest. They sit down with the head sunk in the shoulders or turned back so that the bill rests in the scapular feathers. These postures may be maintained for long periods but the eyes frequently blink open and sleeping birds can become fully alert instantly if disturbed.

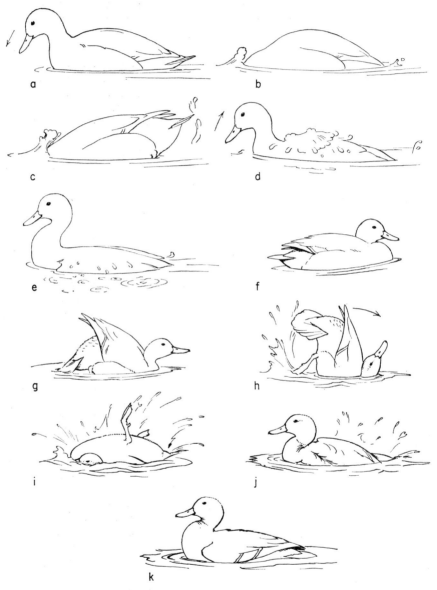

Fig. 108. The sequence of body positions involved in head-dipping (a–e) and somer-saulting (f–k). These movements serve to wet the plumage. (Drawings by Peter Scott from McKinney, 1965b.)

D. Locomotion

Wild Mallards are efficient walkers, surface swimmers, and fliers. The artificial conditions of captivity, with abundant food supply and protection from predators, have led to reduction in locomotory abilities. The heaviest breeds have a more pronounced waddling walk and they have lost the power of flight. The following remarks on locomotion refer to wild birds but they are also appropriate for most game-farm stocks.

In contrast to many species of diving ducks, wild Mallards frequently come out on land to feed, rest or preen. According to the situation, they will walk for long distances (e.g. when feeding in fields or shallow water), run rapidly for short distances (e.g. when chasing another bird), or make single hops (e.g. when jumping on to a log). Forward and back head movements which accompany walking in such species as the Wood Duck (*Aix sponsa*) are normally absent.

Swimming is accomplished by alternate foot movements, the webs being folded as the foot comes forward and expanded during the back stroke. At times, one foot may be shaken in the air and buried in the flank feathers as the other foot makes occasional paddling motions, but Mallards spend less time in this "lazy" swimming activity than Diving Ducks (Aythyini) and Sea Ducks (Mergini) which often sleep on the water.

Pre-flight activities are well marked. The bird faces into the wind, adopts an erect stretched-neck posture, and performs rapid upward *head-thrusting* movements involving the forepart of the body. Repeated lateral *head-shakes* and a variety of other comfort movements (wing-flaps, stretches, etc.) are commonly seen when a bird in flying mood delays take-off for several minutes. These intention movements of flight appear to be important signals in all waterfowl ensuring synchronous take-off by the members of a pair, family or flock (Weidmann, 1956; McKinney, 1965b).

Mallard flight has been analysed from high-speed photographs by Queeny (1946) and the illustrations in his book show clearly the postures adopted during take-off, landing, and a variety of aerial manoeuvres. Changes in the shape and position of wings, tail, head, neck and feet are all involved in accomplishing these feats and the complexity of the co-ordination of body parts can be appreciated only by examining such photographs. The following brief account is drawn from Queeny's conclusions.

Wild Mallards are agile in taking wing, being capable of a vertical rise when alarmed. Such a take-off from the water is achieved by a single vigorous downward stroke, the wings hitting the water. If the bird is on land, it jumps at the same moment and the wings do not make contact with the ground. A more leisurely take-off is less spectacular, but even with a shallow angle of ascent Mallards have no need to patter over the surface with alternate foot thrusts as coots, diving ducks and swans must do to become airborne.

In level flight, the wings are beaten at a rate of about eight per second; during a rapid take-off, the rate rises to ten or twelve a second. During the rapid up-stroke, the primaries are rotated slightly and separated allowing air to pass between them. At the top of the stroke, the primaries overlap tightly and the under surface of the wing becomes slightly concave. The down-stroke provides both lift and propulsion forces, keeping the bird aloft and propelling it forward.

When descending, speed is decreased by braking; the head, tail and feet are brought forward, the long body axis becomes more vertical and the wings beat on a plane close to the horizontal. Spreading and cupping of the tail appears to play an important braking role during the hovering strokes immediately preceding landing. At touch-down, tail and spread feet hit the water simultaneously. Mallards are capable of descending steeply, allowing them to alight on small woodland ponds.

Changes in direction during flight involve complex interactions of all parts of the body. Twisting and fanning the tail, dropping and spreading the webbed feet, and changes in the shape and angle of the wings are all involved. Agility in making rapid turns is put to good use in the course of the aerial pursuits associated with pairing and territorial behaviour.

The feet are normally folded and lie under the tail during flight but in very cold weather they may be brought forward and buried in the flank feathers.

E. Daily Patterns of Activities

According to Raitasuo (1964) who has paid special attention to activity rhythms in wild Finnish Mallards, the day is divided into many periods of activity lasting 45 to 75 minutes with intervening rest periods of 30 to 45 minutes. In autumn and winter, when the birds are living in flocks, rest periods tend to be synchronized and other activities such as bathing appear to be infectious. In the breeding season, each pair follows its own routine and sleeping birds can be observed at all times of the day.

There are wide variations in the routines of different populations at different times of the year. Undoubtedly the distribution of feeding grounds and safe resting places influences the distances travelled each day. In autumn and winter, many Mallard flocks make morning and evening feeding flights to fields but midday trips can occur also. Studies by Bossenmaier & Marshall (1958) and Winner (1959) showed that peak numbers of birds fly to and from the fields around sunrise and sunset. Feeding routines may be disrupted by hunting and birds which are constantly disturbed are said to take to feeding at night.

In general, waterfowl exhibit a basic order in the sequence of four major everyday activities: feeding, bathing, oiling and sleeping. This routine is repeated a number of times each day, but there is considerable variation in the duration of each activity, presumably related to such factors as abundance of food, weather conditions, demands of other activities (e.g. pursuit flights, social courtship, etc.) and amount of disturbance. When food is plentiful more time can be devoted to "loafing".

Weather conditions influence the frequency of many activities. A late cold snap in spring can cause pairs to desert their breeding areas and form flocks again (Sowls, 1955). In wintering flocks at Helsinki, Raitasuo (1964) found that many reproductive activities are less frequent in cold weather and that strong winds have a similar inhibiting effect. Social courtship, pair-formation activities and copulations are all commoner on mild days during winter and spring.

II. SEASONAL ACTIVITIES

While the timing of egg-laying and the duration of the breeding season vary considerably in different latitudes, wild Mallards show a well-defined seasonal cycle. During the winter months, males and females associate in flocks. Social courtship activities are frequent and pairs gradually form. Virtually all Mallards are paired by the time of spring migration but pairs are sociable and migrants arrive on the breeding grounds in flocks. Some populations do not migrate. Soon after their arrival in the breeding habitat pairs spread out and become anti-social. At this time aerial pursuits are frequent.

The male accompanies the female during the selection of the nest-site and he waits for her during her visits to the site for egg-laying. Once incubation has begun the male ranges more widely, the pair bond weakens and usually it breaks before mid-incubation. The males gather in small groups and soon move off to moulting areas. The female usually remains with her brood until they are almost full grown and then she departs for a moulting area. In both sexes the wing-feathers are moulted at one time and the birds are flightless for several weeks. The annual cycle is complete with the autumn migration to the wintering grounds.

Pronounced changes in the gonads coincide with these seasonal changes in behaviour (Höhn, 1947; Johnson, 1961, 1966). In London parks, Höhn found that Mallard testes were small and inactive between July and January. Active spermatogenesis occurred between February and June, coinciding with a marked increase in testes size. The interstitial cells, however, were undeveloped only during the months of July and August. Autumnal courtship activity thus appears to be dependent on the secretory activity of the interstitial cells. Similarly copulation, which is seen frequently throughout the autumn and winter months in this species, apparently is not dependent for its occurrence on sperm production. Ovary and oviduct increase markedly in weight during a shorter period coinciding with egg-laying. Johnson (1961) found considerable individual variation in the stage of gonad development among wild Mallards collected throughout the year in southeastern Washington but in general his findings confirm those of Höhn.

The juvenile Mallard male has a small penis which changes little for the first five to ten months of life (Hochbaum, 1942). By mid-winter (November-January), however, the organ has increased to adult size. Höhn (1960) and Johnson (1961) found that the adult penis undergoes pronounced seasonal changes paralleling those of the testes. In regressed form it is only one-third to one-half the length of the penis of a male in breeding condition and it is reduced correspondingly in diameter.

Both males and females breed in their first year and under normal circumstances only one brood is raised annually. Exceptionally, a juvenile female may lay late in the same season she was hatched (Boyd, 1957).

A. Displays with a Seasonal Distribution

The pre-flight *head-thrusting* and *head-shaking* movements may be seen in every month of the year, though I suspect they are absent or rare while

birds are in the flightless stage. While several other signals, especially some
of the calls, have a wide seasonal distribution, social courtship displays
(*grunt-whistle, head-up-tail-up*, etc.) and displays associated with pair-
formation occur only during the autumn, winter and early spring months.
The following inventory of display movements and calls is compiled from
the accounts of Lorenz (1953), Weidmann (1956), von de Wall (1963) and
Johnsgard (1965) supplemented by my own observations (McKinney,
1965b and unpublished). Most of the names for displays are derived from
Lorenz.

1. FEMALE DISPLAYS

The familiar quacking sound made by female Mallards is given with a
variety of intensities and rhythms, producing several distinctive patterns
which have different seasonal distributions and presumably serve different
functions in communication. While these vocalizations have not yet been
analysed in detail, the major patterns are distinctive.

The *decrescendo call* is heard throughout the autumn and winter but
rarely during the breeding season. It is given by standing or swimming birds
but is not heard from flying birds. It consists of a series of quacks, of variable
number, usually with a strong accent on the first or second syllable, the
remaining notes decreasing in volume and lowering slightly in pitch. Lorenz
(1953) renders it "quaegaegaegaegaegaegaeg" and both Lorenz and Weidmann (1956) state that there are generally six syllables. In the Shoveler
(*Anas clypeata*) and Blue-winged Teal (*A. discors*), the decrescendo is highly
variable in length, form and pitch, and I suspect the same is true in the
Mallard. This variability may be especially important in facilitating individual recognition of females in the wintering flocks. Lorenz notes that this
call is given especially by unmated females and by paired ones when separated from the mate. He also stresses that it is elicited readily by the sight of
other birds flying over. Hunters refer to the decrescendo as the "hail call"
(Hochbaum, 1955), and frequently it is imitated to call passing birds in to
decoys. This is the loudest and most conspicuous call during the autumn
and winter but experimental work is needed to demonstrate its supposed
functions in attracting conspecifics and identifying individuals.

Lorenz (1953) distinguished the *going-away call*, given by females before
flying. Typically this consists of a loud series of short fast quacks. Distinctively sharper and louder *alarm quacks* are heard when a duck is alerted or
alarmed by the presence of a man or bird of prey (Richard Abraham, personal
communication). Judging from the similar call rhythm, head posture, and
situation it is probably this latter call which develops from the "lost peeping"
of the duckling, corresponding to the "raehb" calls of the adult male
(Lorenz, 1953).

Two-syllabled calls ("conversation calls") are distinguished sharply by
Lorenz from the monosyllabic going-away call with the suggestion that these
correspond to the "contentment notes" of the ducklings. He describes these
calls as given by both male and female when the members of a pair come
together after a separation. Weidmann (1956: p. 220) states that the two-
syllabled "quegg-quegg" call of females is seldom heard while the corresponding "rabrab" call of the drake is very frequent.

Persistent quacking is heard from mated females especially on warm quiet mornings and evenings at the beginning of the breeding season—early April on the Canadian prairies (Hochbaum, 1944; Dzubin, 1957). The monotonously repeated "qua, qua, qua, qua . . ." calls are given at a rate of about 16 to 20 per 10-second period and Dzubin associates this calling with a female who is about to begin laying. The female "appears nervous, flying about the countryside from one water area to another or merely swimming or flying about a single slough". This calling makes the presence of a pair very obvious and perhaps it serves the function of advertising intention to settle on a specific nesting area, having a similar "challenge" role to that of many male bird songs (see Three-bird Flights).

Inciting is one of the most important female displays associated with pairing and interactions between paired birds. The female walks or swims beside or behind a chosen male making ritualized threatening movements over one shoulder. The sideways head movements are accompanied by loud, tremulous calls rendered "queggege'ggeggeggeggegg" by Lorenz. A burst of calling accompanies each movement of the head back toward the mate. Often inciting is triggered by the approach of another male but the threatening movements of the female are not always clearly directed toward the stranger (Lorenz, 1958). In many other duck species the homologous display is less highly ritualized, the threatening component being clearly oriented toward the bird being rejected. Inciting appears to serve the function of indicating the female's preference for one male and rejection of another and undoubtedly it is a crucially important signal in resolving disputes during the pair-formation period. It is commonly given on land or water but also in flight (Lebret, 1958b).

The significance of the "tuckata'tuckata'tuckata" calls reported from females in feeding flocks deserves further study. As Hochbaum (1955) notes, this "food call" is imitated widely by hunters to lure birds to the gun. As Collias (1962) points out, this call is similar to that accompanying inciting, and perhaps they are the same. Weidmann (1956) mentions a high, hasty "ku-ku-ku", a mumbling sound, from feeding females, but this does not seem to refer to the "tuckata" call.

Nod-swimming is the most striking movement performed by females during social courtship. The head is stretched forward over the surface of the water and the bird swims rapidly in this posture among the drakes. This movement has a strong stimulating effect on the males, its performance often resulting in a burst of male displays. A low intensity form of nod-swimming in which the head is held high but is moved forward with a single jerky movement is recorded by von de Wall (1963).

"Gestures of repulsion" are highly characteristic behaviour patterns of a broody female pursued by drakes intent on raping her. When given by a bird on land the neck is bent, the head tucked back "in the shoulders", the bill is opened conspicuously by a raising of the upper mandible, the feathers on the back and flanks are ruffled and the tail is fanned, and loud harsh "gaeck" calls are uttered in an irregular pattern. Most of these components are apparent in pursuit flights also, the bent neck and loud calls being especially obvious.

The behaviour of a tame incubating female toward a man or the responses of ducks cornered in a pen have many similarities to this repulsion posture but the loud calls are absent. Sowls (1955) noted that female Pintail (*Anas acuta*) exhibit repulsion behaviour when they are flushed from the nest and he proposed that similar behaviour having a "teasing" function is seen in females which have just lost their eggs and are in need of a mate for a re-nesting attempt. Phillips & van Tienhoven (1962) reached similar conclusions but further observations by Smith (1968) emphasize the complexity of the Pintail's social behaviour. More comparative studies are needed to elucidate the functions of repulsion behaviour in Dabbling Ducks.

The calls given by females with broods will be discussed separately in the discussion of brood behaviour.

2. Male Displays

Male Mallards produce three types of sound: a nasal "raehb" or "rab", a flute-like whistle, and a grunt. A number of distinctive calls involving the nasal sound are the most common vocalizations. Series of evenly-spaced, drawn-out *slow raehbs* are given in situations of mild alarm and after take-off as the bird flies away when flushed. Lorenz (1953) notes that this call develops from the one-syllabled whistle of desertion of the duckling and that it functions as a warning. It is given, for example, by all the drakes on a pond when a strange dog appears on the shore. The necks are stretched, heads are held high, and the birds may swim away at the same time.

In contrast to the slow, even rhythm of the monosyllabic "raehb" calls, rapid double "*rabrab*" calls are characteristic of hostile situations. Lorenz noted that a "*rabrab palaver*" will occur when the members of a pair come together after a long separation or when a flock of birds settle down after being disturbed. Series of these rapid double calls are given as the male tilts his bill up slightly and faces partly toward his mate. While Lorenz did not associate this call with hostility, Weidmann observed that it is characteristic of encounters when attack and escape tendencies are both present. Rabrab calling or threatening movements typically precede fights between males occurring when a drake approaches a pair, when two pairs meet, or when males are disputing during rape attempts. Presumably the rabrab call and accompanying posture function as a threat display.

When a group of males and females forms for social courtship (see below), a number of preliminaries can be detected before a burst of major displays ensues. The males adopt a characteristic posture with the head sunk in the shoulders so that the white neck ring disappears and the head feathers are ruffled. Then *head-shakes, head-flicks*, prolonged *tail-wags*, and *swimming shakes* ("preliminary shaking" of Lorenz, "introductory shake" of Johnsgard) are given. By this time, the observer knows to expect the sudden occurrence of one or another of the major courtship displays (grunt-whistle, head-up-tail-down, down-up).

The *grunt-whistle* begins with a lowering of the bill to the water surface; the bird's body suddenly arches upward and at the peak of the movement the bill is flicked to one side, sending a fine spray of water droplets sideways toward the female (Plate XXIIa). At the same moment a loud whistle is heard, followed by a grunting sound. This is often the first major display to appear in a bout and males seldom give grunt-whistles simultaneously.

The *head-up-tail-down* is a complex display involving the sudden raising of the head with simultaneous vertical cocking of the tail and raising of the closed wings to a position in which the scapulars almost touch the back of the head. The movement is accompanied by a loud whistle. As the wings and tail are lowered, the male turns his head so as to point his bill toward a female (the mate, in the case of paired birds). Then the head is lowered and stretched forward along the water surface as the bird moves forward *nod-swimming* through the group. Finally the sequence is often concluded with a deliberate *turn-back-of-head* to the female, the head being held high.

The *down-up* display involves an equally sudden complex of body actions in which the breast dips deeply into the water, the bill is jerked upward and outward, flipping up a column of water as it goes, and the tail is raised high out of the water (Plate XXIIb). When the head is at the highest level, the drake emits a whistle followed by *rabrab* calls. There is a strong tendency for several males to give courtship displays simultaneously but this usually involves the *head-up-tail-up* and the *down-up*.

Ritualized *jump-flights* are performed by Mallards mostly during social courtship (Lebret, 1958a). The male rises steeply with head held extended horizontal, flies 3 to 8 yards and alights a short distance in front of the female. Immediately after alighting he performs *drinking*.

During social courtship, males (Lorenz, 1953) and also females (von de Wall, 1963) give *gasping* calls without any conspicuous accompanying movements. These calls coincide with the moment when one or more males in the group perform courtship displays.

Bridling, in which the male pulls his head back so that the breast is raised slightly from the water, is normally seen only after copulation in the Mallard but Johnsgard (1960a) records it rarely during social courtship.

Several displays are especially characteristic of situations in which only a single male and female are close together and generally they are associated with birds which are pairing. *Preen-behind-wing* is a conspicuous action in which the male raises the secondaries on one wing and runs his bill along behind them at times producing a rattling noise (Plate XXIf). Other preening movements have also been ritualized as displays, though they are not always very distinct in form from normal preening. *Breast-touching* (Weidmann), *belly-preen, preen-dorsally* and *preen-back-behind-wing* (McKinney, 1965b) are often distinguishable (Plate XXIe). An especially slow and deliberate *wing-flap* is often followed immediately by exaggerated *drinking*. The lateral orientation of the male's body with respect to the female is usually a good clue to the identification of these slightly ritualized movements.

B. Social Courtship and Pair-formation

The earliest observers of Mallard courtship (e.g. Heinroth, 1911; Brock, 1914) assumed that the spectacular activities of groups of birds which are now

20+

collectively known as *social courtship* ("Gesellschaftsspiel") played a direct role in the process of mate selection. Subsequent studies (Weidmann, 1956; Lebret, 1961; Raitasuo, 1964; von de Wall, 1965) suggest that this relationship is not a direct one in the Mallard since pairs are seen before social courtship begins in autumn, paired birds take an active part in social courtship activities, and distinctive behaviour can be identified which seems to be more directly related to pairing. Thus while the "courting parties" of many other duck species appear to result from competitive pairing situations (e.g. *Aythya* species, Hochbaum, 1944; *Somateria*, McKinney, 1961), the biological functions served by Mallard social courtship are still debatable and there are divergent views. At this stage it seems better to discuss pair-formation and social courtship separately until more precise information is available on their relationships.

Weidmann describes *pair-formation* as a gradual process involving inciting by females beside preferred males which respond with drinking and ritualized preening movements. In the early part of the pairing season bonds are temporary and it may be several weeks before firm pairs are apparent. Gradually the two birds spend more and more time swimming close together, engaging in rabrab palavers, and becoming involved in hostile interactions with other birds. Once the bond is strong, the two birds tend to synchronize their daily activities, bathing, preening, and sleeping at the same time. The male behaves aggressively towards other males and the female frequently makes inciting movements beside her mate when unpaired males approach. When the members of the pair become separated the male will respond to the female's decrescendo by giving slow raehb calls and both birds appear to recognize each other individually by voice. On coming together again, drinking, preen-behind-wing, rabrab calling and inciting are commonly seen.

Johnsgard (1960c) has compared pair-formation behaviour in many species of Anatidae and his observations agree in most respects with those of Weidmann. In addition he stresses the frequency of *turn-back-of-head* in response to the female's inciting, comprising the pattern he terms "*leading*". Repeated instances of leading and inciting between the two birds slowly effect a loose pair bond which may be strengthened by copulation, ritualized preening movements and drinking.

Lebret (1961) has made an intensive study of pair-formation in wild Mallards in the Netherlands. While he confirms many of the points made by Weidmann, he suggests that the role of the male in the selection of a partner may be greater than is often believed. He stresses the importance of the male's *jump-flights* in stimulating the interest of a specific female. The occurrence of pursuit-flights involving several males and one female and the observation of trios including two males and one female are interpreted as further evidence for an active role of males in attaching themselves to specific females.

Von de Wall (1965), in a discussion of courtship and pair-formation in Anatini, distinguishes sharply between *directed courtship* ("gerichteten Balz") and *social courtship* ("Gesellschaftsspiel"). Directed courtship appears to correspond to the pair-formation activities described by Weid-

PLATE XXI

a. Both-wings stretch, *b*. scratching, *c*. rubbing head over gland during oiling, *d*. nibbling preening, *e*. preen-back-behind-wing display, *f*. preen-behind-wing display. (Photographs by McKinney.)

PLATE XXII

PLATE XXII

a

b

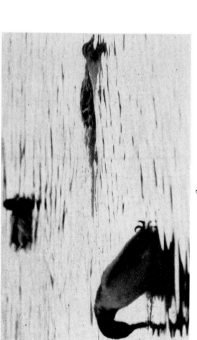

c

d

a. Male performs *grunt-whistle* in response to a female *nod-swimming*; *b.* two males give *down-up*; *c.* male about to mount female in prone pre-copulatory position; *d.* tame incubating Mallard pecking at a man's hand. The latter behaviour is common in park and game-farm stocks; note the fanned tail and pronounced feather erection. For further drawings and photographs of Mallard displays see Lorenz (1953, 1958), Weidmann (1956) and Johnsgard (1965). (Photographs by McKinney.)

mann, characteristically performed by a male and female which show signs of strong mutual attraction. Von de Wall suggests that these inconspicuous and often weakly ritualized patterns (drinking, preening movements, touching of the female's plumage by the male) have an important function in cementing the pair bond.

Pair bonds are established during the late summer, autumn and early winter. Weidmann recorded most pairs forming during September and October but he believed that some pairs which separated during the breeding season re-formed even earlier, in late summer. In the Netherlands, Lebret (1961) records a pair observed as early as 6 July but he notes that the first pairs do not usually appear until the latter part of August. Many form during September and there is then a gradual increase in the number of pairs from October to February. Bezzel (1959) agrees that most pairs are formed by October and Lebret has suggested that pairing after this date concerns only those first-year females which have not yet formed a firm bond. In Finland, Raitasuo (1964) also recorded the first pair-bonds in late August, but he found that most formed in the latter half of February and in March, the last by mid-April. In the exceptionally cold winter of 1962-63, Raitasuo found pair-formation to be greatly delayed. By the end of March, when 75-80 per cent would normally be paired, he found only 12 per cent in pairs. When the weather became mild in early April, pairing occurred very rapidly and by 12 April 95 per cent were paired.

Social courtship is a common activity in wintering flocks. Observers in Europe report slightly different time schedules but the first displays may be seen as early as July (Weidmann, 1956), active groups are widely noted by August and September (Bezzel, 1959; Raitasuo, 1964), peak activity is reported in November (Weidmann) or February-March (Raitasuo) and the last brief subdued bouts are seen in late May or early June.

Von de Wall (1965) reports a daily rhythm of social courtship with peaks in the early morning and late afternoon. He notes, however, that in warm weather the activity ends earlier in the morning and starts later in the afternoon. The inhibiting effect of cold or windy weather has been documented by Raitasuo (1964).

The situations in which social courtship occurs have been studied especially by Weidmann and Lebret (1961). It may begin when there is a sudden mixing of members of the flock or after a period of alert (e.g. caused by a predator). The males adopt the introductory posture and begin shaking. The nod-swimming of a female through a group of males is an especially strong stimulus. Suddenly, one or more males "explode" into one of the three major displays. These bouts of displays may continue at intervals for a quarter of an hour before the group disperses. Paired males have been seen to leave their mates sleeping on shore to swim out and join in such groups. If the mate joins him, the male will orient his displays laterally to her. Observers differ in the amount of hostility between males which they detect in social courtship groups: Weidmann saw little, Lebret saw much pecking between males. But all observers agree that vigorous fighting is absent. There is also unanimous agreement that social courtship is not related in any way to copulation.

Lorenz (1953) suggested that the three major male displays are of "equal value" and it is a matter of chance which one occurs at any given moment. Weidmann (1956) established that this is not the case. He showed that grunt-whistles tend to occur singly, but several males are likely to give head-up-tail-up or down-up simultaneously. He also noted that signs of hostility between males occur only in association with down-up.

Johnsgard (1960a) carried the analysis further in his quantitative comparison of the social courtship displays of Black Duck (*Anas rubripes*) and Mallard. He concluded that the three major male displays represent graded responses to increasingly stimulating female behaviour. In general, an inciting female may produce a grunt-whistle, a female nod-swimming parallel to or away from a male is likely to elicit a head-up-tail-up, while nod-swimming toward a male (the strongest stimulus) usually results in a down-up. But the situation is clearly very complex and analyses of the homologous displays of the Green-winged Teal (*Anas crecca*) (McKinney, unpublished) indicate that the down-up is a response to other males as well as to the female. Johnsgard emphasizes that leading (turn-back-of-head) is especially characteristic of intensive social courtship activity and he believes that this pattern is very important in mate selection.

Johnsgard also showed that the grunt-whistle tends to be the most frequent display during the first few months of pair-formation, but by the time of peak pairing activity down-up has increased in frequency. Johnsgard also confirmed Weidmann's findings, concluding that the grunt-whistle predominates when a single male gives a display. The down-up predominates when four or more males give displays simultaneously and head-up-tail-up usually occurs most often when only two or three males perform.

In summary, the stimulus-response relationships within social courtship groups are still poorly understood, but with each new study it becomes clearer that each display has characteristic situations in which it occurs. The seasonal changes in frequency of displays are presumably linked with changes in hormone levels and these in turn are accompanied by changes in aggressiveness (Etienne & Fischer, 1964). This seems to be the case with down-up in particular, a display which is associated with overt hostility between males.

Many functions have been suggested for social courtship. In spite of the discovery that paired birds participate, many authors still believe with Heinroth (1911) that these performances give females an opportunity to exercise preference for individual mates. The exhibition of specifically distinct male plumage patterns, calls, and display movement repertoires is generally considered to play a role in preventing hybridization with closely related species. In addition, Weidmann suggested that participation in these social activities may have psychosomatic effects in arousing the urge to form a pair in females. Finally, Lebret (1961) has proposed that social courtship may serve as a canalizer of aggression between males during autumn and winter when the birds are living in flocks. These intriguing suggestions remain to be investigated experimentally and the whole subject is still remarkably poorly understood in spite of all the attention it has received.

C. Copulation

As is typical in the family Anatidae, copulation occurs on the water in wild Mallards, but it can be performed on land in domestic breeds. Both male and female make vertical *head-pumping* pre-copulatory movements. In the autumn head-pumping is usually initiated by the female but in spring the male more often begins (Weidmann, 1956). Soon the female lowers her head forward along the water surface (Plate XXIIc), spreads the folded wings slightly to each side and the male mounts, grabbing her crown feathers in his bill. When he has successfully mounted, often after some paddling with his feet presumably necessary to gain a well-balanced position on the female's back, the male passes his tail around the side of the female's cocked tail and after some tail-waggling achieves intromission with a single thrust. The male's tail is usually moved to the left side of the female's tail. The act of copulation is brief, always less than 30 seconds according to Weidmann.

On dismounting, the male immediately pulls his head back into the *bridling* posture often while still holding on to the female's head, emits a single whistle, and then lowers his head forward over the water, *nod-swimming* in a circle around the female. His mate normally bathes without giving any post-copulatory displays.

Weidmann states that paired Mallards copulate once or several times each day starting in the autumn and continuing until incubation begins. Thus copulations are frequent several months before spermatozoa are produced. Copulations can occur at any time of the day. According to Raitasuo (1964) they are most frequent in the middle of the day during winter months, cold or windy weather having an inhibiting effect.

Elder & Weller (1954) investigated the duration of fertility in domestic Mallards after females were separated from drakes. Fertility in a sample of 1,446 eggs remained fairly high during the first week (64 per cent), dropped considerably in the second week (37 per cent) and was very low in the third week (3 per cent).

D. Territorial Behaviour and Aerial Pursuits

Few aspects of waterfowl behaviour have caused so much controversy as the classification of types of pursuit flights and the significance of these flights in pair-spacing. Beginning with the observation of Heinroth (1911) that paired male Mallards attempt to drive off other pairs from their nesting area, a long series of publications has dealt with this and related topics (e.g. Geyr von Schweppenburg, 1924, 1961; Hochbaum, 1944; Sowls, 1955; Lebret, 1955, 1961; Dzubin, 1955, 1957; Weidmann, 1956; Bezzel, 1959; Hori, 1963; McKinney, 1965a; Smith, 1968). Observers are still far from agreeing on many points and continued study is needed if a full understanding of this important aspect of breeding behaviour is to be achieved.

The two most important types of aerial pursuit are now generally labelled *three-bird flights* and *attempted rape flights*. In their characteristic forms the two are quite distinctive. A three-bird flight ("Vertreiben", Geyr von Schweppenburg, 1924; "territorial defence flight", Hochbaum, 1944;

20*

"expulsion flight", Lebret, 1961) is initiated when a paired male takes off from a favourite loafing site on his home range and pursues a pair flying over. The pursuit is directed at the female of the passing pair and the chasing male typically follows her closely, twisting and turning as she changes direction. The male of the passing pair follows some distance behind making little effort to stay close to his mate. After chasing for a distance of several hundred yards, the male breaks off and returns to his starting point.

As Dzubin (1957) points out, the situation is often complicated when two or three drakes launch such a pursuit simultaneously after the same pair. The result is a party of males pursuing a female and it may be difficult to distinguish such a group from an attempted rape flight. Typically, however, the latter includes a larger number of males (up to 20 noted by Dzubin), the pursuit is often more intense and prolonged, and occasionally the female is forced to alight whereupon several males assault her, attempting forced copulation.

The seasonal distribution of these two types of flight is crucially important and more information on individually marked birds of known breeding status is needed. Dzubin (1957) concludes, however, that three-bird flights occur mainly during the period immediately before the mate begins laying and up to early incubation (mid-April to mid-May on the Canadian prairies). At this time the pursuing male has a strong attachment to his mate and this is the period when females are heard giving persistent quacking. Dzubin associates attempted rape flights with the slightly later phase of the breeding cycle when drakes have begun to band together in small groups and pair-bonds have become very weak or have broken completely. By this time, most females are incubating and usually they give repulsion calls while being pursued and frequently adopt the bent-neck posture.

Chasing between males also occurs among Mallards. A male which has settled on a home range will frequently respond to a pair alighting in his area by swimming toward the male of the intruding pair with a characteristic threat posture, head lowered and stretched forward. A chase on water or on land may follow and occasionally fights between males result. Aggressive behaviour between males is rarely seen in flight, however, although contacts of this kind are frequent in some other species of *Anas* notably the Shoveler and Gadwall (*A. strepera*) (Gates, 1962).

A very interesting question arises when we consider what happens if a female loses her clutch after the pair-bond has broken and she requires the service of a male to fertilize a re-nest clutch. Sowls (1955) described pursuit flights in the Pintail (*A. acuta*) similar to attempted rape flights, which he labelled "teasing flights" concluding that "re-nest courtship" was involved. These flights appeared different, however, in that the female did not actively try to escape from the males and she gave the impression of encouraging their pursuit. More attention should be given to this problem in the Mallard, since it is not yet clear whether new pair-bonds are formed, old ones renewed, or whether fertilization for late clutches is achieved by rape as suggested for the Pintail by Smith (1968).

The question of the male's motivational state during three-bird flights has not been settled in spite of much discussion over the last forty years.

It can be argued that the pursuing drake is behaving aggressively. This is suggested by the close temporal association between these pursuit flights and hostility toward males which alight on the water. If we suppose that the male's pursuit of a passing pair is basically hostile in nature, the orientation of the attack toward the female could conceivably be a result of "redirected aggression" (Bastock *et al.*, 1953). On the other hand, the difficulty of distinguishing in all cases between three-bird flights and attempted rape flights suggests that both are expressions of the male's tendency to rape strange females.

Application of the theory of conflicting tendencies as developed by Tinbergen (e.g. 1959) to the interpretation of these pursuit flights may be helpful. The differences between the three-bird flight and the attempted rape flight are understandable if we suppose that the former is an expression of conflict between the male's "tendency to pursue passing females" and his tendency to remain within a small area because of his mate's presence there. As the pair-bond weakens, however, and the male has less attachment to this area, his pursuits become longer and less inhibited resulting in rape attempt flights. If this inverse relationship between the strength of the pair-bond and the increase in rape attempts holds, the theory that the pursuing male in a three-bird flight is motivated predominantly by a tendency to rape is strengthened. The situation may be more complex, however, if aggression toward the female is also present. Possibly the three-bird flight is a complex form of "compromise behaviour" (Andrew, 1956) resulting from the simultaneous presence of all three conflicting tendencies—to attack an intruding pair, to rape a strange female and to return to the mate.

The role of pursuit flights and hostility between males in producing dispersion of pairs over the breeding habitat is also a controversial topic. Suggestions have been made that spacing does occur and has the functions of providing a food supply for the young (Geyr von Schweppenburg, 1924), freedom from interruption during copulation (Hochbaum, 1944) or dispersion of nests as an anti-predator device (Hammond & Mann, 1956; McKinney, 1965a). Other observers (e.g. Bezzel, 1959) suggest that territorial behaviour can have little effect in regulating breeding population densities. The question remains an open and fascinating one.

E. Nest-site Selection, Nest-building and Egg-laying

Migrant Mallards arrive on the breeding grounds in flocks of pairs which break up after a period of days or weeks depending on weather conditions. On mild evenings single pairs are seen flying round over nesting terrain, presumably prospecting for suitable breeding sites. Females home to the areas in which they bred in previous years and some are known to use the same nest-site in successive years (Sowls, 1955). During this period immediately before laying begins *persistent quacking* is frequent (Dzubin, 1957).

Potential sites are inspected by the female, but the male follows her closely waiting nearby as she explores. Mallards will nest in a great variety of situations, for example: on the ground in dense grasses, under bushes or on open fields, over water in emergent vegetation or on muskrat houses, in old crow or hawk nests and in holes in trees. In Dutch decoys and in many parks they take readily to nesting baskets or boxes. Artificial nesting structures of an improbable design, a wire platform on top of a pole placed in open water, have been accepted readily by wild birds in North Dakota in recent years (Lee *et al.*, 1967). Flax straw is placed on the platform and the duck incubates in full view from her safe elevated site. In prairie habitat in southern Manitoba, Sowls (1955) found that most of 123 Mallard nests were situated less than 200 yards from the nearest water, but a few were as far as 400 yards. In south-eastern Alberta, the mean distance for 135 nests was 54 feet (Keith, 1961).

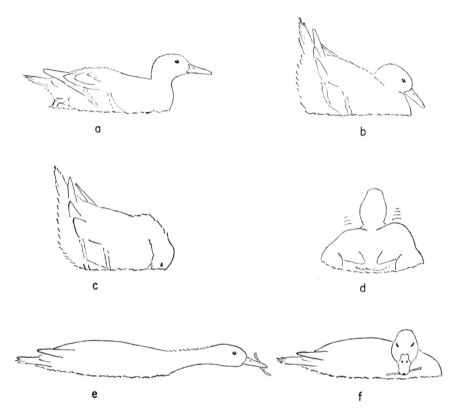

Fig. 109. (a) Scraping the nest bowl, (b) rising to change position during incubation, (c) poking under the eggs, (d) paddling during settling on the eggs, (e) (f) reaching to pull in a piece of grass which is incorporated in the nest by sideways building.

Females probably make several scrapes a few days before the first egg is laid (Sowls, 1955; Coulter & Miller, 1968). These are made by vigorous scraping movements with the feet as the bird turns round on one spot (Fig. 109). Once the site is finally selected and the first egg laid, the bare exposed earth is covered with a thin layer of dead grass, but formation of a substantial nest is a gradual process continuing through the egg-laying and incubation periods. Nest material, mostly in the form of dead grasses and small twigs is gathered solely by means of *sideways building* movements (Fig. 109). The sitting bird leans forward, picks up a piece of grass and places it along one side of her body. From time to time she may rise from the nest, walking off a few feet, continuing to pick up and toss material backward as she goes. Material is never carried back to the nest. Thus the nest gradually accumulates from material that happens to be within a few yards of the site. The nest-rim takes shape as a circle of compressed grass fragments which are patted down into place with the underside of the bird's bill. Pieces of down become detached from the female's breast and are added in the course of incubation. Part of this thick nest-rim of grass and down is drawn over the eggs by the female before she leaves the nest to feed.

Eggs are laid at a rate of one a day, with an occasional day missed (Ogilvie, 1964; Balát, 1967). Laying occurs during the morning (Hochbaum, 1944; Sowls, 1955). Visits to the nest during the egg-laying period vary in duration from 2 to 10 hours but average about 6 hours (Dzubin, *in litt.*). Once the clutch is complete, the female covers the eggs constantly with the exception of one or two short periods off each day.

If the first clutch is destroyed, a second will be laid on a new site. The studies of Sowls (1955), Keith (1961) and Coulter & Miller (1968) suggest that the Mallard is a persistent re-nester as might be expected in a species with such an early breeding season. The period between nest destruction and initiation of the replacement clutch ("re-nesting interval") varies with the stage of incubation at which the nest was destroyed. Sowls found the interval to be 3 days plus 0·62 day for each day of incubation but Coulter & Miller (1968) stress that there is great individual variation.

Precise information on total number of eggs laid by wild birds is difficult to obtain. Single eggs may be "dropped" away from the nest, more than one female may lay in a nest, predators may remove some eggs from a clutch, and females probably remove eggs which become broken in the nest. Bezzel (1966) reviewed the literature on clutch size and pointed out the source of errors in the literature. Extremes of 4 and 18 eggs were recorded by Ogilvie (1964) in England but clutch size varies greatly in samples from various parts of the world. Means range from 7·13 in 252 clutches in Montana (Girard, 1941) to 11·58 for 209 clutches in England (Ogilvie, 1964) but most studies yield means between 8 and 10 (see Bezzel, 1966). Several studies have documented a decrease in clutch size during the breeding season, e.g. from 10·0 to 8·3 (Sowls, 1955), from 10·55 to 6·82 (Bezzel, 1966), suggesting that re-nest clutches tend to be smaller. This has been confirmed by Coulter & Miller (1968) who report that 15 females averaged 10·6 eggs in first nests and 9·6 in their re-nest clutches. These authors also showed that yearling females begin to lay later than older birds.

F. Incubation Behaviour

During her periods on the nest an incubating Mallard spends most of the time sleeping. Periodically she rises from the eggs and changes position. The interval between such active spells varies greatly. Extremes of 3 minutes and 295 minutes were recorded among birds at the Wildfowl Trust in England but the average was between 20 and 40 minutes (McKinney, 1952, 1953). One individual remained virtually immobile on the nest for very long periods (e.g. 120, 169, 213, 295, and 144+ minutes).

At the times of changing position on the nest, the duck performs a series of actions which follow a general pattern with some variations (Fig. 109). The first indication that the bird is about to move is an increased alertness, a turning of the head, or a brief toying with nest-material. Then there may be vigorous tugging or *preening* movements at the breast or flank feathers which often dislodge a few down feathers (*down-pulling*). The bird then rises and moves around on the nest, sometimes making several complete turns. She performs *poking* movements into the nest a number of times, burying her bill deep among the eggs, apparently turning them with motions very similar to those used in egg-retrieval (see below). She then settles down on the eggs again, pushing the breast down and forward while the feet push backward on the edge of the nest. Immediately following *settling* the bird makes vigorous *paddling* movements, apparently involving alternate foot movements, having the effect of rocking the whole body from side to side. I suspect that this activity helps to change the position of some eggs, perhaps arranging the long axes in a horizontal plane in the case of eggs left standing on end after poking.

After these major activities associated with changing position, the bird makes the two kinds of nest-building movements described above, namely *sideways building* and *patting* in the nest-edge with the underside of the bill. While the former movement results in the accumulation of material around the sitting bird, patting apparently serves to arrange the grasses and down in a tightly packed rim. Patting increases in frequency during rain and it probably serves to keep the nest rim tightly pressed against the bird's body.

The final actions in a turning sequence are a lateral *head-shake* and a slight *shuffling* action involving the wings. At times the head-shake serves to dislodge a piece of down or grass which has become attached to the bill, but often no such foreign object can be seen.

In the course of many sequences involving a change in position on the nest, some of the component actions are omitted. Preliminary toying with nest-material occurs in less than 25 per cent of sequences, preening or down-pulling occurs in about 50 per cent, poking in about 65 per cent. Paddling, patting and sideways building are almost always present (80–100 per cent).

Incubating ducks very rarely perform such comfort movements as body-shake, wing-flap, scratching and wing-stretches while they are on the nest. During rain, water is removed from the feathers by delicate dabbling movements of the bill over the surface of the body. In a light shower, these actions successfully remove water droplets from the feathers and they also provide water for drinking, but in heavy rain the duck may become very wet.

If an egg becomes dislodged from the nest bowl on to the rim, it is retrieved by a special *egg-rolling* movement which was first studied in detail by Lorenz & Tinbergen (1938) in the Greylag goose (*Anser anser*) and later in the Mallard and other species by Poulsen (1953). The bird stands up, stretches its head toward the egg, places the tip of its bill beyond the egg and draws it back toward the nest with the underside of the bill. In his experiments with one incubating duck Poulsen showed that eggs as large as a goose egg and as small as a pigeon's egg may be rolled in. Artificially coloured eggs were ignored. In one case, an egg was rolled in when placed as far away as 70 cm from the centre of the nest.

By placing empty egg-shells in the nest of an incubating Shoveler, Sowls (1955) elicited *egg-shell removal* behaviour. The bird lifted each egg-shell in its bill and flew off with it. Observations of wild ducks carrying eggs in the bill (Hochbaum, 1944) probably refer to removal of eggs which have become broken perhaps through fighting on the nest or partial predation of a clutch (McKinney, 1967). There is no evidence that egg-shells are removed at the time of hatching; instead they become buried in the bottom of the nest and broken into small fragments.

When approached on the nest, incubating Mallards often remain frozen in a crouched posture with feathers tightly sleeked and head lowered. Tame birds nesting in close association with humans may refuse to leave the nest even when touched. Such birds will assume an "intimidating" posture with body feathers erected and tail fanned, giving exaggerated deep breathing movements accompanied by hissing and make rapid lunges at the intruder, pecking vigorously at his hand or trouser leg (Plate XXIId). Wild birds are more likely to flush when approached, the distance between intruder and nest varying with the stage of incubation (Weller, 1959). In late incubation, some females will sit tightly until the intruder is close, then flap off the nest with a pronounced *distraction display*. This involves running or swimming with repeated thrashing beats of both wings, which has been likened to "rowing" movements by Stephen (1960).

Where semi-tame Mallards are breeding under crowded conditions females are often molested on the nest by males. Once incubation has begun, females will crouch with head low or stretched forward when strange males fly over or walk nearby. If discovered, they walk or fly off the nest giving *repulsion* behaviour, but they are often pursued relentlessly by the males which attempt to rape them.

Incubating birds leave the nest once, twice or three times each day to feed, drink and bathe. Dzubin (*in litt.*) finds that early in the incubation period, one "recess" per day is usual but two or more periods off the nest are regular in the later stages. Dzubin's study using automatic nest recorders in southern Saskatchewan showed that most morning recesses occur between 03.00 and 04.00, about one half-hour before twilight, while late afternoon departures occurred between 15.00 and 20.00 hours. Morning recesses averaged 47 minutes, afternoon periods 89 minutes. An interesting discovery was that individual females appear to develop short-term rhythms, leaving and returning to the nest at the same time on several consecutive days, then changing to a different schedule.

On rising from the nest, the female draws the nest-rim material over the eggs with backward movements of the bill, often turning round on the nest to draw material from several sides. Then she walks off and flies to her favoured water area, often the closest body of water. On alighting she drinks, bathes, oils, and feeds hurriedly. In the early days of incubation, she usually joins her mate during the off-nest periods. On returning, she alights some distance away and walks to the nest. Some water may be shaken or wiped from her breast before she settles on the eggs again but often some moisture can be seen on her feet and feathers. She sits down on the nest pushing the covering aside with her breast and feet. Within the next few minutes she turns around and paddles until she is again in close contact with the eggs. Finally the nest edge is patted tightly against her body.

The incubation period for wild Mallard eggs hatched in an incubator averaged 22–24 days, with extremes of 21 to 28 days (Hochbaum, 1944). For 33 nests incubated in the wild Balát (1967) gives 24·1 days as the average, varying between 23 and 26, or exceptionally 27 days. In 51 clutches, Ogilvie (1964) found variation between 24 and 32 days with a mean of $27·6 \pm 0·23$ days. Domestic birds of Khaki Campbell and Indian Runner breeds are reported to have 28-day incubation periods (Koecke, 1958; Lamont, 1921).

G. Hatching Behaviour

Three stages in the process of hatching can be distinguished: (1) pipping, (2) emergence and drying, (3) brooding on the nest. The first sign that the eggs are about to hatch is the appearance of a small prominence on the surface of the shell caused by the breaking of the first crack by the duckling's egg-tooth. If the egg is placed against the ear, the duckling can be heard making clicking noises. This stage lasts up to 24 hours or longer and may be said to end when a hole appears in the shell. Emergence of all the ducklings normally takes place over a period of 3 to 8 hours (Bjärvall, 1967). Thereafter the ducklings remain in the nest and are brooded usually for at least 12 to 24 hours, sometimes for as long as 40 hours (Boyd & Fabricius, 1965.)

The duck turns on the nest more frequently during pipping and emergence. Paddling movements become less common but poking in the nest increases in frequency. During brooding, turning components are difficult to identify since the duck changes position frequently in response to movements of the ducklings beneath her.

The female is normally silent while on the nest during incubation but as soon as the eggs start to pip she begins to give very quiet squeaking noises. At first these can scarcely be heard from a distance of a few feet. As hatching proceeds the calls become louder and take the form of series of "quai, quai" or quiet quacking noises.

As the ducklings begin to emerge, the duck frequently raises her back suddenly. Eventually, when several ducklings have hatched, she adopts a half-standing position straddling the nest. As the ducklings become mobile, they poke their heads out at the female's side or under her wing. The duck often attempts to push them back into the nest with her bill with movements similar to patting or poking. Ducklings which have dried off will climb up on to the female's back and this usually seems to prompt the bird to rise, change position and poke into the nest. If pieces of egg shells come to lie on the edge of the nest as a result of the female's poking movements, she will peck at them and may replace them in the nest.

During the brooding stage when the ducklings are moving about in the nest, the female begins to oil. At first the chin is rubbed from side to side across the breast and nibbling of the breast during turning sequences becomes regular. Eventually the oil gland is used, sometimes only once but often three or four times in series. Most of the oil is distributed to the breast, belly and flanks, the wings rarely being touched. There is individual variation in the amount of oiling. I observed one female oiling fourteen times in the course of $5\frac{1}{2}$ hours' observation, the mean of 8 periods between oilings being 26 minutes. This is a much higher rate of oiling than occurs at any other time in a Mallard's life. Apparently the oil is distributed to the ducklings as the Heinroths (1928) suggested.

My observations of normal hatching behaviour suggested that oiling by the duck is released by the presence of ducklings in the nest. I confirmed this by introducing pipped eggs into three nests during late incubation. When the ducklings hatched, all the females showed an increase in preening movements and one bird used its oil gland. The stimuli from the ducklings which induce oiling appear to be tactile and/or auditory but detailed experiments have not been carried out.

The vocalizations of the female gradually increase in frequency and intensity as hatching proceeds. At first, they occur in short bursts and are scarcely audible from a distance of a few feet. Gottlieb (1965a) reports a rate of less than 1 note per 30 seconds at 22 hours before the exodus rising to a peak of 4 notes per second at the time of departure. The female first leads her brood off the nest within 1–2 days of the emergence of the first duckling. She walks off slowly, calling constantly, and the ducklings follow closely in a tight group. According to Gottlieb, the departure never occurs during the night. Captive and tame females may bring the brood back to the nest repeatedly for brooding but this is probably a rare event in the wild.

H. Behaviour of Broods

Beard (1964) has compared the behaviour of broods of several duck species and concludes that Mallard broods are at first surface feeders spending most of their time dabbling or darting and jumping at insects. The ducklings often spread out abreast of each other, but they still maintain contact with the duck and move through the marsh as a unit. Feeding by up-ending is first seen when the ducklings are half grown, at an age of about 4 weeks. Females with broods are constantly alert for signs of danger.

Between spells of feeding in the water broods come out on land to sleep, often choosing such sites as muskrat houses and mud bars. Beard describes how a female deliberately leads her ducklings at a rapid pace toward a favoured resting place. On arrival, the first activity is bathing, then oiling and finally sleeping. Young ducklings are brooded by the duck but as they grow larger they sleep beside the female.

When disturbed, Mallard broods head for the nearest emergent vegetation, the ducklings often skittering across the surface in their haste to retreat. In these situations, females often give repeated loud querulous quacks and may give the wing-flapping distraction display (Bent, 1923;

Munro, 1943; Beard, 1964). Females with broods can be very aggressive toward other birds including ducklings belonging to other females.

As the ducklings grow the bond between parent and young weakens. The brood becomes more dispersed during feeding and the female makes fewer attempts to keep the ducklings close to her by calling. As the ducklings near the flying age at about 7–8 weeks (Dzubin, 1959) they become independent, either by wandering away from the parent and brood mates or through desertion by the hen.

The distance travelled by broods between hatching and flying probably varies greatly depending on local conditions, such as the distribution of favoured feeding areas and the amount of disturbance. Mallard broods have the ability to move considerable distances and in pothole country Evans *et al.* (1952) found that most broods moved from one water area to another. Berg (1956) studied Mallard brood movements in Montana and concluded that there was a tendency to move to large water areas with stable water levels.

I. Duration of the Pair-bond

Dzubin (1955) reported that the drake Mallard ranges more widely once his mate begins to incubate. Gradually he spends less time at his favourite loafing places, so that he is often not present when the female comes off her nest. Apparently the pair bond gradually weakens and probably in most cases the male has lost contact with his mate by the second week of incubation (Oring, 1964; McKinney, 1965a for summary of evidence). Several factors likely influence the precise timing of the break-up of the pair. There must be individual variation but Dzubin (*in litt.*) believes that variations in the timing of the breeding season are especially important. He reports that in the early spring of 1958 in southern Saskatchewan, when Mallard broods began to appear by 15 May, he recorded 15–20 instances of males accompanying broods. In most years, when hatching peaked after 25 May, only a few cases were recorded. Most observers agree that the pair bond in Mallards breaks during the early incubation period; only exceptionally does it survive until the end of incubation and records of males accompanying broods are rare.

III. DEVELOPMENT OF BEHAVIOUR

Embryonic movements have been observed by Gottlieb & Kuo (1965) through a window in the egg. They found that all parts of the embryo are moved to some extent during the first half of the incubation period. At this early stage combined movements involving several parts of the body were the rule. In the later stages of incubation, independent movements of single parts were noted. A few days before hatching the duckling tears the membrane, its head entering the air space, and at this time it begins to give peeping calls and makes "*bill-clapping*" movements. Gottlieb (1965b) demonstrated an increase in the rate of calling and bill-clapping in the foetuses in response to a tape recording of female hatching calls. Thus ducklings and parent are responding to each other's noises before emergence from the shell begins.

The duckling chips the egg shell by repeated *upward nodding* movements of the head which bring the egg-tooth in contact with the shell (Driver, 1960a). The hatching muscle (*M. complexus*) shows a marked increase in relative size immediately before and during the early stages of pipping (Fisher, 1966) and this muscle must play the major role in the process of emergence. The hatching time for each egg, from the appearance of the first small crack to the total emergence of the duckling, appears to vary considerably in Dabbling Ducks. Fisher (1966) records variation between 4 hours and 35·5 hours (mean 9·0 ± 1·8) in the Blue-winged Teal. In the case of these seven teal clutches, the interval between emergence of the first and last ducklings of a clutch (emergence time) was much shorter than the total hatching time for a clutch (11·6 hours as opposed to 31·4 hours). This suggests that emergence from late pipping eggs is speeded up, presumably through stimulation from early pipping eggs as demonstrated in quail by Vince (1964, 1966).

Immediately after emergence, ducklings are relatively inactive until their down has dried off. A few hours later they begin to move around, calling frequently and climbing over the duck. Driver (1960b) reports a "brooding reflex" in newly hatched ducklings. They search for a "feeling of enclosure around the head" by poking the bill into a space where it is in contact with an object on both sides. Driver could produce a state of rest or sleep by holding the duckling's bill between his thumb and finger. He interprets the response as "the mechanism whereby the newly hatched duckling orientates itself to the position of optimum warmth and mechanical protection beneath the female".

Nice (1962) has made detailed observations on the behaviour of newly hatched ducklings. She records nibbling preening movements and jaw-stretching before the bird has completely freed itself from the egg shell. In the first few hours comfort movements such as scratching, tail-wag, wing-flap and wing-and-leg-stretch occurred in some individuals and nibbling and pecking at contrasting objects was frequent. Observations on incubator-hatched Mallard ducklings kept in a small pen (McKinney, 1965b) revealed that oiling preening and bathing involving head-dipping also appear within the first 2 days, but the more complex bathing actions (wing-thrashing, somersaulting, dashing-and-diving) were not seen until the birds were 2 weeks old. At first the young ducklings frequently topple over as they attempt to shake or scratch, but they can soon stand erect and wing-flap quite efficiently in spite of the rudimentary appendages they have to flap.

Ducklings have two distinctively different kinds of vocalizations, commonly referred to as *contentment notes* and *distress calls*. The latter are given when the bird is cold, wet, hungry, alarmed or separated from the parent. They are replaced by contentment notes as soon as their "distressing" situations are overcome. In form, the two types of call are very different: contentment notes being quiet twittering sounds with many short notes while distress calls are loud, strident, evenly spaced and slower, often delivered from an erect posture.

Since the observations of Lorenz (1935) drew attention to the phenomenon of *imprinting* in waterfowl, much research has been carried out on the establishment of filial and sexual preferences in Mallard ducklings. The widely scattered literature reporting studies by comparative psychologists and ethologists has recently been reviewed critically by Sluckin (1964) and Bateson (1966). These publications should be consulted for a full bibliography on this important field of behaviour research in which the Mallard has played a major role.

When ducklings are exposed to a conspicuous object shortly after hatching they become attached to it, following if it moves, sitting close beside it and giving contentment notes when it is stationary and expressing distress through lost peeping calls if it is removed. While such responses would normally be given to the parent, it has been discovered that they can also be elicited by such things as cardboard boxes, wooden duck decoys, human beings, and even flickering lights. The most effective stimuli are those which move and make repetitive sounds similar to the calls of a broody female (Collias & Collias, 1956).

There is a "sensitive period" during which ducklings respond most readily by following models. Ramsey & Hess (1954) found an optimum period between 13 and 16 hours after hatching but Gottlieb (1961) discovered that a more striking peak is obtained if "development age" is calculated, most imprinting occurring on the 27th day from the onset of incubation. After this sensitive period, it becomes more and more difficult to produce following of a model: instead the ducklings tend to show avoidance. Boyd & Fabricius (1965) demonstrated that while most ducklings follow a silent model at 10–20 hours, some still follow when first tested at 240 hours. When auditory stimuli alone were used (rapidly repeated "kom" calls) most birds followed at 40–50 hours but 50 per cent still followed at 240 hours. Readiness to follow is increased greatly when both visual and auditory stimuli are presented simultaneously.

As Lorenz (1935) originally stressed, the early social experiences of ducklings can have marked effects on their later preferences in mate-selection. Recent work by Schutz (1965a) and de Lannoy (1967) has clarified this process of sexual imprinting. Schutz showed that when Mallard drakes are exposed over the first 50 days after hatching to a duckling or foster-mother of another species, many will choose mates belonging to the other species when given a free choice. When female Mallards are raised with members of other species, however, they do not become imprinted but tend to mate with their own species. In the Chilean Teal (*Anas flavirostris*), however, male and female have similar adult plumages and both sexes can be imprinted on other species.

The phenomenon of sexual imprinting has important implications for aviculturists who need to consider the possibilities of producing undesirable "fixations" as a result of the abnormal social environments in which they often raise ducks of many species. Especially unwelcome are the strong homosexual pair bonds that can easily be produced in captivity by keeping groups of male ducklings without females as they develop. Drake Mallards raised under these conditions show no interest in females, directing their courtship displays to males only and forming strong homosexual bonds (Schutz, 1965b).

While mixed-species pairs and resulting hybrids are very frequent in collections of captive waterfowl, the production of hybrids in the wild also occurs on a small scale (Gray, 1958; Johnsgard, 1960b). Sexual imprinting may also be involved in these cases. For example, the accidental attachment of a male Mallard duckling to a Pintail brood may be expected to produce an individual with a preference for a Pintail female as a mate.

REFERENCES

ADDY, C. E. (1964). Atlantic Flyway. In: *Waterfowl Tomorrow*, pp. 167–184. Washington, D.C.: U.S. Government Printing Office.

ANDREW, R. J. (1956). Some remarks on behaviour in conflict situations, with special reference to *Emberiza* spp. *Br. J. Anim. Behav.*, 4, 41–45.

BALÁT, F. (1967). Legefolge und Brutdauer bei der Stockente, *Anas platyrhynchos* L. *Zool. ent. Listy*, 16, 167–172.

BASTOCK, M., MORRIS, D. & MOYNIHAN, M. (1953). Some comments on conflict and thwarting in animals. *Behaviour*, 6, 66–84.

BATESON, P. P. G. (1966). The characteristics and context of imprinting. *Biol. Rev.*, 41, 177–220.

BEARD, E. B. (1964). Duck brood behavior at the Seney National Wildlife Refuge. *J. Wildl. Mgmt*, 28, 492–521.

BENT, A. C. (1923). Life histories of North American wildfowl: order Anseres (Part 1). *U.S. Natl. Museum Bull., Washington, D.C.*, No. 126, 1–250.

BERG, P. F. (1956). A study of waterfowl broods in eastern Montana with special reference to movements and the relationship of reservoir fencing to production. *J. Wildl. Mgmt*, 20, 253–262.

BEZZEL, E. (1959). Beiträge zur Biologie der Geschlecter bei Entenvögeln. *Anz. orn. Ges. Bayern*, 5, 269–355.

BEZZEL, E. (1966). Zur Ermittlung von Gelegegrösse und Schlüpferfolg bei Entenvögeln. *Die Vogelwelt*, 87, 97–106.

BJÄRVALL, A. (1967). The critical period and the interval between hatching and exodus in Mallard ducklings. *Behaviour*, 28, 141–148.

BOSSENMAIER, E. F. & MARSHALL, W. H. (1958). Field-feeding by waterfowl in south-eastern Manitoba. *Wildl. Monogr.*, No. 1, The Wildlife Society, Blacksburg, Va., 32 pp.

BOYD, H. (1957). Early sexual maturity of a female Mallard. *Brit. Birds*, 50, 302–303.

BOYD, H. & FABRICIUS, E. (1965). Observations on the incidence of following of visual and auditory stimuli in naive Mallard ducklings (*Anas platyrhynchos*). *Behaviour*, 25, 1–15.

BRAKHAGE, G. K. (1954). Migration and mortality of ducks hand-reared and wild-trapped at Delta, Manitoba. *J. Wildl. Mgmt*, 17, 465–477.

BROCK, S. E. (1914). The display of the Mallard in relation to pairing. *Scott. Nat.*, 79, 78–86.

COLLIAS, N. E. (1962). The behaviour of ducks. In: *The Behaviour of Domestic Animals*. E. S. E. Hafez (ed.). Baltimore: Williams & Wilkins Co.

COLLIAS, N. E. & COLLIAS, E. C. (1956). Some mechanisms of family integration in ducks. *Auk*, 73, 378–400.

COULTER, M. W. & MILLER, W. R. (1968). Nesting biology of Black Ducks and Mallards in northern New England. *Vermont Fish and Game Dept., Bull.*, No. 68–2. Montpelier, Vermont. 73 pp.

DARWIN, C. (1887). *The Variation of Animals and Plants under Domestication*. 2nd edn, revised (2 vols.). New York: D. Appleton & Co.

DE LANNOY, J. (1967). Zur Prägung von Instinkthandlungen (Untersuchungen an Stockenten *Anas platyrhynchos* L. und Kolbenenten *Netta rufina* Pallas). *Z. Tierpsychol.*, 24, 162–200.

DELACOUR, J. (1956). *The Waterfow lof the World*. Vol. 2. *The Dabbling Ducks*. London: Country Life.

DELACOUR, J. (1964). *The Waterfowl of the World*. Vol. 4. London: Country Life.

DELACOUR, J. & MAYR, E. (1945). The family Anatidae. *Wilson Bull.*, **57**, 3–55.

DRIVER, P. M. (1960a). Field studies on the behaviour of sea-ducklings. *Arctic*, **13**, 201–204.

DRIVER, P. M. (1960b). A possible fundamental in the behaviour of young nidifugous birds. *Nature, Lond.*, **186**, 416.

DZUBIN, A. (1955). Some evidences of home range in waterfowl. *Trans. N. Am. Wildl. Conf.*, **20**, 278–298.

DZUBIN, A. (1957). Pairing display and spring and summer flights of the Mallard. *Blue Jay*, **15**, 10–13.

DZUBIN, A. (1959). Growth and plumage development of wild-trapped juvenile Canvasbacks (*Aythya valisineria*). *J. Wildl. Mgmt*, **23**, 278–290.

ELDER, W. H. (1954). The oil gland of birds. *Wilson Bull.*, **66**, 6–31.

ELDER, W. H. & WELLER, M. W. (1954). Duration of fertility in the domestic Mallard hen after isolation from the drake. *J. Wildl. Mgmt*, **18**, 495–502.

ETIENNE, A. & FISCHER, H. (1964). Untersuchung über das Verhalten kastrierter Stockenten (*Anas platyrhynchos* L.) und dessen Beeinflussung durch Testosteron. *Z. Tierpsychol.*, **21**, 348–358.

EVANS, C. D., HAWKINS, A. S. & MARSHALL, W. H. (1952). Movements of waterfowl broods in Manitoba. *U.S. Fish and Wildlife Service. Spec. Sci. Rept. Wildlife*, No. **16**, 59 pp.

FABRICIUS, E. (1959). What makes plumage waterproof? *Wildfowl Trust 10th Annual Report*, 105–113.

FISHER, H. I. (1966). Hatching and the hatching muscle in some North American ducks. *Trans. Ill. St. Acad. Sci.*, **59**, 305–325.

GATES, J. M. (1962). Breeding biology of the Gadwall in Northern Utah. *Wilson Bull.*, **74**, 43–67.

GEYR VON SCHWEPPENBURG, H. (1924). Zur Sexualethologie der Stockente. *J. Ornithologie*, **72**, 102–108.

GEYR VON SCHWEPPENBERG, H. (1961). Zum Verhalten der Stock- und Schnatterente. *J. Ornithologie*, **102**, 140–148.

GIRARD, G. L. (1941). The Mallard: its management in western Montana. *J. Wildl. Mgmt*, **5**, 233–259.

GOTTLIEB, G. (1961). Development age as a baseline for determination of the critical period in imprinting. *J. comp. physiol. Psych.*, **54**, 422–427.

GOTTLIEB, G. (1965a). Components of recognition in ducklings. *Nat. Hist., N.Y.*, **25**, 12–19.

GOTTLIEB, G. (1965b). Prenatal auditory sensitivity in chickens and ducks. *Science*, **147**, 1596–1598.

GOTTLIEB, G. & Z. Y. KUO (1965). Development of behavior in the duck embryo. *J. comp. physiol. Psych.*, **59**, 183–188.

GRAY, A. P. (1958). Bird hybrids. A check-list with bibliography. *Commonwealth Agric. Bureaux, Farnham Royal, Buckinghamshire*, England. 390 pp.

HAMMOND, M. C. & MANN, G. E. (1956). Waterfowl nesting islands. *J. Wildl. Mgmt*, **20**, 345–352.

HEINROTH, O. (1910). Beobachtungen bei einem Einbürgerungsversuch mit der Brautente (*Lampronessa sponsa* L.). *J. Ornithologie*, **58**, 101–156.

HEINROTH, O. (1911). Beiträge zur Biologie namentlich Ethologie und Psychologie der Anatiden. *Verh. Int. Orn. Kongr.* (Berlin, 1910), pp. 589–702.

HEINROTH, O. & HEINROTH, M. (1928). *Die Vögel Mitteleuropas*. Vol. 3. Berlin: Hugo Bermühler.

HOCHBAUM, H. A. (1942). Sex and age determination of waterfowl by cloacal examination. *Trans. 7th N. Amer. Wildl. Conf.*, pp. 299–307.

HOCHBAUM, H. A. (1944). *The Canvasback on a Prairie Marsh*. Washington, D.C.: Amer. Wildl. Inst.

HOCHBAUM, H. A. (1955). *Travels and Traditions of Waterfowl*. Minneapolis: Univ. Minn. Press.

HÖHN, E. O. (1947). Sexual behaviour and seasonal changes in the gonads and adrenals of the Mallard. *Proc. zool. Soc. Lond.*, **177**, 281–304.

Höhn, E. O. (1960). Seasonal changes in the Mallard's penis and their hormonal control. *Proc. zool. Soc. Lond.*, **134**, 547–555.

Hori, J. (1963). Three-bird flights in the Mallard. *Wildfowl Trust 14th Annual Report*, pp. 124–132.

Hou, H. C. (1928). Studies on the glandula uropygialis of birds. *Chin. J. Physiol.*, **2**, 345–379.

Johnsgard, P. A. (1960a). A quantitative study of sexual behavior of Mallards and Black Ducks. *Wilson Bull.*, **72**, 133–155.

Johnsgard, P. A. (1960b). Hybridization in the Anatidae and its taxonomic implications. *Condor*, **62**, 25–33.

Johnsgard, P. A. (1960c). Pair-formation mechanisms in *Anas* (Anatidae) and related genera. *Ibis*, **102**, 616–618.

Johnsgard, P. A. (1965). *Handbook of Waterfowl Behavior*. Ithaca, N.Y.: Cornell University Press.

Johnson, O. W. (1961). Reproductive cycle of the Mallard duck. *Condor*, **63**, 351–364.

Johnson, O. W. (1966). Quantitative features of spermatogenesis in the Mallard (*Anas platyrhynchos*). *Auk*, **83**, 233–239.

Keith, L. (1961). A study of waterfowl ecology on small impoundments in south-eastern Alberta. *Wildl. Monogr.*, No. **6**, 88 pp.

Koch, E. M. & Koch, F. C. (1941). The pro-vitamin D of the covering tissues of chickens. *Poult. Sci.*, **20**, 33–35.

Koecke, H. (1958). Normalstadien der Embryonalentwicklung bei der Hausente (*Anas boschas domestica*). *Embryologia*, **4**, 55–78.

Lamont, A. (1921). On the development of the feathers of the duck during the incubation period. *Trans. R. Soc. Edinb.*, **53**, 231–241.

Lebret, T. (1948). The "diving-play" of surface-feeding duck. *Brit. Birds*, **41**, 247.

Lebret, T. (1955). Die Verfolgungsflüge der Enten. *J. Ornithologie*, **96**, 43–49.

Lebret, T. (1958a). The "jump-flight" of the Mallard, *Anas platyrhynchos* L., the Teal, *Anas crecca* L. and the Shoveler, *Spatula clypeata* L. *Ardea*, **46**, 68–72.

Lebret, T. (1958b). Inciting ("Hetzen") by flying ducks. *Ardea*, **46**, 73–75.

Lebret, T. (1961). The pair formation in the annual cycle of the Mallard. *Anas platyrhynchos* L. *Ardea*, **49**, 97–158.

Lee, F. B., Kruse, A. D. & Thornsberry, W. H. (1967). An evaluation of certain types of artificial nesting structures for ducks. Paper presented at 29th Midwest Fish and Wildlife Conference, Madison, Wisc., Dec. 10–13, 1967 (mimeo).

Lorenz, K. (1935). Der Kumpan in der Umwelt des Vögels. *J. Ornithologie*, **88**, 137–213, 289–413.

Lorenz, K. (1953). Comparative studies on the behaviour of the Anatinae. *Avicultural Magazine*, London, pp. 1–87.

Lorenz, K. (1958). The evolution of behavior. *Scient. Am.*, **199** (6), 67–78.

Lorenz, K. & Tinbergen, N. (1938). Taxis und Instinkthandlung in der Eirollbewegung der Graugans. I. *Z. Tierpsychol.*, **2**, 1–29.

McKinney, F. (1952). Incubation and hatching behaviour in the Mallard. *Wildfowl Trust 5th Annual Report*, pp. 68–70.

McKinney, F. (1953). Studies on the Behaviour of the Anatidae. Ph.D. thesis, University of Bristol, England.

McKinney, F. (1961). An analysis of the displays of the European Eider *Somateria mollissima mollissima* (Linnaeus) and the Pacific Eider *Somateria mollissima v. nigra* Bonaparte. *Behaviour*, Suppl. VII.

McKinney, F. (1965a). Spacing and chasing in breeding ducks. *Wildfowl Trust 16th Annual Report*, pp. 92–106.

McKinney F. (1965b). The comfort movements of Anatidae. *Behaviour*, **25**, 120–220.

McKinney, F. (1967). Breeding behaviour of captive Shovelers. *Wildfowl Trust 18th Annual Report*, pp. 108–121.

Munro, J. A. (1943). Studies of waterfowl in British Columbia: mallard. *Canad. J. Res.*, **21D**, 223–260.

Nice, M. M. (1962). Development of behaviour in precocial birds. *Trans. Linn. Soc. N.Y.*, **8**, 1–211.

OGILVIE, M. A. (1964). A nesting study of Mallard in Berkeley New Decoy, Slimbridge. *Wildfowl Trust 15th Annual Report*, pp. 84–88.

ORING, L. W. (1964). Behavior and ecology of certain ducks during the postbreeding period. *J. Wildl. Mgmt*, **28**, 223–233.

PHILLIPS, J. C. (1922–26). *A Natural History of the Ducks*. Vols. I–IV, Boston: Houghton Mifflin Co.

PHILLIPS, R. E. & VAN TIENHOVEN, A. (1962). Some physiological correlates of Pintail reproductive behavior. *Condor*, **64**, 291–299.

POULSEN, H. (1953). A study of incubation responses and some other behaviour patterns in birds. *Vidensk. Meddr. dansk naturh. Foren.*, **115**, 1–131.

QUEENY, E. M. (1946). *Prairie Wings*. New York: Ducks Unlimited, Inc.

RAITASUO, K. (1964). Social behaviour of the Mallard, *Anas platyrhynchos*, in the course of the annual cycle. *Helsinki: Papers on Game Research*, **24**, 1–72.

RAMSEY, A. O. & HESS, E. H. (1954). A laboratory approach to the study of imprinting. *Wilson Bull.*, **66**, 196–206.

ROSENBERG, H. R. (1942). *Chemistry and Physiology of the Vitamins*. New York: Interscience Publisher.

ROSENBERG, N. R. (1953). The site and nature of provitamin D in birds. *Archs Biochem. Biophys.*, **42**, 7–11.

RUTSCHKE, E. (1960). Untersuchungen über Wasserfestigkeit und Struktur des Gefieders von Schwimmvögeln. *Zool. Jahrb.*, **87**, 441–506.

SCHUTZ, F. (1965a). Sexuelle Prägung bei Anatiden. *Z. Tierpsychol.*, **22**, 50–103.

SCHUTZ, F. (1965b). Homosexualität und Prägung: eine experimentelle Untersuchung an Enten. *Psychol. Forschung.*, **28**, 439–463.

SLUCKIN, W. (1964). *Imprinting and Early Learning*. London: Methuen.

SMITH, R. I. (1968). The social aspects of reproductive behavior in the Pintail. *Auk*, **85**, 381–396.

SOWLS, L. K. (1955). *Prairie Ducks*. Harrisburg, Penn.: Stackpole Co., and Washington, D.C.: Wildlife Management Inst.

STEPHEN, W. J. D. (1960). Some reactions of the Mallard *Anas platyrhynchos* L. to predators. M.Sc. thesis, University of Toronto, Canada.

TINBERGEN, N. (1959). Comparative studies of the behaviour of gulls (Laridae); a progress report. *Behaviour*, **15**, 1–70.

VINCE, M. A. (1964). Social facilitation of hatching in the bobwhite quail. *Anim. Behav.*, **12**, 531–534.

VINCE, M. A. (1966). Potential stimulation produced by avian embryos. *Anim. Behav.*, **14**, 34–40.

VON DE WALL, W. (1963). Bewegungsstudien an Anatinen. *J. Ornithologie*, **104**, 1–15.

VON DE WALL, W. (1965). "Gesellschaftsspiel" und Balz der Anatini. *J. Ornithologie*, **106**, 65–80.

WEIDMANN, U. (1956). Verhaltensstudien an der Stockente (*Anas platyrhynchos* L). I. Das Aktionssystem. *Z. Tierpsychol.*, **13**, 208–271.

WELLER, M. W. (1959). Parasitic egglaying in the Redhead (*Aythya americana*) and other North American Anatidae. *Ecol. Monogr.*, **29**, 333–365.

WELLER, M. W. (1964). General habits. The reproductive cycle. In: *The Waterfowl of the World* by J. Delacour. Vol. 4, 15–34, 35–79. London: Country Life.

WINNER, R. W. (1959) Field feeding periodicity of Black and Mallard Ducks. *J. Wildl. Mgmt*, **23**, 197–202.

WOOLFENDEN, G. E. (1961). Postcranial osteology of the Waterfowl. *Bull. Florida State Museum, Biological Sciences*, **6**, 1–129.

Conversion Tables

MASS
Imperial

1 ton (2240 lb)	= 1016 kilogrammes
1 hundredweight (112 lb) (cwt)	= 50·80 kilogrammes
1 stone (14 lb) (st.)	= 6·35 kilogrammes
1 pound (avoirdupois) (lb)	= 453·59 grammes
1 ounce (avoirdupois) (oz)	= 28·35 grammes
1 grain (gr)	= 64·799 milligrammes

Metric

1 kilogramme (kg)	= 15·432 grains
	or 35·274 ounces
	or 2·2046 pounds
1 gramme (g)	= 15·432 grains
1 milligramme (mg)	= 0·015432 grains

CAPACITY
Imperial

1 gallon (160 fl. oz) (gal.)	= 4·546 litres
1 U.S. gallon	= 0·8 Imperial gallon
1 pint (pt)	= 568·25 millilitres
	or 0·56825 litre
1 fluid ounce (fl. oz)	= 28·412 millilitres
1 fluid drachm (fl. dr)	= 3·5515 millilitres
1 minim (min.)	= 0·059192 millilitres

Metric

1 litre (l.)	= 1·7598 pints
1 millilitre (ml)	= 16·894 minims

LENGTH
Imperial

1 mile	= 1·609 kilometres
1 yard	= 0·914 metres
1 foot	= 30·48 centimetres
1 inch	= 2·54 centimetres
	or 25·40 millimetres

Metric

1 kilometre (km)	= 0·621 miles
1 metre (m)	= 39·370 inches
1 decimetre (dm)	= 3·9370 inches
1 centimetre (cm)	= 0·39370 inch
1 millimetre (mm)	= 0·039370 inch
1 micron (μ)	= 0·0039370 inch

TEMPERATURE

Centigrade	Fahrenheit	Centigrade	Fahrenheit
110	230	38	100·4
100	212	37·5	99·5
95	203	37	98·6
90	194	36·5	97·7
85	185	36	96·8
80	176	35·5	95·9
75	167	35	95·0
70	158	34	93·2
65	149	33	91·4
60	140	32	89·6
55	131	31	87·8
50	122	30	86
45	113	25	77
44	111·2	20	68
43	109·4	15	59
42	107·6	10	50
41	105·8	+ 5	41
40·5	104·9	0	32
40	104·0	− 5	23
39·5	103·1	− 10	14
39	102·2	− 15	+ 5
38·5	101·3	− 20	− 4

Index

Page references in bold type indicate references to illustrations

PRINTED IN GREAT BRITAIN BY WILLIAM CLOWES AND SONS, LIMITED, LONDON AND BECCLES